281-560-1623

Polypropylene

Polypropylene

An A-Z reference

Edited by

J. Karger-Kocsis

Institute for Composite Materials Ltd.
University of Kaiserslautern
Germany

and

Technical University of Budapest
Hungary

A C.I.P. Catalogue record for this book is available from the Library of Congress

ISBN 0 412 80200 7

Published by Kluwer Academic Publishers,
P.O. Box 17, 3300 AA Dordrecht, The Netherlands

Sold and distributed in North, Central and South America
by Kluwer Academic Publishers,
101 Philip Drive, Norwell, MA 02061, U.S.A.

In all other countries, sold and distributed
by Kluwer Academic Publishers Group,
P.O. Box 322, 3300 AH Dordrecht, The Netherlands.

Printed in Great Britain

Contents

List of contributors

Abdellah Ajji
Industrial Materials Institute, National Research Council Canada, 75 Boulevard de Mortagne, Boucherville, Québec, J4B 6Y4, Canada

S. Al-Malaika
Polymer Processing and Performance Group, Department of Chemical Engineering and Applied Chemistry, Aston University, Aston Triangle, Birmingham B4 7ET, UK

Erik Andreassen
SINTEF Materials Technology, PO Box 124 Blindern, N-0314 Oslo, Norway

György Bánhegyi
Furukawa Electric Institute of Technology, H-1158 Budapest, Késmárk u. 24–28, Hungary

G. Bechtold
Institut für Verbundwerkstoffe GmbH, University of Kaiserslautern, D-67663 Kaiserslautern, Germany

J. Bentham
Borealis Deutschland GmbH, Morsenbroicher Weg 200, D-40474 Düsseldorf, Germany

K. Bernreiter
PCD Polymere GmbH, PO Box 675, A-4021 Linz, Austria

Michael J. Bevis
Wolfson Centre for Materials Processing, Department of Materials Engineering, Brunel University, Uxbridge, Middlesex UB8 3PH, UK

D. Bhattacharyya
Composites Research Group, Department of Mechanical Engineering, University of Auckland, Private Bag 92019, Aukland, New Zealand

Eric Bond
Materials Science and Engineering, College of Engineering, University of Tennessee, Knoxville, Tennessee 37996–2200, USA

P.-E. Bourban
Laboratoire de Technologie des Composites et Polymères, École Polytechnique Federale de Lausanne, CH-1015 Lausanne, Switzerland

Serge Bourbigot
Laboratoire de Chimie Analytique et de Physico-Chimie des Solides, École Nationale Superieure de Chimie de Lille, Université des Sciences et Technologies de Lille, BP 108, F-59652 Villeneuve d'Ascq Cedex, France

C. Brockmann
Institut für Kunststoffverarbeitung, RWTH Aachen, Pontstr. 49, D-52056 Aachen, Germany

W. Brockmann
Werkstoff- und Oberflächentechnik, Universität Kaiserslautern, D-67663 Kaiserslautern, Germany

Witold Brostow
Department of Materials Science, University of North Texas, Denton, TX 76203-5310, USA

S. Brückner
Dipartimento di Scienze e Tecnolgie Chimiche, University degli Studi di Udine, I-33100 Udine, Italy

A. Brunswick
Institut für Kunststoffverarbeitung, RWTH Aachen, Pontstr. 49, D-52062 Aachen, Germany

H. Bucka
PCD Polymere GmbH, PO Box 675, A-4021 Linz, Austria

A. Cervenka
Manchester Materials Science Centre, University of Manchester and UMIST, Grosvenor Street, Manchester, M1 7HS, UK

Chi-Ming Chan
Department of Chemical Engineering, Hong Kong University of Science and Technology, Clear Water Bay, Hong Kong PR China

Kwong Chan
Institutes of Textiles and Clothing, Hong Kong Polytechnic University, Yuk Choi Road, Kowloon, Hong Kong PR China

R. Chatten
Department of Materials Engineering, Brunel University, Uxbridge, Middlesex, UB8 3PH, UK

I. Chodák
Polymer Institute, Slovak Academy of Sciences, SK-842 36 Bratislava, Slovakia

G.R. Christie
Composites Research Group, Department of Mechanical Engineering, University of Auckland, Private Bag 92019, Auckland, New Zealand

K.C. Cole
Industrial Materials Institute, National Research Council Canada, 75 Boulevard de Mortagne, Boucherville, Quebec, J4B 6Y4, Canada

J.-L. Costa
Solvay Polyolefins Europe-Belgium, Rue de Ransbeek 310, B-1120 Brussels, Belgium

T. Czvikovszky
Department of Polymer Engineering and Textile Technology, Faculty of Mechanical Engineering, Technical University of Budapest, H-1111 Budapest, Hungary

A. Dehn
Institut für Verbundwerkstoffe GmbH, Universität of Kaiserslautern, D-67663 Kaiserslautern, Germany

Rene Delobel
Laboratoire de Chimie Analytique et de Physico-Chimie des Solides, Ecole Nationale Superieure de Chimie de Lille, Université des Sciences et Technologies de Lille, BP 108, F-59652 Villeneuve d'Ascq Cedex, France

R. Denzer
Institut für Verbundwerkstoffe GmbH, Universität of Kaiserslautern, D-67663 Kaiserslautern, Germany

B.L. Deopura
Textile Technology Department, Indian Institute of Technology, New Delhi 110016, India

M.M. Dumoulin
ADS Composites Group, 275 North, Monfette Street, Thetford Mines, Quebec, G6G 7H4, Canada

Gottfried W. Ehrenstein
Lehrstuhl für Kunststofftechnik, Universität Erlangen-Nürnberg, Am Weichselgarten 9, D-91058 Erlangen-Tennenlohe, Germany

P. Eyerer
Institut für Polymer Testing (IKP), University of Stuttgart, Pfaffelwaldring 32, D-70569 Stuttgart, Germany

J. Fiebig
PCD Polymere GmbH, PO Box 675, A-4021 Linz, Austria

B. Fisa
Centre de recherche appliquée sur les polymères (CRASP), École Polytechnique de Montréal, PO Box 6079, Station Centre Ville, Montréal, Québec H3C 3A7, Canada

K. Friedrich
Institut für Verbundwerkstoffe GmbH, Universität of Kaiserslautern, D-67663 Kaiserslautern, Germany

Mitsuyoshi Fujiyama
Plastics Research Laboratory, Tokuyama Company, Tokuyama City, Yamaguchi 745, Japan

M. Gahleitner
PCD Polymere AG, PO Box 675, A-4021 Linz, Austria

A. Galeski
Centre of Molecular and Macromolecular Studies, Polish Academy of Sciences, Sienkiewicza 112, PL-90 363 Lodz, Poland

William J. Gauthier
Research and Technology Center, Fina Oil and Chemical Company, PO Box 1200, Deer Park, Texas 77536, USA

G.P. Guidetti
Montell Polyolefins, G. Natta Research Centre, P. le Donegani 12, I-44100 Ferrara, Italy

Archie E. Hamielec
Department of Chemical Engineering, McMaster University, Hamilton, Ontario L8S 4L7, Canada

K. Hammerschmid
PCD Polymere GmbH, PO Box 675, A-4021 Linz, Austria

Eddy W. Hansen
SINTEF Materials Technology, PO Box 124 Blindern, N-0314 Oslo 3, Norway

E. Harkin-Jones
Department of Chemical Engineering, The Queen's University of Belfast, Ashby Building, Stranmillis Road, Belfast BT9 5AH, Northern Ireland

T. Harmia
Institut für Verbundwerkstoffe GmbH, Erwin-Schrödinger-Str. 58, D-67663 Kaiserslautern, Germany

D.R. Hartman
Owens-Corning Science and Technology Center, 2790 Columbus Road, Route 16, Granville, Ohio 43023-1200, USA

Markku T. Heino
NK Cables Ltd, PO Box 419, FIN-00101 Helsinki, Finland

Pavol Hodul
Department of Fibres and Textile Chemistry, Faculty of Chemical Technology, Slovak University of Technology, Radlinského 9, SK-812 37 Bratislava, Slovak Republic

J.A. Horas
Instituto de Matematica Aplicada San Luis, Departamento de Fisica, Facultad de Ciencias Fisico Matematicas y Naturales, Universidad Nacional de San Luis, Ejercito de los Andes 950, 5700 San Luis, Argentina

Martin Jambrich
Department of Fibres and Textile Chemistry, Faculty of Chemical Technology, Slovak University of Technology, Radlinského 9, SK-812 37 Bratislava, Slovak Republic

Just Jansz
Montell Benelux bv, Westelijke Rondwegl, 4791 RS Klundert, The Netherlands

P.K. Järvelä
Tampere University of Technology, Institute of Materials Science and Plastics Technology, PO Box 589, FIN-33101 Tampere, Finland

T.P.A Järvelä
Tampere University of Technology, Institute of Materials Science and Plastics Technology, PO Box 589, FIN-33101 Tampere, Finland

Gürhan Kalay
Wolfson Centre for Materials Processing, Department of Materials Engineering, Brunel University, Uxbridge, Middlesex UB8 3PH, UK

I. Karacan
Zorlu Holding, Korteks, Organize Sanayi Bolgesi, Sari Caddesi No 3, Bursa, Turkey

J. Karger-Kocsis
Institut für Verbundwerkstoffe GmbH, Universität Kaiserslautern, PO Box 3049, D-67653 Kaiserslautern, Germany; Technical University of Budapest, Hungary

S. Kerth
BASF AG, ZEW/BB-L443, D-67056 Ludwigshafen, Germany

R.J. Koopmans
Dow Benelux N.V., PO Box 48, NL-4530 AA Terneuzen, The Netherlands

J. Kressler
Fachbereich Werkstoffwissenschaften, Institut für Werkstoffwissenschaft, Martin-Luther-Universität Halle-Wittenberg, D-06099 Halle (Saale), Germany

Toshio Kunugi
890 Kitamiyaji, Kamiyama-cho, Nirasaki-shi, 407-0042, Japan

Toshio Kurauchi
Toyota Central Research and Development Laboratories Inc., Nagakute-cho, Aichi-gun, Aichi, 480–11, Japan

Francesco Paolo La Mantia
Dipartimento di Ingegneria Chimica dei Processi e dei Materiali, Universita di Palermo, Viale delle Scienze, I-90128 Palermo, Italy

Michel Le Bras
Laboratoire de Genie des Procedes d'Interactions Fluides Reactifs-Materiaux, Ecole Nationale Superieure de Chimie de Lille, BP 108, F-59652 Villeneuve d'Ascq Cedex, France

K. Lederer
Institut für Chemie der Kunststoffe, Montanuniversität Leoben, A-8700 Leoben, Austria

H. Ledwinka
PCD Polymere GmbH, PO Box 675, A-4021 Linz, Austria

N. Legros
Industrial Materials Institute, National Research Council Canada, 75 Boulevard de Mortagne, Boucherville, Québec, J4B 6Y4, Canada

David J. Lohse
Exxon Research and Engineering Co., Route 22 East, Annandale, New Jersey 08801, USA

Bernard Lotz
Institut Charles Sadron, 6 rue Boussingault, F-67083 Strasbourg Cedex, France

A. Luciani
Industrial Materials Institute, National Research Council Canada, 75 Boulevard de Mortagne, Boucherville, Québec J4B 6Y4, Canada; École Polytechnique Fédérale de Lausanne, Laboratoires de Technologie des Composites et Polymères, LTC-DMX-EPFL, CH-1015 Lausanne, Switzerland

A. Lutz
Institut für Verbundwerkstoffe GmbH, Erwin-Schrödinger-Str. 58, D-67663 Kaiserslautern, Germany

N.J. Macauley
Department of Chemical Engineering, The Queen's University of Belfast, Belfast BT7 1NN, Northern Ireland

J.H. Magill
Department of Materials Science and Engineering, School of Engineering, University of Pittsburgh, Pittsburgh, Pennsylvania 15261, USA

M. Maier
Institut für Verbundwerkstoffe GmbH, Universität Kaiserslautern, D-67663 Kaiserslautern, Germany

J.-A E. Månson
Laboratoire de Technologie des Composites et Polymères, École Polytechnique Fèderale de Lausanne, CH-1015 Lausanne, Switzerland

Anton Marcinčin
Department of Fibres and Textile Chemistry, Faculty of Chemical Technology, Slovak University of Technology, Radlinského 9, SK-812 37 Bratislava, Slovak Republic

T.A. Martin
Composites Research Group, Department of Mechanical Engineering, University of Auckland, Private Bag 92019, Auckland, New Zealand

T. Matsuoka
Polymer Processing Laboratory, Toyota Central Research and Development Laboratories Inc., Nagakute-cho, Aichi-gun, Aichi-ken 480–1192, Japan

W. John G. McCulloch
J&M Laboratories, 12 J&M Drive, Dawsonville, GA 30534, USA

A. Meddad
Centre de recherche appliquée sur les polymères (CRASP), École Polytechnique de Montréal, PO Box 6079, Station Centre Ville, Montréal, Québec H3C 3A7, Canada

S.V. Meille
Dipartimento di Chimica, Politecnico di Milano, Via Mancinelli 7-20131 Milano, Italy

D. Meyer
Hohenstaufenstr. 11, D-73349 Wiesensteig, Germany

W. Michaeli
Institut für Kunststoffverarbeitung, RWTH Aachen, Pontstr. 49, D-52062 Aachen, Germany

G.H. Michler
Institut für Werkstoffwissenschaft, Martin-Luther-Universität Halle-Wittenberg, Geusaer Str., D-06217 Merseburg, Germany

Klaus-Peter Mieck
Thüringisches Institut für Textil- und Kunststoff-Forschung e.V., D-07407 Rudolstadt-Schwarza, Germany

Yukio Mizutani
Tokuyama Corp., Fujisawa Research Laboratory, 2023–1 Endo, Fujisawa City, Kanagawa 252, Japan

F. Möller
Institut für Verbundwerkstoffe GmbH, Universität Kaiserslautern, D-67663 Kaiserslautern, Germany

R. Mülhaupt
Freiburger Materialforschungzentrum und Institut für Makromolekulare Chemie der Albert-Ludwigs-Universität Freiburg, Stefan-Meier-Str. 31, D-79104 Freiburg, Germany

W.R. Murphy
Department of Chemical Engineering, The Queen's University of Belfast, Belfast BT7 1NN, Northern Ireland

Ikuo Narisawa
Materials Science and Engineering, Yamagata University, Yonezawa, Yamagata 992, Japan

I. Naundorf
ITT Cannon GmbH, Cannostr. D-71384 Weinstadt, Germany

W. Neißl
PCD Polymere GmbH, PO Box 675, A-4021 Linz, Austria

I. Novak
Polymer Institute, Slovak Academy of Sciences, SK-842 36 Bratislava, Slovakia

U. Panzer
PCD Polymere GmbH, PO Box 675, A-4021 Linz, Austria

B. Pietsch
Ciba Specialty Chemicals Inc., PO Box, CH-4002 Basel, Switzerland

Béla Pukánszky
Department of Plastics and Rubber Technology, Technical University of Budapest, PO Box 92, H-1521 Budapest, Hungary; Central Research Institute for Chemistry, Hungarian Academy of Sciences, PO Box 17, H-1525 Budapest, Hungary

M. Rätzsch
PCD Polymere GmbH, PO Box 675, A-4021 Linz, Austria

Keith Redford
SINTEF Materials Technology, PO Box 124 Blindern, N-0314 Oslo 3, Norway

G.L. Rigosi
Montell Polyolefins, G. Natta Research Centre, P. le Donegani 12, I-44100 Ferrara, Italy

M.G. Rizzotto
Instituto de Matematica Aplicada San Luis, Departamento de Fisica, Facultad de Ciencias Fisico Matematicas y Naturales, Universidad Nacional de San Luis, Ejercito de los Andes 950, 5700 San Luis, Argentina

Norio Sato
Toyota Central Research and Development Laboratories Inc., Nagakute-cho, Aichi-gun, Aichi, 480–11, Japan

K. Schäfer
Barmag AG, PO Box 11 02 40, D-42862 Remscheid, Germany

M.J. Schneider
Freiburger Materialforschungzentrum und Institut für Makromolekulare Chemie der Albert-Ludwigs-Universität Freiburg, Stefan-Meier-Str. 21, D-79104 Freiburg, Germany

Jukka V. Seppäl
Nokia Cables, PO Box 419, FIN-00101 Helsinki, Finland

T.J. Shields
Fire SERT Centre, University of Ulster, 75 Belfast Road, Carrickfergus, Co. Antrim BT38 8PH, Northern Ireland

João B.P. Soares
Department of Chemical Engineering, University of Waterloo, Waterloo, Ontario N2L 3G1, Canada

J.E. Spruiell
Materials Science and Engineering, College of Engineering, University of Tennessee, Knoxville, Tennessee 37996–2200, USA

M. Steiner
Institut für Verbundwerkstoffe GmbH, Universität Kaiserslautern, Erwin-Schrödinger-Str. 58, D-67663 Kaiserslautern, Germany

Tomasz Sterzynski
Department Polymeres-ECPM, Université Louis Pasteur Strasbourg I, 4 rue Boussingault, F-6700 Strasbourg, France; Poznan University of Technology, Institute of Chemical Technology, PL-60–965 Poznan, Poland

J. Suhm
Freiburger Materialforschungzentrum und Institut für Makromolekulare Chemie der Albert-Ludwigs-Universität Freiburg, Stefan-Meier-Str. 21, D-79104 Freiburg, Germany

Hidero Takahashi
Toyota Central Research and Development Laboratories Inc., Nagakute-cho, Aichi-gun, Aichi, 480–11, Japan

Conchita V. Tran
Union Carbide (Europe) SA, 7 Rue du Pré-Bouvier, CH-1217 Meyrin, Switzerland

J.L. Thomason
Owens-Corning Science and Technology Center, 2790 Columbus Road, Route 16, Granville, Ohio 43023–1200, USA

J. Ulcej
Extrusion Dies Inc., 911 Kurth Road, Chippewa Falls, Wisconsin 54729–1443, USA

L.A. Utracki
Industrial Materials Institute, National Research Council Canada, 75 Boulevard de Mortagne, Boucherville, Québec J4B 6Y4, Canada

G.J. Vancso
Materials Science and Technology of Polymers, Faculty of Chemical Technology, University of Twente, PO Box 217, NL-7500 AE Enschede, The Netherlands

József Varga
Department of Plastics and Rubber Technology, Technical University of Budapest, PO Box 92, H-1521 Budapest, Hungary; Lehrstuhl für Kunststofftechnik, Universität Erlangen-Nürnberg, Am Weichselgarten 9, D-91058 Erlangen-Tennenlohe, Germany

D. Vesely
Department of Materials Engineering, Brunel University, Uxbridge, Middlesex, UB8 3PH, UK

M.H. Wagner
Institut für Kunststofftechnologie der Universität Stuttgart, Böblinger Str. 70, D-70199 Stuttgart, Germany

Michael Wehmann
J&M Laboratories, Kolpingstr. 34a, D-63150 Heusenstamm, Germany

J.R. White
Department of Mechanical, Materials and Manufacturing Engineering, University of Newcastle, Newcastle upon Tyne NE1 7RU, UK

Jean Claude Wittmann
Institut Charles Sadron, 6 rue Boussingault, F-67083 Strasbourg Cedex, France

M. Xanthos
Polymer Processing Institute at Stevens Institute of Technology, Castle Point on Hudson, Hoboken, New Jersey 07030, USA; Department of Chemical Engineering, Chemistry and Environmental Science, New Jersey Institute of Technology, University Heights, Newark, New Jersey NJ 07102, USA

J. Zhang
Fire Research Centre, University of Ulster at Jordanstown, Carrickfergus, Co. Antrim BT38 8PH, Northern Ireland

Édesapám emlékének
In memory of my father

Preface

My heart sank when I was approached by Dr Hastings and by Professor Briggs (Senior Editor of Materials Science and Technology and Series Editor of Polymer Science and Technology Series at Chapman & Hall, respectively) to edit a book with the provisional title *Handbook of Polypropylene*. My reluctance was due to the fact that my former book [1] along with that of Moore [2], issued in the meantime, seemed to cover the information demand on polypropylene and related systems. Encouraged, however, by some colleagues (the new generation of scientists and engineers needs a good reference book with easy information retrieval, and the development with metallocene catalysts deserves a new update!), I started on this venture.

Having some experience with polypropylene systems and being aware of the current literature, it was easy to settle the titles for the book chapters and also to select and approach the most suitable potential contributors. Fortunately, many of my first-choice authors accepted the invitation to contribute.

Like all editors of multi-author volumes, I recognize that obtaining contributors follows an S-type curve of asymptotic saturation when the number of willing contributors is plotted as a function of time. The saturation point is, however, never reached and as a consequence, Dear Reader, you will also find some topics of some relevance which are not explicitly treated in this book (but, believe me, I have considered them). On the other hand, I am quite sure that even for those missing themes you will find valuable notes and hints in the many chapters of this book.

During my editing I have had considerable support from some colleagues, whom I want to give credit here: Professor Utracki (CNR, Boucherville, Canada), Dr Neißl (PCD, Linz, Austria) and Professor Mülhaupt (FMF, Freiburg, Germany). They ensured the delivery of several chapters by 'persuading' their staff and coworkers to contribute.

Thanks are, however, due to all contributors for their efficient work. The outcome of all the effort is represented by this book.

The editor wishes to thank David Mackin and the staff at GreenGate Publishing Services for their work in helping to produce the book.

I strongly hope that you, Dear Reader, will be satisfied with it!

A.m.D.g

József Karger-Kocsis
Kaiserslautern and Budapest
October, 1997

REFERENCES

1. Karger-Kocsis, J. (ed.) (1995) *Polypropylene: Structure, Blends and Composites*, Vols. 1–3, Chapman & Hall, London.
2. Moore, Jr., E.P. (ed.) (1996) *Polypropylene Handbook*, Hanser, Munich.

Adhesive bonding of polypropylene

W. Brockmann

INTRODUCTION

Adhesive bonding by definition is the joining of materials with an organic adhesive as an interlayer which adheres on the substrates without macroscopical changing of the material's state.

Adhesion between different materials is created by physical or chemical bonds between the adhesive and the substrate. However, physical or chemical bonds are effective in the nanometer range. As a consequence, good adhesion can only be produced if adhesive and substrate come tight in contact or, in other words, if the adhesive wets the never absolutely flat surface of the substrate as well as possible. Sufficient wetting of a solid substance by a fluid (adhesive) is only possible if the surface energy of the liquid is equal or lower than the surface energy of the substrate [1]. Without discussing details of the theory of adhesion, it must be stated that wetting is a necessary but not sufficient condition for adhesion.

The basic problem of adhesive bonding of polypropylene (PP) is its low surface tension, in the range between 29 and 35 mN/m. An absolute value is not given here, because the surface energy and also the surface state of PP depends, like the surface tension of all materials, on the history of surface creation. Surface energies of adhesives are normally higher than 35 mN/m and can go up on to the surface energy of water (72 mN/m). As a consequence, normally PP in an untreated state is not wettable by adhesives and in an untreated state not bondable. The only

Polypropylene: An A–Z Reference
Edited by J. Karger-Kocsis
Published in 1999 by Kluwer Publishers, Dordrecht. ISBN 0 412 80200 7

exceptions are pressure-sensitive adhesives, whose properties are discussed later.

If surface energy as an average value is divided into dispersed energy and polar energy, PP shows a very low polar part. The reason for this effect is an absence of polar groups in the molecules of PP and thus a low chemical reactivity of PP surfaces. So principally, on a PP surface in an untreated state (also in the case of wetting by the adhesive), only physical bonds as interactions of very low binding energy are created and as a consequence adhesion between adhesive and substrate is very low.

SURFACE TREATMENT OF POLYPROPYLENE

Three ways exist to create improved adhesional properties of a PP surface.

The first way is a particularly developed roughening method which in the literature [2] is described as 'creating skeleton' which can be translated as 'whiskerizing'. In this process, the PP surface is heated up into the softening area near the melting point. Into this surface, a cotton fabric is pressed and shortly after pressing peeled off from the heated surface in a similar way to the peel ply process used for fiber reinforced plastics. The peel-off process produces an extremely whiskerized PP structure which remains bondable for a very long time. It is assumed that in this type of surface, which can be also created on polyethylene (PE) and other plastic materials, the adhesive adheres by micromechanical interlocking and not by physical or chemical interactions. So far as known, this process is the only roughening technique available for improving adhesional properties of PP. All other methods, such as grit-blasting, grinding or high-pressure water spraying, do not lead to significant improvements of the adhesional properties.

The second type of surface treatment is in principle an improving of the chemical reactivity of the substrates by chemical or physical treatment methods. All the systems known lead to oxidizing effects of the nonpolar substrate and create carbonyl, carboxyl and hydroxyl groups at the surface which are polar and partly chemically reactive. The existence of these groups can be measured for example by Fourier transform infrared (FTIR) spectroscopy using the attenuated total reflexion (ATR) technique.

The oldest treatment is an etching process in highly concentrated chromic-sulfuric acid with a temperature near 80°C. Highest adhesional strength between PP and, for example, a two-component epoxy (EP) adhesive is reached after etching times between 120 and 240 s. The recipe of the etching solution is 5078 parts of sulfuric acid (density 1.82 g/ml), 120 parts demineralized water and 75 parts potassium dichromate.

Today, such a process is not recommended because the etching solution contains chromium VI ions which are carcinogenic. Beside this fact, the etching process must be followed by very intensive rinsing and drying processes. The effectiveness of the surface treatment, on the other hand, is very good.

The simplest physical treatment of PP is the so called corona treating process (see the chapter 'Surface treatment of polypropylene by corona discharge and flame' in this book). In this case, the PP part is connected with an electrical conductive carrier. At a distance of some millimeters or centimeters over the surface, another electrode is placed and between electrode and electrical conductive carrier a high frequency field (~ 6 kV) is produced. In this field, electrical discharge effects occur, the air between electrode and the PP surface is ionized and probably some ozone (O_3) is also created. Under these conditions, in times between ½ and 1½ s, the surface is oxidized sufficient for good adhesion. Morphological changes of the surface are not observable. The disadvantage of this surface treatment is that only the surface within range of the electrode can be treated.

Another old treatment method for PP is the flaming process. As a standard gas for the burner, a mixture of propane and air with a mixture ratio of 1–20 or a bit higher air content is used. The flame should not burn or melt parts of the surface. The burner should be guided over the surface at a distance of approximately 20 mm or more with a velocity of 0.12 m/s. Under these conditions with the propane air mixture ratio of 1–20, the best surface properties are produced. Also in this case, the effect is an oxidizing mechanism. With this method only the parts of the surface which are in direct contact with the flame are treated.

Better effects on complicated PP parts are produced by a low-pressure plasma treatment in oxidizing gases, such as oxygen. The disadvantage of the plasma process (pressure 0.01–1 mbar, 10 Mhz, 500 W) is that the plasma chamber after the positioning of the parts must be evacuated and filled with the plasma gas. This process is discontinuous and needs approximately 1.5 min or more for a filled chamber (see also the chapter 'Surface modification of polypropylene by plasma' in this book).

Beside this plasma treatment in a chamber, some variations of plasma treatments using a plasma burner as in a welding process or using a plasma gun are known and described in [2].

Further surface treatment methods for PP such as fluorination exist and will be developed in the future.

The third way of improving the adhesional properties of polypropylene is a coating process. It is called SACO process. In a first stage, the surface is roughened by a grit blasting process using corundum as grit blasting material. In a second step, a grit blasting process is used with silicate coated corundum. From this silicate layer on top of the alumina, some parts adhere to the PP surface very strongly. The binding mechanism of these silicate parts on the polypropylene surface is not well known. In a following step, this silicated PP is coated with a reactive silane as adhesion promoter and on this silanized surface a very good adhesion to adhesives can be achieved. The process is relatively complicated but very efficient. Normally, it is used to treat small parts.

In summary, it can be stated that a large number of different surface treatments exist to improve the adhesional properties of PP. It should be stated here that some manufacturers of cyanoacrylate adhesives recommend special chemical primers which produce adhesion on PP without other treatment. In the experience of the author, this type of primer improves the adhesion but produces bonded joints that are insufficiently durable, especially under humid environmental conditions.

ADHESIVES FOR POLYPROPYLENE

In principle, all physical and chemical curing or hardening types of adhesives, such as contact adhesives, hot melts, acrylics, epoxies, polyurethanes, etc., can be used to join PP in a pretreated state without problems.

One exception of this statement must be noted. These are the so-called pressure-sensitive adhesives which are normally liquids of very high viscosity which do not cure and which are in some cases poorly cross-linked. They adhere on different substrates only by physical bonds and due to the high mobility of their molecules they can 'repair' destroyed bonds in time. This type of adhesion which is called 'dynamic adhesion' meaning a 'centipoid effect' is unique for this class of adhesives and can be used partly on untreated PP for nonstructural bonds. The adhesion between the pressure-sensitive adhesive and the PP surface in an untreated state is not very high but relatively unproblematic.

STRENGTH AND DURABILITY OF BONDED POLYPROPYLENE JOINTS

The strength of bonded PP joints can be measured in accordance with the well-known testing standards such as, for example, DIN 53281, 53282, 53383 or 53289. In some cases, the thickness of the adherends should be different from that used on normal metal parts because the strength of PP is lower. For example, in the case of shear tests with single overlapped joints, the thickness of the PP parts should be 4 mm or more. Also the aging conditions can be in accordance with standards such as DIN 53286 or ASDM B117 (salt spray test). Typical test methods for bonded

Table 1 Tensile shear strength of single overlapped PP joints

Adherent thickness:	4 mm
Overlap length:	10 mm
Glue line thickness:	0.2 mm
Surface treatment:	etched in chromic sulfuric acid
Adhesive:	2 component epoxy (Metallon PA)
Initial shear strength:	1.4 N/mm^2
Shear strength after 1200 h 40/95:	1.2 N/mm^2
Adhesive:	2 components polyurethane (scotch weld 35 32)
Initial shear strength:	2.1 N/mm^2
Shear strength after 1200 h 40/95:	2.1 N/mm^2
Shear strength after 5000 h nat. cl.†:	2.2 N/mm^2

40/95: artificial climate, 40°C and 95% rel. hum. †nat.cl.: natural climate in North Germany.

joints in general are described in [3]. They all can be used in a modified or nonmodified mode for testing bonded PP joints.

A general problem with bonded joints is the moderate durability, particularly in humid environments. This is less pronounced for adhesive bonded PP joints if the surfaces of the substrates are treated accordingly. Table 1 gives some strength values measured with PP joints prepared by using epoxy and polyurethane adhesives (both two-component cold curing systems) respectively. Table 1 lists the initial and the residual strength after different aging procedures. The results show that sensitivity against humidity does not exist.

Other investigations [2], with aging times of 2½ and 3 years under different artificial and natural conditions, show that on the PP side of bonded joints (in that case, PP steel joints) no remarkable changes in the adhesional area occur. So it can be stated that the durability of adhesive bonding of PP is no problem when the surface of the PP is treated as mentioned earlier.

In summary, adhesive bonding of PP is not a particular problem and the design of bonded joints of this material can follow the recommendations given, for example, in [2, 3]. Compared with other joining techniques, such as welding, the advantage of bonding technology is the possibility of joining PP with practically all other materials and creating large joining areas without problems (for example, in sandwich systems, which are gas and watertight and, what is of importance, are free of residual stresses or stress concentrations which can produce problems in design of parts made of thermoplastic polymers). The only problem of bonded joints is that they are difficult to separate and difficult to repair.

REFERENCES

1. Brockmann, W. (1978) *Das Kleben chemisch beständiger Kunststoffe Adhäsion* **22**, 38–44, 80–86, 100–103.
2. Brockmann, W., Dorn, L. and Käufer, H. (1989) *Kleben von Kunststoff mit Metall*, Springer-Verlag, Berlin.
3. Habenicht, G. (1997) *Kleben*, 3rd edn, Springer-Verlag, Berlin.

Keywords: adhesion, adhesive bonding, surface treatment, corona treatment, flame treatment, chemical etching, surface tension, plasma treatment, pressure-sensitive adhesive, adhesive bond strength, adhesive bond durability.

Amorphous or atactic polypropylene

József Karger-Kocsis

INTRODUCTION

Amorphous polypropylene (aPP) is characterized by a random steric orientation of the methyl pendant groups on the tertiary carbon atoms along the molecular chain. The random sequence of these methyl substituents is linked to an atactic configuration. Due to its fully amorphous nature, aPP is easy soluble (even at ambient temperatures) in a great number of aliphatic and aromatic hydrocarbons, esters and other solvents in contrast to the isotactic PP (iPP) of semicrystalline feature.

PRODUCTION [1–2]

Until recently, aPP was obtained as a byproduct during polymerization of propylene for iPP. In the early slurry polymerizations processes, the amount of aPP 'coproduced' was between 2 and 10 wt.% depending on the iPP grade and Ziegler–Natta catalyst type used (see also the chapter 'Industrial polymerization processes'). The removal of aPP from the kerosene or low boiling hydrocarbon diluents (e.g. hexane, heptane) via evaporation made the related processes rather costly. It should be emphasized here that the aPP obtained by this way was never fully amorphous and atactic, but a mixture of PPs with various tacticity and molecular weight (MW). The characteristics and quality of the aPP grades depend mostly on the target iPP polymerized, including also its sensitivity to some changes in the polymerization process.

Polypropylene: An A–Z Reference
Edited by J. Karger-Kocsis
Published in 1999 by Kluwer Publishers, Dordrecht. ISBN 0 412 80200 7

The second- and higher-generation catalysts with improved stereospecificity resulted in much smaller aPP yield so that it should not be removed from the iPP products anymore. Due to this development, the aPP grades offered by the iPP producers at that time (e.g. Epolene® of Eastman, USA; AFAX® of Hercules, USA; Daplen® APP of Chemie Linz, Austria; Vestolen® APP of Veba, Germany; Tipplen® APP A, B and C of Tisza Chemical Works, Hungary) disappeared from the market. In the meantime, however, work was undertaken to find useful and cost efficient uses for aPP and, for some of them, aPP became indispensable. As aPP became less available, the markets with successful aPP use created a considerable demand. So, some companies instead of closing older plants converted them to produce aPP as the main product. However, the history of aPP was not completed by aPP turning from a useless byproduct to a desired, well-selling target polymer. A new age of the aPP history is due to recent R&D activities in the field of metallocene-catalyzed polymerization. This revolutionary polymerization technique allows us to produce aPP types of high molecular weight (HMW) being in the range of several hundreds kg/mol. By contrast, the mean MW of the aPP byproducts is one order of magnitude less (few tens kg/mol; Table 1). The increase in the MW is associated with the appearance of new properties, such as rubberlike elasticity.

PROPERTIES

Commercialized byproducts

At ambient temperatures, aPPs are waxy, slightly tacky solids of white or yellowish color. They become softer and more tacky with increasing temperature. The iPP producers recognized early that the commercial success of aPP depends on whether or not their quality is consistent, i.e the properties can be guaranteed within acceptable limits. The easiest way to fulfil this requirement was to select and offer aPP fractions which were extracted from iPP grades produced in larger quantities. For example, the Tipplen APP A, B and C grades of Tisza Chemical Works using the Hercules slurry technology were byproducts of fiber and injection-moldable homopolymers and injection and extrusion moldable iPP block copolymers, respectively. Despite this product philosophy, the properties of the commercialized aPPs had a rather large scatter. The rationale behind this fact was that the quality control including the necessary feedback was tailored for the main iPP and not for the aPP coproduct. As mentioned earlier, the aPP grades are not fully amorphous. Depending on the polymerization and aPP/iPP separation techniques, the crystalline content of the aPP may reach 15 wt.%. The crystalline

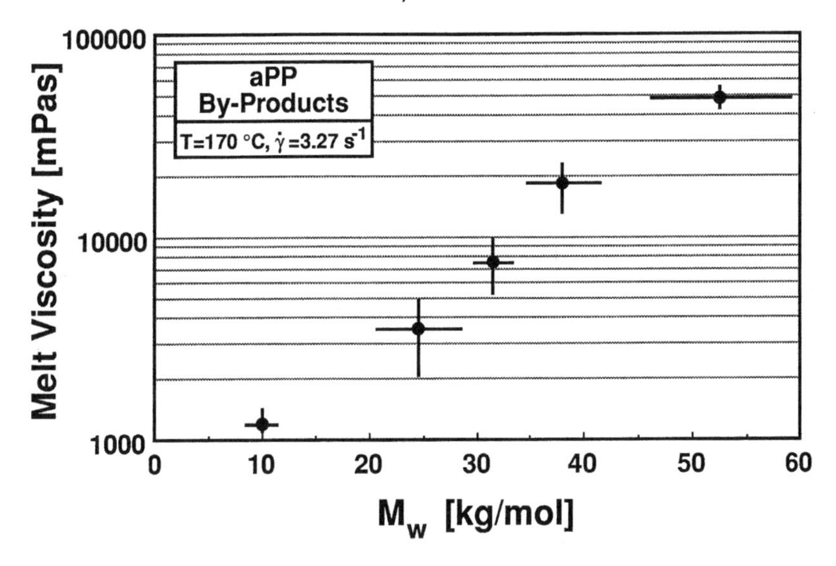

Figure 1 Melt viscosity (rotation viscosimeter, shear rate: $3.27\,s^{-1}$) in function of the weight-average molecular weight (M_w) for various aPP fractions.

content is generally given by the fraction insoluble in boiling n-heptane (C7). Apart from the crystalline content, other impurities (residual solvent, catalyst traces) and additives (e.g. thermooxidative stabilizers) affect the physico-mechanical performance (i.e. flash temperature, softening point, thermal stability) and appearance of the aPP. Due to broad limits on the properties and lacking information on molecular characteristics and crystallinity, general trends in the structure–property relationships are not easy to discern.

Nevertheless, it can be claimed that:

- The melt viscosity increases with increasing MW and in addition, strongly depends on the actual shear rate (γ) (Figure 1).
- The softening point (measured usually by the ring and ball method) is very sensitive to the crystalline fraction of the aPP.
- The tensile strength (lying in the range of 0.2–0.7 MPa) is less dependent, the strain (and also other ductility values), however, strongly depends on the MW and crystallinity.

Due to its tacky consistency, aPP is available mostly in block form. In order to reduce the tackiness and blocking properties, spraying by chalk or talc powders or wrapping in silicon-coated papers are widely used. Pelletized, granulated aPP grades are either of highly crystalline nature or compounded, modified grades.

Table 1 Properties of aPP grades extracted (byproducts) and synthesized (HMW-aPP)

Property	*Test Method or ASTM standard*	*Unit*	*Commercial Byproduct*	*Metallocene-synthesized HMW-aPP*
Density	D1505	gcm^{-3}	>0.86	≈0.85
Shore A hardness	D2240	°	>50	35–45
Intrinsic viscosity	D2857	dl/g	0.4–0.6	1.5–3.5
M_w	GPC	kg/mol	20–80	400–700
Polydispersity (M_w/M_n)	GPC		5–7	2.3–3.3
MFI	D1238	dg/min	>500	<5
T_g	DSC	°C	−25–+5	≈0
T_m	DSC	°C	>140	–
Tensile strength	D638	MPa	<0.7	>1.0
Elongation at break	D638	%	<50	>1000
Compression set (RT)	D395	%	>60	<20

Designations: M_w and M_n, weight- and number-average molecular weight, respectively; MFI, melt flow index (230°C, 2.16 kg); Tg and Tm, glass transition and melting temperature, respectively; GPC, gel permation chromatography; DSC, differential scanning calorimetry.

Metallocene-synthesized aPP

It is worthwhile to collate (Table 1) the basic properties of the aPPs gained as by- (via separation) and main products (HMW-aPP; via metallocene-polymerization) respectively.

The data in Table 1 hint that HMW-aPP exhibits rubbery characteristics. In addition, the synthesized HMW-aPP grades have higher melt viscosity and are less shear rate dependent than the traditional aPP [2].

APPLICATIONS [3–4]

The potential use of aPP was explored in the years 1960–1970. One of the most important applications, that remains until now, is the modification of bitumen. By adding aPP, the softening point is increasing, whereas the ductility decreases. In respect to penetration, the response is strongly affected by the crystalline fraction of the aPP. aPP is much less compatible with blown bitumina having higher polarity. aPP is melt blended also to asphalt and related compounds for the production of various items, such as sound-deadening and roofing sheets, paving compounds. The penetration of aPP into this market is favored by the fact that aPP can be heavily filled. This property is exploited in formulations for self-laying carpets (backing for parquet flooring).

The other main application field of aPP is related to its tackiness. Papers are laminated by roll-coating with aPP-based adhesives. The aPP layer between the papers imparts not only good adhesion but also barrier properties. aPP and related materials can be used to laminate paper to other substrates such as polymer films. aPP may serve also as the basic ingredient of hot-melt adhesives (HMA), even in HMA with some pressure sensitivity.

Many applications are known for aPP in combination with rubbers. aPP has an excellent compatibility with ethylene/propylene and ethylene/propylene/diene rubbers (EPM and EPDM, respectively). Incorporation of aPP in EPM- or EPDM-based rubber recipes is accompanied with the following advantages:

- lowering the Mooney viscosity. This means that less energy is required for mixing the rubber batch;
- improving the extrudability (including die swell properties – see related chapter Die swell or extrudate swell) and surface appearance of the extrudates;
- high uptake of inert and reinforcing fillers.

A thermoplastic dynamic vulcanizate (see also the chapter 'Thermoplastic dynamic vulcanizates') was also developed by using blends of aPP and EPM,EPDM rubber. The product offered earlier (Tauropren® of Taurus Hungarian Rubber Works) was recommended for impact toughening of iPP [5]. In caulking compounds and sealants, aPP usually replaces butyl rubber (IIR). Other notable applications of aPP are as release agents for concretes, dispersing additives for masterbatches (for pigmentation or machinery cleaning), and viscosity modifier for lubricants, oils, inks.

Interestingly, aPP found less use in blends with thermoplastic polymers irrespective of the high number of related patents. The new generation aPP, produced by metallocene synthesis, may change this scenario (see also the chapter 'Elastomeric polypropylene homopolymers using metallocene catalysts'). Blending of HMW-aPP with polyolefins may result in a new mechanical property profile, e.g. rubberlike resilience. Such products may compete with those made of flexible PVC.

The future of aPP and related products seem to depend on the expected breakthrough with the metallocene synthesis.

REFERENCES

1. Liebermann, R.B. and LeNoir, R.T. (1996) Manufacturing, in *Polypropylene Handbook*, Chap. 8, (ed. E.P. Moore), Hanser, Munich, pp. 287–301.
2. Resconi, L. and Silvestri, R. (1996) Polypropyleneatactic, in *Polymeric Materials Encyclopedia*, Vol. 9, (ed. J.C. Salamone), CRC Press, Boca Raton, FL, USA, pp. 6609–6615.

3. Dörrscheidt, W., Hahmann, O., Kehr, H., Nising, W. and Potthoff, P. (1976) Polyolefine, *Kunststoffe-German Plastics*, **66**, 567–574.
4. von Bramer, P.T. (1975) Amorphous polypropylene. Its properties and uses, *Adhesives Age*, July, 15–20.
5. Karger-Kocsis, J., Kozma, B. and Schober, M. (1985) Tauroprene – a new versatile polyolefinic thermoplastic rubber. *Kautschuk Gummi Kunststoffe* **38**, 614–616.

Keywords: amorphous polypropylene (aPP), atactic polypropylene (aPP), additives, hot-melt adhesive, rubber extender, bitumen modifier, metallocene synthesis, molecular weight, properties of aPP, applications of aPP, thermoplastic dynamic vulcanizate (TDV).

Anticorrosion coatings with polypropylene

G.L. Rigosi and G.P. Guidetti

Polypropylene (PP) is utilized as protective coating against corrosion using mainly two technologies: application of powder and extrusion.

APPLICATION OF POWDER

A PP copolymer, modified with grafted polar groups, is used having a high melt flow rate (50–100 dg/min according to ASTM D 1238 conditions L at 230°C, 2.16 kg) in order to obtain a good flow of the melted material during the application process. This PP grade is ground at low temperature using cryogenic mills, obtaining a particle size generally less than 300 μm.

The powder is sprayed on the cleaned metal surface by means of electrostatic guns and then the item is placed in an oven at a temperature of about 200°C to melt the powder. With this technology, coating thickness in the range of 100–200 μm can be obtained. In some cases, epoxy primers are used as a first layer to enhance the adhesion properties.

Alternatively, the metallic item can be heated up in an oven and then dipped in a fluidized bed of modified PP powder; a post-heating step may be included. With the fluidized bed process a coating of 200–600 μm can be applied. The main coating properties are reported in Table 1. As an example of this technology, PP powder is used to coat drums internally, allowing the safe transport of most chemical and foodstuff products.

Polypropylene: An A–Z Reference
Edited by J. Karger-Kocsis
Published in 1999 by Kluwer Publishers, Dordrecht. ISBN 0 412 80200 7

Table 1 Properties of the powder applied coating (epoxy primer and polypropylene) of 0.5 mm of total thickness

Property	*Method*	*Conditions*	*Value*
Cathodic disbonding	NF A 49 711	−1.5 V, 3% NaCl, 23°C, 28 days	220 mm^2
Hot water test	ASTM D 870	95°C, 1000 h, 5.8 g/l NaCl	4 mm
Penetration resistance	NF A 49 711	10 N/mm^2, 24 h, 23°C	0.05 mm
		90°C	0.17 mm
		110°C	0.4 mm
Impact resistance	NF A 49 711	25.4 mm, 10 KV	8 Nm

In the near future a large number of applications are expected to use PP powder coating. This is because of the good mechanical properties, improved adhesion, good barrier effect, excellent resistance to most of the inorganic and organic chemical reagents even at high temperature and relatively low cost of PP. Such applications include external coating of gas tanks, baskets of dishwasher machines, grids of refrigerators, metal items used for appliances and furniture.

EXTRUSION

Since the 1980s, PP has been introduced in pipeline coating due to its excellent ability to protect the steel against corrosion [1, 2]. The pipeline coating is called a three-layer coating because it is composed of a thin layer of epoxy resin, an intermediate layer of modified PP copolymer and a PP outer layer.

The function of the epoxy layer is to ensure a strong bond to the steel interface by interacting with its metal oxides. The epoxy has high resistance to cathodic disbondment and high thermal stability; its melt-flow behavior facilitates the creation of a thin but uniform film sufficient to fill the anchor pattern in the abraded metal surface.

The intermediate layer ensures a powerful bond between the epoxy primer and the external coating. This intermediate, adhesive layer consists of a thin (200–300 μm) coating of specially formulated PP copolymer with polar groups grafted onto the polymer backbone. These polar groups establish bonds with the epoxy layer while the affinity to PP creates a strong bond with the outer layer.

The external coating provides an additional mechanical strength to the system, since the coating is applied at a thickness of 1.5–2.5 mm depending on pipe diameter and pipe wall thickness. The application process of the coating system is a continuous process.

The pipe surface is cleaned by means of an automatic shot-blasting machine to minimum SA 2½ surface finish as per Swedish standard SIS

055900; the obtained surface roughness ranges from 40 to 80 μm. The pipes are than heated using an induction oven up to a temperature of 200–240°C and an epoxy primer is sprayed on the heated pipe in order to cover at least the steel roughness. About 300 μm of adhesive PP is applied by lateral extrusion before the epoxy is fully cured; the melt temperature of the adhesive is generally in the 190–210°C range. The outer coating is than laterally extruded with a number of overlapping layers such that the PP thickness meets the specification requirements. During this process, a pressure roller smooths the coating and avoids entrapping air bubbles and finally the coated pipe is cooled by spraying with water.

Different layouts can also be used to meet the best production speed according to the pipe diameter, applying, for example, the PP adhesive in powder form or using cross head type extruders instead of lateral extruders.

The three-layer PP system offers a high degree of impact resistance, peel strength especially at high temperatures, indentation resistance and flexibility. The main properties of the coating are reported in Table 2.

Using a suitable thermal stabilizer package, the PP coating can withstand a continuous operating temperature up to 120°C. In order to predict the lifetime of the thermal stabilized coating, accelerated tests are performed at 150°C in an oven with forced air and the Arrhenius theoretical equation is used:

$$\log t(x) = \log t(r) + 5540 \times (1/T(x) - 1/423)$$

Table 2 Properties of the three-layer polypropylene coating (−20 +120°C)

Property	*Method*	*Conditions*	*Value*
Peeling	NF A 49 711 DIN 30678	23°C	>40 N/mm
Peeling	NF A 49 711 DIN 30678	120°C	8 N/mm
Impact	NF A 49 711 DIN 30678	23°C	>15 J/mm
Indentation	NF A 49 711 DIN 30678	23°C	0.05 mm
Indentation	NF A 49 711 DIN 30678	110°C	<0.4 mm
Ultraviolet ageing	NF A 49 711	800 h	pass
Thermal ageing	NF A 49 711	1000 h at 150°C	pass
	DIN 30678	2400 h at 140°C	pass
Cathodic disbonding	NF A 49 711	28 days, 23°C	<7 mm

where $t(x)$ is the coating thermal resistance time at temperature $T(x)$ and $T(x)$ is expressed in K, $t(r)$ is the coating thermal resistance time at 150°C (423K).

It is also necessary to ultraviolet stabilize the coating because the pipes are stored outside before laying; in any case, if high continuous service temperatures are expected, it is advisable to reduce the direct sun exposure time and/or to give a coat of paint or to use other kinds of ultraviolet shielding.

PP coating is not affected by environmental stress cracking and provides very low water absorption thus increasing the reliability of the coating over time, especially in the case of sealines. The three-layer PP system shows excellent resistance to cathodic disbondment, well within international standards even at high temperature. The coating system also shows a high degree of resistance to bacterial and fungal attack.

SPECIAL COATINGS

Coatings for cold climates

Where pipelines are laid in very cold climates, the risk of impact damage due to embrittlement of the coating system becomes a key issue.

The recent development of extremely flexible PP grades has enabled the application of coating systems particularly suitable for cold climate [3]. The impact strength and flexibility of the components in these systems is maintained down as far as −45°C, allowing normal handling and laying operations (even reel-barge) at these temperatures. The low-temperature characteristics of these systems are achieved without sacrificing the performance at the other end of the scale; the coatings are capable of withstanding operating temperatures of up to 120°C.

These two special grades of adhesive and top coat can be applied using the standard technology obtaining a coating suitable to meet requirements of the cold regions of the world (Table 3). The peeling test shows a cohesive failure indicating that the adhesion forces to the epoxy are higher than the cohesive forces between the PP matrix phase and the ethylene–propylene rubber (EPR) particles being present as dispersed phase.

Anticorrosion and insulated coatings

Where a temperature drop in the transported fluid could result in phase separation or flow difficulties, the thermal insulation characteristics of the anticorrosion pipeline coatings are also specified.

Table 3 Properties of the low-temperature resistant polypropylene coating

Property	*Method*	*Conditions*	*Value*
Peeling	NF A 49 711 DIN 30678	23°C	15 N/mm
Peeling	NF A 49 711 DIN 30678	110°C	8 N/mm
Impact	NF A 49 711 DIN 30678	−45°C	>10 J/mm
Indentation	NF A 49 711 DIN 30678	23°C	<0.1 mm
Indentation	NF A 49 711 DIN 30678	110°C	<0.6 mm
Ultraviolet ageing	NF A 49 711	800 h	pass
Thermal ageing	NF A 49 711 DIN 30678	1000 h at 150°C 2400 h at 140°C	pass pass
Cathodic disbonding	NF A 49 711	28 days, 23°C	<7 mm

In PP coating systems, this is achieved by incorporating a layer of expanded PP between the intermediate layer and the outer coat [4] (Table 4).

The resultant coating is extremely robust and has a number of advantages:

- The 100% closed cell structure of the PP foam makes the coating extremely water resistant and resistant to hydrostatic pressure.
- The coating is highly flexible and therefore reel-barge laying is possible.
- The coating can be used in fishing areas as it will withstand impacts from trawl-boards.
- Low water absorption and high compression strength (both also at high temperatures) guarantees a constant thermal insulation.

Table 4 Properties of the thermal insulated layer

Property	*Method*	*Conditions*	*Value*
Density	Density bottle	23°C	0.7 g/cm^3
Thermal conductivity	ASTM C 518	23°C	0.16 w/mK
Water absorption	BS 903 A 16	170 h, 95°C, 12 bar	2.5%
Compressive modulus	ASTM D 695	23°C, 5% of deformation	300 MPa
Uniaxial creep	BS 903 A 15	1000 h, 105°C, 30 bar	3%

- Absence of chemical interaction with the sea water and thermal stability provides long coating life.
- The entire coating is applied in a single, in-line extrusion process.

Field joints, repairs, bends and fittings

Where pipelines are laid in sections welded in the field, the area around the welded joint must be protected against corrosion in the same way as the rest of the pipe sections. While coating of these areas must be carried out in the field, it is important that the protection at these sites should be comparable with that of the factory-applied coating, in order not to produce weak spots in the system.

Nowadays, field joints can be coated using a system similar to that applied in the factory [5]. After sand-blasting and heating using an induction heater, the weld area receives a coat of epoxy primer and a layer of powdered adhesive is sprayed on. The two PP layers are applied in form of a coextruded sheet. A small extruder is used to seal the seams between the edges of the joint and the main pipe coating.

Alternatively, a PP copolymer layer can be applied to the epoxy-primed, induction-heated joint by means of flame powder spray guns and thus built up to the required layer thickness. The flame powder spray gun technology is also used to coat curved sections of pipeline and fittings.

In case of damaged coating, where damage is superficial or confined to a small area, it may be repaired by welding using a hot-air gun in conjunction with PP adhesive material in stick form. For major damage, the field joint procedures should be used.

Recommendations, norms and specifications

In December 1988, the CEOCOR (Comite d'Etude de la Corrosion et de la Protection des Canalisations) commission approved a recommendation referred to the PP based coatings, applied by extrusion, as external corrosion protection of steel pipe.

Following the indications of the CEOCOR document and the industrial experiences, national norms have been approved where the main tests to be performed on the coating have been defined. The following national norms have already been issued: DIN 30678 (October 1992), UNI 10416 – Parts 1 and 2, NF A 49–711 (November 1992) and NF A 49–712 (December 1993). At European level the Norm ECISS/TC29/SC4/WG3N4E is under study.

Different companies have their own PP coating specification, among which are ADCO, Agip, BP, Brown & Root Braun, Chevron, Conoco, Elf

Aquitaine, Esso, Gaz de France, Gaz de Sud-Ouest, Lasmo Oil, Shell, Snam, Statoil, Total.

REFERENCES

1. Guidetti, G.P., Locatelli, R., Marzola, R. and Rigosi, G.L. (1987) Heat resistant polypropylene coating for pipelines, in *Proceedings of the 7th International Conference on the Internal and External Protection of Pipes*, (ed. R. Galka), The Fluid Engineering Centre, Cranfield, pp.203–10.
2. Guidetti, G.P., Rigosi, G.L. and Marzola, R. (1996) The use of polypropylene in pipeline coatings, *Prog. Organic Coatings*, **27**, 79–85.
3. Rigosi, G.L., Marzola, R. and Guidetti, G.P. (1995) Polypropylene coating for low temperature environments, in *Proceedings of 4th International Conference on Corrosion Prevention of the European Gas Grid System*, IBC Technical Services, London.
4. Rigosi, G.L., Marzola, R. and Guidetti, G.P. (1995) Polypropylene thermal insulated coating for pipelines, in *Pipeline Protection*, (ed. A. Wilson), Mechanical Engineering Publications, London, pp 297–310.
5. Bond, P.M. and Goff, B.C. (1993) Novel field joint coating techniques match the latest multi-layer polymeric factory applied coatings, in *Pipe Protection*, (ed. A. Wilson), Mechanical Engineering Publications, London, pp 59–89.

Keywords: Coating, anticorrosion, extrusion, powder, adhesion, modified polypropylene, electrostatic spray, fluidized bed, epoxy, insulated coating, repair, field joint

Antistatic and conducting composites of polypropylene

György Bánhegyi

BASIC TERMS AND DEFINITIONS

From the viewpoint of electrical behavior, materials can be divided into two groups: conductors and insulators. Sometimes a third group, that of semiconductors, is also distinguished. The specific resistance (ρ) of a material can be calculated as:

$$\rho = R \frac{A}{d} \tag{1}$$

where R is the resistance measured between parallel electrodes having surface A at a distance of d. Its dimension is ohm m. In certain practical applications, the specific surface resistance (ρ^s) is even more important. It is calculated as:

$$\rho^s = R \frac{l}{d} \tag{2}$$

where R is the resistivity measured between two parallel linear electrodes of length l at a distance of d.

The surface resistivity range of materials is very wide, covering more than 20 orders of magnitude (Figure 1). Pure polypropylene and propylene copolymers are typical insulators with surface resistance in the order of 10^{15}–10^{17} ohm. Materials with such a small conductivity

Polypropylene: An A–Z Reference
Edited by J. Karger-Kocsis
Published in 1999 by Kluwer Publishers, Dordrecht. ISBN 0 412 80200 7

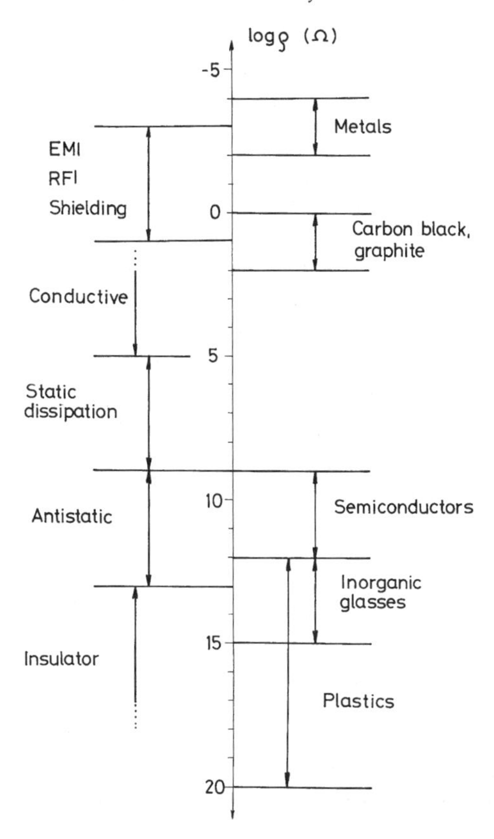

Figure 1 Range of surface resistivity of materials. On the left-hand side, the levels of resistivity are indicated (definitions by various standards are not uniform), while on the right-hand side the resistivity of different groups of materials are shown.

exhibit static charge buildup. This can be an aesthetic problem because of dust collection on plastic furniture, for example, but it can also cause explosion hazard in certain environments. Dust collection can be considerably suppressed if the surface resistance is reduced to 10^9–10^{13} ohm. This can be achieved by adding certain chemicals known as antistatic agents or simply antistats to the PP compound [1]. For a more complete prevention of static charge buildup the composite must exhibit static dissipation, i.e. it must conduct the charge carriers to the ground. This requires a surface resistivity at least in the order of 10^5–10^9 ohm, although in sensitive applications 10^3–10^4 ohm is necessary. Such a resistivity level can be achieved only by adding electrically conductive fillers (carbon black, carbon fiber, metals or metallized fillers) to the composite. An even more stringent condition is if EMI/RFI (electromagnetic or radio frequency interference) shielding is required. Shielding effectiveness (SE, measured in decibels) is defined as:

$$\text{SE(dB)} = 20 \log E_1/E_2 \tag{3}$$

where E_1 denotes the electrical field of a source inside a box made of the shielding material and E_2 is the outer field. *SE* is a sum of reflection (R_1), absorption (A) and internal reflection (R_2) terms:

$$\text{SE(dB)} = R_1 + A + R_2 \tag{4}$$

Attenuation of electrical fields is mainly determined by the conductivity of the wall material, while the attenuation of the magnetic component is determined by the magnetic permeability. Systems with SE < 30 dB are poor EMI shields, while SE > 60–70 dB indicates a very good shielding. The lower range of EMI shielding may be achieved by carbon-filled composites, but for good shielding metals or metallized fillers are necessary. If the attenuation of the magnetic components is also required, the admixture of high-permeability fillers (ferrites) is necessary.

The electrical properties of heterogeneous mixtures can be calculated from the corresponding characteristics of the components if the shape factors of the components are known [2, 3]. If there is a possibility of direct contact between the filler particles, the so-called percolation phenomenon will occur, i.e. at a critical concentration of the conductive filler, the composite conductivity plotted on a logarithmic scale exhibits a rapid change from values typical for the matrix to values characteristic of the conductive filler (Figure 2). This critical concentration is called percolation threshold. The percolation threshold depends on the shape of the filler: it is the highest for ideally spherical particles, lower for flakes (oblate spheroids) and the lowest for fibers or needles (prolate spheroids). The percolation threshold can be significantly reduced for globular particles as well if the elementary particles are agglomerated and can form chains of conductive particles [3]. The percolation threshold is lower for semicrystalline polymers (as PP) and for phase separated block copolymer systems (as rubber toughened PP) than for homogeneous, amorphous polymers. The reason for this is that the conductive particles tend to accumulate at the phase boundaries. As the percolation transition is very sharp, the batch-to-batch reproducibility of the resistivity values especially in the semiconducting range is usually medium to poor.

The conductivity of composites containing conductive fillers is usually field, frequency and temperature dependent. Field dependence can be explained by the nonlinearities of charge transfer between particles separated by thin insulating layers (tunneling, charge injection, microbreakdown). Frequency dependence is caused, among others, by the

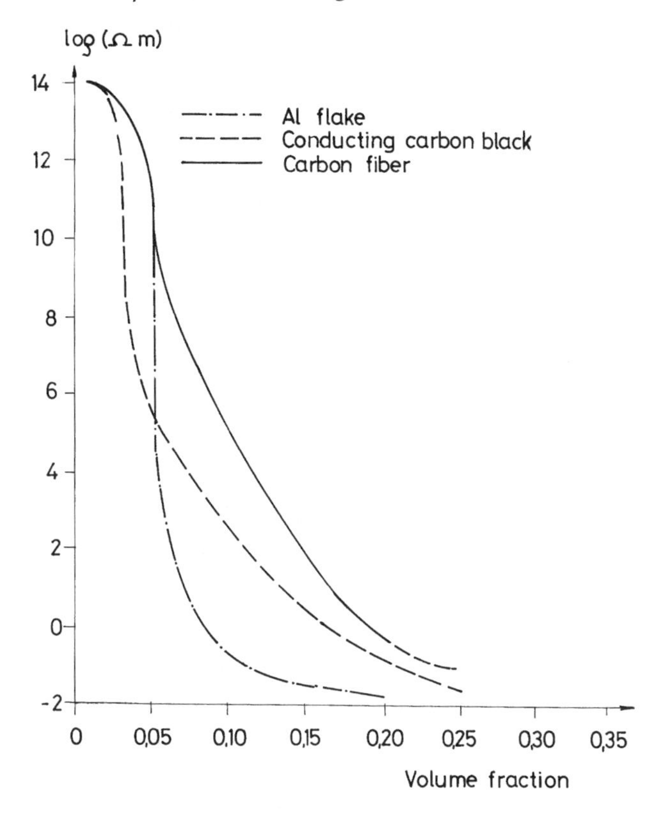

Figure 2 Specific resistivity of various composites made of insulating matrix and conductive fillers.

presence of interfacial polarization between the conducting particles and the insulating matrix. Temperature dependence can be explained by thermal expansion, which also influences the probability of charge transfer between neighboring particles. For the same reason, thermal cycling, processing and mechanical deformation also influence the conductivity level. Thus a great deal of experimentation is necessary to find the best solution for a given application.

COMPOUNDS CONTAINING ANTISTATIC ADDITIVES [1]

Antistats are low molecular or macromolecular compounds which are either sprayed onto the surface of the plastic (external antistats) or migrate from the bulk to the surface (internal antistats). These compounds are in several respects similar to detergents (surfactants), as they contain both nonpolar and polar groups. The latter attract water molecules from the ambient atmosphere and this leads to a reduced surface resistance. Antistats can be cationic (mainly quaternary ammonium salts), anionic (mainly sodium sulphonates or phosphates) or nonionic (as glycerol esters of fatty acids or ethoxylated tertiary amines). They are used typically in 0.1–1 wt.%, and are usually admixed in the form of masterbatches. Proper chemical composition and amount is to be selected empirically, as overdosage may cause exudation and plate-out. Polymeric antistats are growing in importance, as they are less prone to migration and loss.

COMPOSITES CONTAINING CARBON FILLER [4, 5]

There are two kinds of carbon-based fillers used to produce antistatic and conducting PP composites: carbon black and carbon fiber.

Carbon black contains graphite-like microcrystals permanently fused together during the manufacturing process. The most important characteristics of carbon blacks determining their conductivity are particle size, structure, porosity and volatile content. Smaller particles allow denser packing, while high structure (more branching, extended agglomeration) results in a lower percolation threshold, but greater sensitivity to high shear rates during processing. High porosity means more particles per unit weight and smaller interparticle distance. Low volatile content (especially low oxygen content) is required for good conductivity, as chemisorption reduces the number of mobile charge carriers. Surface area is determined by gas absorption (BET surface), structure is characterized by dibutylphthalate (DBP) absorption, while the volatile content can be determined by thermal desorption coupled with weight measurement and gas analysis. Aggregate structure can be determined by electron microscopy. BET surface of conducting carbon blacks varies between 120 and 1500 m^2/g. This technique measures the total area (including micropores), while larger molecules can penetrate into larger pores. DBP absorption characterizing the aggregate structure is typically between 150 and 300 ml/100g. Typical particle size is 15–20 nm, volatile content is 1–2 wt.%.

Carbon black filled antistatic and conducting composites usually contain 10–30 wt.% filler. (Addition of 1–2 wt.% carbon black for improving the weatherability essentially does not change the conductivity, although it somewhat increases the dielectric loss). The addition of conducting carbon black to thermoplastics increases not only the electrical conductivity, but also the modulus, tensile strength, hardness, melt viscosity and the heat distortion temperature of the compound. It reduces, however, the elongation to break and impact properties.

In certain antistatic compositions, graphite or carbon powder is used as conductive filler. In some cases, carbon black is combined with glass

Table 1 Comparison of some properties of PP composites containing carbon-based fillers

Property	*Compound*						
	1	2	3	4	5	6	7
Filler	no	c.b.	c.b.	c.b.	c.b.	c.p.	c.f.
Density (g/cm^3)	0.90	1.07	1.08	1.14	1.08	1.06	1.06
MFI (230°C 10.8 kg)	112	50	n.a.	46	n.a.	n.a.	n.a.
Tensile strength (MPa)	12	18	18	n.a.	29	24	47
Yield strength (MPa)	26	19	22	11	n.a.	n.a.	n.a.
Elongation to break (%)	120	n.a.	31	35	7	22	0.5
Bending modulus (GPa)	0.97	n.a.	n.a.	0.55	2.30	1.10	11.4
Notched impact strength (kJ/m^2)	8	14	5	58	n.a.	n.b.	n.a.
ρ (ohm m)	$>10^{16}$	0.05	0.15	0.24	100	10	1
ρ^s (ohm)	$>10^{16}$	6	400	50	10^5	10_3	100

Abbreviations: no = no filler, c.b. = carbon black, c.p. = carbon powder, c.f. = carbon fiber, n.a. = not available, n.b. = no break.

fiber to improve the mechanical properties (primarily modulus and strength).

Carbon fibers are mainly manufactured from PAN fibers or from tar pitch by high-temperature pyrolysis in an inert atmosphere. Higher graphitization temperature results in higher conductivity and modulus. Surface treatment with donor or acceptor compounds can increase the conductivity of carbon fibers by orders of magnitude. Carbon fibers cause a steeper increase in modulus and strength than carbon black, but the price is higher too. In some cases, carbon fibers are combined with carbon black to achieve the required combination of properties.

Some important properties of carbon black filled and carbon fiber reinforced PP composites are compared with those of nonfilled PP grades in Table 1.

COMPOSITES CONTAINING METAL OR METALLIZED FILLERS [4, 6]

Although an attenuation of 20–50 dB can be achieved by carbon black and 30–50 dB by carbon fibers, if a really good EMI shielding is required, metals must be used. In EMI shielding applications, the parts are usually produced by injection or by compression molding, therefore the viscosity of the compound must not be excessively high. Therefore, although metal powders or metallized glass spheres are used in some applications (mainly in conducting paints and adhesives), anisometric fillers are applied more frequently. Typical anisometric fillers are metal flakes and metal fibers. Metallized platelets (e.g. mica) and fibers (glass or carbon fiber) are also on the market.

Metal flakes are produced by extremely rapid cooling (in the order of 10^6 K/s), which results in very special crystalline structure and mechanical properties. Although several metals can be processed in this way, aluminum alloys are used most frequently. The flakes are strongly anisometric (about 1 mm × 1 mm × 30 μm), therefore the conductivity is significantly decreased at a volume fraction of about 0.1. About 15 vol.% aluminum flake is needed for static dissipation, 20 vol.% for EMI shielding and 25 vol.% for good heat conductivity. In the case of metallized mica, about 4 vol.% is needed for antistatic and 10 vol.% for EMI shielding applications. Due to the specific gravity differences, the weight fractions are higher. Good EMI shielding is, for example, achieved at an aluminum flake content of about 40 wt.%. The shielding efficiency can be well calculated (within 5 dB) using semi-empirical formulae. Dominant terms are R_1 and A (for notation see equation (4)):

$$R_1(\mathrm{dB}) = K_1 - K_2 \log(\rho\mu f) \qquad (5)$$

$$A(\mathrm{dB}) = K_3 t \sqrt{\left(\frac{\mu f}{\rho}\right)} \qquad (6)$$

where μ is the magnetic permeability, ρ is the specific volume resistivity, f is the frequency, and t is the layer thickness. The value of constants K_1, K_2 and K_3 depend on the unit system used. SE values of 30–50 dB can be routinely achieved with aluminum flakes. Aluminum flake filled plastics usually exhibit much higher viscosity than the nonfilled counterparts. These composites must be processed with care; high shear rates, small diameter runners, sharp corners and sudden wall thickness variations should be avoided. In some cases, sandwich structures are molded; the EMI shielding internal layer is covered by an aesthetically attractive nonfilled layer.

Metal fiber filled plastics have become popular, as the high aspect ratio of fibrous fillers allows the lowest loading level for a given conductivity. For example, the same level of conductivity can be achieved by adding 40 wt.% aluminum flake and 7 wt.% stainless steel fiber. There are natural limits for the increase of aspect ratio, however. If the modulus of the filler is high (stiff fibers), above a certain length the fiber breaks into pieces during melt processing or compounding. This is the case, for example, with metallized carbon or glass fibers. Composites containing such fibers have high modulus and strength. If the filler material is tough and ductile (as in the case of metals), the fibers tend to coil and become entangled. Due to the low critical volume fraction, the properties of conductive composites containing metal fillers differ less from those of the nonfilled polymers than in the case of carbon or metallized glass fiber

Table 2 Comparison of some properties of PP composites containing Al flake and stainless steel fillers

Property	*Compound*		
	1	2	3
Filler	no	Al f.	s.s.
Density (g/cm^3)	0.90	1.23	0.95
MFI (230°C 10.8 kg)	112	n.a.	25
Tensile strength (MPa)	12	24	28
Yield strength (MPa)	26	n.a.	n.a.
Elongation to break (%)	120	3	39
Bending modulus (GPa)	0.97	2.75	1.0
Notched impact strength (kJ/m^2)	8	n.a.	n.a.
ρ (ohm m)	$>10^{16}$	1	0.1
ρ^s (ohm)	$>10^{16}$	100	10^4

Abbreviations: no = no filler, Al f. = aluminium flake, s.s. = stainless steel fiber, n.a. = not available.

filled systems. Stainless steel fibers are most popular because of their good mechanical and corrosion resistance properties. Pellets with longer fibers are produced not by compounding the chopped fiber with the melt but by melt coating of infinite sized fiber rovings, followed by pelletization. In these systems, the initial fiber size is comparable to the pellet size. Attenuation levels of 30–40 dB can be achieved by stainless steel fibers and 50–60 dB by nickel-coated carbon fibers. Higher levels of attenuation are required in applications which allow more expensive fillers. Table 2 summarizes the properties of aluminum flake and stainless steel fiber filled PP compounds.

REFERENCES

1. Connor, M. (1997) Antistatic agents, *Modern Plastics Encyclopedia*, C-3.
2. Bánhegyi, G. (1986) Comparison of electrical mixture rules for composites, *Colloid Polym. Sci.*, **264**, 1030–1050.
3. McCullough, R.L. (1985) Generalized combining rules for predicting transport properties of composite materials, *Composites Sci. Technol.*, **22**, 3–21.
4. (a) Whittaker, G. (1986) Antistatic and electrically conductive carbon black filled thermoplastic compounds, in *Fillers, Proceedings of the Joint Conference of the Plastics and Rubber Institute and the British Plastics Federation, March 1986*, Elsevier, Amsterdam. (b) Simon, R. (1986) EMI shielding with aluminum flake filled polymer composites, in *Fillers, Proceedings of the Joint Conference of the Plastics and Rubber Institute and the British Plastics Federation, March 1986*, Elsevier, Amsterdam.
5. Sichel, E.K. (ed) (1982) *Carbon Black – Polymer Composites*, Marcel Dekker, New York.

6. Bhattacharya, S.K. (ed) (1986) *Metal-Filled Polymers; Properties and Applications*, Marcel Dekker, New York.

Keywords: antistatic, conductivity, insulator, electric resistance, electromagnetic shielding, EMI, radio-frequency interference shielding, RFI, percolation, carbon black, carbon fiber, fillers, BET surface, metal-coated fibers, semiconductor, antistatic compounds, conductive fillers.

Appliances

J. Bentham

DEFINITION

Polypropylene (PP) has developed over recent years as the dominant polymer, in volume terms, used by the appliance industry. Before discussing the markets, consumption, applications and properties, it would be worth defining what is included in the appliance industry. The industry can be separated into two product segments.

Household appliances is a generic term used to refer to the 'smaller' appliances which can be bought as an addition to the kitchen, house or garden. Many of these items have developed over the decades as labour-saving devices and more recently as lifestyle products. Within household appliances there are four groupings, small kitchen appliances; floor care appliances; personal care appliances; and garden appliances.

White goods is the collective term for the 'large' appliances generally seen in and around the kitchen. Whereas many household appliances are perceived as optional, most white goods products are regarded as essential to modern day life, particularly in the developed markets. The term 'white goods' is derived from the white paint applied to the products. Three groupings comprise white goods: washing appliances; refrigeration equipment; and cooking appliances.

Polymers, including PP, are widely used in other consumer electrical goods, such as brown goods and the home computer market. These will not be included in this chapter and the focus will be on household appliances and white goods.

Polypropylene: An A–Z Reference
Edited by J. Karger-Kocsis
Published in 1999 by Kluwer Publishers, Dordrecht. ISBN 0 412 80200 7

MARKETS

There are three major world markets for appliances, Europe, North America and Asia, plus some smaller but fast developing areas, such as South America, Africa and the Middle East. There are similarities between the three major markets but also distinct differences in product type and growth rate which thus affects PP and its consumption. The drives which fuel the demand for appliances are numerous but there are several fundamental factors which are common to all markets:

- product saturation: (the percentage of the total number of households which possess the product);
- demographics and social changes;
- gross domestic product (GDP), linked to spending power.

Europe overall is a very saturated market with only a handful of products, such as microwave ovens and dishwashers, having growth potential. The European population is growing slowly, and there is some change in household numbers linked to the rise in single-parent families and vacation home purchases. Growth in GDP is small in Europe and spending confidence remains low. In summary, Europe is a tough market for the appliance industry which interestingly has benefited PP due to the need for continuous cost reduction.

The North American market is very similar to Europe but in some key areas is more advanced in the drive for cost reduction.

If the saturated Japanese market is excluded from the Asian total, then the remainder is probably the most dynamic world area today. Many countries are still developing, so saturation levels are low. Populations are growing rapidly and social changes are increasing the number of households, creating percentage growth figures for most appliance products in double figures. Other regions, such as South America, have the potential of many Asian countries but this will depend upon economic development [1].

POLYPROPYLENE CONSUMPTION

For simplicity, and due to the maturity level, the European market will be used to illustrate the development of PP consumption.

Household appliances

For virtually all units in this segment there is the need for a 'housing' to contain the various electrical components and, in many cases, a foodstuff. Metal was the chosen housing material for many years, but as the polymer producing and conversion industry developed so plastics penetrated this application. The typical water boiling kettle is a classic

Household Appliances (%)

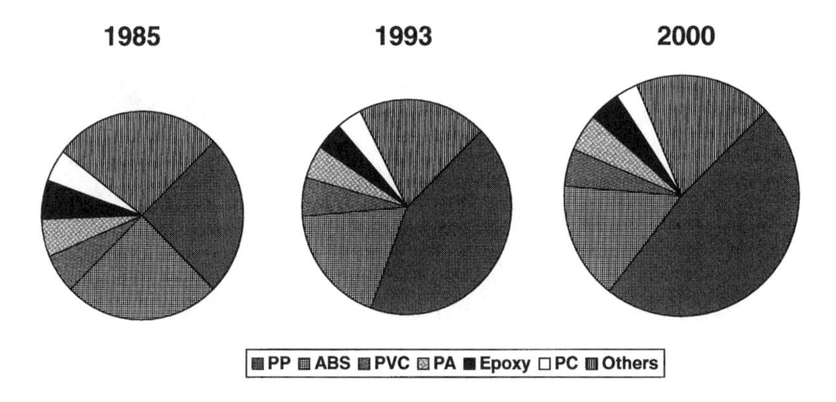

Figure 1 European usage of polymers in household appliances.

example of how this substitution process has worked. Metals were substituted during the 1970s and the 1980s by polymers such as polyoxymethylene (POM) and ABS. Then with the development of PP, married with improved design and the drive to reduce costs, we see that today most kettles have a PP housing. Kitchen appliances, and to some extent floor care appliances, are the two groupings where this has occurred the most. The percentage consumption figures for the segment reflect this development. In the mid 1980s polypropylene was around one-quarter of the total polymer consumption. This grew significantly in the early 1990s and it is probable that by year 2000 PP will account for almost half the polymer consumed in household appliances (Figure 1) [2].

White goods

The role of polymers in this segment varies greatly according to the product grouping. Cooking appliances, not surprisingly, involve high operational temperatures and are therefore dominated still by metal. Refrigeration equipment operates at the opposite end of the temperature scale and hence has different material requirements. Polymers have made inroads for insulation material and for liner materials. PP has some limited applications in refrigeration today, but improved properties should increase consumption in the future.

Washing appliances operate in a temperature range more suited to PP (maximum 95°C) and the presence of fluids creates no problems as PP is not hygroscopic. Many applications have developed for internal components replacing other polymers and metals.

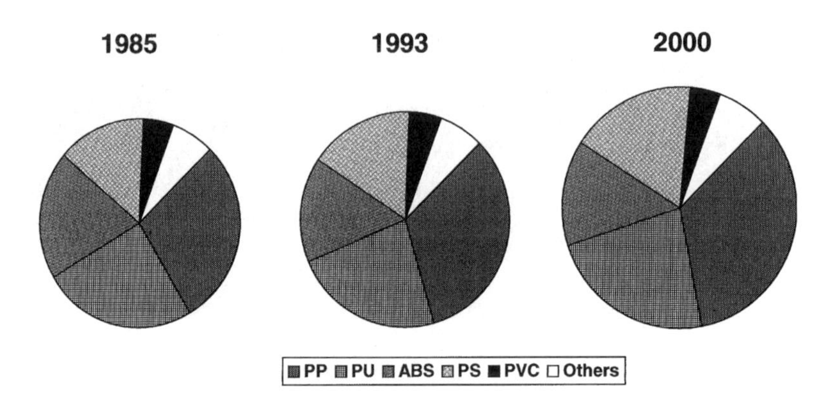

Figure 2 European usage of polymers in white goods.

Combining the three groupings in white goods creates a very different polymer consumption profile than household appliances. The four main polymers, PP, PU, ABS and PS dominate through the years, and although in percentage terms PP continues to grow this is not at the expense of other polymers but at the expense of metals (Figure 2) [3]. In fact, polymers are predicted to account for 24% of all materials consumed in white goods by year 2000 as opposed to 17% in 1985 [4].

APPLICATIONS AND PROPERTY REQUIREMENTS

Household appliances is a very competitive global market which requires PP producers to deliver materials not only to an exacting property specification, but also taking into account design appeal, faster development cycles and cost savings, particularly weight savings [5].

As mentioned earlier, kitchen appliances are where PP is commonly used for housings (Table 1).

Table 1 Typical kitchen appliances and their material requirements

Housings for	*Properties*	*Polypropylene types*
Kettles	Surface finish/gloss	Heat stabilized homopolymers
Coffee machines	Stiffness (flexural modulus)	Mineral reinforced PP
Toasters	Heat resistance (HDT)	High isotacticity PP
Deep fat fryers	Temperature stability	
Irons	Processability (MFI)	
	Colorability	

By utilizing additivation technology, a good high flow PP homopolymer can be stabilized for a coffee machine or kettle housing and offers the most cost-effective solution. Traditionally, if increased heat resistance, heat distortion temperature (HDT) and stiffness is needed then mineral reinforced PP is used, typically 10% or 20% talc reinforced. A more recent development is high isotacticity polypropylene which can achieve the stiffness and heat performance of mineral filled grades with the density of a regular homopolymer. Over the coming years, it is probable that high isotacticity PP will replace mineral reinforced grades in many applications and with improved scratch resistance also challenge ABS, (Figure 3).

Food preparation equipment, such as mixers and blenders, is an area of kitchen appliances where as yet PP has failed to penetrate. This is because the high gloss, stain and scratch resistant requirements are more exacting, although each new PP development brings it closer to the specification. Similarly personal care appliances are still using ABS.

Floor care appliances is a growing area for PP, again as a housing material but also for internal components. Vacuum cleaners dominate

Figure 3 Kettle manufactured from PP with 10% mineral addition.

Table 2 Application of PP in vacuum cleaners

Application	*Cleaner type*	*Properties*	*Polypropylene type*
Housing	Upright	Stiffness	Homopolymer PP
Upper housing	Canister	Aesthetic surface High gloss Colorability	Mineral reinforced PP High isotacticity PP
Lower housing and vacuum head	Canister	Stiffness Impact properties Aesthetic surface	Copolymer PP
Internal, e.g. dust chamber	Upright and canister	Stiffness Heat resistance Temperature stability	Heat stabilized mineral reinforced PP

and there are two classic designs which influence material selection, the upright cleaner and the canister style units. (Table 2)

Today PP is used for housings of uprights and upper housing of canisters on lower range models but for internal components and lower housings of canisters virtually all models utilize PP.

Garden appliances is a developing area for PP. Power tools, like other appliances, require a housing, and metals are being squeezed out by polymers. The products require more durability than, say, a coffee machine which means that mechanical performance is more important. High rigidity (stiffness) and good impact resistance resulted in engineering polymers being used. But the last ten years have seen significant improvements in the performance of glass reinforced PP grades resulting in them substituting other materials.

Lawnmowers and other garden equipment have also experienced the usage of polymers, with PP finding applications in 'hover' style units for example.

White goods units in the European market have high saturation levels which have resulted in a very competitive market place and stagnant, or falling, price levels in real terms. Cost reduction has therefore been at the forefront of many of the material changes in recent years; this has benefitted polymers in general and PP in particular. Cooking appliances are still constructed mainly from metals, with polypropylene being little used. Isolated applications exist for some flame-retarded grades with the only growing area being in internal components for microwave ovens.

The methods used to construct and incorporate white goods appliances into the kitchen environment have developed, with refrigeration equipment often being an integral part of the total storage space. This has enabled redesign, and a recent development has been the use of a large PP base on which the refrigerator is then constructed. Good rigidity (stiffness) is the main property requirement, hence the use of mineral reinforced PP grades.

Applications for PP inside refrigeration equipment are as yet limited. High-impact copolymers have made some inroads, but an interesting development is the use of high-clarity random copolymers. Boxes and trays require a degree of transparency and new PP random copolymers offer an alternative to the appliance producer.

Washing appliances are the biggest application area for PP today, with numerous components manufactured from a variety of grades and the possibility for further developments.

The bases of dishwashers and washing machines, like refrigerators, are typically of a single or two-component design using mineral reinforced PP. External parts, such as control panels, facias and door frames, still utilize polymers such as ABS but for certain models PP grades have penetrated these aesthetic applications.

Internal contact components are where polypropylene's inherent properties make it the ideal material choice. The environment is often hot, up to 95°C, wet and chemically aggressive due to modern detergent solutions. This places high demands on the materials used and, in fact, most appliance manufacturers include in their material specifications requirements for resistance to hot detergent solution.

Meeting these requirements means careful design of stabilization packages to ensure both that the PP is stabilized for long-term heat applications and that the compounds used are chemically resistant to the detergents used. The combination of salt solution and alkali cleaners in dishwashers creates a particularly aggressive environment.

The mechanical needs of the component then determines whether a homopolymer, copolymer, mineral reinforced or glass reinforced grade is used. Typical components are:

Dishwashers – saline solution containers, water distributors, cutlery trays;
Washing machines – pump housings, filter housings, detergent trays;
Tumble dryers – door filter housings, air ducts.

Probably one of the most interesting and technically challenging developments has been the use of glass reinforced PP for washing machine outer tubs. In European design front loading machines, the outer tub is a static cylindrical container which contains the inner clothes drum via a metal shaft from the drive system. During the washing cycle the outer tub experiences a variety of loads and stresses particularly with high spin speed machines (greater than 1000 rpm).

Figure 4 Washing machine outer tub produced from PP with 30% glass fibre reinforcement.

Ten to fifteen years ago, all outer tubs were constructed from metal and a good percentage of models today still use metal. But, following extensive development, washing machines were launched onto the market with a PP outer tub. Due to the high mechanical needs the first tubs utilized a 30% glass reinforced PP grade, whereas more recent models also used mineral reinforced grades (Figure 4).

Total costs for the appliance producer were drastically reduced. The PP tubs are normally injection moulded in two parts with total weight around 5 kg. These can then be assembled relatively easily into the washing machine. Metal outer tubs require more components and several steps of plate work, machining and welding. This is a classic example of how the PP industry, working with converters and appliance designers, can develop a relatively mature product.

CONCLUSION

PP is the polymer of today and tomorrow for the appliance industry. Maturing products and markets require cost effective redesign solutions; PP offers this and more. The property profile and conversion techniques for PP grades have developed significantly and will continue to do so.

REFERENCES

1. Various reports and articles (1995–1996) Appliance, Dana Chase Publications
2. Corporate Development Consultants (1990) *The European Market for Plastics in Home and Garden appliances.*
3. Applied Market Information (1996) *Polypropylene compounds in Western Europe.*
4. SOFRES Conseil for APME (1995) *Information system on plastics in the Electric and Electronic Sector.*
5. Calvin, R. (1996) Performance of PPs make gains in kitchen appliances, *Modern Plastics Int.*, December, 1996.

Keywords: appliances, household appliance, white goods, market, application, property requirement, filled PP, reinforced PP, competition with ABS.

Application of shear-controlled orientation in injection molding of isotactic polypropylene

Gürhan Kalay and Michael J. Bevis

INTRODUCTION

The conventional injection molding of polypropylene is introduced in the related chapter. This chapter extends the subject of injection molding of polypropylene with the addition of shear-controlled orientation in injection molding (SCORIM), which has proved to be an excellent way of managing the morphology and hence the physical properties of polypropylene [1, 2].

SCORIM results in substantial enhancement of the physical properties of iPP. The enhanced properties are related to the formation of shish-kebab morphology, and hence the high levels of molecular orientation that result from the application of macroscopic shears to a solidifying melt in a mold cavity. In this chapter, SCORIM technology is first described, and then relationships between processing conditions and mechanical properties are discussed, including a discussion on shish-kebab morphology, orientation and the occurrence of γ phase in processed iPP.

SHEAR-CONTROLLED ORIENTATION IN INJECTION MOLDING

SCORIM is based on the controlled application of macroscopic shears to a solidifying melt in a mold cavity, and formerly referred to as Multiple Live-Feed Injection Molding (MLFM) [3]. In its simplest form, SCORIM

Polypropylene: An A–Z Reference
Edited by J. Karger-Kocsis
Published in 1999 by Kluwer Publishers, Dordrecht. ISBN 0 412 80200 7

involves the use of a device fitted between the mold cavity and the nozzle of an injection molding machine. The device divides the feed and carries a piston in each of the channels to the mold. The pistons, activated by auxiliary hydraulics and controlled by microprocessor, provide for the control of the frequency and amplitude of their movement during the filling and holding stages of the molding cycle. Recent studies also involve the use of in-mold pistons to achieve a similar effect.

There are three principal operating modes for the SCORIM process. Figure 1 shows a schematic diagram for these operating modes. With mode A, the SCORIM pistons are reciprocated 180° out of phase with each other. This action causes a macroscopic shearing of the solidifying polymer melt as material is displaced from one piston chamber to the other. With mode B, the SCORIM pistons are reciprocated in phase so that the gates and runners to a thick sectioned part can be kept molten by shearing while the material in the cavity continues to solidify. With mode

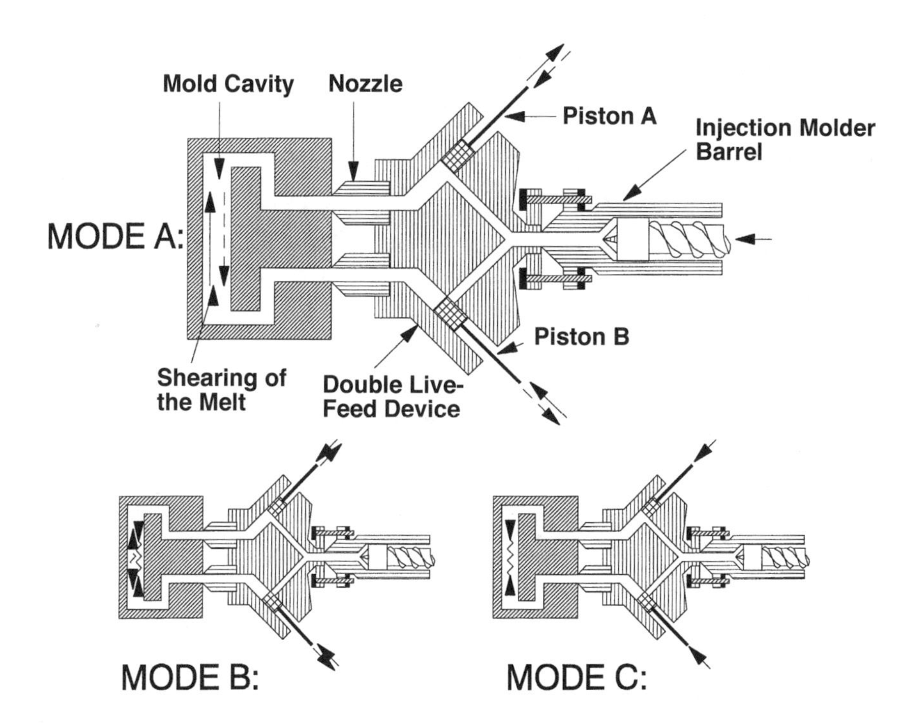

Figure 1 A schematic diagram showing the operatic modes of the SCORIM process: (a) mode A, the pistons are activated 180° out of phase; (b) mode B, the pistons are activated in phase; (c) mode C, the pistons are held down under a constant static pressure.

C, both of the pistons are forced down to impose a static melt pressure in the mold.

The purposeful management of micromorphology in thermoplastics also relates to the use of more than two live-feeds, to produce moldings with quasi-isotropic properties, such as coefficient of thermal expansion, Young's modulus, etc. For example an alternating out-of-phase application of four pistons, pistons 1–3 and 2–4, during the solidification of polymer molded plaques, as illustrated in Figure 2, will produce 0–90° preferred orientations of any alignable constituent contained in the molding compound, i.e fibers, fillers as well as the polymer molecules themselves. The use of SCORIM permits the production of plaques exhibiting 0–90° lamellar morphology, with relative 0° and 90° lamellae

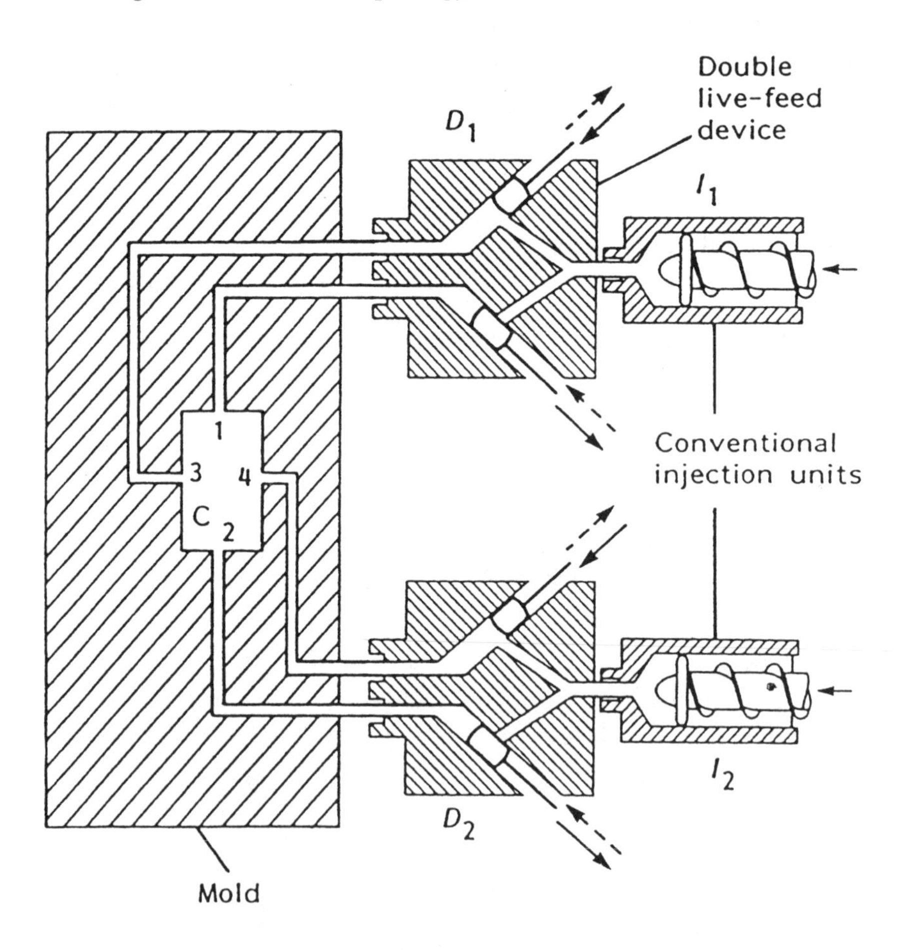

Figure 2 A schematic diagram of a four live-feed arrangement based on twin-barrelled injection molding machine and two double live-feed devices.

thicknesses being determined solely by the relative durations of piston operation.

COMMENTS ON PROCESSING CONDITIONS IN SCORIM

In SCORIM, a high Young's modulus is obtained normally with a low melt temperature in processing. However, as long as controlled solidification is achieved, the use of relatively high melt temperatures should not preclude the attainment of high stiffness; note that the activation of pistons creates shear heating and it is therefore appropriate to keep melt temperatures low. High injection speeds lead to a high Young's modulus in SCORIM. The level of holding pressure is not a determining factor for mechanical properties in SCORIM so long as a high level of orientation is achieved with the SCORIM profile. In SCORIM, the cavity pressure is not critical for the purpose of obtaining high stiffness when the high levels of molecular orientation are not lost during cooling, which is possible to achieve with the right choice of SCORIM profile. The application of excessively high pressures in SCORIM processing may result in mechanical degradation of the polymer melt. The SCORIM operation mode A is the most effective operating mode for the achievement of high stiffness. Further packing with the operation of mode B or mode C may be necessary.

MECHANICAL PROPERTIES

It has been shown that SCORIM may result in more pronounced molecular orientation than conventional injection molding, and results in more than two-fold increase in Young's modulus as recorded for polypropylene – GYM43 (ICI, Welwyn Garden City, UK) – at room temperature. High-temperature tests at 80°C also revealed an 80% increase in Young's modulus and a 65% increase in tensile strength of iPP – KF6100 and GE6100 (Shell, Louvain-la-Neuve, Belgium) – as a consequence of the application of SCORIM. For this grade of iPP, an increase of up to four times in impact strength was achieved with SCORIM as well as the substantial increase in Young's modulus. Table 1 summarizes the mechanical properties of the KF6100 and GE6100 polymers.

PP-copolymers also exhibited the same behavior as iPP homopolymers. Mechanical tests carried out at room temperature and at 80°C with injection moldings produced from HMT6100 (Shell, Louvain-la-Neuve, Belgium) grade PP-copolymer showed enhancement of mechanical properties following the application of SCORIM. At room temperature, conventional moldings of this copolymer exhibit 1.8 GPa Young's modulus and 33.7 MPa ultimate tensile strength (UTS) whereas SCORIM moldings exhibit 2.7 GPa Young's modulus and UTS of 59.3 MPa. At

Table 1 The tensile test results obtained from homopolymers GYM43 and KF 6100, and copolymer HMT6100. CM, conventional molding; SC, SCORIM molding

	Ultimate tensile strength (MPa)	*Yield strain (%)*	*E (MPa)*
CM[GYM43] at 23°C	40.6 (1.5)	–	1540 (90)
SC[GYM43] at 23°C	57.4 (1.4)	–	3910 (340)
CM[KF6100] at 23°C	31.7 (0.5)	11.3 (0.1)	1910 (430)
SC[KF6100] at 23°C	51.2 (0.5)	16.2 (1.2)	3334 (965)
CM[KF6100] at 80°C	18 (0.2)	17.9 (0.9)	607 (378)
SC[KF6100] at 80°C	–	–	934 (42)
CM[GE6100] at 80°C	13	18	501 (19)
SC[GE6100] at 80°C	–	–	922 (59)
CM[HMT6100] at 23°C	33.7 (0.1)	7.4 (0.1)	1806 (119)
SC[HMT6100] at 23°C	59.3 (0.3)	35.9 (1.3)	2719 (176)
CM[HMT6100] at 80°C	16.9 (0.1)	9.4 (0.3)	861 (94)
SC[HMT6100] at 80°C	33.7 (0.5)	20.1 (2.6)	1271 (70)

room temperature, conventional and SCORIM moldings of this copolymer exhibit 7% and 36% strain at peak, respectively, and are representative of a substantial increase in toughness. At a test temperature of 80°C, the mechanical properties of the copolymer processed using SCORIM are also substantially greater than those processed by conventional molding. Table 1 summarizes the room-temperature and high-temperature tensile test data for this copolymer.

SHISH-KEBAB MORPHOLOGY AND MOLECULAR ORIENTATION IN SCORIM MOLDINGS

Figure 3 shows the whole transverse cross-section of a SCORIM molding. There are mainly three regions: the skin, shear-influenced layer and core. Shear bands resulting from SCORIM piston movements can be seen in the shear-influenced region between the skin and the core. The shear-influenced layer exhibits a shish-kebab morphology. A schematic model for the shish-kebab morphology is shown in Figure 4. The reader should refer to reference [2] for TEM micrographs of this morphology. The model involves two main components: the fibrils oriented in the injection direction, and chain-folded lamellae that grow on the fibrils. The fibrils exhibit a preferred *c*-axis orientation parallel to the injection direction. The fibrils are approximately 70 nm in width and the thickness of the chain-folded lamellae is 20 nm. The chain folded lamellae make an angle

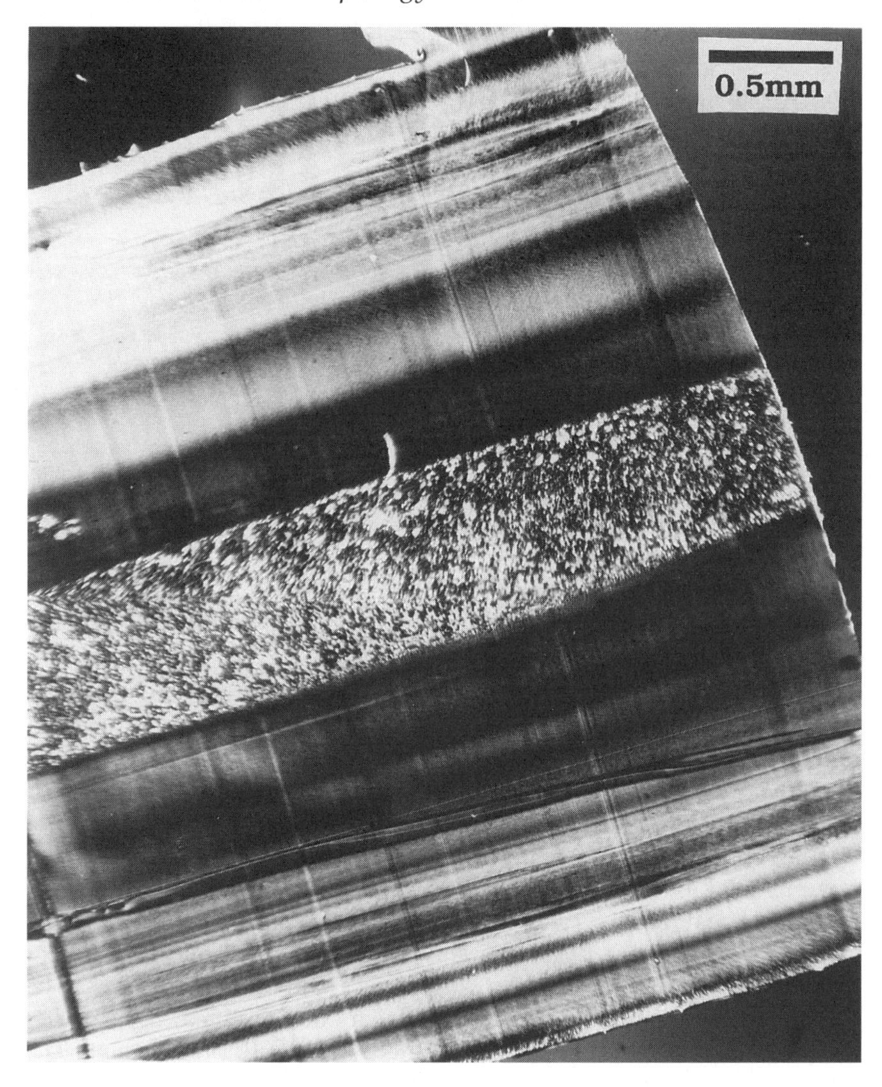

Figure 3 Microstructure of the whole longitudinal cross-section of SCORIM moldings. Shear bands resulting from SCORIM piston movements can be seen in the shear influenced region between the skin and the core.

ϕ (<90°) with the fibrils. This angular relationship is consistent with the epitaxial growth angle of 80° which was determined by Lotz and Wittmann [4]. The assumption that small and nonuniform lamellae, whose a^*-axes are parallel to the injection direction (ID), as proposed by Fujiyama *et al.* [2] was consistent with the $(c + a^*)$-axes orientation recorded on Debye patterns. The main influence on morphology is

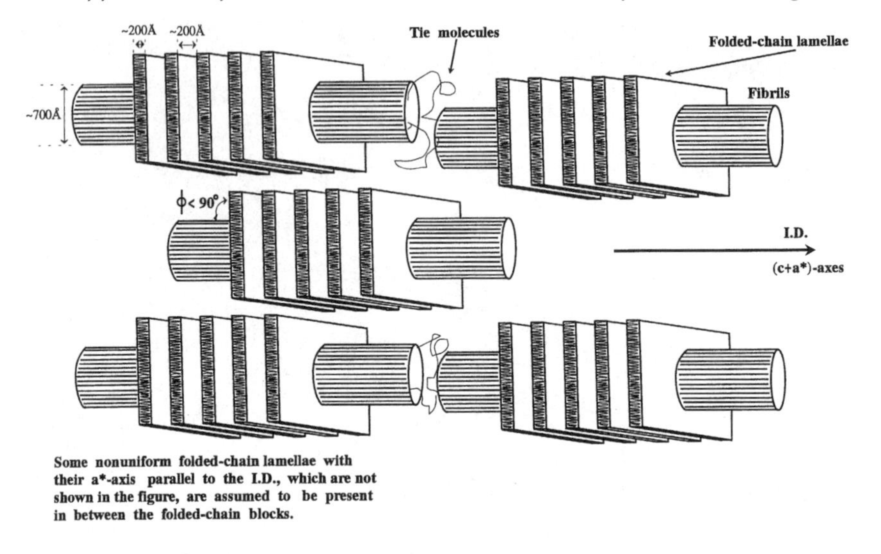

Figure 4 Schematic diagram for the model of the oriented zone morphology in iPP injection moldings [2].

associated with the elongational and shear forces applied to the melt, and subsequently the application of shear forces to the solid at elevated temperatures.

In the molten state, the polymer consists of randomly distributed chains with entanglements. When shear and elongational stresses are applied in SCORIM, the entanglement density decreases and some of the polymer chains are extended. These chains attain a high degree of alignment and they crystallize in the fibrillar form on cooling. The polymer chains which do not initially attain a high degree of alignment crystallize epitaxially on the fibrils forming chain-folded structures. The further reorganization of these crystals in the solidified zone occurs because the cyclic load is applied repeatedly in SCORIM processing. The applied shear generates heat and the mold at 60°C promotes this reorganization, as the thermoplastic is well above 60°C which is sufficient to mobilize the folded chain blocks. These folded chain blocks may by slip and rotation of the lamellae further assume a fibrillar structure, although the main effect is expected to be on the tie-molecules. The tie molecules are stretched and become stiffer thus contributing to the high Young's modulus of the molding. The core region of the SCORIM moldings can be eliminated if gradual cooling is achieved in processing, and if the narrow regions of the flow path, and specially the gates to the mold cavity, are maintained at a temperature that will allow macroscopic shearing to be applied throughout solidification of regions within the molding of particular interest. That is principally where physical property enhancement through control of micromorphology is deemed important.

The reader should refer to Kalay *et al.* (1996) [6] for in depth X-ray diffraction study of SCORIM and conventionally molded iPP. Typical SCORIM moldings mainly exhibit α and γ phases whereas conventional moldings exhibits α and β phases.

γ-PHASE OCCURRENCE IN INJECTION MOLDING

γ-phase can occur in commercially processed isotactic polypropylene, in addition to the α- and β-phases, and all three phases influence the physical properties of the end product.

The occurrence of γ-phase is mainly associated with high molecular alignment and may be used to monitor the level of molecular alignment. The presence of $(117)_{\gamma}$ in X-ray diffraction profiles or Debye patterns obtained for injection moldings indicates the occurrence of γ-phase. The formation of high fractions of α-phase may also indicate degradation during processing. When the intensity of the γ-phase (117) reflection exceeds the intensity of the α-phase (130) reflection then this is an indicator of the deterioration of mechanical properties. The occurrence of γ-phase, when not excessive, is representative of a high molecular alignment, high Young's modulus and high tensile strength. The γ-phase formed in injection molding remains to be identified microscopically. The shish-kebab morphology in the SCORIM moldings, as identified by TEM micrographs of the replicas of etched surfaces, may be associated with the α- and/or the γ-phases, and could only be resolved by selected area electron diffraction studies.

The Debye patterns from the SCORIM moldings which have been subjected to straining showed that the γ-phase formed in injection molding is stable, and does not transform into other phases by deformation. Pole figures from the γ-phase showed that the γ-phase exhibits rotational symmetry about the injection direction as for the α-phase.

CONCLUSION

SCORIM results in substantial improvements in mechanical properties of isotactic polypropylene and provides an effective route for managing the physical properties of injection moldings.

REFERENCES

1. Kalay, G. and Bevis, M.J. (1997) Processing and physical property relationships in injection moulded isotactic polypropylene 1. Mechanical properties. *J. Polymer Sci.: Part B, Polymer Phys.*, **35**, 241–263.

2. Kalay, G. and Bevis, M.J. (1997) Processing and physical property relationships in injection moulded isotactic polypropylene 2. Morphology and crystallinity. *J. Polymer Sci.: Part B, Polymer Phys.*, **35**, 265–291.
3. Allan, P.S. and Bevis, M.J. (1987) British Patent 2170–140-B.
4. Lotz, B., Graff, S., Straupe, C. and Wittmann, J.C. (1991) Single crystals of γ-phase isotactic polypropylene: Combined diffraction and morphological support for a structure with non-parallel chains. *Polymer*, **32**, 2902–2910.
5. Fujiyama, M., Wakino, T. and Kawasaki, Y. (1988) Structure of skin layer in injection moulded polypropylene. *J. Polymer Sci., Part B, Polymer Phys.*, **35**, 29–49.
6. Kalay, G., Zhong, Z, Allan, P and Bevis, M.J. (1996) The occurrence of the γ-phase in injection-moulded polypropylene in relation to the processing conditions. *Polymer*, **37**, 2077–2085.

Keywords: injection molding, SCORIM, structure–property relationships, shear-controlled orientation, molecular orientation, SCORIM pistons, impact strength, Young's modulus, shish-kebab morphology, γ-phase, α-phase, β-phase, X-ray diffraction.

Automation in injection molding

D. Meyer

INTRODUCTION

Automation is a key issue in respect to industrial competitiveness. Robots, together with handling and auxiliary systems and techniques, are gaining acceptance in various plastics molding processes, thanks to improvements in both related hardware and software (Figure 1). Automation of production is triggered by the following goals: more cost-

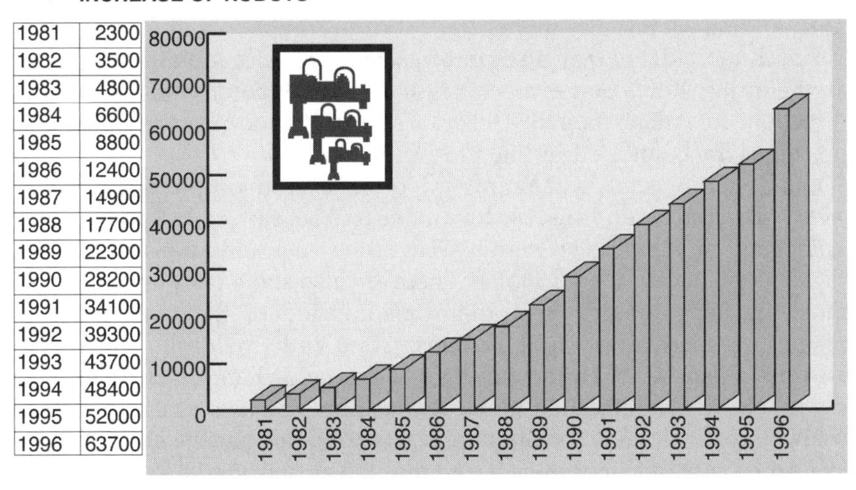

Year	Robots
1981	2300
1982	3500
1983	4800
1984	6600
1985	8800
1986	12400
1987	14900
1988	17700
1989	22300
1990	28200
1991	34100
1992	39300
1993	43700
1994	48400
1995	52000
1996	63700

Figure 1 Increase in robots in Germany between 1981 and 1996.

Polypropylene: An A–Z Reference
Edited by J. Karger-Kocsis
Published in 1999 by Kluwer Publishers, Dordrecht. ISBN 0 412 80200 7

efficient operation; fast return of investment; increase of production quantities; short molding cycles; nonstop production; energy saving; improvement of product quality (to avoid faults and failures) and to free the machine operator from monotonous work. Automation is necessary for the future of plastics molding industry [1].

In processing plastics, there are many possibilities for automation. This extends from automatic drying of materials, through transport to the molding machine, to automatic packaging at the end of the production process. There is no end in sight of this development for processes such as blow molding, rotational molding, extrusion, calendering, thermoforming, foaming, in addition to assembly techniques, such as mounting, welding, glueing and printing. Handling systems and industry robots are prerequisites if we are to realize all the related demands.

HANDLING SYSTEMS

There are various types of industrial robots and handling systems with rotational and/or linear motion [1].

Handling robots can detach runners one-by-one and feed them back into the molding process (called on-line-recycling). They may also pick up moldings and place them onto measuring tools, classify and separate the production into [IO] (accepted) or [NIO] (rejected parts). They may solve tasks related to proper placing, stacking and packing. A widespread application of robots is the placing of metal inserts, inlay decorations and labels into the molds.

Handling systems may also control several plastics molding machines at the same time. Some machines can be interconnected in parallel direction for inlays, quality check, assembly and packaging. Other machines are connected in line in respect to operations.

Quick setup is of paramount importance in injection molding [2]. Mold 'setup changers' are able to remove and set two molds, locking and unlocking in about one minute. The time required depends on the dimension and number of molds. There are also some various systems of mold-fix-clamps which are pre-installed inside of the two clamping platens. Modern and rapid systems work with hydraulic-mechanical clamping. (Figure 2). The actual stage of the mold change is displayed on the monitor of the injection molding machine. The new philosophy is to change molds along with complete plasticizing units at the same time by just one command. In that case, the mold exchange clutch and mold are considered as one unit.

ASSEMBLY

The automation of assembly is a special issue in plastics processing. Several products become economical after automation, for example, the

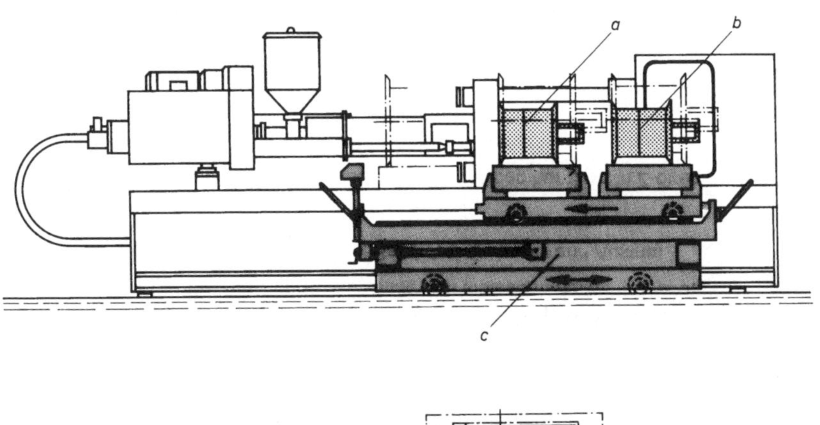

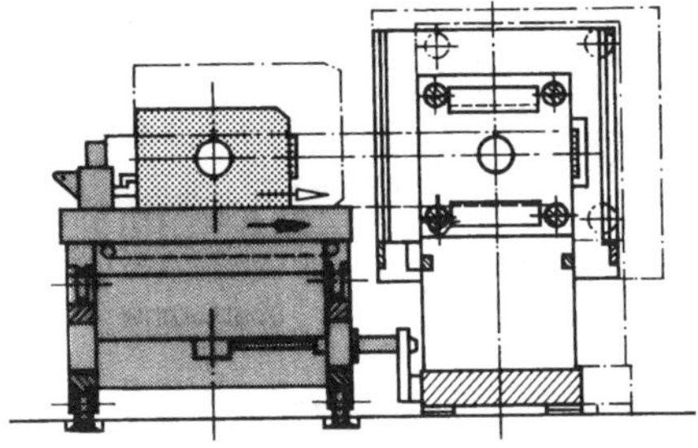

Figure 2 Mold quick exchange from reverse: (a) mold inside the clamping platens; (b) mold in waiting position; (c) mold changing transport car on tracks. (System: Battenfeld, Meinerzhagen, Germany.)

total packaging of audio and phono items. It is known that assembly quality is also affected by human nature. The plastic or 'living' hinge concept (see the chapter " 'Living' or plastic hinges") with polypropylene allows the avoidance of work-intensive product assembly. Recently, a compact disc (CD) case was designed according to this principle. The CD cases are produced in one step, without any assembly.

CONTROL

The control of a handling system is nearly the same as that of a robot. This is due to their similar movement, i.e. top-entry or side-entry handling. Some systems combine these handling methods if necessary. Others have two or more moving heads on the same linear track. The manufacturing cells are complex systems to produce finished products automatically in a sequence of steps. The operation of the handling systems and robots occurs hydraulically, pneumatically, electrically or mechanically. Software programs, for which different languages have been developed, control the operational cycle, including load capacity, speed, range and positioning accuracy. Crucial supplements are set-up reduction systems, as well as holders and sensors, conveyor systems, sorting devices and vision control systems.

REFERENCES

1. Meyer, D. (1995) Kunststoffverarbeitung automatisieren, Hanser, Munich.
2. Menges, G. and Recker, H. (1986) *Automatisierung in der Kunststoffverarbeitung*, Hanser, Munich.

Keywords: automation, handling, manufacturing cells, on-line recycling, robotics, quality assurance (QA), program language.

Beta-modification of isotactic polypropylene

József Varga and Gottfried W. Ehrenstein

Commercial grades of isotactic polypropylene (iPP) crystallize essentially into α-modification (α-iPP) with sporadical occurrence of the β-phase (β-iPP). Crystallization in temperature gradient or in sheared melt encourages the development of the β-phase in commercial, non-nucleated iPP [1]. For preparation of samples rich in β-modification or of pure β-iPP, the introduction of selective β-nucleants is the most reliable method [1]. The known β-nucleating agents are collected in Tables 1 and 2, indicating their advantages and drawbacks. The most widespread high active β-nucleating agent is a γ-quinacridone red pigment. Some two-component compounds obtained by the reactions of certain organic acids with $CaCO_3$, also possess a very high β-nucleating activity. Different calcium and zinc salts of aliphatic and aromatic dicarboxylic acids having high thermal stability, belong to the selective β-nucleants, as well [1]. The β-content of iPP samples – and so the efficiency of β-nucleants and the influence of thermal and mechanical conditions of the crystallization on the polymorphic composition – can be characterized by the *k*-value determined from wide angle X-ray scattering (WAXS), by differential scanning calorimetry (DSC) on non-recooled samples or by polarized light microscopy (PLM) using thin sections of the samples [1]. It should be pointed out that the melting curves of iPP samples cooled to room temperature do not give correct information about the β-content. This statement is due to the βα-recrystallization which takes place during heating of the sample and overlaps with the melting process of the β-

Polypropylene: An A–Z Reference
Edited by J. Karger-Kocsis
Published in 1999 by Kluwer Publishers, Dordrecht. ISBN 0 412 80200 7

Table 1 β-Nucleants for iPP [1]

No	*β-Nucleant*	*Advantage*	*Drawback*	*Literature*
1	γ-Quinacridone	High activity (k=0.8–0.9)	Physical or chemical instability, intense red color	Leugering, 1967 German Patent (DAS) 1188278
2	Aluminium salt of quinizarin sulphonic acid	–	–	Binsbergen, de Lange, 1968
3	Disodium-phtalat	–	Not reproduced	Morrow, 1969
4	Calcium-phtalat	High thermal stability	Moderate activity (k=0.5–0.7)	Hughes, 1970 US Patent 3540979
5	Dihydroquinoacridin-dione and quinacridin-tetrone	–	Moderate activity (β=65%), colored	W. Kathan, 1986 German Patent 3443599
6	Triphenol ditriazine	–	Low activity (k=0.3)	Garbarczyk, Paukszta, 1981
7	Two-component nucleator ($CaCO_3$+organic acids) Ca stearat+pimelic acid	High activity (k=0.8–0.93)	Free stearic acid	Shi *et al.*, 1987
8	Wollastonite (mineral) (Ca silicate)	High thermal stability	Not reproduced	Jingjiang *et al.*, 1990
9	Ca, Zn and Ba salts of polycarboxilic acids	High thermal stability and activity, (k≈1)	–	Varga *et al.*, 1992 Hungarian Patent 209132

Table 2 Some new β-nucleants from scientific and patent literature

No	*β-Nucleant*	*Advantage*	*Drawback*	*Literature*
10	δ-Quinacridone	High activity (k=0.8)	Colored (red)	Filho, Oliviera, 1992
11	Diamides of adipic or/and suberic acids	High activity (β=90%)	–	Ikeda *et al.*, 1992 Japan Patent 92-877559
12	Ca salts of suberic or pimelic acid	High activity (β=90%)	–	Wolfschwenger, 1995 German Patent Appl. 4420989 A1
13	Different types of Indigosol and Cibantine organic pigments	High activity ($k_x \approx 0.9$)	Colored (brown, pink yellow, grey)	M.-R. Huang *et al.*, 1995
14	Quinacridone quinone	High activity	Colored (brown)	Fujiyama, 1996
15	N',N'-dicyclohexil- -2,6-naphthalene dicarboxamide	High activity (β≈80%)	–	N. Ikeda *et al.*, 1996
16	Antraquinon red and bis-azo yellow pigment	Moderate activity	Colored (red, yellow)	A. Marcincin *et al.*, 1996

phase [1, 2]. Unfortunately, the efficiency of β-nucleants was characterized by DSC on cooled samples in many studies. Therefore, the quantitative conclusions of these studies should be treated with caution. Studies on β-nucleated iPP revealed that the formation of pure β-iPP has an upper ($T(\beta\alpha) = 140°C$) and a lower limit temperature ($T(\alpha\beta) = 100–110°C$) [1]. The crystal structure, morphology, melting and recrystallization behavior of β-iPP are discussed in detail in a later chapter 'Spherulitic crystallization and structure'. This section deals with the preparation, properties and application of β-iPP.

PROCESSING OF β-NUCLEATED iPP

For the preparation of β-iPP products, some preconditions should be fulfilled in respect to the processing condition:

- applying of β-nucleating agents with high activity and selectivity and with sufficient thermal stability;
- appropriate selection of the processing parameters for the crystallization in temperature range between $T(\alpha\beta)$ and $T(\beta\alpha)$;
- optimization of the flow and relaxation condition of the melt in order to avoid or to minimize the formation of row-nucleated α-phase;
- avoidance of use of additives (filler, pigment, stabilizers etc.) with α-nucleating activity in β-nucleated iPP compounds.

Compression molding [2]

The simplest way to produce isotropic plaques with different thicknesses is compression molding. The melt squeezed above the melting temperature should be cooled quickly to the crystallization temperature (T_c) lying in a temperature range of 100 – 130°C, in order to avoid any α-crystallization above $T(\beta\alpha)$. Due to the thermal inertia of the melt, the recommended crystallization temperature lies somewhat above $T(\alpha\beta)$, most favorably at about 110–115°C. By an annealing started from T_c, the structural stability of the β-iPP can be markedly enhanced. Annealing of compression molded sheets at 140–145°C reduces their tendency to βα-recrystallization during heating and shifts the melting range of β-iPP to higher temperatures [2]. Sheets of β-iPP based blends with elastomers, with polyethylenes and chalk-filled versions were produced by compression molding under the above thermal conditions [1, 2].

Injection molding [2, 3]

During injection molding, very complex thermal and flow conditions prevail in the cavity. The surface layer of the flowing melt is subjected to extensional stresses, and subsurface layers to shear stresses. This results in a molecular orientation in the melt flow direction. In the central zone, further from the wall, little or no orientation can be detected. That is why injection moldings show so-called skin-core morphology, the characteristics of which strongly depend on the processing parameters, including the heat transfer conditions in the mold. The skin has a complex morphology, but predominantly consists of oriented cylindrites, the growth of which is induced by row nuclei formed during melt shearing. The core has a spherulitic structure indicating a relaxation of the oriented melt prior to the onset of crystallization. So, in the core, the crystallization occurs as in a quiescent melt. For the β-nucleated iPP, the dependence of polymorphic composition on the thermal and mechanical loading conditions makes the scenario even more complicated. This can produce *inhomogenity even in the polymorphic composition*. A preliminary study on β-nucleated iPP showed that no pure β-iPP products can be produced by injection molding [2]. The moldings showed a cylindritic skin consisting of pure or nearly pure α-modification and a spherulitic core, rich in β-modification ($k > 0.9$). The thickness of the skin and the β-content in the spherulitic core depend significantly on processing parameters and β-nucleant content [3]. With increasing melt temperature [3] or injection speed [4], the core thickness is reduced and the β-content in the core may be enhanced. This tendency is more pronounced in case of high molecular weight iPP types nucleated with the calcium salt of pimelic acid [4]. Fujiyama [3] conducted a systematic investigation on the injection molding of various iPP grades nucleated by γ-quinacridone and quinacridone-quinone. In these studies, the melt (cylinder) temperature and the concentration of the β-nucleants were varied in a wide range. The β-content characterized by the k value had a maximum in dependence on the γ-quinacridone concentration of about 10 ppm. A decreasing k value in high concentration ranges refers to a possible, weak α-nucleating activity of this β-nucleant being non-selective. It was also revealed that the nucleating activity of quinacridone-quinone exceeded that of γ-quinacridone [3].

Extrusion [2, 5, and references therein]

The polymorphic composition of extruded β-nucleated iPP products depends significantly on the cooling conditions and on the take-off speed. Sheets of almost pure β-iPP were successfully produced by means of a laboratory extruder (Viskosystem) with wide-gap die and equipped with a calendering unit composed of three superimposed, thermoregulated compression rollers. The temperature of the first two compression rollers should be set between 105–110°C, i.e. in the temperature range between $T(\alpha\beta)$ and $T(\beta\alpha)$ where the formation of pure β-iPP is preferred. In this case, the melt crystallizes mostly in the first roller gap. The next roller can provide an annealing and thus stabilize the structure of the β-iPP [2]. Sheets of β-iPP can be produced at low take-off speed. Both an increased take-off speed and a reduced mass flux (extrusion speed) lead to 'straining' of the melt. This shifts the polymorphic composition of sheets formed toward the α-modification. These pecularities were found in later studies [5, and references therein] performed on the extrusion of films and fibers, too. The cooling medium has also a significant effect on the β-content which is always higher when air-cooling is applied instead of water-cooling [2, 5].

SUPERMOLECULAR STRUCTURE OF MOLDED (PROCESSED) β-iPP

The morphological studies on processed β-nucleated samples by scanning electron microscopy (SEM) revealed that they contain immature spherulites (at least at the nucleant concentration ~0.1%). They are in a transition state of growth between the hedritic and spherulitic structure [4]. As a result, no spherical symmetry exists in their lamellar arrangement. The morphology precursors of β-iPP are hexagonites (see below, in the chapter 'Spherulitic crystallization and structure'). In the early stage of growth, two formations develop symmetrically to the plane of the hexagonite, like flower cups, due to the branches of lamellae on the screw dislocation. The 'flower cups' are located around the central symmetry axis perpendicular to the plane of the hexagonite. Depending on the sterical position of the precursor, two characteristic formations could be observed by SEM on the polished and etched surface of samples: flower cup-like and face-like ones [4]. The former develops from a hexagonal precursor (hexagonit) lying parallel with the plane of cut, the later forms from a hexagonit standing on edge perpendicular to plane of cut (rod-like precursor). These two different views of immature spherulites are seen in Figure. 1. It is worth noting that similar structures were observed by Tjong *et al.* [6, 7], but the authors delivered a doubtful explanation to their origin.

PROPERTIES

Some properties of β-iPP differ significantly from those of α-iPP. In comparison with α-iPP, β-iPP possesses lower crystal density, melting temperature and fusion enthalpy, but a similar glass transition temperature (Table 3). The chemical resistance of β-iPP seems to be lower than that of α-iPP. The former could be selectively extracted with hot toluene from samples of mixed polymorphic composition [8].

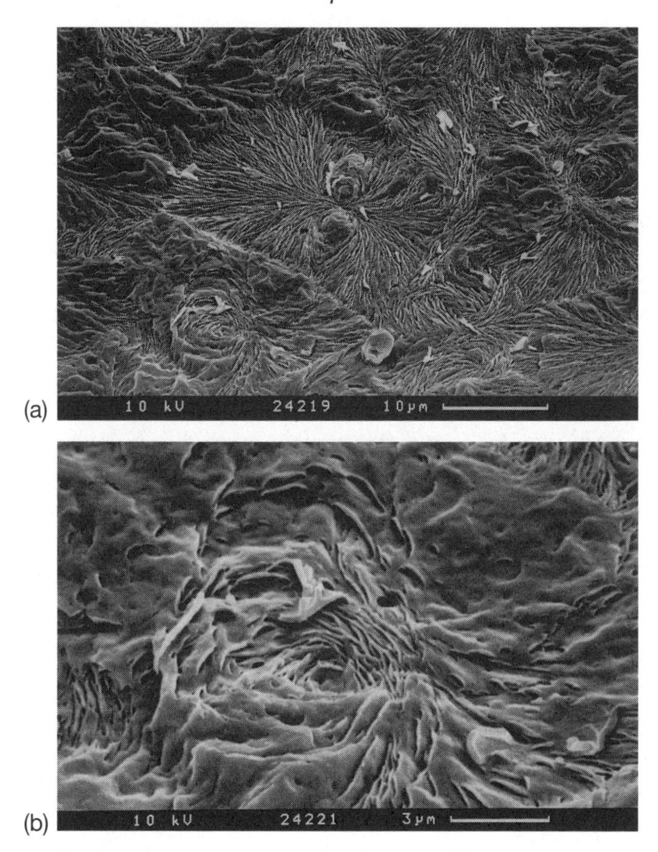

Figure 1 Two different views of immature β-spherulites by SEM: (a) face-like (in centre) and flower-cup like (left bottom) ones; (b) the flower-cup like array at a higher magnification.

Table 3 Some characteristics of α-iPP and β-iPP

Property	*Unit*	*α-iPP*	*β-iPP*	*Reference*
Experimental melting temperature, T_m	°C	~165	~155	[1]
Equilibrium melting temperature, T_m^0	°C	208*	184∓4*	[1]
Glass temperature, T_g	°C	0–25†	0–25†	[9, 10]
Enthalpy of fusion, ΔH_m^0	J/g	148∓10*	113∓11*	[1]
Crystal density, d_c	g/cm³	0.936	0.921	[10]
Amorphous density, d_a	g/cm³	0.858	0.858	[10]

* Contradictory data in literature, see [1].
† Depending on the dynamic conditions of testing.

Mechanical properties [6–10]

In comparison with α-iPP, β-iPP has a lower *E*-modulus and yield stress, but higher ultimate tensile strength and strain. The improvement in the latter might be related to the βα-transition occurring during the necking [1] which leads to the formation of α-phase of enhanced strength (strain-hardening). The impact strength [6, 7, 9, 10] and toughness of β-iPP exceed those of α-iPP [9]. The improvement in toughness and impact strength could be attributed to β to α or (at low deformation rate) to β to smectic transition occurring during the deformation [9, 10]. Another possible reason for superior toughness could be connected to the peculiar morphology of β-iPP discussed above [4]. According to DMTA measurements [8, 10], the intensity of glass transition (T_g), i.e. the value of the loss modulus at T_g, and the storage modulus below T_g increase with the β-content of samples showing mixed polymorphic composition. These findings suggest that β-iPP possesses a lower crystallinity than α-iPP which *per se* contributes to a toughness increase for β-iPP. Nevertheless, direct measurements on the crystallinity contradicted the above assumption, i.e. the crystallinity increases with increasing β-content [8]. To explain this discrepancy, Jacoby *et al.* [8] supposed that the coupling between the crystalline and amorphous phase in β-iPP is weaker than that in α-iPP. This assumption seems to be confirmed by considering the superior damping behavior of β-iPP under impact loading (i.e. the missing load oscillation in the fractograms of β-iPP [10]).

In spite of the lower melting temperature of β-iPP, its heat distortion temperature exceeds that of α-iPP [3]. This could be explained by the βα-recrystallization during heating which generates an α-phase of higher melting temperature [1, 2].

APPLICATION OF β-iPP

The application of β-iPP is favoured in some fields, based on its high impact resistance and toughness. Other application fields are exploiting the micro-void formation and strain-hardening characteristics as well as the βα-recrystallization tendency during partial melting of β-iPP. From the scientific and patent literature, it is known that β-iPP is used for the following: industrial pipeline construction, dielectric capacitors (with roughened surface), paper like films, biaxially drawn microporous film (gas exchange membranes) and porous fibers with improved moisture adsorption.

REFERENCES

1. Varga, J. (1995) Crystallization, melting and supermolecular structure of isotactic polypropylene, in *Polypropylene: Structure, Blends and Composites,*

Vol.1, Structure and Morphology (ed. J. Karger-Kocsis), Chapman & Hall, London, pp. 56–115.
2. Varga, J. (1989) β-Modification of polypropylene and its two-component systems. *J. Thermal Analysis*, **35**, 1891–1912.
3. Fujiyama, M. (1995,1996) Structures and properties of injection moldings of β-crystal nucleator-added polypropylenes, I., II. and III. *Int. Polymer Process.*, **10**, 172–78, 251–54; **11**, 271–74.
4. Varga, J., Mudra, I. and Ehrenstein, G.W. (1988) Morphology and properties of β-nucleated injection-molded isotactic polypropylene. Conference Proceedings, Antec'98 (SPE, Inc. Technical Papers, XLIV. Vol III pp. 3492–3496).
5. Fujiyama, M. (1996) Structure and properties of inflation films of β-phase nucleating agent-added polypropylene. *Int. Polymer Process.*, **11**, 159–66.
6. Tjong, S.C., Shen, J.S. and Li, R.K.Y. (1996) Mechanical behavior of injection molded β-crystalline phase polypropylene. *Polymer Engng Sci.*, **36**, 100–105.
7. Tjong, S.C., Shen, J.S. and Li R.K.Y. (1996) Morphological behavior and instrumented dart impact properties of β-crystalline phase polypropylene. *Polymer*, **37**, 2309–2316.
8. Jacoby, P., Bersted, B.H., Kissel, W.J. and Smith C.E. (1986) Studies on β-crystalline form of isotactic polypropylene. *J. Polymer Sci.: Polymer Phys.*, **24**, 461–91.
9. Karger-Kocsis, J. and Varga, J. (1996) Effects of β–α transformation on the static and dynamic tensile behavior of isotactic polypropylene. *J. Appl. Polymer Sci.*, **62**, 291–300.
10. Karger-Kocsis, J., Varga, J. and Ehrenstein G.W. (1997) Comparison of the fracture and failure behavior of injection molded α- and β-polypropylene in high-speed three-point bending tests. *J. Appl. Polymer Sci.*, **64**, 2057–2066.

Keywords: annealing, application, βα-recrystallization, hexagonit, β-nucleants, compression molding, extrusion, immature spherulites, impact resistance, injection molding, lower and upper limit temperature of formation of β-iPP, mechanical properties, morphology, skin-core structure, strain hardening, toughness.

Biaxially oriented polypropylene (BOPP) processes

Abdellah Ajji and Michel M. Dumoulin

INTRODUCTION

Orientation of polymers enhances many of their properties, particularly the mechanical, impact, barrier and optical properties. Biaxial orientation has the added advantage of allowing this enhancement in two directions, avoiding any weakness in the transverse direction. Biaxial orientation is particularly important in films, where it allows the production of thinner films having superior mechanical, optical and barrier properties and, if required, the ability to shrink when reheated. For example, shrink and barrier properties are particularly important in meat packaging, whereas good mechanical and optical properties and no shrinkage are desired in box overwrap applications of tape cassettes and CDs.

The production of oriented flat films from thermoplastic materials represents a large segment of the polymer industry. Because of its low density (900 kg/m^3) and cost, polypropylene (PP) is very well suited to packaging applications, particularly when it is processed into very thin, biaxially oriented films (12–25 μm) of outstanding mechanical and optical properties. PP has a large share of the packaging sector, right behind polyethylenes (PEs). PP demand forecasts are on the rise: in Western Europe. The demand for PP in the packaging industry was about 1.4 million tons in 1992 and is expected to rise to 2.2 million by the year 1999, an annual growth rate of about 8% [1]. The most widely used biaxial orientation processes for films are tubular film blowing and cast film biaxial orientation (or tentering). Other processes such as blow

Polypropylene: An A–Z Reference
Edited by J. Karger-Kocsis
Published in 1999 by Kluwer Publishers, Dordrecht. ISBN 0 412 80200 7

molding, compression molding and thermoforming involve also some degree of biaxial orientation but will not be discussed here.

BIAXIAL ORIENTATION PROCESSES FOR PP

Tubular film extrusion

The blown film process involves the following steps: extrusion of a tube through an annular die, orientation of the extrudate tube by inflating it into a bubble and pulling it at the same time. There has been little study of tubular film extrusion of PP in the open literature [2]. In the available studies, PP film blowing process was reported as far less stable than for PEs. This is why the process used is mainly the double bubble process. The bubble profile for PP is usually square in shape, and the more square the bubble, the more stable is the operation. More square bubbles also produce films with good optical properties. A large range of production capacities are available, with machines producing bubbles with diameters from a few centimeters to above 2 m.

The double bubble film blowing process is illustrated in Figure 1. The molten PP extruded tube is first cooled in a water bath and then taken to a second stage, the blowing stage, where it is reheated and oriented by blowing while stretched in the machine direction. The material goes through drastic temperature changes. The temperature at this stage is very important as will be discussed later.

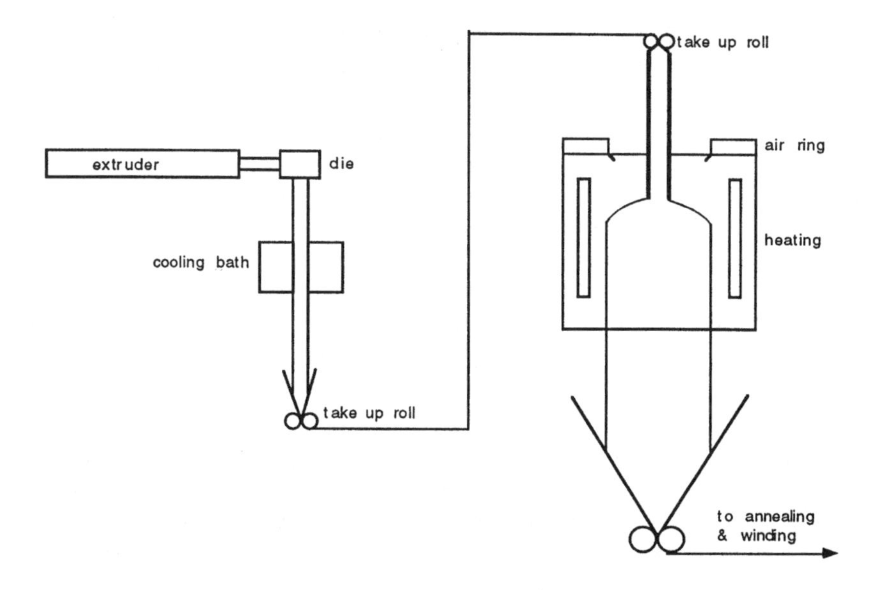

Figure 1 Tubular film extrusion.

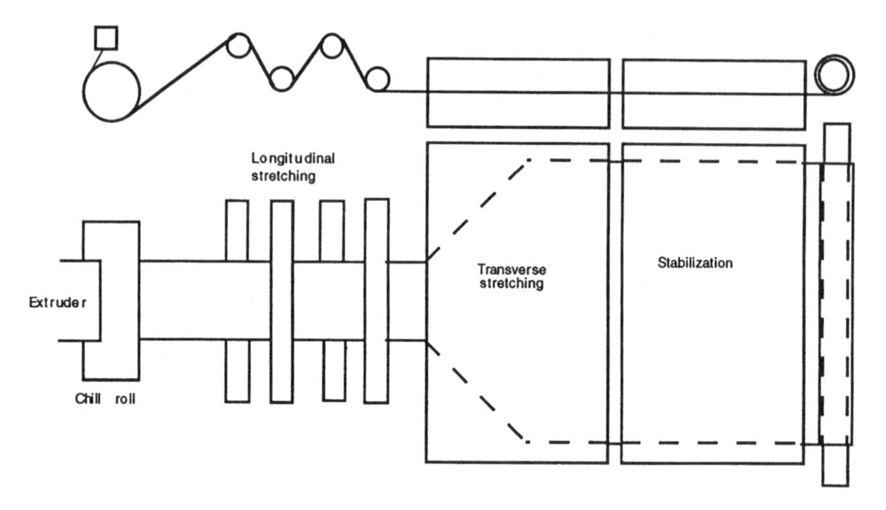

Figure 2 Film tentering process.

Film tentering

The other most important process for biaxial orientation of PP is film tentering, which is illustrated in Figure 2. An extruder delivers the film to the die, with an optional gear pump for optimum control of the overall flow rate. The die is usually of the coat-hanger configuration, with adjustable lips which allow the flow rate distribution at the die exit to be evenly distributed. The molten film is stretched on a short distance in air and quenched on chill roll which rotates at a constant speed. This primary film is stretched again in the longitudinal and transverse directions successively followed by annealing and wind-up. In processing PP to form clear, highly transparent, high-gloss films, the operation of cooling the molten film to form the cast film must be carried out as rapidly as possible. The shorter the cooling time and the lower the film temperature achieved in cooling, the better the optical properties of the film.

The orientation can be done sequentially or simultaneously in the machine and transverse directions. However, clear differences have been observed between the two [3]. The properties (transparency, modulus and strength) were lower for the sequential process. In fact, relaxation processes in the material cause a reduction in properties, since the film is stretched first in the direction of extrusion and can therefore partially relax while it is being stretched in the other direction [3]. Thus, in order to obtain the same properties as for the simultaneous process, the first stretch ratio should be higher and the temperature lowered after this stretching to compensate for and minimize relaxation. Typical data for the extruded film and the chill roll are: molten film 0.5–2 m wide and 0.3–3 mm thick; temperature of the molten film of 250–300°C; chill roll diameter of 0.5–2.5 m; and peripheral speed of 20–60 m/min.

In all these processes and depending on the application of the film, as mentioned above for the shrink properties, an annealing stage can be included or not.

STRUCTURE AND ORIENTATION DEVELOPMENT

Isotactic PP is polymorphic and may crystallize into more than one crystalline form. The most common is the monoclinic form, which is designated as α. The chains in a monoclinic PP have helical conformation and the crystallographic axis is parallel to the axis of the helix. Other crystalline forms have also been observed and depend mainly on cooling and deformation history [2].

PP crystallizes from the melt in the form of spherulitic morphology. The optical and mechanical properties depend to a large extent on the size, perfection and number of the spherulites. In general, the spherulite radius increases with increasing quenching temperature. The increase in radius is also dependent on the type of resin used. Crystallinity increases slightly with annealing time and levels off, whereas the spherulite radius remains mainly constant. It was also found in some studies that annealing of different PP samples at temperatures in the ranges of 120–140°C and 150–170°C yields two melting (thus two structures). Only annealing between 140°C and 150°C yielded a PP with a single melting point [4].

Molecular orientation is the key element in determining the properties of oriented films. Many techniques may be used for the characterization of this molecular orientation developed in PP. The most widely used are birefringence, infrared spectroscopy and wide angle X-ray scattering. Some information on the results obtained using these techniques on oriented PP are presented below and more details on these techniques and their use are presented in other chapters in this book. For a number of polymers (particularly polystyrene and PE), the biaxial orientation function was found to be related to the process stresses in the machine and transverse directions (MD and TD) [2]. In the case of PE, it was found that the relationship between biaxial orientation function factors and stresses were similar to that found in melt spun fibers. It is thought that PP will have a similar behavior.

In biaxial tubular film extrusion, quenching on the mandrel causes metastability and paracrystallinity. This disorder changes on further heat treatment to the well-organized monoclinic crystal form. During bubble forming, this change is increased further by draw in MD and by a force produced by pressure in TD. The *b* axes are randomly distributed about the machine direction at fixed blow-up ratio (BUR) and increasing take-up ratio (TUR). Increasing BUR rotates the *b* axis into a direction normal to the film surface. A similar effect can be found in tubular film operations at fixed BUR, where the TUR is decreased. When TUR equals BUR, the biaxial orientation factors are similar. The same observation was made using birefringence measurements in the MD and TD directions [3].

The use of the three techniques mentioned above for the measurement of orientation leads to much greater confidence in orientation deduced than can be obtained by using any two of the techniques. One-way and two-way (biaxially) drawn PP films, were compared using these techniques [5]. Evolution of the orientation parameters in machine, transverse and normal directions indicate that one-way drawing orient the chain axes of both crystalline and amorphous regions in the draw direction. The crystalline chains are more highly oriented than the amorphous chains and tend to orient towards the plane of the film, whereas the amorphous chains tend to be more uniaxially oriented towards the draw direction. In balanced, simultaneously two-way drawn films the crystalline chains are more highly oriented towards the plane of the film than the noncrystalline chains. For a sequentially, equibiaxially drawn PP film, the orientation of the chain axes of both the crystalline and amorphous regions were found to be higher in the second draw (i.e. transverse) direction than in the first draw direction. The orientation of the crystalline chains was very close to the plane of the film, whereas the amorphous chains were almost uniaxially oriented with respect to the second draw direction. In all the films there is a strong tendency for the *b*-axes of the crystallites to align normal to the plane of the film [5].

Morphological studies determined that the crystalline orientation is not affected by molecular weight distribution (MWD). However, the amorphous orientation, at higher draw temperatures, show a small increase as MWD becomes broader. MWD does not affect the percent crystallinity but the long period spacing (LPS) show a small decrease as MWD broadens.

PROPERTIES OF BIAXIALLY ORIENTED PP FILMS

Optical properties

Haze is a measure of the milkiness of films and is caused by light being scattered by the surface roughness and inhomogeneities in the film. It has been studied carefully and, in general, it has been shown in almost all cases to be associated with scattering from the surface rather than the bulk of films. The magnitude of surface roughness has been found to be primarily related to the crystallization of the polymer. Rough film surfaces and high levels of haze are associated with high levels of crystallinity. Melt elasticity plays a secondary role in determining roughness and haze. In film blowing, it was found that haze increases with increasing frost line height, and decreasing TUR and BUR [2]. Narrowing MWD decreases haze. For films obtained by the tenter process, transparency decreased as the cooling roll temperature increases, and it increased with the stretching velocity and stretch ratio [3]. In general, the larger the orientation, the better (lower) the haze if temperature can be kept low.

Gloss on the other hand is a measure of the 'sparkle' on the film or the ability to reflect incident light specularly and produce a sharp image of any light source. For the same quench conditions, draw temperature has the most significant effect. An increase from 150°C to 160°C can result in a significant fall in gloss. There appears to be an optimum MD and TD stretch ratio of 6 above and below which gloss decreases.

Mechanical properties

The mechanical properties are usually determined from stress-strain tensile curves. Figure 3 shows the tensile strength and modulus in the most highly drawn direction as a function of the effective draw ratio in that direction. The data are taken from different papers involving different types of PP and different processing techniques. In spite of this wide range, the trend can be clearly seen. Undrawn PP has a modulus

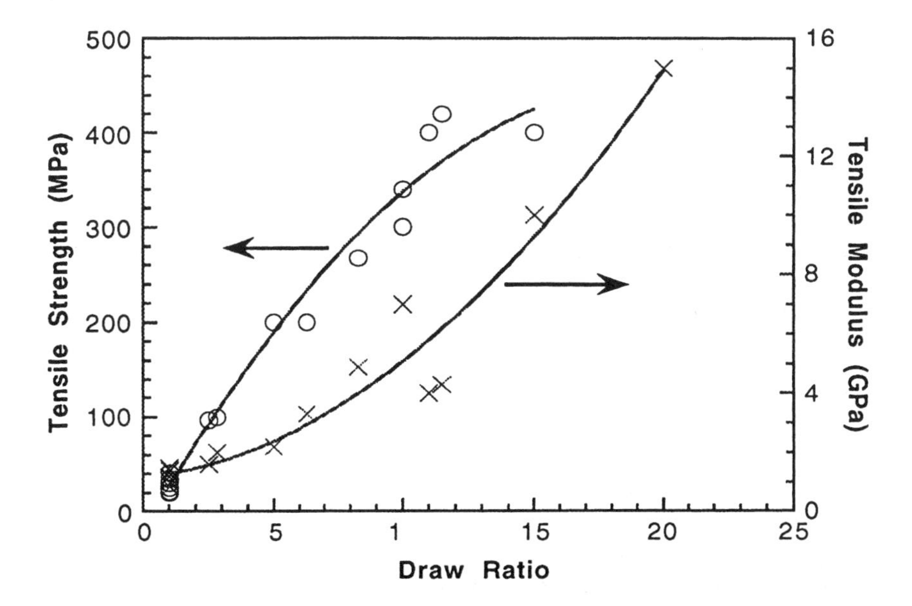

Figure 3 Enhancement of mechanical performance resulting from PP orientation.

around 1.3 GPa and a strength around 30 MPa. Drawing to a ratio of 5 increases these to about 2.4 and 190 respectively. In cases where higher draw ratios can be achieved, the gains are even greater. Yield strength was observed to be the most affected by TD draw ratio. Tensile strength variation with MD and TD shows a maximum at the same draw ratio (around 6). This tensile strength was maximum for a temperature of 145°C. The elongation at break is mostly affected by MD draw ratio. For films obtained by the tentering process, the stress at break was found to have an optimum value for a certain cooling roll temperature (below and above which it decreases), and it increased with stretching velocity and stretch ratio. The elongation at break had a reverse behavior to that observed for the stress at break [3, 4]. As MWD becomes broader, modulus and strength increase slightly, whereas elongation and shrinkage decrease slightly. These effects are attributed to decreases in molecular mobility and increases in the number of tie molecules present in broad MWD resins. Biaxially oriented films show also improved toughness and slow crack propagation.

With respect to recycling, it is to be mentioned that tensile properties of oriented polypropylene tapes can be affected following the addition of recycled process waste at various concentrations. A significant loss in tape strength occurs when recycled or waste PP is added because of thermo-oxidative degradation. Useful tapes with acceptable tenacities can be achieved using up to 25% w/w of some recycled waste PPs in the absence of stabilizers.

Shrinkage

An annealing stage is usually included when the heat shrinkage is to be minimized. For processes where this annealing stage is present, for the same annealing conditions, it was observed that increases in TD draw ratio and draw temperature had the most significant effect on shrinkage. For example, shrinkage can fall from 12% to 5% when the draw temperature increased from 135°C to 160°C. Variation in MD draw ratio in the range 5–7 had little effect. In all cases, shrinkage below 4% is difficult to achieve.

Finally, it is to be mentioned that permeability of oriented films to different gases is significantly reduced compared to unoriented films.

REFERENCES

1. Beer G. (1996) Polypropylene, *Kunststoffe Plast. Europe*, **86**, 14–16.
2. White, J.L. and Cakmak, M. (1988) Orientation, crystallization and haze development in tubular film extrusion, *Adv. Polymer Technol.*, **8**, 27–61, and (1990) Orientation, in *Encyclopedia of Polymer Science and Technology*, (eds F.M. Herman, N.G. Gaylord and N.M. Bikales), Wiley, New York.

3. Nordmeier, K. and Menges, G. (1986) The influence of processing conditions on the optical and mechanical properties of extruded and biaxially stretched PP films. *Adv. Polymer Technol.*, **6**, 59–64.
4. Jabarin, S.A. (1993) Orientation and properties of polypropylene. *J. Reinforced Plastics Composites*, **12**, 480–488.
5. Karacan, I., Taraiya, A.K., Bower, D.I. and Ward, I.M. (1993) Characterization of orientation of one-way and two-way drawn isotactic polypropylene films, *Polymer*, **34**, 2691–2701.

Keywords: biaxial orientation, film blowing, film tentering, structure of oriented PP, mechanical properties of oriented PP, BOPP, packaging, cast film, tubular film, inflation film, amorphous orientation, crystalline orientation, morphology, haze, optical properties, gloss, shrinkage, permeability

Bumper recycling technology

Norio Sato, Hidero Takahashi and Toshio Kurauchi

INTRODUCTION

Background

The necessity to develop an appropriate technology for the recycling of automotive plastic parts has become a key issue in recent years, in terms of environmental protection and preservation of resources. It is obvious that plastics recycling in the automotive industry will focus on bumpers, which are relatively large, and can be dismantled into their constituents easily.

Toyota CRDL and Toyota Motor Company have developed a 'bumper to bumper recycling system' for painted thermoplastics (elastomer modified polypropylene and Toyota Super Olefin Polymer (TSOP), hereafter referred to as TP) bumpers.

Problems for bumper to bumper recycling

Figure 1 compares the physical properties of the virgin and the recycled material obtained by crushing of painted TP bumpers followed by melt pelletizing. In such a simple recycling method, the impact strength and brittleness temperature in particular, which are critical characteristics of bumpers, tend to deteriorate. The cause of the deterioration of physical properties is the relatively large size of paint fragments dispersed in the TP resin. They act as stress concentrators and initiate brittle fracture. In addition, they have a negative effect on the surface appearance of the moldings.

Polypropylene: An A–Z Reference
Edited by J. Karger-Kocsis
Published in 1999 by Kluwer Publishers, Dordrecht. ISBN 0 412 80200 7

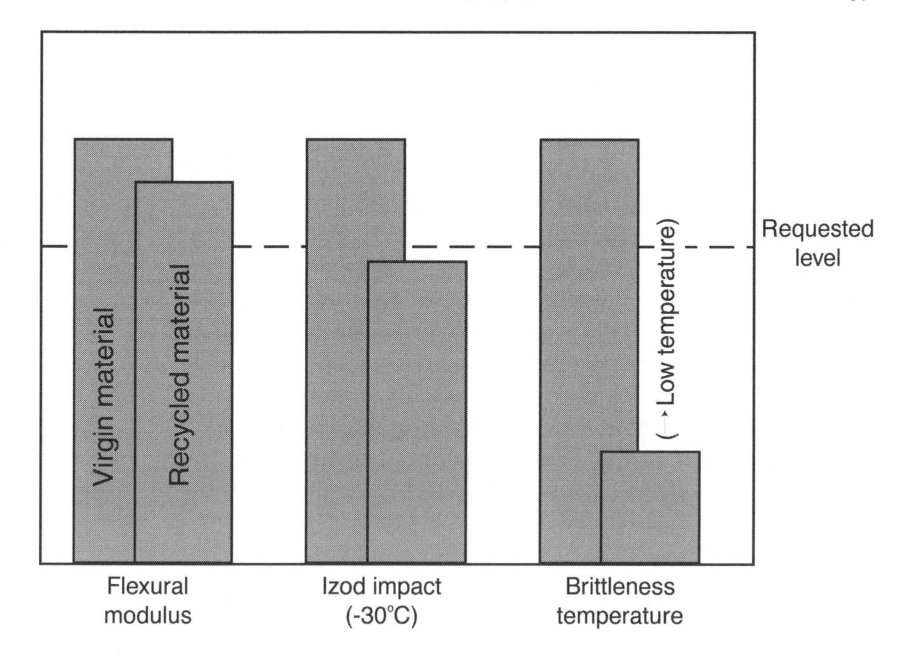

Figure 1 Mechanical properties of virgin and recycled material.

Treatments for elimination of paint film of bumper

Table 1 shows results of various treatments aimed at the elimination of the paint film. The paint separation method specified in (a) is as follows. Making use of the difference in specific gravity between the paint film (1.1–1.6 g/cm^3) and the TP resin (0.96–0.98 g/cm^3), a painted TP bumper is crushed into fine pieces and placed in water in order to sort out material floating up to the surface. This is remelted in an extruder where the paint film is separated by means of a melt filter. This method does not result in adequate paint film separation, yielding a recycled material of poor quality.

Table 1

	Regenerated material	*Quality*	*Productivity*	*Economy*
Paint separation	Separation by specific gravity + filtration	×	○	○
Paint peeling	Shotblasting (waterjet)	○	×	×
	solvent (acid. alkali, metallic salt. etc.)	× – ○	× – ○	×
Paint decomposition	Pressurized hydrolysis	○	○	○

○, Without problem; ×, With problem.

The paint peeling methods specified in (b) are divided into a mechanical peeling method and chemical peeling method. For the mechanical peeling method, shot-blast and waterjet methods were evaluated. The shot-blast method is to spray particulate agents on a painted TP bumper surface by means of compressed air in order to peel off the paint film. The water-jet method uses high pressure water instead of particulate. Both of them are unsuitable for the treatment of such bulky products with complex geometries as bumpers.

The chemical peeling method is to peel off the paint film by immersing the bumper in a solvent of alkali, etc. TP resin may degenerate and the peeling may require a long time, depending on the type of the solvent, which means problems in terms of quality or productivity. Even if there is no problem in quality of productivity, the waste fluid treatments or the solvent itself is too expensive for an industrial line. All of these conventional treatments were proved to be unsuitable for the recycling of bumpers.

PRESSURIZED HYDROLYSIS TECHNOLOGY [1, 2, 3, 4]

With the pressurized hydrolysis technology, water under high temperature and high pressure is applied to hydrolyze and break the crosslinks of the paint film, and then a kneading extruder is used to convert the decomposed film into particles and disperse them in the TP resin. Two types of the processing system have been developed. One is the autoclave reaction and screw extrusion processing system (batch system), and the other is the twin screw reactive extrusion processing system (continuous system).

Autoclave reaction and screw extrusion processing system [1, 2, 3]

Figure 2 shows autoclave reaction and screw extrusion processing. Using an autoclave reactor in hydrolysis, painted TP bumpers are crushed and dipped in the water or exposed to water vapor of 150–160°C under a high pressure for 45–90 min. The pressure of the reactor was the same as the vapor pressure of the water, i.e. 0.48 MPa at 150°C, and 0.82 MPa at 160°C, respectively. After hydrolysis, the specimens were subjected to centrifugal dehydration, and then fed into the kneading extrusion for compounding the TP resin and decomposed paint film and for pelletizing.

Twin screw reactive extrusion processing system [4]

Figure 3 outlines the twin-screw reactive extrusion processing system. The crushed bumper is introduced to a twin-screw extruder whose

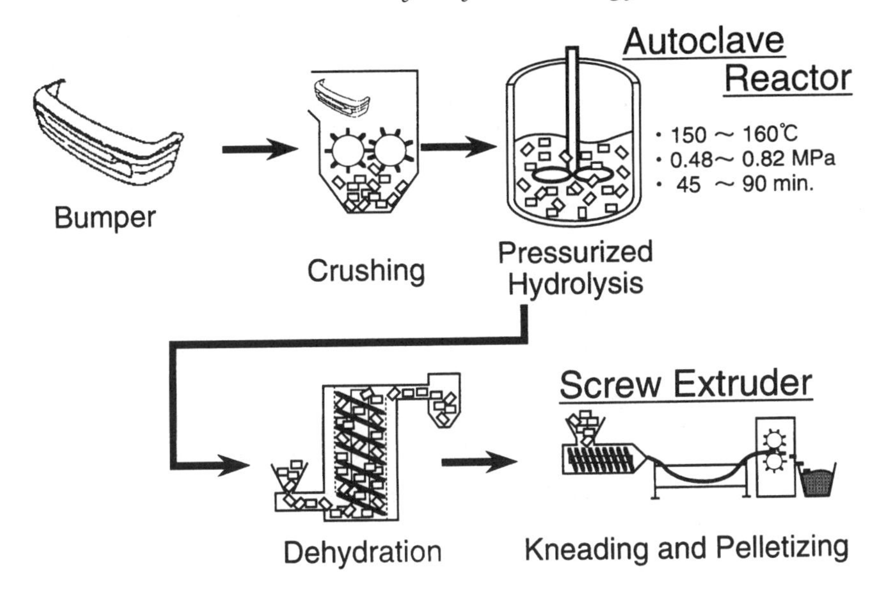

Figure 2 Autoclave reactor and screw extrusion system.

cylinder is divided into three zones: melting, hydrolysis reaction, and mixing. In the melting zone, the crushed TP bumper is melted and kneaded. The paint film on TP resin is fragmented by shear stresses of kneading. In the next hydrolysis reaction zone, water is injected into the cylinder and the vapor is mixed with the melted resin at a temperature of 250°C. The paint film in TP melt is hydrolyzed in a short time, being in intimate contact with the heated and pressurized water vapor. In the next

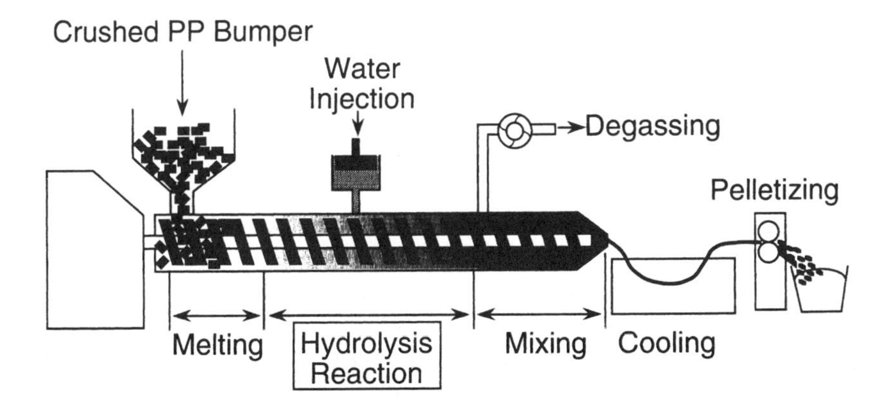

Figure 3 Twin-screw reactive extrusion system.

mixing zone, the hydrolyzed paint film is converted into particulate and dispersed in the TP resin. The volatile fraction of hydrolyzed paint film and the water vapor are evacuated by venting and the compound is pelletized. This method is accompanied by great benefits for both environment protection and economy.

Mechanism of hydrolysis

The major component of the paint is an alkyd-melamine or acrylic-melamine resin. When hydrolysis is completed, the dimethylether bonds ($—CH_2—O—CH_2—$) giving the crosslinks, are broken. This can be easily detected by infrared spectroscopy.

Size of the decomposed paint film in the recycled materials

The size of the paint fragments after mechanical crushing and extrusion with melt filtering is in the range of 200 μm. The size of the paint after the pressured hydrolysis technology is reduced to 15 μm, and the interface with the TP resin is not clearly observed. This indicates that the compatibility and adhesion performance at the interface are improved.

Mechanical properties of recycled materials

The mechanical properties of the recycled materials depends on the conditions of the hydrolysis. The brittleness temperature (JIS K7216) of the recycled materials changed with the hydrolysis time (Figure 4). The

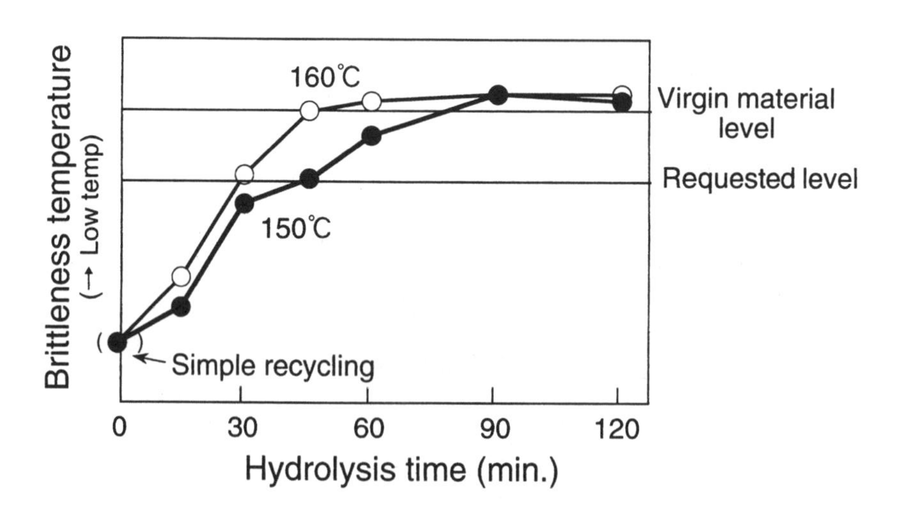

Figure 4 Brittleness temperature of the recycled material.

temperature and duration of the hydrolysis needed to reach the same brittleness temperature as the virgin resin were as follows: at 150°C, 90 min or more; at 160°C, 45 min or more. The other mechanical properties passed the required level under the above hydrolysis condition, also the surface quality did not differ from the virgin material. All of these positive effects are due to the fine particles of the decomposed paint film showing good compatibility to the bare TP resin.

BUMPER DISCRIMINATION TECHNOLOGY

For the post-consumer bumper recycling, the collection of the bumpers from the dealers is done by the Toyota Motor Company. Among the recovered bumpers, bumpers of TP and polyurethane RIM (PU-RIM) are mixed. Furthermore, nonrepaired bumpers and repaired ones are also involved. For bumper to bumper recycling, however, only the nonrepaired TP bumpers are suitable, so that those have to be sorted out. To do this, bumper discrimination methods were developed. One of them is used for the discrimination between TP and PU-RIM bumpers, while the other makes a distinction between nonrepaired and repaired TP ones.

TP Bumper/PU-RIM bumper discrimination

Considering the difference of dielectric constant value between polypropylene (PP) and polyurethane (PU), a dielectric discrimination method with a suitable handy device has been developed. Based on the dielectric constant value obtained by considering thickness and electric capacity of the part, the device indicates a TP bumper or a PU-RIM one with a pilot lamp. Perfect discrimination is achieved by this method.

Nonrepaired TP bumper/repaired TP bumper discrimination [5]

The pressurized hydrolysis technology is available for all TP bumpers of Toyota cars, which are painted with acrylic-melamine resin or alkyd-melamine resin. However, the recovered bumpers contain some repaired ones which are repainted with urethane paint film. The pressurized hydrolysis technology is not available for the urethane paint film because this film is not hydrolyzable with water vapor in such conditions. To maintain the high performance of the recycled material, the repaired bumper must be sorted from the recovered TP bumpers. Considering the difference of the chemical reaction of the paint films with dye agents, a paint film dyeing method was developed. A nonrepaired bumper which is painted with acrylic- or alkyd-melamine resin becomes red and fluorescent color when a mixture of dye agent of Acid Red 52 and a solvent of lactic acid are applied. On the other hand, a repaired bumper which is repainted with urethane paint film hardly takes a color. The time for dyeing is about 1 min, so it is applicable in the recycling process. This method yields a perfect discrimination.

POST-CONSUMER BUMPER RECYCLING SYSTEM

A schematic diagram of post consumer bumper recycling system developed in Toyota CRDL and Toyota Motor Company is shown in Figure 5, which consists of four processes. The first is the TP bumper/PU-RIM bumper discrimination method using a dielectric constant measuring method. The sorted TP bumpers are transferred to the next process step and the rejected PU-RIM bumpers are directed to another PU-RIM bumper recycling line. The next step is the nonrepaired TP bumper/repaired TP bumper discrimination by the paint film dyeing method. The sorted nonrepaired TP bumpers come in the next process and the rejected repaired TP bumpers go to another recycling line. The third step is the mechanical crushing process to cut the bumpers into small pieces which can be fed to the extruder. The last step is the reactive extrusion process involving melting, hydrolysis reaction, kneading and pelletizing in a twin screw extruder. The line is now under operation of Toyota Motor Company and cooperative companies.

CONCLUSIONS

The bumper to bumper recycling system based on the pressurized hydrolysis technology and bumper discrimination technology has been

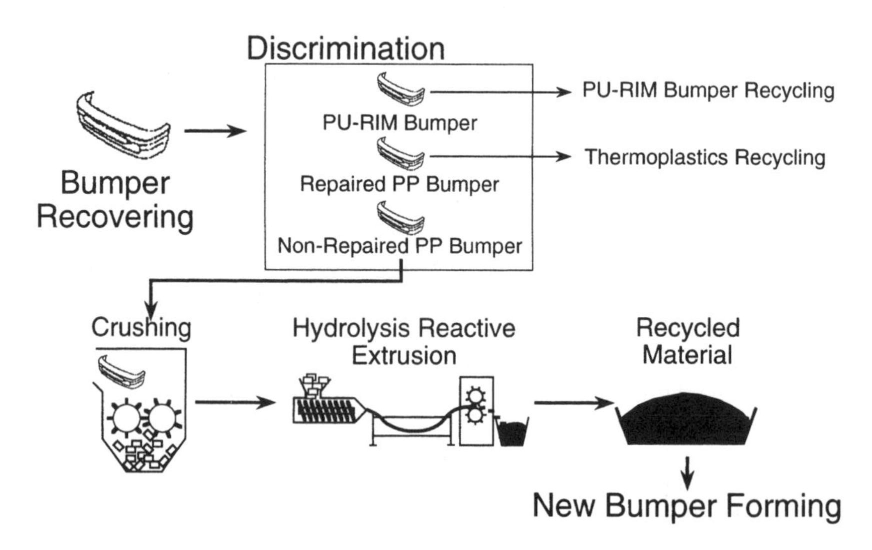

Figure 5 Post-consumer bumper recycling system.

proved to be superior to conventional recycling methods. The quality of the recycled material by this method is equivalent to that of the virgin material. The productivity is also good because the material can be treated in crushed form, without a long processing time. Moreover, a particular solvent and those post-treatments, such as washing, neutralization, etc., are not required by this technology. No waste is generated, and the economical feasibility of this bumper to bumper recycling is hence high.

REFERENCES

1. Sato, N., Takahashi, H. and Kurauchi, T., SAE Technical Paper Series 961029.
2. Ikai, T., Iwai, H., Sakata, I., Ikeda, S. and Sato, N. (1993) *SAE Preprint*, Japan, **931**, 137.
3. Ikai, T., Iwai, H., Sakata, I., Ikeda, S. and Sato, N. (1993) *Trans. Mat. Res. Soc. Japan*, **18A**, 763.
4. Ikeda, S., Kato, T., Inoue, S. and Sato, N. (1994) *SAE Preprints*, Japan, **946**, 205.
5. Handa, J., Iwai, H., Ikeda, S., Suzuki, T., Sato, N., Ohta, T., Matsusita, M. and Sugiyama, S. (1995) SAE Preprints, Japan, **952**, 129.

Keywords: recycling, pressurized hydrolysis, paint decomposition, reactive extrusion, automotive industry, post consumer bumper, IR, elastomer-modified PP.

Calendering of polypropylene

T.P.A. Järvelä and P.K. Järvelä

The most-used thermoplastic material in calendering is PVC, which has been calendered for several years with few difficulties and the process is now well established. Other important calenderable thermoplastics are some styrenic copolymers, such as ABS and ASA, polyurethane, some polyamide grades and nowadays also polyolefins [1, 2] Among polyolefins, polypropylene (PP) is one of the most interesting materials because it has a great potential to replace PVC in existing markets and to penetrate in new ones.

Recent developments in PP quality, together with some ecological credentials of the product and the process, are more and more attracting the attention of manufacturers of calendered products and machinery. The main problem in PP calendering is definitely the lack of experience [3, 4].

Calendering is a versatile processing method to manufacture polymer sheets and coatings. Calendering may be very competitive to various film extrusion techniques specially while manufacturing small batches. Advantages of calendering are, for example, less material loss due to the change of colour and grade, low residual stresses in the product and the possibility to print different three-dimensional patterns on the produced sheet or coating [2, 4].

CALENDERING PROCESS

A calendering line can be either a line for sheet manufacturing or a line for coating a carrier material. The main components of a calendering line in the case of coating are schematically shown in Figure 1. While using

Polypropylene: An A–Z Reference
Edited by J. Karger-Kocsis
Published in 1999 by Kluwer Publishers, Dordrecht. ISBN 0 412 80200 7

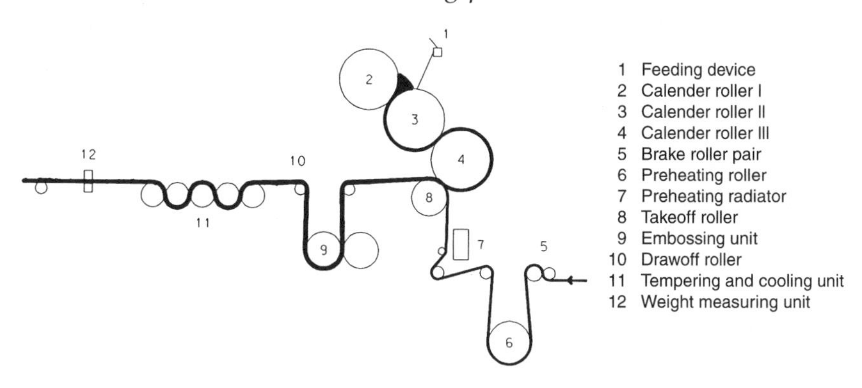

Figure 1 The main components of a coating calender [1].

PVC, premixing of the resin with plasticizers and other components in a dry blender (turbomixer) is required before processing. With PP, there is no need for premixing since the resin (and coloured masterbatches in case of coloured products) can be dosaged directly into the melt mixer. As melt mixers, extruders (planetary or twin-screw) are mostly used. The equipment can be similar to that used with PVC but it must be optimized for each product. The purpose of a melt mixing stage in a PP calendering process is to mix the pigments and fillers into the resin and melt the material. The calender can be feeded directly by the mixing extruder or via an additional feeding device between the extruder and calender [1, 4].

A calender consists of two or more heated rolls with different configurations depending on the product to be manufactured. All different roll configurations (I, L, F, Z, etc. [2]) used for PVC can also be optimized for the processing of PP. The purpose of the heated rolls in the calender is to convert the molten material into a product that has the desired thickness. The thickness decrease is achieved by squeezing the molten plastic through the nips between the heated rolls. The differences in rotation speeds and directions of the rolls drive the molten material to pass the calender [2, 4].

During calender coating, the molten PP film that is covering the heated roll after passing through the nip has to be laminated on a carrier material by using a rubber coated roll. In the sheet manufacturing, the purpose of rubber coated roll is to serve as a take-off roll, i.e. to detach the sheet from the surface of the roll. In both cases, the molten sheet starts to solidify immediately after taking off from the roll. During this phase of the calendering line, an efficient cooling system is needed. A very common and effective cooling system consists of several small diameter rolls of high thermal exchange efficiency [1, 2, 4].

In the calendering line, often a surface treating step such as flame, corona or plasma treatment, is also implemented. It is also possible to cut the produced sheet into pieces as needed. The surface treatment is very important if the calendered PP product is going to be lacquered or printed afterwards. Very often the product is only taken on rolls from which the sheet can be derolled during further processing steps such as lacquering or printing [1, 4].

PROPERTIES OF CALENDERABLE PP

There is a lack of experience on the calendering of PP as opposed to PVC which is a well-accepted calenderable thermoplastic. Therefore it is reasonable to compare PP with PVC when considering the calendering process. The main differences between these two materials are: shape of raw material, structure, heat resistance and formation of toxic vapours during processing [4].

PVC is normally delivered as powder which must be premixed with additives (usually by the processor itself) before calendering. By contrast, PP is sold in form of pellets or granules which are already stabilized and modified by the PP manufacturer [4].

PVC is an amorphous thermoplastic while PP is semicrystalline, which causes differences in their behavior during melting and cooling in calendering. Since PP has a crystalline structure, there is a high density difference between the solid and melt states and the melting temperature range is very narrow compared to PVC. During cooling in the crystallizing PP high stresses generate which can cause defects in the products. Therefore, the control of the cooling process during calendering of PP is the key issue [4]. The processing temperatures for PP are higher than those normally for PVC. The typical working temperature range for PP is between 170°C and 200°C which is about 30°C higher than with PVC. The processing temperature is strongly dependent on the type of the PP. The main difference in heat resistance is that even well stabilized PVC cannot withstand more than 30 min at the processing temperature without degradation, while PP is considerably less sensitive for such thermal loading. The formation of toxic vapours does not take place during calendering PP, in contrast to the processing of PVC where the amount of these vapours is rather high. The main reasons for this difference are that PP does not degrade as easily as PVC, and it does not contain volatile plasticizers (calendered soft PVC grades contain 18wt.% or even more plasticizers) [3, 4].

In respect to the calendered products, the most important differences between PP and PVC are density, rigidity, structure and some ecological factors. Since PP has lower elasticity modulus than hard PVC, a PP film should have a higher thickness in order to reach the same rigidity level.

But, due to its lower density, the PP film can still be lighter than the PVC film at the same stiffness. Due to its crystalline structure the PP film is translucent, while the amorphous PVC film can also be transparent [4].

Processing of PP is free of toxic vapour and also the processing wastes are easier to recycle or dispose than those of PVC. A very important feature of PP compared to PVC is the absence of plasticizers. PVC products lose plasticizers during their service life which may cause problems such as odor, fogging etc. With PP products, no such problems appear. PP products can also be recycled or burned after their life cycle by using already known methods and techniques [4].

The calendering process of PP is faced with a new challenge due to the appearance of metallocene PPs. During the recent years, the development of these materials has resulted in many improved properties compared to conventional PP grades. Also, they are promising materials for calendering. Nowadays the greatest problems with metallocene PPs are their limited availability, associated with the lack of experience with their processing. In the future, it is expected that some metallocene PPs will be tailored in respect to the processability for calendering [2, 5].

APPLICATIONS

Calendered products of soft PP grades can be used as synthetic leathers for various purposes, such as for car interior trims. The semirigid and rigid PP types can be used for tablecloths and shower-tents or they can also be calendered to films having thickness 100–200 μm which can be laminated to sheets or wood boardings used, for example, by the building industry. Other possible applications for calendered PP are, for example, different floor coverings, rigid plates for thermoforming, packaging materials, etc. [1, 4].

Almost all windable webs can be used as a carrier material in the calender coating process. The commonly used carriers are paper, cardboard, fabrics, glass and mineral fiber fabrics. Also, temperature and tension-sensitive carriers, such as polymer-based textile fabrics, can be used as a carrier material. Possible articles produced by calender coating are, for example, different types of wallpapers, book binding materials, coated textiles, multi-layer structures and technical coatings [1].

REFERENCES

1. Schmidt, H. (1988) The roller melt technique in view of new discoveries and developments with the production of coatings and films. *Technical Information*, Saueressig Engineering, April 1988.
2. Kopsch, H. (1978) *Kalandertechnik*, Hanser, Munich.
3. Prentice, P. (1981) Surface irregulations of calendered polypropylene. *Polymer*, **22**, 250–254.

4. Lualdi, R. (1995) Polypropylene calendering on the stage. *Macplas*, February, 39–41.
5. Metallocenes Special Report (1996) *European Plastics News*, April, 26–35

Keywords: calendering, calender coating, film, sheet, PVC replacement application, metallocene PP.

Commingled yarns and their use for composites

K. Friedrich

THE PREFORMS

Commingled yarns are one of the possible preforms used for continuous fiber reinforced thermoplastic composites in order to solve the problem of high melt viscosity during impregnation and consolidation as the required steps for manufacturing of technical components. The preforms can be considered as 'dry' prepregs, in which the solid resin is physically divided and more or less evenly distributed among the reinforcement fibers. Division of the polymer can be made using either powder or the fiber form. Glass fiber (GF) manufacturers (e.g. Vetrotex, France) especially took advantage of their expertise in fiber spinning to develop a commingling process for glass and thermoplastic filaments (Figure 1 (a)) [1]. In their patented one-step process, Vetrotex succeeded in producing commingled yarns with a rather uniform distribution of small glass and thermoplastic fiber bundles, a good consistency of the glass content, a wide range of the GF volume fraction (20–50 vol.%), and the price level of industrial glass composites. Polypropylene (PP) has been their first choice for achieving a good balance between versatility of performance, ease of processing, and price (tradename Twintex®). More recently, a material with PET (polyethylene-terephthalate) fibers was developed for applications needing better thermal resistance and higher strength.

Hoechst AG of Frankfurt am Main, Germany, has been granted a patent (WO 94/20658) for improvements to fiber yarn intermingling which make it possible to produce filaments completely free from broken

Polypropylene: An A–Z Reference
Edited by J. Karger-Kocsis
Published in 1999 by Kluwer Publishers, Dordrecht. ISBN 0 412 80200 7

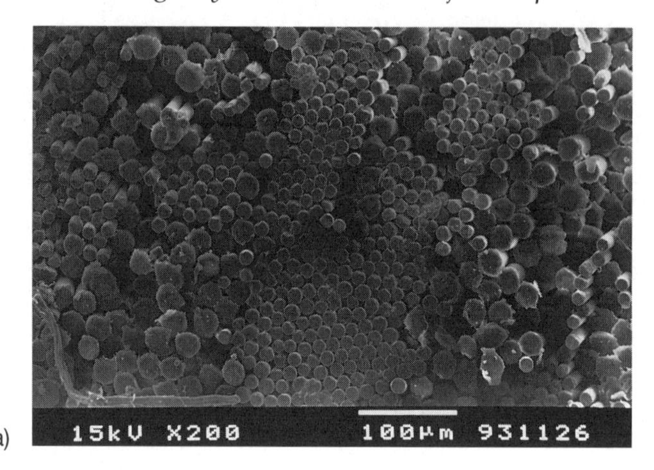

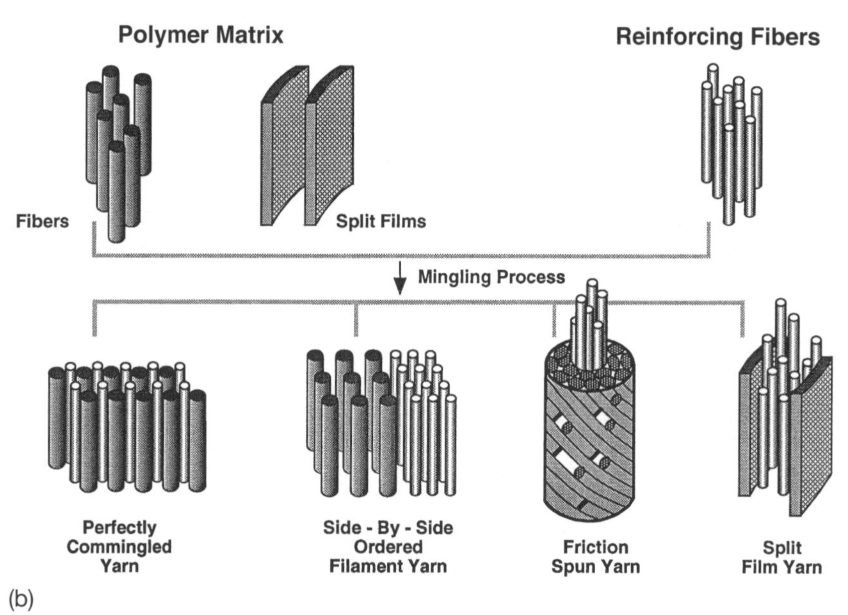

Figure 1 (a) Scanning electron micrograph of an unconsolidated commingled GF/PP bundle (variation 1 with 45 vol.% glass fibers presented in smaller diameter). (b) Schematic illustration of different hybrid yarn structures.

fibers. The process is also suitable for commingling high modulus fibers, such as carbon, with low modulus thermoplastic filaments which form the matrix in subsequent shaping and melting operations for thermoplastic composites. Polyetheretherketone (PEEK), polyetherimide (PEI), polyphenylene sulphide (PPS) and polyamide 66 (PA66) are all examples of thermoplastic polymers that have been used in filament form in the past for the production of dry, flexible, prepreg fabrics [2–4].

The process of commingling different fibers to produce a single multi-filament hybrid yarn has previously been carried out at room temperature using air jets. While this process has been successful on the whole, attrition and stiff, higher modulus fibers leads to fiber breakage and problems of cohesion in the commingled product. The best results were obtained using air jets with jet pressures of 0.2–0.4 MPa. Hoechst claims that yarns produced with this new technique (at elevated temperatures) result in better properties of the final product than cold processed yarns and, because the yarns are smoother and have a higher cohesion, they are more suitable for weaving.

In fact, although the basic product of commingled yarns is a direct roving, it can be easily used as an input for textile operations. The rovings can be woven, knitted or braided and, therefore, can take advantage of the wide range of structures offered by the textile industry [1]. Mäder [5] studied effects of the textile hybrid yarn structure (commingling, friction spun, side-by-side arranged) on the homogeneity of distribution of the reinforcing fibers in the polymer matrix on the mechanical performance of the composites (Figure 1(b)). Different textile structures, such as various knits, wovens and stitch-bondings, were also investigated in comparison with continuous reinforced unidirectional composites.

In material selection and development for PP-based composites, GF with special sizings and both PP-filaments and split-films with and without polymeric coupling agents were used to produce test specimens. These materials were compared to commingled yarns (Twintex®) and different textile preforms produced in pilot studies by various industrial companies (in the framework of a BRITE/EURAM project, No. BE II 7256). Monoaxial and biaxial weft-inserted warp knitted structures carrying both reinforcing GF and matrix material in the form of split-films combine the low cost of split PP-film, the high productivity of knitting and the technical advantages of a highly drapable non-crimp structure. Monoaxial warp-knitted and linen weave structures using Twintex® GF/PP material were also examined. The final aim was to develop a split-film co-knitted structure at low cost and sufficiently high mechanical properties to fulfil the demands of the end-users. Table 1 gives an overview of the mechanical properties obtained so far with textile architectures of warp knits in comparison to a woven fabric. Considering the relatively low fiber volume fraction (23 and 25 vol.% in load direction, respectively), the tensile properties of the split-film structures are rather high. Upscaling with the rule of mixtures, one obtains values for the Young's modulus that are of the order of monoaxial Twintex® material. The biaxial split-film material even seems to be slightly better

Table 1 Properties of monoaxial and biaxial warp knits of split-film and commingled yarns compared to woven fabrics of commingled yarns [5]

Characteristics	*Warp knit split-film*		*Warp knit commingled Twintex®*	*Woven commingled Twintex®*
Fabric structure	*Monoaxial*	*Biaxial*	*Monoaxial*	*Biaxial*
Fiber volume fraction [%]	23	39 (25)	46	37 (37)
Tensile modulus [GPa]	15.2	17.7	32.5	12.7
Tensile strength [MPa]	348.4	447.1	704.0	297.6

than the composite made from the woven Twintex® material in terms of both Young's modulus and strength. This is, however, not surprising, since a non-crimp fabric preform generally gives higher mechanical composite performance than a woven preform.

More fundamental studies with the material under special consideration here, i.e. GF/PP commingled yarns or fabrics, were thoroughly preformed with regard to its isothermal consolidation behavior and to an optimization of its flexural stiffness by Cain *et al.* [6]. In the following, some further details about impregnation, consolidation and resulting properties of GF/PP will be discussed.

IMPREGNATION QUALITY

In the work of Klinkmüller *et al.* [7], different glass fiber (GF)/ polypropylene (PP) commingled yarns (supplied by Vetrotex, France) were investigated. The exact specifications are listed in Table 2. The standard material (800 tex) had a glass fiber content of 19 vol.%, a fiber diameter of the glass and the PP fibers of 14 μm and 20 μm, respectively, and it was well mixed. For the other materials, one of these characteristics was varied, either the glass fiber content (variation 1), or the quality of mixing (variation 2). Layers of unidirectional 'prepreg' sheets were manufactured by winding the yarn onto an aluminum plate, with subsequent welding of the wound yarns at both ends of the plate. Consolidation of the laminates was performed by using a small steel mold and a laboratory heat press. The cold and filled mold was put into the already heated press and set under slight pressure, just to ensure a sufficient contact for the heat transfer. When the processing temperature was reached, the whole pressure was applied immediately. From then on the time of impregnation was measured.

Table 2 Specifications of GF/PP commingled yarn systems studied

Material properties	*Standard*	*Variation 1*	*Variation 2*
Tex of bundle	800	460	800
Tex of gass fibers	320	320	320
Volume fraction of glass, V_g	0.19	0.45	0.19
Densities:			
Glass ρ_g (g/cm^3)	2.56	2.56	2.56
PP ρ_{pp} (g/cm^3)	0.906	0.906	0.906
Radius of fibers:			
Glass r_g [μm]	7	7	7
PP r_{pp} [μm]	10	10	10
Melt flow index (MFI)	20	20	20
Mingling quality	good	good	poor

The following parameters have been varied:

temperature: 155–195°C;
pressure: 0.38–3 MPa;
time: 0.5–10 min;
cooling rate: 17°C/min.

The void content, taken as one characteristic of the consolidation status, was used to define optimum processing conditions. Characterization of mechanical properties as a function of impregnation conditions was carried out by a small three-point bending and a shear test facility.

Based on a rectangular model assumption of the glass fiber/polymer fiber agglomerations, the following equation can be used to calculate the necessary impregnation time (t) in order to achieve a void content X_v, which is lower than a certain limiting value:

$$t = \frac{2\,\eta\,k_{zz}\left(\dfrac{h_0\,(1-V_f) - h_{1(z,0)}\,X_v}{(1-V_f)\,(1-X_v)}\right)^2\left(\dfrac{V_a}{V_f}+1\right)}{r_{gf}^2\,(p_a - p_0)\left(\sqrt{\dfrac{V_a}{V_f}}-1\right)^3}$$

where

η = viscosity of the polymer matrix at a certain processing temperature;
k_{zz} = the permeability constant for matrix flow perpendicular to the glass fibers;
h_0 = half height of a glass fiber agglomeration;
$h_{1(z,0)}$ = half height of total agglomeration (including polymer part) at time 0 s;

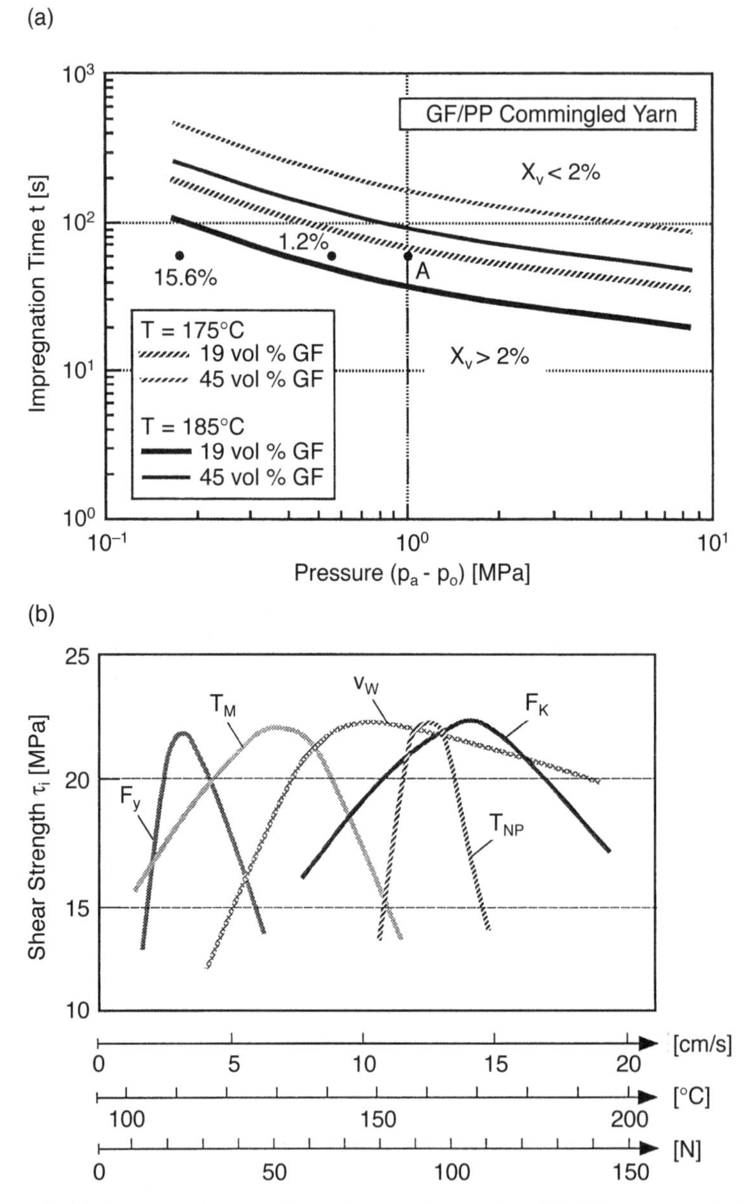

Figure 2 (a) Calculated processing windows for various GF/PP commingled yarn qualities (standard versus variation 1). Point A represents processing parameters (1 MPa; 60 s) under which the standard material has reached a void content of less than 2% when processed at 185°C. The two experimental values given (15.6% and 1.2%) confirm this modelling result. (Major support for these studies came from DFG FR 675/11–2). (b) Interfacial shear strength as a function of filament winding parameters, achieved with GF/PP commingled yarn (800 tex).

V_f = pressure dependent fiber volume content;
V_a = maximum fiber volume content (0.83);
p_a = applied pressure;
p_0 = atmospheric pressure;
r_{gf} = radius of the glass fibers.

This equation leads to predictions of the impregnation times as a function of pressure, temperature and quality of the commingled yarn that agree quite well with the experimentally determined values (Figure 2(a). A void content of 2% as the limiting value was chosen because at this level the mechanical test results had almost reached a maximum plateau (with the shear strength τ_i in the range of 16–22 MPa and the transverse flexural strength in the range of 29–40 MPa). The curves are shifted to higher values of t with increasing size of the agglomerations (i.e. poorer mingling quality), higher fiber volume fractions (lower permeability), higher matrix viscosity (or lower processing temperature), and less optimized fiber sizing.

COMPOSITE INDUSTRY PROCESSES

Various processing routes typical for thermoplastic composites of directionalized reinforcements are also open to commingled yarns or prepregs of them.

Low pressure vacuum or pressure bag consolidation

This type of manufacturing technique takes full advantage of the easy impregnation quality of commingled yarns, as long as the filaments are evenly distributed. With pressure applied for 1 min, a pressure of as low as 4–5 bar is enough to achieve good quality consolidation with the standard material discussed above (i.e. a void content of less than 2%, Figure 2(a)). During this process, however, care must be taken so that the materials used for the mold, bag or bleeder film withstand the melting temperature of the thermoplastic resin. When being used as drapable fabrics, the process is well adopted to low volumes and large parts [1].

Thermoforming

This process has been developed specifically for thermoplastic composite plate materials. But it is also well suited for commingled fabrics, especially when they are pre-consolidated by the use of a double belt press into continuous organic sheets [8]. Further details about the thermoforming procedure are described in the chapter Thermoforming of fiber reinforced composite sheets.

Pultrusion

Thermoplastic pultrusion normally allows only the production of continuous profiles with fibers in the pultrusion direction. How to pultrude commingled yarns into simple profile geometries is outlined in the chapter 'Pultrusion of glass fiber/polypropylene composites' (see also reference [8] in that chapter, dealing in more detail with the impregnation process of GF/PP commingled yarns). One special advantage of commingled yarns is that they can be braided prior to pultrusion, so that also profiles with other than 0° fiber orientation can easily be realized [9].

Roll forming

In roll forming, a sheet of material is formed into some desired shape by feeding it through successive pairs of rolls, arranged normally in tandem. It is demonstrated in the chapter 'Roll forming of composite sheets' that this process is also a very suitable method for producing, e.g. thin-walled U- or Z-profiles out of Twintex® material.

Filament winding

This process has also been proved to be a very cost-efficient method for the manufacturing of axially symmetric components such as pressure vessels, tubes, bearings or nozzles out of thermoplastic composites. Commingled yarns are, thanks to their good flexibility, also suitable for this technology, especially when steep winding angles must be realized. One has to be, however, more careful with the choice of the heating system, because the thin polymer filaments can be burned off more easily than the matrix of thermoplastic composite tapes (see also the chapter 'Impregnation techniques for fiber bundles or tows'). Figure 2(b) illustrates for a GF/PP commingled yarn (800 tex) the qualitative influence of various winding parameters on the quality of rings wound by infrared/hot gas technology [10]. The shear strength, as measured by the Lauke test (see reference in [7]), was used as the quality parameter. Each curve has been constructed as the envelope over the measured shear strength values, achieved after variation of the indicated winding parameters:

F_Y = Pretension on the yarn [N];
F_K = Pressure load of the consolidation roller [N];
V_W = Winding speed [cm/s];
T_M = Temperature of mandrel [°C];
T_{NP} = Nip-point temperature [°C].

It turned out that the optimum processing window for reaching the highest shear strength value (of 23 MPa) of the rings was $F_Y = 25\,N$, $F_K = 105\,N$, $V_W = 10\,cm/s$, $T_M = 130°C$ and $T_{NP} = 160°C$.

REFERENCES

1. Guillon, D. and Saint-John, C. (1995) Twintex®, a material for the composite industry, personal communication at IVW in 1996; *see also*: *Proceedings of the International Conference on Composite Materials ICCM-10*, Whistler, Canada, August 1995, Vol. III, (eds A. Poursartip and K. Street), Woodhead Publishing, Cambridge, UK, pp. 757–764.
2. Hamada, H., Maekawa, Z.I., Ikegawa, N. and Matsuo, T. (1993) Influence of the impregnating property on mechanical properties of commingled yarn composites. *Polymer Composites*, **14**(4), 1993, 308–313.
3. Van West, B.P., Pipes, R.B. and Advani, S.G. (1991) The consolidation of commingled thermoplastic fabrics, *Polymer Composites*, **12**(6), 417–427.
4. Svensson, N., Shishoo, R. and Gilchrist, M. (1998) Manufacturing of thermoplastic composite from commingled yarns – a review, *J. Thermoplast. Composite Mater.*, **11**, 22–56.
5. Mäder E., (1997) Textile preforms for tailored fibre-reinforced thermoplastics, *Wiss. Z. Tech Univers. Dresden*, **46**, 20–26.
6. Cain, T.A., Wakeman, M.D., Brooks, R., Long, A. and Rudd, C.D. (1996) Isothermal consolidation of a co-mingled thermoplastic composite, in *Proceedings of the European Conference on Composite Materials ECCM-7*, London, UK, 1996, (ed. M. Bader), Woodhead Publishing, Cambridge, UK, pp. 57–62 and pp. 221–227.
7. Klinkmüller, V., Um, M.K., Steffens, M., Friedrich, K. and Kim, B.-S. (1995) A new model for impregnation mechanisms in different GF/PP commingled yarns. *Appl. Comp. Mat.*, **1**, 351–371.
8. Ostgathe, M., Breuer, U., Mayer, C. and Neitzel, M. (1996) Fabric reinforced thermoplastic composites – Processing and Manufacturing, in *Proceedings of the European Conference on Composite Materials ECCM-7*, London, UK, 1996, Vol. 1, (ed. M. Bader), Woodhead Publishing, Cambridge, UK, pp. 195–200.
9. Michaeli, W. and Jürs, D. (1996) Thermoplastic pull-braiding: Pultrusion profiles with braided fiber lay-up and thermoplastic matrix system (PP), *Composites*, **27A**(1), 3–7.
10. Neitzel, M., Funck, R., Haupert, F., Friedrich, K. and Schwarz, W. (1994) Filament winding with thermoplastic matrices – current development and equipment, in *Proceedings of the Japan International SAMPE Technical Seminar '94 (JISTES '94)*, Kyoto, Japan, July 14 and 15, 1994, (ed. N. Teranishi), SAMPE, Tokyo, pp. 149–168.

Keywords: commingled yarn, intermingling, Twintex®, hybrid yarn, textile, split-film, impregnation, consolidation, shear strength, modelling, pressure bag, thermoforming, double belt press, pultrusion, roll forming, filament winding, optimum processing window, braiding.

Construction principles of injection molds

D. Meyer

INTRODUCTION

The mold is the 'shaping element' in the process of injection molding with thermoplastics. The construction of the mold has a significant influence on the total molding process and furthermore affects the product quality, the production cycle and the possibility for automation as well [1].

The basic steps for the design of an injection mold are listed in Table 1.

Important for the optimal design of molds is a proper knowledge of the use, assembly and the service conditions of the injection molded items. From the viewpoint of the polymer selection, it is necessary to know the thermal, mechanical, electrical and chemical requirements for the molded part. There is a number of specific design aspects relating to individual tasks; some of them are listed below.

- Automatic ejection (or demolding). The molds are equipped with mechanical or electric driven screw-thread cores for moving parts of the mold.
- High-speed-molds should preferably have a design with a take-out process using horizontal handling systems via the shortest route. In the vertical process, the pieces usually fall down onto conveyor belts (that can be directed by a vertical directed breeze). A special design is called the 'non-open mold concept'. In principle, this concept does not

Polypropylene: An A–Z Reference
Edited by J. Karger-Kocsis
Published in 1999 by Kluwer Publishers, Dordrecht. ISBN 0 412 80200 7

Table 1 Sequence of a mold construction procedure

Step	*Mold characteristics*	*Product quality*	*Injection molding*	
			Productivity	*Automation*
1	Mold dimension	Polymer type	Wall thickness	Demolding flexibility
2	Cavities	Product design	Cycle	Takeout moldings
3	Variable mold inserts	Tolerance	Two-plate mold	Inserts loading
4	Movable cores	Visible surface	Three-plate mold	Assembling
5	Split molds	Surface finish	Sprue-system	Positioning accuracy
6	Mold insulation	Demanding	Setup reduction	Stacking/packaging
7	Injection molding machine	Function	Mold cooling	Robot and handling

require a demolding unit. Article demolding takes place as a parallel operation to 'mold opening', which only requires a very short stroke.

- Placing inserts, in-mold decoration. The mold construction covers also the handling of sheets (decorations, labels, smartcards, etc.), drapery insert (auto-panels, furniture, etc.).
- Production of hollow parts. Gas-assisted injection molding [2] is used to produce hollow bodies with smooth surfaces (auto mirrors, handles).
- Multi-color or multi-component molding. For multi-color molding, two or three colors of the same material are used whereas, for multi-component molding two or three various polymers or the same polymer with different melt flow index (MFI) values are used.

MOLD DESIGN

The mold design starts with deciding the main dimensions (Table 1). Rectangular molds are usually preferred for a better adjustment and fixing at the platen area than round molds. The rectangular molds allow also an easier automatic tool-change. The selection of the cavities is a significant decision that should be based on the amount and price of the product. Three-plate molds (Figure 1) produce twice as much in the same time as two-plate molds at the same clamping force of the injection molding machine. There is also a wide variety of sprues that can be used: stick, point, film, cold and hot runner sprue.

Ejector pins (Figure 1) are responsible for ejecting the molded pieces. They are usually located on the opposite side to the sprue. Round, profile

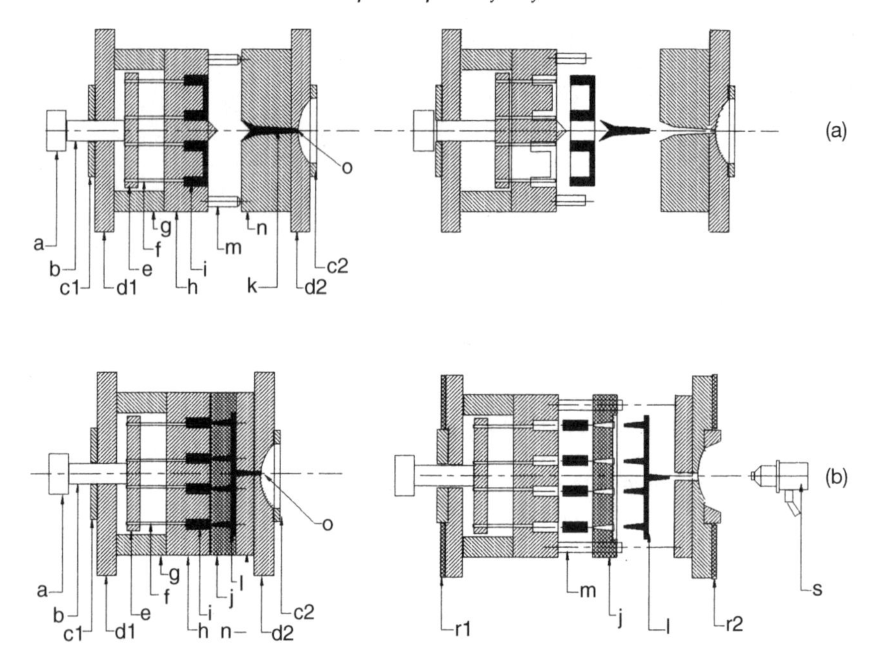

Figure 1 Basic variants of injection molds: (a) two-plate mold (one cavity); (b) three-plate mold (multi-cavity). Designations: a, coupling link; b, ejector rod; c1 + 2, center rings; d1 + 2, clamping plates; e, ejector plate; f, ejector pins; g, distance plate; h, cavity plate; i, cavity; j, transition plate; k, sprue with cone gate; l, runner with 4 sprues; m, center bolt; n, runner plate; o, nozzle; r1 + 2, insulation plates; s, high performance nozzle.

bolts and stripper plates are used as ejectors. When ejector pins are unsuitable, a robot handclasp takes out the parts.

A runner layout of pieces to be demolded by simple ejection should be achieved so that the sprue picker takes the runner easily. A row layout of the mold cavities is more favorable for sprue picking than round or star layout versions. It is often practical to form the upper end of the runner with an extra cone which guarantees an easy take-off.

In three-plate molds generally, a cold runner unit is located between the first and second plates. The cold sprue points here are easy to break during mold opening. The cold runners allow a regular form filling without disturbing the optical appearance of the molded parts.

The hot runner sprue is preferred in high-performance molds due to on-line recycling issues (sprueless injection molding). If this method is not applicable due to the related costs, three-plate mold is the most cost-efficient solution. The sprue picking out of the second mold plate is usually problem free.

Handling systems are not able to pick out parts which shrink on the mold cores at high-shrinking forces. Stripper plates are used more and more for this job. During the drop-out process, the handling clutch takes the piece out of the mold. Core-trains and racketeers require special attention in respect to mold construction.

If mold changes should occur horizontally and automatically across the bars of the injection molding machine, large molds with vertically-oriented hydraulic and pneumatic cylinders and other equipment (such as heating tubes) are sometimes not able to pass. In that case, tiebarless injection molding machines are preferred. They offer a large space to the necessary mold equipment and guarantee the distance needed for the movement of robots and the handling system.

Automatic mold exchange requires a number of preconditions [3]. Special indexing equipments are used to fix the mold in the same position always. They are, for example, center rings and fixing cores, as well as automatic clutches to ensure a fast and safe coupling of ejector pins. Special clutches serve for a proper connection between all the necessary technical equipment. They guarantee a safe transmission of air, hot water, hydraulics, electric and electronics commands. See the EUROMAP norm; this EUROMAP description represents a product-related norm to achieve the use of various molds on different injection molding machines in an identical way. Mold hallmarks are necessary to identify the produced plastic moldings. For identification purpose the following marks are normally introduced: part number with alteration index; production date (year and month); polymer identification; cavity number; producer's name or his identification number or logo. All these marks must be easy to read without affecting the surface quality of the product.

MATERIALS FOR MOLDS

The standard material of injection molds is steel since a great number of parts must be produced without mold erosion [4]. This requires the use of high quality metals. The essential parts of the molds, such as cavity plates, runner and transition plates, as well as ejector pins, are hardened and tempered. The surface of the mold is responsible for the appearance and surface quality of the plastic product. A large number of molds are therefore polished in order to give the products a brilliant surface finish. The modular elements of the mold are normally standard hot work tool steel (CrMoV). The cavity plates, for example, are produced from special case-hardening steel or special fully curing steel (NiCrMo). The insulating plates (Figure 1) are used to prevent the transfer of heat from molds to the machine platen. This insulation reduces the heat and makes the mold temperature more controllable. The usual material for that purpose is glass fiber reinforced resin. Microprocessor controlled equipment keeps the mold temperature constant. High-performance nozzles are used as heated 'extension' of the machine nozzle (Figure 1). They are separately heated in multi-cavity hot runner molds. The polymer flows through the unrestricted channel to the gate.

REFERENCES

1. Gastrow, H. (1990) *Der Spritzgieß-Werkzeugbau in 100 Beispielen*, Hanser, Munich.
2. Eckardt, M. (1996) Gas-assisted injection molding, in *Innovation in Polymer Processing Molding*, Chap 1, (ed. J.F. Stevenson), Hanser, Munich, pp. 1–42.
3. Meyer, D. (1995) *Kunststoffverarbeitung Automatisieren*, Hanser, Munich,.
4. Mennig, G. (1990) *Verschleiß in der Kunststoffverarbeitung*, Hanser, Munich.

Keywords: cavity, demolding, design aspects, EUROMAP, gas assisted injection molding, hot runner, mold construction, mold design, part ejection, robot, runner, sprue, three-plate mold, two-plate mold.

Controlled rheology polypropylene

K. Hammerschmid and M. Gahleitner

MOTIVATIONS FOR PRODUCING CR-PP

The first studies on peroxide-induced degradation of polypropylene (PP) date back to the 1960s. Most of the systematic work in this area was carried out in the 1970s. In 1983, about 5% of the world-wide PP production was finalized in controlled rheology (CR) processes; nowadays, this number is about 20%.

One motivation to produce CR-PP via controlled degradation of a reactor product was the demand for special grades not easily accessible otherwise. Some finalization processes for PP, especially the production of glass-mat reinforced PP (GMT-PP) or of melt-blown fibers, require extremely high flowability, which cannot be achieved in a polymerization process at all or only under extremely uneconomical conditions. Apart from that, problems in the extrusion line as well as in the pelletising process occur with grades having a melt flow index, MFI (230°C/2, 16 kg), higher than 50 g/10 min. CR-PP also shows a number of specific advantages regarding processability and the final material profile in mechanics and optics, which will be discussed in detail below.

Moreover, adjusting the flowability (MFI) of a certain grade in the CR-process allows us to produce a reduced number of reactor grades, thus facilitating production and storage logistics as well as reducing the 'transition' quantities produced when changing the grade in the reactor.

Polypropylene: An A–Z Reference
Edited by J. Karger-Kocsis
Published in 1999 by Kluwer Publishers, Dordrecht. ISBN 0 412 80200 7

This aspect is even more important for large production units and therefore is a factor of growing importance in recent years.

A certain competition may arise for CR-grades from the novel family of metallocene-(MC-)catalyst based PPs, which possess an inherently narrow molar weight distribution (MWD) similar to the characteristics of CR-PP. However, these grades are only just penetrating the market in some applications and it is difficult to judge their actual applicability. The remaining advantage is the rather easily adjustable final shape of the MWD through a variation of the relation between initial and final MFI, a factor which is often referred to as 'crack length'.

PRINCIPLES OF CONTROLLED DEGRADATION

Peroxide-induced degradation of PP is a radical reaction induced by the thermal decomposition of the peroxide, which acts as initiator. The reaction scheme is given in Figure 1. The actual result of a CR-process depends not only on the MWD of the base polymer, but also on its structure. While for PP-homopolymers and most random-copolymers with ethylene as well as for related terpolymers the reaction scheme is straightforward, competing reactions occur in case of high-impact-PP-copolymers with ethylene (termed heterophasic or – falsely – block copolymers). In contrast to the effect on PP, peroxides/radicals promote branching, chain transfer and crosslinking in other polyolefins. Especially relevant in connection with CR-PPs is the effect on polyethylene (PE) and ethylene-propylene-'rubber' (EPR), which leads to problems in degrading (heterophasic) EP-copolymers with high contents of ethylene, where branching and crosslinking becomes dominant in the elastomeric and PE-phase.

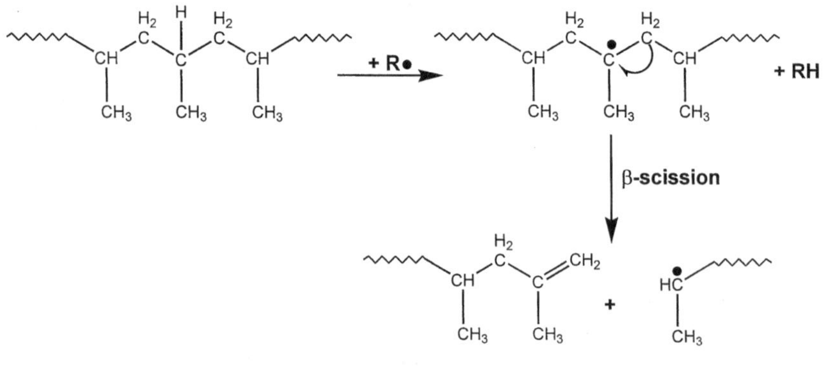

Figure 1 Reaction scheme of peroxide-initiated degradation of PP, with initialization through thermal decomposition of peroxide, and termination by recombination of two radical species.

Different types of peroxides and masterbatches are applied for technical degradation. The main preconditions for applying a peroxide in polymer modifications are: handling and storage security (as a general rule, peroxides must not be stored at temperatures above 30°C); compatibility to polymers; toxicological unobjectionability (of the substance itself as well as its decomposition products – in the case of DHBP, for example, these are methane, ethane, ethylene, acetone and tert.butenol) and desired activity at the melt temperature. For a list of widely applied peroxides, see Table 1. Peroxide masterbatches are supplied by various producers; one example is the Xantrix®-masterbatch marketed by Montell consisting of PP-'reactor granules' with a liquid peroxide adsorbed.

The type and amount of peroxides to be used in the formulation of products for food packaging, medical applications, etc., are subject to regulations in the respective legislation. For example, the German BGA (Bundesgesundheitsamt; federal health office) regulation for materials in contact with foodstuff allows only DIPP, DHPP and DTBP to be applied up to a concentration of 0.1 wt.%.

To be able to predict the result of a specific degradation process, possibilities for modelling are of interest. An easily applicable model for the evolution of the MWD was developed by Tzoganakis *et al.* [1] using a 'quasi-steady state approximation (QSSA)' technique. It yields the evolution of the moments of the MWD as function of peroxide type, concentration and efficiency. If one considers the moments of the MWD as

$$Q_i = \sum_{j=1}^{n} m_j^i p_j \tag{1}$$

with $i = 0, 1, 2, 3, \ldots$; m_j being the number of monomer units and p_j the probability of the molar mass fraction j in a distribution of n fractions then these are related to the number-, weight- and Z-averages of the MWD via the following relations (m_0 being the molar mass of the monomer unit):

$$M_\mathrm{N} = m_0\, Q_1/Q_0 \tag{2}$$

$$M_\mathrm{W} = m_0\, Q_2/Q_1 \tag{3}$$

$$M_\mathrm{Z} = m_0\, Q_3/Q_2 \tag{4}$$

The evolution of the 0th to 3rd moment can then be described as follows:

$$\frac{\mathrm{d}Q_0}{\mathrm{d}t} = 2fk_\mathrm{d}c_\mathrm{P} \tag{5}$$

Table 1 Typical peroxides for controlled degradation of PP

Chemical name	*Type name**	*Form (+23°C)*	*T* ($\tau_{1/2}$=1 *h*) (°C)	*T* ($\tau_{1/2}$=1 *min*) (°C)	*Reaction constant* k_0 [s^{-1}]	*Activation energy* E_a (*kJ/mole*)
2,5-Dimethyl-2,5-bis(tert.butyl-peroxy)hexane (DHBP)	Luperox 101 Trigonox 101	Liquid	142	190	1.7×10^{16}	155.5
2,5-Dimethyl-2,5-bis(tert.butyl-peroxy)hexyne-3 (DYBP)	Trigonox 145 Interox DYBP	Liquid	149	195	1.9×10^{15}	150.7
Dicumyl-peroxide (DCUP)	Perkadox BC	Solid	136	175	9.4×10^{15}	152.7
Di (tert.butyl)-peroxide (DTBP)	Trigonox B Luperox Di	Liquid	146	190	4.2×10^{15}	153.5
Tert.butyl-cumyl-peroxide (BCUP)	Trigonox T	Liquid	138	180	1.17×10^{15}	147.0
Bis(tert.butylperoxy-isopropyl)benzene (DIPP)	Perkadox 14S	Solid	142	190	7.7×10^{15}	152.7

* examples.

$$\frac{dQ_1}{dt} = 0 \tag{6}$$

$$\frac{dQ_2}{dt} = \frac{2fk_d c_P}{3(Q_1 - Q_0)}(Q_1 - Q_3) \tag{7}$$

$$Q_3 = \frac{Q_2(2Q_2Q_0 - Q_1^2)}{Q_1Q_0} \tag{8}$$

with f being the efficiency factor, which is defined by the machine type (normally in the range of 0.6–0.8), k_d the decomposition rate (depending on peroxide type and reaction temperature) and c_P the peroxide concentration.

In principle, any type of mixing equipment can be used for controlled degradation. However, most producers are working with single- or twin-screw extruders, normally coupled directly to the reactor unit. The economics of degradation are then mainly influenced by the following factors: peroxide concentration; MWD of the virgin material; reaction temperature profile; and residence time distribution (RTD) in the extruder. To ensure a homogeneous product, the RTD should be narrow and the mean residence time not above 3–4 times of $\tau_{1/2}$ (half-life) of the peroxide, as no further changes in the MWD will occur afterwards. The homogeneity of the peroxide distribution in the melt is an essential factor for efficiency and product homogeneity. Otherwise, peroxide can be decomposed before melting and mixing is not accomplished sufficiently for the reaction. In an extensive study by Ebner and White [2], twin-screw extruders appeared to be superior to comparable single-screw machines regarding efficiency because of their higher mixing potential and the narrower RTD. Apart from the machine type, the RTD is mainly influenced by the process temperature. Another important influence on process economics comes from other additives, especially antioxidants, which may deactivate a significant portion of the created radicals.

Historically, pure thermomechanical degradation was developed before the application of peroxides for this process. However, due to the high temperature (>260°C) and energy input necessary, it has a much poorer efficiency and is therefore hardly used in the practice. In the case of peroxide-initiated degradation, usual operation temperatures are in the region of 200–250°C. Apart from the applied peroxide, also oxygen either present in the base polymer (powder) or drawn into the extruder at the feeding unit may promote degradation significantly. For an optimum process control as well as for safety reasons, inert conditions (nitrogen atmosphere) should be maintained in the relevant sections.

Feeding the peroxide into the extruder can be accomplished in various ways; solid types (mainly applied for small-scale operations and experimental studies) are mostly premixed with the base polymer while liquid types are either injected directly or adsorbed to a polymer powder in a concentration of 1–10wt.% (the final concentration in the reaction mixture is normally 0.005–0.5 %). Another alternative is masterbatches, which may also contain other additives.

Quality control is generally an important factor in polymer production; in case of the CR-process, it is normally focused on reaching the desired MFI or, more specifically, MWD. The most widely used technique to continuously control and regulate degradation processes is via measuring the rheological properties of the degraded product in an online rheometer (usually a capillary-type) and feeding back the signal into the metering unit for peroxide-addition. An important factor for the performance of such a feedback control system [3] is its response time, which should be as short as possible to minimize quality problems and material losses.

COMPARISON BETWEEN REACTOR AND CR-PP

Basically of interest is the molar weight distribution of the CR-products, particularly its evolution over the degree of degradation or crack length. Rather independent of the original MWD of the polymer, a M_W/M_N-ratio of less than 1.5 cannot be reached in a CR-process. This is below the value of 2 for the 'most probable distribution' according to Flory (see [4]) and confirms that the degradation process follows a Poisson statistics. The 'kurtosis' momentum of M_Z/M_W tends towards 1. The actual evolution in a series of CR-grades can be seen in Table 2. The rheological behavior is in principle connected to the MWD, but not in a simple way for all parameters. The strongest effect normally observed in degraded PPs is a significant reduction of melt elasticity (e.g. extrudate- or die-swell) and extensional viscosity, as these properties are mainly determined by the longest molecules in the MWD. Application of a process model in combination with a 'double-reptation'-model for relating the MWD to the viscoelastic properties [4] confirms this effect. While a log–linear relation is obtained between the MFI and the number of chain scissions, elasticity parameters, such as the steady-state compliance (J_e^0), are changed more strongly. The shape of the viscosity curve (shear-rate dependence) is also altered because of the increasingly narrow MWD. With rising MFI, the shear-thinning effect is reduced and the critical shear rate rises, giving the curve a longer zero shear rate (η_0, Newtonian) region. At high shear rates, all viscosity curves of a degradation series tend to coincide (Figure 2). This shape of the viscosity curve appears to be advantageous, e.g. in fiber spinning processes.

Table 2 Degradation series of PP-homopolymers, development of MWD, rheological and mechanical properties (basic polymer M0 from standard liquid/bulk process, DIPP used as peroxide, degradation in Collin 50 mm twin-screw extruder at 210–220°C; MWD-data from GPC, MFI by ISO 1133 at 230°C/2, 16 kg, flexural modulus by DIN 53452/57, Charpy impact by ISO 179 1eA – V-notch at +23°C; materials as in [5])

Material	c_p *(wt.%)*	M_w *(kg/mol)*	M_w/M_n	*MFI (g/10 min)*	*Flexural modulus (MPa)*	*Charpy impact (kJ/m²)*
M0	0	766	5.5	0.4	1419	7.5
M1	0.026	453	3.5	3.4	1247	4.7
M2	0.051	318	3.1	8.6	1213	3.9
M3	0.108	231	2.8	28	1208	3.0
M4	0.146	181	2.7	51	1175	2.6
M5	0.175	157	2.7	81	1157	2.4
M6	0.24	135	2.5	149	1150	1.9

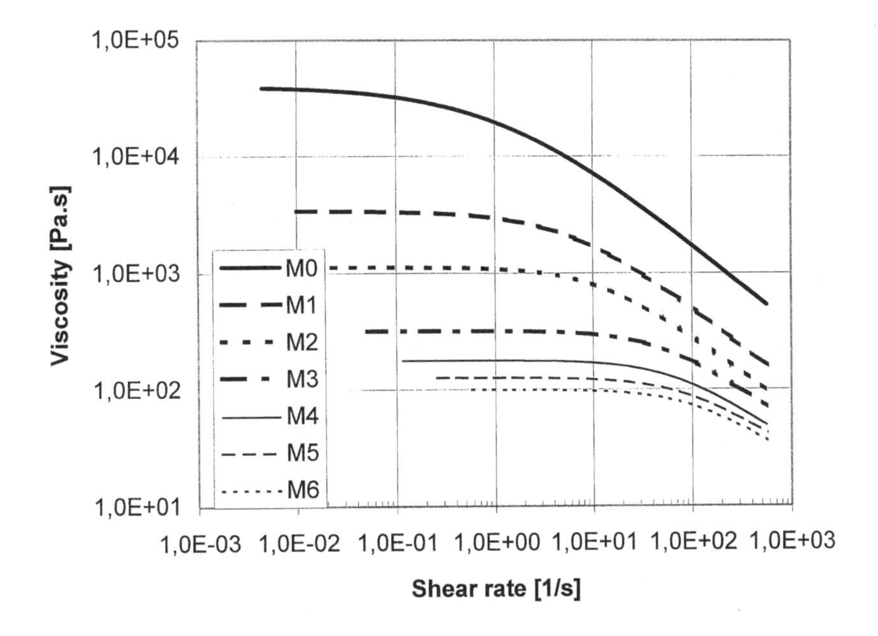

Figure 2 Viscosity curves at 230°C (plate/plate rheometry) for the degradation series of PP included in Table 2.

Rheological effects also determine processability. Especially in high-speed processes, such as film extrusion or paper coating, melt fracture effects are a critical limiting factor. With PP, normally no 'sharkskin'-

phenomena, such as in linear low-density polyethylene (LLDPE), are observed and the material goes with rising velocity or shear stress through a 'pulzation' phase directly to gross melt fracture. CR-products behave advantageously here; as the melt elasticity is reduced, the critical shear stress for melt fracture is increased, showing a linear negative correlation to J_e^0. The same applies to a reduction of draw-resonance phenomena in fiber spinning.

Another important factor defining applicability of a certain grade are crystallization (solidification) and the final mechanical properties. According to the literature [1], as well as our own experience, both the melting and crystallization temperatures (T_M and T_C, respectively) remain unchanged by degradation, while mechanical properties are altered strongly.

Tzoganakis *et al.* [1] showed that, when increasing the MFI of a PP-homopolymer from 3.3 to 126, the flexural modulus is reduced by 10% while the notched impact strength is reduced by 55%. At the same time, the tendency towards 'cold flow', as seen in tensile testing, increased significantly.

The main reason for the modulus changes is the fact that crystallization of CR-grades is hindered through the reduced number of nuclei as compared to reactor (RE) grades with comparable M_W [5]. This effect is even more pronounced in actual processing, especially in injection molding, where shear-induced crystallization effects appear. The latter is dominated by the longest molecules in the material, of which CR-grades contain a lesser fraction. Parallel to that, the flow profile in the mold is changed with the MWD or flow curve, respectively. This results in a significantly reduced stiffness (flexural modulus), while the impact strength is slightly higher than in a comparable RE-grade (Table 2). At the same time, crystallization-induced negative effects, such as shrinkage and warpage after molding, are reduced.

REFERENCES

1. Tzoganakis, C., Vlachopoulos, J., Hamielec, A.E. and Shinozaki, D.M. (1989) Effect of molecular weight distribution on the rheological and mechanical properties of polypropylene. *Polymer Engng Sci.*, **29**, 390–396.
2. Ebner, K. and White, J.L. (1994) Peroxide induced and thermal degradation of polypropylene. *Int. Polymer Processing*, **9**, 233–239.
3. Fritz, H.G. and Stoehrer, B. (1986) Polymer compounding process for controlled peroxide-degradation of polypropylene. *Int. Polymer Processing*, **1**, 31–41.
4. Mead, D.W. (1995) Evolution of the molecular weight distribution and linear viscoelastic rheological properties during the reactive extrusion of polypropylene. *J. Appl. Polymer Sci.*, **57**, 151–173.

5. Gahleitner, M., Wolfschwenger, J., Bachner, C., Bernreitner, K. and Neißl, W. (1996) Crystallinity and mechanical properties of PP-homopolymers as influenced by molecular structure and nucleation. *J. Appl. Polymer Sci.*, **61**, 649–657.

Keywords: peroxide, molar weight distribution (MWD), rheology, crystallization, extrusion, melt flow index (MFI), controlled rheology (CR), peroxide-degradation, residence time distribution (RTD), half-lifetime of peroxides, melt elasticity, die swell, viscosity curve, shear rate, elongational viscosity, melt fracture, heterophasic PP.

Copolymerization

J. Suhm, M.J. Schneider and R. Mülhaupt

INTRODUCTION

Olefin copolymerization and reactor blend formation are important processes to tailor polyolefins. Copolymer properties depend upon the sequence distribution of the comonomers, which is controlled by means of catalyst as well as process technology. Today most copolymers are produced either in solution processes or in solvent-free gas phase polymerization. Recent breakthroughs in catalyst development are stimulating production of a novel range of copolymers, especially of ethylene copolymers. In the past, special catalysts were designed to produce three classes of ethylene copolymers with different comonomer content: (1) high density (HDPE) and linear low density polyethylene (LLDPE) with density ranges of 0.92–0.96 g/cm^3 and 1-olefin content <5 mol.%; (2) polyolefin rubbers, such as ethylene/propylene (EPM) or ethylene/propylene/diene (EPDM) with propylene content varying between 20 and 60 wt.% and content of dienes, such as 1,4-hexadiene, ethylidene norbornene of a few percent; and (3) impact-modified polypropylene containing a few percent of ethylene as comonomer. Special catalyst technology was developed for these three product groups where PP catalyst failed to produce LLDPE or EPM. Moreover, most conventional catalysts were multi-site catalysts containing different catalytically active sites with different reactivity toward comonomer incoporation. As a consequence, a large number of conventional catalysts produced copolymers with broad molecular weight distributions (MWD) and heterogeneous comonomer incoporation. Most traditional catalysts did not tolerate polar comonomers, such as CO or acrylics and produced styrene

Polypropylene: An A–Z Reference
Edited by J. Karger-Kocsis
Published in 1999 by Kluwer Publishers, Dordrecht. ISBN 0 412 80200 7

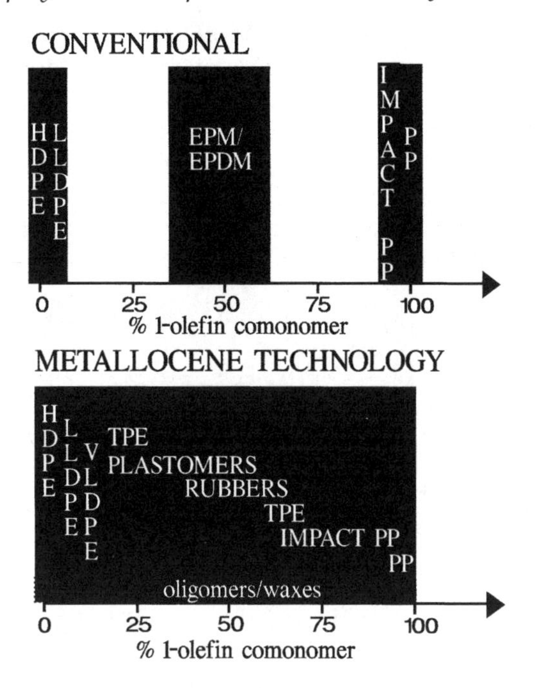

Figure 1 Ethylene/1-olefin copolymers.

homopolymer, in addition to a comonomer with very low styrene incoporation. Breakthroughs in catalysis, especially the development of metallocene catalysts, have eliminated the frontiers between the above-mentioned product classes and enables the production of ethylene copolymers covering the entire feasible composition range as displayed in Figure 1 and including styrene and less reactive long-chain 1-olefins as comonomers. Olefin copolymerization was reviewed by van der Ven [1], and Tait and Berry [2]. Aspects of EPM and EPDM rubber production and opportunities of modern catalyst technology were reviewed by Richter *et al.* [3].

COPOLYMERIZATION PARAMETERS AND CATALYST RANKING

In order to rank performance of catalysts in copolymerization with respect to comonomer incorporation and comonomer sequence distribution, copolymerization parameters have proven to be very useful. They are determined by means of nuclear magnetic resonance spectroscopy (NMR copolymer sequence analysis) taking into account the Markovian statistics of chain growth, as reviewed by Randall [4]. Galimberti and coworkers [5] described the analysis of EPM prepared by means of metallocene catalysts. The analysis and determination of copolymerization parameters of various ethylene copolymers was done by Fink and coworkers [6]. Most copolymerization processes are described by the first-order Markovian statistics, using two copolymerization parameters r_1 and r_2 and taking into account the different reactivity of the comonomers M_1 and M_2 as well as the type of the last inserted monomer unit at the chain end, where the metal alkyl of the catalytically active transition metal center (C*) is located:

$$\text{Pol—M}_1\text{—C}^* + \text{M}_1 \xrightarrow{k_{11}} \text{Pol—M}_1\text{—M}_1\text{—C}^*$$

$$\text{Pol—M}_1\text{—C}^* + \text{M}_2 \xrightarrow{k_{12}} \text{Pol—M}_1\text{—M}_2\text{—C}^*$$

$$\text{Pol—M}_2\text{—C}^* + \text{M}_1 \xrightarrow{k_{21}} \text{Pol—M}_2\text{—M}_1\text{—C}^*$$

$$\text{Pol—M}_2\text{—C}^* + \text{M}_2 \xrightarrow{k_{22}} \text{Pol—M}_2\text{—M}_2\text{—C}^*$$

copolymerization parameters: $r_1 = k_{11}/k_{12}$; $\quad r_2 = k_{22}/k_{21}$

The second-order Markovian statistics for chain growth uses four copolymerization parameters taking into account the last two monomeric units at the chain end. In order to decide which mathematical model to use, copolymerization parameters, determined from ^{13}C NMR spectroscopic sequence analysis, should be compared with calculated distributions using the above mentioned models. In most cases, the first-order Markovian statistics is adequate. However, Herfert, Montag and Fink [6] point out that some metallocene-based copolymers give best fit with the second-order Markovian model.

When propylene (P) is copolymerized with ethylene (E), the copolymerization parameters r_E and r_P, are excellent measures for the olefin sequence distribution as shown below for different copolymer sequences:

$r_E \bullet r_P = 1$	EEEPEPPPEEPPPEE	random copolymer
$r_E \bullet r_P = 0$	EPEPEPEPEPEPEPE	alternating copolymer
$r_E \bullet r_P >> 1$	EEEEEEEEEEEPPEP	block copolymer (homopolymer byproduct is likely)

When comparing *r*-parameters, one should always report which method was applied. Moreover, copolymerization process conditions, e.g. diluent, temperature, pressure, catalyst, conversion, feed control, play a very important role. In the case of metallocenes, it was demonstrated that copolymerization parameters are very sensitive to changes in polymerization temperature, with decreasing 1-octene incorporation at high temperatures [7]. In fact, the metallocene-based copolymer sequence represents a temperature probe which responds to temperature changes. Therefore, adequate temperature control is a prime requirement to exploit the potential of single site metallocene catalysts.

According to Table 1, most conventional $TiCl_3$-based catalysts systems as well as modern Lewis base modified $MgCl_2$-supported $TiCl_4$-based systems polymerize ethylene at much higher rate with respect to 1-olefins. This causes formation of considerable amounts of PE as homopolymer byproduct. Therefore, during the 1960s and 1970s special vanadium catalysts were developed to prepare EPM and EPDM rubber. It is assumed that V^{3+} is the active species and that overreduction during polymerization affords V^{2+}, which is considered to be much less active. In many vanadium-based systems, alkylchlorides, e.g. hexachlorocyclopentadiene, pentachlorocrotonic acid, trichloroacetic acid, and α,α-dichlorotoluene, were added to reactivate the low valent vanadium by means of oxidative addition according to:

$$V(II) + RCl \rightarrow V(III)Cl + R^*$$

Due to moderate catalyst activities and high content of vanadium in the polymer, special purification steps were required to remove catalyst residues. From Table 1 it is apparent that modern metallocene-based catalysts offer attractive opportunities for olefin copolymerization.

Table 1 Copolymerization parameters of ethylene (E)/propylene (P) and ethylene/1-octene (O) copolymerization using various catalyst systems

Catalyst system	r_E	r_P	r_E	r_O	$r_E \times r_{P(O)}$
$TiCl_3/AlEt_2Cl$	25.0	0.10			2.50
$TiCl_4/MgCl_2$/ester/$AlEt_3$	13.4	0.40			5.40
$V(acac)_3/AlEt_2Cl$	15.0	0.04			0.60
$VCl_4/AlEt_3$	10.3				0.25
		0.025			
$VCl_4/AlEt_2Cl$	5.9	0.029			0.14
Cp_2ZrMe_2/MAO	27	0.005			0.14
$(Me_5Cp)_2ZrCl_2$/MAO	250	0.002			0.50
$Me_2SiCp_2ZrCl_2$/MAO	24	0.029			0.70
$Et(Ind)_2ZrCl_2$/MAO	16.6	0.06			0.40
$Me_2C(Cp)(Flu)ZrCl_2$/MAO	1.3	0.20			0.26
Cp_2ZrCl_2/MAO			32.8	0.050	0.17
$Me_2Si(Me_4Cp)(N\text{-}^tBu)TiCl_2$/MAO			4.1	0.290	1.19
$Me_2Si(Ind)_2ZrCl_2$/MAO			18.9	0.014	0.27
$Me_2Si(2\text{-}Me\text{-}Ind)_2ZrCl_2$/MAO			19.5	0.013	0.25
$Me_2Si(Benz\text{-}Ind)_2ZrCl_2$/MAO			10.7	0.076	0.81
$Me_2Si(2\text{-}Me\text{-}Benz\text{-}Ind)_2ZrCl_2$/MAO			10.1	0.118	1.20

METALLOCENE-CATALYZED ETHYLENE COPOLYMERIZATION WITH 1-OLEFINS

The very attractive potential of metallocene-based catalysts was discovered by many groups during the 1980s. Ewen and coworkers [8] demonstrated that metallocene structures played an important role in copolymerization. When Cp is substituted with Me_5Cp ligand, copolymerization parameters of the two metallocene catalysts differ by almost three orders of magnitude in ethylene/1-olefin copolymerization. According to Fink and others, the syndiospecific $Me_2Si(Cp)(Flu)ZrCl_2$/MAO catalyst is more effective with respect to isospecific catalysts, e.g. $Me_2Si(Ind)_2ZrCl_2$/MAO, in ethylene/1-olefin copolymerization. Moreover, Fink's study concluded that stereoregularity was not affected by comonomer insertion. Syndiospecific catalysts produce highly syndiotactic copolymers. Best performance in terms of high 1-olefin incoporation was reported by Soga *et al.* [9], in accord with many observation by other groups, for halfsandwich metallocenes such as $(CpMe_4)SiMe_2N$(tert.Bu)$TiCl_2$/MAO, referred to by Dow Chemical as 'constrained geometry' catalyst. Such ethylene/1-octene copolymers contained appreciable amount of long-chain branching, most likely produced via copolymerization of the vinyl-terminated polymers, which were detected by Soga. When 75 mol.% 1-octene are present in comonomer feed, 44 mol.% are incorporated in the ethylene copolymer with $r_E r_O = 1.19$ typical for random copolymers. As a function of the metallocene structure, it is possible to obtain catalysts which promote either alternating or random copolymerization. The systematic variation of indenyl ligand substitution patterns of $Me_2Si(Ind)_2ZrCl_2$/MAO in ethylene/1-octene [10] and propylene/1-octene [11] polymerization, reported by Schneider and Mülhaupt, revealed that 2-methyl substitution promotes higher copolymer molecular weights without affecting comonomer incorporation, whereas benzannelation improved both randomness and 1-olefin incorporation. In the case of ethylene/1-octene copolymerization, molecular modeling confirmed the unexpected role of benzannelation with respect to lower activation energies of 1-octene insertion subsequent to ethylene as well as 1-octene insertion in comparison to those of the indenyl and 2-methyl-indenyl ligand framework.

As a function of the comonomer incorporation, it is possible to control PE crystallization. Above 20 mol.% 1-octene content, chain folding is not possible and fringed-micelle type nanostructures are formed Basic structure/property relationships of poly(ethylene-co-1-octene) were reported by Minick *et al.* [12].

An important prospect of ethylene copolymerization, reported by Fink and Seppälä [13], is the possibility of incoporating less reactive long-chain 1-olefins, such as hexadecene. Moreover, vinyl-terminated olefin macromonomers are also copolymerized. During ethylene polymerization with halfsandwich metallocene catalysts the vinyl end groups of polyethylene are copolymerized to produce long-chain branching. Long-chain branching influences melt rheology and facilitates processing of polyolefins. Prospects of long-chain branching was demonstrated by Batistini for Dow's Affinity™ and Engage™ ethylene/1-octene copolymers [14].

METALLOCENE-CATALYZED ETHYLENE COPOLYMERIZATION WITH STYRENE AND CYCLOOLEFINS

While most conventional catalysts failed to copolymerize styrene, halfsandwich-metallocene catalysts proved very effective in styrene copolymerization. In addition to random copolymers, specific catalysts were found to produce alternating copolymers. Sernetz *et al.* [15] determined copolymerization parameters of random ethylene/styrene copolymerization using 'single site' $(CpMe_4)SiMe_2N(tert.Bu)TiCl_2$/MAO to be $r_E = 23.4$ and $r_S = 0.015$ with $r_E r_S = 0.15$ in contrast to $r_E = 111$ and $r_S = 0.055$ for a bisphenolate titanium/MAO catalyst, which represents a conventional multi-site catalyst. The single site nature of the halfsandwich metallocene catalyst was confirmed by means of temperature rising elution fractionation of semicrystalline copolymers. Properties of poly(ethylene-co-styrene) depend upon styrene content and vary from semicrystalline polymers to amorphous elastomers above 15 mol.% styrene content. Mechanical properties were reported by Cheung and Guest [16]. Terpolymers of ethylene, 1-olefins and styrene offer attractive potential as blend components.

As reported in the chapter on metallocene catalysis in more detail, cycloolefins are copolymerized with single site metallocene such as the syndiospecific catalyst generations to produce novel cycloaliphatic copolymers [17]. With increasing cycloolefin content, glass transition temperature increases. Above 30 mol.% cycloolefin content, new families of melt processable engineering thermoplastics are obtained. Random, alternating and stereoregular or stereoirregular copolymers are available, depending on the catalyst system. Example of commercial cycloolefin ('COC') copolymer [18] is Topas® of Hoechst, which is considered for applications in optical data storage media and medical packaging. As an alternative route to copolymers containing cyclolefins, a cyclocopolymerization has been introduced where 1,5-hexadiene is cyclized during ethylene copolymerization to produce cyclopentane structural units [19].

When copolymerization using metallocenes is compared with conventional processes, metallocene technology offers several advantages: (1)

metallocenes contain essentially one type of active center ('single site' catalysts) and give very uniform copolymers with respect to both comonomer incorporation and also narrow molecular weight distribution, which is typically reflected by polydispersities of $M_w/M_n = 2$; (2) metallocenes are highly active and do not require removal of catalyst residues ('leave-in' catalysts), which is typical for some V-based systems; (3) comonomer incorporation is independent of molecular weight; (4) uniform copolymers are obtained covering the entire feasible copolymer composition range; (5) no wax-like byproducts are formed due to narrow molecular weight distributions, which is attractive for reducing tackiness and copolymer production in gas phase polymerization; (6) metallocene-based processes are compatible with existing technology ('drop-in' technology); (7) stereoselective metallocene catalysts can produce stereoregular copolymers; (8) less reactive monomers, such as long-chain 1-olefins, e.g. hexadecene, including polypropylene macromonomers, can be incorporated; (9) copolymerization of vinyl end groups affords long-chain branching, which is beneficial to improving polyolefin processing; (10) styrene and cyclolefins are copolymerized to form a great variety of new products; (11) 1-olefin copolymerization is regioselective in contrast to some vanadium catalysts where regiorregular $CH_2CH(CH_3)CH(CH_3)CH_2$ ('head-to-head') units are formed. Moreover, it should be noted that the very clean polymers represent excellent model systems to achieve a better understanding of structure/property correlations. Ethylene copolymers with high comonomer content were the first examples of commercial applications of metallocene technology.

REACTOR BLEND TECHNOLOGY

Cascades of two and more reactors are being used to tailor-make polyolefins. Bimodal polyolefins containing a small fraction of high molecular weight homo- or copolymer, respectively, can be obtained by polymerizing olefin together with comonomer in the absence of hydrogen and then in the presence of hydrogen. Due to the ability of hydrogen to control molecular weight, lower molecular weight polyolefin is produced in the second reactor. As demonstrated by Böhm *et al.* [20] for polyethylene, such high molecular weight 'tie' molecules link together crystallites and account for improved mechanical properties.

In a sequence of gas phase reactors, (see the chapter on Ziegler catalysis) for the Novolen process of BASF AG [21], EPM rubber can be incorporated into PP during the second stage. This reactor blend can be tailored to improve low-temperature impact resistance of PP. Typical PP-based reactor blends contain high rubber contents and exploit the benefits of modern supported catalysts. In comparison to PP extruder blends, PP reactor blends give substantial savings by eliminating the energy intensive compounding step.

In Montell's reactor granule technology (cf. the chapter on Ziegler-Natta catalysis in this book), particle forming catalysts act as templates to produce pellet-sized PP particles. Since deagglomerated catalyst primary particles are homogenously distributed throughout the PP particles, EPM can be incorporated during a second gas phase copolymerization subsequent to PP particle formation in liquid propylene slurry (Montell's Catalloy technology). Control of the PP particle porosity is the key to multiphase PP where monomers such as styrene, styrene/maleic anhydride, acrylics are polymerized by free radical copolymerization inside the PP micropores (Montell's Hivalloy technology). This process was reviewed Galli and coworkers [22]. Development of reactor blend technology benefits from innovative catalyst systems, including catalyst blends (multi-site or hybrid catalysts), and from optimization of reactor configurations and process conditions.

BLOCK COPOLYMERS

For production of copolymers and reactor blends as well as for controlling morphology development during processing, it is very important to produce olefin block copolymers containing alternating blocks of rigid crystallizable PP and flexible amorphous EPM. This is of special interest in improving the toughness/stiffness balance of PP at low temperatures. Since most Ziegler catalysts are not living, special $TiCl_3$-based catalysts and sequenced feed polymerization processes were developed [23]. Such *in situ* formed poly(ethene-block-propene), which are part of a rather complex copolymer mixture, promotes compatibility of PP with HDPE. Gas phase PP technology of the Novolen-type (BASF AG), reported above, produces reactor blends with block copolymer content up to 40%.

Recently, novel living catalysts systems were introduced for 1-olefin polymerization. Brookhart *et al.* [24] used MAO-activated diimine-based catalysts, such as $[ArN{=}C(R){-}C(R){=}NAr]NiBr_2$, to obtain living polypropylene and living polypropylene-*block*-poly(1-hexene) as well as poly(1-octene)-*block*-poly(propene-co-1-octene)-*block*-poly(octene-1). Preferably this living 1-olefin polymerization was performed at temperatures below room temperatures. Van der Linden *et al.* [25] and McConville *et al.* [26] used novel diamide complexes of titanium, e.g. $[RN(CH_2)_3NR]TiMe_2$ with $R{=}2,6\text{-}^iPr_2C_6H_3$ or $R{=}2,6\text{-}Me_2C_6H_3$, activated with equimolar amounts of $B(C_6F_5)_3$ to initiate living nonstereospecific polymerization, such as 1-hexene, 1-octene, and 1-decene, at room temperature in toluene or methylenechloride solution. Molar masses of living polyolefins varied between 4300 and 148 100 g/mol and polydispersities M_w/M_n varied between 1.05 and 1.11. Although ethylene failed to afford living homo- and copolymers, this development of novel initiators for living olefin polymerization will lead to the preparation of novel families of block and star copolymers, impact-resistant polymer blends, thermoplastic elastomers, and rubbers.

FUNCTIONAL ETHYLENE COPOLYMERS

Olefin copolymers are hydrocarbons and cause difficulties in application due to poor adhesion of coatings and adhesives. Therefore, it would be highly desirable to incorporate polar comonomers into the hydrocarbon polyolefin backbone. Most conventional catalysts are severely poisoned, because polar monomers are strong Lewis bases and compete very successfully with nonpolar olefins for vacant Lewis acidic coordination sites, thus causing catalyst poisoning. Three strategies have been applied to overcome this basic problem: (1) reduction of the Lewis basicity of the polar 1-olefin; (2) copolymerization of weak Lewis bases, which are precursors to introduce polar groups during post-polymerization functionalization; and (3) copolymerization using group VIII catalysts, especially Pd and Ni, which are known to better tolerate polar groups.

Following strategy (1) a variety of functionalized copolymers were obtained. For example, Wilén and Näsman [27] prepared polymeric antioxidants by homo- and copolymerization of 6-tert.butyl-[2-(1,1-dimethylhept-6-enyl]-4-methyl-phenol. Most catalyst systems, however, gave rather low catalyst activities at high comonomer content and were severely poisoned by CO. Recently, following strategy (2), Chung [28] prepared a variety of borane-functionalized polyolefins as precursors for functional polymers by means of borane conversion. This was achieved either by polymer-analogous hydroboration of unsaturated polyolefins or by copolymerization of borane-functional 1-olefins. For example, boranes were added to non-conjugated dienes to prouce borane-functional 1-olefins, which were copolymerized, followed by hydrogen peroxide conversion to afford polyolefins with pendant hydroxyalkyl groups.

Strategy (3) was very successful and has led to new families of strictly alternating olefin/CO copolymers and polar branched polyethylenes, equivalent to ethylene/1-olef/polar comonomer copolymers but produced by 'chain walking' polymerization of ethylene in the presence of methylacrylate. Group VIII catalysts tolerate carbon monoxide (CO), which is incorporated to form polyketones being composed of alternating olefin/CO copolymers or the corresponding polyspiroketals, respectively (Figure 2). Recently this remarkable development was reviewed by Drent and Budzelaar [29]. In 1996 Shell started commercialization of

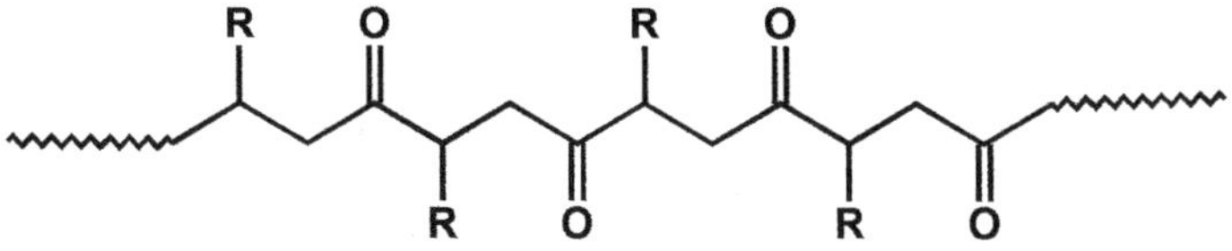

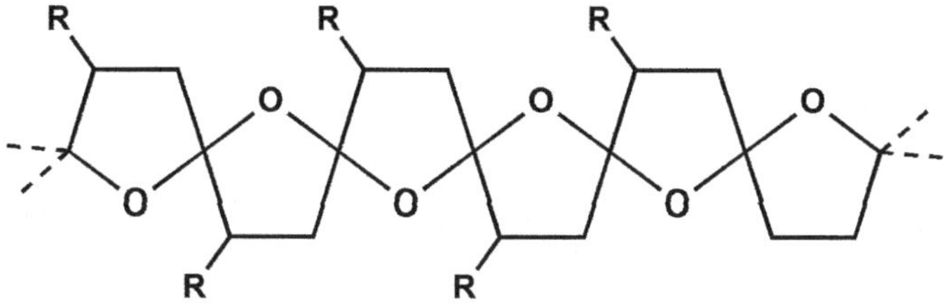

Figure 2 Alternating olefin/CO polymers: polyketones and polyspiroketals.

their Carilon polyketone resins which represent alternating ethylene/CO copolymers containing a few percent of propylene termonomer to lower melting temperature to 225°C in order to facilitate melt processing. Both Shell's Carilon® and BP's Ketonex® alternating ethylene/CO copolymers exhibit high heat distortion temperature, high toughness and excellent barrier properties with respect to hydrocarbon and oxygen permeation. Blends of Ketonex with PVC appear to improve heat distortion temperature of PVC.

Recently, Rieger *et al.* [30] observed that low stereoregular poly(propylene-alt-CO) as well as ethylene/propylene/CO terpolymers afforded elastomers when molecular mass exceeded 100 000 g/mol. Obviously, crystallizable segments afford physical crosslinks of the very flexible atactic polyketone chains, similar to nonpolar elastomeric PP (ELPP – see Elastomeric polypropylene homopolymers using metallocene catalysts) reported above. Polyketones and polyspiroketals as well as the above reported novel alternating copolymers offer attractive potential in polymer synthesis by further exploiting the potential of cheap petrochemical resources such as CO and olefin, cycloolefin and styrene feedstocks.

In a process, recently developed by Brookhart *et al.* [31] ethylene is copolymerized with methylacrylate to form polyethylene containing both long alkyl side chains and ester alkyl side chains, resulting from 'chain walking' (Figure 3). When ethylene is polymerized with diimine catalysts, branched polyethylene is formed due to migration of the transition metal alkyl along the polymer backbone and side chain via repeated β-hydride elimination and reinsertion. When methylacrylate is

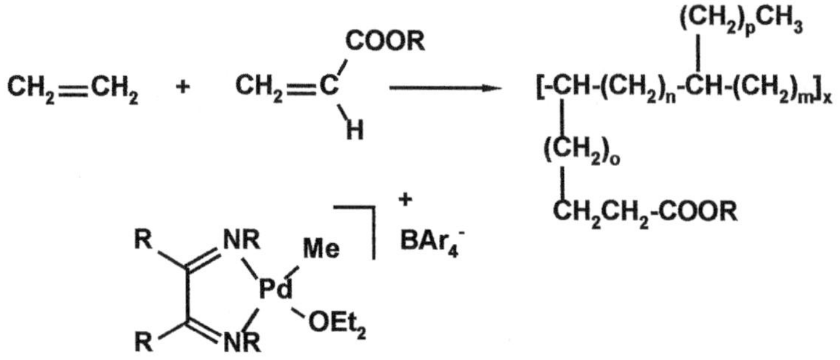

Figure 3 Catalytic low-pressure copolymerization of ethylene with methylacrylate produce polyethylenes with pendant alkyl and esteralkyl chains.

cooligomerized, ester-terminated alkyl side chains are formed. In future, the copolymerization of propylene with polar monomers is expected to become the base of new classes of PP.

REFERENCES

1. Van der Ven (1990) *Polypropylene and Other Polyolefins*, Elsevier, Amsterdam.
2. Tait, P.J.T. and Berry, I.G. (1989) in *Comprehensive Polymer Science*, Vol. 4, (eds G. Allen, J.C. Bevington, G.C. Eastmond, A. Ledwith, S. Russo and P. Sigwalt), Pergamon Press, Oxford, p. 575.
3. Davis, S.C., Von Helens, W., Zahalka, H.A. and Richter, K.-D. (1996) *Polymeric Materials Encyclopedia*, Vol. 3, (ed. J.C. Salamone), CRC Press, Boca Raton, FL, USA, p. 2264.
4. Randall, J.C. (1989) *J. Macromol. Sci. – Macromol. Chem. Phys.*, **C29**, 201.
5. Fan, Z.Q., Locatelli, P., Sacchi, M.C., Camurati, I. and Galimberti, M. (1995) *Macromolecules*, **28**, 3342.
6. Herfert, N., Montag, P. and Fink, G. (1993) *Macromol. Chem. Phys.*, **194**, 3167.
7. Suhm, J., Schneider, M.J. and Mülhaupt, R. (1997) *J. Polym Sci., Part A., Polymer Chem.*, **35**, 735.
8. Ewen, J. (1986) in *Catalytic Polymerizations of Olefins*, (eds T. Keii and K. Soga), Elsevier, Amsterdam.
9. Uozomi, T., Nakamura, S., Toneri, T., Teranishi, T., Sano, T., Arai, T. and Shiono, T. (1996) *Macromol. Chem. Phys.*, **197**, 4237.
10. Schneider, M.J., Suhm, J. and Mülhaupt, R. (1997) *Macromolecules*, **30**, 3164.
11. Schneider, M.J. and Mülhaupt, R. (1997) *Macromol. Chem. Phys.*, **198**, 1124.
12. Minick, I., Moet, A., Hiltner, A., Baer, E. and Chum, S.P. (1995) *J. Appl. Polymer Sci.*, **58**, 1371.
13. Fink, G. and Seppälä, J.V. (1994) *Macromolecules*, **27**, 6254.
14. Batistini, A. (1995) *Macromol. Symp.*, **100**, 137.
15. Senetz, F.G., Mülhaupt, R. and Waymouth, R.M. (1996) *Macromol. Chem. Phys.*, **197**, 1071.
16. Cheung, Y.W., and Guest, M.J. (1996) *ANTEC 96*, 1634.

17. Kaminsky, W., Bark, R., Spiehl, R., Möller-Lindenhof, N. and Niedoba, S. (1988) in *Transition Metals and Organometallics as Catalysts for Olefin Polymerization*, (eds W. Kaminsky and H. Sinn), Springer-Verlag, Berlin, p. 291.
18. Cherdron, H., Brekner, M.-J. and Osan, F. (1994) *Angew. Makromol. Chem.*, **223**, 121.
19. Sernetz, F.G., Mülhaupt, R. and Waymouth, R.M. (1997) *Polym. Bull.*, **38**, 141.
20. Böhm, L.L., Bilda, D., Breuers, W., Enderle, H.F. and Lecht R., in *Ziegler Catalysts*, (eds G. Fink, R. Mülhaupt and H.H. Brintzinger), Springer-Verlag, Berlin, p. 387.
21. Hungenberg, K.D., Kerth, J., Langhauser, F., Marcinke, B. and Schlund, R. (1995) in *Ziegler Catalysts*, (eds G. Fink, R. Mülhaupt and H.H. Brintzinger), Springer-Verlag, Berlin, p. 363.
22. Galli, P., Haylock, J.C., and Simonazzi, T. (1995) in *Polypropylene: Structure, Blends and Composites*, (ed. J. Karger-Kocsis), Chapman & Hall, London.
23. Chen, C.M. and Ray, W.H. (1993) *J. Appl. Polymer Sci.*, **49**, 1573.
24. Killian, C.M., Tempel, D.J., Johnson, L.K. and Brookhart, M. (1996) *J. Am. Chem. Soc.* **118**, 11664.
25. Van der Linden, A., Schaverien, C.J., Meijboom, N., Ganter, C. and Orpen, A.G. (1995) *J. Am. Chem. Soc.*, **117**, 3008.
26. Scollard, J.D. and McConville, D.H. (1996) *J. Am. Chem. Soc.*, **118**, 10008.
27. Wilén, C.-E. and Näsman, J.H. (1994) *Macromolecules*, **27**, 4051.
28. Chung, T.C. (1997) in *Reactive Modifiers for Polymers*, (ed. S. Al-Malaika), Blackie Academic & Professional, Glasgow, p. 303.
29. Drent, E. and Budzelaar, P.H.M. (1996) *Chem. Rev.*, **96**, 663.
30. Abu-Surrah, A.S., Eckert, G., Pechhold, W., Wilke, W. and Rieger, B. (1996) *Macromol. Rapid Commun.*, **17**, 559.
31. Johnson, L.K., Mecking, S. and Brookhart, M. (1996) *J. Am. Chem. Soc.*, **118**, 267.

Keywords: copolymerization, reactor blend, comonomer distribution, ethylene/propylene rubber (EPM), ethylene/propylene/diene rubber (EPDM), multi-site catalyst, single site catalyst, Markovian statistic, metallocenes, random copolymer, alternating copolymer, block copolymer, chain folding, fringed micelle, Affinity®, Engage®, Topas®, polydispersity, stereoregularity, regioregularity, drop-in technology, Novolen®, functional copolymers, polyketones

Crash performance of glass fiber reinforced polypropylene tubes

S. Kerth, A. Dehn and M. Maier

INTRODUCTION

It has been found that the energy absorption capacity per unit weight of composite structural parts under axial crush load is much better than that of steel or aluminium [1, 4]. For that reason, they are increasingly used in crash-loaded structures of any kind of vehicles, especially in automobiles. Beside special composite crash elements, complex composite structures, such as longitudinal girder, cross members, mudguards, under bodies and even complete car bodies, are subjects of research. The investigations on the crash behavior of composites focused mainly on crash elements with simple geometry, e.g. tubes with cylindrical and quadratic cross-sections and cones [1–3, 5, 6]. Nowadays the good crash performance and energy absorption capacity of composite tubes with thermoplastic matrix systems has to be pointed out [2, 3]. This can be attributed to higher fracture toughness and ultimate strain of thermoplastics resulting in better mechanical damping and less tendency for crack propagation. So composites with thermoplastic matrix systems exhibit high damage tolerance. This, together with an industrial-scale manufacturing technique and benefits due to easy recycling, means that thermoplastic composite structural parts have great potential for future applications [4].

Polypropylene: An A–Z Reference
Edited by J. Karger-Kocsis
Published in 1999 by Kluwer Publishers, Dordrecht. ISBN 0 412 80200 7

ENERGY ABSORPTION CAPACITY

The outstanding energy absorption capacity of composites is an important aspect in the development of lightweight structures. Several studies have shown that the energy absorption of composite structures is superior to structures of steel or aluminium. The energy dissipation during crash of metallic structural elements is mainly due to plastic deformation. Yield strength and fracture toughness are the only material parameters influencing the crush behavior with respect to a given geometry. The crushing and energy consumption process in continuous fiber reinforced composites is much more complex and thus affected by numerous parameters, e.g. fiber and matrix properties (strength, stiffness, failure strain), fiber volume fraction, reinforcement system (e.g. short fiber, continuous fiber, woven fabric, knitted fabric), lay-up and manufacturing process. The reason for the improved energy absorption is the 'controlled' propagation of microcracks in the crash front where the material undergoes various failure mechanisms. These mechanisms are fiber and matrix cracking under tension, compression and shear, as well as delamination, fiber pull-out, interface failure between fiber and matrix and friction effects between fragmented material and the loading surface [5]. Therefore it is hard to quantify the influence of a single failure event on the crash energy absorption. Furthermore the geometric shape, the properties of fiber and matrix material, their interface, fiber volume fraction and fiber orientation have strong influences on the crash behavior.

Farley and Jones [5] identified three unique crushing modes which occur in continuous-fiber-reinforced composite tubes, namely transverse shearing, laminae bending, and local buckling. Most composite materials with brittle fibers crush in a combination of transverse shearing and laminae bending referred to as the brittle crushing-fracturing mode [6]. This fracture scenario includes interlaminar, intralaminar, and in-plane crack initiation and propagation, and local fracture of laminate bundles. Whereas composites with brittle fibers, such as glass or carbon, fail in a splaying mode, an aramide reinforcement leads to failure by local buckling (comparable to metals), so that post-crush integrity of the structure is excellent. However, the folding mode is less efficient for polymer composites, because less material volume is destroyed.

CRASH PERFORMANCE TESTS

The most common way to perform crash tests on components is to use a servo-hydraulical or electrical driven universal testing machine. The typical speed of these testing devices reaches up to 1 m/s. Tests at higher velocities require a drop tower, where a defined weight is dropped from

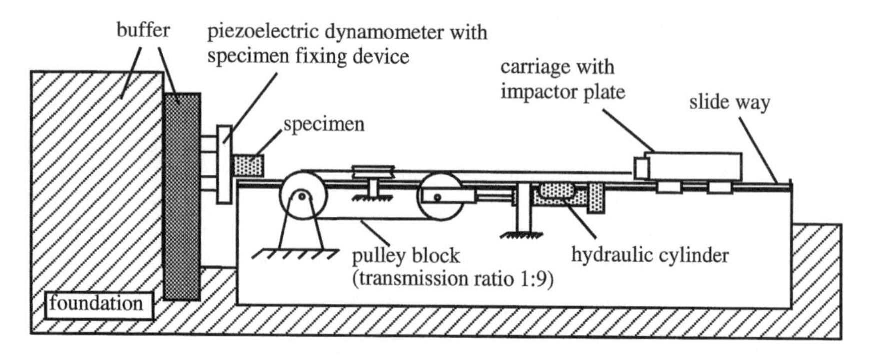

Figure 1 Principle scheme of a horizontal catapult crash test rig.

a certain height upon the specimen, or a horizontal catapult test rig. With these test facilities, velocities of up to 40 m/s can be reached. Figure 1 shows a scheme of the mechanical assembly of a horizontal catapult based on the acceleration of a carriage with variable load by a hydraulic cylinder. The pulley block with 1:9 transmission extends the carriage speed to 9 times cylinder speed. Before hitting the specimen, which is fixed to a piezoelectric measurement plate, the carriage is detached from the traction rope. In any case, the crash event has to be recorded by measuring at least force or acceleration versus time. By numerical integration of the acceleration, signal velocity and displacement are calculated. The forces are obtained by dividing the acceleration by the striker mass. More exact results are obtained when force and displacement are measured separately by two independent measurement systems [7].

The experimental investigations on the crush behavior on filament wound glass fiber reinforced tubes were performed at a crash test rig, as shown in Figure 1, which was designed for experiments on larger specimens and structural parts [8]. A striker which can attain a mass scalable between 45 and 220 kg is accelerated horizontally from 1 m/s up to 24 m/s and impacts the specimen which is clamped to the force measurement plate equipped with four piezoelectric discs (sampling rate 200 kHz). Other measured parameters are the acceleration and the displacement of the carriage as well as the initial velocity at the onset of the crush event. The crash event can be filmed with up to 10 000 pictures per second. All crash tests were performed with a striker mass of 96.6 kg at an initial velocity of 8.3 m/s (30 km/h). The kinetic energy was 3.34 kJ.

For characterization of crash worthiness the maximum (peak) force and the mean force are considered. The ratio of peak and mean force (load uniformity) for ideal absorbers is considered to equal 1. The mass specific energy absorption E_s, expressing the efficiency of the specimen concerning energy absorption capabilities, is calculated from the absorbed energy E, which is given as the area under the load–displacement curve, and the mass of the crushed material m_{CR}. The specific crash stress σ_s gives information about the stress level where the crash takes place. The value σ_s is defined as the mean crash load F_m divided by the hit area A. If σ_s is divided by the density ρ of the material one gets the specific sustained crash stress which equals E_s. Typical values of E_s for steel tubes are 12 kJ/kg, for aluminium 20 kJ/kg [1].

CRASH BEHAVIOR OF GF/PP TUBES

The axially crashed tubes were produced from polypropylene powder-impregnated glass fiber roving with the flame assisted thermoplastic winding technology. By using this flame torch speeds up to 30 m/min were reached using infrared/hot gas technology processing powder impregnated fiber bundles. During the consolidation process, matrix flow occurs until 'interlaminar welding' is achieved. The efficiency of this process mainly depends on temperature, pressure and time.

The fabricated tubes had an internal diameter of 70 mm, a wall thickness of 4 mm for ±85° winding angle, 4.5 mm for ±45° winding angle and 5 mm for ±45° with outer 90° hoop layers. The winding angle is counted with respect to the length of the tube. The tube length was 150 mm, respectively 200 mm for the ±85° tubes. The specimens have been triggered by a 45° chamfer in order to assure a progressive failure with a defined onset location. It is known that composite structural parts fail by progressive crushing when the end of the specimen is chamfered, which acts as a local imperfection. A total of three specimens per variation of process parameters have been tested. The scatter of data was within 10% of the mean value. The macroscopic outlook of the crashed specimen was identical for the test series.

TEST RESULTS

Table 1 gives an overview of specimen geometry and recorded crash data. Figure 2 shows typical load–displacement curves.

The specimens with the ±45°-lay-up fail by internal and external splaying of fiber bundles separated by a debris wedge leading to a centre delamination propagating progressively down the specimen. The specimens show a very good load uniformity (maximum peak load: 26 kN; mean crash load: 24 kN). The application of outside hoop layers to the specimens raised the mean crush load by almost 20 kN and the specific energy absorption increased from 17 to 28 kJ/kg. The reason for higher

Table 1 Test results of GF/PP tubes

Reference	*Lay-up* (°)	*Carriage Velocity* (m/s)	*Specimen mass* (g)	*Crashed length* (mm)	*Mean load* (kN)	*Specific energy* (kJ/kg)
Gfpp4500	±45	10.58	215.76	97.81	24.57	17.39
Gfpp4501	±45	7.68	214.80	79.63	25.16	17.71
Gfpp4502	±45	7.88	215.79	82.63	24.11	17.02
Gfpp4503	±45/90	7.85	242.55	45.39	44.79	28.19
Gfpp4504	±45/90	7.64	237.79	45.90	43.81	28.93
Gfpp8500	±85	10.58	223.65	92.87	26.71	24.11
Gfpp8501	±85	10.56	241.56	90.20	26.14	22.82

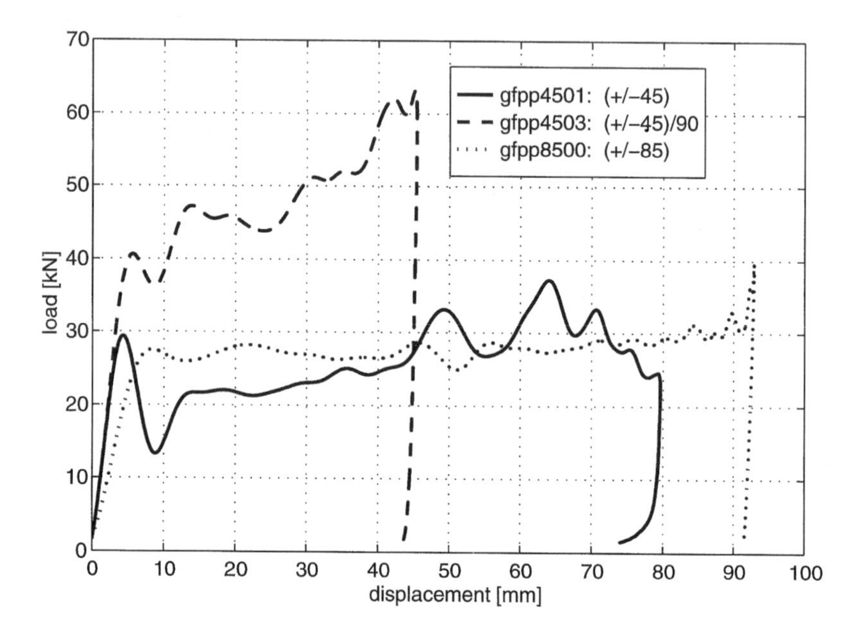

Figure 2 Typical load–displacement curves for GF/PP tubes.

specific energy absorption values is an additional energy consumption is needed for tearing the hoop layers. The smaller crash radius forced by the hoop layers and consequently the smaller bending radius for the ±45°-layers also contribute to higher energy dissipation. The tubes with the ±85°-lay-up show as well a smaller bending radius. In this case, the almost circumferential fibers are torn. The lack of fibers aligned in the crash direction leads to lower compressive strength of the tube resulting

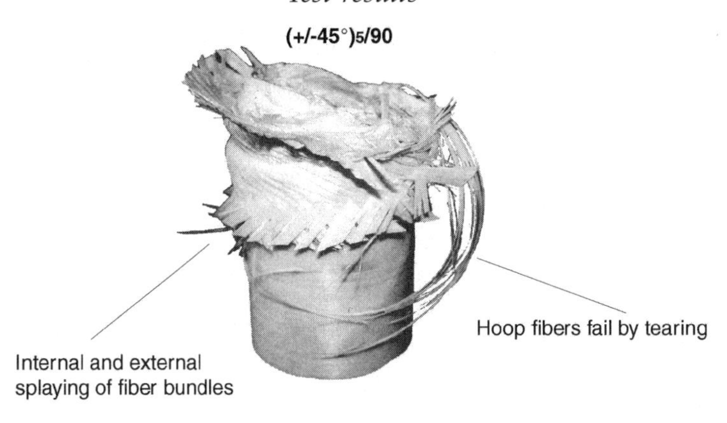

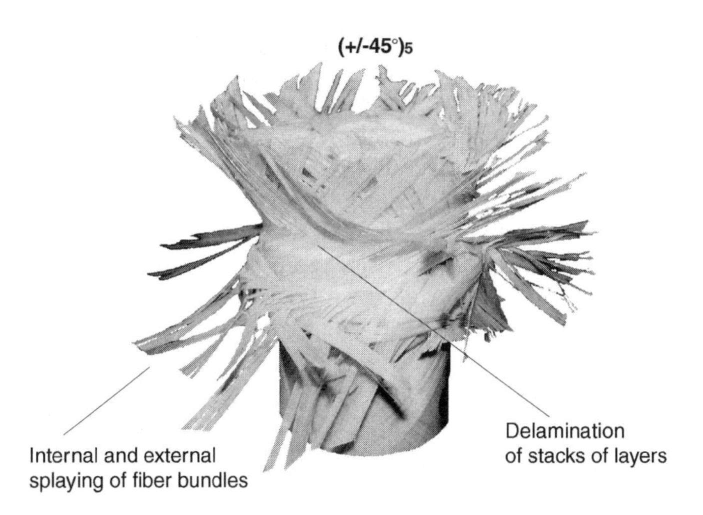

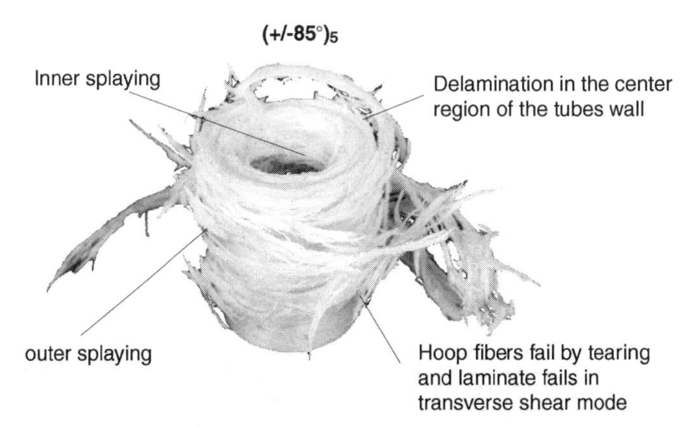

Figure 3 Failure mechanisms of thermoplastic filament wound GF/PP tubes.

in a lower mean crash force. The laminate fails by transverse shearing. The failure modes are depicted in Figure 3.

CONCLUSION

Considerable research into crash behavior of composites has been done in the last decades. However, this research has been focused on high-performance composites. In contrast, polypropylene matrix composite tubes are relatively inexpensive and offer, together with a flexible manufacturing technique and material advantages, such as high volume capability or recycling aspects, a great potential for crash applications in the future. Future research into crashworthiness should be extended to complex structural parts, such as sandwich parts or spherically shaped (e.g. stamp formed) organic sheets. Another aspect is the strong need to perform research and development mainly by computer-aided engineering (CAE) to reduce the costs for prototyping and experimental investigations as well as to shorten the development time of, for example, new generations of cars. Actually, there is an enormous lack of reliable and efficient tools to analyze the failure behavior of composite structural parts especially with respect to ductile material behavior and rate dependency. The possibility of performing an exact CAE simulation of crash behavior is of vital importance and will be of major interest in research.

REFERENCES

1. Maier, M., (1990) Experimentelle Untersuchung und numerische Simulation des Crashverhaltens von Faserverbundwerkstoffen, Dissertation Universität Kaiserslautern.
2. Hamada, H., Coppola, J.C., Maekawa, Z. and Sato, H. (1992) Comparison of energy absorption of carbon/epoxy and carbon/PEEK composite tubes, *Composites*, **23**(4), 245–252.
3. Kerth, S., Dehn, A., Ostgathe, M. and Maier, M. (1996) Experimental Investigation and Numerical Simulation of the Crush Behaviour of Composite Structural Parts, *Proceedings of the 41st SAMPE Symposium and Exhibition*, 1996, Anaheim, USA, (eds G. Schmidtt *et al.*), SAMPE, Covina, CA, USA, pp. 1397–1408.
4. Breuer, U., Ostgathe, M., Kerth, S. and Neitzel, M. (1996) Fabric Reinforced Thermoplastic Composites – A Challenge for Automotive Applications, Proceedings of the XXVI Congress FISITA 96, 17–21.06.1996, Prague, Plecknowledge International, New York.
5. Farley, G.L. and Jones, R.M. (1992) Prediction of the energy-absorption capability of composite tubes, *Composite Mat.*, **26**(3), 388–404.
6. Hull, D. (1991) A unified approach to progressive crushing of fibre-reinforced composite tubes, *Composite Sci. Technol.*, **40**, 377–421.
7. Hanefi, E. H. and Wierzbicki, T. (1995) *Calibration of Impact Rigs for Dynamic Crash Testing*, Final Report, EU Project ERB 4050 PL 930800.

8. Himmel, N., Kerth, S. and Maier, M. (1994) Rechnergestützte Bauteilprüfung, *Kunststoffe*, **84**, 12.

Keywords: crash, crush, failure mode, energy absorption, filament wound tube, thermoplastic filament winding, axial crash of tubes, GF/PP composite tubes.

Crazing and shear yielding in polypropylene

Ikuo Narisawa

Crazing and shear yielding are essentially the two main modes of deformation which are responsible for brittle and ductile fractures of all polymers, respectively. However, craze-like features, as those in amorphous polymers, have been observed in crystalline polymers for a long time, but crazing in polypropylene (PP) is still not completely understood. One basic problem of studying crazing in crystalline polymers, such as PP, is the fact that they are normally organized into various microstructures, such as spherulites, consisting of lamella-type crystal and interlamellar amorphous regions (see 'Spherulitic crystallization and structure' in this book). Another basic problem is that, at room temperature, crystalline polymers are above their glass transition point and they tend to deform and to yield without forming crazes. Such typical deformation is frequently found in plane stress tensile tests at room temperature. Hence, it has not been believed for a long time that crazing can be a dominant deformation mode in PP.

Nevertheless, craze-like features are often observed in deformed PP especially at lower temperatures or well-crystallized PP and they were assigned different names by different investigators [1, 2]. The appearance of these craze-like features is rather less evident because of the lack of transparency. However, their characteristics are quite in agreement with those of amorphous glassy polymers. As shown in Figure 1, although the fibril thickness are more than 10 times larger and much thicker and more randomly oriented than those typically observed in amorphous glassy polymers, they contain stretched microfibrils that span the space

Polypropylene: An A–Z Reference
Edited by J. Karger-Kocsis
Published in 1999 by Kluwer Publishers, Dordrecht. ISBN 0 412 80200 7

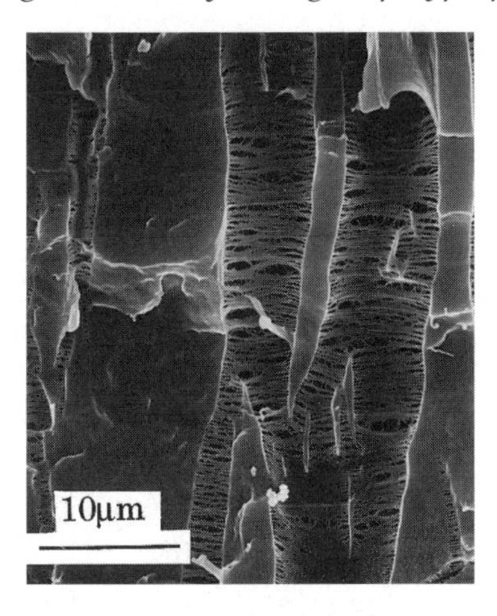

Figure 1 Craze structure in PP.

between the walls of what would otherwise be called microcracks. The boundary between the craze structure and surrounding uncrazed material is very sharp. Small pieces are separated from the undeformed matrix near the craze interfaces. The separated matrix pieces are further broken to form fibrils by drawing and, at the same time, a new craze structure initiates at the gap between matrix and separated pieces. Finally, the craze thickens drastically by coalescence of the 'old' and 'young' craze. A craze-thickening mechanism in which the new undeformed polymer is drawn into the craze interfaces maintaining the extension ratio of the fibril constant is generally called a surface drawing mechanism [2]. Craze thickening in PP mentioned above occurs by this surface drawing mechanism as usually encountered in amorphous glassy polymers.

Furthermore, the crazes in PP show other similar characteristics to those of amorphous polymers. They grow apparently normal to the direction of major tensile stress which somewhat deviates from the tensile direction because of spherulitic structure. There are similar environmental effects on craze initiation (see also 'Environmental stress cracking of polypropylene' in this book). Crazing is also an important source of toughness in toughened PP alloy systems such as propylene-ethylene block copolymers.

With respect to general criteria for craze initiation and growth, important contributions have been made for amorphous glassy polymers by many investigators during past years. Recognizing that crazing is also inherently a cavitational process for crystalline polymers leads to the criterion that craze initiation involves a dilatational stress component. In other words, crazing never occurs when the stress field is negative or compressive. This is expressed as:

$$\sigma_p = \frac{\sigma_1 + \sigma_2 + \sigma_3}{3} = \sigma_c \tag{1}$$

where σ_1, σ_2 and σ_3 are the maximum, medium and minimum principal stresses, respectively, while the criteria for shear yielding are generally characterized by the following Tresca or von Mises yield condition as:

$$\tau = \frac{\sigma_1 - \sigma_2}{2} = \tau_T \tag{2}$$

$$(\sigma_1 - \sigma_2)^2 + (\sigma_2 - \sigma_3)^2 + (\sigma_3 - \sigma_1)^2 = 6\tau_M^2 \tag{3}$$

where τ_T and τ_M are the material constant. The critical stress for craze nucleation can be estimated from tensile tests or bending tests using a thick specimen with a blunt notch [2]. When the thickness is sufficient to maintain a plane strain state, a tensile force which resists the contraction due to Poisson's effect appears in the thickness direction. For the same reason, a tensile stress is generated in the width direction. Thus, the triaxial tensile stress (dilatational stress) shows the maximum at the base of the notch. On the other hand, the shear stress is also maximum at the notch tip. If the tensile load is further increased, shear yielding will start at the base of the notch. This shear yielding will propagate toward the inside of the specimen as the applied stress increases. The slip line theory of an elastic-perfectly-plastic material can predict that the maximum dilatational stress is again produced at the tip of the shear yielded region. The maximum triaxial stress increases with an increasing size of the shear yielded region. The maximum stress is limited to a certain value because the shear yielded region has a maximum size which is determined only by the notch geometry. This dilatational stress is sufficient to form crazes at the shear yielded region. This is a critical stress σ_c for void nucleation in the craze structure as defined by equation (1).

In amorphous glassy polymers, the critical stress for nucleation of a craze is generally high and the strength of fibrils in the craze is close to the critical stress for craze nucleation (void initiation) because the cohesive strength has a more narrow distribution if compared with crystalline polymers. And loading of PP, which has inhomogenous structure, easily results in the nucleation of porosity under a plane strain tensile condition due to microrupture in a region with a weak molecular cohesive strength (such as an amorphous layer connecting adjacent crystallites). Therefore, the craze initiation occurs at a low stress relative to the strength of fibrils (i.e. strength of craze) and, consequently, the propagation of a craze in PP is rather stable compared with that of the amorphous region. Thus, the critical stress for craze nucleation in PP is firmly influenced by the fine structure. The strength of craze is close to the stress of fracture strength evaluated from the strength of the oriented region formed by necking during uniaxial extension and it strongly depends on the number of tie molecules between crystalline lamellae [3].

On the other hand, PP shows shear yielding without craze formation and yielding behavior (i.e. necking) with a plateau in a plane stress state. The stress for yielding is also affected by the density of tie molecules because the tie molecules are pulled out from the lamellar fragment during the yielding process [4].

REFERENCES

1. Friedrich, K. (1983) *Advanced Polymer Science*, (ed. H.-H. Kausch) 52/53, Springer-Verlag, Berlin, pp. 225–274.
2. Narisawa, I. and Ishikawa, M. (1990), *Advanced Polymer Science*, (ed. H.-H. Kausch) 91/92, Springer-Verlag, Berlin, pp. 353–391.
3. Ishikawa, M., Ushui, K., Kondo, Y., Hatada, K. and Gima, S. (1996) *Polymer*, **37**, 5375–79.
4. Takayanagim, M. and Nitta, K. (1997) *Macromol. Theor. Simul.*, **6**, 181–195.

Keywords: crazing, shear yielding, dilatational stress, Tresca criterion, von Mises criterion, necking, void nucleation, tie molecules, yielding.

Crosslinking of polypropylene

Ivan Chodák

Crosslinking is a widely used method for modification of polymer properties. The process consists in a formation of a tri-dimensional network structure, a gel, causing substantial changes in the material properties. Vulcanized rubbers and thermosetting resins can be mentioned as the most common examples of crosslinked systems. However, crosslinked polyolefins are of significant interest as well.

Various procedures may be used for initiation of polyolefin crosslinking. All commercially used processes are based on a formation of polyalkene macroradicals in certain stage of the process. Common ways of crosslinking initiation consist in macroradical formation via thermal decomposition of organic peroxides, high energy irradiation (gamma or electron beam), ultraviolet radiation in the presence of ultraviolet-sensitizer, and grafting of silane groups onto the polyolefin backbone with subsequent crosslink formation via hydrolysis of the silanole moieties. Other procedures of initiating polyolefin crosslinking are rarely used or are only investigated in the laboratory. A detailed description of various initiation possibilities together with mechanisms of the processes and specific features of each procedure has been given recently in a comprehensive review by Lazár *et al.* [1].

Although general features of the crosslinking procedure are the same for all polyolefins, substantial differences exist regarding the efficiency and mechanism of the process depending on the chemical structure and physical state of the material. Since basically all common procedures leading to polyolefin crosslinking are based on a free radical formation and subsequent recombination of macroradicals, it is important to be

Polypropylene: An A–Z Reference
Edited by J. Karger-Kocsis
Published in 1999 by Kluwer Publishers, Dordrecht. ISBN 0 412 80200 7

aware of peculiarities of the free radical transfer and macroradical decay in polypropylene (PP), especially when compared to polyethylene (PE).

The formation of PP macroradicals is an easy process initiated by more or less any radical initiator. It occurs spontaneously in oxidative processes. Alkyl radicals (except for methyl) are usually not reactive enough to initiate an efficient macroradical formation in PP. Oxyl radicals, formed by a thermal decomposition of peroxides, are the most convenient species for crosslinking initiation. The transfer of the radical centre to PP is selective to a certain extent. At temperatures usual for peroxide decomposition, the ratio of the rate constants of the abstraction of hydrogen from primary, secondary, and tertiary carbon by the oxyl radical is approximately 1:3:10 [2].

The PP macroradicals can decay by several ways (Figure 1).

When discussing crosslinking, recombination is the only useful process. Disproportionation leads to a formation of unsaturation (double

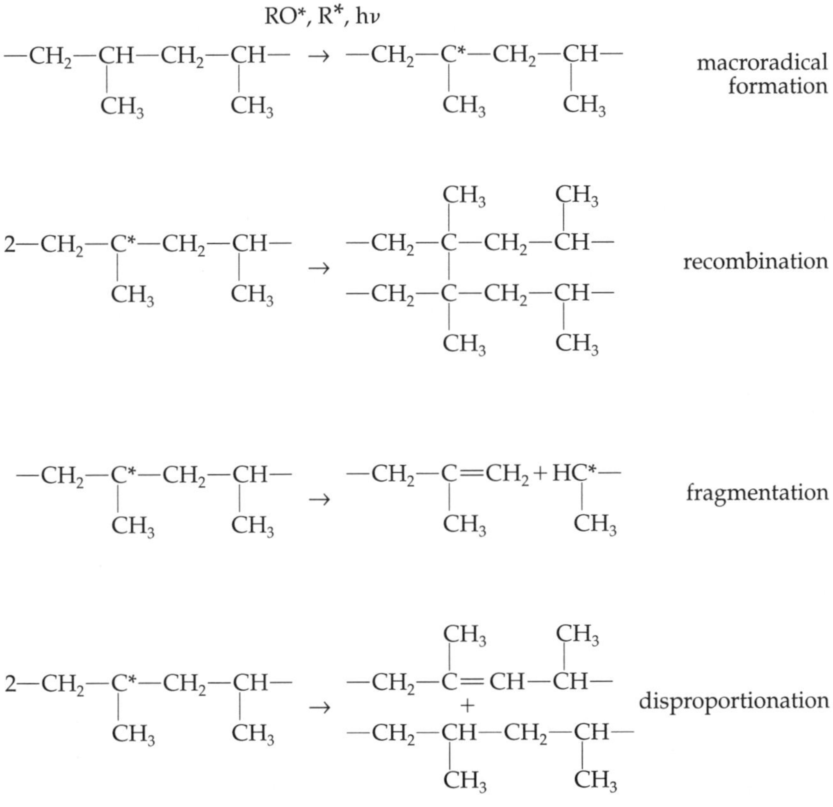

$$-CH_2-\underset{CH_3}{\underset{|}{CH}}-CH_2-\underset{CH_3}{\underset{|}{CH}}- \xrightarrow{RO^*,\ R^*,\ h\nu} -CH_2-\underset{CH_3}{\underset{|}{C^*}}-CH_2-\underset{CH_3}{\underset{|}{CH}}- \qquad \text{macroradical formation}$$

$$2-CH_2-\underset{CH_3}{\underset{|}{C^*}}-CH_2-\underset{CH_3}{\underset{|}{CH}}- \rightarrow \begin{matrix} -CH_2-\overset{CH_3}{\overset{|}{C}}-CH_2-\overset{CH_3}{\overset{|}{CH}}- \\ | \\ -CH_2-\underset{CH_3}{\underset{|}{C}}-CH_2-\underset{CH_3}{\underset{|}{CH}}- \end{matrix} \qquad \text{recombination}$$

$$-CH_2-\underset{CH_3}{\underset{|}{C^*}}-CH_2-\underset{CH_3}{\underset{|}{CH}}- \rightarrow -CH_2-\underset{CH_3}{\underset{|}{C}}=CH_2+\underset{CH_3}{\underset{|}{HC^*}}- \qquad \text{fragmentation}$$

$$2-CH_2-\underset{CH_3}{\underset{|}{C^*}}-CH_2-\underset{CH_3}{\underset{|}{CH}}- \rightarrow \begin{matrix} -CH_2-\overset{CH_3}{\overset{|}{C}}=CH-\overset{CH_3}{\overset{|}{CH}}- \\ + \\ -CH_2-\underset{CH_3}{\underset{|}{CH}}-CH_2-\underset{CH_3}{\underset{|}{CH}}- \end{matrix} \qquad \text{disproportionation}$$

Figure 1

bonds), resulting in an inefficient decay of initiator. Therefore a decrease in the crosslinking efficiency is observed. From the point of view of crosslinking, fragmentation of macroradicals is the most detrimental reaction. The ratio of macroradicals decaying by recombination and fragmentation depends on several factors. Since the activation energy of fragmentation is higher than that of recombination, a temperature rise leads to an increase of the fragmentation probability. On the other hand, an increase in the macroradical concentration results in a higher rate of bimolecular recombination, compared to competing monomolecular scission. Considering all these factors, the increase of initiator concentration or dose rate would lead to an increase in crosslinked portion. It was demonstrated that the parameters of crosslinking, e.g. gel point, can be enhanced by using peroxides with higher decomposition rates. A rough calculation based on experimental data published [3] indicates that peroxide-initiated process at around 160°C (this is the decomposition temperature of many important peroxides) results in a ratio of recombination/scission ≈ 1.0. Less than one-half of PP macroradicals formed decay by processes leading to crosslinking, since the disproportionation process also has to be considered. From a practical point of view this means that, although it is possible to obtain a high degree of crosslinking (up to 80%) by using a high initiator concentration or irradiation dose, the high degree of fragmentation will lead to excessive degradation. Initiating crosslinking of PP by simple methods leads to detrimental deterioration of the properties, and thus to a product of no industrial importance.

An increase in the crosslinking efficiency can be achieved by addition of so-called coagents of crosslinking. These are species which can react with PP macroradicals by addition reaction before fragmentation occurs. By such a way the macroradical is transformed to a more stable form which does not undergo scission. Efficient coagents possess two or more active groups in the molecule so that they can react with two or more PP macroradicals. Thus, the crosslinks are formed by the coagent molecules which are bound into the network.

The most common coagents described in the literature are polyfunctional monomers, e.g. bi-, tri-, or tetraacrylates, allylics, vinyl compounds, etc. Either peroxides or high-energy irradiation (gamma, electron beam, or ultraviolet) can be used as the initiators for macroradical formation [4]. Addition of these species results in an increase of the crosslinking efficiency so that a high crosslinked proportion can be produced with reasonable irradiation dose or amount of the peroxidic initiator. Very efficient coagents are those containing acryloyloxy moieties, e.g. tetramethylolmethane tetraacrylate or trimethylolpropane triacrylate. For crosslinking of PP to rather high degree, the required radiation dose is ten times lower than if other coagents (e.g. triallyl cyanurate) are used. However, with any polyfunctional monomer, at too high concentration of the coagent, too many double bonds decay by polymerization reaction and not by addition with PP macroradicals. Therefore, a high degree of the crosslinking can be reached without extensive degradation but with the irradiation dose or peroxide content too high to allow an industrial application of the process.

Many other species have been tested as possible coagents for PP crosslinking. Sulfur and some sulfur-containing chemicals, e.g. thiourea, are among the most effective species. Oligomeric polybutadiene or other unsaturated rubbers are also effective and can be considered as special types of polyfunctional monomers. Acetylene belongs to the same group. Hydroquinone and benzoquinone have been reported to be extremely efficient coagents, superior to other systems described in literature [5]. However, these coagents are only effective in combination with organic peroxides and act as inhibitors of radical processes if the reaction is initiated by radiation.

In crosslinked PP [6], the change in the molecular weight has a direct effect on the behavior of the material. The composition of the initiating system, especially the presence or absence of the coagent, is important in respect to the degradation:crosslinking ratio. Generally, crosslinking leads to an increase in viscosity of polymer melt; however, with small initiator amount, when degradation and branching are dominant, a decrease in PP viscosity is common. This feature is used for the production of PP with controlled rheology (CR, see as cross-reference Controlled rheology polypropylene). CR-PP grades have lower viscosity and better processability due to a certain decrease in molecular weight and narrow molecular weight distribution and are used mainly for production of PP fibres.

The change in crystallinity due to crosslinking is another fundamental feature influencing other properties. From this point of view, it is important whether crosslinking is initiated in the melt (e.g. by thermal decomposition of peroxides) or in solid state (by high-energy irradiation, such as gamma or electron beam). Crosslinking in the melt leads to homogeneous distribution of junctions while the crosslinking processes in solid state proceed mostly in amorphous region only or at boundaries of crystallites. This difference results in certain differences in the behavior of crosslinked material. It was reported that irradiation up to 1500 kGray does not lead to a substantial changes of crystallinity measured by density and X-ray diffraction. The changes at molecular level due to gamma-irradiation, such as conformational, destructive, crosslinking, oxidation, and post-irradiation effects, did not affect the supermolecular smectic structure. On the other hand, peroxide-initiated crosslinking results in a dramatic drop of crystallinity if no coagents are present. The size of spherulites is diminished to one third of their original dimensions. Differential scanning calorimetry (DSC) measurement showed that after crosslinking the crystalline portion fell from an initial 55–60% down to 10–15%. In the presence of an efficient coagent, such as hydroquinone, the decrease recorded was from 120 J/g for the virgin PP to about 100 J/g for the highly crosslinked PP. Obviously, in the absence of a coagent, a decrease of crystallinity is caused mainly by defects formed by branching. In the presence of a coagent, scission and subsequent branching is suppressed and the network junctions are the prevailing structural defects influencing the crystallinity degree. The crystallinity drop is comparable with that of chemically crosslinked PE where almost no scission is supposed to occur.

Mechanical properties depend directly on the change in molecular weight and crystallinity. Irradiation of PP in absence of an effective coagent results in a substantial decrease of the molecular weight accompanied by a drop in the tensile strength. At an absorbed dose above 100 kGray the material becomes extremely brittle. Similar effects are also caused by PP macroradical reactions initiated via thermal decomposition of peroxides.

It has to be pointed out that substantial numbers of free macroradicals remain in the polymer after irradiation. These macroradicals, formed in crystalline region, contribute substantially to a long term degradation of mechanical properties. The rate of decay of the residual radicals was found to depend strongly on the crystalline portion, i.e. it can be influenced by processing and molding procedures.

Peroxide-initiated crosslinking results in similar changes as caused by irradiation. In the absence of a coagent, a decrease in molecular weight is observed accompanied by a decrease of both modulus and yield strength. The decrease of the impact strength is substantial. The opposite tendency was found for crosslinking in the presence of a coagent. Superior efficiency was reported for peroxide-initiated crosslinking if benzoquinone or hydroquinone were used as coagents. Since beta-scission of PP macroradicals is suppressed significantly, the mechanical properties are less affected by degradation compared to the other crosslinking initiating systems. An improvement in the impact resistance at low temperature and lower temperature of brittle–ductile transition was observed. Significant enhancement in the temperature resistance, even above 300°C, was also demonstrated. The yield strength depended on the crystallinity as expected. Somewhat surprisingly, the same dependence was found for tensile strength, although modulus and elongation values drop steadily with increasing crosslinking degree [6].

Ionic crosslinking of polypropylene can also contribute to better properties. PP with grafted maleic anhydride can be crosslinked by addition of zinc diacetate or sodium acetate. It is surprising that besides bivalent zinc monovalent sodium can also act as a crosslinking agent indicating that the crosslinking proceeds via a somewhat complex mechanism in this case.

Improvement of creep behavior and increased resistance to crack growth can be considered as beneficial effects of crosslinking as well. The shrinkage stress of oriented PP is also affected. Similar to PE, the shrinkage stress is enhanced if crosslinking is performed prior to drawing, while a very small effect was observed if a drawn material was crosslinked.

Various applications are targeted by crosslinked PP. Crosslinking to a gel content of 55% proved to be beneficial for cable insulation. Both crosslinking via grafting of silanes initiated by peroxide decomposition and radiation in the presence of polyfunctional monomers result in a substantial increase of the resistance to electrical breakage in high-voltage tests.

Maintaining a certain level of mechanical properties even above the melting temperature can be exploited in the production of foamed materials. Twenty-fold foaming can be reached using common foaming agents. The thermal stability of foamed PP is higher than that of foamed PE which is understandable considering the difference in melting temperatures of both polymers.

Crosslinking was also investigated as an alternative for compatibilization in blends containing PP [7]. The blends of PP with low density polyethylene possess very high impact resistance at low temperature accompanied by only marginal drop in tensile strength and modulus, when crosslinked with peroxide in the presence of an effective coagent. The same blends behave as rather brittle material if no crosslinking is applied. The effect is attributed to an *in situ* formation of cocrosslinked PE and PP chains which act as an efficient compatibilizer within the blend.

REFERENCES

1. Lazár, M., Rado, R. and Rychlý, J. (1990): Crosslinking of polyolefins, *Adv. Polymer Sci.*, **95**, 149–197.
2. Denisov, E.T. (1985) Reactivity of polyfunctional compounds in radical reactions (in Russian), *Usp. Khimii*, **54**, 1466.
3. Henman, T.J. (1983) Controlled crosslinking and degradation, in *Degradation and Stabilisation of Polyolefins*, Chapter 2, (ed. N.S. Allen), Applied Science, London, 29–62.
4. Sawasaki, T. and Nojiri, A. (1988) Radiation crosslinking of polypropylene, *Radiat. Phys. Chem.*, **31**, 877–886.
5. Chodák, I. and Lazár, M. (1986) Peroxide-initiated crosslinking of polypropylene in the presence of p-benzoquinone, *J. Appl. Polymer Sci.*, **32**, 5431–5437.
6. Chodák, I. (1995) Properties of crosslinked polyolefin-based materials, *Prog. Polymer Sci.*, **20**, 1165–1199.

7. Chodák, I., Repin, H., Bruls, W. and Janigová, I. (1996) *Macromol. Symp.* **112**, 159–166.

Keywords: crosslinking, crosslinking mechanism, crosslinking coagent, macroradical, peroxide initiation, irradiation, degradation, controlled rheology PP, foams, compatibilization.

Crystallization

A. Galeski

Every crystallization in polymers is induced by crystallization nuclei. Nuclei give rise to lamellar crystal growth. Lamellae radiate from the central primary nuclei and by sufficiently frequent branching and cross-hatching form crystalline aggregates called spherulites.

Crystallization of polypropylene (PP) is accompanied by the release of latent heat of fusion of nearly 210 J/g of crystal. In metals and other low-molecular substances, the crystallization front, at which the transformation occurs, follows the isotherm of melting point. The controlling factor is then the dissipation of latent heat of fusion which leads to square-root time dependence of the growth rate of crystals. There is a substantial difference in crystallization of polymers and metals. Crystallization of polymers proceeds at large undercoolings; solidified polymer is well below its melting point; the crystallization growth rate is usually constant in time and controlled by the secondary nucleation mechanism rather than by the dissipation of latent heat of fusion. In isotactic polypropylene (iPP), the growth rate of crystals is constant with time for a fixed temperature in thin films. The release of latent heat of fusion increases the temperature and slows the crystal growth in thicker samples due to insufficient heat subtraction from the interior of the sample limited by low thermal conductivity of PP. Nonisothermal crystallization with the cooling rate about 5000°C/min shows that the release of latent heat of fusion maintains the growth rate of spherulites at the level of 8–9 μm/s and a constant temperature at the interface, where the growth occurs, even in thin samples.

The growth of crystals occurs on primary nuclei by a series of secondary nucleation events and spreading of a new crystalline layer on

Polypropylene: An A–Z Reference
Edited by J. Karger-Kocsis
Published in 1999 by Kluwer Publishers, Dordrecht. ISBN 0 412 80200 7

the growth face. The kinetic nucleation theory with chain folding now provides the best general tool for understanding the nucleation and the growth of polymer crystals at isothermal conditions from unstrained melt [1]. The reptation concept is also adapted for the description of chain motion and transport in the melt. The secondary nucleation of new crystalline layers on an existing crystal is the controlling factor of the crystal growth. The observable crystal growth is the result of two processes, the first being the nucleation of initiating stems on the surface of the crystal, and the second being the coverage of the surface by new stems beginning at the initial stem. The crystal grows macroscopically in the direction normal to its surface while on the molecular level the elementary growth mechanism is the growth along the crystal surface. After completion of a layer on the surface of the crystal a new surface secondary nucleus must be created. The temperature dependence of the

Table 1 Basic parameters for the description of nucleation and crystallization of isotactic and syndiotactic polypropylene

Properties	*Isotactic polypropylene α form*	*Isotactic polypropylene β form*	*Syndiotactic polypropylene*
Equilibrium melting point, T_m^0	208°C±8°C	176°C	158–161°C
Glass transition, T_g	−10°C	−10°C	0°C
Enthalpy of fusion, Δh_f	209.3±29.9 J/g	177 J/cm^3	190 J/g
Chain conformation	3_1 helix	3_1 helix	two-fold helix S(211)2
Unit cell parameters	Monoclinic,	Hexagonal	Orthorombic
Space group	C_{2R}^6-C2/c	D_3^4-P$3_1$21	C 222_1, Ibca
a	0.665 nm	1.274 nm	1.46 nm, 1.45 nm
b	2.095 nm	1.274 nm	0.56 nm, 1.12 nm
c	0.650 nm	0.635 nm	0.74 nm, 0.74 nm
β	99.3°	120°	90°
The main growth direction	(010)	(300)	
a_0	0.549 nm	0.636 nm	
b_0	0.626 nm	0.551 nm	
Regime I/II transition	155°C		
Regime II/III transition	137°C	123–129.5°C	
L in regime I	≈0.11 μm		
L in regime III	$\approx 2a_0$=0.91 nm		
Fold surface energy, σ_e	122 erg/cm^2	48–55 erg/cm^2	49.9 erg/cm^2
Work of chain folding, q	6.6 kcal/mole		5.8 kcal/mole
Lateral surface free energy, σ	9.2–11.5 erg/cm^2		11–12 erg/cm^2
Activation energy for reptation, Q_D^*	6276 J/mole		

secondary nucleation rate at low and moderately high undercoolings is determined by the free enthalpy of the formation of nucleus of the critical size on the crystal face which is proportional to $T_m^0/(T\Delta T)$. The appropriate expression for the secondary nucleation process predicted by the kinetic nucleation theory with reptation is as follows:

$$I = (N_0\beta_g p_i)/(a_0 n_s)\exp[-4b_0\sigma\sigma_e T_m^0/(\Delta T \Delta h_f kT)] \quad (1)$$

where N_0 is the number of reacting species at the growth front, $\beta_g = (K/n)(kT/h)\exp[-Q_D^*/RT]$, n is the number of macromolecule segments in the melt, n_s is the number of stems of width a_0, h is the Planck constant, and Q_D^* is the activation energy for reptation. K is a constant usually of the order of unity as determined from the experiments where T_m^0 is the equilibrium melting temperature, T is the temperature of crystallization and ΔT is the undercooling. The appropriate crystallization parameters for PP are listed in Table 1. After the formation of a stable secondary nucleus, the layer will be completed with new stems by the attachment-detachment mechanism. The completion rate, as the layer spreads in the direction parallel to the surface of the crystal, expressed as a difference of attachment and detachment rates of stems to the nucleus on the surface of the substrate, is not a strongly dependent function of temperature.The expression for the completion rate is as follows:

$$g = a_0 Q\beta_g \exp[-2a_0 b_0 \sigma_e(1-\psi)/kT] \quad (2)$$

ψ being the fraction of the free energy of fusion for the forward reaction and Q being a factor of the order of unity.

There is a competition during crystallization between secondary nucleation and the completion of the layer on the substrate [3]. Three cases can be distinguished:

- *Regime I.* The secondary nucleation process is slow allowing for completion of the nucleated layer before the next event of the secondary nucleation.
- *Regime II.* The secondary nucleation events occur before the completion of the nucleated layer allowing for multiple nucleation on the substrate.
- *Regime III.* The secondary nucleation occurs so often that it does not allow for the completion of the nucleated layer with new stems. The crystallization occurs mainly by the intense nucleation of new stems on the substrate rather than the completion of layers across the surface of the substrate.

These three cases are the reason for the existence of three regimes of crystallization and the respective changes in polymer morphology. The transitions to different regimes are possible because the completion rate depends on temperature but much less than the nucleation rate. The necessary data for the description of processes involved in crystallization of iPP are collected in Table 1.

The observable growth rate, G, is a combination of the secondary nucleation rate, I, and the size of crystal growth face and is described by basic relationships:

- for regime I, with low nucleation rate allowing a rapid completion of the entire crystal growth face $G_1 = b_0IL$
- for regime II, with the completion rate of the layers, g, allowing for multiple nucleation on the substrate $G_2 = b_0(Ig)^{1/2}$
- for regime III, for which the crystallization occurs mainly by the intense nucleation of new stems on the substrate rather than the completion of layers across the surface of the substrate $G_3 = b_0IL'$.

Here L is the length of a growth layer of thickness b_0, g is the completion rate of a layer and could be identified in terms of the nucleation process $g = a(A - B)$, A and B being attachment and detachment rates, respectively, L' is the effective length of a niche between neigbouring secondary nuclei on the growth face. In general, L' is considerably smaller than L. The spherulite growth rate versus temperature and regime transitions is illustrated in Figure 1. iPP shows the

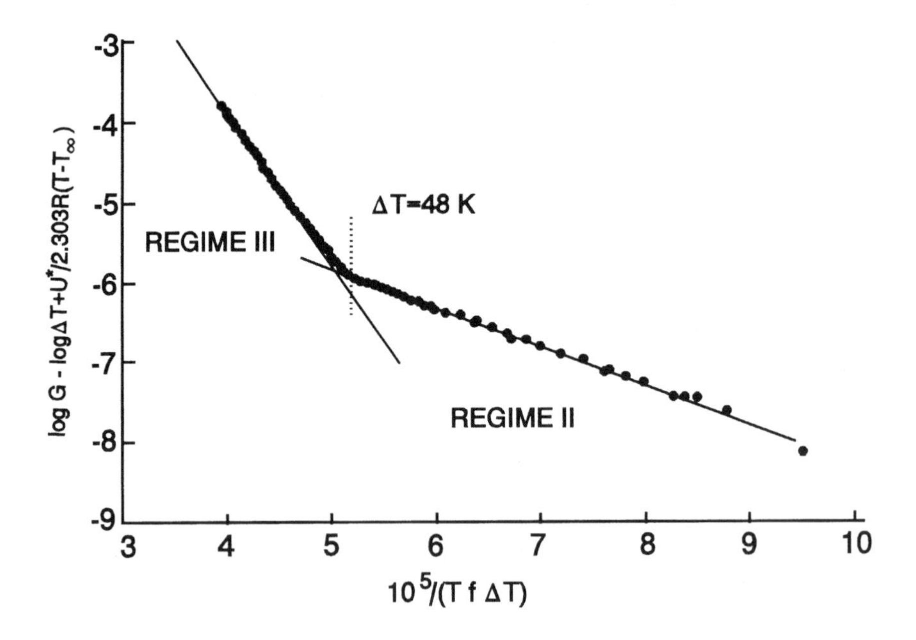

Figure 1 Growth rate of α-sperulites in isotactic polyproplene versus temperature in isothermal conditions (Cheng, S.Z.D., Janimak, J.J., Zhang, A., Cheng, H.N. (1990) *Macromolecules*, **23**, 298).

transition temperatures: regime III/II around 138°C and regime II/I shows up around 155°C only for low-molecular fractions of PP.

The secondary nucleation can be easily determined from the measurements of the spherulite growth rate based on the knowledge of the basic crystallographic and thermodynamic parameters characteristic for a PP shown in Table 1.

PP crystals grow with certain crystallographic faces depending on the thermodynamic parameters. The (110) plane of monoclinic α crystals is the growth face at low supercooling while at higher supercooling also other planes including the most probable (010) planes. At a temperature below 155°C, α crystalline modification exhibits a cross-hatched texture; the tangential lamellae inclined at about 100°C are nucleated on (010) planes of lamellae radiating from the primary nucleus [2]. The cross-hatching density decreases with increasing crystallization temperature. At low temperature, the tangential lamellae develop at almost the same time as the radial leading lamellae. The thicknesses of radial and tangential lamellae are thus similar in the range of 20–25 nm. At high temperature, the cross-hatched lamellae grow after the radial lamellae, and therefore, are thinner. Radial lamellae are ≈ 40 nm thick while tangential lamellae only 25–30 nm thick. The cross-hatching density decreases with decreasing isotacticity which is due to increasing irregularity of basal (010) planes of radial lamellae. However, the thickness of radial and tangential lamellae increases with decreasing isotacticity. Cross-hatched texture has not been observed in other crystalline modifications of iPP. In β crystalline modification the lamellae thickness ranges from 21 to 30 nm.

The description of crystallization of PP requires consideration of movements of macromolecules in a molten polymer. The displacements of macromolecules called 'reptation' are possible along a virtual tube circumscribed around a macromolecular chain. Other movements are more difficult since they require extensive cooperation of other chains in the vicinity. Entanglements with neighboring chains, beside the friction of chains against the tube, are additional impediments for the chain movements. Entanglement knots constitute a hindrance for chain transport in the course of crystallization. The chain displacement during crystallization is enforced by crystallization forces after the secondary nucleation event. Calculations show that in the case of the disentangled chains, the crystallization forces easily overcome the chain friction against the tube. Reptation motions of other macromolecules are so slow in this time scale that the chains can be considered as motionless. As the supercooling increases, the secondary nucleation rate increases. Competition between the secondary nucleation rate and the completion rate of layer on the crystal growth face results in three regimes of crystallization. Macromolecules are reeled out of the melt mainly in regime I and regime II of crystallization which is connected with the completion of layers on the crystal face in chain folded manner. In regime III the chains are displaced only for short distances as required by the number of chain folds in secondary nuclei. Crystallization in regime III of PP is preferred for improved mechanical strength because of many chain links between neighboring crystals and cilia. Since the regime transition is controlled by the relative rates of secondary nucleation and layer completion, any factor that affects either of these rates will alter the temperature at which the transition occurs. Elimination of reptation by crosslinking or by grafting with relatively short side groups moves the crystallization inevitably to regime III because the completion of the layer on the growth face requires adjacent re-entry folding and transport of macromolecular chains over large distances. The incorporation of nucleation agents shifts the crystallization to the range of higher temperatures. However, for a cooling rate of 10°C/min the most active nucleation agents can shift the crystallization peak of isotactic polypropylene at most to 132°C which is still in regime III.

The overall crystallization rate is used to follow the course of solidification of iPP. Differential scanning calorimetry (DSC), dilatometry, dynamic X-ray diffraction and light depolarization microscopy are then the most useful methods. The overall crystallization rate depends on the nucleation rate, $I(t)$ and the growth rate of spherulites, $G(t)$. The probabilistic approach to the description of spherulite patterns provides a convenient tool for the description of the conversion of melt to spherulites. The conversion of melt to spherulites in the most general case of nonisothermal crystallization is described by the Avrami equation:

$$\alpha(t) = 1 - \exp\left\{-\pi \int_0^t I(T)\left[\int_T^t G(s)\mathrm{d}s\right]^2 \mathrm{d}T\right\} \tag{3a}$$

for films and

$$\alpha(t) = 1 - \exp\left\{-(4\pi/3) \int_0^t I(T)\left[\int_T^t G(s)\mathrm{d}s\right]^3 \mathrm{d}T\right\} \tag{3b}$$

for bulk samples.

The analysis of the above equations is often applied for obtaining the nucleation data from isothermal and nonisothermal crystallization experiments. Several simplifications of the equations are developed and used for isothermal crystallization (with instantaneous or spontaneous nucleations only) and nonisothermal processes with a constant cooling rate. It was found that the crystallization of iPP follows the dependence $\log\,[1 - \alpha(t)] \sim t^n$ where n is around three for relatively low supercoolings which indicates instantaneous character of primary nucleation.

To characterize the spherulitic nucleation during nonisothermal crystallization, the Ozawa equation is applied, which could be obtained by integrating twice by parts the Avrami equation and assuming cooling at the constant rate, a. The slope of the plot $\ln\{-\ln[1 - \alpha(T)]\}$ versus $\ln(a)$ equals two or three for instantaneous nucleation, three or four for nucleation prolonged in time, in two- and three-dimensional crystallization, respectively. The values from three to four, depending on temperature range were obtained for iPP from DSC nonisothermal crystallization [4].

Recently, the statistical approach was developed [5] for the description of the kinetics of conversion of melt to spherulites and the kinetics of formation of spherulitic pattern during both isothermal and non-isothermal crystallizations. The final spherulitic pattern can also be described. The rates of formation of spherulitic interiors and boundaries (boundary lines, surfaces and points) as well as the their final amounts could be predicted if spherulite growth and nucleation rates are known. Applied to iPP crystallized during cooling with various rates, the approach allowed for the predictions of tendencies in the kinetics of formation of spherulitic structure and its final form.

REFERENCES

1. Clark, E.J. and Hoffman, J.D. (1984) *Macromolecules*, **17**, 878.
2. Bassett, D.C. and Olley, R.H. (1984) *Polymer*, **25**, 935.
3. Hoffman, J.D., Davies, G.T. and Lauritzen Jr., J.I. (1976) The rate of crystallization of linear polymers with chain folding, in *Treatise on Solid State Chemistry*, Vol. 3, Chap. 7, (ed. N.B. Hannay), Plenum, New York, pp. 497–614.
4. Monasse, B. and Haudin, J.M. (1986) *Colloid. Polymer Sci.*, **264**, 117.
5. Piorkowska, E. *J. Phys. Chem.*, (1995) **99**, 14007; *ibid.* (1995) **99**, 14016; *ibid.* (1995) **99**, 14024.

Keywords: entanglement, disentanglement, cross-hatching, lamellae, crystallization, nucleation, reptation, nucleation (crystallization) regimes, nucleation agents, nucleation rate, spherulitic growth rate, Avrami-equation, Ozawa-equation, isothermal crystallization, nonisothermal crystallization, secondary nucleation, supercooling.

Crystallization of syndiotactic polypropylene

J. Kressler

The modern story of syndiotactic polypropylene (sPP) starts with the introduction of metallocene catalysts allowing for a large scale production of this polymer. This initiated immediately a large interest in the crystallization behavior of this polymer. Three different chain conformations were reported for crystallized sPP in contrast to iPP where exclusively a 3_1-helix is observed. A planar zig-zag form $(TT)_n$ occurs in the melt or in quenched samples [1]. A second possible chain conformation of a $(T_6G_2T_2G_2)_n$ type of helix is formed when quenched samples are exposed to solvent vapour [2]. A stable helical conformation with $(TTGG)_n$ (s(2/1)2) symmetry is usually formed by highly stereoregular sPP when isothermally crystallized from the melt or when moderately cooled. For this helical conformation, three orthorhombic packing modes were described; a C-centred cell I with the space group $C222_1$, a primitive cell II with antichiral packing of chains along the *a*-axis (space group Pcaa) and finally a body centred cell III with fully antichiral packing of the chains (space group Ibca) [3] (Figure 1).

The orthorhombic type I cell was already described by Corradini *et al.* and is formed by helices of one hand only (all right-hand or all left-hand). The cell types I to III (Figure 1(a–c)) are based on the $(TTGG)_n$ helix. The dimensions of the type I cell are, $a = 1.44$ nm, $b = 0.56$ nm and $c = 0.74$ nm. The orthorhombic type II cell has the same dimensions as the type I cell but is formed by layers of helices with different hands. Type I and II cells are formed when sPP is rapidly cooled at least 40°C below the equilibrium melting point which is usually lower compared to

Polypropylene: An A–Z Reference
Edited by J. Karger-Kocsis
Published in 1999 by Kluwer Publishers, Dordrecht. ISBN 0 412 80200 7

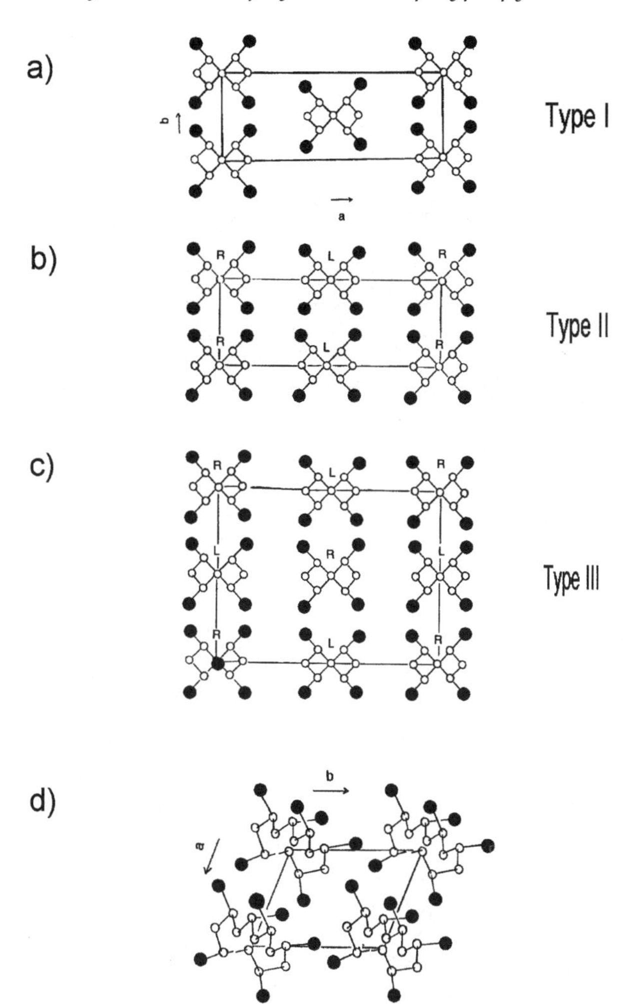

Figure 1 The three modifications of sPP formed $(TTGG)_n$ according to Lovinger *et al.* (a)–(c) [1]. Another cell type is formed by $(T_6G_2T_2G_2)_n$ type helices (d) [2]. The dark circles symbolize the methyl groups.

iPP. Type II crystallization of sPP is most frequently found and can be obtained as a neat modification. The type I has never been found as a neat modification. The type III crystallization leads to a C-centred orthorhombic cell. The helices are alternating left-handed and right-handed packed along the *a*- and *b*-axis. The dimensions are the double of the type II cell in *b* direction, $a = 1.44\,\text{nm}$, $b = 1.12\,\text{nm}$ and $c = 0.74\,\text{nm}$. This modification can be obtained by crystallization at low supercoolings or by rearrangement of the type II crystals at temperatures above 130°C. It should be mentioned that an orthorhombic unit cell is also formed in quenched and elongated sPP samples. The methyl groups do not have the maximum distance from each other but they are located at the 'same

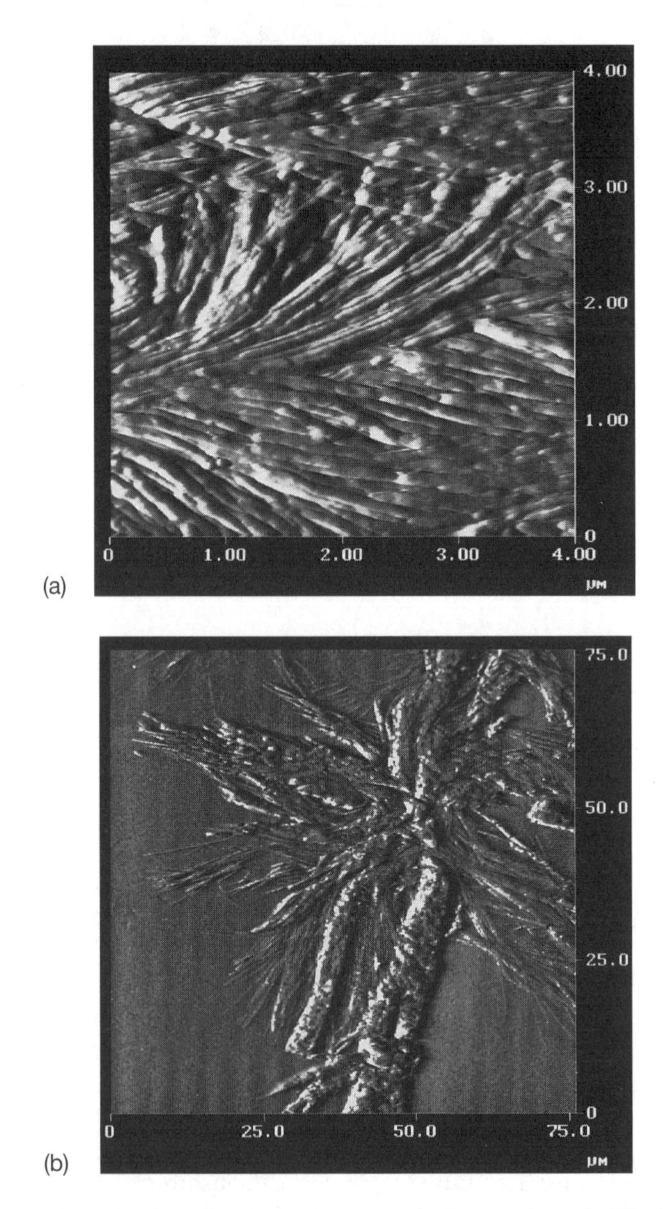

Figure 2 Atomic force microscopy on etched samples of sPP isothermally crystallized at 135°C (a) and at 143°C (b).

side′ of the helix. The cell dimensions are $a = 0.522$ nm, $b = 1.117$ nm and $c = 0.506$ nm. The triclinic cell formed by $(T_6G_2T_2G_2)_n$ type helices is shown in Figure 1(d). This helix conformation can be assumed as an intermediate state between planar zig-zag and the normal $(TTGG)_n$ conformation. The cell has the dimensions of $a = 0.572$ nm, $b = 0.764$ nm and $c = 1.16$ nm. The angles are $\alpha = 73.1°$, $\beta = 88.8°$ and $\gamma = 112°$. This is also a metastable modification.

The supermolecular crystalline morphologies of sPP are very different from those of iPP. The formation of large spherulites in highly stereoregular sPP is rather an exception. In polarized light microscopy, a very disordered crystalline morphology appears even at very low crystallization rates. Atomic force microscopy (AFM) is able to show these morphologies in more detail after permanganic etching [4]. In Figure 2(a), typical bundles of lamellae (fibrils) can be seen in a sample space filling crystallized at 135°C. More open aggregates of bundle structures and single-crystal like entities can be seen in Figure 2(b). This sample was isothermally crystallized at 143°C. Obviously, the cross-hatching, i.e. a kind of homoepitaxial growth where daughter lamellae grow onto original lamellae in iPP, does not occur in sPP. Thus always more extended and less spherical morphologies are formed.

Another characteristic feature of sPP is the formation of large rectangular single crystals in thin layers or in bulk when crystallized at low

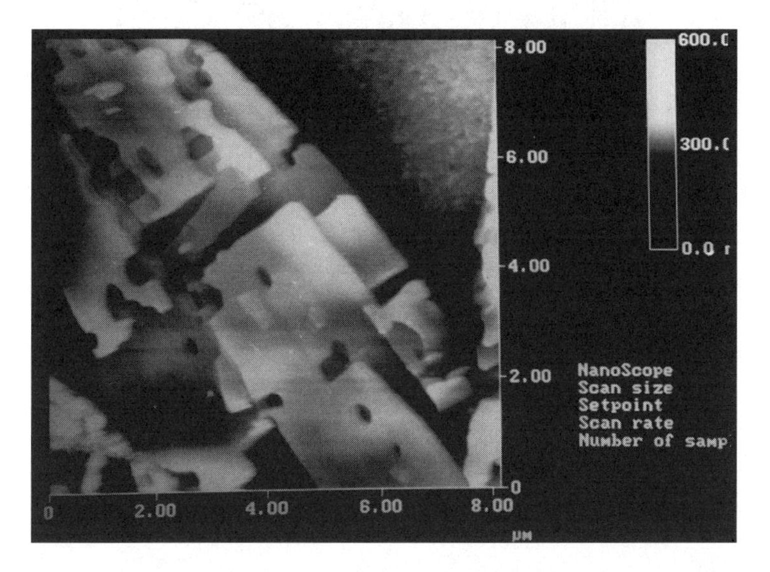

Figure 3 Fractured single crystal of sPP after isothermal crystallization at 139°C in bulk. The photograph is obtained by atomic force microscopy on etched samples.

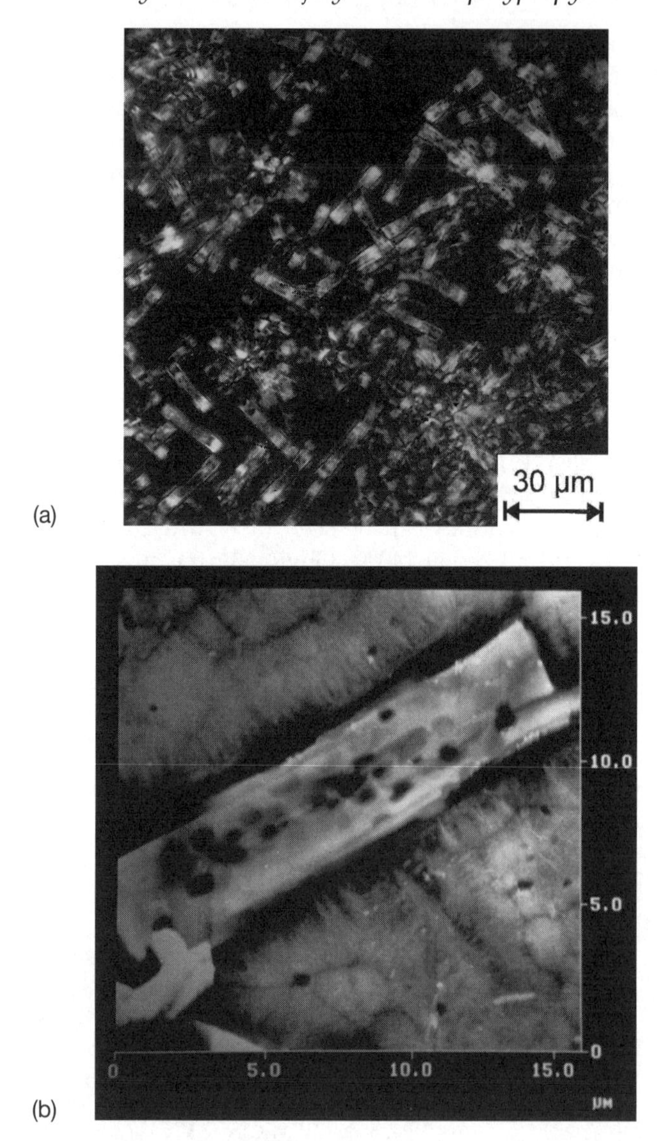

Figure 4 (a) Polarized light micrograph of a syndiotactic polypropylene sample which contains 4 wt.% 1-octene. The sample was isothermally crystallized at 132°C for 24 days. (b) Atomic force micrograph of the same sample as described in (a) after permanganic etching.

supercoolings [5]. These single crystals show transverse fractures and ripples which are caused by anisotropic thermal expansion coefficients in the single crystal. The thermal expansion coefficient along the *b*-axis (long axis of the rectangular crystal) is almost one order of magnitude larger than the linear thermal expansion coefficient along the *a*-axis. Also the thermal expansion coefficient of the chain-fold region has to be taken into account [5]. A fractured single crystal of sPP isothermally crystallized in bulk at 139°C can be seen in Figure 3. This is an AFM photograph taken after permanganic etching.

Copolymerization of small amounts of 1-octene into sPP is able to prevent the occurrence of fractures in single crystals [5]. Figure 4(a) displays a polarized light micrograph of a syndiotactic propylene copolymer containing 4 wt.% 1-octene. The sample was isothermally crystallized for 24 days at 132°C. These single crystals appear very similar to that of neat sPP. But AFM of the same sample (Figure 4(b)) reveals after etching that the large transverse fractures are absent. Only small holes appear which are the result of screw dislocation in the crystal.

REFERENCES

1. Lotz, B., Wittmann, J.C. and Lovinger, A.J. (1996) Structure and morphology of poly(propylenes): a molecular analysis. *Polymer*, **37**, 4979–92.
2. Chatani, Y., Maruyama, H., Asanuma, T. and Shiomura, T. (1991) Structure of a new crystalline phase of syndiotactic polypropylene. *J. Polymer Sci., Polymer Phys.*, **29**, 1649–52.
3. Stocker, W., Schumacher, M., Graff, S., Lang, J., Wittmann, J.C., Lovinger, A.J. and Lotz, B. (1994) Direct observation of right and left helical hands of syndiotactic polypropylene by atomic force microscopy. *Macromolecules*, **27**, 6948–55.
4. Thomann, R., Wang, Ch., Kressler, J., Jüngling, S. and Mülhaupt, R. (1995) Morphology of syndiotactic polypropylene. *Polymer*, **36**, 3795–801.
5. Thomann, R., Kressler, J. and Mülhaupt, R. (1997) Single crystals of syndiotactic poly[propene-co-(1-octene)] and syndiotactic polypropene crystallized in bulk. *Macromol. Chem. Phys.*, **198**, 1271–79.

Keywords: crystallization, syndiotactic PP (sPP), conformation, unit cell, supramolecular structure, single crystal, copolymer, dislocation, thermal expansion coefficient.

Designing properties of polypropylene

K. Bernreitner and K. Hammerschmid

MOTIVATIONS FOR TAILORING OF PROPERTIES

The strongest impact for tailoring properties by polypropylene (PP) suppliers comes from the high growth rate of the plastic materials market. Because of its outstanding cost-to-performance ratio, PP has become the fastest developing member of the thermoplastics family throughout the world. The rapid advancement of PP into application segments of high value resins (polyamides, linear polyesters or ABS), e.g. in automotive or electrics applications, is essential for holding this position. Furthermore, backmarket, converters and machine producers are more and more demanding materials with high output rates accompanied by a constantly high quality assured for different lots.

PRINCIPLES OF TAILORING

Concerning the design principles PP possesses the following attractive features: low density of 0.9 g/cm^3, high crystallite melting point (about 165°C), chemical resistance, high surface gloss and orientability [1]. The three main tools for further property improvements are molecular design, post-reactor modification and variation of the processing conditions.

Molecular design during the polymerization process

Catalyst system and polymerization technology are primary contributing factors for building up the polymer chain. The main structural factors

Polypropylene: An A–Z Reference
Edited by J. Karger-Kocsis
Published in 1999 by Kluwer Publishers, Dordrecht. ISBN 0 412 80200 7

affecting the basic properties of conventional PP-homopolymers are chain-stereoregularity, molecular weight and molecular weight distribution (MWD). These controlling parameters significantly affect the degree of crystallinity, which furthermore determines the processability and material performance. Heterophasic PP-copolymers basically have at least a two-phase structure, consisting of a matrix and a disperse elastomer phase (most common: ethylene-propylene-rubber, abbreviated as EPR). In normal high-impact PP reactor blends, the matrix consists of a homopolymer while in special cases random copolymers of propene with other comonomers, such as ethene, butene, hexene and so on, form the continuous phase. In each case, the molecular structure determines the mechanical property profile. The main goal is to achieve a balance between rigidity and resistance to heat deflection under load on one hand and impact strength and low temperature resistance on the other.

Post-reactor modification

Because the virgin polymer does not meet the needs of end-users, it is necessary to do some finishing work. Further driving forces for this modification are to broaden the property range of PP or to tailor its property profile considering the special demands of particular applications. The most common techniques for the required improvements are:

- *Stabilization*. The chemical industry, in cooperation with the PP manufacturers, succeeded in developing effective organic systems for process, long-term and special stabilization. These mainly involve sterically hindered phenols and, for the first two purposes, thioethers, phosphites and phosphonites. The concentration of the stabilizers will range from a few parts per million (ppm) to several percent, highly depending on the end-use requirements.
- *Nucleation*. Nucleating offers an interesting way of PP modification and a high number of PP-grades modified in such a way is offered by most PP manufacturers. The main reason for nucleation is to obtain high clarity, but also the improvement of the mechanical properties (especially stiffness enhancement) and the reduction of cycle time are of paramount interest.
- *Degradation*. Whilst in the initial phase, degradation was achieved by thermal means, for economic reasons the industry had switched to the chemical degradation of PP by peroxides by the early seventies. Today, CR (controlled rheology) technology is the state of the art at all PP producers. The first use of CR-PP was to produce high-tenacity fibers. Soon after, the benefits of degraded PP became evident in the optical properties like transparency and gloss. As a consequence, high-quality film grades were introduced in the market. Through degradation the

stiffness-toughness ratio is shifted towards toughness, and the lower warpage tendency makes CR-PP ideally suited for complex injection molding items, especially in the high melt flow index (MFI) range.

- *Filling or reinforcing.* Inorganic reinforcements mainly produce an improvement in stiffness, accompanied by a certain increase in heat distortion resistance [2]. The property profile can be tailored by incorporating inorganic fillers via their quantity geometry (aspect ratio), fineness and adhesion to the matrix (Figure 1). Compounds based on organic fillers, such as jute-, hemp-, wood-fibers or wood-flour, are used for ecological (recycling) reasons. Starches are of interest as fillers for formulating biodegradable compounds.
- *Colouring.* Pure PP is a translucent material with a whitish color. For a wide field of applications, it is desirable to pigment the resin to a specific color. Generally, there is a trend to substitute toxic substances like cadmium-based pigments by food and drug approved dyestuffs.
- *Impact modification.* Post-reactor impact modification commonly will be done by the addition of EPR, ethylene-propylene-diene rubber (EPDM) or special resins like ethene-vinyl acetate copolymers (EVA), but also with the new generation of linear low density polyethylene (LLDPE) produced by metallocene catalysts.

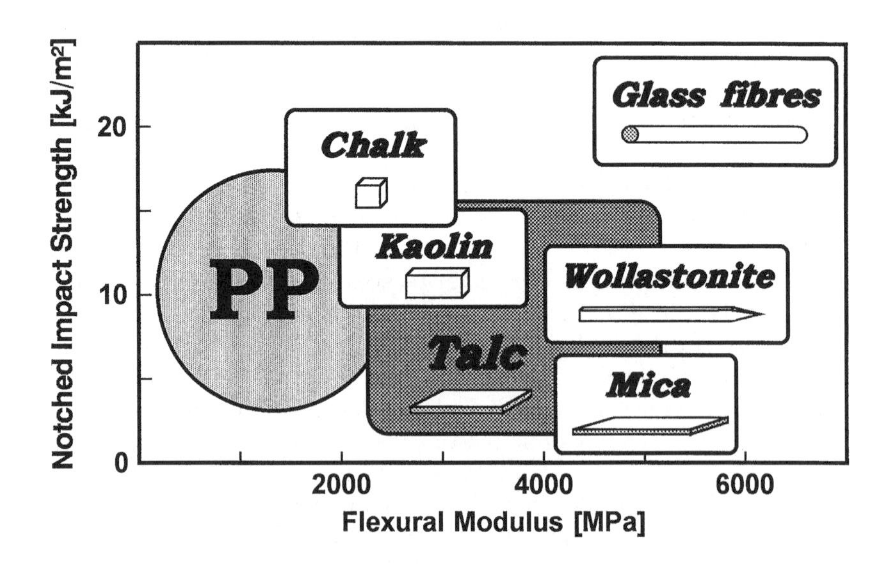

Figure 1 Influences of the aspect ratio of different fillers on the property profile.

Influence of processing parameters

For a proper fine-tuning of the basic characteristics of the PP-material, the processors have to keep in mind how the processing influences the end-use performance of the moldings. The most important processing conditions are the degree of melt shearing and the cooling rate, respectively the annealing history. For example, compression molding with slow cooling will result in phase segregation due to coalescence of dispersed elastomer phase, while on the other hand, injection molding at high melt shearing, yields a fine and well dispersed morphology. In general, all alterations of processing conditions may significantly influence crystallinity and morphology, and with them also the final properties of the material.

MECHANICAL PROPERTIES

An overview of the most important mechanical, thermal and surface data are given in Table 1. The data of common PP-grades were arranged according to the respective stiffness ranges from 'supersoft' to 'ultrastiff' materials. In addition, detailed information can be obtained from the databases, such as CAMPUS, containing relevant polymer characteristics of several producers. Most of the national testing methods (e.g. DIN) will continuously be replaced by the global ISO standards (already used in the CAMPUS database). In the future, data sheets or material specifications will be based mostly on ISO standards.

Stiffness/softness

Standard PP covers a flexural modulus range of 700–1500 MPa and thus is used in commodity applications. Modern trends, such as downgauging, to save material costs create a demand for increased material stiffness. Further advantages of this development are improved barrier performance to water vapor and the potential to substitute high-priced engineering thermoplastics, such as ABS or polyamides. Benefits of filled systems are for example low shrinkage and low thermal expansion (e.g. 'zero gap' bumpers). A possible loss of impact strength can be retarded by adding impact modifiers. A tailored decrease of the modulus allows PP to penetrate in the class of supersoft materials, such as plasticized PVC, EVA-copolymers or thermoplastic olefinic elastomers (TPO). Their typical applications are sealing layers, roofing membranes, sanitary films and cable compounds.

Testing methods

The determination of the modulus of elasticity can be done by the tensile test, compressive test and flexural test (e.g. DIN 53457/ISO 527/ASTM D

Table 1 Selected mechanical, thermal and surface property-ranges of natural PP and PP-compounds

		Natural PP-Grades					*Compounds*		
		Supersoft (1)	*Soft (2)*	*Standard*	*Stiff (3)*	*Highstiff (4)*	*Talkum 20–40%*	*Glass-fibers 20–40%*	*Filled impact*
Density (g/cm^3)	DIN 53479	0.890	0.900	0.905	0.905	0.905	1.05–1.25	1.04–1.22	0.95–1.15
Flexural modulus (MPa)	DIN 53457	50–300	300–700	700–1500	1500–2000	>2000	2500–5000	5000–9000	1000–2000
3.5% Flexural stress (MPa)	DIN 53452	3–15	10–25	25–35	35–40	>40	40–55	100–200	20–30
Yield stress (MPa)	DIN 53455	4–10	15–25	25–35	35–40	40–50	25–35	70–140	15–30
Elongation at yield (%)	DIN 53455	0–100	15–25	5–15	5–10	4–7	2–4	2–4	5–10
Elongation at break (%)	DIN 53455	>600	>600	20–800	20–200	10–30	15–30	2–4	20–400
Impact strength (kJ/m^2)	DIN 53453								
+23°C		n.b.	n.b.	70–n.b.	50–n.b.	30–n.b.	15–n.b.	25–35	n.b.
−20°C		30–n.b.	20–n.b.	10–n.b.	10–80	10–40	5–20	20–30	20–n.b.
Notched impact strength (kJ/m^2)	DN 53453								
+23°C		n.b.	10–n.b.	3–40	3–10	2–4	2–10	7–15	15–30
−20°C		5–n.b.	2–20	2–10	1–5	1–2	1–3	4–10	5–20
Vicat-softening point (°C)	DIN 53460								
Vicat-A		50–100	80–120	100–140	140–150	150–160	150–160	140–160	100–130
Vicat-B		<50	30–60	50–90	70–100	100–120	80–100	100–130	40–60
Heat distortion temperature (°C)	DIN 53461								
HDT-A		<30	35–45	40–60	50–70	60–80	60–90	130–150	40–60
HDT-B		<60	50–70	70–100	100–120	110–130	120–140	150–160	70–100
Shrinkage (%)									
longitudinal		<1.0	<1.0	1.0–1.5	1.0–1.5	1.0–2.0	0.6–1.1	0.1–0.3	0.4–0.8
transverse		<1.0	<1.0	1.0–1.5	1.0–1.5	1.0–2.0	0.6–1.1	1.0–1.1	0.5–0.9
Thermal expansion (μm/mK)		n.a.	n.a.	110–140	90–110	70–90	60–80	20–30	40–80
Gloss at 20° (%)	DIN 67530	20–50	30–60	60–80	70–80	70–80	5–30	20–40	20–40
Ball indentation hardness (N/mm^2)	DIN 53456	<20	20–40	30–80	80–90	>90	80–90	100–120	40–60

(1), Soft and elastic PP-materials; (2), heterophasic- and random-heterophasic-PP; (3), homo– and heterophasic-PP; (4), homopolymers; n.b. – no break; n.a. – not available.

638 and D 790). Yield stress, elongation at yield, break and tensile strength will be measured according to DIN 53455/ISO 527/ASTM D 638. Flexural modulus and strength, stress at 3.5% strain and strain of the outer fiber can be determined by DIN 53452/ISO 178/ASTM D 790 standards.

Design principles

Enhanced crystallinity raises stiffness, service temperatures, surface hardness, as well as chemical resistance and barrier properties. However, in every case, the low-temperature impact strength is reduced. With mineral filled systems, the stiffness can further be improved up to 5000 MPa. To achieve superior values of stiffness (up to 10 000 MPa) the best choice is the reinforcement with coupled glass fibers. In principle, a higher degree of stiffness can be obtained by broadening the MWD, by increasing the stereoregularity of the polymer chain and through the addition of nucleating agents suitable for the formation of α-spherulites. If softness is the target, incorporation of comonomers, such as ethene or butene, in the polypropylene chain is the best way. By blending PP with soft components during the polymerization stage or in a compounding step, the 'softness' can be adjusted exactly to the desired levels (moduli down to 50 MPa are achievable).

Impact strength

The impact range of PP spans from very low values (e.g. homo-PP) to an extremely high impact resistance where notched specimens don't break even at low temperatures down to −20°C (e.g. impact-modified PP, rubbers or TPOs). Super-high impact grades are gaining in importance in the automotive sector, especially in exterior applications. The most important uses are bumpers, spoilers and protective side moldings [3]. Materials with good impact resistance while retaining an attractive transparency, are a result of a special phase design during the polymerization process (Figure 2). They are suitable for applications in the household sector, e.g. flexible lids for freezer containers and films for food packaging.

Testing methods

The impact tests indicate the energy required to break notched or unnotched specimens under standard conditions. For PP two groups of testing methods have been established, Izod and Charpy, which mainly differ in the way of fixing the specimen for the test. The most common variations for the Izod group are ISO 180 1C unnotched, ISO 180 1A

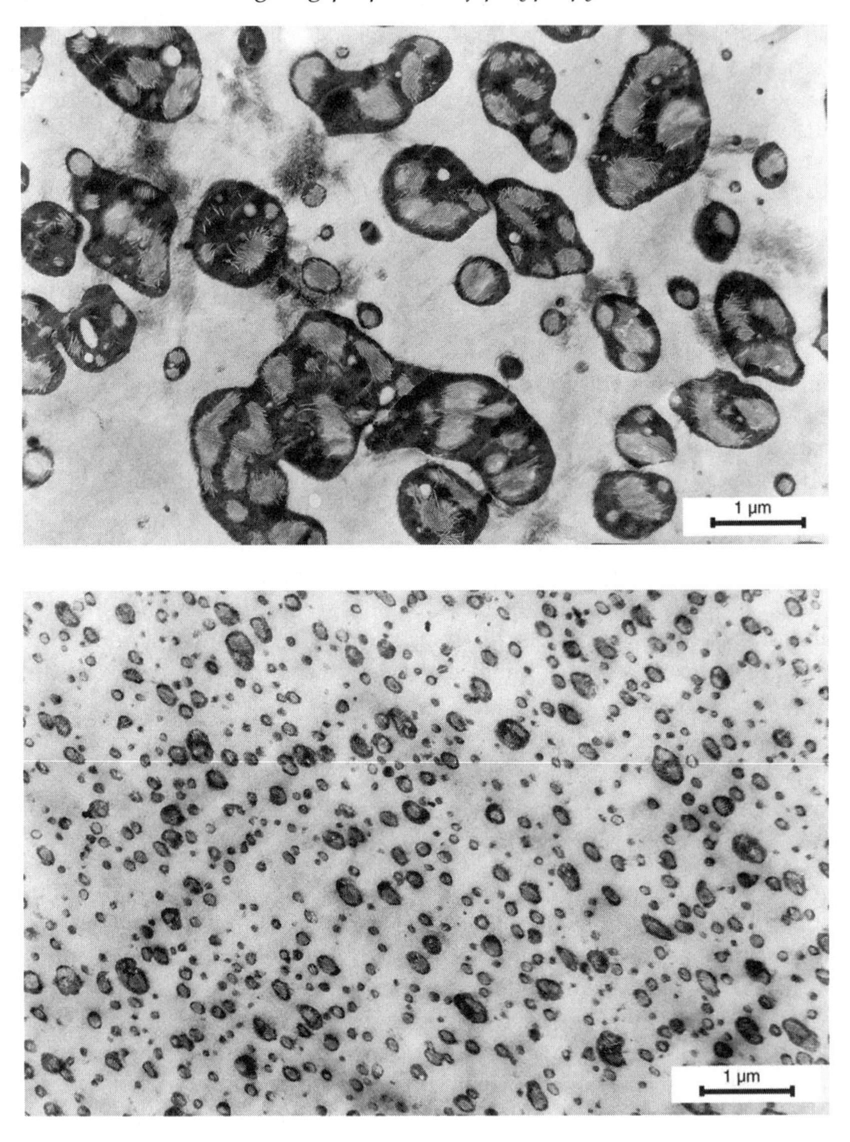

Figure 2 Phase morphologies of heterophasic copolymers with different matrices. (a) homopolymer; (b) random copolymer.

notched or ASTM D 256/A, and for the Charpy group ISO 179 1eA, 1eU, DIN 53453, ASTM 256/B are relevant. The trend is to concentrate the impact tests on ISO 179 (Charpy). In special cases (e.g. testing of very soft and elastomeric PP-based TPOs) the determination of the tensile impact strength by DIN 53448/ISO 8256 seems to be useful.

Design principles

The most important controlling factors to influence the impact properties are:

- molar masses of the EPR and the matrix, which determine primarily the phase structure (particle size of the rubber) and the strength of each of the two phases;
- volume fraction of the EPR phase and the resultant ratio between the diameter and spacing of the elastomer particles;
- structure of the elastomer phase: the monomer distribution (ethene/propene and/or other comonomers) determines both the compatibility with the matrix and also the inner structure of the particles;
- structure of the matrix phase: type and content of comonomer determine the compatibility with the elastomer phase and, via crystallinity, the strength of the matrix.

As mentioned above, the composition of the rubber phase is controlled by the monomer ratio in the copolymerization step. Therefore the crystallinity is adjustable and subsequently the impact-strength/stiffness ratio; when the EPR quantity is kept constant, the impact strength increases with the propene content; the stiffness displays an opposite trend. To get a high-impact PP grade with high transparency a very fine distribution of the EPR particles is required. This morphology, seen in Figure 2(b), is achieved by matching the viscosities of the matrix and elastomer phase and ensuring a low interfacial tension between the phases [4, 5]. Through the incorporation of ethene in the PP matrix, the molecular design of the matrix resembles that of the ethene–propene elastomer. As a consequence, the deformation mechanism of the material is altered which is accompanied by an additional benefit: the susceptibility to stress whitening can nearly be eliminated.

By the addition of fillers to impact-modified PP, formulations with a well balanced combination of stiffness and toughness can be achieved ('filled-impact' materials, Table 1).

THERMAL PROPERTIES

Efficient tools to predict the service temperature range of final parts are methods for the determination of the heat deflection temperature under load (HDT), the Vicat softening temperature (VST) and the short- (oxidation induction temperature/time; OIT) and long term thermal resistance (oven aging test; OT). Articles made of common PP may be used up to temperatures of 100°C. With high crystallinity PP (HCPP) this threshold can be raised up to 120°C. The application range can be widened up to 150°C with filled or reinforced material, which is close to the melting point of the neat polymer. A very important property combination in respect of the substitution of ABS is the proper balance of high impact strength and high softening point. Good sealability, which is often required in film applications (food packaging), is attained by lowering the melting point of PP.

Testing methods

Melting point and glass transition temperature are measured by differential scanning calorimetry (DSC) or differential thermal analysis (DTA) (ISO 3146/method C and IEC 10006-A). By running two or three heat- and cool-cycles during DSC-measurement, some further information about the crystallization behavior (e.g. content of α- or β-modification) can be received. The HDT, measured by ISO 75-1/2/3, reflects that temperature at which an arbitrary selected deflection occurs under a given load. The Vicat test (ISO 306) indicates the softening temperature of the material surface. Methods such as OIT or OT (ISO 4577/DIN 53383) are used to observe the changes of selected properties (e.g. brittleness, discoloration) under heat aging conditions.

Design principles

The necessity of adding greater quantities of a soft rubber phase to the PP-matrix in order to get high impact strength, results in an undesirable 'softening' of the material. To overcome this problem, enhancement of crystallinity and optimization of molecular design of the rubber phase are the most efficient methods. Nucleation or the addition of fillers is also very helpful for the improvement of HDT and the Vicat softening point. To improve short and long term thermal resistance, well known basic stabilization packages have been developed by several additive suppliers. The melting and the glass transition temperatures are reduced if PP is randomly copolymerized with ethene or butene. A random copolymer with 10 mol.% ethene shows a melting point of approximately 135°C.

SURFACE PROPERTIES

Generally, the term 'surface properties' covers end-user aspects (e.g. gloss, scratch resistance, ball indentation hardness), converter and processor-related needs (e.g. blocking, slipping, fogging and blooming) and the suitability for surface treatments (e.g. painting, printing, coating). Determination of above-mentioned properties is important if they are foreseen for application fields such as housings, automotive exterior or interior parts and food packaging.

Testing methods

The determination of blocking resistance will be made according to DIN 53366 or ISO 11502. Slipping behavior is characterized by the coefficient of friction (DIN 53375/ISO 8295), while fogging and blooming usually will be detected by user-specific techniques. These methods are closely connected to film applications. Important procedures for the characterization of housings or other visible parts, such as kitchen appliances, cosmetic packaging or automotive interior trims, are the measurements of gloss (DIN 67530), ball intendation hardness (ISO 2039) and scratch resistance. The visibility of scratches can only be evaluated by miscellaneous methods tailored for different customer needs.

Design principles

Blocking and slipping properties are mainly influenced by the addition of substances like silica (antiblocking) and lubricating agents such as primary and secondary amides. For slipping agents, migration speed to the surface is the most crucial criterion. Fogging and blooming depend on the chemical nature of the additives and the degree of polymer crystallinity. Gloss and scratch resistance will be reduced by increasing the amount of filler. The loss of stiffness due to filler reduction for enhancement of gloss and scratch resistance can be compensated by increasing the crystallinity of PP. To get good paintability, the surface has to be modified by flame and plasma treatment or by adding polar substances such as maleic anhydride grafted polyolefins.

REFERENCES

1. Neißl, W. and Ledwinka, H. (1993) Polypropylen – Die Zukunft hat gerade begonnen. *Kunststoffe*, **83**, 577–583.
2. Katz, H.S. and Milewski, J.V. (1987) *Handbook of Fillers For Plastics*, Van Nostrand Reinhold, New York.
3. Moore, E.P. (ed.) (1996) *Polypropylene Handbook*, Hanser, Munich.
4. Paulik, Ch., Gahleitner, M. and Neißl, W. (1996) Weiche, zähelastische PP-Copolymere. *Kunststoffe*, **86**, 1144–1147.
5. Martuscelli E. (1995) Structure and properties of polypropylene–elastomer blends, in *Polypropylene – Structure, Blends and Composites*, 1st edn, (ed. J. Karger-Kocsis), Chapman & Hall, London, Vol. 2, pp. 95–140.

Keywords: property design, stiffness, impact strength, thermal properties, surface properties, morphology design, molecular design, testing methods, modification, CAMPUS, standardization, gloss, crystallinity, fogging, homopolymer, copolymer, heat deflection temperature, HDT, Vicat, oxidation induction time, OIT, composite, filled systems, thermoplastic olefinic elastomers, TPO, nucleation, degradation, stabilization, ethylene-propylene rubber, EPR, melt flow index, MFI.

Die swell or extrudate swell

R.J. Koopmans

INTRODUCTION

Die swell is a complex rheological phenomenon [1]. It can be observed as an extrudate with a cross-section (D_{ex}) which is greater than the die cross-section (D_o). This effect, also known as extrudate swell, Barus effect, or % memory, is defined as the ratio $D_{ex}/D_0 = B$ and is a feature of polymer melt flow. Die swell is associated with the viscoelastic nature of polymer melts as it exceeds the swelling of constant viscosity (Newtonian) fluids. Accordingly, for laminar flow situations, the swelling due to velocity profile rearrangements or mass balance considerations accounts for only 10–20% and cannot explain the 50–300% increase in extrudate cross-section of the polymer emerging out of a die.

Understanding and determining the quantitative swelling features of polymers is very import for a number of polypropylene (PP) processing applications such as extrusion blow molding and fiber spinning. However, die swell depends on many variables related to the molecular composition of the polymer as well as the processing conditions of the melt, making research into its nature a complex endeavour.

A PHENOMENOLOGICAL UNDERSTANDING

One way to explain die swell is to consider the ability of the polymer melt to remember its flow history. The idea is to imagine a fluid element moving from the reservoir into a capillary die as a short, fat cylinder getting squeezed into a long, slender cylinder. If the residence time of the fluid element in the die is shorter than the time of its fading memory, it will try to return to its original shape and produce the die swell effect.

Polypropylene: An A–Z Reference
Edited by J. Karger-Kocsis
Published in 1999 by Kluwer Publishers, Dordrecht. ISBN 0 412 80200 7

This explanation implies the importance of die geometry, flow rate and a characteristic polymer time. More detailed explanations are needed to arrive at a quantitative definition for die swell. In this respect, the recoverable strain and the first normal stress difference are critical rheological variables.

MEASURING DIE SWELL

Die swell is easily measurable on a qualitative basis by way of any standardized test method. Capillary die rheometers are much preferred for such an exercise. The data thus obtained can be very useful for relative comparison purposes but will fail for more detailed needs. This may be associated with the time-dependent viscoelastic nature of die swell. Under isothermal conditions and omitting the influence of gravity, die swell develops gradually over time and may reach its maximum value only after several hours. These ideal experimental conditions are not always achieved in which case cooling and gravity will tend to reduce die swell. Such complicating factors make experimentation difficult. Furthermore, by using capillary dies extrapolating the results to actual processing behavior of the various commercial polymers may prove to be difficult.

In order to approach the processing reality of, for example, extrusion blow molding several techniques have been developed to measure die swell of annular dies. Even sophisticated on-line parison die swell measurement devices using fiber optics and real time video analysis have been designed. For these dies two independent swell ratios can be defined, i.e. the diameter swell B_1 and the thickness swell B_2. The diameter swell B_1 is the ratio of the parison diameter (d_p) and the die diameter (d_0). The thickness swell B_2 is the ratio of the parison thickness (h_p) and the die gap (h_0). Another ratio which is often referred to is the weight swell (B_w), defined as the ratio of the weight of a parison of length L to the weight of the same length of sample having the inner and outer diameters of the die. Under isothermal conditions, the weight swell is equal to the area swell (B_A). If the die gap of the annular die is rather small compared to its diameter, a slit approximation can be applied which makes the area swell, defined as the product of B_1 and B_2, approximately equal to the weight swell.

The ultimate aim for these measurement approaches is to obtain reliable experimental data which may lead to quantitative models, able to describe very complex processing operations. The key feature to capture is not so much the maximum swell or equilibrium swell (B_∞) but the time evolution of the swelling process.

VARIABLES INFLUENCING DIE SWELL

As the response of the flow towards a deformation defines die swell behavior, the property invariably depends on the nature of the molecules constituting the melt [2]. The effects of molecular weight (MW) and molecular weight distribution (MWD) have been studied most extensively for polyethylene (PE). However, the same general conclusions seem to apply to PP. Accordingly, die swell will increase with increasing molecular weight and molecular weight distribution. However, as the time scale of the experiment in relation to the memory or relaxation time scale of the polymer may be smaller or larger, experimental evidence can yield a maximum in this relationship, i.e. very high molecular weight polymers with broad molecular weight distributions may show a lower die swell than expected from a linear extrapolation.

The importance of experimental time scales is reflected in the dependence of die swell on processing variables. Longer dies reduce die swell as well as more gradual narrowing of constrictions. Higher flow rates or shorter die residence times will increase die swell. At a fixed flow rate, die swell decreases with increasing temperature. Experiments at fixed shear stress show little dependence on temperature of die swell.

Even though many experimental studies for various thermoplastic polymers are available, it is not yet possible to propose a single model capturing the many influences on a quantitative level of the variables relating to die swell.

MODELLING DIE SWELL

A number of mathematical expressions have been proposed to calculate die swell in relation to some measurable rheological variable. Most of these expressions relate to the maximum swell or equilibrium swell (B_∞), assuming fully developed laminar flow in long dies. One of the first simple expressions, relating die swell as an 'elastic recoil' mechanism to the first normal stress difference $\tau_{11} - \tau_{22}$ and the shear stress τ_{21}, can be expressed as follows [3]:

$$B = 0.1 + \left[1 + 0.5\left(\frac{\tau_{11} - \tau_{22}}{2\tau_{21}}\right)^2\right]^{1/6} \tag{1}$$

An alternative expression relates die swell to the recoverable shear strain at the die wall (γ_R) which may be measured via creep experiments and the creep recovery function [4]:

$$B^2 = \frac{2}{3}\gamma_R\left[\left(1 + \frac{1}{\gamma_R^2}\right)^{3/2} - \frac{1}{\gamma_R^3}\right] \tag{2}$$

where γ_R can be expressed in terms of the steady shear compliance (J_s), the zero shear viscosity (η_0) and the shear rate ($\dot{\gamma}$):

$$\gamma_R = J_s \eta_0 \dot{\gamma} \tag{3}$$

In both equations the shear rate dependence of die swell is directly evident. The possibility of die swell going through a maximum by changing the MW and MWD as described above is evident from the steady shear compliance (J_s). This value is known to show a maximum for polymers – each having different MW and narrow MWD – as a function of the blend ratio. Most commercial polymers can also be considered as blends of narrowly distributed molecular weight fractions.

In order to capture the time evolution of die swell, some empirical equations have been proposed for annular die geometries. The diameter (B_1) and thickness swell (B_2) are made time dependent via an exponential function by including a 'characteristic time' (ζ). This leads to expressions of the form:

$$B_1(t) = B_{1,\infty} - (B_{1,\infty} - B_{1,0})e^{-t/\zeta_1} \tag{4}$$

$$B_2(t) = B_{2,\infty} - (B_{2,\infty} - B_{2,0})e^{-t/\zeta_2} \tag{5}$$

B_∞ is the maximum swell and B_0 is the respective diameter (B_1) and thickness (B_2) swell at zero time.

More equations have been reported in literature but all are variations of the above and should only be applied for qualitative purposes.

Alternatively, much attention has been devoted in literature to calculating die swell via proper constitutive equations. Using finite-element methods (FEM), the most successful results for quantitative die swell predictions have been obtained by using integral-type constitutive equations of the Kaye-Bernstein-Kearsley-Zapas class in combination with a spectrum of relaxation times. Quantitative correspondence between calculations using the proper rheological data and experimental die swell have been obtained mainly for linear polymers. However, obtaining all the relevant rheological data is not a trivial exercise. Moreover, a complete rheological data set does not promise success especially for rheologically complex polymer systems (e.g. blends). The challenge relates to defining new and improved constitutive equations.

REFERENCES

1. Dealy, J.M. and Wissbrun, K.F. (1990) *Melt Rheology and its Role in Plastics Processing*, Van Nostrand Reinhold, New York.
2. Minoshima, W., White, J.L. and Spruiell, J.E. (1980) Experimental investigation of the influence of molecular weight distribution on the rheological properties of polypropylene melts. *Polymer Eng. Sci.*, **20**, 1166–1176.

3. Cogswell, F.N. (1981) *Polymer Melt Rheology.* George Godwin, London.
4. Tanner, R.I. (1985) *Engineering Rheology.* Clarendon Press, Oxford.

Keywords: die swell, extrudate swell, blow molding, modelling of die swell, rheology, finite element modelling (FEM), creep fiber spinning.

Dielectric relaxation and dielectric strength of polypropylene and its composites

György Bánhegyi

DIELECTRIC RELAXATION

Dielectric relaxation means the adjustment of dielectric displacement (D) or polarization (P) to the time-dependent electrical field (E). *Relative permittivity* (ε) characterizes the capacitance ratio of a condenser filled with an insulating material and with vacuum. If the field is sinusoidal, the permittivity becomes a complex number:

$$\tilde{\varepsilon} = \varepsilon' - i\varepsilon'' \qquad (1)$$

where ε' is the real part (proportional to the capacitive current) and ε'' is the imaginary part (proportional to the resistive current). The complex permittivity is the imaginary Laplace transform of the so-called *response function*, ϕ (t) or *dipole autocorrelation function* [1].

The so-called relaxation strength, the difference between the relaxed and unrelaxed permittivities can be calculated as [1]:

$$\varepsilon_R - \varepsilon_U = \frac{3\varepsilon_R}{2\varepsilon_R + \varepsilon_U} \frac{N_r}{3\varepsilon_0 kT} \left(\frac{\varepsilon_U + 2}{3}\right)^2 g_r \mu_0^2 \qquad (2)$$

where N_r is the concentration of the dipoles, ε_0 is the vacuum permittivity, k is the Boltzmann constant, T is the absolute temperature, μ_0 is the gas phase dipole moment of the dipolar molecules and g_r is the so-called Kirkwood factor, which takes into account the possible orientational

Polypropylene: An A–Z Reference
Edited by J. Karger-Kocsis
Published in 1999 by Kluwer Publishers, Dordrecht. ISBN 0 412 80200 7

correlation of the dipoles. The inherent dipoles in polypropylene are very small, the dipole moment of the methyl group is in the order of 0.4 debye, therefore very small inherent relaxation strength is expected. Due to this low polarity the contribution of polar additives (as e.g. antioxidants) may become dominant in spite of their low concentration.

In the simplest case (when the interaction between the dipoles is negligible) the autocorrelation function can be described by an exponential. In this case the complex permittivity, after the separation of the real and complex variables can be described as:

$$\varepsilon'(\omega) = \varepsilon_U + \frac{\varepsilon_R - \varepsilon_U}{1 + \omega^2\tau^2} \quad (3)$$

$$\varepsilon''(\omega) = \frac{(\varepsilon_R - \varepsilon_U)\omega\tau}{1 + \omega^2\tau^2} \quad (4)$$

where τ is the so called *relaxation time*, which is determined by the local hindrance of the molecular reorientation (microviscosity) and ω is the angular frequency ($2\pi\nu$). The total loss contains not only the dielectric loss, but also the contribution of the ohmic conductivity:

$$\varepsilon''_{total}(\omega) = \varepsilon''(\omega) + \frac{\sigma_R}{\varepsilon_0\omega} \quad (5)$$

where σ_R is the low-frequency limiting conductivity. If plotted against $\log\omega$, ε' exhibits an inflection and ε'' a maximum at $\omega\tau = 1$. In practical cases, usually several relaxation processes exist simultaneously (distribution of relaxation times). As the relaxation time depends on the temperature, it is possible to find $\omega\tau = 1$ not only by varying the frequency, but also by the temperature. In the first approximation the temperature dependence of τ can be described by the Arrhenius relation:

$$\tau(T) = \tau_0 \exp\left(\frac{E_a}{kT}\right) \quad (6)$$

where τ_0 is the pre-exponential factor and E_a is the activation energy. If the relaxation is determined not by local motions but by the cooperative movement of several molecular segments (glass–rubber transition), the temperature dependence is better described by the so-called *Williams–Landel–Ferry (WLF) equation* [1].

Relaxation spectra can be measured in time, frequency and temperature domains. The first two can be converted into each other by Laplace transformation, while the interpretation of the last is more complicated, due to the temperature dependence not only of τ, but also of the relaxation strength. Nevertheless, this last mode is used most frequently, as temperature can be varied more easily than frequency, and by scanning the temperature all relaxation regions can be covered relatively easily and can be compared with other thermal analysis data (as e.g. differential scanning calorimetry or dynamic mechanical spectroscopy).

In the case of composites, consisting of at least two solid components (which is a very important subclass of PP based plastic compounds), the effective unrelaxed permittivity and the relaxed conductivity of the composite can be described as follows [1, 2]:

$$\bar{x} = f(x_1, x_2, v_2, A) \tag{7}$$

where the overline denotes the effective parameters of the composite, and x is either ε_U or σ_R. In our case, index 1 denotes the matrix (PP), 2 the inclusion (filler), v_2 is the volume fraction of the filler, A is the so-called shape factor, which is 0.33 for spheres. For the functional forms of f and A, see the cited literature. If $\varepsilon_1\sigma_2 \neq \varepsilon_2\sigma_1$, a new relaxation process appears, which is called *Maxwell–Wagner relaxation*. It can be ascribed to the charge accumulation at the interface of two, dissimilar materials, therefore it is also called *interfacial relaxation*. If the concentration of the fillers is low, the interfacial relaxation can be described by equations. (3)–(5). In this case, the effective relaxed permittivity of the composite (ε_R) and the relaxation time of the interfacial relaxation (τ) depend on both the permittivities and conductivities of the components. If the concentration of the fillers is high, equations (3)–(5) cannot be applied, but equation (7) remains valid, but the complex permittivities of the components must be inserted.

The presence of surface treatments (coupling agents, stearic acid derivatives, fiber sizings, elastomeric additives) and of adsorbed water complicates the dielectric behavior, as the new interfaces may give rise to additional interfacial and ionic and dipolar relaxation processes (see the examples).

DIELECTRIC RELAXATION IN NONFILLED ISOTACTIC PP

In the case of semicrystalline polymers, usually the following relaxation regions can be distinguished (in the order of increasing temperature) [1]:

- local motions in the glassy state (T_{gg} relaxations or γ relaxation). It appears around -100°C.
- glass-rubber transition (T_g or β relaxation) of the amorphous phase. It appears around 0°C, the corresponding activation energy is about 200 kJ/mole.
- premelting or crystalline transition (T_c or α relaxation), which is attributed to the crystalline–amorphous interface or to the crystalline

defects. This process can be observed between 40 and 90°C, depending on the measuring conditions and sample pre-treatment; the activation energy is around 400 kJ/mole.

Reference [3] describes a careful examination of dielectric relaxation in isotactic PP films, which takes into account the effect of antioxidants and other additives. The temperature range (from −40 to +120°C) covers only the α and β relaxation regions and a doubling of the β relaxation is observed. The properties of the samples and the observed relaxation properties are summarized in Table 1. Due to its good heat resistance, low permittivity and loss, PP is frequently used in power capacitors, therefore samples A and B were selected from these grades. Morphological properties, such as density, crystallinity and isotactic content, are practically identical. Type 2 antioxidant concentration is also similar, but sample B contains three to four times more of antioxidant 1, and sample C contains a considerable amount of nonidentified extractable impurities. Figure 1 shows the temperature-dependent dielectric permittivity and loss values measured at 110 Hz for the three samples. The permittivity exhibits negligible dispersion (frequency dependence) and it begins to decrease rapidly with temperature in the α relaxation range. This is partly due to the decrease of the density (equation (2)) in the premelting range, but in part it is an artefact due to dimensional changes in the sample thickness. The doubling of the β relaxation can be observed in all three samples, but it appears most clearly in sample A. The α relaxation process can be well observed in sample A only, where the ohmic loss (the second part of equation (5)) is negligible in comparison to the dipolar loss. Samples B and C, which contain larger amount of antioxidant 1 and

Table 1 Properties of three isotactic PP samples taken from [3]

	Sample A	*Sample B*	*Sample C*
Type	Impregnated capacitor	Dry capacitor	General grade
Thickness (μm)	18	20	47
Density (g/cm^3)	0.907	0.907	0.907
Crystallinity	57%	57%	57%
Isotacticity	96%	96%	96%
Antioxidant 1 (ppm)	1500	5500	1410
Antioxidant 2 (ppm)	450	560	650
Other impurity	small	small	large
Residual catalyst (ppm)	14	n.m.	n.m.
E_a (α) (kJ/mol)	385	n.m.	381
E_a (β_1) (kJ/mol)	180	193	201
E_a (β_2) (kJ/mol)	306	331	318

n.m. = not measured or not measurable.

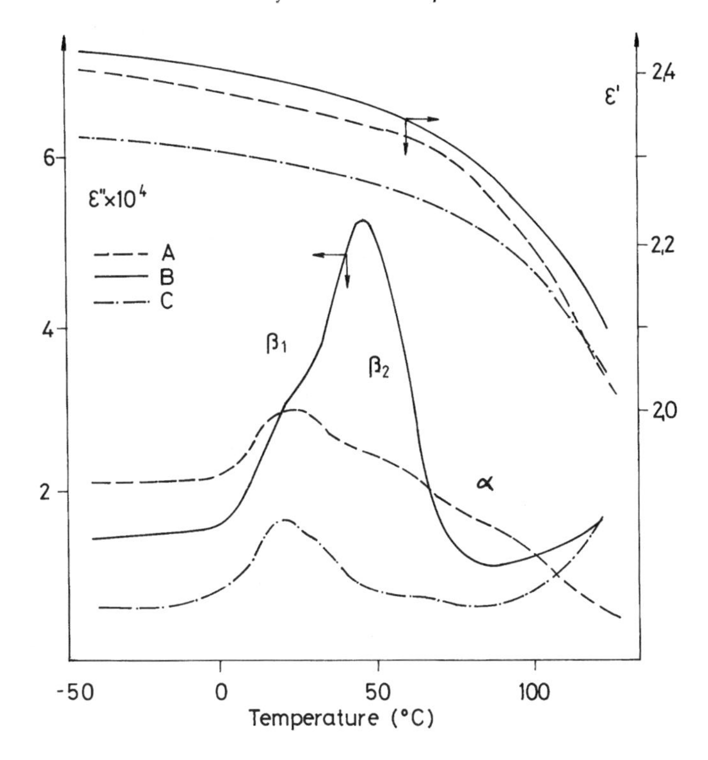

Figure 1 Dielectric permittivity and loss of three different PP grades (see Table 1) measured at 110 Hz. (After [3]).

unspecified impurities respectively, exhibit higher conductivity as well. It is suspected that in sample B the β_2 process and the increased conductivity is mainly caused by antioxidant 1. This was proven by the comparison of the dielectric properties of virgin and acetone extracted grade B samples. It is concluded that β_1 is due to the slightly oxidized atactic-rich amorphous phase, while β_2 to the antioxidants which are also concentrated in the amorphous phase. The activation energy of β_2 is much closer to that of α than to β_1, which proves that the β_2 relaxation belongs to an isotactic-rich amorphous phase with more restricted mobility. The crystalline relaxation (α) in nonoxidized PP under AC conditions is relatively unimportant.

DIELECTRIC RELAXATION IN FILLED PP COMPOSITES AND IN PP BLENDS

Due to the limited space only two examples will be mentioned. One is $CaCO_3$ filled PP composite [4], the other is PP/polyurethane blend [5]. Figure 2 shows the dielectric permittivity and loss of two PP/$CaCO_3$

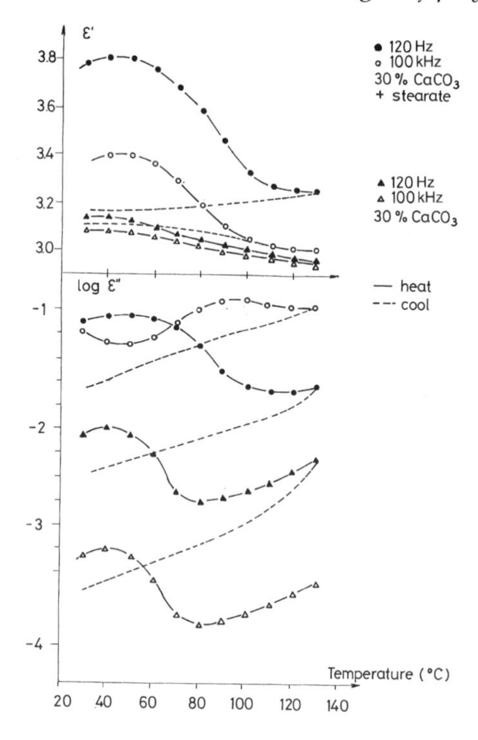

Figure 2 Dielectric permittivity and loss of two, $CaCO_3$-filled PP samples measured at 100 kHz and 120 Hz during heating and cooling. Sample A contains 30 wt.% nontreated filler, while Sample B contains 30 wt.% stearate treated filler. (In part after [4] and unpublished data).

composite samples filled with 30 wt.% filler, in one case without any surface treatment, in the other case with stearate treatment. The curves are shown for both samples at two frequencies (100 kHz and 120 Hz) during heating after ambient storage. On cooling the frequency dependence of the permittivity becomes smaller and the loss maxima disappear. The following features can be observed: (1) The high frequency permittivity of the composite is higher than that of the PP matrix. This can be explained by equation (7) and by the fact that the dielectric permittivity of the filler ($\varepsilon_2 = 8$) is higher than that of PP. In the case of the stearate treatment, there is an additional increase ascribed to the dipolar relaxation of the stearate component; (2) Heating causes a drastic change in the dielectric spectrum, which, however, recovers if the samples are stored under ambient conditions. This can be explained by the gradual desorption of the water accumulated at the filler/matrix interface. The loss peaks with frequency-dependent maxima, which disappear on cooling, are the result of two opposing effects: the increasing mobility of water molecules and desorption. In the case of stearate treatment, it is further complicated by the dielectric activity of the stearate component. This picture is further corroborated by the fact that the dielectric loss peak measured on samples filled with $CaCO_3$ particles of different average diameters increases with the specific surface area of the filler. The cooling curves after desorption are, however, similar for all nonstearate-treated samples. There is no detectable interfacial relaxation for the dried, nonstearate-treated composites. The residual loss in the stearate-treated composite after drying can be attributed to a combination of dipolar and interfacial loss processes.

The dielectric properties of an incompatible PP/polyurethane (PUR) blend containing 20 wt.% PUR component [5] is shown in Figure 3. A comparison with Figure 1 clearly shows that the dielectric properties are dominated by the more polar minority component. The two relaxation regions can be assigned to the soft and hard segments of the PUR component respectively. The dielectric properties of the nonblended PUR component can be easily calculated using equation (7), which is in good

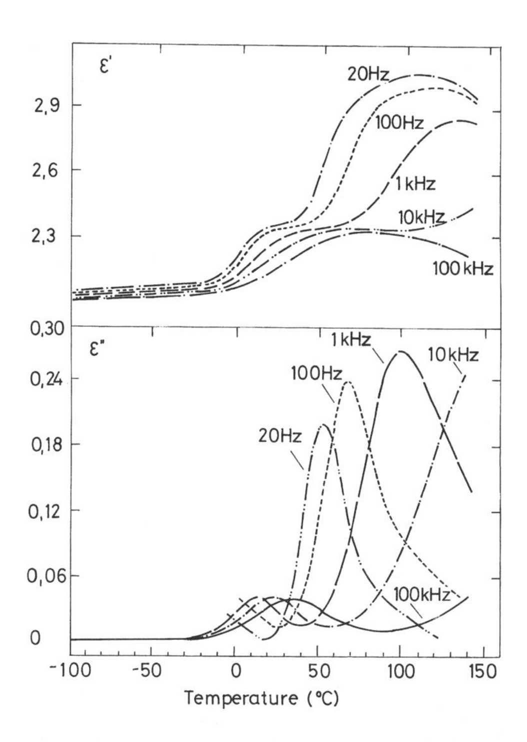

Figure 3 Dielectric permittivity and loss of an incompatible PP/polyurethane blend containing 20 wt% PUR component. (After [5]).

Table 2 Dielectric strength values on PP samples of different thickness

Sample thickness	*Dielectric strength (kV/mm)*	*Reference*
3 mm	20–25	*Modern Plastics Encyclopedia*, 1996
0.5 mm	30–35	C. Brinkman, *Die Isolierstoffe der Elektrotechnik*, Springer, Berlin, 1975
10 μm	300	C. Brinkman, *Die Isolierstoffe der Elektrotechnik*, Springer, Berlin, 1975
6–8 nm	1000–1500	T. Ogawa *et al. Proc. 4th Int. Conf. on Properties and Applications of Dielectric Materials*, Vol. 1., pp. 211–14, 1994

agreement with other measurements. No ohmic interfacial relaxation is observed in this system.

DIELECTRIC STRENGTH OF PP

Dielectric (or breakdown) strength is the maximum field which a given material can survive without breakdown [6]. It can be measured either by gradually increasing the voltage (short-term dielectric strength) or at a constant voltage for a long time (lifetime testing). The first is used for material comparison and development, but the latter is more important in practice. The long-term maximum strength is usually much lower than the short-term value, and the two are related by empirical equations [6]. Dielectric strength is an ultimate property, determined by the weakest link, therefore it is not really a material constant. It depends on electrode geometry (field homogeneity), on the applied voltage (AC, DC or impulse), temperature and, most notably, on the sample thickness. Some representative dielectric strength data measured on samples of different thickness are listed in Table 2. Most investigations were performed on PP films, as they are used in power capacitors and to replace kraft paper in high-voltage oil-filled cables. It has been shown that crystallinity, orientation and molecular weight influence dielectric and mechanical strength, and stability in insulating oil in very high-voltage oil-filled PP-laminated paper (PPLP) cables [7]. PPLP cables have about 10% higher dielectric strength and about 0.25 loss as compared to conventional paper insulated cables.

REFERENCES

1. Hedvig, P. (1977) *Dielectric Spectroscopy of Polymers*, Adam Hilger, Bristol.
2. Bánhegyi, G., Hedvig, P. *et al.* (1991) Applied dielectric spectroscopy of polymeric composites. *Polymer Plast. Technol. Engng*, **30**, 183–225.

3. Umemura, T., Suzuki, T. *et al.* (1982) Impurity effect of the dielectric properties of isotactic polypropylene. *IEEE Trans. El. Ins.*, **EI-17**, 300–305.
4. Bánhegyi, G., Karasz, F.E. *et al.* (1990) AC and DC dielectric properties of some polypropylene/calcium carbonate composites, *Polymer Engng Sci.*, **30**, 374–383.
5. Bánhegyi, G., Karasz, F.E. *et al.* (1990) Dielectric relaxation properties in polypropylene–polyurethane composites, *J. Appl. Polymer* Sci., **40**, 435–452.
6. Dissado, L.A. and Fothergill, J.C. (1992) *Electrical Degradation and Breakdown in Polymers*, Peter Peregrinus, London.
7. Gao, L.Y., Tu, D.M. *et al.* (1990), The influence of morphology on the electrical breakdown strength of polypropylene film, *IEEE Trans. El. Ins.*, **EI-25**, 374–383.

Keywords: dielectric relaxation, dielectric strength permittivity, dipole moment, polarization, relaxation, conductivity, relaxation time distribution, activation energy, Arrhenius equation, WLF-equation, Maxwell–Wagner polarization.

Dyeing of polypropylene fibers

Anton Marcinčin

INTRODUCTION

Polypropylene (PP) fibers having a nonpolar paraffinic character are generally undyeable by the classical bath-dyeing method and therefore the substantial part of the PP fiber production is colored with pigments (mass dyed fibers). Only a small part of PP fibers is dyed after preliminary modification.

EXHAUST DYEING PROCESS

The dyeing of unmodified PP fibers is very rarely applied in industry. The classical bath-dyeing procedure utilizes some structures of disperse dyes with a long linear hydrocarbon chain, e.g. anthraquinone dye structure with alkyl chain beyond C_{10}–C_{12} units.

Dyeing of polypropylene modified fibers

Chemical modification. From among a number of procedures, the method of graft copolymerization of fibers by monomers, such as acrylic acid, vinylpyridine, acrylonitrile, styrene and vinyl acetate was introduced into pilot production.

Physical modification of PP found its industrial application. It uses addition of the low-molecular and polymer compounds containing functional groups which are able to bind dyes. The following groups are usually used [1, 2]:

1. Organometallic compounds of Al, Cu, Zn, but mainly of Ni in the form of salts of organic acids, alcoholates, phenolates, amines, triazols,

Polypropylene: An A–Z Reference
Edited by J. Karger-Kocsis
Published in 1999 by Kluwer Publishers, Dordrecht. ISBN 0 412 80200 7

etc. Special dyes were developed for this modification capable of creating metal complexes. The Ni compounds found wider industrial application.

2. Oligomer and polymer additives with a primary or secondary amino group or with tertiary nitrogen able to bind water-soluble dyes containing acid groups, assortment palette of which is very wide. This function is fulfilled, e.g. by ethylene copolymers with alkyl amino-acrylates, vinylpyridine, N-vinylcarbazole and acrylamide, further acrylate copolymers, copolyamides with the derivatives of piperazine, polyaminotriazoles, polyureas and styrene–amine resins. They are added to PP before spinning in an amount up to 10 wt.%
3. Oligomer and polymer additives able to bind disperse dyes. This group of modifiers covers mainly ethylene and vinyl acetate copolymers, polyethyleneterephthalate and its copolymers, polyamides, etc. Polymer additives form in PP fibers a polyfibrillar mixture with microfibers of dispersed phase oriented in the direction of the fiber axis. During dyeing of this mixture, the diffusion and anchoring of the dyes take place mostly at the matrix–fibrils interfaces which have a higher surface energy than the bulk phase. Modification of PP with anionic additives did not find any application.

PP fibers modified with organometallic compounds, substances containing nitrogen and polymer additives are dyeable by both bath solution process and printing. The outcome of dyeing depends on modification as well as on the kinetics of the dye diffusion from a solution into the fiber. The dye diffusion is negatively affected by the hydrophobic character of PP and the high crystalline portion. The hydrophobic finish of the dye (alkyl chains), microporous structure of the fiber as a result of the deformation at lower temperature or addition of other additives, (e.g. based on polyoxyalkylene oxides, their copolymers, esters with carboxylic acids and ester of aminoalcohols with organic acids) show the beneficial effect.

PIGMENTATION–MASS COLORATION

The substantial part of colored PP fibers are prepared as spun dyed fibres. Mass coloration includes dispergation and homogenization of pigments or dyes in a polymer before extrusion and spinning. The dyes are color substances dissolved in polymer. Pigments are solid color particles insoluble in polymer, usually also water-insoluble. They are divided according to their origin into organic and inorganic. The PP melt is, in principle, not colored by the dyes and pigments directly (besides some special developed processes) but by means of concentrated color dispersions called also masterbatches which are usually divided into:

- liquid pastes, where the carriers are liquid low-molecular compounds, oligomers or polymers;
- solid pastes where the carrier is a solid compound at room temperature;
- color concentrate where the carrier is the same polymer for which the dyeing concentrate is determined;
- predispersed (flushed) dispersions, where pigment particles or their agglomerates are covered by additive with a dispersing effect.

Dyes for polypropylene

Dyes have not practically been used in the industry because of their low light fastness (mainly due to the action of degradation products, like peroxides and aldehydes) low washing and dry-cleaning fastness.

Inorganic pigments

The most important are oxides, sulfides, selenides, and chromate compounds of metals, mainly Zn, Ti, Cd, Cr, Ni, Co, Pb, Mo. Carbon black is also often included into this group. They have high thermostability and good dispersibility in PP.

Organic pigments

Besides carbon black and TiO_2 organic pigments are most often used for dyeing of PP fibers. They have to fulfil following requirement: high thermostability; high light stability; compatibility with other additives; they should not negatively influence the efficiency of light stabilizers; they are not allowed to migrate on the fiber surface; and reduce the mechanical and physical properties of fibers. Their thermostability is, however, limited. Some of them are thermostable, above 300°C for several hours (anthraquinone, phthalocyanine), others to 260°C for only several minutes (monoazopigments) [1, 3]:

- Monoazopigments with one azo group are sometimes less stable at high temperatures, similarly they may have lower rubbing fastness and higher migration (e.g. C.I. Pigment Orange 36).
- Disazopigments are derived from benzidine or they are condensates of monoazopigments; they are more stable (e.g. C.I. Pigment Yellow 95).
- Isoindolinones are derivatives of tetrachlorophthalic imides. They are easy to disperse but have lower color strength.
- Perylene pigments are derived from pentenecarboxylic acid, have high thermostability, but are difficult to disperse in PP (e.g. C.I. Pigment Orange 43).
- Anthraquinone pigments have high thermostability but they are difficult to disperse (e.g. C.I. Pigment Red 177).

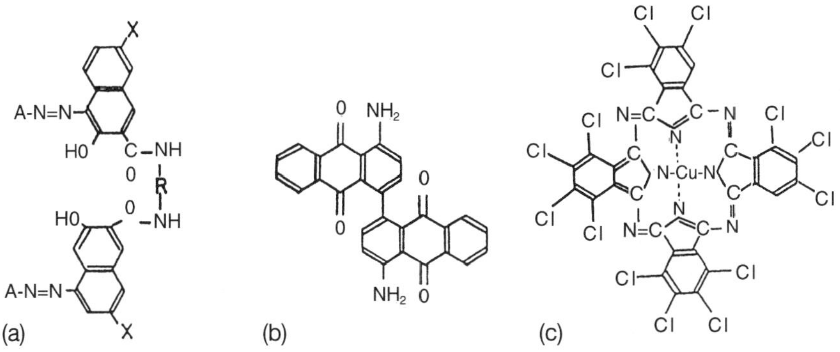

Figure 1 Pigments for spun dyed polypropylene fibres: (a) C.I.Pigment Red 144; (b) C.I. Pigment Red 177; (c) C.I.Pigment Green 7.

- Quinacridone pigments show good fastness properties.
- Phthalocyanine pigments have high fastness, high thermostability (e.g. C.I. Pigment Green 7).
- Nacreous and luminescence pigments are special types. Fluorescing ones absorb light of short wavelength and emit a light of higher wavelength. Pigments with phosphorescence activity emit the transformed light with a time delay.

PREPARATION OF PIGMENT CONCENTRATES

In the preparation of color concentrates, a compatibilizer-dispersant is used in addition to the carrier which is usually polypropylene (although some concentrates are polyethylene-based). Concentrated dispersion of pigment in PP with a high degree of dispersity cannot be prepared without efficient dispersant, which can be a low molecular or oligomeric substance. The role of dispersants is to increase the wettability of pigments by the polymer, particle separation within agglomerates of pigments by the formation of the so-called 'liquid wedge', improvement of the rheological properties of the dispersion and improving of the stability of the heterogeneous system. Dispersants can be of multifunctional character, e.g. antistatic agents, substances with thermostabilizing effect, etc., at the same time. With regard to the paraffinic character of PP, a dispersant must create an interface between PP and pigment with lower surface energy. They are added in an amount from 5 to 40% by weight of pigment.

Polyoxyalkyleneglycols, their copolymers and mixtures, esters, and ethers with carboxylic acids and alcohols belong to very efficient dispersants for pigments with polar groups like disazopigments (C.I. Pigment Yellow 95, C.I. Pigment Red 144), esters of glycerine and carboxylic acids, derivatives of poly(propylene) oxide and copolymer of ethylenoxide–(propylene) oxide also PP oil for pigments with lower polarity of functional groups (C.I. Pigment Yellow 147, Red 177). The third group of pigments (e.g. phthalocyanine C.I. Pigment Green 7) does not require any strict selection of dispersants from the point of view of polarity [4, 5]. For carbon black and TiO_2, dispersants based on oligomer polyolefins and waxes are suitable. Highly efficient dispersing systems contain esters of polyoxypropyleneglycol and carboxylic acids, ionomers, and modified polyethylene and PP waxes. The increase of the degree of pigment dispersity is achieved by using 'hyperdispersants' and by surface treatment of pigments before drying (flushed pigments). The concentration of inorganic pigments in concentrates is up to 70 wt.%, for organic pigments to 50 wt.%. Color concentrates are prepared by homogenization of powder polymer, pigment and dispersant prior to melting and granulation in a twin-screw extruder.

PREPARATION OF SPUN DYED FIBERS

Concentrated dispersions are added to the basic polymer in two ways:

- by volume or weight dosers directly to the extruder feeder;
- after melting and homogenization in a separate extruder the melt of the concentrate is dosed by injecting into the melt of the basic polymer. The melt mixture is then homogenized in a static or dynamic mixer.

The content of the organic pigment usually varies from 0.01 to 1.5 wt.% in the colored melt, according to the required color strength. A greater amount of inorganic pigment, but smaller amount of soluble dye is needed to reach the same color strength. At this pigment concentration (incl. additives), the rheological properties of the melt do not change significantly. The colored melt is extruded through the spinneret holes.

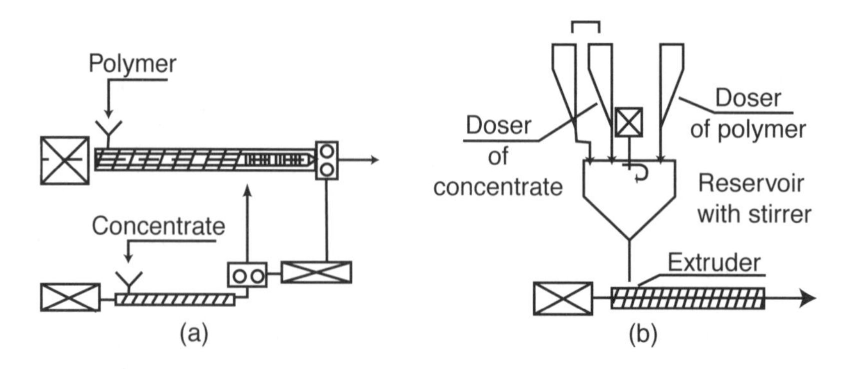

Figure 2 Scheme of mass coloration: (a) by the injection method; (b) homogenization before the melt process – direct coloring.

The spinning and the following operations are identical with the preparation of uncolored fiber according to the type and the use.

Some pigments act as nucleation agents and significantly accelerate the crystallization of the polymer stream below the spinneret, e.g. phthalocyanine pigments. Others, such as carbon black and TiO_2, have less effect on the kinetics of crystallization.

The color properties of the spun dyed fibres as color strength, color shade, and others can be evaluated using a color coordinates, e.g. CIELab system (approved by Commission Internationale de l'Eclairage in Paris 1976). Color fastness evaluation usually expressed by color difference between original and threaded samples can be used for various quality tests of fibers or textile material. The light fastness, rubing fastness, dry-cleaning and wash fastness are most frequently estimated (ISO 105). The majority of pigments give the highest fastness values.

REFERENCES

1. Ahmed, M. (1982) *Polypropylene Fibers – Science and Technology,* Elsevier, Amsterdam.
2. Ondrejmiška, K. (1994) Recent trends in dyeing of synthetic fibres. *Vlákna a textil – Fibers and Textiles,* **1**(1) 11–16.
3. Kaul, B., Ripke, C. and Sandri, M. (1996) Technical aspects of the mass-dyeing of polyolefin fibres with organic pigments. *Chemical Fibers Int. (CFI),* **46**(April), 126–129.
4. Marcinčin, A., Zemanová, E. and Marcinčinová, T. (1993) Role of dispersant during spin dyeing of synthetic fibres (Rus). *Tekstilnaja Chimija,* **1**(3) 32–43.
5. Marcinčin, A. and Krištofič, M. (1994) Some aspects of dyeing synthetic fibres in mass. *Fibers Text. East. Europe,* **2**(3) 38–42.

Keywords: anthraquinone pigment, bath-dyeing, color masterbatch, color properties, concentrate, dispersant, dry-cleaning fastness, dry rubbing, dyeability fastness, dyeing, exhaust process, mass-coloration, phthalocyanine pigment, pigmentation.

Elastomeric polypropylene homopolymers using metallocene catalysts

W.J. Gauthier

INTRODUCTION

A thermoplastic elastomer (TPE) can be generally described as a material that exhibits classical elastomeric properties (such as low tensile set and high ultimate elongation) and typically comprises hard (crystalline or glassy) domains and soft (amorphous) domains. Elastomeric polypropylene (ELPP) represents a relatively new type of polyolefin-based TPE which is of commercial interest. Recent improvements in supported catalyst technologies to efficiently prepare ELPP and the discovery of metallocene catalysts capable of producing this material has sparked renewed interest in ELPP as an economically viable TPE. The following is a brief overview of the preparation of semicrystalline ELPPs produced by the homopolymerization of propylene and their properties and potential applications.

SYNTHETIC METHODS TO PRODUCE ELPP [1]

DuPont heterogeneous technology

One of the most successful methods to produce ELPP directly has been reported extensively by DuPont researchers and involves the use of heterogeneous catalysts comprised of tetra-alkyl complexes of Group IVB transition metals of the general formula (MR_4) on metal oxide sup-

Polypropylene: An A–Z Reference
Edited by J. Karger-Kocsis
Published in 1999 by Kluwer Publishers, Dordrecht. ISBN 0 412 80200 7

ports where the alkyl substituent lacks beta hydrogen atoms [2]. Bisarene complexes of Group IVB transition metals, such as M(mesitylene)$_2$ on metal oxide supports have also been reported to produce ELPP by DuPont. The DuPont ELPP process produces polymer with high melting temperature (125–165°C) and good elastic recovery. Commercial production of ELPP using DuPont's technology (or based upon) is believed to commence in the near future and will likely employ the Zr(neophenyl)$_4$ complex supported on alumina as reported by researchers from PCD [3]. Process improvements for the continuous production of ELPP and improvements in the preparation of these catalysts by both Eastman [4] and PCD makes it increasingly likely that these materials will become commercially available in the near future. The ELPP produced using DuPont technology, however, is compositionally impure in that the polymer molecular weight distribution is broad ($M_w/M_n >> 2$), the melting endotherm (by DSC) is broad and multimodal, and the polymer can be fractionated into compositionally different materials using low boiling solvents such as diethylether and hexane. These properties are indicative of multiple active sites being present in the heterogeneous catalyst. The key attributes of catalysts for the preparation of ELPP was believed to be the ability to prepare high molecular weight polymer with an intermediate quantity of crystallizable segments (high enough to crystallize and prevent cold flow but low enough to obtain an elastomeric matrix with no yielding or permanent set).

Up until recently, there was essentially no well-defined, single site catalytic method to directly manufacture ELPP. Several research groups have explored the use of well-defined single site metallocene catalysts to prepare compositionally pure ELPP with limited success. Generally speaking, the metallocene catalysts explored to date can be subdivided into two groups: those which are conformationally rigid (by virtue of containing a chelating bridging group between the two ancillary cyclopentadienyl moieties thus restricting the conformation about the transition metal) and those which are unbridged and fluxional. Since these represent two distinct approaches to obtaining compositionally pure ELPP, they will be described separately.

Conformationally rigid metallocenes

The use of *ansa* (or bridged)-metallocenes has proven to be a successful route to the preparation of highly stereoregular polypropylene with catalyst activities, polymer melting temperatures and polymer molecular weights approaching that of commercial heterogeneous systems.

In 1990, Chien *et al.* [5] discovered the nonsymmetrical metallocene: rac-[ethylidene(1-η^5-tetramethylcyclopentadienyl)(1-η^5-indenyl)] titanium(IV)dichloride (**1**) with methylaluminoxane (MAO) as cocatalyst

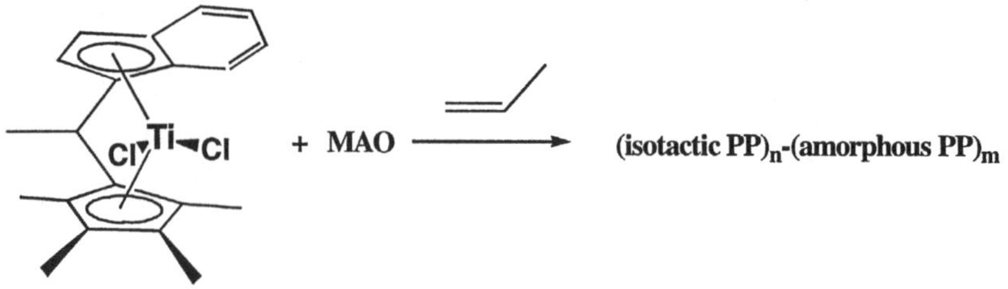

Figure 1 ELPP produced using Chien/Rausch titanocene catalyst [1].

could produce homopolymers of propylene which were compositionally pure (polydispersity index $(M_w/M_n) \sim 2$ and could not be fractionated) with elastic recoveries in excess of 90%. The melting temperature of the polymer was low (~70°C) but the percent elongation to break was high at 1300% while having a tensile strength of 12 MPa. This significant discovery was the first example of a single site catalyst generating a compositionally pure ELPP and it sparked renewed interest and research in this field. Chien and Rausch proposed that the microstructure of the ELPP contained crystalline and amorphous blocks which were produced by multiple insertions at an isospecific site followed (presumably) by chain isomerization and multiple insertions at an aspecific site, thus providing polymer which contained crystallizable and noncrystallizable segments in any given polymer chain (Figure 1). This catalyst system, however, displays low overall activity towards propylene polymerization and the rate of polymerization appeared to diminish with time possibly due to catalyst deactivation by reducing the oxidation state of Ti (IV to III).

To further explore the utility and critical requirements of other nonsymmetrical metallocenes for the production of ELPP, Gauthier and Collins [6] studied the polymerization behavior of $Me_2C(1\text{-}\eta^5\text{-Cp})(1\text{-}\eta^5\text{-indenyl})MCl_2$ (**2**) and its analogous dimethyl silylene-bridged metallocenes $Me_2Si(1\text{-}\eta^5\text{-Cp})(1\text{-}\eta^5\text{-indenyl})MCl_2$ (**3**) where M = Ti (**a**), Z (**b**), and Hf (**c**) – Figure 2. They found that ELPP could be prepared using the dimethyl-silylene-bridged hafnocene catalyst $Me_{12}Si(1\text{-}\eta^5\text{-Cp})(1\text{-}\eta^5\text{-indenyl})HfCl_2$ (**3c**). The polymer produced using this catalyst also displayed high elastic recovery from strain (low tensile set) with values in excess of 90%. Low overall catalyst activity and the cost of the ultrapure $HfCl_4$ (essentially free of zirconium), however, appeared to be prohibitive to the commercialization of this catalyst system.

In both instances, the microstructure of ELPP produced using stereorigid metallocenes could not be fully accounted for using a simple migratory insertion model of propagation. Collins proposed a chain

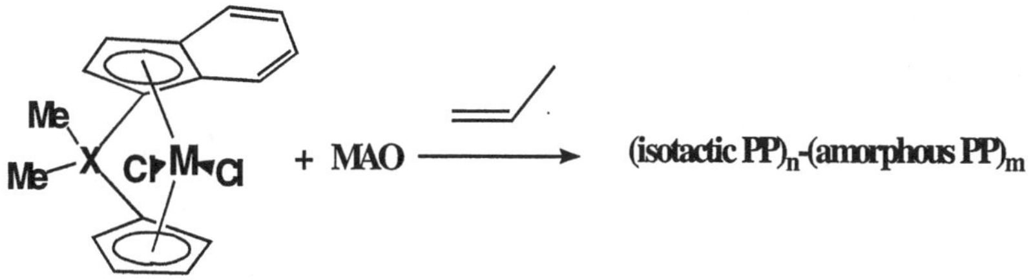

Me₂X(Cp)(Ind)MCl₂ + MAO → $(isotactic\ PP)_n$-$(amorphous\ PP)_m$

Figure 2 ELPP produced using bridged Cp-indenyl catalysts (**2** or **3**).

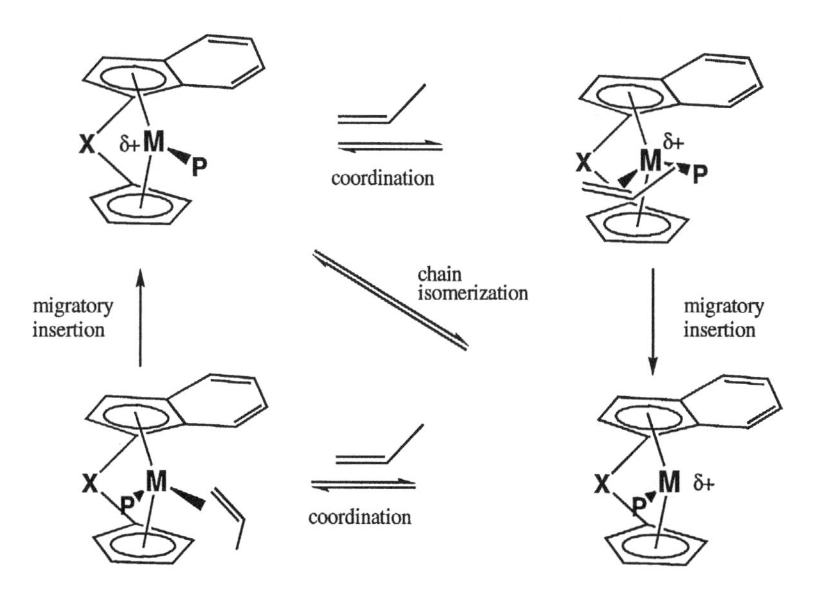

Figure 3 Propagation mechanisms for nonsymmetrical metallocene catalysts.

isomerization mechanism to account for the formation of ELPP with stereorigid catalysts as shown in Figure 3.

One limitation of the polymers produced using catalysts such as **1** or **3c** was that the ELPP produced using these catalysts exhibited rather low melting temperatures ($T_m < 100°C$), crystallinities were low, and the crystallization rate was fairly slow. Since the upper usage temperature of the ELPP is typically governed by the melting temperature of the polymer, this meant that the ELPP could only be used in applications in a very narrow temperature range. A more successful approach employing unbridged fluxional metallocenes appears to have extended the upper usage temperature of ELPP into the realm of DuPont's higher melting ELPPs.

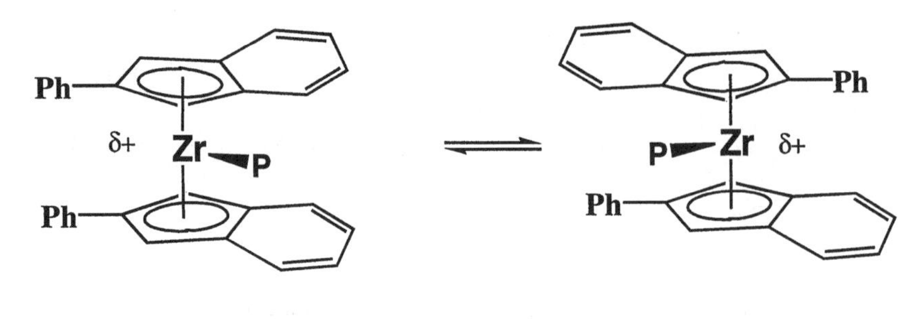

Figure 4 Interconversion of $(2\text{-Ph-Ind})_2ZrCl_2$ (**4**) for ELPP production.

Fluxional metallocenes

In 1994, Waymouth *et al.* [7] discovered the unbridged $(2\text{-Ph-Ind})_2ZrCl_2$ (**4**) catalyst produces highly elastic PP with high melting temperature ($T_m > 130°C$). The key element to the generation of ELPP with this catalyst system appears to be the isomerization or interconversion (Figure 4) of the ligand environment about the transition metal from an isospecific configuration (generating isotactic PP segments) to one with an aspecific configuration (generating atactic PP segments) during growth of a single polymer chain. The isotactic sequence length and regioregularity of the polymer is reportedly large giving rise to higher melting ELPP with larger crystallites compared to that produced using catalysts **1**–**3**.

The origin of the stereoselectivity of this catalyst (**4**) is believed to involve pseudofixation of the racemic or meso conformer, likely due to a π- stacking (phenyl to phenyl or phenyl to indenyl) type process as proposed by Pietsch and Rappe [8]. This oscillating process between isospecific and aspecific conformers partially accounts for the higher melting temperatures of the polymers obtained (Figure 5) compared to other single-site catalytic systems. There are still some unanswered questions concerning the microstructure and compositional purity of the ELPP produced from catalysts such as (**4**) which will undoubtedly be addressed in the near future.

PROPERTIES OF ELPP HOMOPOLYMERS

The most distinguishing feature of ELPP is its good elastomeric performance with high elastic recovery (typically $> 90\%$), high elongation to break (1000–2000%), good tensile strength (3–34 MPa) and low modulus. The usage temperature of ELPP is limited by the melting temperature (upper limit) and the T_g ($\sim -5°C$ for propylene homopolymers). The usage temperature range, however, could be broadened by copolymerization or by blending with other resins to extend the lower temperature limit (using low T_g rubbers for instance) or higher melting resins such as isotactic PP to extend the upper usage temperature. The modulus, tensile strength and elastomeric performance of ELPP depends primarily on the mass fraction crystallinity with increasing crystalline fraction leading to higher modulus and tensile strength. If the crystallinity is too high, however, plastic yield (a non-recoverable deformation) occurs and hence a decrease in elastic recovery is observed. Figure 5 depicts the relationship between polymer isotacticity and polymer melting temperature for various metallocene resins and blends. This figure shows that the resins produced using the stereorigid metallocenes such as **1–3** and other C_2-symmetric metallocenes give polymers which show similar trends in % isotacticity (by carbon-13 nuclear magnetic resonance spectroscopy, ^{13}C NMR) versus T_m (%mmmm $\sim 0.6\,T_m$) whereas the polymer blends of amorphous PP with isotactic PP (reported by Exxon [9] or Rexene [10]) display melting temperatures which, as expected, are much less sensitive to the percent meso pentad (%mmmm) of the blend. The ELPP homopolymers with longer isotactic block lengths (using catalysts such as **4**) possess much higher melting temperatures than one

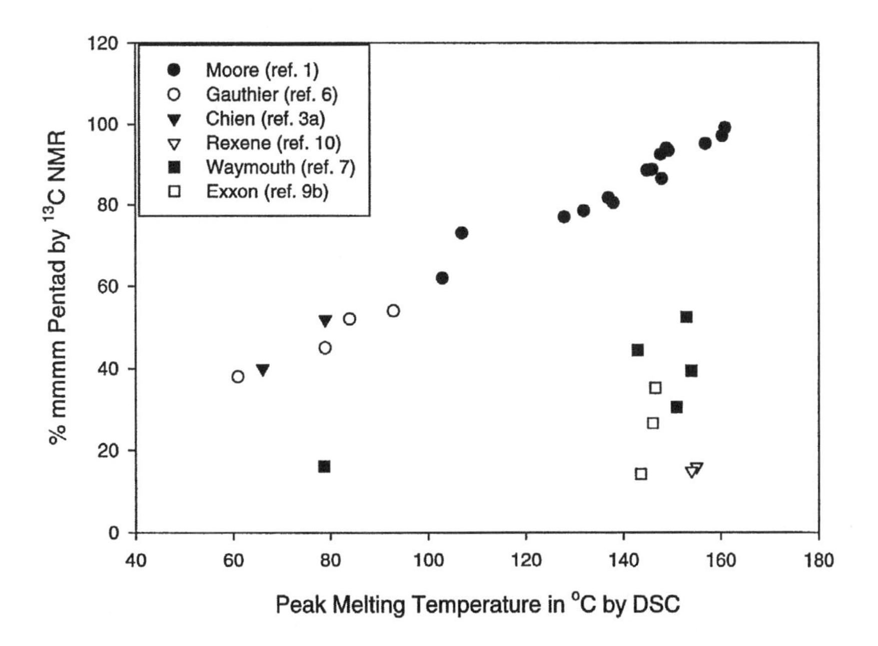

Figure 5 Relationship between PP isotacticity and melting temperature [1].

would predict based on a simple single site catalytic process with a more random distribution of isotactic block lengths.

It might be expected that an ELPP with a block-type microstructure would display a relationship between %mmmm pentad intensity and polymer melting temperature somewhat intermediate between the blended samples and the compositionally pure, pseudorandom microstructures produced using stereorigid metallocenes. This finding supports Waymouth's contention that a very different propagation process is occurring with their catalyst system.

POTENTIAL APPLICATIONS OF ELPP

ELPP can be considered as a general-purpose thermoplastic elastomer and hence could be a potential candidate to replace traditional TPEs, such as plasticized polyvinylchloride (PVC), styrene–butadiene block copolymers (SB, SBS) or styrene–isoprene copolymers (SI) block copolymers (SIS) in a variety of applications. It is reported that ELPP can be converted to fibers, sheet, film, foam, wire and cable, tubing, soft bottles/containers as well as car bumpers, medical contacts, intravenous bags, grafting stock, adhesives, sealants, caulks, gaskets, damping elements for furniture and flexible flooring. It is further reported that ELPP can be successfully employed as a modifier for various thermoplastics (such as PE or PP) as well as bitumen, as a compatibilizing agent for ethylene-propylene rubbers (EPR and EPDM)/PP blends. Ultimately, the successful integration of ELPP into the TPE marketplace will depend on its cost and performance attributes (which have yet to be determined) compared to other TPEs.

REFERENCES

1. For an elegant overview of propylene polymerizations, see Albizzati, E., Giannini, U., Collina, G., Noristi, L., Resconi, L. (1996) *Polypropylene Handbook* (ed. E.P. Moore), Hanser, Munich.
2. For an overview of Du Ponts' ELPP technology, see the following and references therein: (a) Tullock, C.W., Tebbe, F.N., Mulhaupt, R., Ovenall, D.W., Setterquist, R.A. and Ittel, S.D. (1989) *J. Polymer Sci. Polymer Chem.*, **27**, 3063. (b) Collette, J.W., Tullock, C.W., MacDonald, R.N., Buck, W.H., Su, A.C.L., Harrell, J.R., Mülhaupt, R. and Anderson, B.C. (1989) *Macromolecules*, **22**, 3851. (c) Collette, J.W., Ovenall, D.W., Buck, W.H., Ferguson, R.C. (1989) *Macromolecules*, **22**, 3858.
3. Gahleitner, M., Ledwinka, H., Hafner, N. and Neißl, W. (1996) *Proceedings of the Sixth International Business Forum on Specialty Polyolefins*, Houston, TX, p. 251.
4. Ames, W.A., Holliday, R.E., McKeon, T.J., Pagan, L.A., Scott, J.H., Seeger, H.K., Slemons, G.T., Statman, M. and Vanderbilt, J.J. (1996) WO 9616996.
5. For an overview of this work see Llinas, G.H., Dong, S., Mallin, D.T., Rausch, M.D., Lin, G.-Y., Winter, H.H. and Chien, J.C.W. (1992) *Macromolecules*, **25**,

1242 and references therein. More recent efforts to prepare ELPP via homogeneous binary (mixed metallocene) catalyst system have also been reported. For details see Chien, J.C.W., Iwamoto, Y., Rausch, M.D., Wedler, W. and Winter, H.H. (1997) *Macromolecules*, **30**, 3447 and references therein.
6. (a) Gauthier, W.J. and Collins, S. (1995) *Macromol. Symp. 98*, **223**. (b) Gauthier, W.J., Corrigan, J.F., Taylor, N.J. and Collins, S. (1995) *Macromolecules*, **28**, 3771. (c) Gauthier, W.J. and Collins, W.J. (1995) *Macromolecules*, **28**, 3779.
7. Waymouth, R.M., Coates, G.W. and Hauptman, E.M. (1997) US patent 5 594 080 and PCT Int. Appl. WO 9525757. See also (a) Coates, G.W. and Waymouth, R.M. (1995) *Science*, **267**, 217. (b) Wagener, K.B. (1995) *Science*, **267**, 191. (c) Kravchenko, R., Masood, A. and Waymouth, R.M. (1997) *Organometallics*, **16**, 3635 and references therein.
8. Pietsch, M.A. and Rappe, A.K. (1996) *J. Am. Chem. Soc.*, **118**, 10908.
9. (a) Canich, J.A.M., Yang, H.W. and Licciardi, G.F. (1996) US Patent 5 516 848. (b) Yang, H.W., Canich, J.A.M. and Licciardi, G.F. (1996) US Patent 5 539 056.
10. Pellon, B.J. and Allen, G.C. (1992) European Patent Application 475 306 A1.

Keywords: elastomeric polypropylene (ELPP), thermoplastic elastomeric polypropylene, TPE, metallocene catalysts, physical properties, applications, tacticity, elasticity, regioregularity, stereoselectivity.

Electron microscopy

G.H. Michler

In many fields of research in science, engineering and medicine, electron microscopy as a method for directly imaging submicroscopic structures has become increasingly important in recent decades. Electron microscopy (EM) includes several different techniques: conventional transmission electron microscopy (TEM), high-resolution electron microscopy (HREM), high-voltage electron microscopy (HVEM), scanning electron microscopy (SEM), analytical electron microscopy (AEM), and others. In the past the central aim of using electron microscopy was structure determination, but recently it has been of growing importance for also investigating different processes, i.e. changes in materials by interaction with several influentional factors (e.g. heat, electric or magnetic fields, mechanical loading). Of particular interest is the study of the micromechanical processes of deformation and fracture. Therefore, electron microscopy is a very powerful tool for materials science.

TECHNIQUES OF ELECTRON MICROSCOPY

Electron microscopy (EM) can be divided into the techniques of transmission electron microscopy and of direct imaging of surfaces. There are a number of reviews and monographs on the different techniques of EM [1, 2].

Transmission electron microscopy

In transmission electron microscopy, the specimen is traversed by an electron beam, typically in an energy range between 80 and 200 kV. Electrons with these energies can penetrate specimens only up to a

Polypropylene: An A–Z Reference
Edited by J. Karger-Kocsis
Published in 1999 by Kluwer Publishers, Dordrecht. ISBN 0 412 80200 7

maximum thickness of some tenths of a micrometer, therefore, ultrathin specimen foils must be produced from bulk material.

Contrast formation in TEM depends on the physical interaction processes between electrons and specimen, mainly on scattering and absorption of penetrating electrons. For an amorphous object, the electrons are scattered more intensively at specimen places with a higher atomic number and a greater thickness. These electrons scattered through larger angles are eliminated by the objective aperture and do not contribute to the image. This so-called scattering-absorption contrast is similar to the absorption contrast in light-optical microscopy. Beside this contrast in the usual bright-field image, dark-field imaging, phase contrast, and – for crystalline objects – diffraction patterns can be used.

As in optical microscopy, the *resolution limit* is defined by the wavelength of electrons and the numeric aperture of the optical system. Because of the very small wavelengths of electrons, the achievable resolution limit in high-performance TEM with typical operating voltages of 100–200 kV is about 0.2–0.3 nm. There are special high-resolution electron microscopes operating at voltages from 300 kV up to about 1.2 MV with attainable resolution in the 0.1–0.2 nm range.

Accelerating voltages well beyond 200 kV (mostly of 1000–1200 kV) define the range of high-voltage electron microscopy (HVEM). Due to the enlarged penetration depth of the high energetic electrons, thicker polymeric specimens up to several micrometers thickness can be investigated.

Direct imaging of surfaces

The main technique for direct surface investigations is scanning electron microscopy (SEM). The principle of SEM differs substantially from TEM. A focused electron beam with accelerating voltage from 0.1 to 50 kV is scanned line by line across the specimen surface. At the incident point of the primary electron beam, secondary electrons are emitted (besides several other electron beam–specimen interactions, see below). The intensity of secondary electrons depends on surface topography, i.e. on the angle between the direction of the primary electron beam and the surface. The number of the secondary electrons modulates the brightness of a display screen, which is controlled by the same scan generator as the primary electron beam. For example, areas with a higher intensity of secondary electrons appear brighter on the display screen than others. This yields a good contrast of SEM images with a very good spatial visibility of details and a high depth of focus (a factor of about 100 greater than in light-optical imaging).

The *magnification* is attained from the ratio of the size of the display screen to the size of the scanned surface region on the sample, and it can be changed easily between about 5 to 200 000. Therefore, the resolving power of SEM lies between light optical microscopy and TEM, as is illustrated in the scheme of Figure 1 together with some morphological details in polymers.

There is an essential advantage of SEM that, in general, no special preparations are necessary to perform morphological investigations of specimens. However, for nonconducting materials such as polymers the deposition of a thin conducting layer of metal or carbon (by vacuum evaporation or sputtering) is necessary.

Due to the interaction between the primary electron beam and speci-

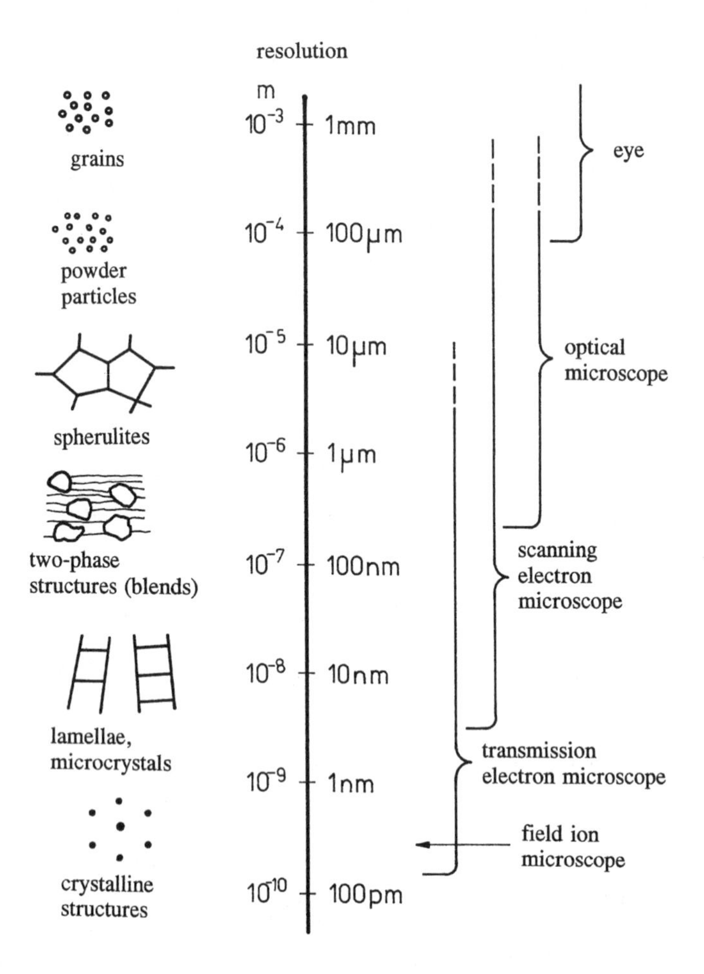

Figure 1 Scheme of attainable resolutions with different microscopic techniques together with some morphological details in polymers.

men, a number of other signals beside the emission of secondary electrons may arise and yield specific information on the specimen. Most important is determination of characteristic X-rays, enabling qualitative and quantitative analysis of elements (EDX – energy dispersive analysis of X-rays).

Using additional signals and information (backscattered electrons, absorbed electrons, energy loss of transmitted electrons) is the field of analytical electron microscopy (AEM).

SPECIMEN PREPARATION METHODS

In general, direct investigations of polymers at higher magnifications by transmission electron microscopy involve three problems:

1. preparing ultrathin specimens from bulk polymers is often difficult;
2. polymers are particularly sensitive to electron beam irradiation;
3. contrast between structural details is often very low because they consist only of light elements.

To overcome these difficulties, several preparation techniques have been developed, which can be summarized in two general directions:

1. preparation of surfaces and investigation by means of replicas in the TEM or directly in the SEM;
2. preparation of thin sections by ultramicrotomy, generally after special fixation or staining procedures and investigation by TEM or HVEM.

Modifications of these basic techniques are necessary, depending on whether study of morphology or of micromechanical processes is the centre of interest.

Investigation of morphology

An overview on main directions of preparation techniques often used to investigate the morphology is shown in Figure 2.

Investigation of surfaces

The external surfaces of polymeric solids only in some cases yield information on the morphology because they are often strongly modified by the processing and manufacturing conditions. Therefore, special surfaces from the interior have to be prepared.

The traditional method to study heterogeneous materials is to produce brittle fracture surfaces. This is usually done at low temperatures to avoid plastic deformation, which would hide the morphology. The fracture path occasionally follows structural details so that they become

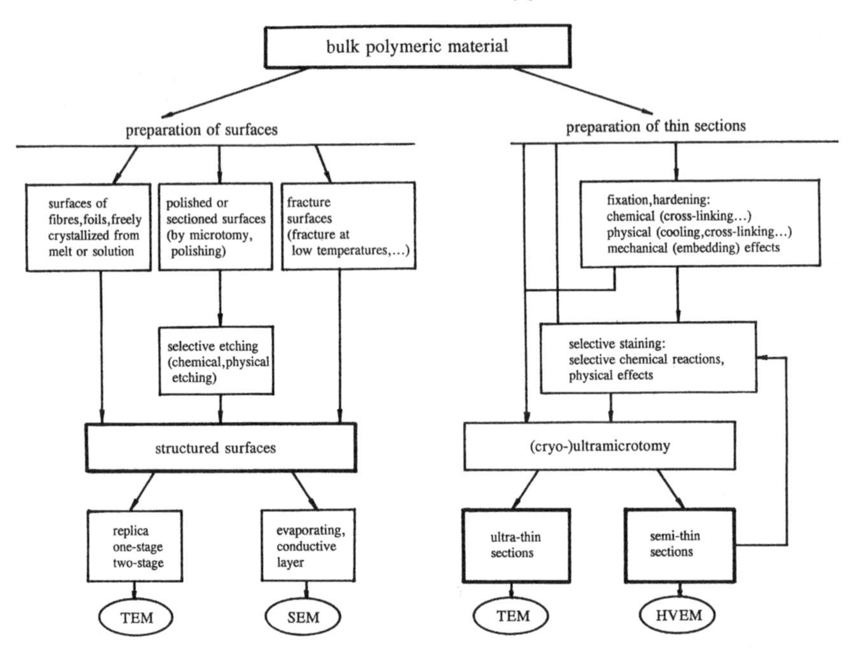

Figure 2 Overview on preparation techniques and electron microscopic methods for investigating the morphology.

visible, e.g. large spherulites or inorganic filler particles. More recently, the 'soft matrix fracture', a technique to produce a fracture surface at higher temperatures (in the melting range of the polymer), has been developed, to avoid the brittle fracture features on the surface [3]. By using this technique, hard particles in the matrix are more clearly detectable.

The second possibility of producing internal surfaces consists in polishing or sectioning them in a microtome. Selective etching of these surfaces may lead to the 'development' of a structure. Etching can be performed chemically by solvents (usually potassium permanganate [4], Figure 4) and physically by ion etching or etching in activated oxygen [5]. If the etching rate of several components (crystalline and amorphous parts, several polymer phases, inorganic particles) is different, these components appear on the surfaces.

Surfaces can be investigated by TEM by the replica technique. Since it is difficult to remove a replica from a polymer surface, two-stage techniques are usually necessary. The gold decoration technique to mark surface structures [2] has been tested, but the difficulties of removing the replica films from the surfaces are too great for wide application. SEM has been increasingly used because it is a very easy way to investigate surfaces (only a coating with a conductive layer – C or C/Au – is necessary). The disadvantage is that the resolution is not better than about 10 nm for polymeric materials.

Investigation of thin sections

The second principal method of determining the morphology of bulk polymeric material is TEM investigation of ultrathin sections. Several improvements in instrumentation and sample preparation during the past decades have made ultramicrotomy an almost universally applicable method. It allows the production of thin specimens from bulk polymers and direct investigation by TEM. Ultramicrotomy is a mechanical cutting technique whose success depends essentially on the compressibility and plastic deformation of polymeric materials in terms of cutting parameters (e.g. glass or diamond knifes, knife angle, cutting velocity) [6]. The possibility of artefacts being produced by local plastic deformation (smearing, chatter, folds, scratches) should always be taken into considerations.

To avoid artefacts during cutting of PP samples, they have to be fixed or hardened before being cut. In general, there are two methods for doing this:

1. *Chemical fixation or hardening*. Bulk samples are treated with chemical agents, mostly with chlorosulfonic acid and uranyl acetate or osmium tetroxide or with ruthenium oxide [7]. Several reactions cause a fixation or a hardening of the material. For the most part, these reactions are highly selective, therefore, a selective staining of structural details takes place as a positive secondary effect (cf. Figure 5).
2. *Physical fixation*. Using cryo-ultramicrotomy, polymeric samples are cooled below their glass transition temperature and cut at these temperatures. Additional selective chemical staining of the ultrathin sections is helpful to enhance contrast.

Investigation of micromechanical processes

A survey of successful methods is given in Figure 3, demonstrating three main techniques:

a) investigation of fracture surfaces by SEM (or TEM using replicas):
b) deformation of bulk material, followed by investigating the changes at the surfaces by SEM or, after replication, by TEM. Changes inside the bulk material are studied by preparing ultrathin sections using ultramicrotomy (occasionally after chemical staining) and by TEM investigations.

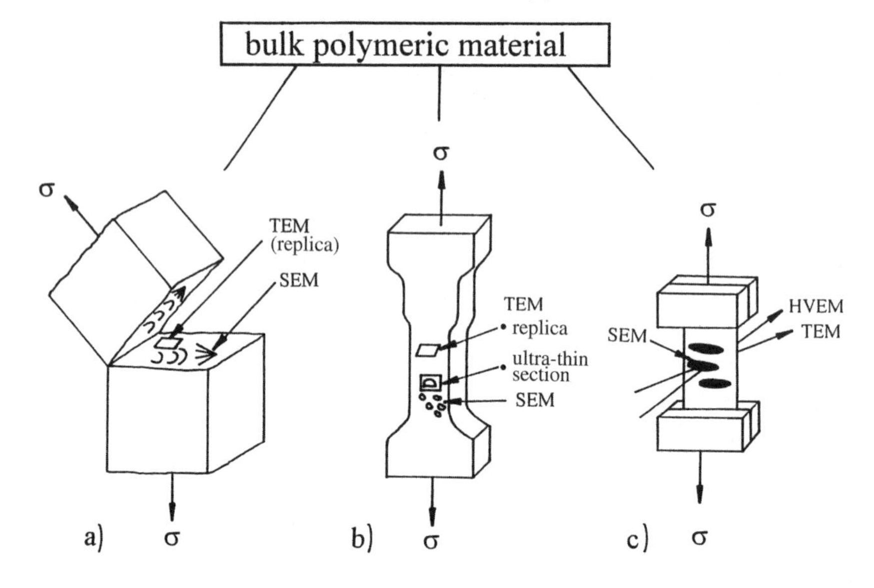

Figure 3 Survey of electron microscope methods for investigating micromechanical processes in polymers (a, b, c, see text).

c) Deformation of thin films or semithin sections in a tensile device and direct investigation of the deformed sample by SEM, HVEM, or TEM.

Changing from technique a) to techniques b) and c), reveals more and more details of the micromechanical processes in their dependence on real morphology [7]. The microfractographic analysis of fracture surfaces by SEM (microfractography, technique a) yields information about the processes of crack initiation and crack propagation up to final fracture. The influence of structural heterogeneities ('defects') on the initiation as well as on the propagation of cracks can be studied. Using SEM in technique b) visualizes several processes of crack initiation. Technique c) enables the investigation of both morphology and deformation processes at a high resolution. A particularly advantageous method is the investigation of semithin sections by HVEM [2, 7].

MORPHOLOGY OF PP

Investigation of surfaces by SEM

Sectioning of bulk samples and permanganic etching very easily reveals the spherulitic morphology of PP. The lower magnification of Figure 4(a)

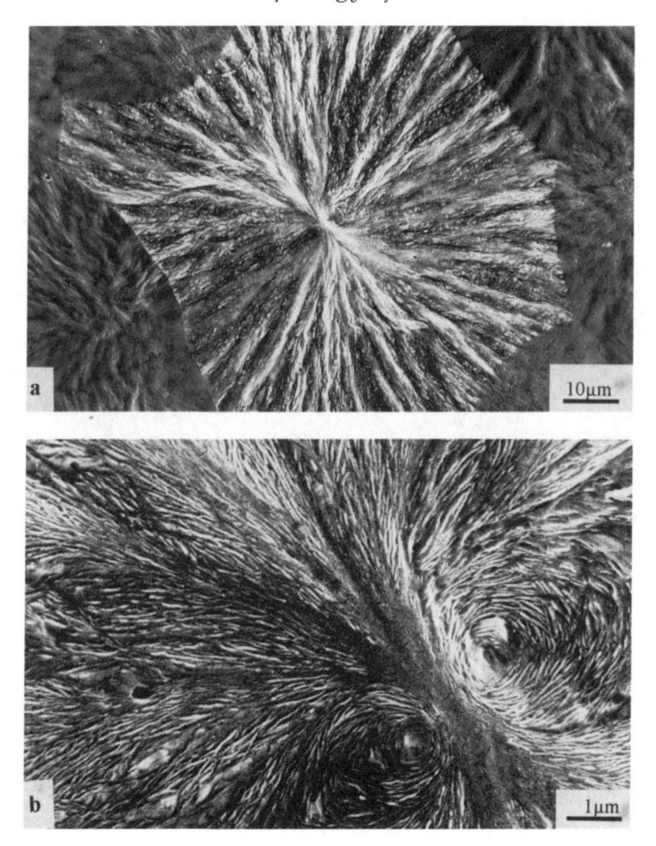

Figure 4 Sperulitic morphology of bulk PP on a microtomed surface after chemical etching, SEM micrographs: (a) lower magnification with a whole spherulite; (b) higher magnification of the central part of a spherulite.

shows a well-developed spherulite, limited by several adjacent ones. The diameter of these spherulites is in the range of some tenths up to 100 μm. From the centre of the spherulite lamellae are arranged in radial directions. The higher magnification in Figure 4(b) reveals the central part of a spherulite with the parallel arrangement of lamellae in the very centre and radial growth direction of lamellae. On both sides near the spherulitic centre, lamellae are bent and form the so-called 'eyes' of spherulites. As result of permanganic etching parts of amorphous material between lamellae are removed, revealing crystalline material at the surface. However, the bright lines with thicknesses of about 100 nm are not individual lamellae, but sticks of lamellae in parallel order. More information about the spherulitic structure of PP are to be found in the chapter Spherulitic crystallization and structure.

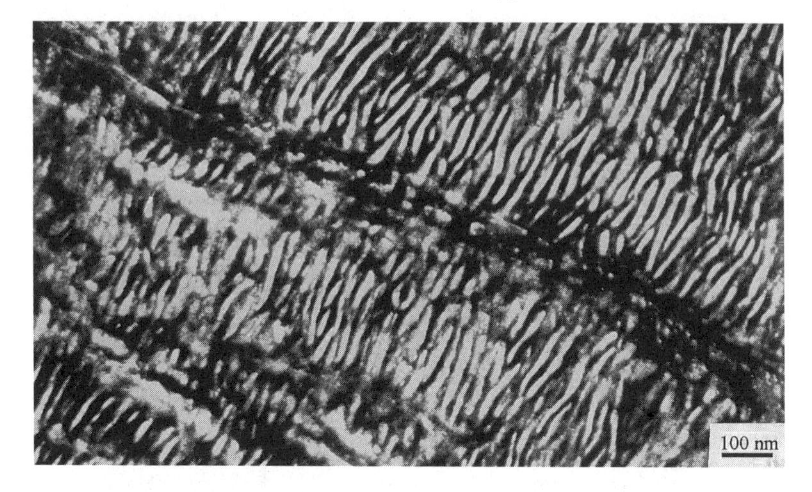

Figure 5 Shish-kebab morphology in an oriented PP foil, short pieces of lamellae are arranged in parallel order perpendicularly to a backbone; chemically stained ultrathin section, TEM micrograph.

Investigation of thin sections by TEM

Due to crystallization in an oriented melt with a pronounced molecular orientation the special morphology of 'shish-kebabs' may appear (Figure 5). Perpendicularly to a backbone structure short pieces of lamellae are arranged in parallel order like in a ladder structure. The molecular arrangement is in parallel direction to the backbone (the 'shish') and perpendicular to the short lamellae.

MICROMECHANICAL PROCESSES

Deformation of PP

At room temperature, PP is a tough material with expanded plastic deformations in form of shear yielding. These processes are connected with drastic rearrangements on a lamellar and spherulitic level. However, a brittle fracture can be initiated in materials with large spherulites and a reduced interspherulitic bonding [8]. Lower temperatures, high strain rates, and high stress concentrations at crack tips favour the formation of crazes (overview in [9]).

Toughened PP

It is of particular interest for many applications to produce PP with a better toughness at lower temperatures. To improve the toughness, various modifier particles with different physical properties can be

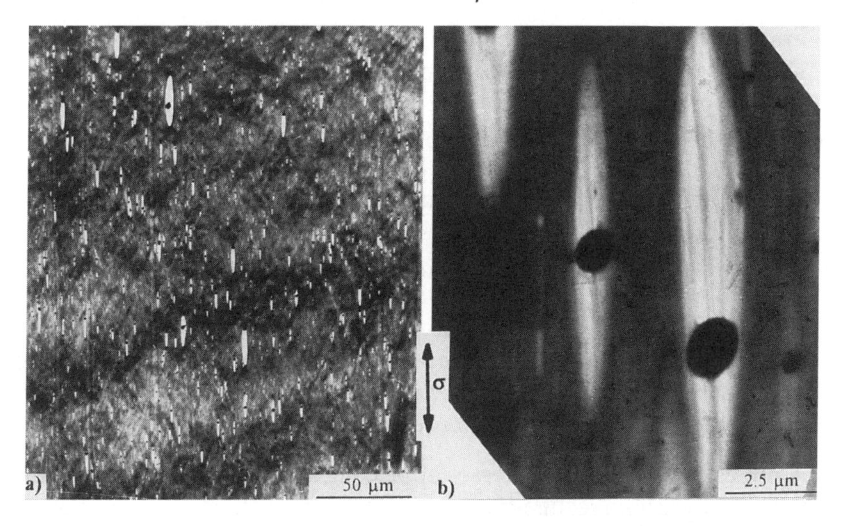

Figure 6 Initiation of high local plastic deformation by cavitated core-shell particles in modified PP; semithin section, deformed and investigated in a 1000 kV HVEM. (a), (b) lower and higher magnification.

added to the PP matrix. The function of the modifier particles is to act as stress concentrators and to initiate plastic deformation processes (shear yielding) of the matrix strands between the particles. An important precondition is void formation or cavitation within the particles or at the interface [10]. Up to now, there is some controversy regarding the detailed role of the size of modifier particles, the interparticle distance, and others.

Deformation structures of PP modified with 20 wt.% EPR-particles are shown in Figure 6. In the early stage of the deformation process, particles deform together with the matrix, and void formation (cavitation) appears suddenly in the deformed EP rubbery shell. Then the voids grow gradually with increasing strain of the material. Together with void formation and particle elongation, intense shear deformation appears between particles/voids, visible as diffuse, bright deformation bands in micrograph (a) in Figure 6. The higher magnification of micrograph (b) shows cavitated and strongly elongated shells of the core-shell particles together with highly deformed adjacent matrix strands (the elongation reaches up to 900%).

Results of investigating the micromechanical processes by EM are summarized in a three-stage mechanism, described in more detail in [10]:

1. *Stress concentration*. Under external stress, stress concentrations or stresses additionally increased by superposition of local stress fields

are built up between the modifier particles. In places of a maximum shear stress component, weak shear bands form between the particles at an angle of roughly 45° to the load direction.

2. *Void formation*. Owing to stress concentrations higher hydrostatic stresses are built up inside the particles, causing particles to crack and microvoids to form inside (cf. Figure 6), yielding a higher local stress concentration between the particles.
3. *Induced shear deformation*. The high local stress concentration initiates shear processes in the matrix bridges between the particles/voids. Shear deformation proceeds at numerous adjacent matrix bridges simultaneously, thus taking place in fairly large polymeric volumes (cf. Figure 6a). This is the deformation step during which energy is mainly absorbed (main contribution to toughness).

A direct simulation of processes appearing at low temperature loading is possible by EM investigations using cryotensile stages. An example of rubber-toughened PP (with 20% EPDM particles) is shown in Figure 7. Owing to deformation at −40°C the intense plastic shear yielding of the matrix strands is limited onto narrow deformation bands, aligned perpendicular to the loading direction. Besides the diffuse shear bands some fibrillated crazes are visible (see also the related chapter Crazing and shear yielding in polypropylene). The very bright areas are from cavitated particles.

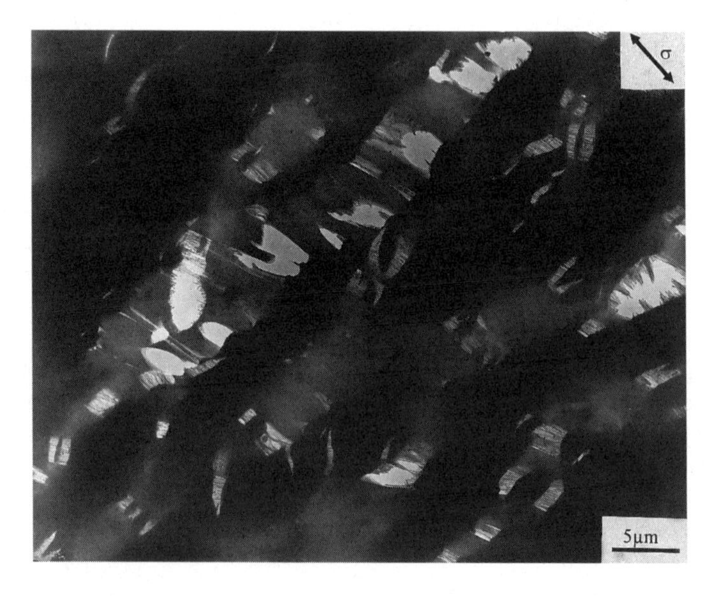

Figure 7 Deformation structures in rubber toughened PP at −40°C; ultrathin section, deformed and investigated in a 200 kV TEM.

Particle-filled PP

Filling of polymers with inorganic particles are often used to improve stiffness and form-stability of materials (see the related chapter Fillers for polypopylene). Beside the size and spatial distribution of particles, adhesion or interfacial strength between particles and PP-matrix is important for mechanical behavior.

It is interesting to note that there is a similarity of the deformation mechanisms in toughened PP and particle-filled PP. By the effect of locally increased deformability such particle-filled polymers show not only a better stiffness, but also a relatively good toughness.

REFERENCES

1. Glauert, A.M. (1973–1991) *Practical Methods in Electron Microscopy,* Vols. 1–13, Elsevier Science, Amsterdam.
2. Bethge, H. and Heydenreich, J. (eds) (1987) *Electron Microscopy in Solid State Physics,* Elsevier, Amsterdam.
3. Lednicky, F. and Michler, G.H. (1990) Soft matrix fracture surface as a means to reveal the morphology of multi-phase polymer systems, *J. Mat. Sci.,* **25**, 4549–4554.
4. Bassett, D.C. (1984) Electron microscopy and spherulitic organization in polymers, *CRC Crit. Rev. Solid State Mat. Sci.,* **12**, 97–163.
5. Spit, B.J. (1963) Gas discharge etching as a new approach in electron microscopy research into high polymers, *Polymer,* **4**, 109–117.
6. Reid, N. and Bessley, J.E. (1991) Sectioning and Cryosectioning for Electron Microscopy, in *Practical Methods in Electron Microscopy,* Vol. 13, Elsevier Science, Amsterdam.
7. Michler, G.H. (1992) *Kunstoff-Mikromechanik Morphologie, Deformations- und Bruch-mechanismen,* Carl Hanser Verlag, München.
8. Friedrich, K. (1978) Analysis of crack propagation in isotactic polypropylene with different morphology, *Progr. Colloid Polymer Sci.,* **64**, 103–112.
9. Jang, B.Z., Uhlmann, D.R. and van der Sande, J.B. (1985) Crazing in polypropylene, *Polymer Engng Sci.,* **25**, 98–104.
10. Michler, G.H. (1993) The role of interparticle distance in maximizing the toughness of high-impact thermoplastics, *Acta Polymerica,* **44**, 113–124.

Keywords: electron microscopy, crazing, shear yielding, transmission electron microscopy (TEM), resolution, surface imaging, scanning electron microscopy (SEM), contrasting methods, etching, failure mechanisms, morphology, lamellae, spherulites, deformation, toughening, cavitation.

Elongational viscosity and its meaning for the praxis

M.H. Wagner

INTRODUCTION

Since the invention of the rotary clamp rheometer (Figure 1(a)) by Meissner [1], the elongational viscosity of linear and branched polyethylene (PE) melts has been studied extensively. For polypropylene (PP), only a few studies have been published. The measurement of elongational viscosity at constant strain rate or constant tensile stress is of great importance for characterizing the structure of the polymer melt.

For processing behavior, however, the so-called Rheotens test, where an extruded filament is subjected to elongational deformation under the action of a tensile force (Figure 1(b)), is of more direct relevance, as it measures elongational behavior as influenced by a prototype industrial flow. In a Rheotens test, the draw ratio is increased until the filament ruptures, and usually the maximum drawdown force ('melt strength') and the maximum draw ratio ('extensibility') are reported. Of course, elongational viscosity and the force/draw-ratio relationship measured in a Rheotens test are interconnected; however, Rheotens behavior is influenced by the viscoelastic prehistory of the melt in the extrusion die, while elongational viscosity is measured starting from the stress-free, isotropic state. Again, while melt strength of PE has been studied extensively, and melt strength is as important for PP as for PE, only few studies on melt strength of PP have been published.

Processing operations where elongational viscosity and melt strength play an important role include fiber spinning, blow molding and foam

Polypropylene: An A–Z Reference
Edited by J. Karger-Kocsis
Published in 1999 by Kluwer Publishers, Dordrecht. ISBN 0 412 80200 7

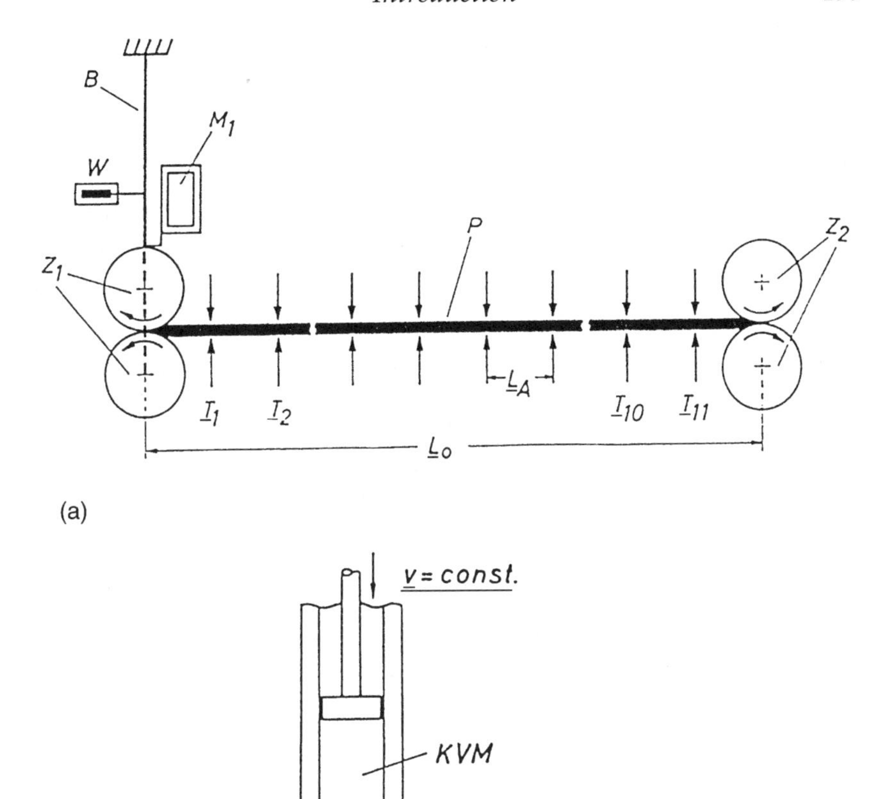

Figure 1 (a) Rotary clamp rheometer. (b) Rheotens test. For details see Meissner [1]. (Reprinted from *Rheol. Acta* (1971), by permission of Steinkopff Verlag.)

extrusion. In fiber spinning, an insufficient melt strength can lead to insufficient spinline stability and fiber breakages. In blow molding, a poor melt strength is related to sagging of the parison. In foam extrusion, a low melt strength results in cell wall rupture and a nonuniform cell structure.

ELONGATIONAL VISCOSITY

The relationship between elongational viscosity and molecular structure is summarized in Figure 2 [2]. The elongational viscosity of long-chain branched PE (LDPE IUPAC A) shows strong 'strain hardening', i.e. at Hencky strains $\varepsilon > 1$, the elongational viscosity increases dramatically above the linear-viscoelastic start-up curve. Fewer long-chain branches (LDPE III) lead to a smaller strain-hardening effect. For polymer melts with no long-chain branches (HDPE I, PS I), the elongational viscosity increases only slightly above the *V*-linear-viscoelastic limit; as the steady-state elongational viscosity (if ever attained in a constant strain-rate experiment) lies below the zero-deformation-rate viscosity plateau, elongational behavior of polymer *V*-linear polymer melts is sometimes described as 'strain thinning'.

In this sense, linear PP melts are extremely 'strain thinning' [3]: their elongational viscosity at large strain increases hardly above the linear-viscoelastic start-up curve (Figure 3). This behavior is only weakly influenced by the molecular weight distribution. By introducing long-chain branches into PP (through electron irradiation in the presence of polyfunctional monomers or through the use of metallocene catalysts), high melt strength PP can be produced with a strain-hardening elongational viscosity comparable to long-chain branched PE. For these materials, improved behavior in blow molding, extrusion coating and foam extrusion is reported [4, 5].

The meaning of 'strain hardening' and 'strain thinning' is more clearly seen, when the effects of the linear-viscoelastic spectrum of relaxation times and the nonlinear strain measure **Q** on the elongational viscosity are separated. In the tube model, the strain measure can be represented by the second rank orientation tensor (describing the orientation of tube segments) and a molecular stress function f [6],

$$\mathbf{Q} = f^2 \left\langle \frac{\mathbf{u}'\,\mathbf{u}'}{u'^2} \right\rangle$$

where $\mathbf{u}'$ describes a deformed unit vector $\mathbf{u}$, u' its length, and $\langle\ \rangle$ indicates an average over all orientations.

Figure 4 shows the square of the molecular stress function, f^2, as a function of the average deformation $\langle u' \rangle$ for linear and long-chain branched polymer melts and for crosslinked rubbers (NR, PDMS). It is clear from Figure 4 that the amount of molecular stress which can be induced in linear macromolecules by deformation in the melt state, is

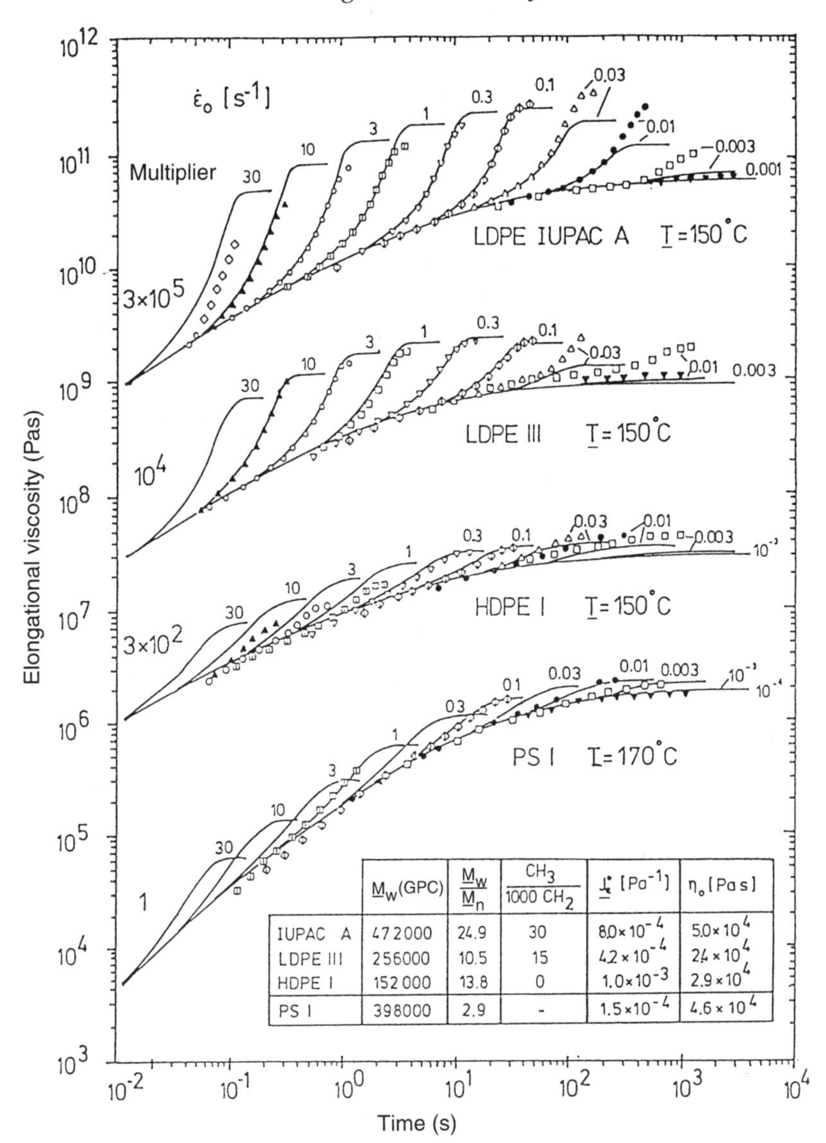

	M_w(GPC)	M_w/M_n	$CH_3/1000\ CH_2$	J_e^0 [Pa⁻¹]	η_0 [Pa s]
IUPAC A	472000	24.9	30	8.0×10^{-4}	5.0×10^{4}
LDPE III	256000	10.5	15	4.2×10^{-4}	2.4×10^{4}
HDPE I	152000	13.8	0	1.0×10^{-3}	2.9×10^{4}
PS I	398000	2.9	-	1.5×10^{-4}	4.6×10^{4}

Figure 2 Time-dependent elongational viscosity $\mu(t,\dot{\varepsilon}_0)$ measured at constant elongation rate $\dot{\varepsilon}_0$ for long-chain branched polyethylene (LDPE IUPAC A, LDPE III), linear polyethylene (HDPE I) and polystyrene (PS I). The number of long-chain branches is indicated by the number of CH_3 groups per 1000 CH_2 groups ($CH_3/1000CH_2$). (From Laun [2], by permission of Plenum Publishing.)

limited. This is especially so for linear PP melts, probably due to the high flexibility of PP. Long-chain branching increases the molecular stress

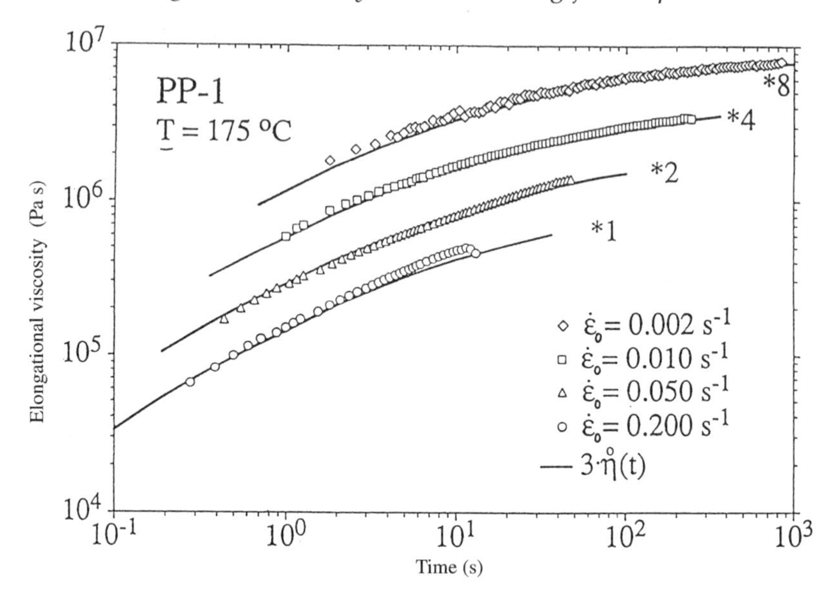

Figure 3 Time-dependent elongational viscosity $\mu(t,\dot{\varepsilon}_0)$ measured at constant elongation rate $\dot{\varepsilon}_0$ for a linear polypropylene (PP-1, $M_w = 9.1 \times 10^5$ g/mol, $M_w/M_n = 4.2$). The solid lines indicate the linear-viscoelastic start-up curve (From Hingmann and Marczinke [3], reprinted with permission of the American Institute of Physics).

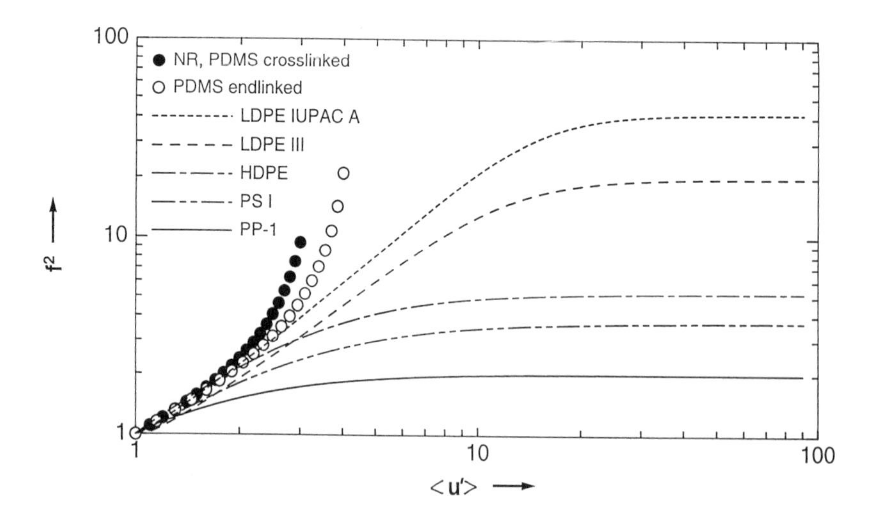

Figure 4 The square of the molecular stress function, f^2, as a function of average deformation <*u*'>.

plateau, which is reached at large deformations. Crosslinked systems (being solids) show finite extensibility, i.e. the molecular stress diverges at a finite extension.

MELT STRENGTH

The melt strength behavior of 18 (linear) reactor grade and controlled rheology PP melts was investigated by Ghijsels and De Clippeleir [7]. They found for all PP melts tested with a Rheotens apparatus a unique relationship between increasing zero-shear viscosity and increasing melt strength. This implies that the melt strength of linear PP is solely governed by the weight average molecular weight, independent of the width of the distribution.

For PP as for other polyolefins, melt strength decreases and melt extensibility increases with increasing temperature. Figure 5 shows Arrhenius plots for the temperature dependence of the melt strength for a number of PP grades. All grades follow a linear relationship at high

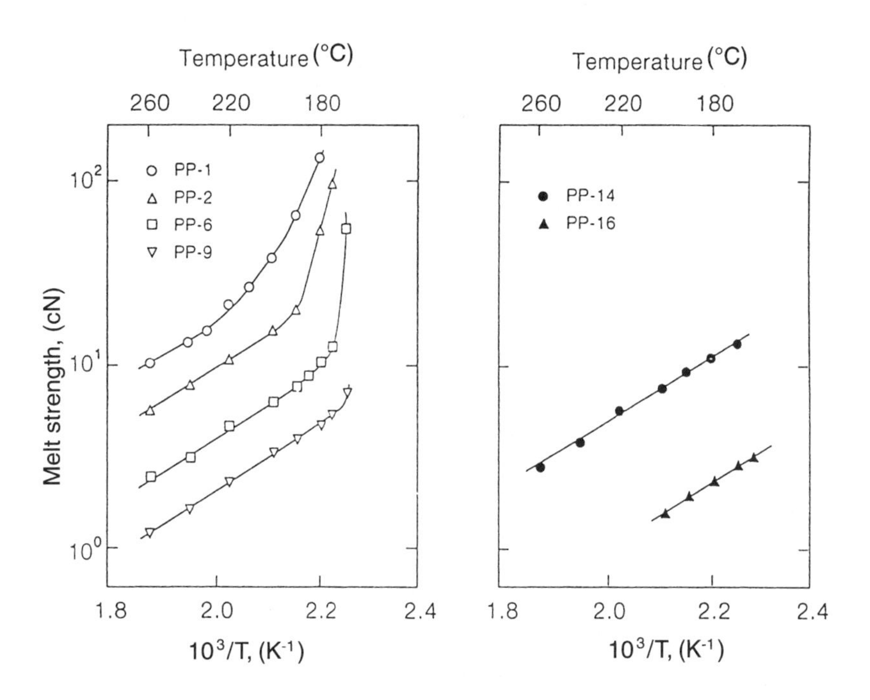

Figure 5 Effect of temperature on the melt strength of reactor (left) and controlled rheology (right) polypropylenes. M_w in 10^3 g/mol: 478 (PP-1), 440 (PP-2), 292 (PP-6), 233 (PP-9), 299 (PP-14), 188 (PP-16). (From Ghijsels and De Clippeleir [7], reprinted by permission of Carl Hanser Verlag.)

melt temperature, while large upward deviations from the linear relationship are observed at lower test temperatures for the high molecular weight grades. The onset temperature for this inflection point increases with increasing molecular weight. The non-Arrhenius increase in the melt strength is caused by flow-induced crystallization, which can already occur at the entry region upstream of the die or within the die itself.

So far, no melt strength data have been reported for branched PP melts. However, from experience with PE melts, it can be expected that melt strength is increased by long-chain branching [8].

MEANING FOR THE PRAXIS

Elongational viscosity of polymer melts can be categorized as 'strain thinning' or 'strain hardening', and is related to macromolecular structure and the amount of 'molecular stress' that can be induced into a polymer melt by deformation. Linear PP melts show 'strain thinning' (low molecular stress), while long-chain branching leads to high molecular stress and 'strain hardening'.

For linear PP, melt strength, as defined by maximum drawdown force in a tensile experiment, increases with increasing zero-shear viscosity and increasing weight average molecular weight. At lower melt temperatures, melt strength can be enhanced by flow induced crystallization. Long-chain branched PP is expected to have a superior melt strength relative to a linear PP of comparable zero-shear viscosity due to the 'strain-hardening' effect.

Elongational viscosity and melt strength play an important role in all processing operations where extensional deformations occur.

REFERENCES

1. Meissner, J. (1971) Dehnungsverhalten von Polyäthylen-Schmelzen, *Rheol. Acta*, **10**, 230–242.
2. Laun, H.M. (1980) Stresses and recoverable strains of stretched polymer melts and their prediction by means of a single integral constitutive equation, *Rheology*, Vol.2, (ed. G. Astarita *et al.*), Plenum, New York, pp. 419–425.
3. Hingmann, R. and Marczinke, B.L. (1994) Shear and elongational flow properties of polypropylene melts, *J. Rheol.*, **38**, 573–578.
4. Phillips, E.M., McHugh, K.E., Ogale, K. and Bradley, M.B. (1992) Polypropylen mit hoher Schmelzestabilität, *Kunststoffe*, **82**, 671–676.
5. Lesca, C. and Pohl, M. (1996) Extrudierter PP-Partikelschaum, *Kunststoffe*, **86**, 831–836.
6. Wagner, M.H. and Schaeffer, J. (1993) Rubbers and polymer melts: Universal aspects of nonlinear stress–strain relations, *J. Rheol.*, **37**, 643–661.
7. Ghijsels, A. and De Clippeleir, J. (1994) Melt strength behavior of polypropylenes, *Int. Polymer Process.*, **9**, 252–257.

8. Wagner, M.H., Schulze, V. and Göttfert, A. (1996) Rheotens-mastercurves and drawability of polymer melts, *Polymer Engng Sci.*, **36**, 925–935.

Keywords: elongational viscosity, melt strength, extensibility, strain hardening, strain thinning, molecular stress function, flow induced crystallization, fiber spinning, blow molding, foam extrusion.

Environmental stress cracking of polypropylene

R. Chatten and D. Vesely

Environmental stress cracking (ESC) is the premature cracking of a polymer due to the combined action of a stress and a fluid. It is associated with the phenomenon of crazing and solvent plasticization of the polymer. The embrittlement by oxidative or other chemical degradation is not included under ESC, but is classed as corrosion stress cracking (CSC).

It has been observed that fracture of material, which is accelerated by the environment, requires crack initiation in the crazed region and subsequent fast crack propagation, and therefore a brittle failure mode. The transition from a ductile to a brittle failure mode is accelerated by aging, temperature, tensile stress, cyclic loading, stress concentrations and contact with aggressive fluids. In isotactic polypropylene (iPP), the transition of failure mode is affected by morphology through the change of fracture toughness. Increasing crystallinity decreases the fracture toughness, whilst spherulite size has been found to have little or no effect. Crystallinity hinders craze formation, and in the presence of a solvent, voiding and solvent-induced crystallization are competitive. Stress-whitening is associated with the formation of craze-like structures [1].

Data collected from commercial extrusion grades of iPP shows that at ambient temperature the principal plastic deformation is crazing [2]. The plastic deformation mode is, however, dependent upon temperature. A transition is observed whereby the mode of plastic deformation changes

Polypropylene: An A–Z Reference
Edited by J. Karger-Kocsis
Published in 1999 by Kluwer Publishers, Dordrecht. ISBN 0 412 80200 7

from shear yielding at high temperatures to crazing at lower temperatures. The temperature range over which this transition occurs is dependent upon grade and thermal history. Crazing has also been observed for blends of iPP and ethylene-propylene block copolymer where microcrazes of different structure to those seen in the pure homopolymer are observed. Studies of glass-fiber reinforced iPP show that the fibers induce crazing within the matrix. Standard adhesion fibers cause craze formation at low stress levels whereas higher adhesion fibers elevate the process to higher stress levels.

YIELDING

Crazing can be considered as a type of inhomogenous localized yielding. A major factor affecting ESC is yield strength σ_y, and this can be expressed in terms of thermally activated shear flow, as described by Eyring's equation:

$$\frac{\sigma_y}{T} = \frac{E}{VT} + \left(\frac{k}{V}\right)\frac{\ln\dot{\varepsilon}}{A}$$

where E is the activation energy, V is activation volume, k is the Boltzmann constant, A is the materials constant and $\dot{\varepsilon}$ is the deformation rate.

The localization of yield can be explained in terms of density fluctuations. On cooling from the melt, low density regions are frozen in, the volume of these regions increasing with cooling rate. Also, in semicrystalline polymers, such as iPP, the relative amounts of different crystal phases and the stresses between phases need to be considered when calculating bulk properties.

From the thermal activation and density fluctuation treatments, it can be seen that yield stress increases with decreasing temperature, increasing strain rate, increasing hydrostatic pressure and annealing below T_g. An accurate predictive model for yielding requires an integrated approach using both flow and density considerations.

CRAZING

Crazing is the mechanical separation of entangled groups of polymer chains under a tensile stress. Crazes are well-defined regions filled with crazed matter. The principal features of a craze are illustrated in Figure 1. The crazed material consists of primary fibrils oriented to the tensile stress, and connected by secondary fibrils interspersed with voids. The

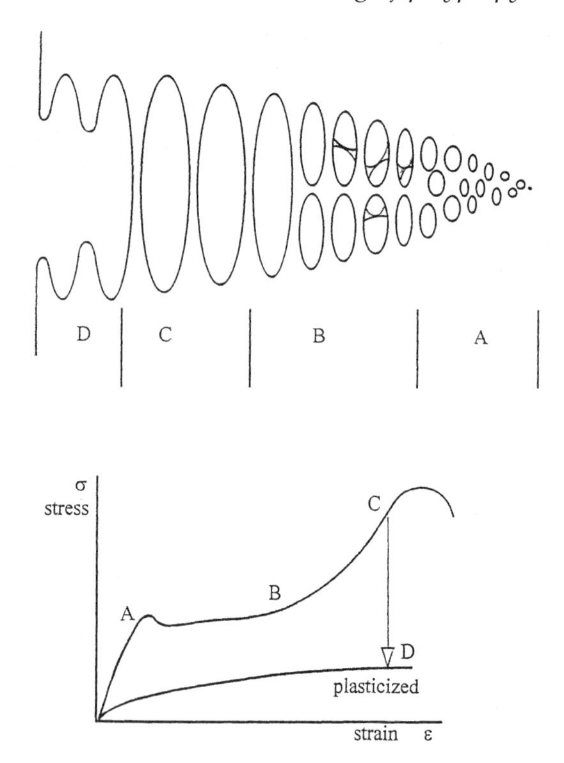

Figure 1 Schematic two-dimensional drawing of major features of a craze: A, microvoiding; B, formation of fibrils; C, extension of fibrils; D, fracture of fibrils. The stress–strain curves indicate the load-bearing capability of the material at different stages of craze formation and the effect of plasticization.

polymer volume fraction in a craze depends upon time and level of stress, temperature and degree of plasticization. An increase in any of these variables causes increased fibril rupture, resulting in coarser structures which are of lower polymer content and therefore weaker. These regions, typically of density 40–60 vol.% of that of the matrix, but as low as 10 vol.%, distinguishes this phenomenon from cracking. Crazes can sustain tensile stresses at near bulk-yield-stress values although the associated strains are much greater. This load-bearing quality gives rise to very large deformation energies for highly crazed systems [3]. Typical crazing behavior of iPP is shown in Figure 2.

Crazing does not occur where material orientation coincides with the draw direction, and the tensile behavior of crazes is an extension of the formation process. The crazing of a given material is dependent on the dynamics of the loading process: At low strain-rates, the craze can

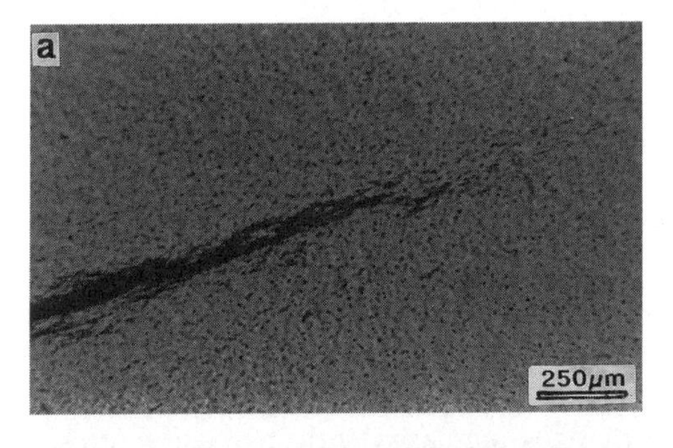

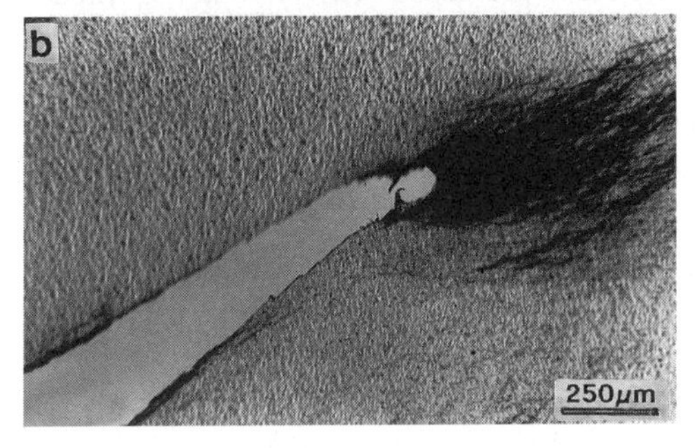

Figure 2 (a) Light micrograph of a crazed region formed in a thin film (~ 25 μm thick) of iPP in tension. The individual crazes are small but merging into a large craze. (b) The same sample area after a crack has propagated. Broken fibrils show the localization of the crazed region. In the middle part of the crack, the absence of the fibrils indicate that the crack propagated in a brittle manner, before being stopped by further crazing which has blunted the crack tip.

grow laterally by both fibrillating the surrounding material and elongating the existing fibrils, whereas at higher rates of loading the craze does not draw new material into itself. An extension of this behavior is that as the craze grows, near-tip fibrils are thicker than those more orientated fibrils, which are found near the craze-initiation site. Systems that craze exhibit an incubation time, which is a delay between loading and the appearance of crazes. An increase in stress causes a decrease in this incubation time.

Both dry crazes and those formed in weakly plasticizing fluids exhibit a tip region, where the stress is at a maximum, and a body where the stress is lower and constant, and the craze–bulk interfaces are parallel. The length of the tip region is dependent upon the stress distribution, which is in turn dependent upon plasticization. A more highly plasticizing fluid induces higher stress concentrations at the tip, reducing strength.

CRAZE INITIATION

There is no firm agreement as to the mechanism of craze initiation but it is known that both initiation and propagation are stress-biased, thermally activated processes, dependent upon the material properties.

In general the following rules can be applied to the craze initiation process:

1. The incubation time decreases with increasing stress.
2. The strain at the onset of crazing decreases with decreasing stress level.
3. The density of crazes increases with increasing stress.
4. The crazing stress at a given time decreases with increasing temperature.
5. Craze initiation is accelerated by dilational stress.

Once initiated, crazes can grow in all three dimensions. Generally a system requiring high strain for craze initiation results in smaller crazes. Small crazes are also connected with systems of molecular orientation normal to the applied stress. The effect of molecular orientation on crazing can be explained in terms of the number of chain segments aligned in the direction of stress. A greater density of aligned chains increases strain hardening which is a key factor in stabilizing craze microstructure. The strain-hardening rate can be related to the amount of chain entanglements. Systems which can strain harden to high levels give high craze strength.

Initiation is thought to be a three-stage process. In the first stage, decohesion of molecular coils leads to the development of regions of microporosity. In the second stage, these agglomerate into stable voids. The third stage is an extension process whereby the voided areas form the characteristic craze geometry.

CRAZE GROWTH

Taking into consideration the formation of many crazes simultaneously or sequentially, a particular craze grows either by displacement of the craze tip into uncavitated material by formation of voids ahead of the craze, or by widening of the craze whilst extending molecular strands, which may rupture or disentangle giving a crack.

The propagation of crazes can be split into three considerations: kinetics, interfacial stress and breakdown. Growth is explained in terms of the Taylor meniscus instability model, where the polymer at the growing craze tip becomes less viscous due to the action of stress. The velocity of the craze tip through the material can then be calculated from material properties, the state of the stress–strain field and environmental variables since they affect the viscosity boundary at the tip. Variants of this model explain the fibrillation of the matrix material at the craze–bulk interface.

In a growing craze, molecular strands are formed and extended. Growth is slowed or halted if the craze hits an obstacle preventing fibrillation, if the craze has grown so that the initiating flaw becomes uncritical, or if the craze interacts with another craze or the environment causing growth conditions to change. Whilst subjected to a tensile load the molecular strands in a craze degrade and eventually fail, causing transition from a craze to a crack by either slippage and disentanglement of molecular coils leading to fibril rupture, breakage of entangled chains and fibril rupture, or fibril–matrix separation.

CRAZE FRACTURE

Crazes act as initiation sites for cracks. The resistance to craze breakdown is dependent upon time, stress, temperature, molecular weight, chain structure, environment and thermal history. The local temperature rise caused by opening and breakdown can be large enough to both melt and thermally degrade a polymer. Cross-tie fibrils which exist in iPP crazes are able to transfer stress between broken and unbroken primary fibrils, and it has been observed that craze fracture could be accelerated by the failure of these secondary fibrils. An increase in tacticity or number average molecular weight has the effect of increasing fibril strength thereby increasing the stress to craze fracture.

Craze morphology and breakdown are highly dependent upon molecular weight. Craze width and length increase with molecular weight up to a limiting weight where dimensions remain stable. The width of a craze can far exceed the length of the constituent coils and therefore each fibril consists of many thousands of entangled, elongated coils. At low molecular weights, the interpenetration and entanglement of coils is so small that fibril formation is extremely difficult, whereas at the critical molecular weight, where craze dimensions become stable, the coils are larger and the absolute number of entanglements per coil increases, giving stronger fibrils that are more resistant to slippage and disentanglement. At higher molecular weights, the fibril strength becomes determined by the material strength and not fibril failure. The kinetics of fibril failure and hence craze breakdown can therefore be modelled by Eyring's equation for yielding.

Analysis of extruded and injection molded iPP shows that a skin-core structure is formed due to high shear and thermal gradients upon cooling. These frozen-in tensile stresses in the skin can act as the stress component in the environmental stress cracking process [4]. In injection moldings, molecular weight has a secondary effect. This effect is that residual orientation in the direction of flow increases with higher molecular weight, leading to higher growth resistance when flow direction and stress direction coincide. Oscillating loads can enhance the disentanglement process so that craze dimensions and critical weights change with test frequency.

THE EFFECT OF THE ENVIRONMENT

The reason why certain fluids act as stress cracking agents is not fully understood. Whilst any fluid that is significantly absorbed by a polymer is a potential ESC agent for that polymer, high solubility is not a prerequisite since some systems can give very low absorption but greatly accelerated cracking. The environmental stress cracking resistance of a polymer/fluid pair can often be evaluated by microhardness or creep testing.

The effect of the environment within ESC is best understood in terms of how it modifies the material and failure processes. Aggressive fluids may be corrosion stress cracking agents that chemically attack and degrade a polymer (e.g. nitric acid for iPP), or environmental stress cracking agents that do not chemically modify or degrade the surface but are absorbed into the bulk of the polymer. Craze initiation stress decreases with decreasing solvent chain length showing that a slight swelling of a microscopic surface area is both necessary and sufficient to initiate a craze.

Liquid uptake may have several consequences that affect ESC. Swelling causes the formation of compressive stresses at the surface, whilst plasticization increases the ease of void formation within a pre-immersed layer. Craze initiation, propagation and breakdown criteria are all affected as a consequence.

Solvent crazing has been studied extensively with respect to the solubility parameters of both solvent and polymer [5]. In increasing chain mobility, the first and second stages of craze initiation are aided, although the formation of craze precursors is independent of the environment. Solvent-induced crazes grow more quickly and to greater dimensions than those in inert environments. The limiting speed for solvent-induced crazes is thought to be the penetration velocity of the fluid, and since the plasticized bulk is more easily fibrilated, resultant fibrils are softer and weaker.

The surface absorption of a fluid (and subsequent plasticization of the polymer) is generally held to be the primary influencing factor as far as the effect of the fluid is concerned. The thermodynamics of the mixing process have been studied extensively in attempts to construct a material property-based predictive model. These attempts have not as yet been successful but some important relationships have been established (Hildebrand, Flory and Huggins etc.). A more general and accurate approximation might be found by considering Hansen partial solubility parameters, where the cohesive energies for the three van der Waals forces are treated separately.

Attempts have been made to explain how the severity of an ESC agent can be predicted from the solubility parameters. If the solubility parameters of the polymer and fluid are very different then negligible interaction will occur. If the parameters are equal, the absorption will be complicated by dissociative reactions and the fluid will not plasticize the polymer strongly and may well act as a cracking agent. Considering these two extremes, it can be seen that a fluid is potentially most likely to be an ESC agent when the solubility parameters are close but not equal.

Examples of solubility parameters ($J^{1/2}\,cm^{-3/2}$) are:

Polypropylene	18.8–19.2	Dodecanol	20.1
Acetone	20.3	Naphthalene	20.3
Benzene	18.8	Propylene oxide	18.8
Carbon tetrachloride	17.6	Toluene	18.2
Chloroform	19.0	Water	47.9
Cyclohexane	16.8	Xylene	18.0

The previous treatment shows whether a particular system will mix or not, but in order to predict the equilibrium solubility, the thermodynamic potential must be calculated. The equilibrium solubility is at the point of minimum thermodynamic potential, and this point can be found by using treatments such as that of Flory and Huggins. The temperature effect on solubility is critical, as most solvents will not solvate iPP below 80°C.

TESTING STANDARDS

Several standards have been devised in order to evaluate specific ESC systems. Since this phenomenon has not previously been reported in polypropylene, the standards pertaining to polyethylene may be used with suitable adjustments in parameters such as load and temperature.

Generic standards include:

- ISO 4599: 1986 Determination of Resistance to Environmental Stress Cracking – Bent Strip Method (see also DIN 53499 Part 3).
- ISO 4600: 1981 Determination of Environmental Stress Cracking – Ball or Pin Impression (see also DIN 53499 Part 1).
- ISO 6252: 1981 Determination of Environmental Stress Cracking – Constant Tensile Stress Method

Polyethylene specific standards include:

ASTM D 1693: Standard Test Method for Environmental Stress Cracking of Ethylene Plastics.
ASTM D 2561: Standard Test Method for Environmental Stress Cracking Resistance of Blow Molded Polyethylene Containers.
ASTM F 1248: Environmental Stress Cracking Resistance of Polyethylene

CONCLUSION

iPP, as other polymers, can fail by environmental stress cracking. In order to minimize this effect, the specific polymer/solvent pair needs to be studied in terms of yield strength, failure mode transition and craze initiation, propagation and breakdown. Generally, improved barrier properties will give improved ESC resistance, increasing the lifetime of components. It should be noted, however, that iPP has excellent resistance to ESC and that failure of this type is much less frequent than in case of glassy polymers.

REFERENCES

1. Liu, Y. and Truss, R. (1994) A study of yielding of isotactic polypropylene, *J. Polymer Sci., Part B – Polymer Physics*, **32**, 2037–2047.
2. Frontini, P. and Fave, A. (1995) The effect of annealing temperature on the fracture performance of isotactic polypropylene, *J. Mat. Sci.*, **30**, 2446–2454.
3. Kambour, R. (1987) in *Polymers – An Encyclopaedic Sourcebook of Engineering Properties*, (ed. J. Kroschwitz), Wiley, New York, pp. 152–176.
4. Karger-Kocsis, J. and Friedrich, K. (1989) Effect of skin–core morphology on fatigue crack-propagation in injection molded polypropylene homopolymer, *Int. J. Fatigue*, **11**, 161–168.
5. Andrews, E. and Bevan, L. (1972) Environmental crazing in a polymeric glass, *Polymer*, **13**, 337–346.

Keywords: crazing, craze initiation, craze growth, yielding, craze fracture, morphology, disentaglement, Eyring equation, swelling, plasticization, solvent-induced crazing, environmental stress cracking (ESC), corrosion stress cracking (CSC), solubility parameter.

Epitaxial crystallization of isotactic and syndiotactic polypropylene

Bernard Lotz and Jean Claude Wittmann

Epitaxy is defined as the growth of one crystalline phase (the guest crystal, here the polymer) on the surface of a crystal of another phase (the host crystal, here the substrate) in one or more strictly defined crystallographic orientations. The interactions imply a structural analogy between the two species in their contact planes, either in two, or sometimes only one direction. The details of the molecular interactions may be difficult to reach, and, as a consequence, the epitaxy is often defined in terms of a geometrical concordance of matching unit-cell dimensions: the disregistries should remain $< 10\%$.

Polymer/substrate epitaxial interactions are of interest in order [1]:

1. to produce well-oriented polymer thin films well adapted for morphology and structural analyses by electron microscopy and electron diffraction;
2. to produce well-defined exposed crystalline surfaces highly appropriate for combined electron diffraction-atomic force microscopy (AFM) molecular modelling studies;
3. to understand the mechanisms of action and the efficiency of nucleating agents for polymers, and therefore establish guidelines for the search and design of new, or more effective, nucleating additives.

A major problem in the processing of crystalline polymers is indeed to achieve high concentration of crystalline nuclei which initiate crystal

Polypropylene: An A–Z Reference
Edited by J. Karger-Kocsis
Published in 1999 by Kluwer Publishers, Dordrecht. ISBN 0 412 80200 7

growth, and thus to reduce molding cycles while improving properties (mechanical, optical). The number of 'spontaneous' nuclei in isotactic and syndiotactic polypropylenes (iPP and sPP) is, under normal conditions, in the range of $10^6/cm^3$, giving rise to spherulites of $\approx 100\,\mu m$ diameter. Processing constraints would make it desirable to increase this number, sometimes up to $> 10^{12}/cm^3$ (spherulite size $< \approx 1\,\mu m$), i.e. a size for which spherulites barely scatter light, and samples become less hazy. To this end, nucleating agents are added to the polymer. As a rule, they are minerals or low molecular weight organic materials, sometimes other polymers, which are crystalline at the crystallization temperature of the polymer; they act as a *crystalline substrate* for the crystallization of the polymer. Their overall activity depends on their degree of dispersion and the 'quality' of the interactions between the substrate and the polymer.

Although a number of nucleating agents are used for 'simple' polymers, such as polyethylene (PE), and for iPP and sPP, the real nature of the interactions has been worked out only in the last ten years or so. These interactions are essentially of *epitaxial character*, i.e. do not involve chemical reactions. For polymers with a 'linear' envelope (such as PE), the epitaxial relationship involves, as a rule, the chains lying 'flat' on the substrate (with the helix axis parallel to the substrate). The major dimensional match involves the interchain distance in the contact plane; the latter may differ for different substrates. As a consequence, the polymer lamellae stand 'edge on' on the substrate [1].

EPITAXY OF ISOTACTIC POLYPROPYLENE

The structural correspondence between iPP (in its stable α phase) and the substrate is more complex than for linear chains. Indeed, iPP crystallizes in a three-fold helix in which the helical *path* rather than the helix axis (which is not materialized) interacts with the substrate when the chain lies on the substrate. Since this helical path is at a significant angle to the helix axis, and has opposite tilts for right- and left-handed helices, two orientations of the chains (and therefore the lamellae) are generated [1]; these are illustrated in Figure 1. They are very reminiscent of the so-called 'cross-hatched' morphology of iPP in its α phase [2]. Technically, the *ac* face of the iPP unit-cell is in contact with the substrate (in this case, benzoic acid, used as a model for the well-known nucleating additive sodium benzoate). The major dimensional match involves the αiPP interstrand distance (distance between two successive helical turns) of $\approx 0.5\,nm$, and a substrate near 0.5 nm periodicity. Note that this $\approx 0.5\,nm$ distance corresponds also to a standard interchain distance in PE. This feature explains that sodium benzoate is a nucleating agent for both αiPP

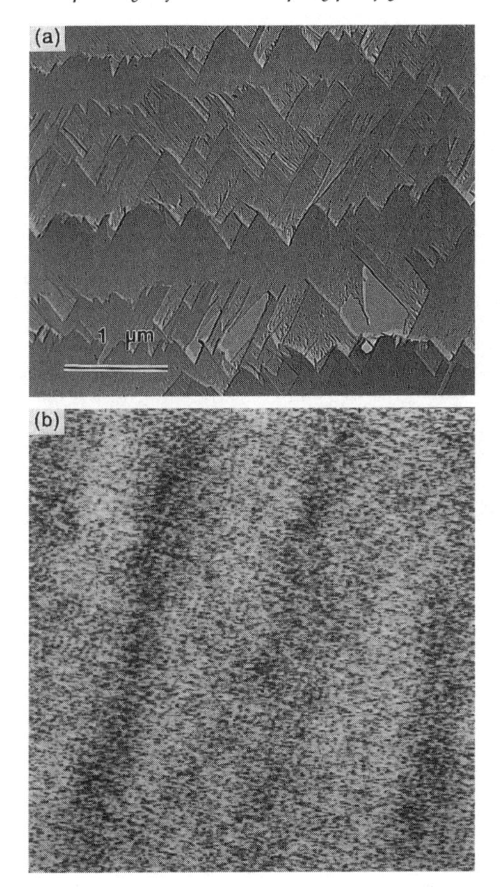

Figure 1 (a) Lamellar morphology of epitaxially crystallized isotactic propylene in the α phase onto benzoic acid. The latter substrate was located on top of the thin film of iPP. After cocrystallization of the polymer and substrate, the latter was dissolved away, revealing the polymer morphology exposed. Note the two orientations of the lamellae, due to the different orientations of helical paths in right- and left-handed helices (see text). Electron micrograph, platinum–carbon shadowing. (Reproduced from [3a] with permission.) (b) AFM image of the contact face in (a). The three dark stripes oriented at one o'clock are lamellar edges. The smaller scale, nearly square pattern is due to the organization of methyl groups, 0.65 nm apart in the exposed (010) contact face of αiPP. (Reproduced from [3b] with permission.)

and PE, and further that, despite different chain conformations and cell geometries, αiPP and PE nucleate each other's crystallization [1].

The structural correspondance between polymer and substrate could be analyzed one step further for the epitaxy illustrated in Figure 1a.

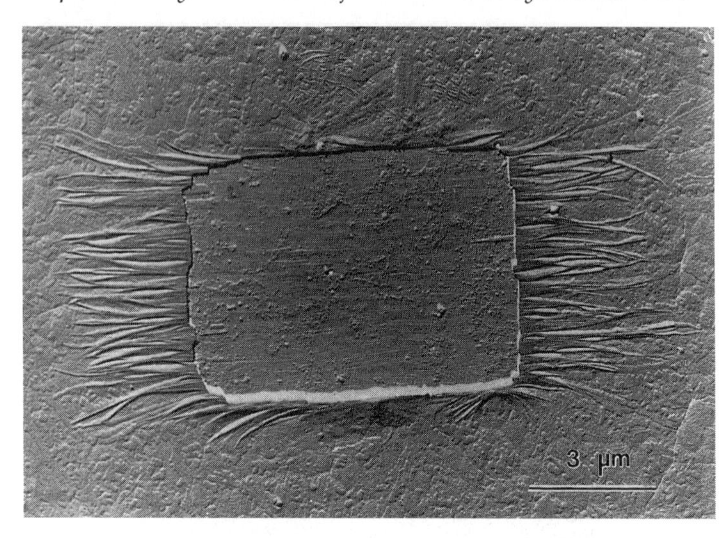

Figure 2 Lamellar morphology of epitaxially crystallized iPP in the β phase onto a crystal of dicyclohexylterephtalamide. As in Figure 1(a), the rectangular substrate crystal has been dissolved away, leaving the bulk, which is the basis of the nucleation efficiency of this additive. Electron micrograph, platinum–carbon shadowing. (Reproduced from [6], with permission.)

Indeed, whereas diffraction evidence indicates that the *ac* plane contacts the substrate, there exist two such planes in αiPP which differ by the density of the exposed methyl groups. Discriminating between these two planes becomes possible when resorting to atomic force microscopy; as shown in Figure 1b, AFM indicates that the plane with the lowest density of methyl groups is involved in the epitaxy [3, 4].

The above analysis illustrates the most common epitaxial relationship which generates αiPP. At least one other epitaxy has been documented, which involves the αiPP (110) contact face, but the molecular analysis has not yet been performed in similar detail.

iPP exists also in a metastable modification, the β phase, which melts at ≈ 150–155°C, has a faster growth rate than αiPP (between 141°C and 100°C), but nucleates much less profusely; iPP normally crystallizes in the α phase. Generation of the β phase rests therefore critically on the development of appropriate, β phase-specific nucleating agents. Early work has indicated that γ-chinacridone is such a nucleating agent [5]. Recent patents have disclosed a number of other agents. The structural relationships between these substrates and βiPP have been elucidated recently [6, 7]. βiPP interacts with the substrate with its dense (110) plane of its trigonal unit cell. The epitaxy seems to rest on a matching of a ≈ 0.65 nm periodicity in the substrate with the chain axis repeat distance of the β phase. Further, the orthogonal geometry of the two unit cells in contact seems an important feature. Again, AFM has been used to visualize the structure of the βiPP contact face. Interestingly, the images reveal [6, 7] the unorthodox character of the βiPP crystal structure, which can be described as a frustration [4]; two helices in the contact plane have different setting angles, and therefore different crystallographic environments in the unit cell.

EPITAXY OF SYNDIOTACTIC POLYPROPYLENE

The development of metallocene catalysts has made it possible to synthesize syndiotactic polypropylene (sPP). sPP has nearly the same melting temperature as αiPP, but its industrial development seems hindered by its lower overall crystallization rate. As a consequence, the search for appropriate nucleating agents may be essential. sPP has been shown to interact with low molecular weight organics and polymers, and notably αiPP.

The most thorough analysis of sPP epitaxy was conducted on *p*-terphenyl as a substrate [6, 8]. The end surface of *p*-terphenyl provides a quasi-perfect two-dimensional match with the *bc* face of the sPP unit-cell. Further, the helix conformation of sPP is very similar to a rectangular staircase, with three 'steps' made of a CH_3, a backbone CH_2 and a CH_3 exposed in that *bc* face. AFM with methyl group resolution makes it possible to visualize these 'steps' in the contact plane. The resulting pictures [8] demonstrate that in this plane, left- and right-handed helices alternate regularly, in agreement with a recent reinterpretation of the crystal structure of sPP (see chapter on 'Morphology and nanostructure of polypropylenes by atomic force microscopy' in this book). They are also the first instance, in polymer science, where a definite helical hand could be assigned to every helical stem embedded in its crystallographic environment [6, 8].

sPP is often blended with iPP which acts as a nucleating agent. The epitaxial relationship between the two PPs could be worked out [9] by using appropriately oriented sPP samples first obtained by epitaxial crystallization. The interacting crystallographic planes are those already considered in the other epitaxies: *ac* for iPP (in its α modification); *bc* for sPP. The epitaxial relationship rests on the parallelism of the sPP 'flight of stairs' (CH_3, CH_2, CH_3) and the αiPP rows of CH_3. Since these flights and rows are both at significant angles to the respective helix axes, the epitaxial relationships involve complicated (at first sight) orientations of the interacting iPP and sPP.

To summarize, the epitaxial crystallization of iPP and sPP onto crystals of small organic molecules as well as onto each other rests on a lattice and structural match of submolecular features (arrays of methyl groups, rows of CH_3, CH_2, CH_3) in the contact planes. These interactions are now understood in great detail, thanks in part to the development of atomic force microscopy, which is able to probe the surface of the PPs with methyl group resolution. These structural analyses of epitaxial interactions are a necessary step in the understanding of the efficiency of nucleating agents, and a prerequisite for the design of new, more effective ones.

REFERENCES

1. Wittmann, J.C. and Lotz, B. (1990) Epitaxial crystallization of polymers on organic and polymeric substrates. *Prog. Polymer Sci.*, **15**, 909–48.
2. Lotz, B. and Wittmann, J.C. (1986) The molecular origin of lamellar branching in the α (monoclinic) form of isotactic polypropylene. *J. Polymer Sci., Part B, Polymer Phys.*, **24**, 1541–58.
3. (a) Stocker, W., Magonov, S.N., Cantow, H.J. *et al.* (1993) Contact faces of epitaxially crystallized α and γ phase isotactic polypropylene observed by atomic force microscopy. *Macromolecules*, **26**, 5915–23. Correction: (1994) **27**, 6690–94 (b) Stocker, W., Graff, S., Lang, J. *et al.* (1994) Contact faces of epitaxially crystallized α phase isotactic polypropylene: AFM imaging with a 'liquid cell'. *Macromolecules*, **27**, 6677–78.
4. Lotz, B., Wittmann, J.C. and Lovinger, A.J. (1996) Structure and morphology of poly(propylenes): a molecular analysis. *Polymer*, **37**, 4979–92.
5. Leugering, H.J. (1967) Einfluss der Kristallstruktur und der Überstruktur auf einige Eigenschaften von Polypropylen. *Makromol. Chem.*, **109**, 204–16.
6. Stocker, W., Lovinger, A.J., Schumacher, M. *et al.* Atomic Force Investigations of epitaxially crystallized tactic poly(propylenes). *ACS Symposium Series*, in press.
7. Stocker, W., Schumacher, M., Graff, S. *et al.* (1998) Epitaxial crystallization and AFM investigation of a frustrated polymer structure: isotactic poly(propylene), β phase. *Macromolecules*, **31**, 807–14.
8. Stocker, W., Schumacher, M., Graff, S. *et al.* (1994) Direct observation of right and left helical hands of syndiotactic polypropylene by atomic force microscopy. *Macromolecules*, **27**, 6948–55.
9. Lovinger, A.J., Davis, D.D. and Lotz, B. (1991) Temperature dependence of structure and morphology of syndiotactic polypropylene and epitaxial relationship with isotactic polypropylene. *Macromolecules*, **24**, 552–60.

Keywords: epitaxy, isotactic polypropylene, syndiotactic polypropylene, crystal structure, crystal polymorphism, helical hand, nucleating agents, AFM, α-phase, β-phase, lamellae.

Extrusion die design guidelines for polypropylene

J. Ulcej

PRODUCTS/PRODUCTION/TYPES OF DIES

The first consideration in the design of a flat extrusion die system to convert polypropylene (PP) feed stocks is to determine the design and processing parameters for the final extruded product(s). The basic parameters that have to be defined are the final product width if the final product is a monolayer or coextruded structure, the thickness range, the polymer(s) rheology, the processing temperature range and the desired through-put range.

To produce monolayer film, sheet or coating, it is most common to use a single manifold die design. However, if a multilayered coextruded structure is required, a single manifold die design used in conjunction with a coextrusion feedblock, or a multimanifold die, which may also be used in conjunction with a coextrusion feedblock, may be employed. In coextrusion applications, the decision to use a single or multimanifold die design is based on the final product structure specifications. Typically a single manifold die and feedblock approach can produce coextruded structures with a layer to layer thickness tolerance that is in the range of ±15–20%; a multimanifold die design is capable of producing tolerances that are in the range of ±5–10%. Additionally, it is almost always recommended that a multimanifold die design be used for applications that require skin layer thickness that are less that 5% of the total product thickness, or in applications where the various resins require a major difference in processing temperatures and/or viscosity characteristics.

Polypropylene: An A–Z Reference
Edited by J. Karger-Kocsis
Published in 1999 by Kluwer Publishers, Dordrecht. ISBN 0 412 80200 7

A common multilayer PP product is bi-axially oriented polypropylene (BOPP), produced on a line including a machine direction orientation unit, and a tenter frame for transverse orientation. This process commonly uses a three-manifold die that allows thin skin layers of copolymer to be applied on either side of a homopolymer base layer. The copolymer is used in this product so the film can be sealed, which is a requirement in packaging applications. Typically, the skin layer thickness is 5% of the total film thickness. The adhesive property of the copolymer makes it difficult for the film to be processed through a tenter frame, because the film would stick to the tenter frame clips. The multimanifold die can be designed with narrower width flow channels for the skin layers, to keep the adhesive out of contact with the tenter clips, allowing the product to be extruded successfully.

MANIFOLD DESIGN

The most important feature of an extrusion die is the flow channel design. As plastic processing becomes specialized, the die design becomes more tailored to match the process. Using the rheological results from samples of the polymers that the processor intends to run, the design can begin. A die intended to run a myriad of materials has to be designed with a generic manifold that may not be ideal for any of the individual materials, but could produce adequate results with some operator intervention. If a processor focuses on producing a specific, specialized product, then the die, and particularly the manifold in the die can be custom designed for the specific polymer at a known output rate.

The flow channel in the die is designed to uniformly distribute the melted polymer to a desired width. It has four parts, the manifold, preland, secondary manifold, and final land. The manifold can be viewed as a pipe, which carries the material to the desired width. The preland is used to vary the resistance of the flow paths within the die. The secondary manifold is used to reduce the pressure drop within the die. The final land establishes the gauge, or thickness of the melt leaving the die.

The standard manifold design for a PP until a few years ago would have been a coat hanger design. This manifold relies on a delta shaped preland to control the distribution within the die. The manifold design is typically done using a one-dimensional set of design equations. Pressure drops across various sections of the die are calculated, and a good design will balance them so that the theoretical exit flow is usually within 5%.

A basic problem with the coat hanger design is that it causes the distance from the lip exit to the back of the manifold to be longest at the center of the die, and this distance decreases to the ends of the die. When a simple deflection analysis is done, the manifold area of the die is viewed as a cantilevered beam. If the deflection is taken at cross-sections from the center to end of the die, differing results are usually obtained. Typically, a greater force is applied by the polymer at the center of the die where the wetted length is also longest. This results in a phenomena called 'clamshelling' or greater deflection at the center, and less deflection at the ends. [1] The design distribution of the die is affected, and the die distribution tends to vary as the output rate is changed. Additionally, coathanger manifolds can cause 'M' or 'W' patterns to occur. This type of pattern is the result of the changing shear rates within the die, and the shear thinning of the PP. As the viscosity of the material lowers, the localized flow rate increases, causing poor die distribution.

A new development in flow channel design is the patented hybrid manifold [2]. This design addresses some of the shortfalls of the coat hanger design. It clamshells uniformly because the back of the manifold is kept in a straight line across the width of the die. The hybrid design performs better as a coextrusion manifold, because the aspect ratio in the manifold is controlled, reducing the high shear region that exists at the coextrusion interface of a coathanger style manifold. The preland is also shaped to eliminate the 'M' and 'W' patterns associated with a coat hanger manifold. Finally, the hybrid manifold design relies on the generation of a finite-element mesh, and the solution of three-dimensional flow equations; this type of rigorous analysis provides a much more accurate solution to the manifold's design. Previously, the design of a manifold would rely on solving one- or two-dimensional sets of equations. Figure 1 depicts the top view of a hybrid manifold.

Restrictor bars are used in most manifold designs to aid in flow tuning if the lip opening of the die is greater that 2 or 3 mm. These bars can be either fixed or adjustable, and are either profiled or bent with external adjustments to change the distribution in the flowstream.

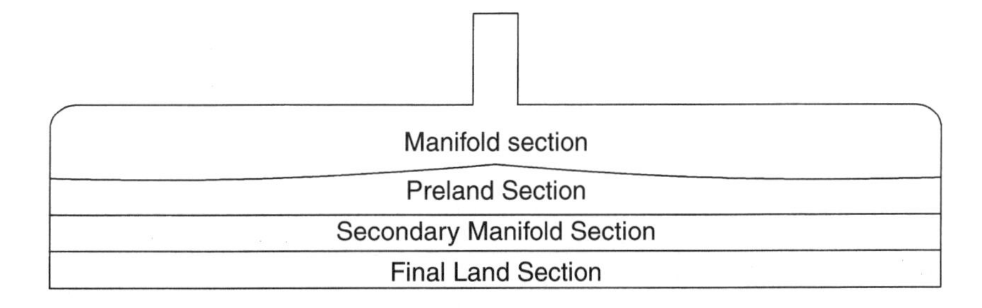

Figure 1 Top view of a hybrid manifold.

MECHANICAL DESIGN

A typical PP die is manufactured from a P20 tool-steel billet. The flow channel is polished to 2–4 Ra, and hard-chrome plated. The exterior of the die is flash chrome plated to make die clean-up easier, and also to prevent corrosion during shipping and storage. Dies can also be built from stainless steel, which is more costly that a chrome-plated tool-steel die, and of no significant advantage, because PP is not a corrosive material.

The following conditions must be addressed to insure a successful design.

Operating pressures within the die are typically in the range of 100 bar. The die's cross-section must be strong enough to keep distortion to a minimum, but it must also fit within the extrusion line. It is desirable to keep the distance from the die exit to the take-away system to a minimum. Some extrusion line layouts allow for smaller clearance angles. Subsequently the die can be made stronger, although in some cases a trade-off occurs between the optimal die position and the die's strength.

Large fasteners, under high torque, up to 1700 Nm, are closely spaced to keep the die from leaking. The die body halves are precisely ground to finishes under 10 Ra. The metal-to-metal sealing surfaces must be kept clean and nick free to prevent leakage. Metal-to-metal sealing surfaces are used in lieu of gaskets to insure a precise match of die parts where they are joined in the flow channel.

PP extrusion temperatures average 260°C. Dies must be heated, either with cartridge heaters fit into holes drilled into the die body or plate heaters fastened to the exterior of the die. These are electric resistance heaters, wired in zones, and controlled via a thermocouple feeding back to an instrument, or a programable logic control (PLC). Temperature control is usually within 1°C. In applications where air knives or vacuum boxes are also used, it is advisable to heat the die lips.

OPTIONS FOR INCREASED PERFORMANCE

All dies should have some type of flexible lip and an adjustment system that is used to fine-tune the distribution of the polymer as it exits the die. The simplest form of adjustment is a set of push screws, usually with fine threads to give the operator better control.

A differential thread system, which takes advantage of the difference in screw thread pitches to reduce the axial motion of the adjustment per revolution of the fastener, is a refined manual adjustment system. The differential system gives finer control to the operator, increasing the adjustment resolution by about a factor of four. Differential adjustment systems can also be designed to push and pull against the die lip, so that low spots in the sheet or film can be adjusted by pulling open the die lip, and high spots can be adjusted out by closing down the lip.

Autoflex™ (trade mark of Extrusion Dies, Inc.) dies take the adjustment system's level of sophistication one step further by closing the loop between the gauging system and the die. These systems use thermal expansion and contraction to open and close the die lips. An electric resistance heater with air cooling applied to each adjustment bolt on the die provides the thermal energy required to expand and contract the actuator. A scanning head downstream of the die measures the thickness of the material. Through control software and a microprocessor, the measurements are analyzed. The controller cycles the heaters on and off as required to open or close the die lip. Closing the die lip will restrict the flow of material it that area, and the downstream result is thinner gauge. Cooling the autobolt will open the die lip, allowing more material flow, and increase the material's thickness in that area. These closed loop systems are highly accurate, and in many applications gauge control of less that 1% can be obtained.

One of the drawbacks associated with earlier Autoflex™ dies was the adjustment range, ±0.2 mm. This was often less than the movement required to make enough of a change in pressure on a die with a large lip opening, 2.5 mm and over, to influence the final product. This range has

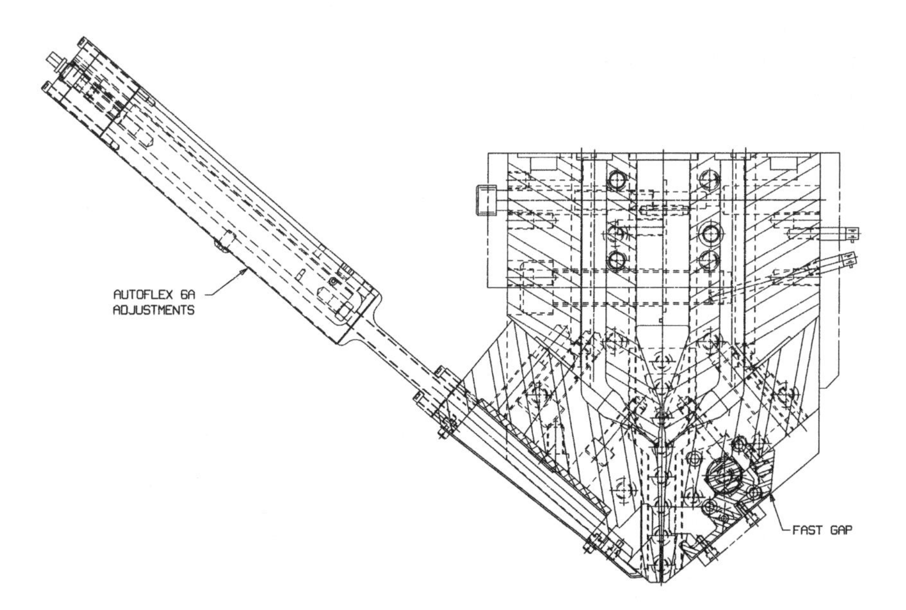

Figure 2 Typical cross-section of a three-manifold BOPP die.

recently been increased to ±0.4 mm, by re-engineering the adjustment system, making auto control practical in BOPP applications.

Dies can be manufactured with removable lower lips. Their primary advantage is to make the die more versatile. The thickness of the final product being produced is related to the thickness of the material exiting the die. A removable lower lip increases the lip opening range of a die. The disadvantage of a removable lip is the parting line in the flow surface.

A Fast-Gap™ system (trade mark of Extrusion Dies, Inc.) on a die extends the adjustment range, and eliminates the parting line of a removable lower lip. This device provides an additional 5 mm of adjustment (±2.5 mm), by simply turning a hex headed power screw on the end of the die. The operator can now also flex open the lip to clean it, and then close it back to the run position, saving time and reducing scrap.

Figure 2 depicts a three manifold BOPP die, with hybrid flow channels, an extended range Autoflex™ system, and Fast-Gap™. This brief overview is intended to highlight some of the considerations in the specification and design of a high-performance PP extrusion die.

REFERENCES

1. Michaeli, W. (1992) *Extrusion Dies for Plastic and Rubber*, Hanser, Munich, pp. 147–148.
2. US Patent Number 5 494 429 Granted to Extrusion Dies, Inc.

Keywords: extrusion die, die design, biaxially-oriented PP (BOPP), heat sealing, manifold, coextrusion, extrusion parameter, process control, flow.

Fatigue performance of polypropylene and related composites

József Karger-Kocsis

INTRODUCTION [1]

Polypropylene (PP), without and with fillers and reinforcements, is widespreadly used in applications where a high resistance to repeated loading, i.e. to fatigue, is required. The fatigue performance can be characterized either by the endurance limit (the maximum allowed stress causing no damage after 10^6 fatigue cycles) or by the description of the fatigue crack propagation (FCP). The latter approach, based on the fracture mechanics, is straightforward when the PP material is inhomogeneous, i.e. when inherent flaws of various origin and type are present. Such flaws may act as initial cracks and grow under the prevailing external loading conditions. In that case, the life expectation of the material or construction part is determined by the FCP behavior. On the other hand, if the controlling step is the development of cracks or more generally damage, the target is to determine the fatigue endurance limit and describe the related curve (i.e. maximum stress versus cycle number, *S–N*; Wöhler curve). From the viewpoint of the loading mode, a distinction is generally made between static and cyclic (sometimes falsely termed dynamic) loading.

Polypropylene: An A–Z Reference
Edited by J. Karger-Kocsis
Published in 1999 by Kluwer Publishers, Dordrecht. ISBN 0 412 80200 7

FATIGUE CRACK PROPAGATION

Cyclic loading

This fracture mechanical approach has been successfully applied for both neat [2–3] and short (SGF) and long glass fiber (LGF) reinforced PP [3–4]. It was found that the stable FCP behavior, which is a part of the da/dN versus ΔK curves, can be described by the well-known Paris–Erdogan relationship:

$$\frac{\mathrm{d}a}{\mathrm{d}N} = A(\Delta K)^m \tag{1}$$

where da/dN is the FCP rate, A is a pre-exponential factor, ΔK is the stress intensity amplitude and m is the exponential term. It is worth noting that ΔK is a function of the load amplitude and specimen configuration. The main effects caused by the reinforcing GF, including reinforcement volume fraction (V_f), molding-induced layer structuring (transverse, T, or longitudinal, L, notching of the injection-molded plaques in respect to the mold filling direction, MFD) and fiber aspect ratio (l/d) are depicted schematically in Figure 1.

The scheme displayed in Figure 1 is qualitatively similar for all injection-molded chopped fiber reinforced composites. It should be emphasized here that the main prerequisite of adopting this approach is that the crack advance of the crack tip can be well resolved and followed in course of the fatigue test. In many cases, this requirement is not met, e.g. for neat PP with high molecular weight and/or fine spherulitic structure. Instead of a sharp crack travelling through the free ligament, here only a change in the shape and size of the damage zone can be detected. For the description of such a diffuse, extended damage zone the crack layer theory [5–6] can be adopted though the physical meaning of some of its constituting terms are not yet fully clear.

In FCP studies performed on neat and discontinuous GF reinforced PP composites, it was found that the stable acceleration range is often preceded by a stable deceleration one [2–4]. The appearance of this range is due to a 'blunting' process in which different matrix- and fiber-related events are involved.

Static loading [3]

For this case, a similar description to equation (1) is used:

$$\frac{\mathrm{d}a}{\mathrm{d}t} = A'K_I^{m'} \tag{2}$$

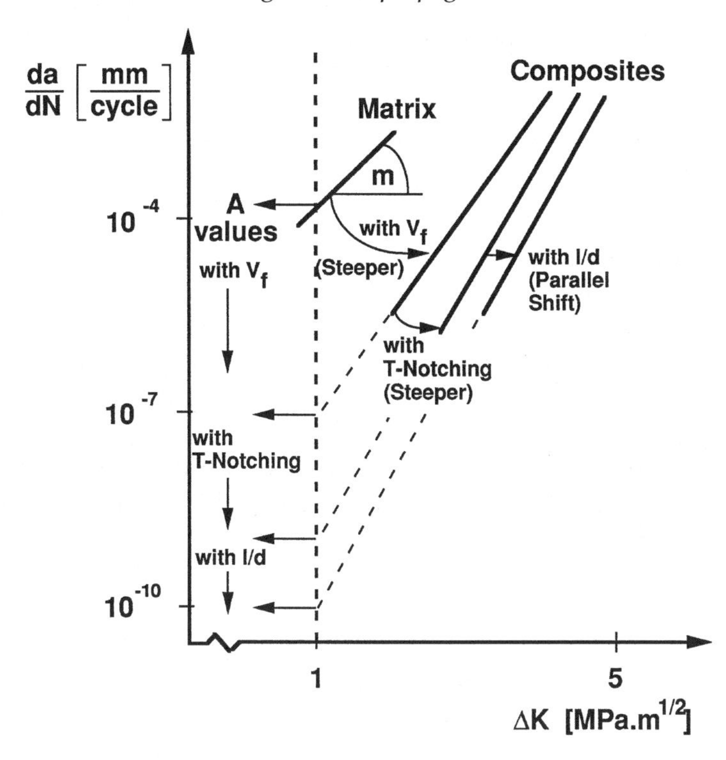

Figure 1 FCP response as a function of microstructural parameters for discontinuous GF-reinforced PP composites [3]. Note: A and m are the parameters of the Paris–Erdogan power law function (see Equation (1)), respectively.

where $\mathrm{d}a/\mathrm{d}t$ is the crack growth in function of time, K_I is the initial stress intensity factor under mode I (i.e. crack opening) condition, and A' and m' are the pre-exponential and exponential constants, respectively. Though the shape of the $\mathrm{d}a/\mathrm{d}t$ versus K_I curve may be very peculiar, especially when the test is performed in different environments (environmental stress corrosion cracking, ESC; see the chapter Environmental stress cracking of polypropylene), a part of the curve can always be approached by a linear relationship when the $\mathrm{d}a/\mathrm{d}t$ versus K_I data pairs are represented in double logarithmic scale. Comparing the onset of the stable acceleration range in cyclic and static fatigue, the lower threshold stress intensity value is considerably higher for the latter. In addition, the slope of the related curve section becomes very steep. This indicates an enlarged damage zone (a more pronounced 'blunting'), which is, however, less resistant to crack advance (steep slope due to fast crack growth).

It should be noted here that the primary crack growth results in static fatigue can also be represented in another form: initial K_I versus time to failure (t_f). By extrapolation of the related curve, the threshold K_I value can be determined below which no crack growth occurs. This description is favored in ESC studies [1].

WÖHLER DESCRIPTION

Cyclic loading

As mentioned above, the fatigue behavior with damage development should be assessed by the Wöhler curve and the fatigue endurance limit. The latter value is the most important parameter for design and construction purposes. For the description of the Wöhler curves, different methods can be selected according to the plotting (semi- or double logarithmic scale).

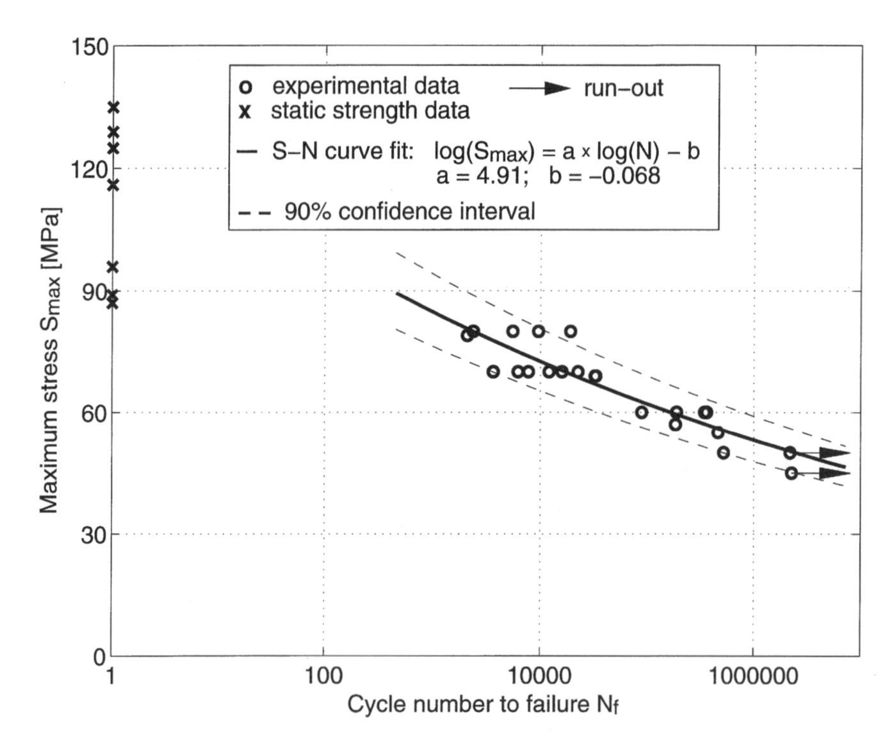

Figure 2 Wöhler curve (S_{max}-N_f) for a unilaterally compression-molded GMT-PP with 40 wt.% glass swirl mat reinforcement. Note: the following testing conditions were set: $R = 0.1$, $f = 5$ Hz, $T = 30$°C (DFG Ne 546/5–1, [7]).

The best demonstration example of this behavior is the fatigue performance of glass mat reinforced PP (GMT-PP; see the chapter Glass mat reinforced thermoplastic polypropylene). The fatigue performance of this composite is controlled by damage development and not by crack growth (though the latter has also been claimed). For the damage development and growth, the continuous swirl mat reinforcement is responsible which affects both stress transfer and stress redistribution processes under external loading. Figure 2 shows the maximum stress (S_{max}) versus cycles to failure (N_f) curve of a tension–tension ($R = 0.1$) loaded GMT-PP produced by restricted flow (flow in one direction). The specimens were cut and loaded along the flow direction [7]. Figure 2 also demonstrates the usual large scatter in the static tensile strength of molded GMT-PP products.

Static loading or creep

Information about the creep behavior can mostly be found for neat PPs. Creep tests at different temperatures are widely used for estimation of the service life of the material (see also the chapter 'Long-term properties and lifetime prediction for PP' in this book). During this test (performed above the glass transition temperature, T_g, of PP) the polymer may undergo considerable morphological changes in which both the amorphous and crystalline regions are involved. Though several phenomenological descriptions are recommended to describe the creep behavior, its controlling terms in molecular, supermolecular levels are still less understood. The lack of knowledge is even higher for PP-composites. Even the existence of the endurance limit is a topic of dispute.

REFERENCES

1. Karger-Kocsis, J. (1991) Structure and Fracture Mechanics of Injection-molded Composites, in *International Encyclopedia of Composites*, Vol. 5, (ed. S.M. Lee), VCH, New York, pp. 337–356.
2. Karger-Kocsis, J. and Friedrich, K. (1989) Effect of skin-core morphology on fatigue crack propagation in injection moulded polypropylene homopolymer, *Int. J. Fatigue*, **11**, 161–168.
3. Karger-Kocsis, J. (1995) Microstructural Aspects of Fracture in Polypropylene and its Filled, Chopped Fiber and Fiber Mat Reinforced Composites, in *Polypropylene: Structure, Blends and Composites*, Vol. 3, Chap. 4, (ed. J. Karger-Kocsis), Chapman & Hall, London, pp. 142–201.
4. Karger-Kocsis, J., Friedrich, K. and Bailey, R.S. (1991) Fatigue crack propagation in short and long glass fiber reinforced injection-molded polypropylene composites, *Adv. Composite Mat.*, **1**, 103–121.
5. Chudnovsky, A., Moet, A., Bankert, R.J. and Takemori, M.T. (1983) Effect of damage dissemination on crack propagation in polypropylene, *J. Appl. Phys.*, **54**, 5562–5567.

6. Dolgopolsky, A. and Botsis, J. (1989) The crack layer approach to polymers and composites, in *Application of Fracture Mechanics to Composite Materials*, Chap. 7, (ed. K. Friedrich), Elsevier, Amsterdam, pp. 249–271.
7. Fischbach, F. and Himmel, N. (1997) Manufacturing influences on static and fatigue properties of glass mat reinforced polypropylene (GMT), in Proceedings of the International Conference on Fatigue of Composites (ICFC-1), Paris, 3–5 June, 1997, pp. 462–470.

Keywords: creep, fatigue, fatigue endurance limit, fatigue crack propagation (FCP), Paris–Erdogan relationship, Wöhler curve, short and long fiber reinforced PP, glass mat reinforced thermoplastic PP (GMT-PP), cyclic fatigue, crack layer theory.

Fiber orientation due to processing and its prediction

T. Matsuoka

INTRODUCTION

Fiber reinforced polypropylene (FRPP) is used as a structural material because of its high strength and stiffness to weight and cost performance. FRPP parts are usually manufactured by a conventional polymer processing method, such as injection molding. In injection molding, polymer melt is injected into mold cavities by applying high pressure. Fiber orientation is induced by polymer melt flow during mold filling. Fiber orientation causes the anisotropy of mechanical properties in molded parts. For example, when all fibers align in a given direction, the elastic modulus in the fiber direction will be much larger than that in the perpendicular direction. Also, fiber orientation is a main reason of warpage of FRPP molded parts. Therefore, the prediction of fiber orientation is very important to control fiber orientation and prevent warpage in a design stage. Computer simulation can be used to predict fiber orientation in molded parts with three-dimensional complex geometry qualitatively in injection molding CAE(Computer Aided Engineering).

FIBER MOTION IN FLOW

Motions of a rigid rod-like fiber in polymer melt flow are translation, rotation and spin (rotation around its axis). The motion of spin can be neglected because of no change of fiber orientation. Figure 1 shows schematic representations of basic flows for predicting fiber orientation.

Polypropylene: An A–Z Reference
Edited by J. Karger-Kocsis
Published in 1999 by Kluwer Publishers, Dordrecht. ISBN 0 412 80200 7

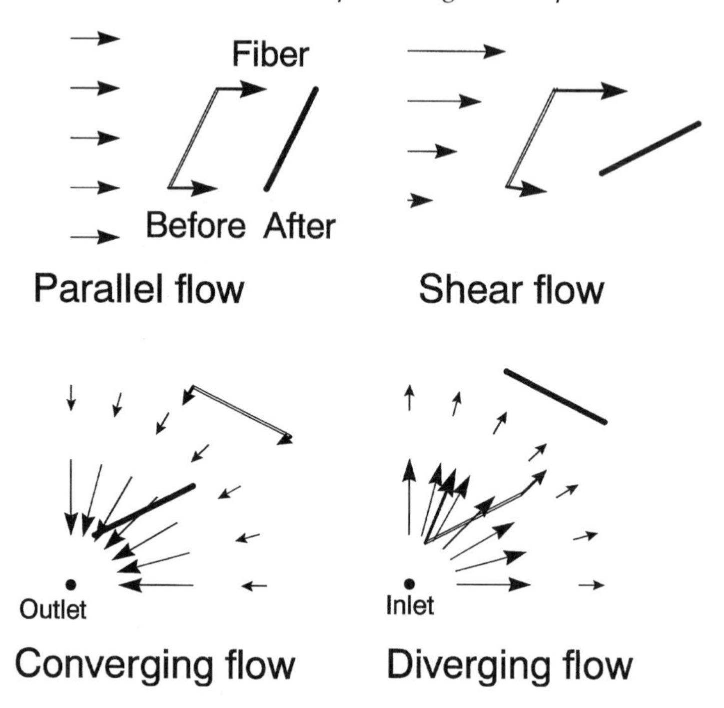

Figure 1 Schematic representation of basic flows and related fiber motions.

There are four kinds of basic flows, namely, parallel, shear, converging and diverging flows. For the parallel flow, since the velocity profile is uniform, the fiber only moves downstream without rotation. Then, the fiber orientation caused at the upstream will be kept in this flow. In shear flow, the fiber will begin to rotate clockwise in the figure because the velocity of its upper end is faster than that of the lower end. The rotational velocity of the fiber is the fastest when the fiber takes its position perpendicular to the flow and is the lowest when it lies in the flow direction. Therefore, the probability that the fiber orientates in the flow direction is very strong in the shear flow. When a fiber is exposed to a converging flow, in which the fluid velocity increases as it approaches the outlet, it will point to the outlet. Finally, in the case of diverging flow, the fiber orientation is peculiar. The fiber will turn perpendicularly to the radial flow direction or orient in the direction of the circumference elongational flow.

Flow-induced fiber orientation in complex moldings can be quantitatively predicted by considering a combination of four types of basic flows along streamlines [1]. Fibers will tend to align in the streamline direction at regions of shear and converging flows, but across the streamline at a region of diverging flow. Fiber orientation in the parallel flow will be the same as that at the upstream.

FIBER ORIENTATION IN INJECTION MOLDING

Figure 2 shows an X-ray plane photograph of brass fibers in a square plate (100 × 100 × 3 mm) injection-molded of 20 wt.% glass fiber reinforced PP [2]. Brass fibers of 2.5 mm long and 0.15 mm diameter were filled in FRPP at a fraction of 1 wt.% as a tracer. Brass fibers and FRPP were directly mixed in the injection cylinder and were injected into the cavity from a side gate located at the center of the left-hand side. A segment denotes an individual brass fiber. It is clearly observed that brass fibers align in the circumferential direction around the gate and in the flow direction toward right-hand upper and lower corners. The observation reveals that not all fibers align in streamlines from the gate to two right-hand corners. The observed fiber orientation is qualitatively explained by considering the above-mentioned fiber motion because the planar flow is characterized by a diverging flow around the gate and the converging flow toward the corners.

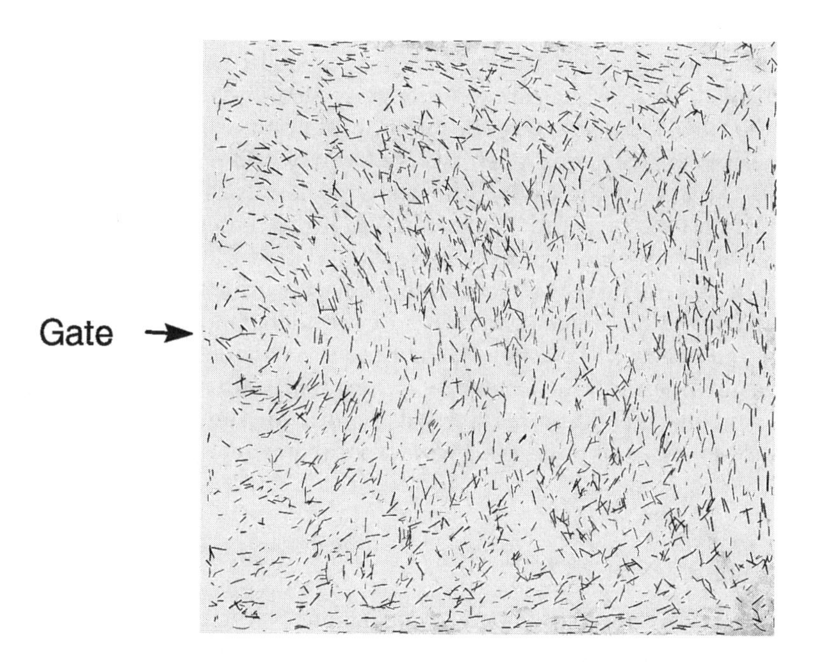

Figure 2 Photograph of X-ray observation of brass fiber tracers in an injection molded square plate (100 × 100 × 3 mm) of glass fiber reinforced polypropylene [2, 5]. A gate is located at the center of the left-hand side.

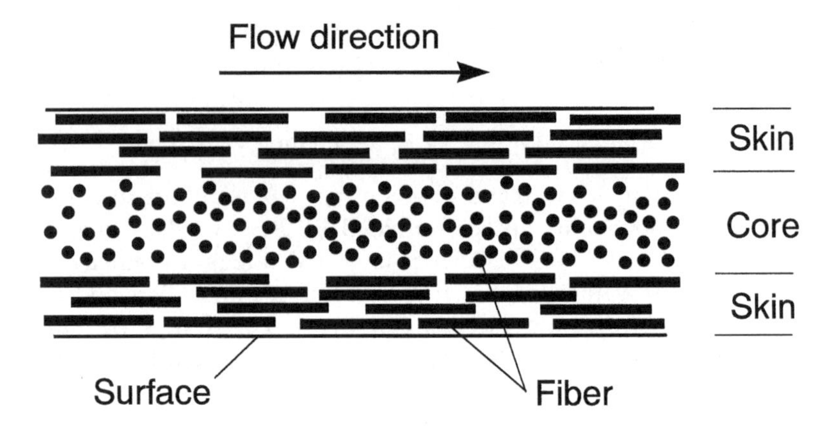

Figure 3 Schematic representation of layered structure of fiber orientation in a cross-section of an injection-molded part.

Injection-molded parts have a layered structure of fiber orientation in its cross-section through thickness as shown in Figure 3. The flow direction is from left to right. The cross-section is divided into three layers, upper and lower skin layers and core layer. Fibers align in the flow direction in the skin layers, but lie perpendicularly in the core. The mechanism of this interesting structure is also understood by considering the flow type. The flow in injection molding consists of a gapwise flow between narrow gapped walls of cavity and a planar flow in the broad expanse of the cavity. The gapwise flow is always a shear flow, which is strong near the wall and weak at the center because of a plug type velocity profile due to non-Newtonian behavior of polymer melt. Then, fibers near the wall will align in the flow direction, as in the skin layer. On the other hand, the type of the planar flow depends on the cavity geometry. Where the planar flow is a diverging flow, as around the gate, fibers at the center will turn perpendicularly to the flow direction and then the core layer will be formed. The layered structure is developed by combining the gapwise shear flow and the planar diverging flow.

FIBER SUSPENSION

There are three types of fiber suspensions for predicting fiber orientation due to flow. They are dilute, semiconcentrated and concentrated suspensions defined as follows: $f_v < (d/l)^2$, dilute; $(d/l)^2 \leqq f_v \leqq d/l$, semiconcentrated; $d/l < f_v$, concentrated. Here d is the diameter of fibers and l is the length of fibers, and f_v is the volume fraction of fibers.

For the dilute suspension, fibers are free to rotate without interactions between fibers. The volume necessary for the free rotation of a fiber must be greater than the volume of a sphere, the diameter of which equals the fiber length. On the other hand, fibers interact with other fibers in the concentrated region. Fibers interfere via their rotation, when the average space occupied by a fiber is less than the volume of a cylinder whose diameter and height equal the fiber length and the fiber diameter, respectively. In the semiconcentrated region, interactions between fibers are frequent.

The diameter of glass fibers is generally about 13 μm and their length is about 250 μm. Thus, the aspect ratio, l/d, is 19. According to the above classification, the volume fraction is less than 0.27% for a dilute suspension and is greater than 5.2% for a concentrated suspension. The volume fraction of fibers f_v is converted to the weight fraction of fibers f_w by the equation:

$$f_w = \rho_f f_v / \{\rho_f f_v + \rho_m (1 - f_v)\},$$

where ρ_f is the density of the fiber and ρ_m is the density of the polymer matrix. When $\rho_f = 2.5$ and $\rho_m = 1$ gcm^{-3}, the weight fraction becomes 0.6 wt.% for a dilute suspension and 12 wt.% for a concentrated one. Glass fibers are usually present as 10–30 wt.% in commercially available FRPP. Therefore, the melt of FRPP must be treated as a concentrated suspension for predicting fiber orientation.

FIBER ORIENTATION MODEL

Motion of a fiber in flow is described by Jeffery's model [3]. It is assumed that the fiber is a single rigid ellipsoidal particle suspended in a viscous fluid, the flow is a creeping flow of a Newtonian and incompressible fluid, and Brownian motion and inertia terms of the fiber are neglected. Jeffery's model was used for prediction of fiber orientation in the early period of injection molding CAE. Since it is, however, for dilute suspension, the model is replaced with the Folgar–Tucker model for concentrated suspension.

The Folgar and Tucker model describes planar behaviors of rigid fibers using a statistical approach instead of considering an individual fiber in concentrated suspensions [4]. A distribution function φ is defined for fiber orientation as a function of orientation angle ϕ. The integral of the orientation distribution function from ϕ_1 to ϕ_2 is the probability that a fiber will be oriented between them. The function must be periodic with period π because a fiber at ϕ is not distinguished from one at $\phi + \pi$, $\varphi(\phi + \pi) = \varphi(\phi)$. Since all fibers orient from 0 to π, the integral of φ from 0 to π is equal to 1. The Folgar–Tucker equation is:

$$\partial\varphi/\partial t + u(\partial\varphi/\partial x) + v(\partial\varphi/\partial y) = C_I\dot{\gamma}(\partial^2\varphi/\partial\phi^2) - (\partial\Psi/\partial\phi) \qquad (1)$$

$$\Psi = \varphi[-\sin\phi\cos\phi\,(\partial u/\partial x) - \sin^2\phi\,(\partial u/\partial y) + \cos^2\phi\,(\partial v/\partial x) + \sin\phi\cos\phi\,(\partial v/\partial y)] \quad (2)$$

where t is the time, u and v are velocities in the x and y direction, respectively. $\dot{\gamma}$ is the magnitude of the strain rate tensor, C_1 is the interaction coefficient determined experimentally. The first term on the right-hand side expresses the interaction effect between fibers. The second term describes the hydrodynamic effect and is the same as the two-dimensional Jeffery's model for a fiber with infinite aspect ratio.

PREDICTION OF FIBER ORIENTATION

In injection molding CAE, fiber orientation analysis is carried out combining with mold filling analysis. Using velocity distributions obtained from the mold filling analysis, the fiber orientation equation is numerically solved by a finite-difference method for predicting fiber orientation. The orientation distribution function is calculated and is applied to the estimation of mechanical properties for predicting warpage of molded parts.

Figure 4 shows predicted results of fiber distribution functions on skin and core layers of the above-mentioned square plate molded of glass fiber reinforced PP. An elliptical symbol is drawn with 36 segments, which are directed from 0 to 180° every 5° and indicate values of fiber distribution functions by their lengths. Therefore, a circle-like shape means that fiber orientation is nearly random and an ellipsoidal shape denotes that fibers orient strongly in the longitudinal direction. In the skin layer, fibers orient radially from the gate in the flow direction owing to the gapwise shearing flow, while, in the core layer, fibers align

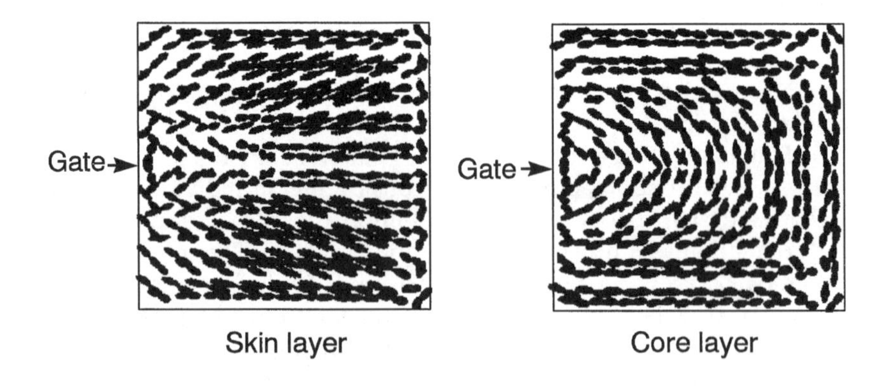

Figure 4 Predicted fiber distribution functions in skin and core layers of a square plate (100 × 100 × 3 mm) molded of glass fiber reinforced polypropylene [5].

perpendicularly against the flow direction around the gate. The circumferential orientation agrees qualitatively with the results of the X-ray observation. The simulated results demonstrate the difference in fiber orientation between skin and core layers, or the layered structure of fiber orientation.

REFERENCES

1. Lockett, F.J. (1989) Prediction of fibre orientation in moulded components. *Plastics Rubber Processing*, **5**, 85–94.
2. Matsuoka, T., Takabatake, J., Inoue, Y. and Takahashi, H. (1990) Prediction of fiber orientation in injection molded parts of short-fiber-reinforced thermoplastics. *Polym. Eng. Sci.*, **30**, 957–966.
3. Jeffery, G.B. (1922) The motion of ellipsoidal particles immersed in a viscous fluid. *Proc. Roy. Soc.*, **A102**, 161–179.
4. Folgar, F. and Tucker III, C.L. (1984) Orientation behavior of fibers in concentrated suspensions. *J. Reinforced Plastics Composites*, **3**, 98–119.
5. Matsuoka, T. (1995) Fiber orientation prediction in injection molding, in *Polypropylene: Structure, Blends and Composites* Vol. 3, Chap 3, (ed. J. Karger-Kocsis), Chapman & Hall, London, pp. 113–141.

Keywords: computer aided engineering (CAE), computer simulation, fiber orientation prediction, fiber orientation modelling, fiber reinforced thermoplastics, finite difference method, finite element method (FEM), flow, glass fiber reinforcement, injection molding, mold filling analysis, warpage.

Fillers for polypropylene

Béla Pukánszky

INTRODUCTION

Behind PVC, particulate fillers are used in the second largest quantity in polypropylene (PP). Table 1 gives an estimate of the amount of fillers used in Western Europe and about the increase of their consumption. Three fillers are applied in considerable quantities: talc, $CaCO_3$, glass fibers (GF). Mostly technical and technological reasons justify the application of fillers, since a separate compounding step increases the price of the composites considerably. Combination of fillers with PP results in a new material with changed properties; some of these changes are advantageous (improved stiffness, higher heat deflection temperature (HDT), better heat conductivity), while others are less favorable (decreased deformability and impact, wear of the equipment). An optimum of

Table 1 Use of fillers for PP in Western Europe (1000 tons)

Filler type	*1987*	*1990**	*1993**
Calcium carbonate	34.0	48.0	72.0
Talc	43.0	55.3	74.0
Glass fibers	7.0	8.5	11.0
Silica	Small	Small	Small
Wollastonite	0.2	0.4	0.5
Mica	1.9	2.4	2.8
Glass beads	0.3	0.4	0.5
Wood flour and fibers	0.1	0.2	0.2

* estimated.

Polypropylene: An A–Z Reference
Edited by J. Karger-Kocsis
Published in 1999 by Kluwer Publishers, Dordrecht. ISBN 0 412 80200 7

properties must be found during the development of a new material and selection of the proper type and grade of filler is always an important part of the process.

FILLERS

The most important information about the most important particulate fillers of PP are listed in the following paragraphs. For more detailed information refer to handbooks [1] and monographs [2].

$CaCO_3$

Calcium carbonate exists in three crystalline modifications (calcite, aragonite, vaterite), but only calcite has practical importance. It can be found in large quantities all over the world, but fillers mined at different locations differ considerably in purity, size of the crystals and origin, which all influence their use as fillers. In nature, it can be found in three different forms: limestone, chalk and marble. Limestone is a consolidated sedimentary rock formed by the deposition of shells and skeletons of marine organisms, chalk is soft-textured limestone laid down in the cretaceous period and consists of nanofossils; marble is metamorphic limestone formed under high pressures and temperatures. $CaCO_3$ occupies second place in usage behind talc in PP.

Talc

The chemical composition of pure talc is $Mg_3(Si_4O_{10})(OH)_2$; it is a secondary metamorphic mineral. The filler consists of platelets. Layers of the particles are bonded only by secondary van der Waals forces, cleavage takes place very easily also under the conditions of usual thermoplastic processing technologies. Talc is the softest mineral, its Mohs hardness is 1 (diamond has a hardness of 10 at the other end of the scale). Its surface is rather hydrophobic, active hydroxyl groups can be found only on broken edges and surfaces, surface treatment of the filler is difficult. It reinforces polymers due to its platelet-like geometry and has a strong nucleation effect on PP further increasing its stiffening effect.

Glass fiber

This reinforcement is used in two forms in PP. Glass mat reinforced PP (GMT-PP) is produced in small quantities for the automotive industry. Composites containing short GF are also used in this area, mostly for under-hood applications. The properties of GF reinforced PP are in the

Table 2 Composition and properties of particulate fillers

Filler	*Composition*	*Density (g/cm³)*	*Mohs hardness*	*Shape*
Calcium carbonate	$CaCO_3$	2.7	3	Sphere
Talc	$Mg_3(Si_4O_{10})(OH)_2$	2.8	1	Plate
Kaolin	$Al_2O_2 \cdot 2SiO_2 \cdot 2H_2O$	2.6	2.5–3.0	Plate
Wollastonite	$CaSiO_3$	2.9	4.5	Needle
Mica	$KM(AlSi_3O_{10})(OH)_2$	2.8	2.0–2.5	Plate
Barite	$BaSO_4$	4.5	3.5	Plate

range of engineering thermoplastics. Proper surface treatment of the fiber is crucial in these composites.

Other fillers

Kaolin is used only in small quantities in PP. The anisotropic particle geometry of wollastonite and mica reinforces the polymer. The latter is applied more extensively in USA and Canada due to the closer location of mining facilities. Barite-filled PP has excellent vibration damping properties due to the high density of this filler. Wood flour filled PP is occasionally applied in the automotive industry for the preparation of door panels. Water absorption of this filler creates problems in processing and application. The most important characteristics of mineral fillers are collected in Table 2.

SURFACE TREATMENT

Surface treatment of fillers is an important issue of their application. Nonreactive treatment decreases aggregation, improves processability and surface appearance of the product, while reactive treatment increases strength, but decreases deformability and impact resistance.

$CaCO_3$

Mostly nonreactive treatment is used; the majority of commercial grades are treated with stearic acid. Other organic acids can be used as well; these form ionic bonds with the surface of the filler. Reactive treatment of $CaCO_3$ for PP is difficult; certain aminosilanes may couple the components with covalent bonds. Usually the improvement in properties is counteracted by the cost of the treatment.

Talc

Efficient surface treatment of talc is practically impossible due to its inactive surface and low surface energy. Occasionally, its surface is covered by phenolic resins to decrease the effect of heavy metal contamination. Reactive treatment is applied sometimes, but it is superfluous and not cost-effective.

Glass fibers

Reactive treatment is achieved by the combined effect of silanes attached to the active hydroxyl groups of the surface and the introduction of functionalized PP (maleic anhydride or acrylic acid) into the composite.

EFFECT OF FILLER CHARACTERISTICS ON COMPOSITE PROPERTIES

The chemical composition of fillers, which is usually supplied by the producer as relevant information, is not sufficient for their characterization. Further physical, mostly particle characteristics are needed to forecast their performance in a composite for any application [3, 4].

Chemical composition

Chemical composition and especially purity of the filler has both a direct and an indirect effect on its application possibilities and performance. Traces of heavy metal contamination decrease the stability of PP. Insufficient purity leads to discoloration of the product. High-purity $CaCO_3$ has the advantage of white color, while the grey shade of talc-filled composites excludes them from some applications. Chemical composition of the filler surface has some influence on its nucleation effect.

Particle size, particle size distribution

Mechanical properties of PP composites containing nontreated fillers are determined mainly by their particle characteristics. Part of the basic information supplied by the manufacturer is the average particle size of the filler, which has a pronounced influence on composite properties. The strength (sometimes also the modulus) increases, the deformability and impact strength, decrease with decreasing particle size. Particle size in itself, however, is not sufficient for the characterization of any filler; knowledge of the particle size distribution is equally important. Large particles – top cut – beside changing abrasion and appearance of the product, drastically alter, usually deteriorate, deformation and failure characteristics of the composites. The volume in which stress concentration is effective is said to increase with particle size and the matrix/filler adhesion also depends on it. This latter factor is especially important in PP composites; because of the low polarity of this polymer, it adheres weakly to fillers. Under external load, this weak adhesion leads to the separation of matrix/filler interface (debonding) and to the catastrophic failure of the composite. The other end of the particle size distribution, i.e. the amount of small particles, is equally important. The aggregation tendency of fillers increases with decreasing particle size. Extensive aggregation, however, leads to insufficient homogeneity, rigidity and lower impact strength. Aggregated filler particles act as crack initiation sites under dynamic loading conditions.

Specific surface area

Specific surface area of fillers is closely related to their particle size distribution. Furthermore, it has direct impact on composite properties, as well. Adsorption of both small molecular weight additives, and also that of the polymer, is proportional to the size of matrix/filler interface. Adsorption of additives may change stability, while matrix/filler interaction significantly influences mechanical properties, mainly yield stress, tensile strength and impact resistance.

Shape

Reinforcement increases with anisotropy of the particle. In fact, fillers and reinforcements are very often differentiated by their degree of anisotropy (aspect ratio). Plate-like fillers, such as talc and mica, reinforce PP more than spherical fillers and the influence of GF is even stronger. The effect is demonstrated in Figure 1. Anisotropic fillers orientate during processing, which enhances their reinforcing effect. In fact, the reinforcing effect depends very much on the distribution of orientation. Orientation and orientation distribution is determined by the flow pattern during processing, and also by the filler content of the composite.

Hardness

Hardness of the filler has a strong effect on the wear of the processing equipment, but this is influenced also by the size and shape of particles, composition, viscosity, speed of processing etc.

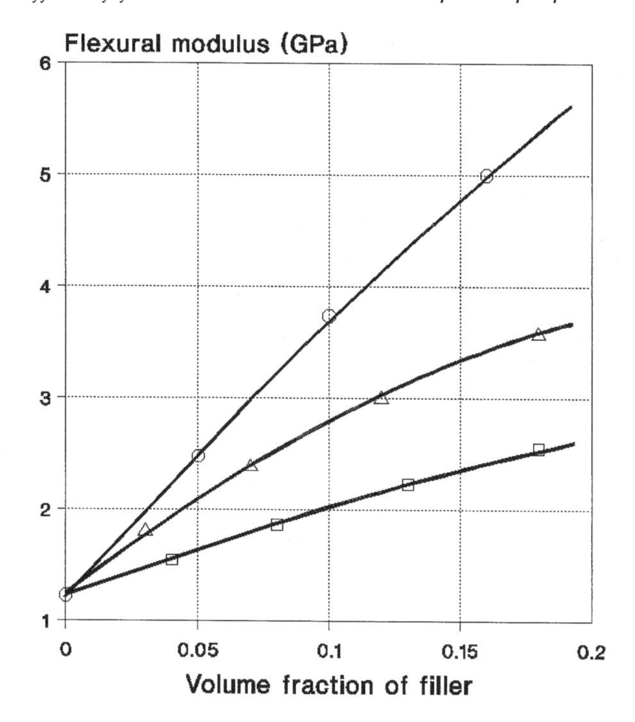

Figure 1 Effect of filler anisotropy on the flexural modulus of PP composites; (○) glass fiber, (△) talc, (□) $CaCO_3$.

Surface free energy

Surface free energy (surface tension) of the fillers determines both matrix/filler and particle/particle interaction. The former has a pronounced effect on the mechanical properties, the latter determines aggregation. Both interactions can be modified by surface treatment.

Thermal properties

Thermal properties of fillers differ significantly from those of PP. This has a beneficial effect on productivity and processing. Decreased heat capacity and increased heat conductivity reduce cooling time. Changing thermal properties of the composites result in a change of the crystalline skin-core morphology and thus properties of injection-molded parts as well. Large differences in the thermal properties of the components, on the other hand, lead to the development of thermal stresses, which also influence the performance of the composite under external load.

Filler content of PP is usually considerable, in the case of both extenders and functional fillers. Composite properties, however, depend strongly on composition. The dependence is usually a function of particle characteristics. Performance of a filler in any composite, therefore, must be determined always as a function of filler content.

REFERENCES

1. Katz, H.S. and Milewski, J.V. (1978) *Handbook of Fillers and Reinforcements for Plastics*, Van Nostrand, New York.
2. Rothon, R. (1995) *Particulate-Filled Polymer Composites*, Longman, Harlow.
3. Schlumpf, H.-P. (1983) *Kunststoffe*, **73**, 511.
4. Pukánszky, B. (1995) Particulate filled polypropylene: structure and properties, in *Polypropylene: Structure, Blends and Composites*, Vol. 3., (ed. J. Karger-Kocsis), Chapman & Hall, London, p. 1.

Keywords: filler, aspect ratio, particle shape, calcium carbonate, talc, platelets, reinforcement, glass fiber, kaolin, particle size, particle size distribution, chemical composition, adhesion, interface, aggregation, specific surface area, flow-induced orientation, hardness, surface free energy, surface tension, surface treatment, mechanical properties, thermal properties.

Fire hazard with polypropylene

T.J. Shields and J. Zhang

INTRODUCTION

Uncontrolled fires cause much loss of human life and property damage every year. Fire in simple terms is the consequence of complex chemical and physical reactions induced by heating materials to their respective ignition temperatures. The growth and development of a fire can be represented by a number of stages or states, i.e. ignition, fire growth, fully developed fire, decay and finally extinction. At each stage in the progress of a fire, many hazards usually in the form of heat energy and combustion products are generated which may pose a threat to life and property. Generally, the fire hazards associated with material are referred to as ignitability, flame spread, heat release, smoke production, toxicity and corrosivity. However, the fire hazard associated with a material is not only dependent on the nature of the materials but is also a function of the environment in which the fire occurs, i.e. ventilation rates and geometry of the enclosed space as well as the orientation of the material. Different end-use applications of the same material or component may create different fire hazards.

FIRE HAZARD

Ignitability

Ignitability may be considered in terms of the ease with which the material or its pyrolysed products may be ignited under given conditions of temperature, pressure and oxygen concentration. In terms of fire hazard, if a material can be ignited more easily than another material,

Polypropylene: An A–Z Reference
Edited by J. Karger-Kocsis
Published in 1999 by Kluwer Publishers, Dordrecht. ISBN 0 412 80200 7

Table 1 Cone calorimeter test results for thermoplastics [2]

Polymers (non flame retarded)	*Thickness (mm)*	*Heat flux* (kWm^{-2})	*Time to ignition (s)*	*Heat release rate* (kWm^{-2})		*SEA** (m^2kg^{-1})
				Peak	*Overall*	
HDPE	6	20	403	453	265	326
	6	30	171	866	396	396
	6	40	91	944	453	367
	6	50	58	1133	478	320
PP	3	20	120	377	192	465
	3	30	63	693	199	447
	3	40	35	1095	199	455
	3	50	27	1304	187	476
ABS	6	20	299	683	447	1017
	6	30	130	947	615	1052
	6	40	68	994	634	1053
	6	50	43	1147	794	1090
Rigid	3	20	–	–	–	–
PVC	3	30	487	90	42	1088
	3	40	276	107	48	1019
	3	50	82	155	54	861
Plasticized	3	20	140	126	81	928
PVC	3	30	61	148	83	999
	3	40	41	240	105	967
	3	50	25	250	102	1073

* specific extinction area.

given all other factors are equivalent, such a material is more likely to contribute to the initiation and propagation of fire. Polypropylene (PP) will start to decompose at approximately 500K and ignite when its surface temperature is at approximately 610K [1]. PP has a relatively lower time to ignition than some other thermoplastics as shown in Table 1 [2] but exhibits a higher ignition time (45 s) than PMMA (22 s) when tested under the cone calorimeter protocol which employs an incident radiant flux of 50 kW m^{-2} [3]. Since ignitability can be affected by minor variations in the ingredients in polymer compounds and test environments the ignition time data reported for PP should therefore be treated as indicative.

Flame propagation and spread

Direct contact with flame or exposure to increasing levels of incident heat radiation can create a serious threat to life and property. Flame propagation also contributes significantly to the spread of fire to unburnt regions and other fuel sources. Thus the rate of flame spread is considered an important parameter in the growth, development and spread of fire. When ignited, depending on the ignition power of the ignition source, PP burns slowly at first with fairly gentle flaming but burns vigorously with large flames when exposed to a high-temperature environment. A PP specimen with a diameter of 100 mm and a thickness of 25 mm can, when exposed to an irradiant heat flux of 50 kW m^{-2}, generate flames as high as 1.5 m when the boiling pool of liquid polymer has formed [1]. The rate of flame spread on the surface of a PP sheet depends on the orientation of the material and the intensity of the heat load [3].

Heat release

The rate of heat release generated by a combusting material is considered to be a most important single parameter, i.e. it is a measure of the size of a growing fire and is related to the production rate of smoke and toxic effluents.

The net heat of combustion of PP is approximately 43.4 MJ kg^{-1} [1] which is higher than most other plastics and thus has a potentially higher rate of heat release when involved in fire. Although there is some scatter in the reported heat release rate data for PP, the measured rates of heat release for PP are generally higher than those for most other thermoplastics as shown in Table 1.

Smoke emission

Smoke generated from combusting materials consists essentially of air-borne solid and particulate, aerosols and gases. Smoke is directly responsible for the majority of deaths and injuries in fires when compared with exposure to flames and radiation. Polymers which contain aromatic rings or halogen compositions usually emit more smoke than simple hydrocarbon polymers. However, smoke emission is also related to heat release rates, e.g. in a real fire situation, a material with a lower smoke emission rate but having higher heat release rate is not necessarily better than one with a higher smoke emission rate but lower heat release rate. The former may burn up completely and thus release more smoke while the latter may self extinguish at an earlier stage and thus produce less smoke.

Compared with polymers containing halogen, such as PVC, and polymers consisting of large amounts of aromatic ring structures, such as polystyrene, PP falls into the category of polymers with low smoke emission as shown in Table 1. However, considering the high peak heat release rate, PP will, in real fire situation, produce large quantities of smoke very quickly thus representing a considerable hazard.

Toxicity

Carbon monoxide has been identified as the primary cause of death of most fire victims. Apart from carbon monoxide and carbon dioxide, the evolved products from PP in a fire include pyrolysis products, which are primarily saturated aliphatics, and combustion products, which are mainly a mixture of simple and complex aldehydes and ketones. However, carbon monoxide is generally considered to be the principal resulting toxicant from burning PP. The LC_{50} value (for definition, see later) measured for PP is $17\,mg\,l^{-1}$ and the severity of toxicity associated with PP is considered similar to PE and no worse than wood [4]. It is generally accepted that PP is one of the polymers with least toxic potential in a fire situation because of its simple hydrocarbon structure [1].

Corrosivity

In fires, corrosive effluents are generated from burning materials which may be deposited on and contaminate the surface of materials exposed within the environment. Although all fire effluents may be corrosive to some degree, acid gases, such as HCl and H_2SO_4, are particularly corrosive on metals. PP, however, is a simple hydrocarbon polymer which is not expected to generate such corrosive gases. The results obtained from an intensive corrosivity study on polymer materials using four existing testing methods showed that an intumescent (phosphorus flame retarded) PP has much lower corrosivity than PVC [5]. For instance, the metal loss during combustion for the intumescent PP is 66.3 nm, for PE (non-flame retarded) 14.5 nm and for PVC more than 250 nm (measured by ASTM E05.21.70); the pH value for the PP is 5.58, for the PE 3.58 and for the PVC 1.75 (measured by DIN 57 472) [5].

FIRE HAZARD ASSESSMENT

There are various fire hazard test methods associated with PP and its products, including international and national standard test methods. Basically two groups of tests are used, i.e. small scale and large scale tests. The small scale tests rely on relatively simple fundamental reaction-to-fire parameters while large scale tests generate more realistic fire performance or behavior of the materials under end-use conditions. Assessment methods for PP are generally included in those for polymers. For derived end-use products of PP, such as textiles, cables or building products, additional specific testing methods are required. Some of the important assessment methods are discussed in this section.

Ignitability

Ignitability tests are designed to assess the ease of ignition of pyrolysis products of a material under given conditions of temperature, pressure and oxygen concentration. Generally, two ignition sources are used, i.e. the direct flame impingement or contact with a heated object and exposure to electric resistance elements or arcs. *ISO 5657–1986 (E), Fire Test – Reaction to Fire – Ignitability of Building Products* employs a horizontal specimen with a diameter of 150 mm which is exposed to a radiant heat flux from a cone-shaped heater and uses a small pilot flame to ignite the specimen. A similar method but using a spark igniter and 100 mm^2 sample is the cone calorimeter method as used in *ISO 5660, Fire Tests – Reaction to Fire – Rate of Heat Release from Building Products (Cone Calorimeter Method)* and *ASTM E1354, Test Method for Heat and Visible Smoke Release Rates for Materials and Products Using Oxygen Consumption Calorimeter*. In addition, *ASTM D 1929, Standard Test Method for Ignition Properties of Plastics* uses a hot-air furnace capable of reaching 750°C to test for the flash ignition and self-ignition temperature of plastics.

Flame spread

Flame spread may be defined as the rate of travel of a flame front under given conditions of burning. Directions of flame spread may be upward, downward or lateral depending on the orientation of sample tested and test conditions. The mechanism for flame spread is either concurrent (wind-aided), i.e. air flow direction is same as flame spread direction, or opposed, i.e. air flow is opposed against flame spread direction. *ISO 12992, 1995 Plastics – Vertical Flame Spread Determination for Film and Sheet* is designed to vertically test plastics in thickness 3 mm or less, subject to a small ignition flame. There are various test methods which accommodate inclined specimens, e.g. *ASTM E 648* for 0°; *ASTM D 1230* for 45° and *ASTM D 3014* for 90°, and also, for different end-use products such as *ASTM D 2859* test for floor coverings, *ASTM D 568* test for measuring the upward flame spread with specimen 450 by 25 mm using a 25 mm flame and *ASTM E 108* test for roof coverings.

Heat release

There are various standard methods available to assess the heat release rate. Widely used bench scale methods include the cone calorimeter which is based on oxygen consumption principle, such as *ISO 5660* and *ASTM E1354*, and the Ohio State University (OSU) heat release apparatus (accepted for *ASTM E906 Test Method for Heat and Visible Smoke Release Rates for Materials and Products*) which measures heat release rate by the sensible enthalpy rise method. The large scale heat release tests using oxygen consumption technique, such as the room/corner test (ASTM proposal) and furniture calorimeter (NBSIR 82–2604) have been developed for end-use tests [1].

Smoke emission

Most smoke emission tests measure the optical density of smoke released from burning materials. There are various standard testing methods available for assessing smoke emission. *ASTM E662 Specific Optical Density of Smoke Generation by Solid Materials* (NBS smoke chamber) uses a completely closed smoke chamber and measures the optical density of smoke accumulated in the enclosure. *ISO 5659–2, Plastics – Smoke Generation Part 2: Single Chamber Static Method* improves on the traditional NBS chamber and was published recently. The cone calorimeter is used for measuring dynamic smoke production, such as *ASTM E1354* and ISO proposal *ISO 5659–3 Plastics – Smoke generation Part 3: Determination of Dynamic Optical Density.*

Toxicity

Combustion toxicity is assessed by either identifying and analyzing the chemical compounds in the gaseous combustion products or studying the effects of gaseous combustion products on laboratory animals. Testing methods for analysis of gases evolved during pyrolysis and combustion include *ASTM E 800 – Measurement of Gases Present or Generated during Fires* and the French *NF X 70–101 Test Chamber Method, Analysis of Gases from Combustion or Pyrolysis.*

The laboratory animal tests are to analyze the number and cause of deaths and incapacitation of small animals like rats or mice when the animals are exposed to the gaseous products of decomposition and combustion under carefully controlled conditions. Lethality parameters usually used include LD_{50} (dose lethal for 50% of the test animals), LC_{50} (concentration lethal for 50% of the test animals) and LT_{50} (time lethal for 50% of the test animals). Various such toxicity tests are currently used such as *NBS Cup Furnace Method* and *DIN 53436 Tube Furnace Method* although none of these have been as yet accepted by standards bodies such as ISO and ASTM.

Corrosivity

Traditional standard corrosion tests assess the effects of corrosive gases on metallic surfaces in terms of the pH and conductivity of aqueous solutions of combustion products, the yield of soluble metal ions from corrosion of metal, or change in resistance of a circuit board exposed to corrosive gases. *IEC 754–1 Tests on Gases Evolved during Combustion of Electronic Cable Components* and *DIN 57 472 Part 813 Testing of Cables, Wires and Flexible Cords; Corrosivity of Combustion Gases* measure changes in pH and conductivity of aqueous solutions used to adsorb combustion gases. However, combustion corrosivity is not only associated with acids such as halogenated gases but all combustion effluents. Recently published *ISO 11907–2 1995, Plastics – Smoke Generation – Determination of Corrosivity of Fire Effluents, Part 2, Static Method* (based on French *CNET* corrosivity test) involves measuring in a printed wiring board exposed to combustion products. Direct and dynamic test methods which measure overall effects of combustion effluents on metal surface are under development, such as *ASTM D09.21.04 Fire Response Standard for Determining the Corrosive Effect of Combustion Products Using a Cone Corrosimeter* (also *ISO DIS 11907–4 Dynamic Determination Method Using a Conical Radiant Heater*) which uses the cone calorimeter to measure corrosion by introducing a special sampling system [5].

FIRE PROTECTION

Modern fire protection techniques have developed through multidiscipline activities in science and engineering and involve careful selection of materials, end-use product design and manufacture, and fire performance and risk assessment. Reducing the flammability of materials is still an essential primary fire safety consideration. PP is an inherently flammable material and therefore flame retardant treatment of the polymer is an essential consideration in relation to fire safety. For material scientists, however, it is also very important to be aware of developments in fire safety science and engineering including the transition from prescriptive to functional regulation of fire safety in many countries.

REFERENCES

1. *The SFPE Handbook of Fire Protection Engineering*, (1995) 2nd edn, National Fire Protection Association, Quincy, MA, USA.
2. Scudamore, M.J., Briggs, P.J. and Prager, F.H. (1991) *Fire and Materials*, **15**, 65–84.
3. Zhang, J., Shields, T.J. and Silcock, G.W.H. (1996) Proceedings of Interflam '96, Interscience Communications, London, pp. 853–857.
4. Gad, S.C. and Anderson, R.C. (eds) (1990) Combustion Toxicology, CRC Press, Boca Raton, FL.
5. Kessel, S.L., Rogers, C.E. and Bennett Jr, J.G. (1994) *J. Fire Sci.*, **12**, 197–233.

Keywords: corrosion, fire, fire hazard, flame spread, heat release, ignition, smoke emission, toxicity.

Flame-retardant polypropylene compositions

S. Bourbigot, M. Le Bras and R. Delobel

INTRODUCTION

Economical manufacturing methods of mass production and improvements to the properties of finished products have in many applications greatly helped to replace traditional materials, such as metals or wood, with plastics and rubbers. In particular, polypropylene (PP) is the fastest growing commodity plastic world-wide. It has found its place in many sectors such as building, transportation (automotive, railways, etc.), electrical engineering (electrical/household appliances, housings, etc.) or paper industry.

The use of organic polymer systems, such as PP, which are flammable, leads to greater fire risks and thus to the growing importance of flame retardancy. Recent events have shown the need to improve fire resistance. One can notice as an example, the PP fire at the BASF plant (March 1995 in Teeside, UK) which has been described as one of the largest fires ever seen in the UK during peacetime (more than 10 000 tonnes of PP were consumed) [1]. This clearly indicates the need for a comprehensive appraisal of all aspects of the combustion of polymers and the ways to prevent it. Reductions in the propensity of organic materials to ignite or emit dense and/or toxic fumes are equally important.

Ignition of polymers occurs either spontaneously or from an external source if the concentration of volatile combustible products evolved by pyrolysis or thermo-oxidative degradation of the polymer is within the flammability limits. A self-sustained combustion cycle is then triggered.

Polypropylene: An A–Z Reference
Edited by J. Karger-Kocsis
Published in 1999 by Kluwer Publishers, Dordrecht. ISBN 0 412 80200 7

This is driven by the heat of the flame promoting the pyrolysis of the polymer. The process will continue as long as the heat transmitted to the polymer is sufficient to keep its rate of thermal degradation above the level required to feed the flame, otherwise it will extinguish. The self-sustained combustion cycle occurs in both the condensed and gas phase. This means that in order to extinguish the flame by depressing the rate of chemical and/or physical processes taking place in one or both phases, polymers have to contain a variety of additives that may act as fire retardants or they have to be modified (chemical or physical modifications) to resist fire [2].

There are therefore three main techniques to create flame retarded polymers:

- synthesis of polymers with low flammability;
- chemical and physical modification of polymers;
- application of flame retardants.

This chapter will review various flame-retardant technologies which are applicable in PP matrices. Halogen and nonhalogen flame retardants will be examined.

MECHANISMS OF FLAME RETARDANCY

The high carbon and hydrogen content of PP (like many other synthetic polymers) makes it combustible. To enable PP to be used safely, flame retardants are added in applications where regulation or specifications require an enhanced flame-retardant capability over and above the unmodified polymer. The flame retardant will not make the resulting compound non-combustible. It will however make the material more ignition resistant. By adding flame retardants, the flame spread time during burning is considerably increased compared to the unmodified PP.

Flame retardants exert many different modes of action as a function of the chemical nature of the polymer–flame-retardant systems and of the interactions between the components. It is considered that inhibition of burning is achieved by modification of either the condensed phase or the dispersed or gaseous phase in a physical and/or chemical mode. In the condensed phase, the following may occur [3]:

- dilution of the combustible content;
- dissipation of heat, shielding of surface by charring or forming of a glassy coating or gas or liquid barrier reducing the heat transfer;
- alteration of the distribution of the condensed pyrolyzates;
- lowering of substrate's temperature (due to the endothermic decomposition of the flame retardant);

- influence on the combustion mechanism in both condensed and gaseous phases (flame retardant interrupts the chain reaction by radical acceptor species), while in the dispersed phase, by generation of noncombustible gases that dilute the fuel gas and tend to exclude oxygen from the polymer surface.

TESTING METHODS

As shown above, burning implies a great number of parameters of mutual influence. In such a complex system, it is impossible to measure all variables and the attempts to standardize the testing methods have led to the development of specialized norms. It is therefore good practice to describe fire performance of plastics by code numbers or letters related to a special test rather than by perhaps misleading descriptions such as 'self-extinguishing' or 'nonburning'. If needed, statements such as 'flammable with difficulty' (by a flame) or 'ignitable' ('by radiation' 'by this test' might be used). In the following sections, we present the usual tests used to determine the fire proofing performances of flame retarded PP.

Underwriters' laboratory (UL 94)

UL 94 is used to measure burning rate and characteristics based on standard samples. The ratings are HB, V-2, V-1 and V-0. 'HB' means that once ignited the sample will continue to burn but at a controlled rate. 'V-0' is the best rating and is often required for flame retarded polymers in many sectors. PP is HB.

Oxygen index (ASTM D2863)

The oxygen index is the minimum percentage of oxygen it takes in an air-like gas mixture to support flaming combustion. In their unmodified state, i.e. no flame retardants, polymers vary greatly in their ability to support combustion at normal atmospheric conditions. The relative resistance to burning is based on the chemical composition of each polymer. The oxygen index is a method of ranking the polymers based on combustion. As the normal amount of oxygen in air is approximately 21%, polymers with indexes below 21 (commonly around 18% like PP) will burn continuously at normal conditions.

Flammability and flame spread tests

A flame is applied for a short time to the edges and/or on the surface of plastics specimens. Criteria for grading include extinction, afterflame, afterglow time and/or mass burning rate and area of burned material. The articles are tested in a horizontal position (ASTM D 635; ISO 1210) or more severely, vertically (ASTM D 568 for flexible plastics and ASTM D 3713 ignition response index (IRI)). These standards should be consulted for the different test conditions and coding of results. For surface flame action on sheets which are horizontal or in the 45° position, see ASTM D 1433; VDE 0340. Ignition by radiation (ASTM E 162) is applied in the internationally used 'Flooring Radiant-Panel Test' (ASTM E 648). ASTM 1929; ISO 881 give standard test methods for the flash and self-ignition temperatures of plastics.

Glow wire test

The Schramm–Zebrowski incandescent rod resistance method (ASTM D 757; ISO/R 181; DIN/VDE 0318 part 2) has been adopted internationally for solid electrical insulating materials. A glowing silicon carbide rod (950°C) is pressed against the specimen for three minutes. BH1 indicates nonignition and less than 5 mm penetration by the rod. BH2 indicates >5 mm penetration in three minutes. If the rod penetrates more than 95 mm in the test time, then BH3 gives the burning rate in mm/min. PP is BH3-20 mm/min.

Cone calorimeter (ASTM E 1354)

The 'Cone Calorimeter' was based on a design by Vytenis Babrauskas (Fire Research, National Bureau of Standards) in the early 1980s [4]. This apparatus allows simultaneous and continuous determinations to be made of heat release rate, smoke production rate, mass loss rate and concentration of the various combustion gases formed. It is capable, at the same time, of being used to obtain ignition, heat of combustion and soot production data for the materials tested. As an example, PP has a rate of heat release maximum of 1800 kW/m^2 (heat flux = 50 kW/m^2, thickness of the sheet = 3 mm).

CLASSIFICATION OF FLAME RETARDANTS [5]

Despite the large number of different flame-retardant additives, most commercial flame retardants currently belong one of the following six chemical classes: hydrates of aluminium or magnesium, organobromine compounds, organophosphorus (including halogenated phosphorus) compounds, antimony oxides and boron compounds.

Many inert fillers and additives, such as talc, $CaCO_3$ and clays, can act also as flame retardants or thermal sinks. At high loading (>40 wt.%), they lead to mass dilution and slow down heat generation, promoting charring and reducing flammability and smoke generation.

Table 1 Flame retardancy in condensed and gas phases

Condensed phase	
Mode of action	*Flame retardant*
Endothermic degradation	$Al(OH)_3$, $Mg(OH)_2$, ...
Dilution	Inert fillers : talc, $CaCO_3$, ...
Thermal shield	Intumescent systems
Gas phase	
Mode of action	*Flame retardant*
Radical inhibition	Halogenated compounds, phosphorus species, Sb_2O_3/Halogenated compounds, ...
Dilution leading to the decrease of the flame temperature	Products may evolve : CO_2, H_2O, HX, ...

Generation of volatile, combustible gases can also be prevented by protecting the surface against the heat of the flame by a thermal barrier such as an intumescent coating.

Smoke suppressants are also important additives in fire retardancy. Resulting from an incomplete combustion, opaque smoke may evolve that leads to panic and slow the rescuers' progress. Additives, such as antimony oxides, metal borates, hydrates of magnesium or aluminium and magnesium oxychlorides, are used as fillers and flame retardants, but they are also good smoke reducers.

Flame retardants can act at different stages of the combustion process; hence in many cases a combination of fire retardants are employed. Sometimes a synergistic effect can be observed. For example, antimony oxides serve as synergists and strengthen the effect of organic halogenated compound. Other commonly used synergists are boron compounds, titanium oxides, molybdenum oxides, zinc oxides, silicates, phosphorus-halogens, etc. The synergists are used for economic reasons or for obtaining the properties of interest of the plastics.

To resume this classification, Table 1 gives the different modes of action of the flame retardants in both condensed and gas phases.

FLAME RETARDANTS FOR PP

Hydrates of aluminium and magnesium [6]

Inhibition in the condensed phase implies mainly physical processes. The strong endothermic degradation of the hydrate slows down the rate of pyrolysis. The evolved water dilutes the combustible gases issuing from the degradation of the plastic and cools the flame, while the residue constituted in Al_2O_3 or MgO acts as a thermal barrier on the polymer surface (reactions 1 and 2) and eventually promotes charring.

$$2\ Al(OH)_3 \rightarrow Al_2O_3 + 3\ H_2O \qquad \Delta H = 1247\ kJ/mol \qquad (1)$$

$$Mg(OH)_2 \rightarrow MgO + H_2O \qquad \Delta H = 1590\ kJ/mol \qquad (2)$$

The decompositions of these hydrates are in the temperature range 180–240°C in the case of $Al(OH)_3$ and in the temperature range 330–460°C in the case of $Mg(OH)_2$. This implies therefore that $Al(OH)_3$ is little used in PP because of its processing temperature (around 220°C).

The use of these hydrates has also an important drawback: 40–65 wt.% must be used in PP to provide the fire performances of interest. This is associated with material stiffening and embrittlement. Nevertheless some grades of magnesium hydroxide have been developed by producers such as Dead Sea Bromine, Morton Thiokol or Solem which can be used in PP at high loading (65 wt.%) and which may keep the rheological properties of PP. These grades are generally surface treated by compounds such as fatty acid, terpolymer with carboxylic groups or modified polyolefins; the products are fine powders with average particle size of around 1–2 μm. The performances that can be obtained with such additives are summarized in Table 2. It can be noted that in developing a flame retarded PP compound with magnesium hydroxide one should select a grade which gives the desired flame retardancy rating at minimized influence on the key mechanical properties. The formulator should be aware of specialty treated grades available in order to meet these demands.

Table 2 Performances of flame retarded PP loaded at 65 wt.% by some grades of $Mg(OH)_2$ or $Al(OH)_3$

Formulation	*UL 94 V Rating (3.2 mm)*	*Heat Distortion Temperature (°C)*	*Melt Flow Index (MFI) (5 kg/230°C) (g/10 min)*	*Notched Izod Impact (ASTM D 256) (J/m)*
PP	No rating	56	20	34
PP-$Mg(OH)_2$ treated by fatty acid	V-0	62	3	27
PP-$Mg(OH)_2$ treated by terpolymer with carboxylic source	V-0	80	4	19
PP-$Mg(OH)_2$ treated by modified polyolefins	V-0	67	19	25
PP-$Al(OH)_3$	V-0	–	–	–

Halogen-containing compounds [7]

Mode of action by radical transfer

Due to the release of halogenated acid during decomposition, halogen-containing compounds interrupt the chain reaction of combustion by replacing the highly reactive $OH^{\bullet}$ and $H^{\bullet}$ radicals by the less reactive halogen $X^{\bullet}$ according the following reactions (3–5) where RX and PH represent respectively the flame retardant and the polymer:

$$R—X + P—H \rightarrow H—X + R—P \qquad (3)$$

$$H—X + H^{\bullet} \rightarrow H_2 + X^{\bullet} \qquad (4)$$

$$H—X + OH^{\bullet} \rightarrow H_2O + X^{\bullet} \qquad (5)$$

By dissipating the energy of the $OH^{\bullet}$ radicals by trapping, the thermal balance is modified which reduces strongly the combustion rate.

Mode of action by inhibition

This process occurs when the combination of antimony trioxide and a halogenated compound is used. These two additives act in synergy and the system is very efficient. In the mechanism of action it is assumed that the flame retardancy is due to SbX_3 according to the following reactions (6–10).

$$R—X + P—H \rightarrow H—X + R—P \qquad (6)$$

$$Sb_2O_3 + 2\ HX \rightarrow 2\ SbOX + H_2O \qquad (7)$$

$$5\ SbOX \rightarrow Sb_4O_5X_2 + SbX_3 \qquad (8)$$

$$4\ Sb_4O_5X_2 \rightarrow 5\ Sb_3O_4X + SbX_3 \qquad (9)$$

$$3\ Sb_3O_4X \rightarrow 4\ Sb_2O_3 + SbX_3 \qquad (10)$$

Among all species formed, SbX_3, HX and H_2O can be evolved in the conditions of a combustion. These are therefore those compounds that play the role of inhibitor in the gas phase.

Examples

Table 3 shows typical examples of using halogen-containing compounds in PP. Aliphatic-type flame retardants are excellent to achieve V-2 rating. The thermal stability of these compounds compared to PP's pyrolysis temperature is satisfactory for delivering HBr in sufficient concentration during burning. But when the processing temperatures of PP are over 230°C, aromatic brominated flame retardants are recommended for PP because of their better thermal stability. Organic bromine is more reactive

Table 3 Performances of flame-retarded PP loaded with halogen-containing compounds

Brominated aliphatic flame retardants

Ingredients	*Percentage (wt.%)*		
PP	100	97.15	58
HBCD		2.15	
EBTPI			22
Sb_2O_3		0.7	6
Talc			14
Properties			
UL 94 V Rating (1.6 mm)	NR	V-2	V-0
Dripping	Yes	Yes	No

Brominated aromatic flame retardants

Ingredients	*Percentage (wt.%)*						
PP	100	91.1	93.0	90.9	88.0	58	64.3
TBBA		5.2					
DBDPO			4.8				23.8
INDAN				5.8			
PBB-PA					8.5		
Sb_2O_3		1.7	1.6	2.6	2.8		11.9
Additives		2.1	0.6	0.7	0.6		
Properties							
MFI (g/10 min)	6	7	4	–	5	–	–
UL 94 V Rating (3.2 mm)	NR	V-2	V-2	V-2	V-2	V-0	V-0(1.6 mm)
Dripping	Yes	Yes	Yes	Yes	Yes	No	No
Notched Izod Impact (J/m)	21.5	26.7	21.5	21.5	21.5	–	21.5
Oxygen Index (%)	18	–	–	–	–	–	22

Chlorinated flame retardants

Ingredients	*Percentage (wt.%)*				
PP	100	55	61	60	63
Dechlorane		35	30	27	
Chlorinated paraffin					27
Sb_2O_3		4	6	13	10
Zinc borate		6	4		
Properties					
UL 94 V Rating (1.6 mm)	NR	V-0	V-0	V-0	V-0
UL 94 V Rating (3.2 mm)	NR	V-0	V-0	V-0	V-0
Oxygen Index (%)	18	–	–	–	27

NR, No Rating; HBCD, hexabromocyclododecane; EBTPI, ethylene bis(tetrabromophthalimide); TBBA, tetrabromobisphenol-A; DBDPO, decabromodiphenyloxide; INDAN, octabromotrimethylphenylindan; PBB-PA, poly(pentabromobenzylacrylate); Dechlorane, 2 mol hexachlorocyclopentadiene + 1 mol cyclopentadiene.

than chlorine and thus this is a more effective flame retardant. Nevertheless brominated compounds are generally more expensive than chlorinated ones. Chlorinated flame retardants are preferred when not too high fire proofing performances or well-defined rheological properties (due to their higher amount) are required.

Phosphorus [6–7]

Phosphorus is the most flame-retardant element known; typically 5 wt.% P in PP can provide fire-proofing properties. Organic or mineral phosphorus may be used as flame retardant but generally the performance even at high loading ($>$ 30 wt.%) remains poor and the weak compatibility between the additives and the polymeric matrix leads to a strong decrease of the mechanical properties of PP.

Moreover, organic halogen–phosphorus combinations are synergistic, producing flame retardancy at much lower concentrations. Typically where 5% P or 40% Cl are required for good performance, 2.5% P + 9% Cl work equally well; and where 5% P or 20% Br are required, 0.5% P + 7% Br work equally well. In many cases, 3–6 wt.% of organic bromine–phosphorus compounds is sufficient.

Intumescents [8]

In recent times, intumescent flame retardant systems have been very carefully studied for their use in flame-retarded plastics. The intumescent effect is achieved by using an acid source, then the acid formed initiates the dehydration and subsequent charring of the carbonific compound which is an oxygen-containing component (i.e. cellulose, an oxygen-containing plastic or pentaerythritol). A spumific compound is added which evolves a non-combustible gas and which leads to the formation of the carbonaceous foamed layer over the substrate. The molten resin itself or a resin binder covers the foam with a skin which prevents the gases from escaping. For further details about typical intumescent formulations and mechanisms of action, the reader could refer to the chapter in this book entitled Intumescent fire retardent polypopylene formulations.

CONCLUSION

In this article different actual strategies to flame-retarded PP are surveyed. The choice of the flame retardant depends on applications, specifications and/or legislation. As an example, the European searches for nonhalogen replacements because of the potential for halogenated compounds to form toxic dioxins and furans. The formulators have therefore to carefully balance material, processing and economic needs. Various flame retardants are available to meet those needs.

REFERENCES

1. Carty, P. (1996) *Fire Mat.*, **20**(3), 158.
2. Cullis, C.F. and Hirshler, M.M. (1981) *The Combustion of Organic Polymer*, Clarendon, Oxford.
3. Troitzsch, J. (1990) *International Plastics Flammability Handbook*, 2nd edn, Hanser, Munich.
4. Babrauskas, V. and Grayson, S.J. (1995) *Heat Release in Fires*, Chapman & Hall, London.
5. *Handbook of Flame Retardant Chemicals and Fire Testing Services* (1988) Technomic, Lancaster, PA.
6. Green, J. (1989) *Thermoplastic Polymer Additives*, (ed. J.T. Lutz), Marcel Dekker, New York.
7. Green, J. (1982) *Flame Retardant Polymeric Materials*, Vol. 3, (eds M. Lewin and E.M. Pearce), Plenum Press, New York.
8. Vandersall, H.L. (1971) *J. Fire Flamm.*, **2**, 97.

Keywords: flame retardancy, flame retardants, flammability, ignition, oxygen index, UL-94, cone calorimeter, burning rate, smoke suppression, intumescent coating, fire proofing, char formation, rate of heat release.

From quality control to quality assurance in injection molding

D. Meyer

Modern injection-molding machines are currently able to work according to deliberately chosen processing parameters. It is possible to set tolerance values, for example, with respect to injection pressure, injection time, injection speed, injection rate, material heating zones, cavity pressure, cooling time, cycle time and others.

Parts molded outside of the chosen tolerance are sorted out. This happens either by separation tools on the bottom of the mold or by suitable conveyors. These 'NIO' parts (rejected parts) hold information about the cause of a quality problem. By statistical analysis, the reasons for the deviations from the set limits can be discovered.

Abbreviations used in this field are:

- AQP Advanced Quality Planning
- CAQ Computer Aided Quality
- CPC Continuous Process Control
- SPC Statistical Process Control
- SQC Statistical Quality Control
- TIA Totally Integrated Automation
- TQM Total Quality Management

The starting point of TQM is an audit of personnel organization and related job descriptions. AQP, on the other hand, starts with a design and feasibility review. Key issues in quality control (QC) and quality assurance (QA) are failure mode effects and analysis (FMEA), control plans with SPC and inspections by audit and customer [1].

Polypropylene: An A–Z Reference
Edited by J. Karger-Kocsis
Published in 1999 by Kluwer Publishers, Dordrecht. ISBN 0 412 80200 7

By using portable measuring equipment, the QC measurements can be performed alongside the injection-molding machine. The equipment carried on trolley is usually an efficient PC, printer, digital calipers, micrometers, scales and contact-less optically measuring tools. It is more straightforward to transmit all the measured data by a datalogger or multiplexer over a row of inputs from inductive measuring tools. All tools, equipped with analogue, BCD and IEC exits, can be connected. The programs allow a flexible and efficient use of the equipment.

The software program is tailored for each molded piece. An operator, led by a menu, performs all the indicated measurements. All sampling, the period between the statistical analysis of the data, remarks of the inspector are recorded under the corresponding design number of the part on the data-disc. The results of the measurements appear on the monitor along with the actual and index values and tolerance limits.

The software package covers the recent requests of the ISO 9000 standard and may be displayed on a Local Area Network (LAN) of a data acquisition and storing system. In this way, the customers' inspection forms and long term statistics can be drawn and supplier validations can be stored. The program supports the printing of reports and histograms requested by the customer.

QA, with the help of automatic handling systems, has become very popular for the molding of thermoplastics. To equip the production machine with a handling system and measuring tools supports the quality assurance *per se*. It is also possible to instruct the handling system via CAQ on how to select parts produced according to the set tolerance. For example, a precisely defined scale can be used to measure whether a part has completely filled all requirements, or it can be checked whether or not the molded items contains all inserts. Every 'IO' part is placed by the handling system on the correct conveyor belt or goes into the box. The 'NIO' parts land on a separate conveyor belt and the supervisor can intervene immediately. He decides also how to bring the production back to the required tolerance.

A zero-defect production of injection-molded parts is guaranteed when internal and external quality controls work properly together. In this case a QA protocol can be sent to the customer. All ranges of protection are specially significant. The automated quality control for injection molding is illustrated in the following example.

The handling system or robot gives a signal to quality assurance when the product is placed on the measuring tool. According to the result of the quality check, the robot receives an 'IO' or 'NIO' signal. At 'IO', the plastic parts are placed on a conveyor or directly into a box. At 'NIO', the part is placed in the rejected line of the conveyor belt. No-one is required to check the quality of inserts before they are placed in the injection mold. The decision on whether the produced parts are 'IO' or 'NIO' is made by measuring tools. Finally, a robot transports the parts with acceptable quality to the next process step. The sampling time of the quality check is controlled by the computer program.

REFERENCE

1. Keating, M. (1995) *How to Assure Quality in Plastics*, Hanser, Munich.

Keywords: ISO 9000, quality assurance (QA), quality control (QC), zero defect

Gamma-phase of isotactic polypropylene

J. Kressler

There are basically three crystal modifications known for isotactic polypropylene (iPP) which forms in the crystalline state exclusively a 3_1-helix caused by the energy minimum at this conformation [1]. The α-modification is the common form of iPP and well characterized already by Natta and coworkers. The β-modification occurs in a certain interval of crystallization temperatures and is usually induced by nucleation agents. Turner-Jones *et al.* assigned a triclinic unit cell to the γ-modification of iPP [2]. Later studies applying a Rietveld analysis showed that the γ-modification of iPP forms an orthorhombic unit cell. The original triclinic unit cell can be considered as a part of the orthorhombic unit cell. The main difference of the new model is the nonparallel chain packing in the crystal of the γ-modification which is unique in the field of synthetic polymers and was introduced by Meille *et al.* [3]. The orthorhombic unit cell of the γ-modification is formed by bilayers composed of two parallel helices. The direction of the chain-axis in adjacent bilayers is tilted with an angle of 80°. This is a unique packing arrangement for polymers but has been known e.g. for fatty acids. The dimensions of the unit cell are $a = 0.854$ nm, $b = 0.993$ nm, $c = 4.241$ nm. Lotz *et al.* have made a remarkable work to support this nonparallel chain packing model [1]. They pointed out that the electron diffraction experiments on a flat-on γ-phase single crystal resulted in the same patterns after clockwise 40° rotation and anticlockwise 40° rotation, which corresponds to an angle of 80° between the double layers of

Polypropylene: An A–Z Reference
Edited by J. Karger-Kocsis
Published in 1999 by Kluwer Publishers, Dordrecht. ISBN 0 412 80200 7

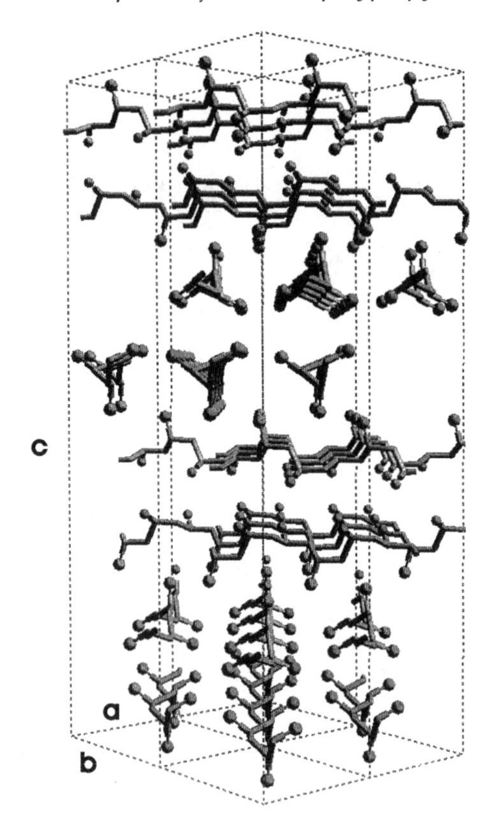

Figure 1 Four unit cells of the γ-modification of iPP according to the nonparallel chain packing model of Meille *et al.* [3]. The spherical entities represent the methyl groups.

helices arranged in the unit cell. Figure 1 shows four unit cells of the γ-modification formed by 3_1-helices of iPP.

Calculations of the packing energies of the γ- and α-modifications show that the γ-modification should be slightly more stable than the α-modification. There are several ways to influence the formation of different crystal modifications of iPP. The content of the γ-modification in iPP can be increased in the case of [4]:

1. using iPP prepared with metallocene catalyst systems having a relatively large number of stereo- and regioirregularities in the chain or by using very low-molar mass samples;
2. application of high pressure during the crystallization process;

3. using random copolymers of propene with other 1-olefins with 1-olefin contents between 2.5 and 20 wt.%;
4. crystallization in shear fields.

The ratio of the γ-modification to the α-modification is usually determined by wide angle X-ray (WAXS) measurements using the method of Turner-Jones. It is calculated directly from the ratio of height of the (130)-reflection of the α-modification (h_α) and of the height of the (117)-reflection of the γ-modification (h_γ). The content of the γ-modification, X_γ, is then given by

$$X_\gamma = h_\gamma \,/\, (h_\gamma + h_\alpha) \quad (1)$$

Figure 2 shows the WAXS traces of as-prepared metallocene iPP, and of the same sample isothermally crystallized at different temperatures.

The as-prepared iPP shows mainly the α-modification. The samples isothermally crystallized from the melt have increasing amounts of the γ-modification with increasing crystallization temperature up to an amount of nearly 100%. This iPP sample is synthesized using the homogeneous catalyst system rac.-ethylenbis(4,5,6,7-tetrahydroindenyl) zirconium-(IV)dichloride/MAO. This iPP has an M_w value of 181000 g/mol and M_w/M_n is 1.6. The sample contains 81.5% of mmmm pentads, 8.8% of mmmr pentads, 0.1% of rmmr pentads, 6.5% of mmrr pentads, 0.2% rrrm pentads and 2.9% of mrrm pentads, determined by nuclear magnetic resonance (^{13}C-NMR). The average length of isotactic chain segments is about 22.4 monomeric units, i.e. a length of 4.85 nm when arranged in a 3_1-helix. Obviously the microstructure of different iPPs has a large influence on the formation of different crystalline iPP modifications. The formation of the γ-phase can closely be related to the difficulties of chain folding in random copolymers or polymer chains with a large number of defects as stereoirregularities. The γ-modification has 4.39 chains per nm^2 emerging from the crystal surface into the amorphous phase. In contrast, the α-modification has 5.73 chains per nm^2. Therefore, the chains of the α-modification must backfold in order to guarantee that the amorphous phase has a lower density compared to the crystalline phase. It is reasonable that the sharp backfolding is hampered by chain irregularities or copolymer units because they need to be excluded from crystalline regions. This is different for the γ-modification. Because the chains have a lower density at the crystal surface, the backfolding is not necessary. Thus polymers with problems during the sharp backfolding tend to form the γ-modification.

It should be mentioned that the supermolecular morphologies formed by the γ-phase are very different from the typical spherulitic structures of α- or β-phase iPP [5]. Figure 3 shows four polarized light micrographs taken after different crystallization times of the same iPP sample as discussed above.

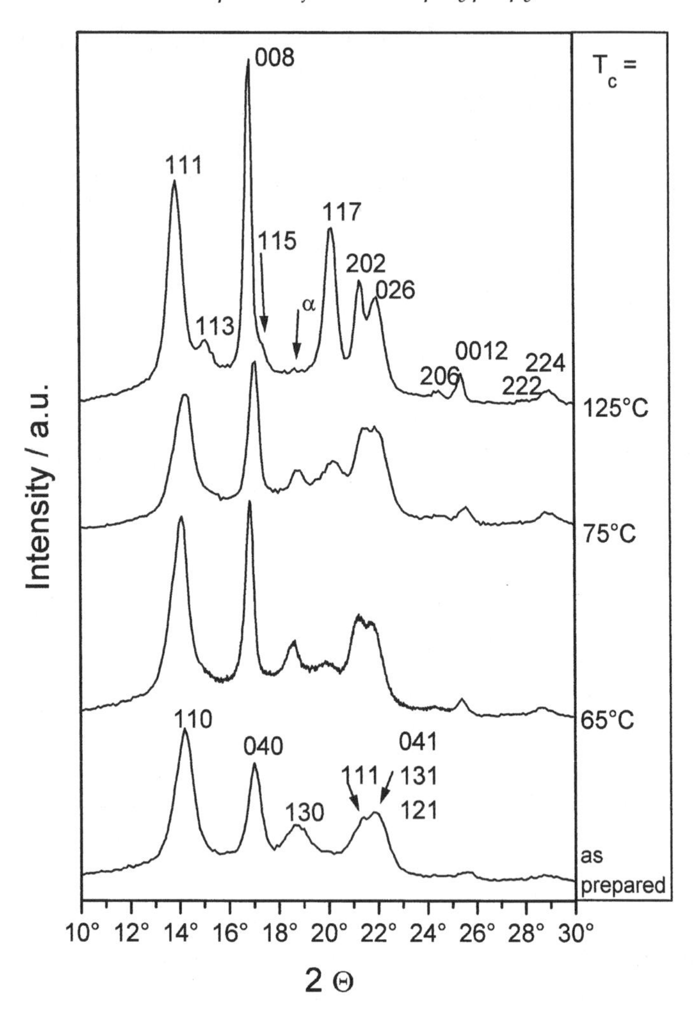

Figure 2 WAXS traces of as-prepared iPP (bottom) and of the same sample isothermally crystallized at 65°C, 75°C and 125°C. The peak assignment of the γ-phase iPP (top trace) is due to Brückner and Meille and of the α-phase iPP is due to Natta *et al.* The patterns of the as-prepared polymer and the sample isothermally crystallized at 65°C show exclusively or mainly the α-modification. With increasing crystallization temperature, the γ-modification is promoted and at $T_c = 125°C$ the γ-modification is formed nearly exclusively.

The micrograph taken after $t_c = 1{:}50\,h$ (t_c is the crystallization time) at a crystallization temperature of 120°C (upper left) shows elongated entities that are typical for the early stage of crystallization of α-phase lathes with γ-phase ongrowth similar to that reported for low molecular

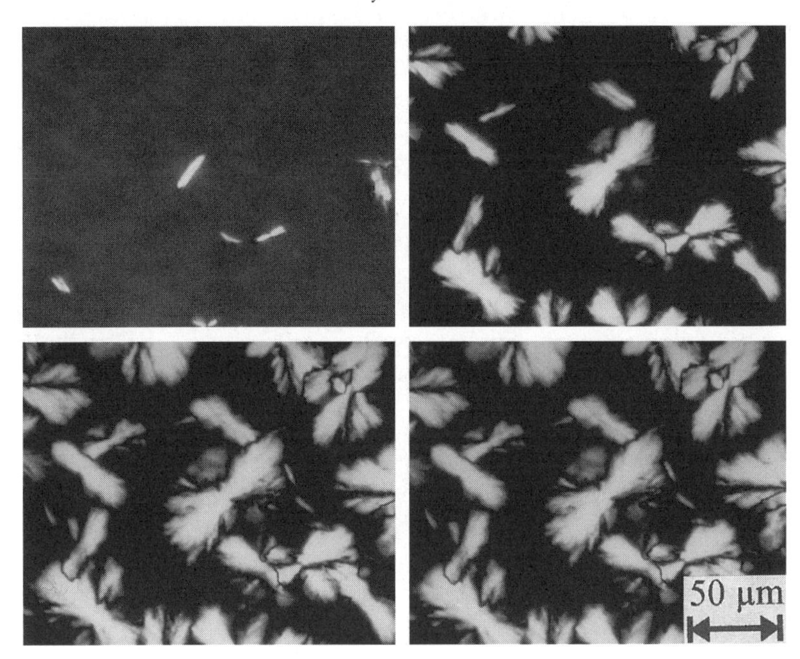

Figure 3 Polarized light micrographs taken during isothermal crystallization of a metallocene iPP sample (see text) at 120°C after different crystallization times: 1:50 h (upper left), 4:30 h (upper right), 23 h (lower left), 43 h (lower right).

weight samples. The micrograph taken at t_c = 4:30 h (upper right) shows more bundle-like entities. The micrographs taken after 23 and 43 h (lower left and lower right, respectively) do not show any significant difference indicating that the growth of the bundle-like entities is finished long before the bundles become space filling. After that, only diffuse structures between the needle-like entities appear, not visible in the polarized light micrographs. This material is crystallized completely in the γ-modification without any contributions of the α-modification.

Real spherulites are formed in samples where the α and γ-modification appear simultaneously and the amount of the α-modification exceeds 30 %. It is also possible to prepare single crystals of the γ-modification iPP. Also in this case an ongrowth of γ-phase iPP on α-phase lathes can be observed [1, 5]. Single crystals of the neat γ-phase have a triangular appearance [5].

REFERENCES

1. Lotz, B., Wittmann, J.C. and Lovinger, A.J. (1996) Structure and morphology of poly(propylenes): a molecular analysis. *Polymer*, **37**, 4979–92.

2. Turner-Jones, A., Aizlewood, J.M. and Beckett, D.R. (1964) Crystalline forms of isotactic polypropylene. *Makromolekulare Chemie*, **75**, 134–58.
3. Meille. S.V., Brückner, S. and Porzio, W. (1990) γ-isotactic polypropylene. A structure with nonparallel chain axes. *Macromolecules*, **23**, 4114–21.
4. Brückner, S., Philips, P.J., Mezghani, K. and Meille, S.V. (1997) On the crystallization of γ-isotactic polypropylene: a high pressure study. *Macromol. Rapid Comm.*, **18**, 1–7.
5. Thomann, R., Wang, Ch., Kressler, J. and Mülhaupt, R. (1996) On the γ-phase of isotactic polypropylene. *Macromolecules*, **29**, 8425–34.

Keywords: gamma phase, γ-modification crystallography, morphology, X-ray scattering, unit cell, single crystal, metallocene-synthesized, chain packing, stereoregularity, regioregularity.

Gas diffusion in and through polypropylene

J.A. Horas and M.G. Rizzotto

INTRODUCTION

A better understanding of transport phenomena in polymers is encouraged by the fact that a number of important practical applications depend wholly or in part on such phenomena. These applications include various protective coatings, selective barriers for the separation of gas and liquid mixtures, biomedical devices, packaging materials for foods and beverages, etc. Small molecules can be used as very sensitive 'probes' to explore a polymer matrix because their movement is strongly dependent on polymer structure and morphology.

Polypropylene (PP) is the most used commercially tactic polymer. The packaging industry uses about 25% of the PP output, mainly as films. This is due to its remarkable barrier characteristics. Its notable resistance against heat and solvents is mainly due to the fact that it is a partially crystalline polymer, which is dense and rigid, fuses near 165°C and has a low glass transition temperature ($T_g = -18$°C).

The main qualitative features concerning the transport of gases through polymer films can be described as the condensation and solution of gas at the one surface followed by diffusion through and evaporation to the gaseous state at the other surface.

The nature of the penetrant and the nature of the polymer are the main factors affecting the diffusivity or permeability of gases. Thus if the operating temperature is lower than or equal to the critical temperature of the penetrant vapour, T_c, there appears a strong concentration and

Polypropylene: An A–Z Reference
Edited by J. Karger-Kocsis
Published in 1999 by Kluwer Publishers, Dordrecht. ISBN 0 412 80200 7

pressure dependence of the diffusion and sorption coefficients in the rubbery state ($T > T_g$) and in the behavior of advancing fronts in the glassy state ($T < T_g$).

If the polymer is above its glass transition tempetature, T_g, it responds rapidly to changes in its physical condition and we have Fickian or Case I diffusion. This is the simplest case, and for $T > T_c$ Henry's law is valid for sorption and the diffusion coefficient is a constant (ideal Fickian diffusion). Its temperature dependence is well approximated by a simple Arrhenius expression with a constant activation energy.

In contrast, at temperatures below T_g, we have the so-called Case II and Super Case II transport, the other extreme, in which diffusion is very rapid compared with simultaneous relaxation processes. Sorption processes may be complicated by a strong dependence on swelling kinetics. Finally we have anomalous diffusion, which occurs when the diffusion and relaxation rates are comparable.

From a modelistic viewpoint we can divide the contributions in two groups:

1. *The phenomenological or free volume models.* These models are macroscopic or thermodynamic and consider that molecular transport is caused by a redistribution of the free volume, resulting from random fluctuations in the local density. Some of these models are those of Fujita, Vrentas and Dudda, Kreituss and Frisch, and others. These models were reviewed by Frisch and Stern [1] and also by Comyn [2].
2. *Molecular or detailed models.* These models are based on statistical mechanics analysis of specified motions of penetrant molecules and polymer chains, and take into account the pertinent intermolecular forces. Diffusion is visualized as a thermally activated process. The main goal of this type of model is the calculation of the activation energy for diffusion. Some of these models are the Barrer activated zone theory, the models of Brandt, DiBenedetto and Paul, Pace and Datyner and others, also reviewed by Frisch and Stern [1] and Comyn [2].

MEASUREMENTS AND MODEL APPLICATIONS TO RUBBERY CRYSTALLINE FILMS

In a tactic polymer such as polypropylene, the regularity of the lateral groups allows a compact arrangement of the chains, producing regions with a complex crystalline packing. Structures called spherulites are formed, with their sizes reaching some 150 μm, depending on the crystallization conditions and on the material. The amorphous phase, consisting of chains without any special organization, fills the interlamellar and interspherulitic space.

From the transport point of view, crystallinity has at least two effects on sorption and diffusion. First, at temperatures well below the melting point, crystalline regions are generally inaccessible to most penetrants. Hence they act as excluded volume for sorption and as impermeable barriers for the diffusion. Second, crystalline domains act as giant cross-linking regions with respect to those chains which enter and leave those regions from the surrounding noncrystalline matrix in which sorption and diffusion takes place.

The resulting decrease in diffusion (D) and sorption (S) characteristics can then be related to the total amount of crystalline material in the polymer, and so diffusivity can be expressed as:

$$D = D_a/\tau\beta \tag{1}$$

where D_a is the diffusion coefficient in the amorphous region, τ is a tortuosity factor expressing the necessity for the diffuser to avoid crystallites, and β is the immobilization factor, which is an empirical correction. β and τ increase with crystallinity. A further complication appears by medium inhomogeneities and the fact that the local diffusion coefficient is a strongly varying function of the penetrant concentration, ϕ_1.

Several models of the phenomenological or free volume type have been proposed to account for this concentration dependence of the diffusion coefficient. These models use the general expression

$$D^{\text{eff}} = D_a f(\phi_c) \exp[-B/v_f(\phi_1, \phi_c, T)] \tag{2}$$

where $f(\phi_c)$ is a function of the crystallinity fraction, ϕ_c, depending on model specifications, B is a constant factor reflecting the necessity of a smallest cavity size for a penetrant. The concentration dependence is considered through the free volume fraction $v_f(\phi_1, \phi_c, T)$ which is also explicitly dependent on crystallinity ϕ_c and temperature T. Thus a chain immobilization factor was included in a modified version of a Fujita free volume fraction and has been explicitly obtained using some simplifications and statistical considerations. The tortuosity effect and the one related to crystallite size, shape and its distribution in the polymer matrix are also included through the exponent appearing in a power law expression of the function $f(\phi_c)$ in equation (2).

This model is compared with experimental data of CCl_4 in ethylene–propylene copolymers of different composition, obtaining good agreement for the diffusion coefficient concentration dependence [3]. The same experimental data are contrasted by Kreituss and Frisch [4] using another free volume model.

The temperature dependence of the corresponding diffusion coefficient in PP and most rubbery semicrystalline polymers obeys an Arrhenius dependence. Molecular or detailed models are used to account for this temperature dependence, the goal being activation energy calculation.

As crystallinity affects the chain flexibility, the chain separation may be either symmetric or nonsymmetric for a diffusive step and any approach must take into account both separation types to properly describe the diffusive process. So in [5] a molecular model is developed, obtaining an expression of the activation energy needed for diffusion, in which both separations are considered. The chain's rigidity is accounted for by means of the flexibility modulus. The latter is related to the average chain immobilization factor, β, which is obtained from a free volume model. This model fits experimental data of activation energy with good agreement for diverse gases and vapors in polypropylene and other polymers.

REFERENCES

1. Frisch, H.L. and Stern, S.A. (1983) Diffusion of small molecules in polymers. *CRC Crit. Rev. Sol. Stat. Mat. Sci.*, **11**(2), 123.
2. Comyn, J. (ed) (1985) *Polymer Permeability*, Elsevier, London.
3. Horas, J.A. and Rizzotto, M.G. (1996) Gas diffusion in partially crystalline polymers. Part I: concentration dependence. *J. Polymer Sci., Polymer Phys.*, **34**, 1541.
4. Kreituss, A. and Frisch, H.L. (1981) Free volume estimates in heterogeneous polymer systems. I. Diffusion in crystalline ethylene-propylene copolymers. *J. Polymer Sci., Polymer Phys.*, **19**, 889.
5. Horas, J.A. and Rizzotto, M.G. (1996) Gas diffusion in partially crystalline polymers, Part II: temperature dependence. *J. Polymer Sci., Polymer Phys.*, **34**, 1547.

Keywords: diffusion through polymeric films, diffusion coefficient: temperature and concentration dependence, modelling of diffusion, sorption, crystallinity.

Geotextiles and geomembranes

Kwong Chan

DEFINITION OF GEOTEXTILES AND GEOMEMBRANES

A geotextile is any textile, or related product, permeable to water, used within a foundation, soil, rock, earth or any geotechnical engineering – the material is an integral part of a man-made project, structure, or system [1]. The geotextile can include a coating deposit on one or both sides so as to consolidate the fabric or to protect it against external attacks, such as various environmental degradation. If the coating makes the geotextile impermeable then it is referred to as a geomembrane.

FIBERS FOR GEOTEXTILES

Natural fibers and the majority of regenerated fibers are seldom used to make geotextiles and geomembranes because they are biodegradable. Because of ease of manufacture, useful properties, and general economy, the most common synthetic fibers used in the manufacture of geotextiles are polypropylene (PP), polyethylene (PE), polyester (PET) and polyamide (PA).

PP is a semicrystalline thermoplastic with a melting point of 165°C and a density of around 0.9 g/cm^3 which makes it the lightest polymer used in geotextiles [2]. Since the density of PP is less than that of water, PP geotextiles float and therefore have to be weighted if placed underwater. Their ultimate tensile strength depends upon the degree of orientation of their constituents (fiber, yarn) and falls typically in the range of 400–700 MN/m^2. The maximum working temperature of PP is 100°C.

Polypropylene: An A–Z Reference
Edited by J. Karger-Kocsis
Published in 1999 by Kluwer Publishers, Dordrecht. ISBN 0 412 80200 7

The resistance of PP to microorganisms and ultraviolet light is good and fair to good, respectively.

Additives required to facilitate processing include various lubricants and plasticizers, whilst those required to enhance durability during service include thermal stabilizers and carbon black which enhance the resistance to ultraviolet attack. PP is more prone to oxidation than PE and therefore the incorporation of antioxidants is an important issue.

PRODUCTION METHODS OF GEOTEXTILES

The production methods of geotextiles are shown in Figure 1. The first step involves the transformation of the raw polymer from solid to liquid form by heating the polymer until it is molten. The liquid polymer is then extruded through a spinneret, a device consisting of many fine holes, and solidified into continuous strand (filaments) by cooling. While the above fiber-forming technique generally results in filaments of circular or elliptic cross-section, a slightly different technique is sometimes used for the production of PP fibers. First, a thin sheet is extruded, which is subsequently cooled and slit to tape-like filaments (split-film tapes). Once the filaments are formed, they are uniaxially drawn. This increases greatly the orientation of the molecular chains within the filament structure, and thus improves the physical property profile.

Although the majority of synthetic filaments consist of a single polymer, it is possible to manufacture bicomponent fibers consisting of two polymers, called also heterofilaments. They combine the beneficial properties of both polymers and may also provide an easier and novel mean of producing a geotextile. Common bicomponent combinations are PP/PE and PP/PA. Further processing involves a yarn-forming operation whereby many short fibers or continuous filaments are combined to form a yarn. From the yarns, consisting of filaments and tapes, various geotextiles can be produced which are grouped in five major groups:

- *Woven geotextiles* are manufactured by interlacing two sets of monofilaments, tapes or yarns perpendicular to each other.
- *Nonwoven geotextiles* first form a loose web or layers of fibers or filaments, arranged in either an oriented or random pattern, prior to fixing them by bonding (by an adhesive or thermal fixation) and in this way a cohesive planar structure is developed.
- *Knitted geotextiles* are produced by mechanically interlooping yarns or monofilaments.
- *Stitch-bonded geotextiles* are produced by stitching yarns or monofilaments through a fibrous web, thus obtaining a cohesive fabric. Alternatively, cross-laid yarns may be stitched together.
- *Combination geotextiles* combine two of the above major fabric-forming processes.

Outline of Steps in the Production of Geotextiles

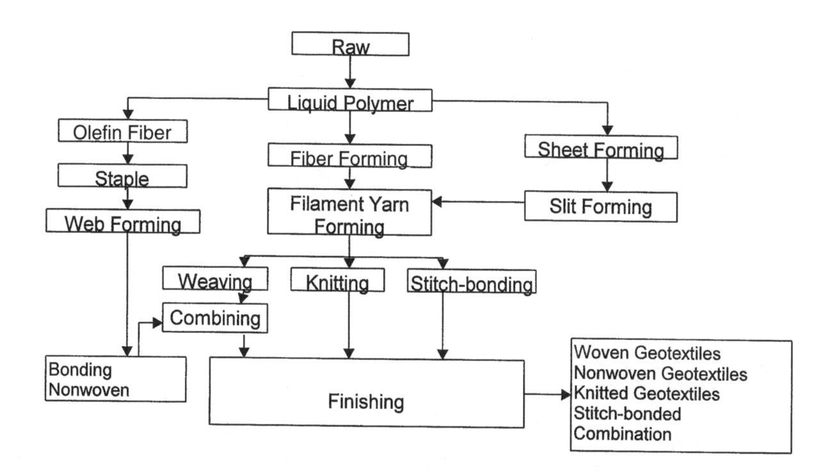

Outlines of steps in the Production of Geomembranes

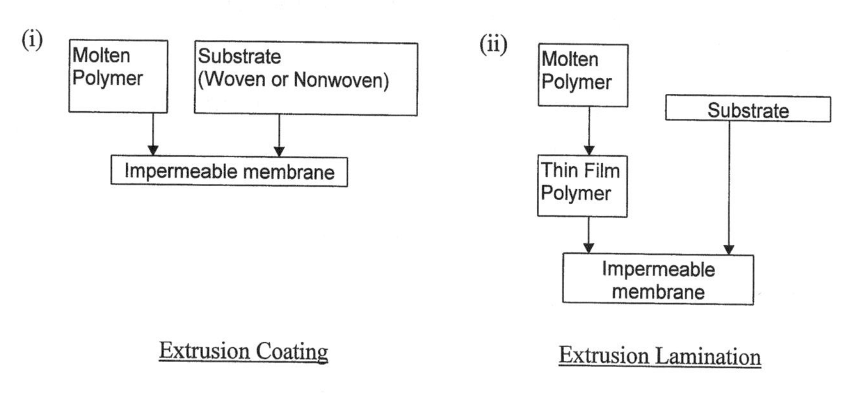

Figure 1 Production routes of geotextiles and geomembranes.

The architectures of woven, nonwoven, knitted and stitch-bonded geotextiles are shown schematically in Figure 2.

PRODUCTION METHODS OF GEOMEMBRANES

The process of applying a thin layer of material upon a substrate is called coating. For the coating process resulting in an impermeable end-product, polymeric fluids (including melts) or powders (consolidated in

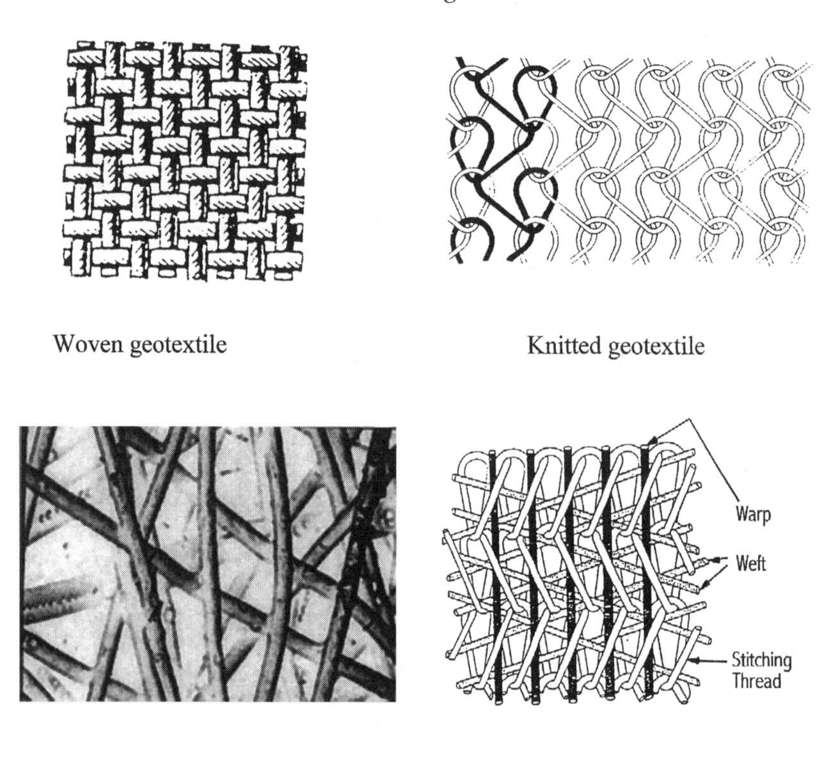

Woven geotextile Knitted geotextile

Nonwoven geotextile Stitch-bonded geotextile

Figure 2 Various forms of geotextiles.

a further step) are used. The production methods of geomembranes are shown also in Figure 1.

The substrate, for example a woven or nonwoven fabric, serves as support carrier to the end product, while the coating seals it. In the melt impregnation process, the polymer melt is extruded into a thin smooth film, laid down onto the substrate. The impregnation of the latter is achieved by pressure. The product is then cooled and wound up by a winder. An impermeable polymer film can also be formed between two textile fabrics. Besides sealing, the film also takes the role of an adhesive layer between the two fabrics.

GEOTEXTILES APPLICATION BY SECTOR AND FUNCTION

The applications of geotextiles are the following: transportation, environmental protection, water resources and coastal stabilization [3, 4]. The functions of geotextiles can be varied according to their applications.

- *Stabilization.* Geotextiles serve to reinforce the soil, thereby creating the internal stability required (fabric-reinforced soil). Moreover, geotextiles can be placed between two different soil layers with a separation function, preventing the intermixing of the two soil layers due to dynamic loading.
- *Erosion protection.* Geotextiles are used to cover civil engineering constructions of earth. Their function here is to reduce the kinetic energy of raindrops, thereby impeding the washing out of soil particles. Geotextiles act as a filter in contact with soil, allowing water to seep through the soil but holding back most soil particles, carried by the water current.
- *Pond underliner.* It is placed between two materials to act as a cushion via stress and strain relief for the layers.
- *Asphalt overlay.* It is placed onto a roadway on which a hot asphalt is poured (fabric-reinforced pavement).
- *Drainage.* Geotextiles collect liquids or gases and convey them towards an outlet.

MARKET FOR GEOTEXTILES

A total of 309 million square metres of geotextiles was consumed in 1990, in which more than half of that total were used by the USA construction market. The geotextiles industry is growing at 10–12% per year and is expected to continue at that rate well into the next decade [5]. It is therefore believed that approximately 800–840 million square metres of geotextiles should be consumed annually by year 2000. A market breakdown of the end-use of geotextiles is given in Figure 3.

GEOMEMBRANES APPLICATION BY SECTOR AND FUNCTION

Unlike geotextiles which have multiple functions, geomembranes serve only as a barrier to contain or exclude a fluid. For example:

- Geomembranes are used to exclude water or other fluids, to prevent the penetration of toxic gases or vapors into inhabited buildings, etc. In the majority of such applications, geomembranes are used in combination with geotextiles which act as cushioning and thus prevent the puncture of the geomembrane by sharp obstacles.
- Geomembranes are used as walls, linings of containers, canals and water reservoirs to contain fluids.

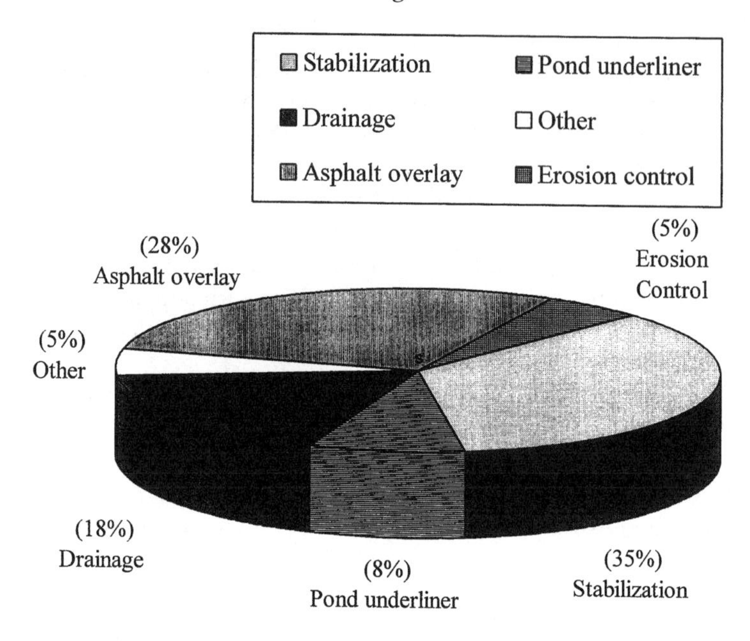

Figure 3 Geotextile market in 1990 [5]. Total use 309 Mm^2.

GENERAL TESTING METHOD FOR GEOTEXTILES AND GEOMEMBRANES

Although an International Organization for Standardization (ISO) exists, there are very few ISO standards which apply to geotextiles and geomembranes. In the USA, there is a comprehensive set of national

Table 1 The common ASTM test methods used for geotextiles and geomembranes

Test method	*ASTM standard*
Grab tensile strength	ASTM D4632
Fiber/yarn tensile properties	ASTM D3822
Puncture resistance	ASTM D4833
Trapezoidal tear strength	ASTM D4533
Burst resistance (Mullen)	ASTM D3786
Permittivity (the quotient of the coefficient of permeability and thickness)	ASTM D4491
Apparent opening size (permeability)	ASTM D4751
Strip tensile strength	ASTM D751
Wide width tensile strength	ASTM D4595
Hydraulic transmissivity (permeability)	ASTM D4716

standards produced under the auspices of the American Society for Testing and Materials (ASTM). The common test methods together with their ASTM test number for Geotextiles and Geomembranes are listed in Table 1.

REFERENCES

1. Van Santvoort, G.P.T.M. (1994) *Geotextiles and Geomembranes in Civil Engineering*, Balkema, Rotterdam.
2. Ingold, T.S. (1994) *The Geotextiles and Geomembranes Manual*, Elsevier, Amsterdam.
3. Chan, K. and Kan, M.M.K (1993) Engineering ingenuity with fibres and fabrics for geotextiles in civil engineering, in *Proceedings of the 2nd Asian Textile Conference*, Vol. 1, 1993, pp. 358–363.
4. Chan, K., Kan, M.M.K. and Wan, W.C. (1995) Survivability assessment of site damages on geotextiles, in *Proceedings of the 3rd Asian Textile Conference*, Vol. 2, 1995, pp. 1187–1194.
5. Stephens, T. (1988) Geotextiles: A down-to-earth market that's lifting off, in *Proceedings of the 75th Annual Convention of the Industrial Fabrics Association International*.

Keywords: geotextile, geomembrane, woven fabric, nonwoven fabric, knitted fabric, stitch bonding, fabric architecture, coating, application of geotextiles and membranes.

Glass mat reinforced thermoplastic polypropylene

József Karger-Kocsis

INTRODUCTION

Glass mat reinforced thermoplastic (GMT) sheets were first introduced in the early 1970s [1] in order to produce large parts by hot pressing techniques economically. Their earlier but still-used designation 'organic sheets' clearly indicate that the main target with GMTs was to replace stampable metallic sheets especially in the automotive industry. Considering the cost-related mechanical performance, GMTs are located between injection-moldable discontinuous (short or long) glass fiber (GF) reinforced and advanced thermoplastic composite with various fiber and fabric architecture. Their market acceptance is triggered by the following:

- high innovation potential in respect to material properties, production, processing and ecological demands;
- design freedom (production of complex parts by integrating several functional elements and thus reducing assembly costs);
- low density associated with weight saving when other construction materials are replaced;
- high productivity of both manufacturing and processing;
- easy recycling.

Although GMTs are available with different polymeric matrices, the market is dominated by polypropylene (PP) based composites (GMT-PP) due to their low price.

Polypropylene: An A–Z Reference
Edited by J. Karger-Kocsis
Published in 1999 by Kluwer Publishers, Dordrecht. ISBN 0 412 80200 7

MANUFACTURING

Manufacturing and forming of GMT-PP occur mostly separately. Due to economic reasons efforts are, however, being undertaken to combine the production and shaping on-line.

In respect to the GMT-PP sheet production, it is reasonable to distinguish between dry- and wet-laid methods by regarding how the GF mat is placed [2]. The majority of the present production lines work according to the dry-laid principle using various methods of melt impregnation (Figure 1). In practice, GF mats are impregnated and laminated in a double-belt press yielding compacted (consolidated) sheets of 2–4 mm thickness. A key issue of this process is the wetting-out of the GF mat. For stampable GMT-PP types, chemically bonded swirl mats composed of endless GF rovings are used. They exhibit restricted flow and deformation behavior during shaping. On the contrary, the GF mats used for hot-flowing grades are produced by needling of endless or chopped long GFs. As good deformability ('flow') of the GF mat and its layers is a prerequisite for molding complex parts (containing ribs, thickness changes, metallic inserts etc.), the R&D work is mostly focused on this aspect. Hot-flowing GMT-PP grades dominate the market at present. The melt impregnation technique is used by several firms including e.g. General Electric (Azdel®), Symalit (Symalit®), Elastogran (Elastopreg®) and PCD (Daplen TCF®).

A recent development of the dry-laid techniques can be ascribed to the preformed GMT produced by a fluidized bed process [3]. In this case, fiber rovings are cut and mixed with thermoplastic powder, conveyed and deposited by air stream onto a mold. The loose web formed on this mold is heated by hot gas stream prior to being transferred to a matched die compression mold. So, in this case, the web (i.e. the GMT preform) is consolidated and formed simultaneously in the compression molding

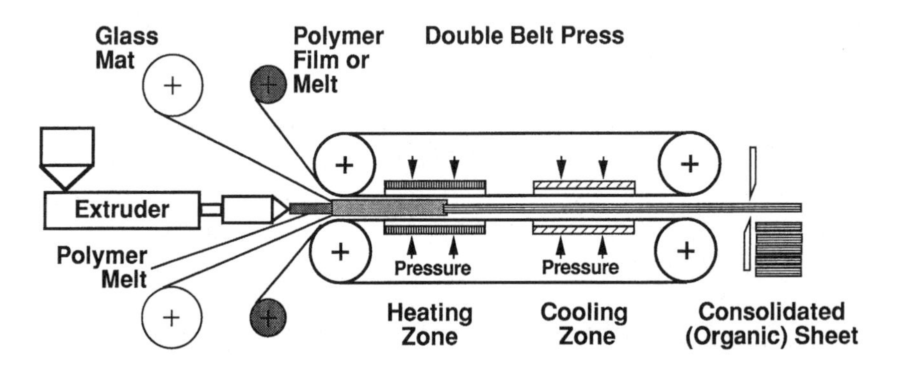

Figure 1 Dry-laid melt impregnation techniques of GMT sheet manufacturing.

step. There are further possible dry-laid techniques of producing GMTs or GMT-like materials: coneedling of GF mats and nonwoven PP layers, coknitting or coweaving of PP fibers (also split fibers) with the reinforcing GFs into various textile assemblies (see the chapter 'Commingled yarns and their use for composites' in this book). These textile techniques are gaining acceptance in the manufacturing of natural fiber mat-reinforced PP (NMT; see the chapter 'Natural Fiber/polypropylene composites' of this book).

Among the wet-laid techniques, the slurry and foam deposition should be mentioned (Figure 2). Since these processes were originally developed by companies producing paper, these variants are often referred to as papermaking processes. They result in thermoplastic composite sheets with long discontinuous (chopped) GF reinforcement [1, 2]. The main task in the manufacturing is to avoid segregation between the components due to a large difference in the density (PP $\approx$ 0.9 g/cm^3, whereas GF $\approx$ 2.6 g/cm^3). Segregation can be overcome either by setting a proper consistency or using binders which hold the GF and the polymer (in powder or fiber form) together (e.g. latex binder). The length of the GF incorporated varies usually between 10 and 30 mm. Generally several dried webs are combined prior to the consolidation on a double-belt press. Alternatively, the consolidation may be done simultaneously with the shaping that occurs via hot pressing. The long discontinuous GF (mat) containing GMT-PP exhibits excellent flowability which allows the use of 'low pressure' molding techniques. Improved flow characteristics are achieved, however, at costs of the mechanical properties (especially toughness). Further problems may be related to the foam or slurry ingredients (stabilizer, binder etc.) which

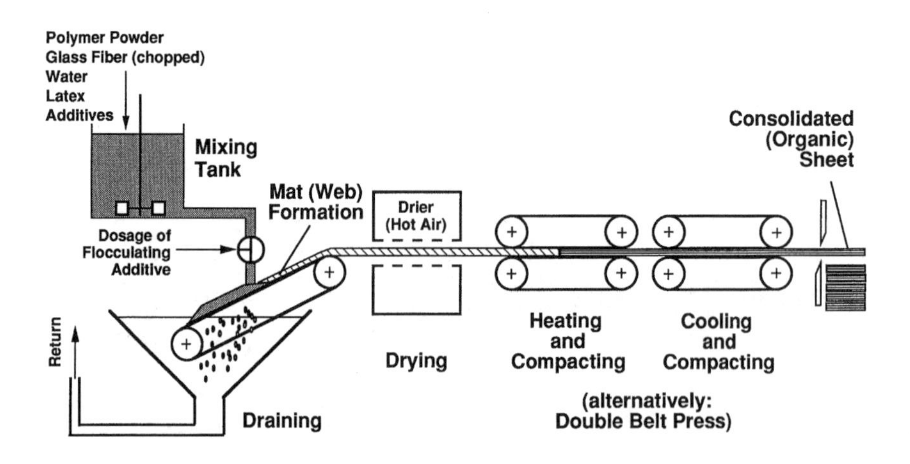

Figure 2 Wet-laid (papermaking) technology of GMT sheet manufacturing.

may affect especially the long term mechanical response of GMT-PP. Wet-laid GMTs are offered among others by General Electric (Radlite®), Exxon (Taffen SC®), Ahlström (RTC®) and DuPont (MCS®).

PROPERTIES

The reinforcement content in GMTs varies between 20 and 45 wt.% (in GMT-PP this range corresponds to 8–22 vol.%). The basic mechanical and thermal parameters for different GMT grades can be taken from references [1–3] and from the chapter 'Mechanical and thermal properties of long glass fiber reinforced polypropylene' of this book. Here only the excellent toughness performance along with its explanation will be briefly treated. The high toughness of GMT-PP is related to a peculiar failure mode occurring in a large damage zone [4, 5]. The basic difference in the failure mode between discontinuous and mat reinforced PP is that the apparent mesh (network) structure of the mat triggers an efficient stress transfer and redistribution process which is completely missing for discontinuous fiber reinforced composites (Figure 3). The large damage zone observed in GMT-PP should be considered when the determination of reliable mechanical properties is undertaken. Needless to say the mechanical response of GMTs may show a large scatter due to the local microstructure of the mat (which obviously affects the stress transfer and distribution within). The mesh-related deformability of the mat may also influence the long term mechanical behavior (creep,fatigue); they are

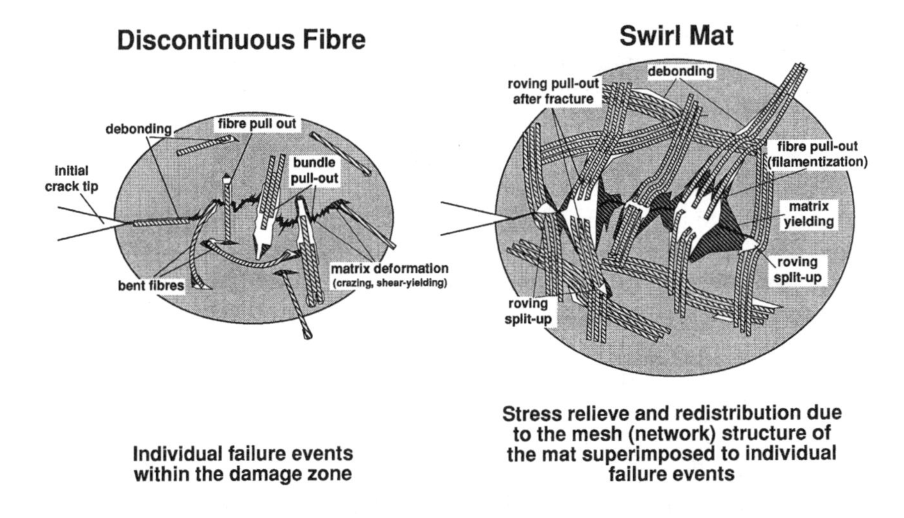

Figure 3 Difference in the damage zone and related failure mechanisms between discontinuous fiber and mat reinforced PP composites (from the project DFG Ne 546/5–1).

superior to the chopped short and long GF reinforced thermoplastic composites. The fatigue endurance limit of GMT-PP at 10^6 fatigue cycles can be given by 25–30% of the static tensile strength.

PROCESSING

As mentioned earlier, hot stampable GMT sheets having a thickness of about 2 mm contain a swirl GF mat with continuous, generally chemically bonded GFs. Their processing occurs by melt phase stamping (being a variant of thermoforming [6]). Hot blanks having a similar overall surface as that of the mold are deposited in the mold prior to closing the press. The pressure used is between 5 and 15 MPa, the temperature of the blanks is between 180 and 220°C and that of the mold usually between 20 and 70°C. The thickness of the final part agrees with that of the blank, so no complex parts can be produced by this way. Stampable GMT-PP grades are losing their importance with the exception of natural fiber reinforced types.

GMTs with semicrystalline thermoplastic matrices can be formed also by solid-phase thermoforming [1]. The temperature of shaping is selected between the glass (T_g, in case of PP at about 0°C) and melting temperature (T_m). Shaping occurs by fast pressing the clamped blank associated with thickness reduction (since the clamped surface agrees with that of the projected one of the mold cavity). Forming is due to morphology rearrangement in the matrix (see the chapter 'Thermoforming of polypropylene' in this book). The achievable part complexity is higher than for stampable parts, but lower then for flow-moldable GMTs. The industrial acceptance of this technique is very limited.

Complex structural parts can be produced from flow-moldable GMTs containing GF mat composed of long GFs and rovings. The blanks are thicker (initial thickness 3–4 mm) but smaller in area than the final part, so their compression flow molding by short stroke presses results in complex-shaped parts. Heating-up and conveying of the blanks to the press unit are similar to the stampable GMT-PP. It is obvious that both the pressure required (10-30 MPa) and pressing cycle time (about 60 s) are higher than in case of stampable grades.

Heating of the precut blanks in the aforementioned processes may occur by radiation (infrared), convection (hot air or gas) or conduction (contact). Preheating by hot air or infrared radiation takes 3–5 min. During heating the thickness of the consolidated GMT increases ('lofting') which influences the heat transfer.

The recent trend with GMT-PP is to combine the manufacturing and shaping process and thus avoid the cooling (during sheet consolidation) and reheating (prior to forming, see above) steps. In-line compounding and molding with long GF is now offered by Composite Products (USA).

Kannegiesser (Germany) has developed an extrusion deposition techniques for long discontinuous GF-reinforced systems, as well. These developments have further benefits to energy savings: both the clamping pressure and the flow length required from the machinery and compound, respectively, can be reduced. It is worth noting here, that these composites should be referred as GMT-like systems or press-molding compounds due to their reinforcement structuring. The surface appearance of GMTs is inferior to SMC and steel sheets. This is due to the crystallization of the matrix accompanied with some shrinkage that makes the GF mat structure visible beneath the surface ('readout' or 'readthrough' effect). The surface appearance can be improved by surface embossing (pattern given by the mold surface), in-mold decoration or textile covering.

MARKET

GMTs are increasingly replacing steel and sheet-molding compounds (SMC) in the automotive industry. They are widely used for the production of seat shells, bumpers and construction front ends. Further applications that may be suitable for GMTs are inner door panels, spare wheel covers and the like (see also the chapter 'Polypropylene in automotive applications' in this book). Another possible large use for GMTs is handling and transportation.

RECYCLING

The recycling philosophy with GMT-PP follows various ways. Production waste and scrap can be on-line refeeded up to about 20 wt.% without penalty in the mechanical property profile. Cutouts, trims and larger parts may be directly hot pressed for less demanding applications or converted in injection-moldable short fiber reinforced PP-granules. It is worth noting that there is hardly any drop in the stiffness and toughness properties if the mesh-like structuring of the mat is 'reconstructed' in the secondary molding operation (which is obviously linked to some limiting size of the GMT-PP waste)

REFERENCES

1. Bigg, D.M. (1995) Manufacturing methods for long fiber reinforced polypropylene sheets and laminates, in *Polypropylene: Structure, Blends and Composites*, Vol. 3, Chap. 7, (ed. J. Karger-Kocsis), Chapman & Hall, London, pp. 263–292
2. Karger-Kocsis, J. (1996) Glass mat reinforced thermoplastics in *Polymeric Materials Encyclopedia*, Vol. 4, (ed. J.C. Salamone), CRC Press, Boca Raton, FL, USA, pp. 2761–2766.

3. Berglund, L.A. and Ericson, M.L. (1995) Glass mat reinforced polypropylene, in *Polypropylene: Structure, Blends and Composites*, Vol. 3, Chap. 5, (ed. J. Karger-Kocsis), Chapman & Hall, London, pp. 202–227
4. Karger-Kocsis, J., Harmia, T. and Czigány, T. (1995) Comparison of the fracture and failure behavior of polypropylene composites reinforced by long glass fibers and by glass mats, *Composites Sci. Technol.*, **54**, 287–298
5. Karger-Kocsis, J. and Fejes-Kozma, Zs. (1994) Failure mode and damage zone development in a GMT-PP by acoustic emission and thermography, *J. Reinf. Plast. Composites*, **13**, 768–792
6. Throne, J.L. (1987) *Thermoforming*, Hanser, Munich.

Keywords: glass mat, glass mat reinforced thermoplastics, GMT-PP, swirl mat, chopped glass fiber mat, thermoforming, hot stamping, flow molding, lofting, melt impregnation, slurry impregnation, papermaking technology, solid phase pressure forming (SPPF), dry-laid techniques, wet-laid techniques, in-line compounding and forming, in-mold decoration.

Hard-elastic or 'springy' polypropylene

Ivan Chodák

In the 1960s it was discovered that polypropylene (PP), as well as several other highly crystalline polymers, can be extruded in the form of elastic fibers and films. These materials were termed as 'hard' elastic to distinguish them from common elastomers. The principle of the formation of hard-elastic structure consists in a crystallization under stress resulting in a superstructure of stacked lamellar aggregates. When the stress is very high and the superstructure is perfected by annealing, the resulting material displays an unusual mechanical behavior [1].

Originally hard-elastic PP was made from crystalline lamellar materials and was processed via melt spinning and crystallization under stress, followed by annealing under tension. The structure of the material consists of stacked crystalline lamellae (5–40 nm thick) with fold planes normal to the fiber direction. Between the lamellae, microfibrils oriented parallel to the draw direction are located. Under load, the lamellae tend to separate and voids bridged by fibrils appear. The void volume is initially about 18% at zero strain which increases to about 65% at 15% strain. No further increase in void volume was observed above 15% deformation.

Hard-elastic polypropylene has an interesting combination of mechanical and physical properties [2] including:

1. initial Hookean elasticity;
2. high recoverability from large strains;
3. energetic retractive forces;

Polypropylene: An A–Z Reference
Edited by J. Karger-Kocsis
Published in 1999 by Kluwer Publishers, Dordrecht. ISBN 0 412 80200 7

4. healing after work softening;
5. constant cross-sectional area during deformation.

Other properties of interest include a reversible reduction of density with an extensive increase of pore volume on stretching and a negative temperature coefficient of retractive force effect on stretched material. It is surprising that high deformability with good and rapid recovery is observed even at liquid nitrogen temperature. When these 'hard' elastic fibers are stretched, there is little or no decrease in diameter. The decreased density is associated with a formation of large, surface-connected voids which can be detected by mercury porosimetry. The unique elastic properties of hard-elastic 'springy' PP fibers deteriorate when the deformation exceeds 200–300%.

Several models explaining the behavior have been suggested [3]. Sprague proposed a leaf-spring model based on stacks of lamellae which bend during deformation, with interlamellar fibrils limiting the extent of bending deformation. The elastic restoring forces are energy – and not entropy – driven as in rubbers [4]. Chou's model [2] explains the hard elasticity in terms of the effect of increased surface stress on microfibrils during extension. Calculated values of the fibril diameter as a function of strain show that stress-induced subfibrillation is the mechanism responsible for the gradual increase in the surface stress. Subfibrils are stable only under an imposed stress; the smallest fibril diameter obtained with this material is approximately 2 nm. Other models which should be mentioned are reversible shear of lamellae between fixed tie points [5], a general model based on a change in entropy in the intermolecular layer and an increase in surface energy during extension [6], and a combined structural model that attributes the stress for extension to the pulling of fibrils from lamellae and the retractive force resulting in the particular elastic properties [7]. All the models are based on certain experimental observations and can help to understand the behavior of the material. However, none of the models describes all features of the 'springy' PP behavior in the whole range of conditions, and, therefore, no model exists which can be considered as generally valid. A rather simple but instructive model of Samuels [1] is shown in Figure 1 as an example. In the unstretched material (a), the crystal lamellae are *c*-axis oriented. After room temperature stretching, the lamellae are deformed and noncrystalline chains between the lamellae become strained (b). If this material is annealed, then both lamellae and noncrystalline chains in interlamellar space relax, while the noncrystalline chains outside of lamellae undergo an orientation (c). If this mechanism occurs together with microfibrillation, it can serve as a simple but rather instructive explanation of the hard- elastic PP behavior.

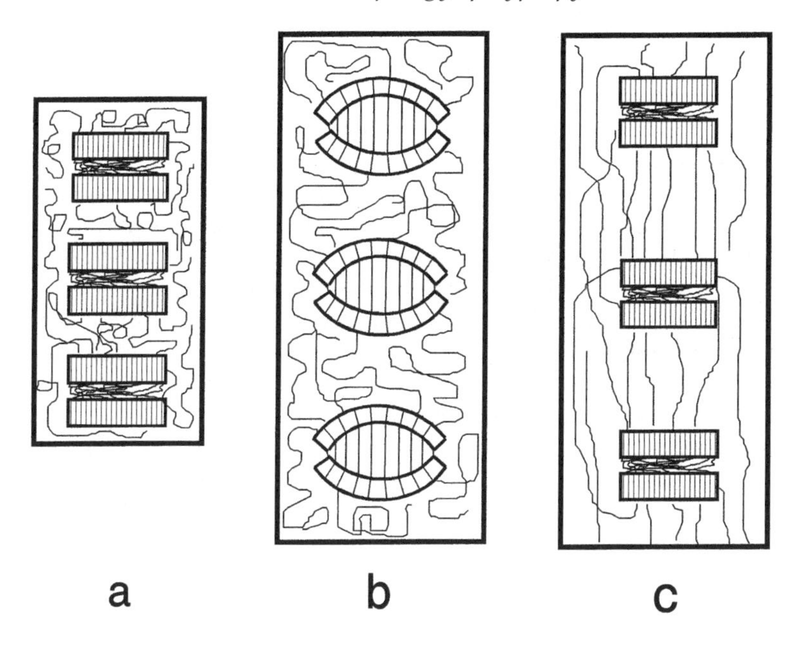

Figure 1 Schematic representation of the formation of high strength elastic fiber structure. Unstretched material (a), room temperature stretched (b), high temperature annealed under tension (c) (from Samuels [1], with permission).

Almost all the models proposed involve a bending or shearing of row structure lamellae. Thus, crystallization seems to be the essential criterion for hard-elastic behavior. The crystallization process of polymer melt under strain is quite different from that in relaxed state. In mechanically strained polymer, crystallization occurs at higher temperature and proceeds much faster than in unconstrained melt. The morphology is completely different. Instead of spherulites growing more or less isotropically, extremely anisotropic growth patterns have been registered. Highly oriented cylindrites compressed in the strain direction and having the primary nuclei more or less aligned in rows parallel to the strain form the rows of lamellae oriented perpendicular to the direction of deformation. This morphology is characteristic of hard-elastic PP.

Relaxation and cyclic deformation behavior of the hard-elastic polypropylene is of particular interest [2]. If the fiber is allowed to relax, the stress needed to continue the deformation is different from the initial stress. The stress increment can be divided into a permanent and transient part. As the recovery properties of the fibers begin to deteriorate, the permanent increment changes from a negative to a positive value and the transient component decreases to zero. Stress relaxation of the fibers can be described by an exponential law at shorter times and a power law for longer times.

The cyclic stress–strain behavior of hard-elastic PP is characterized by gradually decreasing modulus and high recovery upon unloading in the first cycle. In the second cycle, in which the same sample is deformed immediately after the first cycle, the initial modulus is lower and the area of the hysteresis loop is much smaller. Further cycles are similar to the second one. Time-dependent effects associated with deformation indicate that a substantial portion of viscous or disordered elements must be present. Hysteresis curves are very different if made after short time of storage (few hours), or longer storage (days or months) after the first cycle. The longer the storage time, the closer is the shape of the second stress–strain curve to the first run [2].

Liquid-induced stress depression was also observed for hard-elastic PP. The specimen was stretched to 10% elongation under vacuum. After one hour relaxation, ethanol vapor was introduced leading to a rapid drop in equilibrium stress to a new level. The process is completely reversible, i.e. after removing the vapor, the stress returned to the original relaxed value as measured *in vacuo* [2].

Some attempts have been done to use the hard-elastic PP as a component in a blend [3]. Films with improved mechanical properties have been obtained by blending a small portion of extrusion-grade PP with fiber-grade PP. The addition of a small amount of high density polyethylene (HDPE) results in a substantial increase of the hard elasticity of the filaments.

REFERENCES

1. Samuels, R.J. (1979) High strength elastic polypropylene, *J. Polymer Sci., Polymer Phys.*, **B 17**, 535–568.
2. Chou, C.J., Hiltner, A. and Baer, E. (1986) The role of surface stresses in the deformation of hard elastic polypropylene, *Polymer*, **27**, 369–376.
3. Du, Q. and Wang, L. (1989) The hard elasticity of polypropylene and its blends with polyethylene, *J. Polymer Sci. Polymer Phys.*, **B27**, 581–588.
4. Sprague, B.S. (1973) Relationship of structure and morphology to properties of hard elastic fibers and films, *J. Macromol. Sci. Phys.*, **B8**, 157–187.
5. Clark, E.S. (1973) Mechanism of energy-driven elasticity on crystalline polymers, in *Structure and Properties of Polymer Films*, (eds R.W. Lenz and R.S. Stein), Plenum, New York, pp. 267–282.
6. Göritź, D. and Müller, F.H. (1974) *Colloid Polymer Sci.*, **252**, 862.
7. Miles, M., Petermann, J. and Gleitner, H. (1976) Structure and deformation of polyethylene hard elastic fibers, *J. Macromol. Sci. Phys.*, **12**, 523–534

Keywords: hard elastic, springy PP, microstructure, lamellae, crystallization under stress, recovery, relaxation, cyclic deformation.

High-modulus and high-strength polypropylene fibers and films

Toshio Kunugi

Many studies on improvements of mechanical properties in isotactic polypropylene (iPP) fibers and films have been performed by researchers since 1964. The ultimate values of modulus and strength are the crystal modulus along molecular chains and theoretical strength calculated on the basis of the bond energies and the cross-section of a molecular chain. The high-modulus and high-strength fiber has to possess a high molecular orientation and a high crystallinity. Fortunately, iPP can be easily drawn and crystallized up to fairly high degrees of orientation and crystallinity. The maximum values of Young's modulus and tensile strength reported up to date are 36–40 GPa and 1.5 GPa, respectively. Further, the polymer has desirable characteristics, such as a low density, chemical resistance and hydrophobic properties. Therefore, the high-modulus and high-strength iPP fibers and films are highly ranked among those of flexible chain polymers.

CRYSTAL MODULUS AND THEORETICAL STRENGTH

The molecular chain in crystal lattice of the monoclinic α form has a helical conformation with three units per turn to avoid the steric hindrance of the bulky methyl groups and a large cross-section of 0.344 nm^2. The crystal modulus along molecular chains therefore is low, which is estimated to be 41.2 GPa by X-ray diffraction measurement or 88.2 GPa by Raman spectroscopy. The theoretical tensile strength is calculated to be 18.2 GPa.

Polypropylene: An A–Z Reference
Edited by J. Karger-Kocsis
Published in 1999 by Kluwer Publishers, Dordrecht. ISBN 0 412 80200 7

DRAWING BEHAVIOR

The drawing behavior of iPP fiber produced by melt-spinning is strongly influenced by drawing conditions such as drawing temperature and strain rate, molecular weight, molecular entanglement, and initial morphology of the starting material. When iPP fiber with a common molecular weight is drawn at a low temperature below 100°C, strain hardening occurs and the draw ratio is limited to 7–9, which is called the 'natural draw ratio'. Therefore, in order to attain a higher draw ratio, drawing temperatures of 110°–170°C are well set. Since as-spun original fibers or as-extruded films have a fairly high crystallinity of 50–60%, a high drawing temperature near to the melting point is useful to soften the lamellae and to transform them to extended-chain crystals. During the drawing, the fiber structure alters from a lamellar into a fibrillar one [1]. Gel films and single crystal mats having an ultrahigh molecular weight can be easily drawn up to very high draw ratios, because the density of the molecular entanglements is reduced by the use of extremely dilute solutions.

SUPERSTRUCTURE

The superstructure of the fiber controls the mechanical properties. The superstructure of semicrystalline polymer consists of crystalline and amorphous portions. The crystallites in semicrystalline polymer fibers are easily oriented by drawing at low draw ratio, but amorphous chains can hardly be fully oriented and extended. Furthermore, it is necessary that the amorphous regions contain many taut tie chains connecting two or more crystallites. In order to obtain excellent mechanical properties, the high-modulus fiber should have a network structure composed of tie molecules and extended-chain crystals. In such a structure, the movement of the amorphous chains is strongly inhibited. The dynamic loss modulus (E")-temperature curve of an iPP fiber shows generally three peaks, i.e. α, β, and γ relaxations. The β relaxation peak is at −13–0°C whereas the α relaxation peak is found between 120–150°C. The β relaxation is attributed to micro-Brownian motion of molecular chains in the amorphous region and this corresponds to the glass transition (T_g). The α relaxation is attributed to molecular motion within the crystal regions. As the orientation and crystallinity increase, the β relaxation peak decreases in height and finally disappears. On the contrary, the α relaxation peak increases and shifts toward higher temperatures. This means that the amorphous chains are strongly restrained and the crystallites becomes rigid. A variety of techniques using X-ray, differential scanning calorimetry (DSC), nuclear magnetic resonance (NMR), infrared (IR), optical and electric microscopes and so on, as well as the dynamic viscoelastic measurements, are utilized for characterizing the superstructure of high-modulus and high-strength fibers. The maximum values of birefringence and crystallinity reported in the literature [2] are 44×10^{-3} and 74%, respectively.

MECHANICAL PROPERTIES

The helical conformation and the large molecular cross-section ($0.344\,nm^2$) in iPP are disadvantageous in respect to high-modulus and high-strength. In polyethylene, the molecular chain in crystal lattice takes a planar zigzag molecular conformation, and the cross-section is much smaller ($0.182\,nm^2$). Therefore, the crystal modulus and theoretical strength are as high as 240 GPa and 32 GPa, respectively. Also, the values in poly(vinyl alcohol), polyamide 6, and poly(ethylene terephthalate) are 250, 165, and 108 GPa in the crystal modulus, and 26, 32, and 28 GPa in the theoretical strength, respectively. These values are much higher than those of iPP, because these polymers take the planar zigzag molecular conformation and have smaller cross-sections (0.228, 0.192, and $0.217\,nm^2$, respectively).

However, the ratios of a maximum modulus attained to the crystal modulus (E_{max}/E_c) for iPP and polyethylene, are very high, 0.971 and 0.967, respectively. The ratios for polyamide 6 and poly(ethylene terephthalate) are 0.12 and 0.34, because the maximum moduli reached are 20 and 37 GPa, respectively. The maximum moduli are lower than or comparable to that of iPP which is in the range of 37–40 GPa. Therefore, the high-modulus and high-strength fiber of iPP has a special position among the fibers produced of other flexible chain polymers.

PREPARATION

Table 1 indicates main methods presented for the preparation of high-modulus iPP fibers. The methods are divided into two groups according to molecular weight of the used materials. There are hydrostatic extrusion, two-step drawing, die drawing, zone-drawing, and zone-drawing/zone-annealing methods in the first group, and the superdrawing methods of gel-film and single crystal mat in the second group. A few examples in each group will be briefly described below.

Common molecular weight grade

The hydrostatic extrusion [3] is carried out by extruding a billet through a conical die at an optimum temperature. The pressure required for extrusion is given to the billet through a transmitting fluid (Figure 1). The modulus of the extruded rod is 17 GPa. Die-drawing is a process in

Table 1 Various methods for preparation of high-modulus iPP fibers

Molecular weight	*Year*	*Preparation method*	*Maximum modulus (GPa)*
Common molecular weight	1964	Multistep drawing	14
	1964	Hot-drawing	10
	1973	Hydrostatic extrusion	17
	1973	Hot-drawing	16
	1976	Super-drawing	19
	1978	Two-step drawing	22
	1979	Zone-drawing	15
	1979	Die-drawing	20
	1983	Zone-drawing	21
	1987	Die-drawing	20
	1991	Coextrusion/zone-annealing	17
	1996	Multistep zone-drawing/zone-annealing	27
Ultrahigh molecular weight	1984	Dried gel film/hot-drawing	36
	1984	Single crystal mat/coextrusion/hot-drawing	33
	1986	Dried gel film/two-step hot-drawing	40

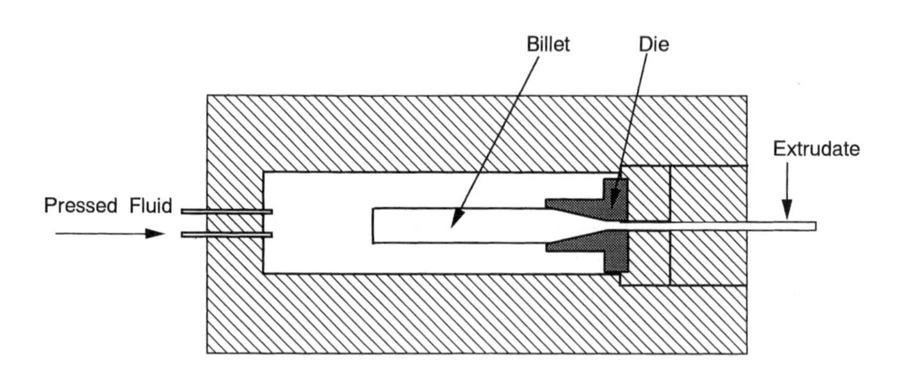

Figure 1 Apparatus for hydrostatic extrusion method.

which the highly oriented iPP rod is continuously produced by pulling out a heated billet through a conical die. The maximum modulus of 20 GPa is obtained by drawing at a draw temperature of 140°C up to a draw ratio of 23.2. The two-step drawing process works as follows. In the first step, an as-extruded billet is rapidly drawn to the natural draw ratio of seven in a hot silicone oil bath. In the second step, the drawn fiber is

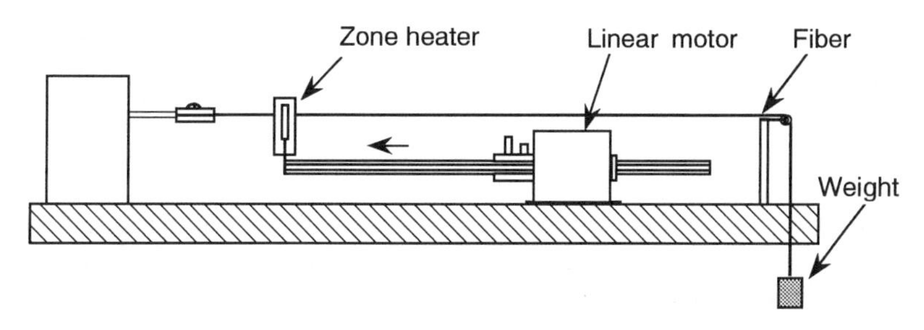

Figure 2 Apparatus for zone-drawing and zone-annealing method.

subsequently drawn at a slow draw rate of 4%/min and at a carefully controlled temperature of 130°C. Young's modulus and tensile strength obtained by this way are 22 and 0.93 GPa, respectively. Zone-drawing is carried out by heating the narrow zone of fiber under a constant tension. The heating zone can be moved along the fiber axis at a constant speed (Figure 2). The heating temperature and tension are decided so that the drawing and annealing are most effectively performed. In the multistep zone-drawing/zone-annealing [2], the zone-drawing was repeated five times at 150°C increasing the tension from 2 to 120 MPa stepwise, and subsequently the zone-annealing was repeated four times at 170°C increasing the tension from 120 to 210 MPa. The maximum modulus and draw ratio are 27 GPa and 25, respectively.

Ultrahigh molecular weight grade

The initial morphology of a starting material strongly influences the drawing behavior. By applying the force in the direction parallel to the lamella surface, the folded molecular chains in lamellae can be easily pulled out from the lamellae, because of the weak intramolecular forces. This technique which was established for ultrahigh molecular weight polyethylene (UHMW PE) can be applied to iPP with an ultrahigh molecular weight of 10^6 g/mol range. The dried gel films and single crystal mats have a similar superstructure to each other which consists of roughly stacking lamellae. The gels are generated by pouring a hot dilute solution of the polymer (e.g. 0.5–1.5 g/100 ml in decahydronaphthalene) into an aluminum tray kept at 0 to −25°C. The solvent is allowed to evaporate from the gel during several days. Strips of the dried gel film can be drawn at 140–170°C up to a maximum draw ratio of 40–100, and the superdrawn films have outstanding mechanical properties: Young's modulus of 36–40 GPa and a tensile strength of 1–1.5 GPa. The single crystal mat [4] is produced by the following process. First, the dilute iPP solution (concentration: 0.2% in xylene) is kept at 55°C for 20 h. Thereby, crystals grow isothermally and are suspended in the solution. The deposited crystals are slowly filtered and then dried at room temperature in vacuum for one week. The single crystal mats thus prepared are drawn by a two-stage drawing technique which consists of a solid-state coextrusion at 130°C and a hot drawing at 155°C. The solid-state coextrusion is carried out by extrusing the single crystal mat sandwiched between two split billet halves of a more ductile polymer (PE or POM). The mats can be drawn up to a total draw ratio of 66, and the superdrawn films have a modulus of 34–37 GPa and a tensile strength of 1.5 GPa.

APPLICATION

iPP is a very light polymer having a density of 0.91 g/cm^3. On the other hand, the high crystallinity imparts to it excellent mechanical properties, such as high modulus, strength, stiffness, and hardness. The modulus and strength in high-performance fibers are often compared by means of specific modulus and specific strength which are given by the ratio of their maximum values and specific density. The specific modulus and strength of iPP are beneficial in many applications when compared to metallic or ceramic high-performance fibers. Today, light and strong materials are required in various field of applications. The high-modulus and high-strength iPP fibers are widely used as ropes, nets, and cables.

REFERENCES

1. Peterlin, A. (1977) Mechanical properties of fibrous structure, in *Ultra-High Modulus Polymers*, (eds A. Ciferri and I.M. Ward), Applied Science., London, pp. 279–320.
2. Kunugi, T. (1996) Preparation of high oriented fibers or films with excellent mechanical properties by the zone-drawing/zone-annealing method, in *Oriented Polymer Materials*, (ed S. Fakirov), Hüthig & Wepf, Heidelberg, pp. 394–421.
3. Capaccio, G., Gibson, A.G. and Ward, I.M. (1979) Drawing and hydrostatic extrusion of ultra-high modulus polymers, in *Ultra-High Modulus Polymers*, (eds A. Ciferri and I.M. Ward), Applied Science, London, pp. 1–76.
4. Porter, R.S. and Wang, L. (1995) Uniaxial extension and order development in flexible chain polymers. *JMS Rev. Macromol. Chem. Phys.*, **C35**(1), 63–115.

Keywords: iPP fibers and films, high modulus, high strength, crystal modulus, theoretical strength, mechanical properties, drawing, draw ratio, molecular weight, superstructure, lamella, tie molecule, relaxation transitions, ultrastrong fibers and films.

Impregnation techniques for fiber bundles or tows

A. Lutz and T. Harmia

A composite part which is made of a polymer matrix and fibrous reinforcement shows better mechanical properties than the unreinforced version. Prerequisite for the superior mechanical behavior is that the fibers are wetted by the matrix and a good adhesion between fibers and matrix is established. This means that a void-free impregnation of the very slender fiber filaments is the key issue when the production of high-quality composite materials or parts are targeted. In case of thermoset resins, the fiber impregnation can easily be carried out due to the low viscosity of the uncured resin. In contrast, well-impregnated composite parts using highly viscous and glutinous thermoplastic melts are much more difficult to realize. The viscosity of the thermoplastic melt (100–1000 Pa) is generally more than two or three orders of magnitude higher than that of the uncured thermoset resins (1–10 Pa). Additionally, the 'handling' of the hot impregnation unit, which is necessary for processing the thermoplastic polymers, is less comfortable. Therefore, the impregnation of fiber tows with thermoplastics has been an important research and development topic. Many different routes are followed with the ultimative goal to overcome the problems associated with impregnation of fibers by thermoplastic polymer melts.

The impregnation of thermoplastic tow composites is often supported by intermediates which is an important processing step in manufacturing thermoplastic composite parts. In an overview, the totally impregnated intermediates, such as tapes or fiber fabrics, and the semi-impregnated intermediates, such as powder impregnated fiber bundles,

Polypropylene: An A–Z Reference
Edited by J. Karger-Kocsis
Published in 1999 by Kluwer Publishers, Dordrecht. ISBN 0 412 80200 7

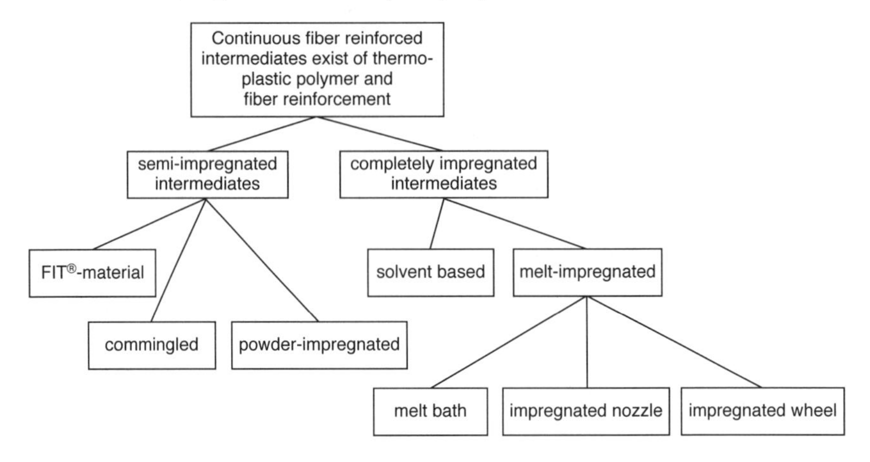

Figure 1 Fiber bundle impregnation techniques.

commingled fiber bundles or cowoven mats, are divided into two separate categories (Figure 1). The main difference between these categories is that in the first group of materials (tapes) the impregnation and consolidation step is completed in contrast to the semi-impregnated intermediates. In the latter case, fibers and matrix material are mixed in such a manner that only short ways of melt flow are necessary to impregnate the individual filaments in the tow when the matrix is molten.

The impregnated tows can be used for many applications which are generally subdivided into those where the fiber reinforcement is continuous (e.g. filament winding or pultrusion) and those where the tow is cut and the unidirectionally reinforced pellets (discontinuous long fiber reinforced thermoplastics) are then further processed into parts (e.g. in extrusion or injection molding).

The goal of advanced impregnation techniques is to carry out the wetting as effectively as possible. The most significant parameters that influence the flow of the molten polymer through an aligned fiber bundle are the applied pressure, the temperature and the shear rate of the melt (the latter is due to the non-Newtonian flow behavior of the melt). In addition, the duration during which the melt is squeezed between and through the reinforcing filaments is a very important factor.

A wide array of fiber tows (each of which consists of hundreds of filaments) are used for composite tape manufacturing. The diameter of filaments are usually in a range from 6 to 8 μm for carbon fibers (CF) and from 10 to 17 μm for glass fibers (GF). Also, aramid fiber (AF) tows or other polymeric fibers in this range of diameters are available. The filaments themselves are often inhomogeneously distributed. Usually the filaments are sized in order to improve the fiber–matrix adhesion in thermoplastic composite applications.

SEMI-IMPREGNATED INTERMEDIATES

Powder impregnation

The impregnation of fiber bundles with thermoplastic powders occurs continuously. The fiber bundles are pulled through a whirl bed of pulverized thermoplastic polymer. A homogeneous distribution of the polymer powder in the fiber tow is achieved by air whirling and spreading the tow by guiding it around pins in the powder bed. Additionally, electrostatic charging of the tow or powder can be used to increase the adhesion between the components. After this step, the polymer powder is melted by an infrared heater so that the fibers are partly covered by the matrix. If a definite geometry of the powder impregnated fiber bundle is needed, it can be shaped on a calender system. Powder impregnation is an economic route for the prefabrication of thermoplastic composites, provided that the polymer is available in powder form and can be directly used in the powder impregnation process.

FIT® material

Alternatively, the powder impregnated fiber tows can be jacketed, i.e. covered in an extruded thermoplastic sheath. The purpose of the sheath is to prevent the loss of powder during handling of the semi-impregnated fiber tows. The thermoplastic sheath is of the same polymer as the impregnation powder. This kind of intermediate is called FIT® (Fiber Impregnée Thermoplastique) and developed for filament winding processes [1].

Commingled yarn

A homogeneous distribution between fibers and polymer can also be achieved by commingling reinforcing and polymer fibers where the polymer fibers after melting form the matrix [2]. Commingled yarns are often used in filament winding and pultrusion processes. The fiber volume content is determined by the ratio of polymeric to reinforcing filaments. The diameter of modern polymer filaments is close to that of the reinforcing fibers. Therefore, good impregnation is achieved after a short distance of flow when the polymer fibers are molten. The commingled yarns show a more homogeneous distribution of polymer and reinforcing fibers than the powder impregnated intermediates.

The main drawback of semi-impregnated intermediates is that their impregnation and consolidation do not occur prior to the processing. As a consequence, the degree of impregnation cannot be determined before the last manufacturing step occurs. Since the impregnation and consolidation with thermoplastics, even with optimized distribution of fibers and polymer, is time consuming, the manufacturing of thermoplastic composite part from semi-impregnated intermediates is less effective than using fully impregnated tapes.

COMPLETELY IMPREGNATED INTERMEDIATES

Solvent impregnation

This technique uses an additional solvent in order to reduce the 'impregnation viscosity' of the thermoplastic resin. The fiber tow is pulled through a bath containing the thermoplastic polymer solved. Usually amorphous polymers are used in this process since semicrystalline polymers are less soluble. In order to improve the degree of impregnation, pins are situated in the solvent bath leading the fiber bundle. After the impregnation step, the excess solvent has to be removed from the tow (e.g. in a heating oven) and refed into the process after condensation. The difficulty of removing and condensing the solvent makes this impregnation process complex, cost-intensive and less friendly to the environment.

Melt impregnation techniques

The main advantage of this technique is that the introduction of the thermoplastic matrix occurs via simple melting. This has led to the development of various types of melt impregnation techniques, which are surveyed below.

Impregnation bath

The most widespread melt impregnation technique for fiber rovings today is the melt bath technique [3]. The impregnation is carried out by carrying fiber bundles alongside a number of pins which are located in a heated bath of molten polymer matrix (Figure 2). After the impregnation step, the fiber bundles are 'calibrated' by pulling them through a wipe-off die positioned at the end of the matrix bath.

The 'melt bath' technique has been studied by several research groups and both qualitative and quantitative models exist to describe the tension build up in the fiber bundles. Irrespective of its success, this process has some drawbacks. Since a high fiber tension is needed in order to spread the tows in the bath, the process is very sensitive.

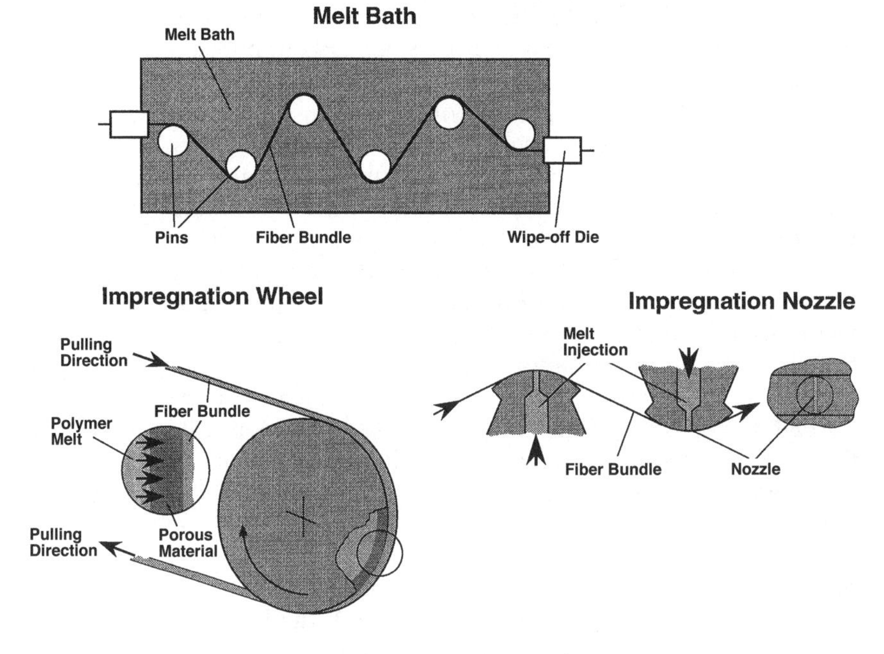

Figure 2 Melt impregnation techniques.

Furthermore, effective impregnation takes place in the contact zone of the tow with the pins which results in rather short impregnation times. The size of the melt bath is large and its closed design makes an easy set up (change of fiber or matrix grade) difficult.

Impregnation nozzle

This melt impregnation process uses one or more nozzles where the melt is fed through and upon which the fiber bundle is situated [4] (Figure 2). In this process, the impregnation takes place in a short time interval during which the fiber bundle passes the nozzle. A high fiber tension must be generated in order to keep the fiber bundle in contact with the nozzles. In this way, the melt can penetrate transverse through the bundle. The benefit of this technique is that high pressure can be applied to overcome the flow resistance of the fiber bundle. Based on the scheme in Figure 2, it is obvious that the impregnation time is very short (i.e. a function of the pulling speed).

Impregnation wheel

The restriction in output of the above-mentioned melt impregnation techniques is governed by the time during which the effective impregnation takes place (at the pins or the nozzle). Irrespective of their continuous operation, the tow impregnation occurs in a discontinuous manner.

Using the impregnation wheel and by feeding melt through a porous body, a continuous impregnation of a fiber bundle can be realized [5] (Figure 2). The impregnation wheel consists of a porous ring, which allows the molten thermoplastic matrix to penetrate through and impregnate the fiber bundle. The fiber bundle is in intimate contact with the outer surface of the ring along half of its perimeter. The advantage of this technique is that by increasing the diameter of the impregnation wheel, the duration of the impregnation step can be prolonged. This means that the effective impregnation time is much longer compared to the processes described before and it takes place continuously. Because of the high shear rate while the molten polymer matrix penetrates through the ring, the viscosity is reduced to a minimum value. Therefore, the impregnation pressure can be reduced (compared to the nozzle technique), so that only a low fiber bundle tension is needed. This reduces the fiber damage during impregnation and allows a safe operation. This design guarantees high output based on the adjustable impregnation time. Additionally, the fiber bundle tension can be kept low, because of the low diversion of the tows (by the large diameter of the wheel). Further benefits are short set-up times, and easy handling (no dies through which the fiber bundle has to be pulled are present). The fiber volume fraction with this technique can be adjusted by controlling the amount of molten polymer penetrating through the impregnation wheel per unit time.

REFERENCES

1. Atochem, Ganga, R.A. (1986) US Patent 4 614 678.
2. John, C. St. (1995) Co-mingled thermoplastic prepregs industrial applications, in Proceedings of the 10th International Conference on Composite Materials, ICCM-10, Whistler, Canada, 14–18 August, 1995, eds: A Poursatip and K. Street, Vol. III pp. 757–769, Woodhead Publ., Abington.
3. ICI (1982) New European Patent Specification, 82 300 150.8
4. I. Du Pont de Nemours (1987) US Patent 4 690 861.
5. Institut für Verbundwerkstoffe GmbH, Lutz, A. (1995) European Patent Application, PCT/EP95/01476.

Keywords: commingled yarn, composite, composite tape, discontinuous fiber reinforcement, fiber bundle, filament winding, FIT®, impregnation, long fiber, melt impregnation, powder impregnation, pultrusion, solvent impregnation, thermoplastic filament winding, thermoplastic pultrusion, tow impregnation, wetting.

In-situ reinforced polypropylene blends

Markku T. Heino and Jukka V. Seppälä

INTRODUCTION

Blending with high-performance engineering polymers is a feasible way of upgrading the properties of commodity thermoplastics. The aim of blending is often to improve the mechanical strength or toughness or, for example, the thermal resistance of the matrix polymer. Addition of conventional thermoplastics such as polyamides or polyesters, to polyolefins improves the properties, however, only to a relatively small extent. Completely different phenomena are instead achieved when thermotropic main-chain liquid crystalline polymers (LCPs) are used.

ABOUT LCPs

Classification of different liquid crystalline polymers, their phase behavior, processing and general properties are extensively reviewed in [1, 2]. Here, the focus is on the so-called thermotropic main-chain liquid crystalline polymers. These consist of mostly linear, rigid rod-like molecules that are capable of aligning to a very high degree during melt flow, and thus form a highly ordered melt phase in a certain temperature range just above their melting temperature. Owing to the relatively long relaxation time, the orientation is retained when the LCP melt is cooled, resulting in a highly ordered fibrous structure and anisotropic properties in the solid state. The excellent mechanical strength and dimensional and thermal stability of thermotropic LCPs are based on this fibrous structure. Because of the highly oriented melt phase, thermotropic LCPs

Polypropylene: An A–Z Reference
Edited by J. Karger-Kocsis
Published in 1999 by Kluwer Publishers, Dordrecht. ISBN 0 412 80200 7

exhibit significantly lower viscosities and are typically more shear thinning than isotropic polymers. The fibrillar structure of LCPs can be considerably affected by product design and processing conditions, especially by applying high shear or elongational forces to the molten polymer. The close relationship between the orientation during processing and the flow-induced morphology makes the processing of LCPs particularly important [1–5].

Several chemically different thermotropic main-chain LCPs have been synthesized and reported in the literature [1, 2]. However, most commercial polymers are polyesters, polyamides or their copolymers. The first thermotropic LCPs made by Du Pont in 1972 were high melting point aromatic copolyesters consisting of p-hydroxybenzoic acid (HBA), terephthalic acid (TA) and 4,4' dioxydiphenol (DODP). Later these polymers formed the family of commercial LCPs presently known under the trade name Xydar® (Amoco; $T_m > 400°C$, rigid, HDT about 300°C). Among the most important commercial LCPs at the moment are those based on HBA and 6-hydroxy-2-naphthoic acid (HNA) developed by Hoechst Celanese (Vectra®; $T_m = 280°C$, rigid, HDT about 200°C). The copolyesters based on polyethylene terephthalate (PET) and HBA developed first by Tennessee Eastman Chemical Co. (code name X7G) and currently produced by Unitika Ltd. (Rodrun®; T_m about 220°C, semirigid, HDT about 100–120°C), are also well-known. Other commercial LCPs are e.g. Granlar® (Granmont), Zenite® (Du Pont), Novacor® (Mitsubishi), Econol® (Sumitomo) and Rhodester® (Rhone-Poulenc). KU (Bayer), Ultrax (BASF) and Victrex (ICI) have recently disappeared from the market [2, 3].

WHY PP/LCP BLENDS?

Materials with totally new property combinations may be achieved by blending two or more polymers together. Through blending of thermotropic main-chain LCPs with thermoplastics, the highly ordered fibrous structure and good properties of LCPs can be transferred to the more flexible matrix polymer. The main aim is often to reinforce the matrix polymer or to improve its dimensional stability, but LCP addition may modify several other important properties, as well, e.g. thermal stability, gas barrier, chemical resistance and flame retardance. Owing to the relatively low melt viscosity of thermotropic LCPs, often a small amount of LCP decreases the blend viscosity significantly and renders the matrix thermoplastics easier to process [1–5].

Blends of LCPs and thermoplastics show some analogy to fiber reinforced composites, though many differences exist, as well. Compared with solid-fiber reinforced thermoplastics, LCP blends have several advantages. Because of their low melt viscosity, they are easily processed and molded, even into thin complex shapes. Solid fibers, in contrast, tend to increase the melt viscosity of thermoplastics, which leads to higher torque and energy consumption. Since LCP is totally molten during processing and the fibers are formed only during cooling, the blends do not wear the processing equipment as the solid fibers do. Nor do the problems related to fiber breakage during processing exist in LCP blends. A separate compounding step, e.g. in a twin-screw extruder, is not necessarily required, especially if the blended polymers are carefully chosen. And finally the quality of the surfaces of molded LCP blends is good. Hence, *in-situ* composites, i.e. LCP blends, may be thought of as supramolecular analogs of short-fiber composites, but exceptionally they provide a rare combination of processability and performance [4].

Generally, it can be stated that at small concentrations (0–15%), LCP acts in blends as a processing aid by reducing the viscosity whereas, at higher LCP contents (15–85%) it produces *in-situ* composites and mechanical reinforcement. If the LCP content is very high (85–100%), the blend can be considered a modified LCP.

Mechanical properties of some injection-molded PP/LCP blends are shown in Table 1 and the effects of elongational drawing on corresponding extruded ones in Figure 1. Since the properties of LCPs and LCP blends depend significantly on processing and, for example, the thickness of the sample, these results should be taken only as examples. In general, LCP addition provides mechanical reinforcement which is seen as almost linearly increased strength and stiffness with increasing LCP content. In extrusion, the fibrillation and orientation of LCP domains are further enhanced by additional drawing. In this way, significant improvements in strength and stiffness are achieved with LCP contents of 20–30 wt.% even with rather small draw ratios (Figure 1). Higher draw ratios than used here would lead to even higher stiffness. Strain at break and impact strength are, however, strongly reduced after LCP addition.

Table 1 Example of mechanical properties of some injection-molded uncompatibilized blends consisting of PP and LCP (Vectra A950, Hoechst Celanese) [4]

Material PP/LCP	*Tensile strength (MPa)*	*Elastic modulus (MPa)*	*Elongation at break (%)*	*Charpy impact strength* (kJ/m²)*
PP/0	34 (1)	1805 (20)	588 (14)	not broken
PP/5	32 (1)	2072 (26)	13.2 (2.4)	30 (2)
PP/10	34 (1)	2063 (27)	9.6 (2.4)	23 (2)
PP/20	39 (1)	2438 (73)	4.8 (0.3)	14 (1)
PP/30	44 (1)	2937 (38)	3.6 (0.2)	11 (1)
PP/50	51 (1)	4730 (92)	2.5 (0.4)	7.6 (0.7)

* unnotched samples; standard deviations in parenthesis.

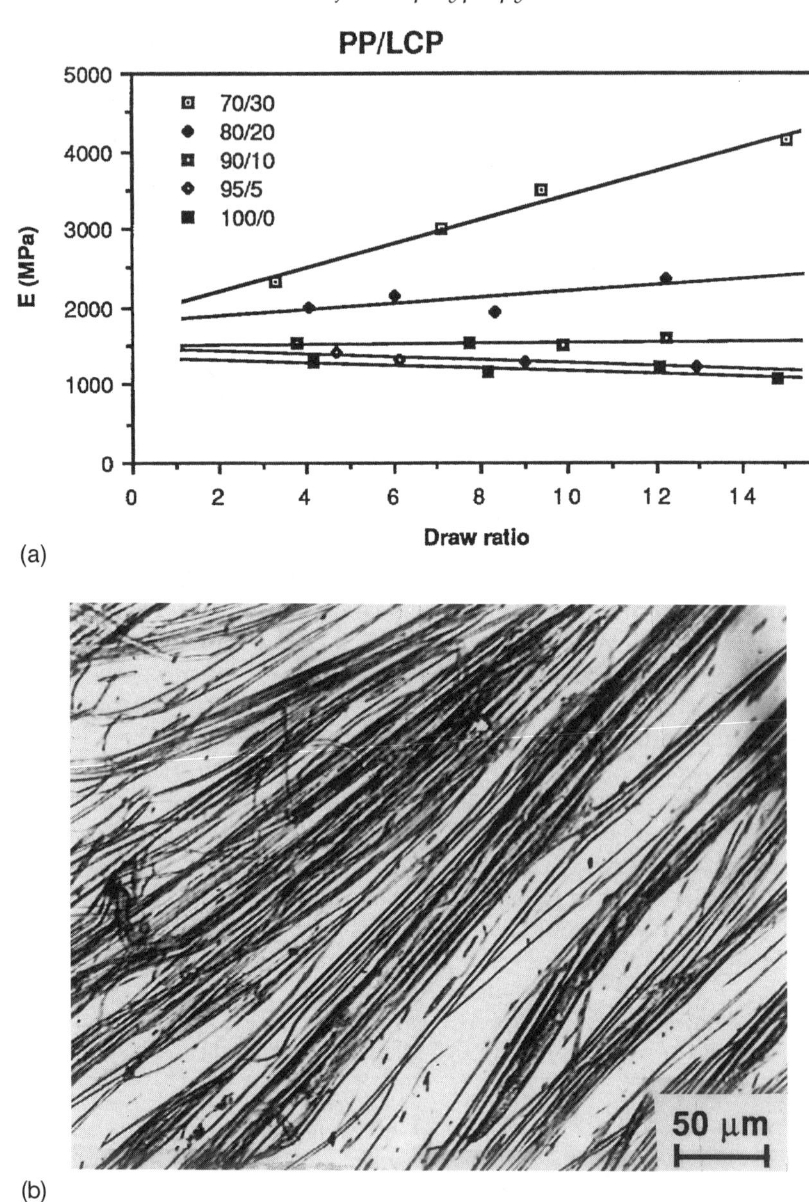

Figure 1 (a) Effect of elongational drawing on the elastic modulus (E) of extruded PP/LCP blends. (b) Morphology of highly drawn PP/LCP 80/20 blend (LCP = Vectra A950, Hoechst Celanese) [4].

The blends are thus stiff in the fiber direction but brittle in the transversal direction. While this is partly due to the anisotropic blend structure, probably the main reason is the poor interfacial adhesion between the blend components. The latter can be improved by proper compatibilization (see below) [4].

In addition to mechanical reinforcement, LCP addition improved the dimensional stability and thermal resistance of the PP matrix. The former appeared as remarkably reduced post-shrinkage and lower linear thermal expansion coefficient, the latter as significantly increased heat deflection temperature (HDT). The HDT values increased from the level of 52°C for neat PP to 78°C for PP/LCP 70/30 and 110°C for 50/50 blends. Even greater improvements were found for blends with another, copolyesteramide-type LCP (Vectra B950): the HDT value for PP/30 wt.% B950 was as high as 132°C [4].

Studies with other LCPs, such as Vectra B950 (Hoechst Celanese) or Rodrun LC-3000 (Unitika), have shown qualitatively similar properties, but the level seems to depend a lot on the performance of the LCP. Hence, for example, the former leads to higher stiffness, but the latter allows the use of rather low processing temperature. Other flexible semirigid LCPs exhibiting a melting temperature closer to that of PP have been blended with PP as well [2, p.138].

BLEND MORPHOLOGY

Mechanical properties of the blends are closely related to their morphology. Due to the difference in chemical nature, the PP and LCPs discussed here are immiscible to each other. Generally, LCP forms a composite-like fiber structure in the PP matrix, with thin oriented LCP fibers in the skin region and spherical or ellipsoidal LCP domains in the core (skin-core morphology). The size of the LCP domains tends to increase with the LCP content, and the formation of LCP fibrils is enhanced with higher LCP content and draw ratio. For example, at 20–30 wt.% of LCP, fibril-like LCP domains existed even in the core of the extruded strand (Figure 1). The particular morphology created depends on material characteristics of the blend components (in particular, the viscosity ratio) but also on the conditions selected for blending and processing.

Our studies showed the optimal viscosity ratio (η_{LCP}/η_{PP}) for formation of small LCP domains and their spontaneous fibrillation to range from about 0.5 to 1. At lower viscosity ratios, the fiber structure was coarser, while at viscosity ratios above unity the LCP domains tended to be spherical or cluster-like. As the temperature and shear rate dependencies of LCP and the matrix polymer may well be different, it is extremely important to know the viscosity ratio under the actual processing conditions. Its effect on spontaneous fibrillation of the LCP domains forms the basis for material selection, but the final processing conditions determine how well the LCP phases will be fibrillated and oriented in the product. Other properties, such as melt elasticity and interfacial tension of the blended polymers, affect the morphology as well [4, 5].

COMPATIBILIZATION

As mentioned above, the interfacial adhesion between PP and LCPs is generally poor which reflects, for example, on the impact properties of the blends. Special efforts are often needed to increase the interfacial adhesion and overcome the brittleness. The objective in compatibilizing LCP blends is to stabilize the fibrous blend morphology and enhance the interaction and interfacial adhesion between the two distinct phases. It should be noted that the aim is not to achieve miscible blends or very fine dispersions, since the good properties of LCP blends are based on the composite-like fiber morphology formed *in situ*. Toughness of PP/LCP blends can be improved by adding small amounts of suitable copolymers such as functionalized polyolefins, e.g. maleic anhydride grafted polypropylene. We have achieved good results with an ethylene-based terpolymer functionalized with glycidyl methacrylate. The compatibilizing effect of this elastomeric polymer was based on good mixing with the PP matrix and strong interactions, involving a chemical reaction, with the LCP phase. The problem with such reactive compatibilizers is however their tendency to disperse the LCP phases to tiny droplets and to prevent the fibrillation [4].

Successful compatibilization depends on the chemical suitability of the compatibilizer for the system, its amount and distribution, and the viscosity ratios of the blend components. An effective blending, preferably in a twin-screw extruder, is usually required. The morphology of ternary systems is more complex and perhaps more difficult to control; in particular, the accommodation of small amounts of compatibilizer precisely at the interfaces may be difficult to achieve. Proper compatibilization and processing are thus the key factors in producing optimal LCP blends.

PROCESSING AND APPLICATIONS

PP/LCP blends can be processed in the molten state by all conventional melt processing techniques, such as extrusion, melt spinning, injection or compression molding. Because of the rheological behavior of thermotropic LCPs, extrusion and melt spinning are advantageous methods of achieving good fibrillation and orientation of the LCP in the matrix. In injection molding, controlling the LCP orientation and achieving good fibrillation is slightly more complicated. Thin-wall products are favourable. In any case, the final processing of the blends is extremely important, since it determines the blend morphology and resultant properties.

Naturally, the anisotropic nature of LCP blends should be taken into account in the product design. The effect of weld lines, which are the weak points for all composites as well, may be diminished with special processing techniques such as push-pull or multiple-live-feed (MLF) injection molding. Moreover, the orientation field of LCP fibers in extrusion of tubular products may be controlled by using a counter-rotating annular die [2–4].

In addition to melt processing, we have successfully processed pre-blended PP/LCP blends without remelting the LCP fibers. The resulting composites exhibited significantly higher impact strength than conventional uncompatibilized melt blends. This is based on highly fibrillar morphology throughout the sample, instead of skin-core structure typical for melt blends [5]. By a similar technique it is also possible to make prepregs which may be subsequently laminated together in desired positions to obtain more isotropic properties [3].

For the moment, thermotropic LCPs are not yet widely used as engineering plastics, in part no doubt due to their high price which, in turn, is at least partly due to the relatively small production volumes. Blending of LCPs with thermoplastics may lead to useful combinations of material properties and new applications, thanks to the several exceptional properties of LCPs. In addition, blending could be a means to reduce the price of LCP.

REFERENCES

1. Brostow, W. (1990) Properties of polymer liquid crystals: choosing molecular structures and blending. *Polymer*, **31**, 979–995.
2. La Mantia, F.P. (ed.) (1993) *Thermotropic Liquid Crystal Polymer Blends*, Technomic, Lancaster, USA.
3. Isayev, A. (1996) Self-reinforced composites involving liquid-crystalline polymers: Overview of development and applications, in *Liquid-Crystalline Polymer Systems: Technological Advances*, (eds A.I. Isayev, T. Kyu and S.Z.D. Cheng) ACS Symposium Series 632, American Chemical Society, Washington, DC, pp. 1–20.
4. Heino, M. (1994) Blends of thermotropic main-chain liquid crystalline polymers and thermoplastics. *Acta Polytechnica Scandinavica, Chemical Technology and Metallurgy Series*, No. 220, 1–49 (+ 7 appended publications).
5. Heino, M. and Vainio, T. (1996) Effect of viscosity ratio and processing conditions on the morphology of blends of liquid crystalline polymer and polypropylene, in *Handbook of Applied Polymer Processing Technology*, (eds N.P. Cheremisinoff and P.N. Cheremisinoff), Marcel Dekker, New York, pp. 233–263.

Keywords: polymer blend, liquid crystalline polymer (LCP), morphology, compatibilization, processing, polymer reinforcement, *in situ* composites.

Industrial polymerization processes

H. Ledwinka and W. Neißl

HISTORICAL OVERVIEW

In 1954 G. Natta polymerized propylene in the Montecatini laboratories by means of a modified Ziegler catalyst and obtained a blend of isotactic and atactic polypropylene (PP). For this pioneering invention he and K. Ziegler received the Nobel price for chemistry in 1963.

Montecatini started the production in 1957 based on a very complicated process in a small scale of only a few thousand tons per year. Since then the production and catalyst technologies as well as the consumption of PP have developed tremendously. World-wide consumption of PP in 1996 amounted to around 20 million tons. This figure is all the more remarkable when it is remembered that in 1968 worldwide consumption had only reached 1 million tons. By the year 2000, this number will be at least 25 million tons.

CORRELATION OF CATALYST AND PROCESS DEVELOPMENT

Since the beginning of PP production, the fourth generation of so-called Ziegler–Natta (ZN) catalysts has been reached [1]. In the case of PP, the catalyst system is the decisive element in respect to both product properties and polymerization technologies. Also metallocene catalysts have just gained commercial interest, allowing the design of PPs with property combinations not realizable up to now.

The above-mentioned first generation catalysts developed by Natta consisted of δ-$TiCl_3$ with aluminium diethylchloride as activator. PP

Polypropylene: An A–Z Reference
Edited by J. Karger-Kocsis
Published in 1999 by Kluwer Publishers, Dordrecht. ISBN 0 412 80200 7

polymerization in suspension has largely established itself for this system. The process was very elaborate, since the large quantity of atactic PP (aPP) obtained had to be removed. Furthermore, the low level of activity of the catalyst necessitated deactivation and washing-out of the latter.

The second generation was developed by Solvay. Compared with the first, it was distinguished by a level of activity about four times as high and by a high stereospecifity, obtained through the inclusion of an electron donor in the system. From the technological aspect, this enabled separation of the aPP fraction of the product, but not yet the washing-out of the catalyst.

The decisive breakthrough to the third generation was achieved in 1975 by Montedison and Mitsui Petrochemical. They developed a highly active catalyst system with good stereospecifity. This advance was made possible by supporting the titanium component on magnesium chloride. It permits a significant simplification of the polymerization technology by eliminating the washing-out of the catalyst. As the expensive catalyst removal (addition, separation and recovery of a suitable decomposition agent like alcohols) was eliminated, this resulted in a reduction of at least 20% in the investment costs and major savings in both energy and maintenance. Furthermore, due to this improvement the PP process became an extremely environment-friendly technique.

The further improvement to the fourth generation in the 1980s was characterized by developing catalysts which retain their spherical and porous morphology during the polymerization process. Himont/Montell has conducted these studies intensively in recent years calling the concept, in which each polymer granule becomes a 'minireactor', Reactor Granule Technology leading to developments such as Addipol, Catalloy and Hivalloy [2].

MODERN POLYMERIZATION PROCESSES

General comments

The medium for polymerization can be liquid or gaseous propylene or an inert hydrocarbon diluent such as hexane. Today about 25% of PP is produced in inert hydrocarbon slurry processes. Their relative proportion will steadily decrease because practically all new plants and investments are based on gas phase, bulk or combined bulk/gas phase technologies.

The combined progress in catalysis and technology has lead to a complete realization of the experience curve: In the 1960s, Ziegler–Natta catalysts yielded about 1 ton PP/kg catalyst in plants with capacities of some thousand tons per year. Today the yield is as high as 50–100 tons PP/kg catalyst at a production capacity of 2–300 000 tons per year.

Next, the most widely used processes will be briefly described.

Bulk processes

Montell

By far the most commonly used PP-process is Montell's 'Spheripol Process'. The first reaction stage consists of one or two tubular loop reactors where bulk polymerization of homopolymers, random and terpolymers is carried out in liquid propylene. The prepolymerized catalyst, liquid propylene, hydrogen for controlling molecular weight and eventually comonomers are continuously fed into the reactor in which polymerization takes place at temperatures of 60–80°C and pressures of 35–40 bar. The tubular configuration enables a perfect heat transfer and control of the reaction temperature.

The PP circulating in liquid propylene inside the loops is continuously discharged to a separation cyclone. Unreacted propylene is recycled to the reaction while the polymer is transferred to one or two gas phase reactors where ethylene, propylene and hydrogen are added to produce heterophasic impact copolymers. The granules are discharged to the monomer flashing and recovery section and sent to a monomer stripping system using moist nitrogen. In case of homopolymers and random co- and terpolymers, PP is directly fed to the flashing drum bypassing the gas phase stage.

After the drying unit, the granular resin is conveyed to an extrusion system for stabilization, eventual modification and pelletization. In Addipol-technology, pelletization is omitted.

Mitsui Chemicals

Up to now, about twenty PP plants in Asia and America have been built with Mitsui's Hypol process.

As already mentioned, Mitsui and Montedison are the pioneers of the supported Ziegler–Natta catalysts. Their independently developed polymerization processes are based on improved catalyst generations of the joint Ziegler-Natta development started in 1975. One of the major differences between the two processes is the concept of the homopolymerization stage. Instead of tubular loop reactors, Mitsui uses stirred tank vessels.

Similar to the Spheripol process, the Hypol process generates spheric granules, making possible the development of a 'nonpelletizing' process.

Gas phase processes

Amoco – Chisso

This gas phase PP process is a joint development of these two companies. The main feature is the innovative reactor design [3]. The polymerization takes place in one or two horizontal stirred vessels according to the polymerization mode. The reactors are subdivided into individually gas-composition-controllable and polymerization-temperature-controllable polymerization compartments. With continuous catalyst injection, essentially at one end of the reactor, and powder removal at the other end, the residence time distribution approaches that of plug flow. In this way the batch polymerization is 'reproduced', however with the good economics of a continuous process. Further characteristics of this technique are: rapid grade change and a high flexibility in tailoring the product properties. For the production of homopolymers and random co- and terpolymers, one reactor is used followed by a second reactor in series for heterophasic impact copolymers.

BASF

The BASF gas phase Novolen process was the first advanced process technology that come into commercial operation. The polymerization takes place in vertical stirred gas phase reactors. As usual, the production of heterophasic copolymers requires two reactors. Homopolymers, random co- and terpolymers can be produced in a single reactor or two reactors in a cascade. This results in a high output rate and allows to manufacture grades of bimodal molecular weight distribution. Cooling takes place by vaporizing liquid propylene which is injected into the reactor. Continous further development has led to an output of about 230 000 tons per year in a 75 m^3 reactor. Downstream a desorption step gives the possibility of producing types with a very low level of odor.

Union Carbide/Shell

The Unipol PP process combines the Union Carbide fluid bed technology developed for polyethylene with the Shell supported high yield catalysts (SHAC).

The polymerization system for homopolymers and random co- and terpolymers consists of a fluidized bed reactor, single-stage centrifugal cycle gas compressor, cycle gas cooler, catalyst feeder and product removal system. This compressor circulates reaction gas upward through the bed, providing the agitation required for fluidization, backmixing and heat removal. No mechanical stirrers or agitators are needed in Unipol process reactors. For impact copolymers a smaller replica of the first reaction system is operating in series.

The granular resin is conveyed from the polymerization system to a bin for purging with nitrogen to remove residual hydrocarbons and then to pelleting.

OUTLOOK

Further developments in metallocene- and Ziegler–Natta catalysts as well as innovative improvements in polymerization technology are guarantees for further product diversifications using the existing processes and for improvements in reactor productivities.

For metallocenes, two aspects are relevant. Firstly, this catalyst family enables the creation of PPs with property profiles not accessible with Ziegler–Natta systems (e.g. combination of stiffness of homopolymers with melting ranges of random copolymers). Furthermore, the production of syndiotactic PP (sPP) in commercial scale can be realized for the first time. In the future, totally new PP-materials will be synthesized by the use of comonomers, such as polar compounds not polymerizable by Ziegler–Natta systems.

Secondly, metallocenes do not require new polymerization processes. They can be used in existing technologies. This drop-in characteristic will definitely accelerate the break-through of this new generation of catalysts [4].

With Ziegler–Natta systems, ongoing developments target further product improvements (e.g. control of molecular mass distribution or copolymerization characteristics) and the increase of activity retaining at least the already achieved levels of stereospecificity and morphology. Enhanced activities will enable a flexible operation of high capacity plants due to technically feasible reduced residence times (i.e. short transitions at product changes).

A very interesting and promising technology is the polymerization of propylene under supercritical conditions. On the one hand, this polymerization mode enhances the heat removal capabilities drastically. On the other hand, this technology combines the advantages of both the gas phase and bulk processes in respect to polymer solubility, comonomer and hydrogen miscibility and space–time yield. A precondition of this method is to find catalysts which work properly at temperatures of around 100°C.

REFERENCES

1. Neißl, W. and Ledwinka, H. (1993) Polypropylen – die Zukunft hat gerade begonnen. *Kunststoffe*, **83**, 577–583.
2. Galli, P. (1996) The reactor polymer alloys. The shifting of the frontier of polymer research. *Macromolecular Symp.*, **112**, 1–16.
3. Shepard, J.W. *et al.* (1976) Divided horizontal reactor for the vapor phase polymerization of monomers at different hydrogen levels. US Patent 3 957 448.
4. Beer, G. (1996) Polypropylen (PP) *Kunststoffe*, **86**, 1460–1463.

Keywords: polymerization, Ziegler–Natta catalysts, metallocene-catalysts, supported catalysts, gas phase polymerization, slurry polymerization, reactor granule technology, Sheripol, Hypol, Novolen, Unipol.

Infrared and Raman spectroscopy of polypropylene

Erik Andreassen

GENERAL

Vibrational spectroscopy is one of the most versatile methods of polymer characterization [1]. Other methods may be better in a certain area, but the advantage of vibrational spectroscopy is that it supplies several types of information, while being a rapid and inexpensive method. Infrared (IR) and Raman spectroscopy probe the vibrational spectrum by absorption and inelastic scattering, respectively. In many cases these two techniques are complementary. For a noncentrosymmetric molecule, such as isotactic PP (iPP) in a 3_1 helix, bands which are strong with one technique are often weak with the other. From a practical point of view, the requirements for sample preparation are different. IR spectroscopy, being an absorption method, is mostly performed on thin films in transmission mode. However, reflection methods or photoacoustic detection can be utilized to analyze thick or opaque samples, with a range of penetration depths. In principle, no sample preparation is required for Raman spectroscopy. Fluorescence has been a problem when analyzing polymers with this technique. However, this problem is almost eliminated with Fourier transform (FT) Raman instruments, which excite in the near infrared (NIR) range ($\sim$ 4000–11000 cm^{-1}) and analyze the scattered light in a modified FT-IR instrument.

The literature on IR and Raman spectroscopy applied to PP is vast [2]. It goes from fundamental studies in polymer chemistry and physics to applications, such as forensic analysis and identification systems for

Polypropylene: An A–Z Reference
Edited by J. Karger-Kocsis
Published in 1999 by Kluwer Publishers, Dordrecht. ISBN 0 412 80200 7

recycling processes. A range of specialties, such as surface analysis, microspectroscopy, on-line analysis and diffusion studies, are represented. The following sections summarize the basic elements in spectroscopic characterization of PP. These methods extract information from the vibrational spectra regarding molecular configurations, conformations and orientations, as well as deformation and degradation mechanisms.

CONFIGURATIONS, CONFORMATIONS, CHAIN PACKING AND LAMELLAR STRUCTURE

Vibrations in a polymer crystal can be excited at widely different levels, ranging from dimensions corresponding to a bond length up to the thickness of a lamella. There are two types of modes in a polymer crystal: Lattice modes and internal modes. In the former, whole chains move relative to each other. However, apart from a few special cases, vibrational spectroscopy only utilizes the internal modes, with molecular-level information. Internal modes can further be divided into modes with high and low degrees of intramolecular coupling. The latter type is not sensitive to the chemical environment of the active group. Hence, the frequencies associated with such 'localized' modes (often referred to as group frequencies) are roughly the same for different molecules containing the same group.

The spectra of atactic PP (aPP) and iPP melts contain some well-known group frequencies (involving CH_2 and CH_3 groups), while other bands are due to coupled modes. By applying various theoretical and experimental methods, the distribution of conformations can be deduced from these bands. Several additional bands appear in the spectrum of iPP in the solid state, especially in the 800–1200 cm^{-1} range (Table 1). These bands are attributed to the 3_1 helix, which is the regular conformation in all iPP polymorphs, as well as in the mesomorphic phase. Two commonly used 'helix bands', at 998 and 841 cm^{-1}, only appear for helix segments with at least ~ 11 and ~ 14 repeat units, respectively. The band at 973 cm^{-1} is attributed to shorter helix segments, and is also observed with iPP melts, although with reduced intensity and a small frequency shift. For solid samples, it is sometimes resolved into two peaks, attributed to amorphous and crystalline domains, respectively. With aPP, this peak is usually only a shoulder.

Because the ability to form regular helices depends on the degree of isotacticity, the latter can be estimated indirectly using the helix bands. Among the various band combinations used as isotacticity indices, the peak area ratio $A(998\,cm^{-1})/A(973\,cm^{-1})$ is one of the most common. Raman and IR measures of isotacticity have been shown to correlate well with direct measurements, such as nuclear magnetic resonance (NMR),

Table 1 The most used bands in the vibrational spectrum of semicrystalline iPP (based on [2] with some modifications). Note that the actual frequencies depend on factors such as instrument type, analysis method and sample type.

Raman frequency (cm^{-1})	*Infrared frequency* (cm^{-1})	*Main active group vibrations*
–	2956vvs	νCH_3 asym.
2952m	2953vvs	νCH_3 asym.
2920m	2921vvs	νCH_2 asym.
2905m	2907sh	νCH
2883s	2877vs	νCH_3 sym.
2871w	2869vs	νCH_2 sym.
2840m	2840vs	νCH_2 sym.
1458vs	1460s	δCH_3 asym., δCH_2
1435w	1434m	δCH_3 asym.
1371sh	1370s	δCH_3 sym., ϖCH_2, δCH, νCC_b
1360s	1357m	δCH_3 sym., δCH
1330vs	1326vw	δCH, τCH_2
1306vw	1305w	ϖCH_2, τCH_2
1296vw	1296vw	ϖCH_2, δCH, τCH_2
1257w	1255w	δCH, τCH_2, ρCH_3
1219s	1220vw	τCH_2, δCH, νCC_b
1167sh	1164m	νCC_b, ρCH_3, δCH
1152vs	1154w	νCC_b, $\nu C\text{-}CH_3$, δCH, ρCH_3
1102w	1101vw	νCC_b, ρCH_3 ϖCH_2, τCH, δCH
1040s	1045vw	$\nu C\text{-}CH_3$, νCC_b, δCH
998m	998m	ρCH_3, δCH, ϖCH_2
973s	973m	ρCH_3, νCC_b
941m	940vw	ρCH_3, νCC_b
900m	900w	ρCH_3, ρCH_2, δCH
841vs	841m	ρCH_2, νCC_b, $\nu C\text{-}CH_3$, ρCH_3
809vs	809w	ρCH_2, νCC_b, $\nu C\text{-}CH_3$
530m	528w	ϖCH_2, $\nu C\text{-}CH_3$, ρCH_2
458m	456vw	ϖCH_2
398s	396vvw	ϖCH_2, δCH
321m	320vvw	ϖCH_2
252m	248vvw	ϖCH_2, δCH

Abbreviations: b = backbone, m = medium, s = strong, sh = shoulder, v = very, w = weak, δ = bending, ν = stretching, ρ = rocking, τ = twisting, ϖ = wagging.

and calibration curves have been constructed. It is important that the samples have experienced the same solidification conditions, and annealing procedures may be necessary in some cases. On the other hand, the same band combinations can also be used to estimate the degree of crystallinity as a function of both material and processing parameters. Although these measurements are only based on one-dimensional order, they are highly correlated with crystallinity data from other sources, as well as density measurements. Bands corresponding to irregular conformations can be used to characterize the amorphous phase. However, few pure 'amorphous' bands have been identified. It has been suggested that the band at 1154 cm^{-1} mainly originates from the amorphous phase in iPP.

Similar correlations exist for syndiotactic PP (sPP), but the picture is complicated by the fact that at least three different regular conformations have been observed in crystalline sPP. These are all different from the regular conformation in iPP, which makes it easy to distinguish between iPP and sPP. The two most common regular sPP conformations, the 2_1 helix (tggttg'g't) and the planar zig-zag, can be identified by characteristic bands at 977 and 962 cm^{-1}, respectively. A band at 867 cm^{-1} exists for both conformations and is often used as a syndiotacticity index, in various combinations.

In addition to tacticity, the molecular configuration is also affected by chemical defects. Head-to-head addition gives rise to peaks at 755 cm^{-1} (—$(CH_2)_2$—) and 1030 cm^{-1} (—$CH(CH_3)$—$CH(CH_3)$—).

Although iPP does not have any bands representing three-dimensional crystalline order, intermolecular coupling may lead to band splitting, due to phase differences between internal modes in adjacent chains, and non-equivalent molecular sites in the unit cell. With iPP, this usually appears as peak broadening, because the splitting cannot be resolved. However, some multiplets have been resolved in low-temperature studies. Differences in the spectra of the α, β, γ and mesomorphic phases of iPP have been attributed to differences in chain packing, i.e. different unit cells and packing defects. The positions and intensities of the CH_3 stretching bands are especially sensitive to chain packing. These bands have, for instance, been used to study the β-α transition.

At the extreme low-frequency end of the Raman spectrum, the so-called longitudinal acoustic modes (LAM), probe the lamellar structure, including the fold surfaces. These lattice modes correspond to accordion-like deformations of the chain. The lamella thickness is related to the frequency of these modes. A quantitative expression has not been found for helical PP, but an inverse relation has been proved. LAM bands in the range 10–20 cm^{-1} have been used to compare PP samples in terms of lamella thicknesses. A similar mode exists in the amorphous phase; the so-called disoriented LAM (DLAM). A DLAM doublet observed near 220 cm^{-1} in iPP melts has been associated with 3_1 helix segments.

MOLECULAR ORIENTATION AND DEFORMATION

Molecular orientation is a key parameter when studying relationships between processing parameters and mechanical properties. It can be measured by both IR and Raman spectroscopy. The magnitude of the IR absorption depends on the angle between the incident electrical field and the transition moment. Using polarized incident light, this so-called dichroic effect can be used to assess the molecular orientation. The orientation parameter obtained by IR dichroism is essentially the average of $\cos^2 \theta$, where θ is the angle between a reference axis (usually the main axis of deformation) and the chain axis. The Raman scattering process is more complicated than IR absorption. This leads to some problems, but also extended possibilities. As an example, Raman scattering gives a more detailed description of the orientation by assessing the average of both $\cos^2 \theta$ and $\cos^4 \theta$.

The angle between the transition moment vector and the helix axis must be known in order to quantify the IR orientation data. Transition moment angles for PP bands are usually assumed to be 0 or $\pi/2$, but this is probably not correct. The large variation in orientation factors calculated from different bands is partly due to different transition moment angles (not taken into account) and partly due to actual differences in orientation (since different bands may represent different structural states). As mentioned above, the bands in IR and Raman spectra of PP are unresolved multiplets. Most multiplets have the same transition moment angle, but for the Raman scattering the picture is more complicated. This reduces the number of Raman bands which can be used for orientation measurements.

A large number of IR bands have been used in orientation studies. Bands at 809, 841, 973 and 1154 cm^{-1} have been used in both IR and Raman studies, with various phase assignments. IR and Raman orientation factors are correlated with data from other methods, such as wide-angle X-ray scattering (WAXS) (orientation of crystallites) and birefringence (average of all phases). The IR orientation factors have in some cases been calibrated by adjusting the transition moment angle. The 1460 cm^{-1} IR band has been reported to be insensitive to orientation, because it is due to two overlapping bands with opposite transition moments. Hence, this peak may be used as a reference.

For a system with uniaxial orientation, only one orientation factor is needed. However, a bimodal orientation (autoepitaxy) may develop in iPP solidified under high stress in a uniaxial stress field. Due to this effect, some IR studies, e.g. of melt spun fibers, are flawed by assuming a uniaxial orientation for all samples. The orientation along three orthogonal axes have been measured for films (uniaxially and biaxially oriented) and injection-molded samples, using reflection and tilting methods. A simple IR transmission method without tilting has also been demonstrated for iPP, based on the 841 and 809 cm^{-1} peaks. These bands originate from vibrations of the same regular helix conformation, and the transition moments are close to 0 and $\pi/2$. The orientation along the three orthogonal axes can be calculated from three measurements of the ratio $A(841\,\text{cm}^{-1})/A(809\,\text{cm}^{-1})$, with polarization parallel and perpendicular to a reference direction in the plane, and no polarization, respectively. This method is also well suited for samples with a very high degree of uniaxial orientation.

Orientation measurements have also been implemented for *in-situ* analyses of PP. The orientation and relaxation kinetics of melts in shear flows have been studied by IR spectroscopy in a rheometer. In another example, the orientation in the transition front of the neck during tensile deformation has been analyzed by Raman microspectroscopy. This last example is a natural link to studies of samples subjected to mechanical stress.

Several IR and Raman bands are sensitive to stress, i.e. molecular deformation. Stress generally affects both the position and the profile of the peaks. The latter effect is taken as evidence of an uneven stress distribution, but usually only the former effect (a frequency shift) is utilized. The bands with the highest stress sensitivity are those with large contributions from axial stretching of C—C bonds along the backbone. The shift of the band at $1164\,\text{cm}^{-1}$ is usually reported to be the most sensitive probe, with a stress sensitivity factor on the order of $-20\,\text{cm}^{-1}/\text{GPa}$. The sensitivity factors of vibrational bands vary with temperature, morphology and stress level. Most factors are negative, but a few positive shifts have also been reported for PP. Observed sensitivity factors have been confirmed by calculations.

Raman and IR spectroscopy have been used in a number of studies of molecular load distributions and deformation mechanisms in PP, usually in combination with a mechanical loading device (tension or compression). The topics studied include true loads on atomic bonds, chain scission under stress, stress relaxation and creep, residual stresses, and stresses along aramid fibers in a PP matrix during pull-out testing.

A number of new analysis techniques have been developed recently, combining polarized IR spectroscopy and periodic mechanical perturbation. If the strain amplitude is small, the spectral response (due to molecular orientation and deformation) is linear and can be analyzed in terms of in-phase and out-of-phase components. This technique is often referred to as dynamic IR linear dicroism (DIRLD). DIRLD evolved into the 2D IR technique, in which a 2D spectrum, $I(\nu_1, \nu_2)$, is obtained by considering the relative phase differences between IR absorptions at different frequencies, ν_1 and ν_2. With these techniques, species with different response to the external perturbation can be distinguished. As an example, peaks with contributions from both crystalline and amorphous domains can be resolved. Some initial studies of PP with these techniques have been reported.

COPOLYMERS AND BLENDS

Ethylene–propylene copolymers are of high commercial interest, and IR spectroscopy is an excellent tool for analyzing these systems [3]. Isolated ethylene units (—$(CH_2)_3$— sequences), typically found at low ethylene concentrations, give rise to a band at 733 cm^{-1}. It has been shown that ethylene units can be accommodated inside PP crystals without reducing the degree of crystallinity, but the average helix length, as measured by IR helix bands, decreases, due to the ethylene interruption. A band at 721 cm^{-1} appears when there are two or more consecutive ethylene units between propylene units. PE crystallites can be identified by a pure 'crystalline' band at 730 cm^{-1}. The 721/730 cm^{-1} pair is a doublet due to intermolecular coupling. The band at 730 cm^{-1} may be confused with the band at 733 cm^{-1}, but the former is sharper and changes with temperature. At low propylene concentrations (or with special catalysts), isolated propylene units can be identified by a band at 935 cm^{-1}. Sequences of more than about three units give rise to a band at 973 cm^{-1}, which was discussed above.

Various band combinations have been used to characterize the ethylene and propylene distributions in copolymers, depending on the copolymer type (statistical or block) and the ethylene content. It must be noted that additives often contain methylene sequences comparable to those incorporated in the copolymer chain. Copolymers can also be analyzed in the NIR range, which is well suited for on-line implementation and nondestructive analysis. The analysis and identification of copolymers have gained much in sensitivity and reliability by applying statistical methods to large parts of the spectrum. Studies of size exclusion chromatography (SEC) fractions have given detailed information on the performance of catalyst systems. Comonomer distributions obtained with IR spectroscopy agree well with NMR data.

Several other propylene-based copolymers and blends have been analyzed by vibrational spectroscopy. The EPDM (ethylene-propylene-diene-monomer) content in PP/EPDM (low temperature impact) blends is a linear function of the ratio $A(2850\,cm^{-1})/A(2920\,cm^{-1})$ up to 80% EPDM. Functionalized PP (compatibilizer) for use in PP blends is often characterized by IR spectroscopy. These compatibilizers typically consist of PP grafted with anhydrides or acids, and may be analyzed in terms of chemical details and graft content. Intermolecular interaction between groups on the grafted side chain and the non-PP component in the blend has been characterized by band shifts and band broadening.

Vibrational spectroscopy makes it possible to assess morphological parameters (e.g. order and orientation) of the blend constituents separately. This has, for instance, been demonstrated with PP/polyamide melt spun fibers. The composition and morphology of microdomains in PP based blends have been analyzed by IR and Raman microspectroscopy. The micro-Raman imaging technique, which is based on confocal laser line scanning, offers the best spatial resolution (typically 1 μm laterally and 3 μm in depth).

DEGRADATION

IR spectroscopy is one of the most common methods for studying the degradation and stabilization of PP under various conditions [4]. It is a versatile tool for identifying the chemical species formed in the degradation process and obtaining kinetic parameters. IR spectroscopy also offers some special possibilities: orientation, stress and degradation can be measured simultaneously, in order to study the complicated relationship between degradation kinetics and molecular orientation and stress. The difference between bulk and surface degradation can be assessed. Microspectroscopy can reveal local variations in degradation, e.g. due to inhomogeneous distributions of antioxidants. IR emission spectroscopy has been demonstrated as a tool for studying the thermal degradation of PP.

The exact degradation mechanism depends on the conditions (temperature, stress, UV light, γ irradiation, plasma treatment, electrical field, etc.), the type of PP and the morphology. However, the same chemical species are involved, containing hydroxyl (OH) groups, carbonyl (C=O) groups and unsaturated (C=C) groups (Raman spectroscopy is more sensitive to unsaturated groups than IR spectroscopy). Hydrogen-bonded hydroxyl groups give rise to a broad peak around 3400 cm^{-1}, while multiple overlapping carbonyl peaks appear somewhat later, typically centered around 1720 cm^{-1}. The degree of degradation is often characterized by a carbonyl index, which is the area of the major carbonyl peak relative to an internal standard, such as the 1166 cm^{-1} band. From the hydroxyl and carbonyl groups, degradation products such as peroxides, alcohols, carboxylic acids, ketones, aldehydes, esters and γ-lactones can be identified by performing derivatization reactions and consulting handbooks on group frequencies of organic molecules. An analysis based on carbonyl peaks is the most reliable.

Bands due to unsaturated groups are less studied. In the absence of oxygen, these are the only new groups that appear. Thermal degradation in an inert atmosphere typically leads to the formation of vinylidene end groups (with characteristic bands at 889 and 1648 cm^{-1}) at temperatures above 300°C, followed by vinyl end groups (910 cm^{-1}) and trans-vinylene (965 cm^{-1}) at higher temperatures. After prolonged treatment at elevated temperatures all PP bands disappear – only an unsaturated hydrocarbon liquid is left.

REFERENCES

1. Bower, D.I. and Maddams, W.F. (1992), *The Vibrational Spectroscopy of Polymers*, Cambridge University Press, Cambridge.
2. Arruebarrena de Báez, M., Hendra, P.J. and Judkins, M. (1995) The Raman spectra of oriented isotactic polypropylene. *Spectrochim. Acta*, **A51**, 2117–24.
3. van der Ven, S. (1990), *Polypropylene and Other Polyolefins – Polymerization and Characterization*, Elsevier, Amsterdam.
4. LaCoste, J., Vaillant, D. and Carlsson, D.J. (1993), Gamma-initiated, photo-initiated, and thermally-initiated oxidation of isotactic polypropylene. *J. Polymer Sci. Polymer Chem.*, **31**, 715–22.

Keywords: infrared spectroscopy, Raman spectroscopy, vibrational spectroscopy, conformation, configuration, near infrared (NIR), surface analysis, on-line analysis, microspectroscopy, forensic analysis, chain packing, lamellar structure, polymorphism, group frequencies, coupled vibation modes, isotacticity index, syndiotacticity index, lattice modes, internal modes, degree of crystallinity, head-to-head addition, longitudinal acoustic mode (LAM), molecular orientation, molecular deformation, stress-induced frequency shifts, IR dichroism, 2D IR, ethylene-propylene copolymers, blends, ethylene-propylene-diene-monomer (EPDM), functionalized PP, compatibilizer, degradation, oxidation, carbonyl index, carbonyl groups, hydroxyl groups, unsaturated groups.

Injection molding of isotactic polypropylene

Gürhan Kalay and Michael J. Bevis

INTRODUCTION

Injection molding is one of the widely used methods in the processing of isotactic polypropylene (iPP), where a screw in a heated barrel rotates to transport, melt and pressurize the polymer in the barrel, and then reciprocation of the screw injects the melt into the mold. Processing parameters have a very strong influence on the morphology and hence the physical properties of the molded product. This section discusses the effect of main processing parameters on the morphology and properties of conventionally injection-molded isotactic polypropylene. The application of shear controlled orientation in injection molding (SCORIM) for converting iPP into molded products is the subject of another chapter in this book, which presents the morphological changes induced in SCORIM with the objective of enhancing physical properties. These chapters together present the fundamental characteristics of injection-molded polypropylene which may serve as a guide to the more informed processing of iPP, and to the realisation of optimum physical properties.

EFFECT OF PROCESSING PARAMETERS ON THE PROPERTIES OF MOLDED iPP

Mold temperature

The optimum mold temperature for processing iPP is 60°C. The use of mold temperatures lower than this value result in a thick skin which

Polypropylene: An A–Z Reference
Edited by J. Karger-Kocsis
Published in 1999 by Kluwer Publishers, Dordrecht. ISBN 0 412 80200 7

exhibits a layer of premature spherulites. The term 'skin' defines the region showing premature spherulites including a very thin layer exhibiting an oriented morphology resulting from extentional flow described in Tadmor's [1] model for the filling of a mold cavity, which is in any case difficult to identify by light microscopy. The core of a molding defines the region with mainly spherulitic morphology in conventionally molded iPP. There is a transition region between the skin and the core which exhibits spherulites with increasing diameter towards the core. Moldings produced using narrow section and/or converging geometry cavities may exhibit an oriented region between skin and core. This oriented region exhibits a shish-kebab morphology. High mold temperatures result in an increase in cycle times, and also an increase in the β spherulite content.

Melt temperature

Melt temperatures are normally set in the range 190°C and 250°C. In conventional injection molding, low melt temperatures may lead to a low Young's modulus due to the occurrence of β spherulites. However, in thin section moldings, this effect may be dominated by high molecular orientation attributed to fast cooling. Therefore a low melt temperature may be preferable for achieving high stiffness in thin section moldings. Low melt temperatures may also result in high impact resistance in thick section moldings produced at high mold temperatures which tend to exhibit a high β phase content.

Injection speed

In conventional injection molding, a very high injection speed results in a decrease in Young's modulus (E) due to high shear heating. A very low injection speed may result in a low level of molecular orientation which is associated with low Young's modulus. Therefore an optimum injection speed has to be set for achieving high stiffness.

Holding pressure

In conventional injection molding, high holding pressures result in high stiffness. High stiffness is the result of increased levels of molecular orientation which can be measured by X-ray diffraction studies. High pressures may result in formation of γ phase in injection-molded samples. It is worth emphasizing that the general view that processed polypropylene exhibits only α phase with sporadic β phase is suspect because injection-molded iPP may contain substantial amounts of γ phase [2–3].

MORPHOLOGY

Injection-molded iPP [2–5] shows clear skin and core morphologies when cross-sections are observed with a polarizing microscope. The thickness of the skin layer varies widely with the grade of resin and molding conditions, and physical properties such as Young's modulus, yield strength, and mold shrinkage, systematically change accordingly. Figure 1 shows the transverse cross-section of a conventionally molded thick section iPP molding produced using a cylindrical cavity. The skin-core structure is evident, and is mainly composed of α and β spherulites.

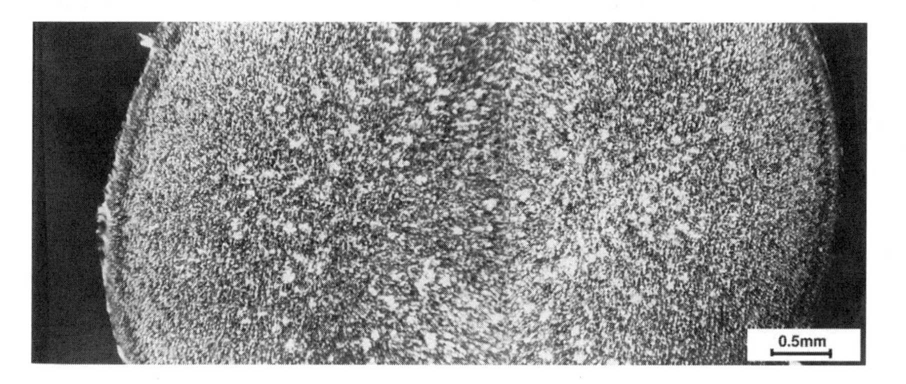

Figure 1 Microstructure of the whole cross-section of a thick section iPP molding exhibiting α and β spherulites.

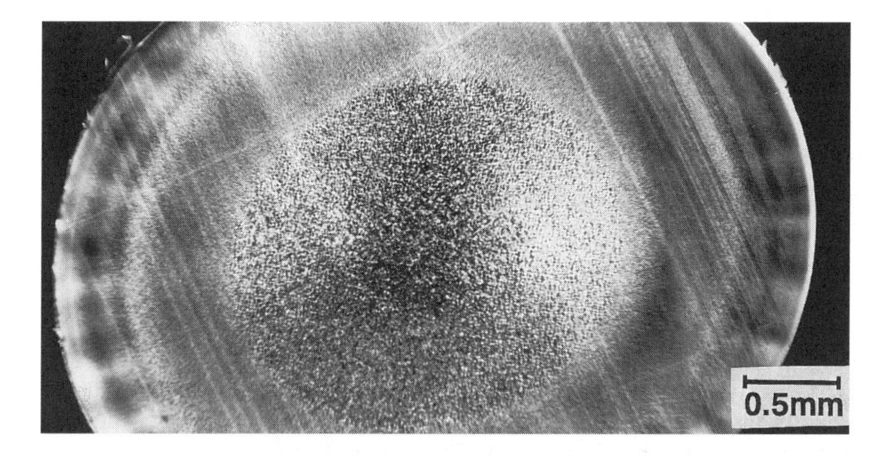

Figure 2 Microstructure of the whole cross-section of a narrow section iPP molding exhibiting an oriented zone between the skin and the spherulitic core.

Table 1 Tensile test data for thick and thin section conventional iPP injection moldings. Standard deviations are shown in parentheses

	Yield strain (%)	*Yield stress (MPa)*	*E (MPa)*
S1 (thick section)	10.6 (1.3)	31.6 (0.6)	1195 (113)
S2 (thin section)	9.7 (1.2)	44.3 (1.0)	2401 (114)

Spherulite size increases from the edges to the core. Figure 2 shows the whole transverse cross-section of a thin section iPP molding produced using a converging die, and exhibits an oriented region between skin and core. Figures 1 and 2 illustrate the marked influence of mold cavity geometry and processing conditions on the morphology of molded iPP. The morphology of the oriented region is not revealed under light microscopy because of insufficient resolution, but may be identified by transmission elecron microscopy studies which reveal a shish-kebab morphology (see the chapter entitled Application of shear-controlled orientation in injection molding for more details).

MECHANICAL PROPERTIES

Table 1 summarizes the tensile test data for conventionally molded thick and thin section moldings of a typical commercial iPP. Thin section moldings exhibit an oriented region as shown in Figure 2, which contributes to the enhancement of mechanical properties.

Charpy flexural impact testing showed that the thin section moldings that contained an oriented zone exhibit greater impact resistance than thick section moldings.

MOLECULAR ORIENTATION AND CRYSTALLINITY

Wide angle X-ray diffraction [3] is a reliable way of assessing the molecular orientation and phase relationships in injection-molded iPP. WAXD scans from conventionally injection-molded iPP may exhibit a reflection at 2θ of 16.07° which corresponds to the β phase, in addition to α-phase reflections. However when iPP is processed under high cavity pressures γ phase reflection at 2θ of 20.06° may be detected in WAXD scans.

Application of high pressures in injection molding also leads to higher crystallinity iPP moldings. The β phase is associated with low modulus and tensile yield strength values. The occurrence of β phase in conventional injection moldings may be prevented through application of high pressures.

The intensity of the (111) and (041) α-phase reflections, with *d*-spacings of 0.4192 and 0.4064 nm respectively, have lower intensities for moldings exhibiting lower levels of molecular orientation than those moldings with greater levels of molecular orientation. This is attributed to the higher α phase orientation in oriented moldings. (110), (040) and (130) α phase reflections are representative of the crystalline orientation parallel to the surface of the molding whereas this is not the case for (111) and (041) reflections. As the intensity of the (111) and (041) α peaks decreases the α-orientation index increases. This doublet exhibits a low intensity for the γ phase containing moldings. This also confirms the association of the γ phase with the high molecular alignment.

Debye patterns reveal orientation characteristics in moldings. For conventional moldings (040), (130) and (060) reflections show preferred orientation in the longitudinal sections. The preferred orientation decreases from the edges to the centre of the molding. The level of orientation is greatest at the skin region of the moldings and the orientation is bimodal *c*- and *a**-axes orientation. Skin layers, i.e. the layer including shish-kebab morphology, show a higher melting point than the core region of the moldings because of the higher level of crystallinity in the skin layer.

CONCLUSION

In conclusion it can be said that iPP is a very readily injection-moldable thermoplastic, though it exhibits very complex morphology, phase relations and orientation characteristics which are very closely related to processing conditions. This implies that it should be possible to control properties by informed control of processing parameters. The chapter on page 38 introduces the concept of shear-controlled orientation in injection molding for the conversion of iPP into molded products, and in a way that provides for the management of micromorphology and hence physical properties.

REFERENCES

1. Tadmor, Z. (1974) Molecular orientation in injection moulding. *J. Appl. Polymer Sci.*, **18**, 1753–1772.
2. Kalay, G. and Bevis, M.J. (1997) Processing and physical property relationships in injection moulded isotactic polypropylene 1. Mechanical properties. *J. Polymer Sci., Part B, Polymer Phys.*, **35**, 241–263.
3. Kalay, G. and Bevis, M.J. (1997) Processing and physical property relationships in injection moulded isotactic polypropylene 2. Morphology and crystallinity. *J. Polymer Sci., Part B, Polymer Phys.*, **35**, 265–291.
4. Trotignon, J.P., Lebrun, J.L. and Verdu, J. (1982) Crystalline polymorphism and orientation in injection moulded polypropylene. *Plastics Rubber Processing Applic.* **2**, 247–251.

5. Fitchmun, D.R. and Mencik, Z. (1973) Morphology of injection moulded polypropylene. *J. Polymer Sci., Part B, Polymer Phys.*, **11**, 951–971.

Keywords: injection molding, processing parameters, mold temperature, melt temperature, injection speed, holding pressure, skin-core morphology, spherulites, molecular orientation, α phase, β phase, Young's modulus, structure–property relationships, SCORIM, polymorphism.

Injection molding: various techniques

W. Michaeli, C. Brockmann and A. Brunswick

INTRODUCTION

Injection molding is one of the most important processes for the production of plastic parts. The advantages of this process are:

- the direct conversion of the raw material into a finished product;
- no or only little refinishing of the molding;
- the ability of complete automation; and
- high reproducibility in production.

In recent years, the demands on the injection-molding process increased and the diversity of products has increased. Especially, the processing of polypropylene (PP) has changed from standard (e.g. packaging) to 'high-tech' injection-molded parts. Typical examples of complex PP-parts can recently be found in the automotive industry (e.g. door panels). To satisfy the demands of such parts, special processes of injection molding (IM) were developed, such as injection-compression molding, cascade injection molding and gas-assisted injection molding. In the following sections, these processes will be explained and typical applications will be described.

INJECTION-COMPRESSION MOLDING

The injection-compression molding (ICM) process starts, as shown in Figure 1, with the closing of the mold until the compression gap is reached. At this point, the mold is already sealed, because it is equipped

Polypropylene: An A–Z Reference
Edited by J. Karger-Kocsis
Published in 1999 by Kluwer Publishers, Dordrecht. ISBN 0 412 80200 7

$t = t_1$ injection phase

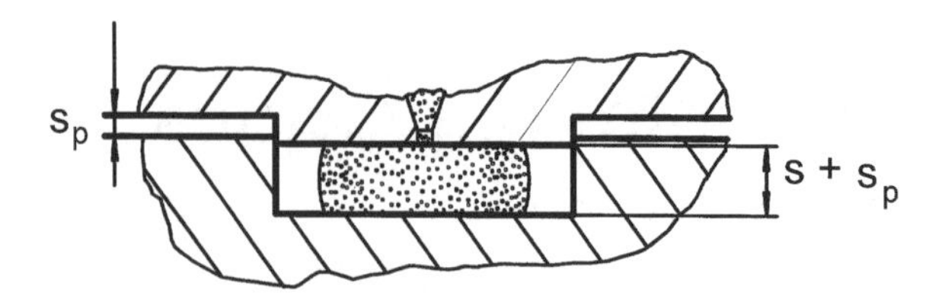

$t = t_2 > t_1$ compression phase

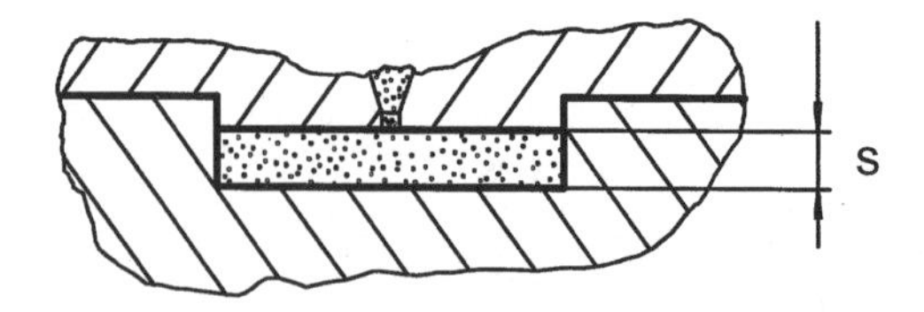

s_p = height of the compression gap

s = part thickness

Figure 1 Injection-compression molding.

with shear edges. In the next step, the melt volume, which is needed to fill the molding, is injected into the cavity and the shut-off nozzle is closed afterwards. In the last step, the mold is closed completely by the compression movement. Due to this movement, the melt is spread throughout the cavity until it is filled.

Finally the melt is compressed, because the injected melt volume is usually greater than that of the cavity. This is necessary to compensate the volume shrinkage. In conventional IM, this is done by the packing phase, which is only exceptionally used in ICM.

In most cases, the compression movement is done by the clamping unit of the injection-molding machine. Modern machines with hydraulic or toggle clamping units are able control the position and the velocities during the compression phase very exactly. In some cases, also, movable cores in the mold are used for the melt compression.

The possible pressure reduction is the reason for using ICM for the production of decorated moldings, such as used in the automotive industry (e.g. in-mold decorated door panels) [1, 2]. Another application of this process are thin-walled moldings with wall thicknesses below 0.5 mm and a flow path/wall thickness ratio below 150. Finally, there are parts for optical applications which benefit from the reduced shear rates and the more uniform molecular orientation along the flow path, using injection-compression molding.

CASCADE INJECTION MOLDING

Cascade injection molding can be used for parts with long flow paths, if the mold is equipped with more than one gate. As shown in Figure 2, the melt is injected over the first valve and the second valve is opened when

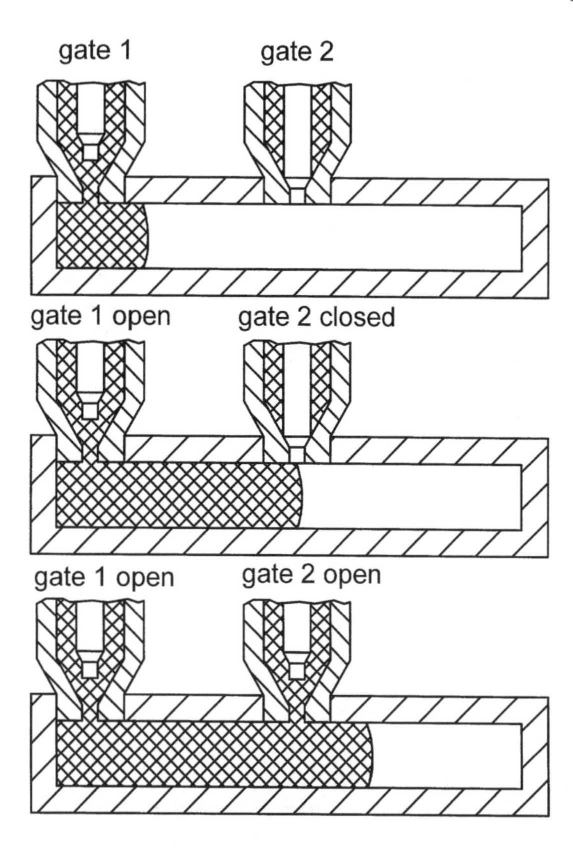

Figure 2 Cascade injection molding.

the melt front has already passed. Using this process, the cavity pressures can be reduced close to the values of simultaneous injection over both valves and, at the same time, the formation of weld lines can be avoided.

The only problem with this process are the shut-off valves in the mold, which are operated by a hydraulic system. Modern injection-molding machines supply an internal control for these valves but it is also possible to use an external hydraulic unit. The opening and closing of the valves can be controlled in dependence of time (e.g. starting from the beginning of injection) or of the screw position. Usually the reproducibility is higher when the screw position is used for control purposes.

Typical applications for this process are bumpers or A-, B- and C-pillar covers [1, 3]. In connection with in-mold surface-decoration, cascade injection may improve the quality of the molded parts. Such moldings are not only damaged by high pressures but also there may be wrinkles at the locations of weld lines. Both problems can be avoided using cascade injection molding. But this process is not limited on decorated moldings. Often this process is also adopted to produce visible parts for the automotive industry (e.g. bumpers).

GAS-ASSISTED INJECTION MOLDING (GAIM)

In this process, gas (nitrogen) is injected into the molten polymer to form hollow sections inside a part. The gas can be injected via the machine nozzle or through mold nozzles. GAIM is widely employed in the automotive, consumer and furniture industry, especially for parts with thick sections. Typical products are bumpers, dashboards, pillar covers, pedals, rear guidances, automotive door panels and arm rests, handles, copier doors, housings and covers, TV and monitor cabinets, CD-ROM trays, crates, chair legs, etc. [4].

A large number of patents have been issued for different GAIM process variants and hence processors are advised to clarify the legal situation before using these processes. One of these variants, the standard GAIM process (short-shot process) is described in the following paragraph.

The short-shot process starts with the injection of the melt into the cavity (Figure 3). Once the cavity is partly filled with a short-shot to between 50% and 95%, the gas is injected. The gas is best conducted into areas where there are large accumulations of melt. The gas helps to fill the cavity by pushing the melt front further forwards. By this way, hollow cores can be generated in the molded part. The gas is kept under pressure until the molded part becomes dimensionally stable. During this phase, a constant pressure acts over the entire gas channel. After the melt is solidified, the gas pressure is removed, either by letting the gas

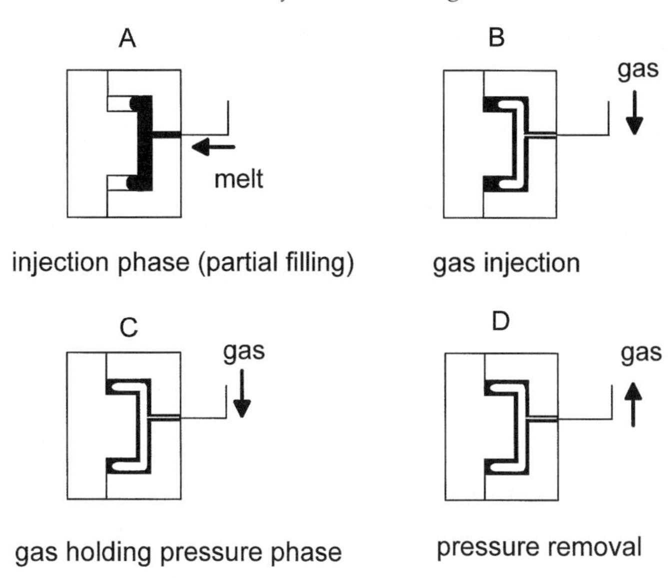

Figure 3 Standard-GAIM process.

escape into the environment or by recovering a certain percentage of it (up to 90%) via the machine or mold nozzles.

Todays variety of GAIM products shows that this process is very versatile and not specific for the manufacturing of thick rod-shaped parts (to reduce the amount of material, cycle time and sink marks). The main recent challenges are to combine different functional components into a single design having thick and thin cross sections and to decrease distortion and clamping force [5]. For example, enlarged ribs, which can be used in GAIM act as melt flow supporting channels or 'gates on the part' during melt injection and are later 'cored out' by the gas. Easier filling of the cavity (e.g. of large flat parts) requires less injection pressure and therefore a substantially reduced clamping force. The warpage and shrinkage are kept at a minimum due to the constant gas pressure. In addition, the stiffness of the part is increased. Therefore, cheaper materials, such as PP, can be used instead of more expensive low-shrink or engineering polymers to achieve high part qualities.

The advantages of utilizing GAIM depend strongly on the part design, but can be summarized as follows:

- greater freedom in design during mold layout (combination of thin and thick part sections);
- more uniform shrinkage, lower inherent stresses;

- reduced raw material costs, through the use of less material or the use of less expensive materials, such as PP instead of engineering polymers;
- integration of functional components;
- reduction of sink marks and warpage;
- reduced clamping force;
- shorter cycle times for rod-shaped and thick-walled moldings;
- increased mechanical rigidity at the same weight;
- lower demolding forces;
- feasibility of long flow paths;
- lower mold costs (simpler tooling, avoidance of cores and slides);
- the possibility to use gas channels for the transport of fluids.

Although, GAIM is increasingly used because of the number of advantages and growing process know-how, the plastics processor has to consider that there are many products for which minor or no benefits might be obtained by using GAIM. In these cases, it is more reasonable to use standard injection molding. It should be born in mind that GAIM also has some disadvantages, such as difficulties during the production set-up for complex parts and higher costs for quality assurance.

REFERENCES

1. Galuschka, S. (1994) In-Mould Surface-Decoration – Manufacturing of Textile-Covered Injection Mouldings, RWTH, Aachen. Dissertation.
2. Michaeli, W. and Brockmann, C. (1977) Injection Compression Moulding – A Low Pressure Process for Manufacturing Textile-Covered Mouldings. *Proceedings of the SPE Annual Technical Conference of Plastic Engineers (ANTEC)*, Toronto.
3. Michaeli, W. and Brockmann, C. (1995) Process Simulation Increases the Productivity of Special Processes in Injection Moulding in *Proceedings of the 28th International Symposium on Automotive Technology and Automation* (ISATA), Stuttgart.
4. Findeisen, H. (1997) Formation of the Residual Wall Thickness and Process Simulation Using the Gas Injection Technique, RWTH Aachen. Dissertation.
5. N.N. Plug and Play *Automobil Industrie* (1996) **3**, 46–47.

Keywords: injection molding, injection-compression molding, cascade injection molding, gas-assisted injection molding, process control, application.

Integrated manufacturing

P.-E. Bourban, and J.-A.E. Månson

INTRODUCTION

Polymers and polymer composites have the advantages of weight savings, high specific mechanical properties and good corrosion resistance which make them outstanding materials. On the downside, manufacturing costs can be very high, they can reach up to 60% of the total cost. Consequently, new economic manufacturing concepts will be increasingly appreciated in the future. As illustrated in Figure 1, several

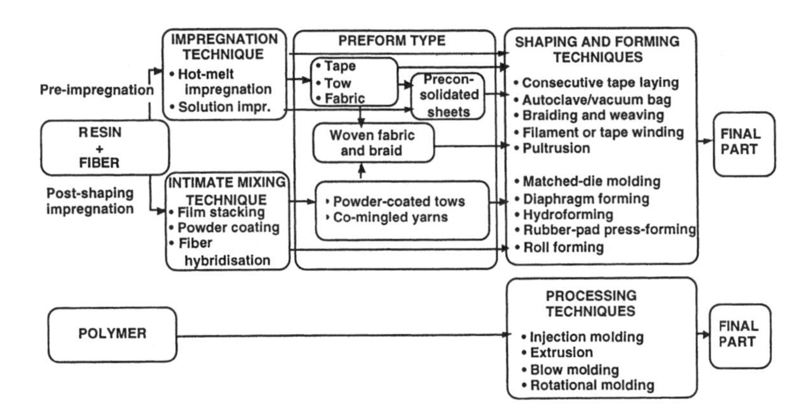

Figure 1 Alternative manufacturing routes for thermoplastic-based composites and thermoplastic polymers (modified from [1]).

Polypropylene: An A–Z Reference
Edited by J. Karger-Kocsis
Published in 1999 by Kluwer Publishers, Dordrecht. ISBN 0 412 80200 7

alternative routes and novel material preform types are being investigated for manufacture of thermoplastic-based parts [1, 2].

Furthermore, the production of complex shaped parts is still a challenge for the composite industry. Parts with relatively simple geometries are commonplace today for reinforced thermoplastic composites; the pultrusion of simple sections and the compression molding of parts with simple curvature are examples of well-developed technologies. But the production of complex three-dimensional parts usually requires injection or compression molding. Assembly technologies used to obtain complex geometries are sometimes inefficient and costly.

Therefore, the development of a multiprocess device could open interesting perspectives for design freedom and economical production of complex shaped parts.

In an attempt to reduce manufacturing costs by the suppression of intermediate processing and assembly stages, the primary approach is to develop a novel processing technique where neat polymers and composites are combined in a single operation. This is achieved by combining several processes into a single step or as a sequence of steps in rapid succession. Hence, integrated processing is defined here as the combination of processing techniques and of materials.

PROCESS INTEGRATION

The option of having various processing techniques assembled in one equipment is very interesting for the development of fast, flexible and cost-effective manufacturing. A combination of different processing techniques, such as compression molding, forming, multi-material injection molding, and extrusion, has been investigated. As a result, a unique equipment has been developed. A schematic illustration of the integrated processing machine is presented in Figure 2. Additional impregnation lines provide preconsolidated preforms to be integrated by the robot into the molds of the main unit. All assembled techniques can be used individually, furthermore the choice and the control of a large number of automatic sequences is made from a control desk. Each sequence combines different processes and material transformation steps (compression–injection–extrusion, etc.) which are executed in the same mold. The mold geometry defines the final external shape of the integrated part. The association of six moving half-molds increases the diversity of available complex shapes. Drapable preforms placed in existing molds increase the possible geometries further. Postforming and thus final shaping is an extra possibility using the compression facility of the present machine.

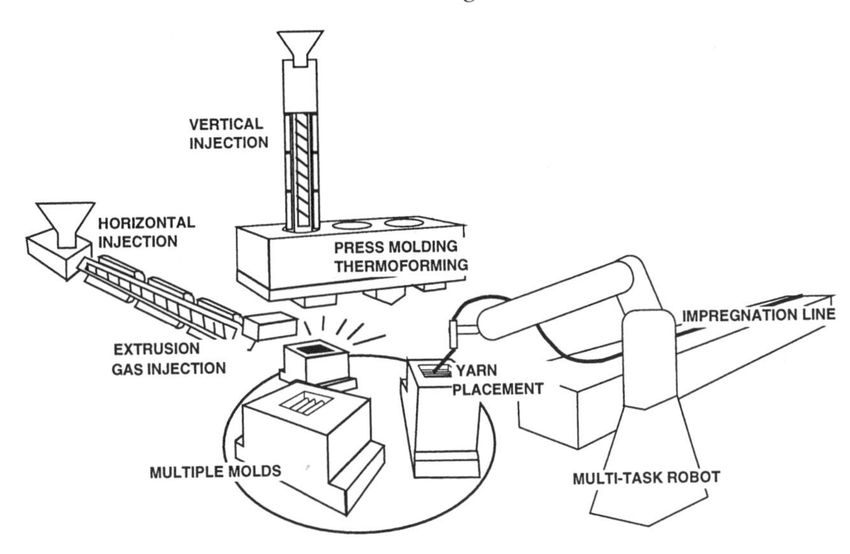

Figure 2 Schematic representation of the so far developed integrated processing equipment.

Flexibility in shape production is an interesting added value not only for achieving a required geometry but also for improving mechanical efficiency. For structural components, shape factors are defined to characterize the efficiency of a section shape in a given mode of loading [3]. For example, in the elastic bending of beams, rectangular and I-sections have respective shape factors of two and ten. By means of the previously described shaping facility, all components of an integrated part can be successively processed into the adequate optimum section shape.

Each sequence of molding and forming steps is optimized also in terms of energy consumption and time savings. For example, the manufacture of a two-component part usually consists of three main operations: processing of each component and then bonding. Part transfers and logistics as well as finishing operations are often required. With the integrated processing, where only one sequence and one heating cycle is needed, the time to manufacture the final part is drastically reduced.

MATERIAL INTEGRATION

The performance of a structural part does not depend only on geometric parameters. For composites more than for other materials, material properties and functional requirements are part of any optimum design. With the available integration of various material types into complex

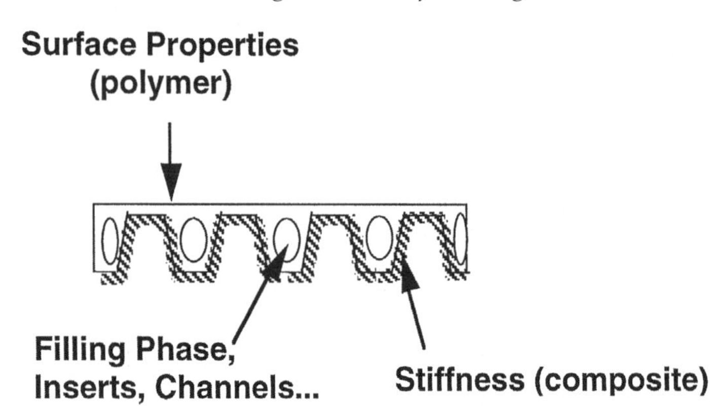

Figure 3 Schematic representation of the section of a part, illustrating the integration of functions achieved during one operation of the integrated processing.

shaped parts, the integrated processing concept introduces therefore new perspectives in part design.

Up to now, combinations of several polymers have been studied by numerous industries for different applications. For example, multimaterial injection is used for multicolor lenses or food containers. Electronics and hygiene/cosmetics industries are other application sectors. Coextrusion is another example of a technique successfully developed for multilayer packaging.

This idea of material integration is appealing and one can envision combining composite preforms and reinforced polymers partially or entirely embedded in neat polymers. By a judicious disposition of tailor-made intrinsically stiff composite preforms, the load transmission is optimized. Then nonstructural components are molded to keep the composite in place and to fulfil additional requirements, such as surface properties (finish or wear resistance). Different types of materials can be integrated, such as neat and modified polymers, particles and fiber reinforced polymers, composites, metals. Subsequently, as illustrated in Figure 3, various functions can be integrated in one part: load transmission, connections using inserts, complex geometry, wear and corrosion resistance, heat insulation or transfer, integration of sub-structures (channels, doors, fastening elements, etc.). Consequently, via material integration, multifunctionality and fine tailoring of properties can be obtained by the integrated processing.

The combination of materials with different rheological, thermal and mechanical properties requires the study of phenomena related to multimaterial structures, such as optimization of *in-situ* consolidation and

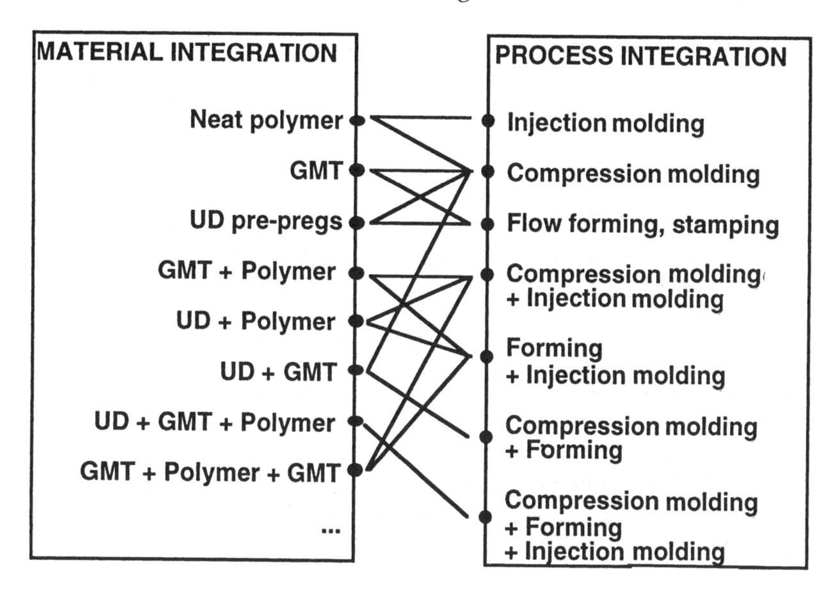

Figure 4 Combinations of PP-based preforms so far investigated and integrated processes applied.

control of interfaces between integrated materials, management of internal stresses and dimensional stability. Compatible thermoplastic polymers are quite easy to bond, numerous 'healing' studies have explained the autohesion mechanisms, especially for amorphous polymers. A similar approach has been developed to determine process windows available for an optimum adhesion between semicrystalline polypropylene (PP) adherends [4]. It was shown that, under specific nonisothermal conditions, bonds can be processed in shorter times than under isothermal conditions. These results are used to optimize the interface processing stage and to achieve good bond strengths during the integrated processing. Currently different parts are realized with the integrated processing equipment combining various PP-based materials. Neat PP is used with semiproducts, such as glass mat thermoplastic (GMT) and unidirectional glass fibers prepregs (UD). One of several processes available on the integrated processing machine is sequentially selected in function of the material preform type and of the desired part shape. Figure 4 presents the current investigated combinations of material and process techniques. Demonstration parts are manufactured considering the processing parameters determined to achieve optimum interfacial adhesion between reinforced PP components. For example, optimum nonisothermal interfacial temperature distribution was applied during the injection of PP polymer onto GMT composites (Figure 5). It

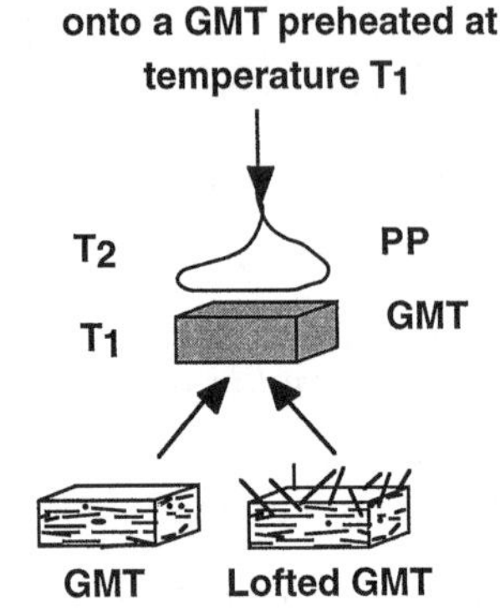

Bonding conditions	Fracture Energy [kJ/m^2]
PP on 'as received' GMT	0.35 ± 0.05
PP on lofted GMT	0.56 ± 0.07
PP on preheated GMT	1.24 ± 0.18
PP on lofted and preheated GMT	1.78 ± 0.23

Figure 5 Influence of interfacial temperature (via GMT preheating) and superficial composite lofting on fracture energy of PP/GMT bonds obtained during nonisothermal integrated processing. Fracture energy was measured using double cantilever beam (DCB) tests.

was shown that preheating of the GMT is necessary in order to expose the interface at temperatures promoting 'healing' and interfacial crystallization. Furthermore, a superficial lofting of the GMT, obtained by scanning its surface with hot air, improves the fracture energy of the bonds by creating fiber bridging at the interface (Figure 5). Several other different combinations and parts have been realized to illustrate the feasibility of the integrated processing [5].

Transportation industry, mechanical industries and medical engineering have manifested their interest in the development of both the integration of materials and the integration of processes because of the manufacturing flexibility and cost-efficiency.

CONCLUSION

Integration of materials and of manufacturing processes are considered in order to optimize manufacturing costs and to add value to parts made from thermoplastic polymers and composites. The approach involves the design of the part and the selection of the appropriate sequence of processing operations on the integrated equipment to yield the best cycle time, rather than forcing the mold or the part design to accommodate conventional machine limitations. Various types of polymers and composite preforms are combined. The use of flexible manufacturing techniques is currently essential to satisfy critical application requirements. A higher degree of design freedom, shape complexity, mechanical efficiency, and multi-functionality are some added values provided by the integrated manufacturing approach (Figure 6).

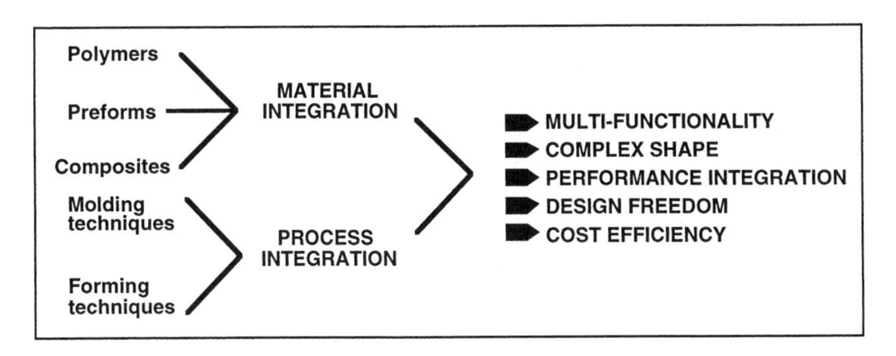

Figure 6 Summary of the integrated processing approach and the added values provided to polymer and composite manufacturing.

REFERENCES

1. Månson, J.-A.E. (1993) Processing of thermoplastic-based advanced composites, in *Advanced Thermoplastic Composites Characterisation and Processing*, (ed. H.H. Kausch), Hanser, Munich, p. 273.
2. Gibson, A.G. and Månson, J.-A.E. (1992) *Composites Manufacturing*, **3**, 223–233.
3. Ashby, M.F. (1993) *Materials Selection in Mechanical Design*, Pergamon Press, Oxford.
4. Smith, G.D., Bourban, P.E. and Månson, J.-A.E., The fusion bonding of neat polypropylene, submitted to *Polymer.*
5. Bourban, P.E., Bögli, A., Bonjour, F. and Månson, J.-A.E. (1998) Integrated processing of thermoplastic composites, *Composite Sci. Technol.*, **58**, 633–637.

Keywords: integrated processing, thermoplastic polymers, composites, cost-efficiency, multifunctionality, interfaces, integrated materials, GMT-PP.

Interfacial morphology and its effects in polypropylene composites

J. Karger-Kocsis and J. Varga

INTRODUCTION [1–3]

Certain fillers (e.g. talc, mica) and reinforcements (fibers of cellulose, polyamides, polyesters, carbon etc.) may act as heterogeneous nucleants in polypropylene (PP) composites and produce a peculiar interfacial superstructure termed a transcrystalline layer. An essential prerequisite of transcrystallization is the presence of active nuclei on the surface of the substrate (viz, fillers and reinforcements) in high density. The closely spaced nuclei hinder the lateral extension of spherulites which are then forced to grow in one direction namely perpendicular to the substrate's surface (oriented crystallization or oriented growth). Since the density of the nuclei on the substrate's surface is higher than in the bulk polymer, a columnar morphology appears (Figure 1). The development of a columnar structure is a necessary but not sufficient feature of transcrystallization (see below). Though transcrystallization was first reported for PP in 1952, its exact mechanism is not yet fully understood. Among the factors believed to induce transcrystallization, similarities and dissimilarities in both chemical and physical characteristics (such as analogous chemical composition, topology, differences in the surface energetics, matches in the crystalline structure, mismatches in the thermal coefficients, diffusion and sorption phenomena, temperature gradient at the interface etc.) have all been reported. According to the present state of knowledge, epitaxy

Polypropylene: An A–Z Reference
Edited by J. Karger-Kocsis
Published in 1999 by Kluwer Publishers, Dordrecht. ISBN 0 412 80200 7

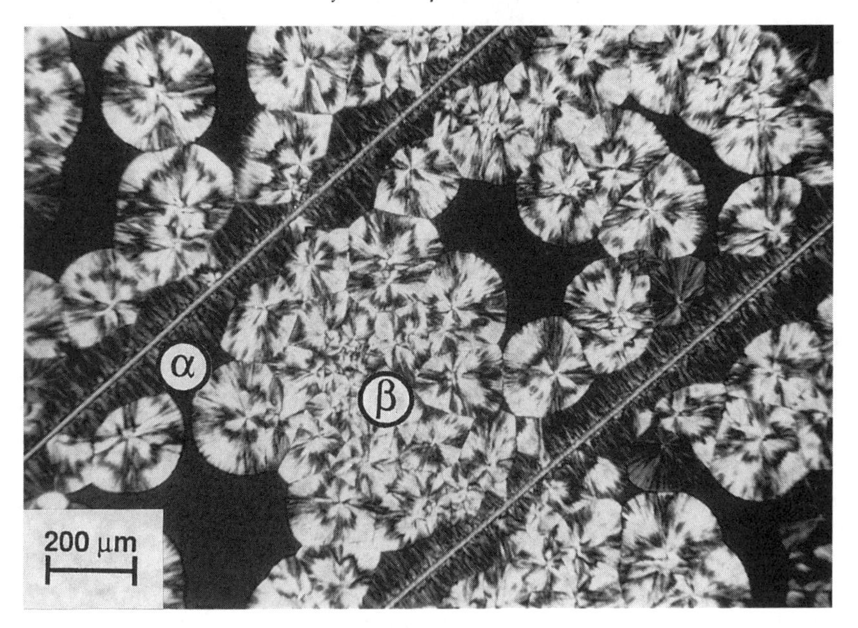

Figure 1 α transcrystalline layer induced by a polyethylene terephthalate (PET) fiber in a quiescent β-nucleated PP melt under isothermal conditions (crystallization temperature and time $T_c = 124°C$ and $t_c = 30$ min, respectively).

seems to play a decisive role in transcrystallization (see also the chapter 'Epitaxial crystallization of isotactic and syndiotactic polypropylenes' in this book). Why are extensive R&D activities still being devoted to transcrystallization phenomena? This can be understood by the assumption that the heterogeneous nucleation on the surface (i.e. interfacial morphology) is accompanied with an improved adhesion between reinforcement and PP matrix. Consequently, via tailored transcrystallization, semicrystalline thermoplastic matrix-based composites of improved mechanical performance can be produced. It should be borne in mind that in advanced composites, containing unidirectionally (UD) oriented endless fibers in about 60 vol.%, the whole matrix may be of transcrystalline nature.

INTERFACIAL SUPERSTRUCTURES

Transcrystallization

Transcrystallization was reported to occur during cooling of a quiescent PP melt being in intimate contact with various organic and inorganic substrates having high nucleation ability. In this case, the nucleation rate on the surface is much more higher than the polymer melt (see also the chapter on 'Nucleation' in this book). The attribute quiescent means that no relative movement between the melt and substrate takes place during crystallization.

Cylindritic crystallization

During processing of fiber reinforced isotactic PPs, the melt velocity may differ from that of the reinforcement which results in a local shear field. This may generate row-nuclei that initiate cylindritic growth. Characteristics of cylindritic crystallization of isotactic PP were studied in detail by the authors of this chapter using the fiber pulling technique [4–5]. They showed that the polymorphic nature of cylindrites developed on a sheared layer along the pulled fiber strongly depends on both crystallization and shearing temperatures (T_c and T_{pull}, respectively) and also on the PP type, but is practically unaffected by the quality of the fiber used. The sheared region is very rich in β-modification if the melt is sheared and crystallized between 100 and 140°C, where the growth rate of the β-form is higher than that of the α-form. By increasing T_{pull} in the temperature range beyond 140°C the proportion of the β-phase decreases and even diminishes irrespectively that the crystallization was performed under $T_c < 140$°C. Cylindrites formed above 140°C consist of pure α-modification.

By exploiting the polymorphism and the difference in the melting behavior between the α- and β-form of PP (see also 'Beta-modification of isotactic polypropylene' in this book), Varga and Karger-Kocsis demonstrated [4–5] that shear-induced crystallization results in cylindritic instead of transcrystalline superstructure. What is the basic difference between trans- and cylindritic crystallization? Figure 2 shows the demonstration route of the authors cited. Figure 2(a) dislays the supermolecular formations after isothermal crystallization for a quiescent (A) and pulled (B) polyacrylonitrile-based high-modulus carbon fiber (CF). Based on the spherulites in the view field, one can state that the columnar morphology along the unperturbed CF consists of α-, whereas at the pulled CF of β-PP. Along the CF (B), a banded negative birefringent transcrystalline-like structure can be resolved. The cylindritic feature of the latter can well be demonstrated by melting the β-phase at T = 157°C (Figure 2(b)). After selective melting of the β-form along the surface of the pulled CF (B), a saw-teeth-resembling α structure becomes visible. This implies that, during the melt shearing, first α-row nuclei were formed, on the surface of which an αβ-secondary nucleation took place [5]. The secondary β-nuclei induced the formation of β-cylindritic segments. Since in the case of transcrystallization no change in the

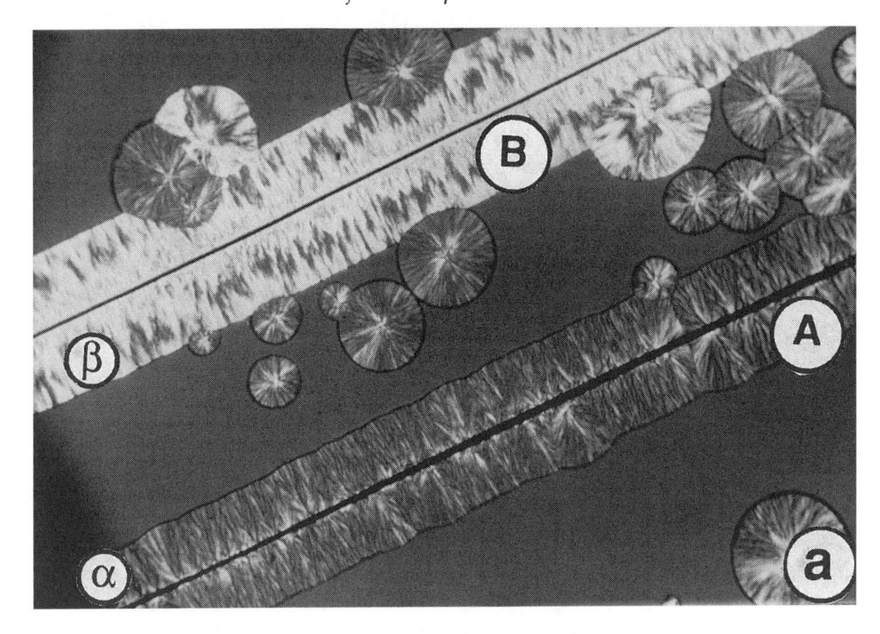

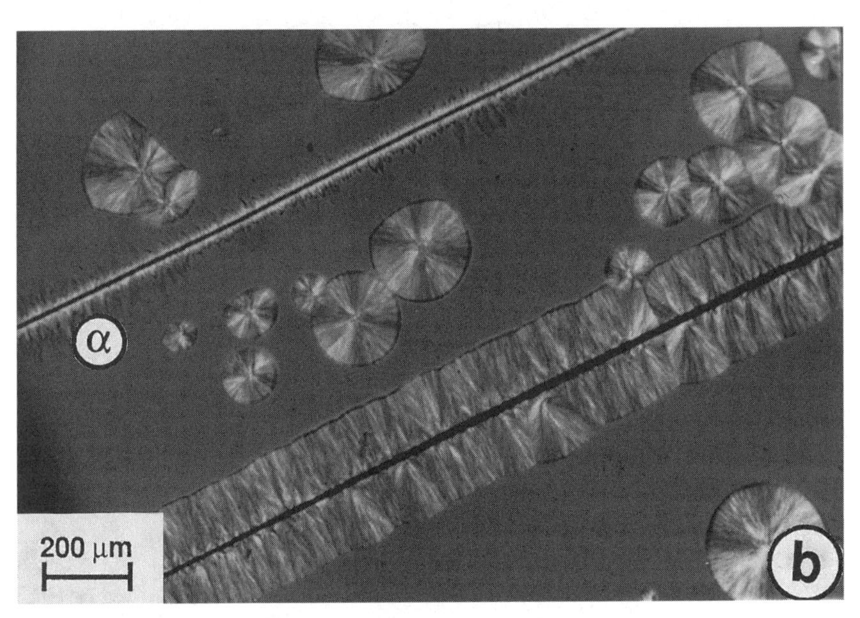

Figure 2 (a) Interfacial crystallization along unperturbed (A) and pulled (B) carbon fibers (polyacrylonitrile-based CF) at $T_c = T_{pull} = 133°C$ ($t_c \approx 90$ min). (b) Morphology showing the presence of the α-form after selective melting of the β-PP phase at $T = 157°C$. (Note: the α-phase grown on the shear-generated α-row nuclei in form of saw-teeth is clearly visible in case of fiber B in Figure 2(b)).

crystalline modification may occur, this structure should be distinguished from the transcrystalline one. The term cylindrite is credited to Peterlin [4–5]. The difference between a transcrystalline and cylindritic structure is depicted schematically in Figure 3. Figure 3 emphasizes that cylindrites are generated by α row-nuclei, produced by melt shearing in the vicinity of the substrate. The aforementioned build-up of the cylindrites was confirmed by atomic force microscopy (see the chapter on 'Morphology and nanostructure of polypropylenes by atomic force microscopy' in this book). The substantial difference between transcrystallization and cylindritic crystallization has been overlooked or

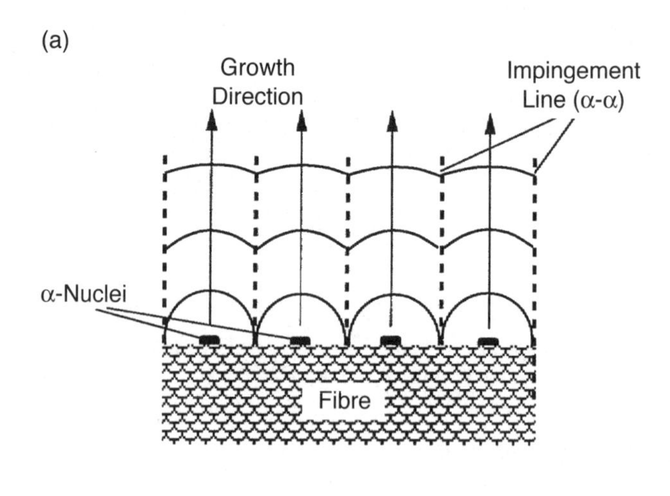

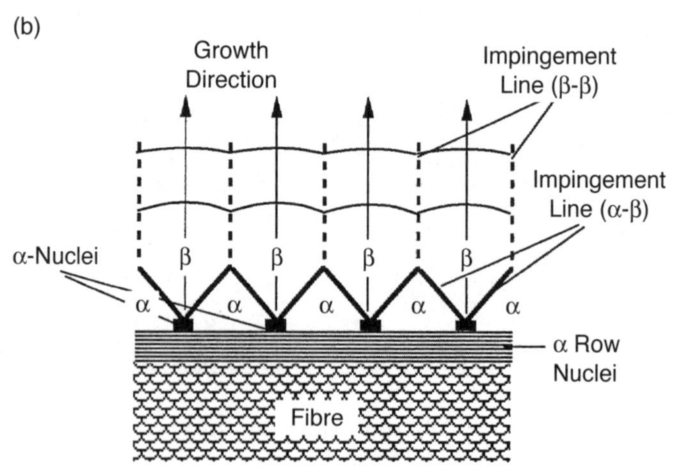

Figure 3 Scheme showing the difference between transcrystallization (a) and cylindritic crystallization (b).

disregarded by many researchers. It can be claimed that shear or elongational type deformation of the PP melt along the surface of reinforcements and fillers (or more generally a relative movement between the crystallizing melt and substrate) favors cylindritic growth. Therefore the findings of most papers indicating for 'transcrystallization' effects in samples or specimens produced by injection, extrusion molding, hot stamping and the like, should be revisited and interpreted as effects of cylindritic superstructures.

EXPERIMENTAL TECHNIQUES FOR STUDYING INTERFACIAL CRYSTALLIZATION [2]

The most widely used technique for morphology studies is optical microscopy. Since the thickness of the transcrystallization layer is in the range of few tens of micrometers and the superstructures are highly birefringent, their development can well be followed under crossed polarizers (polarized light microscopy, PLM; Figures 1 and 2). In order to gain insight into the nucleation process and sites, scanning electron microscopy (SEM) and AFM are used for specimens with various preparation. X-ray diffraction is one of the oldest techniques adopted for studying transcrystallization. Although reflected X-rays seemed to work well for detecting transcrystallinity, this method failed to deliver the expected results until recently. It can be predicted that the recently available microbeam techniques (e.g. microfocus wide-angle X-ray scattering using synchrotron radiation) will revitalize the research with X-ray diffraction. The Fourier transform infrared spectroscopy (FTIR), especially in the attenuated total reflexion (ATR) mode can also help to distinguish between surface and bulk regions in respect to orientation and crystalline characteristics. The use of differential scanning calorimetry (DSC) is very straightforward since this technique allows us to study the nucleation process, crystallization kinetics, crystallinity, etc. – all of which are of great importance to understand the mechanism of interfacial crystallization. Features of the interfacial morphology can also be deduced from dynamic and thermomechanical analysis (DMA and TMA, respectively), especially when suitable model specimens (e.g. transcrystalline strips) are tested where the influence of the bulk matrix is negligible. Further valuable information is expected from the use of stretching calorimetry [6]. The advantage of this technique yielding both elastic (Young's modulus) and thermal characteristics (linear expansion coefficient, change in the internal energy as a function of strain) is that it can be adopted for plain (e.g. transcrystalline strips) PP and single or multiple fiber reinforced model composites, too.

EFFECTS ON THE INTERFACIAL SHEAR STRENGTH

As mentioned earlier, the interfacial morphology in semicrystalline polymer matrix composites may affect the macroscopic response. In the literature, many examples can be found that the transcrystalline layer improves, has no effect on or even reduces some selected mechanical properties. The authors believe that this uncertainity is mostly linked to the structure of the reinforcement; effects such as imperfect fiber alignment, presence of voids, etc., termed nowadays as 'mesostructure' (caused by manufacturing and/or processing) 'overwrite' the influence of the interfacial morphology. That is the main reason why researchers are favoring tests on single-fiber composites (SFC) using preferentially fiber pull-out, microdebond or microdroplet pull-off and fragmentation tests. In case of cylindritic structure, it was found that they do not affect the interfacial shear strength (ISS) [7]. There is also an increasing number of reports stating that transcrystallinity has also a marginal effect on ISS. It is more likely that the transcrystalline sheath affects the frictional stresses than those of the debonding (used for computing ISS) [8]. Recall that the fibers stay under radial residual stresses since the thermal expansion coefficient of the matrix is higher than that of the fiber, i.e. the matrix is shrinked on the reinforcing fiber. The assessment of the residual thermal stresses is of paramount importance. For this purpose the micro-Raman spectroscopy (MRS), developed for composite materials by Galiotis, works well. MRS is a direct measuring technique for the axial compression behavior of crystalline fibers (such as aramid [AF], CF) of suitable Raman activity. The fiber in the SFC acts as an embedded strain gauge. If the shift in the corresponding Raman band of the fiber under both tension and compression is known, the strain data (mapped with a resolution of about 5 μm) can be converted for the locally accommodated stresses. By this MRS technique it was found that CF and AF are under axial compressive stresses when surrounded by a transcrystalline or spherulitic layer [9–10]. This is likely to be the case in radial direction, as well. In respect to ISS, slightly higher data were measured in SFC of AF/PP for transcrystalline than for spherulitic PP interface [9]. It is worthwhile mentioning that ISS is computed by adopting different models for the measured data, which may affect the outcome. In a recent pioneering work, Nielsen and Pyrz [10] followed the strain development in an SFC composed of CF and PP during cooling. Their main finding is that the linear thermoelastic approach, widely used, overestimates the residual thermal strains and thus stress values. To overcome this difficulty, the authors introduced a thermorheological description by which a better fit with the experimental data was achieved. This is a clear hint that the viscoelastic nature of the crystallizing melt strongly affects the ISS. This result will definitely trigger research activities on model systems of various viscoelasticity and thus some 'puzzling' effects attributed to the interfacial morphology may be elucidated. A further open question is whether or not the arrangement and stacking of lamellae in these supermolecular formations (e.g. the banded structure caused by rotation of the lamellae in case of the pulled CF (B) in Figure 2) influence the ISS.

CONCLUSIONS

Transcrystallization is caused by heterogeneous nucleation in quiescent melt. According to the authors' knowledge all substrates have an inherent α-nucleating capacity (Figure 1). The effect of the transcrystalline layer on the load transfer from the matrix to the reinforcement is still not conclusive and likely has a small effect, at least in PP. The cylindritic structure generated by processing-induced row-nuclei seems to affect the interfacial shear strength between fiber and matrix also marginally. It is expected that the micro-Raman spectroscopy on well-defined composite models will contribute to shed light on the open questions related to the transcrystalline layer.

REFERENCES

1. Folkes, M.J. (1995) Interfacial crystallization of polypropylene in composites, in *Polypropylene: Structure, Blends and Composites*, Vol. 3, Chap. 10, (ed. J. Karger-Kocsis), Chapman & Hall, London, pp. 340–370
2. Ishida, H. and Bussi, P. (1993) Morphology control in polymer composites, in *Materials Science and Technology*, Vol. 13, Chap. 8, (ed. T.-W. Chou), VCH, Weinheim, pp.339–379
3. Devaux, E. and Chabert, B. (1993) Origin and development of transcrystallinity in polypropylene in the presence of active substrates: role of the surface energy, *Sci. Engng Composite Mat.*, **3**, 135–144.
4. Varga, J. and Karger-Kocsis, J. (1995) Interfacial morphologies in carbon fibre-reinforced polypropylene microcomposites, *Polymer*, **36**, 4877–4881.
5. Varga, J. and Karger-Kocsis, J. (1996) Rules of supermolecular structure formation in sheared isotactic polypropylene melts, *J. Polymer Sci., Part B, Polymer Phys.*, **34**, 657–670.
6. Tregub, A., Kilian, H.-G. and Karger-Kocsis, J. (1996) Thermodynamics of deformation of micro composites, *Composite Int.*, **3**, 333–341.
7. Hoecker, F. and Karger-Kocsis, J. (1995) On the effects of processing conditions and interphase of modification on the fiber/matrix load transfer in single fiber polypropylene composites, *J. Adhesion*, **52**, 81–100.
8. Wang, C. and Hwang, L.M. (1996) Transcrystallization of PTFE fiber/PP composites II. Effect of transcrystallinity on the interfacial strength, *J. Polymer Sci., Part B, Polymer Phys.*, **34**, 1435–1442.
9. Heppenstall-Butler, M., Bannister, D.J. and Young, R.J. (1996) A study of transcrystalline polypropylene/single-aramid-fibre pull-out behaviour using Raman spectroscopy, *Composites A*, **27A**, 833–838.

10. Nielsen, A.S. and Pyrz, R. (1997) In-situ observation of thermal residual strains in carbon/thermoplastic microcomposites using Raman spectroscopy, *Polymer Polymer Compos.*, **5**, 245–257.

Keywords: transcrystallization, heterogeneous nucleation, row-nuclei, cylindritic crystallization, interfacial shear strength (ISS), single fiber composites (SFC), mesostructure, micro-Raman spectroscopy (MRS), α-PP, β-PP, residual thermal stresses, composites.

Intumescent fire retardant polypropylene formulations

Michel Le Bras and Serge Bourbigot

Applications of polyolefinic materials (transports, electrical cables and wires, upholstered furniture, synthetic fibers) require fire retardant (FR) properties using low additives content (≤ 30 wt.% loading) to preserve the original mechanical properties of the materials.

First of all, halogen-based fire retardant additives (especially brominated compounds associated with the antimony trioxide) are widely used. These systems release obscuring, corrosive and toxic smoke when they perform their fire retardant action. More, some of them release super toxic compounds ('dioxins' and polybrominated dibenzofurans) when exposed to heat during manufacturing or in fire. A continuous trend is the development of polymeric materials with reduced fire hazard, in order to meet the requirement of the international regulations (5th OECD Draft Status report (04/1993) and UN Environmental Program 1st Draft Report (01/1993)).

Among the additives which may be used as substitutes for these systems, intumescent additives seems to be particularly attractive. Fire protection of flammable materials (in particular, paints, varnishes and cellulose-based materials) by an intumescence process has been known for several years [1]. Intumescent technology has more recently found a place in polymer science as a method of providing flame retardancy to polymer formulations, especially polypropylene (PP) formulations [2]. These systems interrupt the self-sustained combustion of the polymer at its earliest stage, i.e. the thermal degradation with evolution of the gaseous fuels.

Polypropylene: An A–Z Reference
Edited by J. Karger-Kocsis
Published in 1999 by Kluwer Publishers, Dordrecht. ISBN 0 412 80200 7

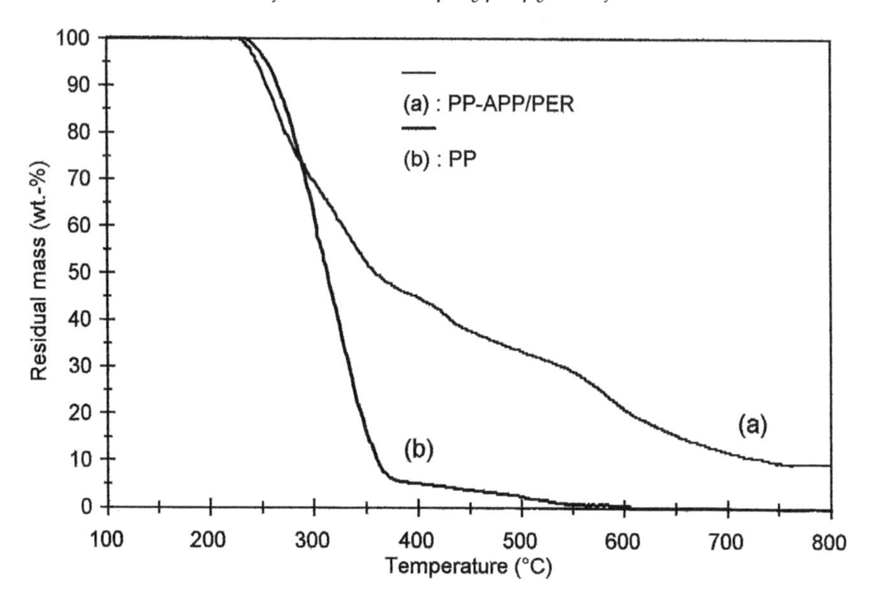

Figure 1 TG curve of PP and PP-APP/PER under air.

PRESENTATION OF INTUMESCENCE

Intumescent polymeric systems decompose on heating with formation of large amounts of thermally stable carbonaceous residues. TGA shows that this residue forms from the standard PP-APP/PER system (intumescent additives: ammonium polyphosphate (APP: $(NH_4PO_3)_n$, $n = 700$)/ pentaerythritol (PER) = 3/1 at 30 wt.% loading) at about 300°C and that its amount is maximum between 350 and 430°C (Figure 1). Intumescent 'char' result from a combination of charring and foaming of the surface of the burning polymer (observed with PP-APP/PER heat treated between 280°C and 430°C under air). The resulting foamed cellular charred layer (whose density decreases from 1.05 at 300°C to 0.30 at 430°C) protects the underlying material from the action of the heat flux or the flame.

PROTECTIVE ACTION OF AN INTUMESCENT COATING

The proposed mechanism is based on the charred layer acting as a physical barrier which slows down heat and mass transfer between the gas and the condensed phases (Figure 2). More, the layer inhibits the evolution of volatile fuels via an 'encapsulation' process. Finally, it limits the diffusion of oxygen to the polymer bulk.

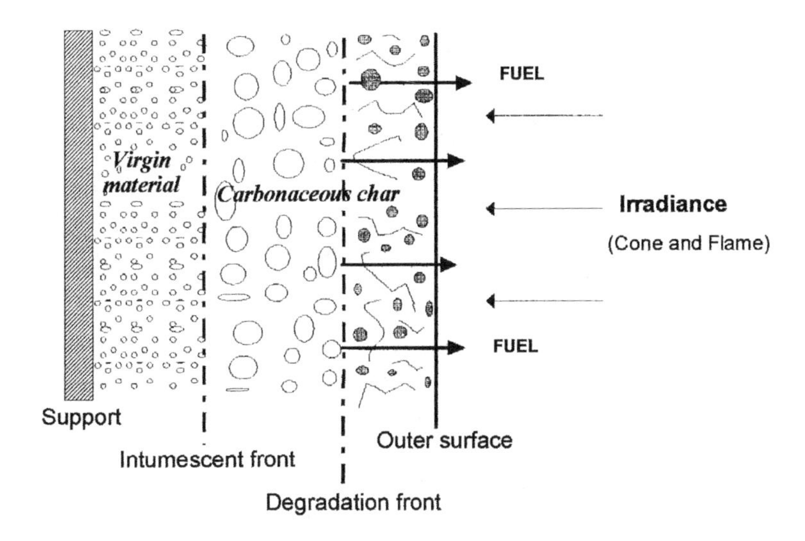

Figure 2 One-dimensional intumescence model.

The limitation of the heat transfer (thermal insulation) by an intumescent coating is illustrated by the temperature profile in a PP-APP/PER sheet (under a heat flux of $50\,kW/m^2$) versus time and depth, which presents three successive plateaux (Figure 3).

The first plateau at about 190°C may be ascribed to both polymer melting and formation of esters (reaction between the two additives with evolution of water and ammonia), the second one at 280°C to the carbonisation process [3]. The material shows then a constant temperature 350°C which is the temperature at which the foamed coating forms [3]. A high rate of the thermooxidative degradation of PP is observed at 310°C in the experimental conditions (horizontal plane map in Figure 3). So, the degradation of the polymeric material is only a surface process during the first 900 s of the treatment and about 50% of the material is preserved after 1800 s thermal treatment.

The stability of the intumescent material consequently limits the formation of fuels and leads to self-extinction in standard conditions. Oxygen consumption calorimetry in a cone calorimeter (according to NBS-IR 82:2604) confirms the low rate of the degradation related to the presence of a surface intumescent material. As consequence, the rate of heat release (rhr) $\leq 300\,kW/m^2$ during 180 s with PP-APP/PER [4]. The maximum of rhr ($rhr_{max} = 450\,kW/m^2$, four times lower than rhr_{max} of PP) is observed (at 250 s) when the intumescent coating degrades (Figure 4).

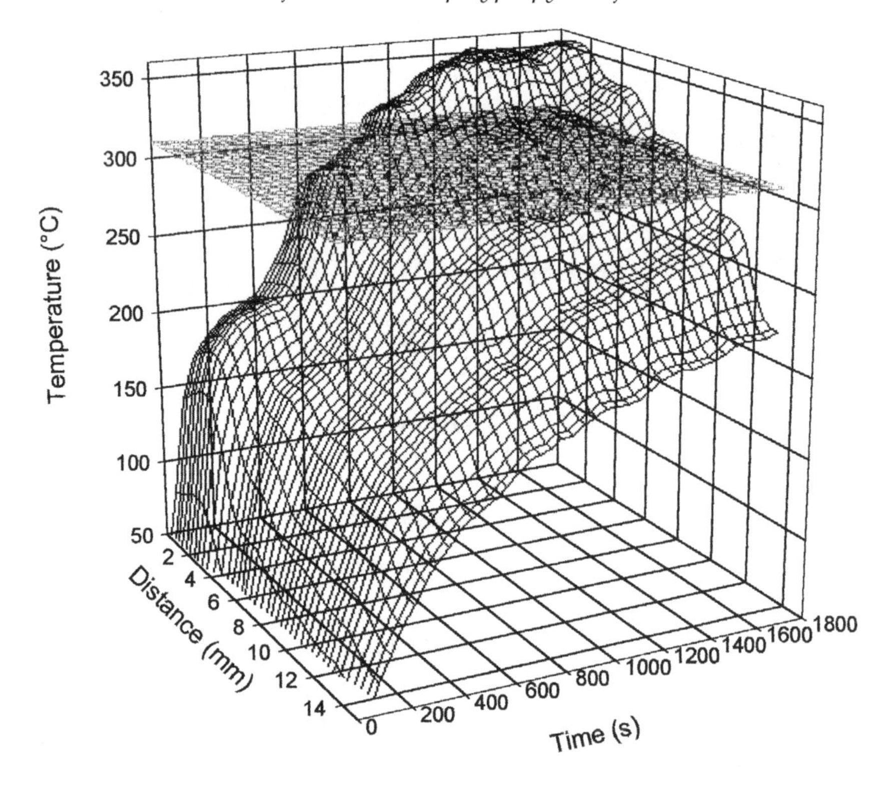

Figure 3 Temperatures in a PP-AP/PER sheet under a 50 kW/m^2 heat flux versus time and depth.

CHEMISTRY OF INTUMESCENT POLYPROPYLENE FORMULATIONS

Generally, intumescent formulations contain three active ingredients: an acid source (precursor for catalytic acid species), a carbonific (or polyhydric) compound and a spumific agent. In a first stage ($T < 280$°C), the reaction of the acidic species with the carbonization agent takes place with formation of esters mixtures. Then carbonization process follows (via a Friedel–Craft and/or a free radical process). In a second step, the spumific agent decomposes to yield gaseous products which cause the char to swell ($280 \leq T \leq 350$°C). The intumescent material decomposes then at highest temperatures and loses its foamed character at about 430°C.

The intumescent structure consists of carbonaceous and polyaromatic species whose carbon organization in stacks is characteristic of a pre-graphitization stage. This 'carbon', formed from the additives, plays two different chemical parts in the fire retardancy process:

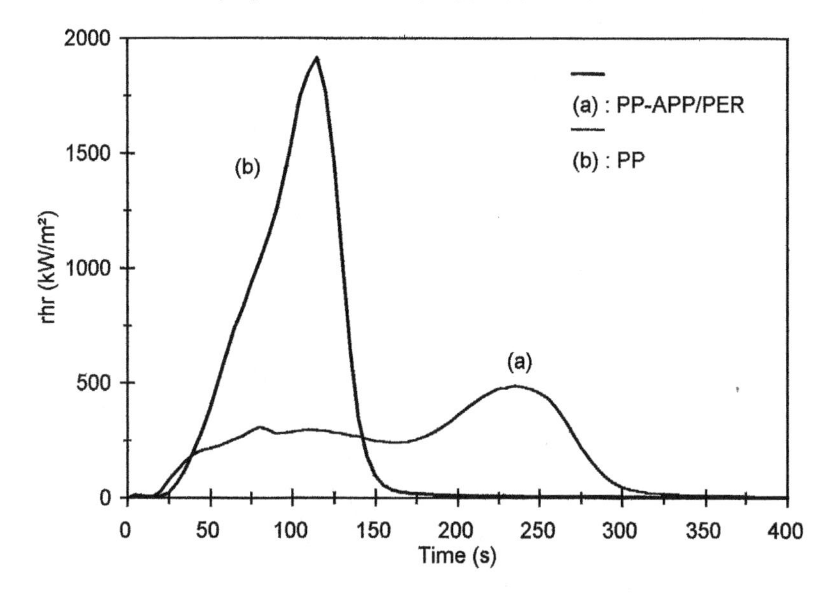

Figure 4 Rates of heat release (rhr) of PP and PP-APP/PER under an irradiance of 50 kW/m^2.

- It contains free radical species which react with the gaseous free radical products formed during the degradation of PP. These species may also play a part in termination steps in the free radical reaction scheme of the pyrolysis of the PP and of the degradation of the protective material in the condensed phase,
- It is a support for acidic catalytic species which react with the oxidized products formed via the thermooxidative degradation of the material. So, the intumescent material consists of polyaromatic stacks bridged by polymer links and phosphate (poly-, di- or orthophosphate) groups, as presented in Figure 5.

Synergistic agents may be added to the three additive components. They may play different parts: they increase the amount of stable carbon formed (dextrine derivatives), increase the amount of stable 'high temperature' acidic species (clays, zeolite or 'functionalized' co-polymers) or increase the number of polymer chains trapped in the intumescent shield (aluminosilicates).

It should be noted that antagonistic effects are often observed using FR additives in association with antioxidants or anti-ultraviolet additives (which prevent the free radical process) or fillers (large solids particles in

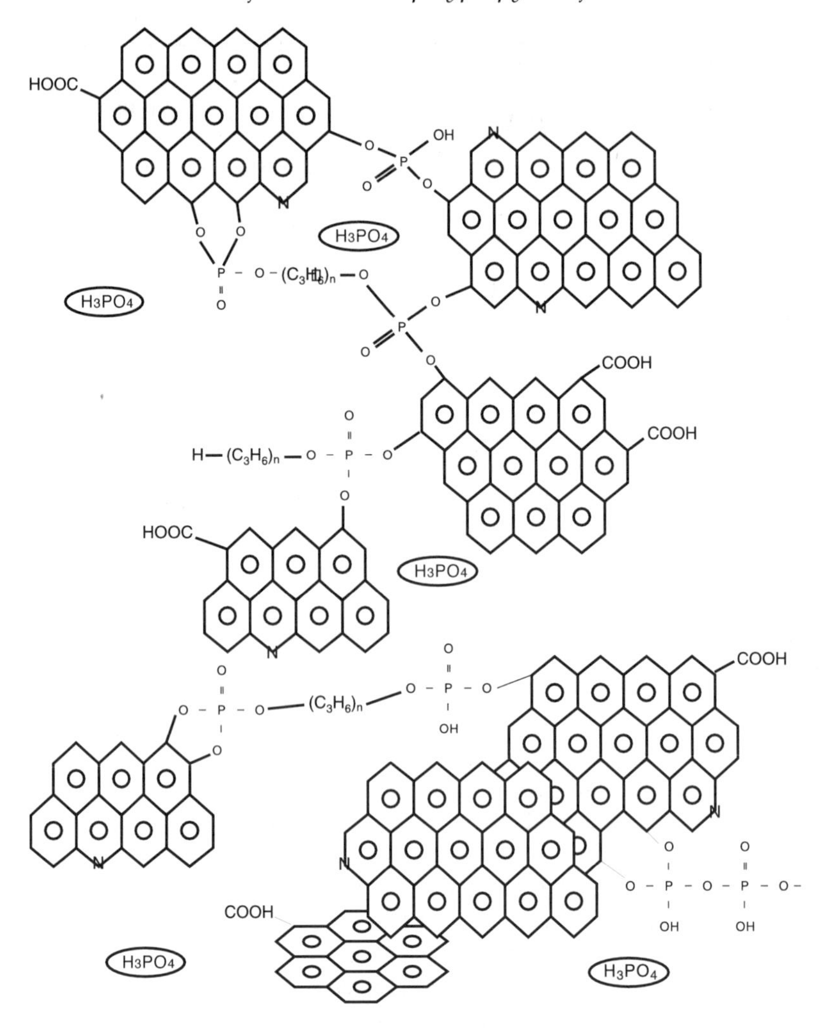

Figure 5 Intumescent coating result from PP-APP/PER heat treated at 350°C.

the intumescent materials lead to the formation of cracks and to the related loss of the protective character).

Introduction of processing and lifetime improving additives in PP may lead to the rejection of these FR compounds (migration, exudation) out of the polymeric matrix so that suitable interfacial agents have to be selected. As examples, tetraethyl-siloxane (TEOS-) or methyl-siloxane (MEOS)-based resins or functionalized co-polymers have been developed as compatibility agents.

Table 1 Compositions and FR performances in typical intumescent polypropylene formulations

Composition				FR performance			
Acidic agent	*Carbonization agent*	*Spumific*	*Synergistic agent*	*LOI volume (%)*	*UL-94*	*rhr_{max} (kW/m^2)*	
APP: exolit 422 (Hoechst)	Pentaerythritol			30	V0	400	*,†
APP: exolit 422 (Hoechst)	Pentaerythritol		Clays	25 to 35	Not classed to V0 depending on the clays		*,†
APP: exolit 422 (Hoechst)	Pentaerythritol		Zeolite 4A	42	V0	300	*,†
APP: exolit 422 (Hoechst)	Pentaerythritol	Urea		42	V0	300	?
APP: exolit 422 (Hoechst)	Pentaerythritol	Melamine		42	V0	300	?
$P_2O_7(NH_4)_2$	Pentaerythritol			32	V0		*,‡
$P_2O_7(NH_4)_2$	Pentaerythritol		Zeolite 4A	51	V0		*,‡
$P_2O_7)NH_4)_2$	Xylitol			24	V0		*,‡
$P_2O_7(NH_4)_2$	Sorbitol			24	Not classed		*,‡
$P_2O_7(NH_4)_2$	Mannitol			24	V0		*,‡
$(NH_4)_2B_{10}O_{16}$, $8H_2O$	Pentaerythritol			24	Not classed		*,‡
APP: exolit 750 (Hoechst)	Unknown embedding			37	V0	250	
APP spinflam MF82 (Himont)	Pyrazole type Resin			32	V0	155	
APP: exolit 422 (Hoechst)	PA-6		Functionalized polymer	31	V0	300	
APP (Himont)	Poly(ethyleneurea-formaldehyde)	Melamine	Hydrated silica or alumina	34			

Exolit is the old denomination of hostaflam; * exudation; † reactive during processing at 210°C; ‡ reactive during processing at 190°C.

INTUMESCENT FR POLYPROPYLENE FORMULATIONS

This section concerns only formulations of unmodified PP. Laboratory works on intumescent PP formulations via specific copolymerization and/or grafting of chemical groups on the PP chain have not been reported. Main strategies for fire retardancy of PP by intumescence for direct industrial applications are addition of additives systems during the polymer processing. The additives or their mixtures have to be selected using two main criteria:

- each additive is thermally stable in the temperature range 190–250°C corresponding to processing of PP;
- no reaction between the additives (or an additive and PP) occurs in this temperature range.

Four kinds of additives may be considered:

- esters of a mineral acids (such as melaminium,2,6,7-trioxa-1-phosphabicyclo[2.2.2]octane-4-methoxy sulphates) and, more generally, phosphorus compounds with a phosphospirane or a phosphabicyclo structure (such as 3,9-dihydroxy-2,4,8,10-tetraoxa-3,9-diphosphaspiro[5.5]undecane-3,9-dioxide and 4-hydroxymethyl-2,6,7-trioxa-1-phosphabicyclo[2.2.2]octane-1-oxide and their derivatives);
- mixtures of a precursor for catalytic acid species, a polyhydric compound, a spumific agent and a synergistic agent. Typical examples of additives and their FR performances (oxygen index; LOI (ASTM D2863–77), UL-94 test (ANSI/ASTM D635–77) and rhr) are compared for a 30 wt.% loading in Table 1;
- a single additive, precursor for catalytic acid species (APP-based additives are more particularly developed) embedded in a self-carbonizing resin. Commercial additives are given in Table 1;
- addition of an acidic agent in PP-based blends. The other polymers in the blends are self-carbonizing resins and/or give char via a reaction with the acidic agent. Typical examples are reported in Table 1.

The last two typical additive formulations seem to be the more promising. Some of them have already be used in PP-based materials for electric, flooring or house applications.

REFERENCES

1. Vandersall, H.L. (1971) Intumescent coating systems, their development and chemistry, *J. Fire Flammability*, **2**, 97–140.
2. Camino, G., Costa, L. and Trossarelli, L. (1984) Study of the mechanism of intumescence in fire retardant polymers: Part II Mechanism of action in polypropylene-ammonium phosphate-pentaerythritol mixtures. *Polymer Degrad. Stability*, **7**, 25–31.
3. Delobel, R., Le Bras, M., *et al.* (1990) Thermal behaviours of ammonium polyphosphate-pentaerythritol and ammonium pyrophosphate-penta-

erythritol intumescent additives in polypropylene formulations. *J. Fire Sci.*, **8**(3–4), 85–108.
4. Bourbigot, S., Le Bras, M. and Delobel, R. (1995) Fire degradation of an intumescent flame retardant polypropylene using the cone calorimeter, *J. Fire Sci.*, **13**(1–2), 3–22.

Keywords: intumescent coating, intumescent formulations, fire retardance, fire protection, char formation, heat flux, thermal decomposition, carbonization, pyrolisis, antagonistic effect, UL–94, LOI, spumification.

Joining: methods and techniques for polypropylene composites

M. Steiner

During the last five decades, much research was devoted to polymers and polymeric composites, starting with material development via chemical and physical modification (e.g. fiber reinforcement) and ending with new processing technologies. In the field of long or endless fiber reinforced thermoplastic composites (TPC) with polypropylene (PP) matrix the glass mat-reinforced (GMT; see also 'Glass mat reinforced thermoplastic polypropylene (GMT-PP)' in this book) and commingled yarn-based systems (see 'Commingled yarns and their use for composites' in this book) have achieved considerable market share.

In order to close the gap between possible technical application of endless fiber reinforced composite material and the actual practice, limited by consolidation and forming processes, various joining techniques were adapted and developed for glass fiber reinforced PP (GF/PP).

The production of a finished complex engineering part is technologically and economically rarely viable. Nevertheless, considerable efforts are made to integrate some manufacturing and processing steps in line. The interested reader is advised to read the chapter 'Integrated manufacturing' in this book. It is more straightforward to produce simply designed and shaped components which are assembled to a complex part afterwards (Figure 1). For that purpose, different methods of joining exist, which are widely used in polymer and metal processing, and which can also be adapted to fiber reinforced polymeric composites. Figure 2 gives a general overview of the different methods.

Polypropylene: An A–Z Reference
Edited by J. Karger-Kocsis
Published in 1999 by Kluwer Publishers, Dordrecht. ISBN 0 412 80200 7

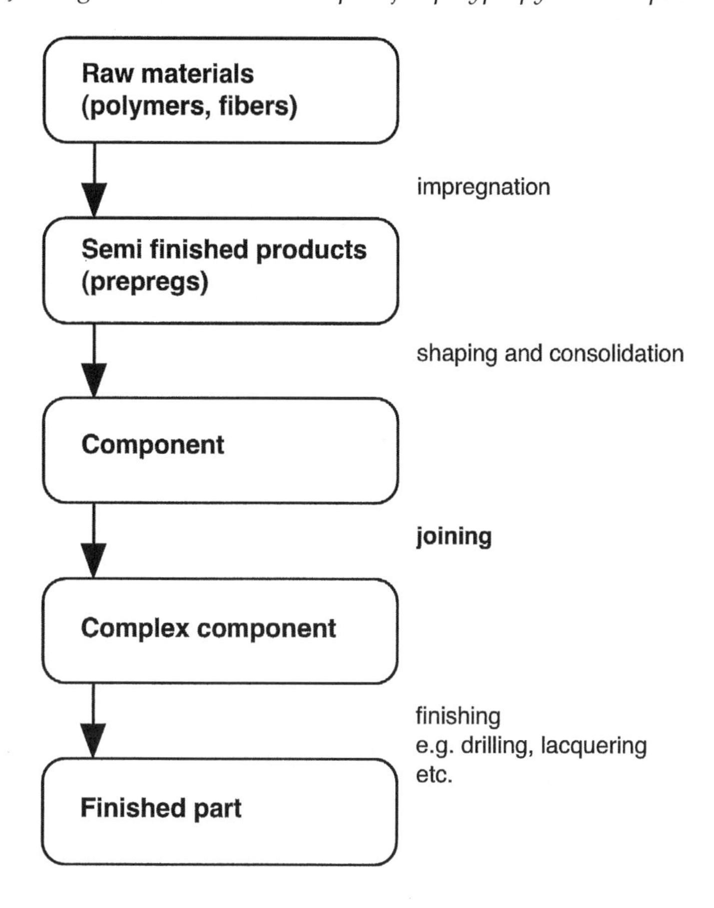

Figure 1 Processing steps for composite components.

The joining techniques (except mechanical fastening) used at present are suitable only for fiber reinforced polymeric composites with a maximum fiber volume content of 35%. If the fiber volume content is higher, joining of the fiber reinforced composites may result in a drastic decrease in the seam strength (especially in case of welding). For joining of advanced composites (i.e. with very high reinforcement content), mechanical fastening is preferred. The use of elements, such as screws and rivets etc., not only leads to additional weight, but also weakens the material through notch effects (holes, cut-outs which reduce the load bearing capacity acting as stress concentrator sites).

Gluing, adhesion bonding (see also 'Adhesive bonding of polypropylene' in this book) has to be ruled out, because in most cases the bond strength of glued polymeric composites is not sufficient or very large

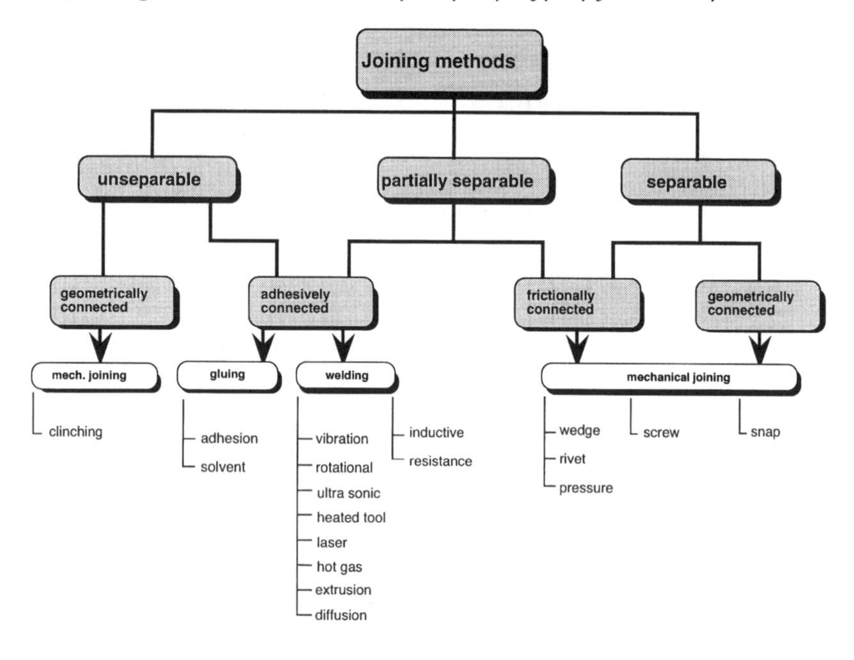

Figure 2 Joining technologies for fiber reinforced composites.

joint surfaces would be required. Furthermore, adhesives tend to creep, their fatigue strength is also limited and preparation of the adhesive joint requires considerable skill. Despite the above limitations, adhesion bonding is used in different fields.

According to investigations considering economic aspects as well, the following joining methods proved to be promising for the joining of fiber reinforced PP composites with a high fiber volume fraction:

- vibration welding;
- heated tool welding;
- hot air welding;
- inductive welding;
- resistance welding; and
- clinching.

Other methods (e.g. ultrasonic welding) are too expensive due to high investment costs. In addition, they do not allow a flexible production. To apply the above methods for GF/PP composites with high GF volume fraction successfully and on industrial scale, it is still necessary to carry out some fundamental studies.

VIBRATION WELDING

During vibration welding, the two components which have to be joined are pressed together and through a vibrationally oscillating movement they are moved in relation to each other. The frictional energy produces the heat, which is necessary for melting the polymer in the welding joint. The variable parameters of the process are the oscillating amplitude, pressure during the welding (constant or variable) and vibration time as well as time of repressing. For most types of machines, the frequency of oscillation is a constant parameter of 100 Hz or 200 Hz.

The remarkable features of this process are the very short cycle times and the relatively low energy requirements. Reinforcing materials have a positive effect since they prevent the bending of the component, and with that it is possible to convert the total vibration energy into frictional heat. Due to the fact that the welding procedure occurs with the exclusion of oxygen, it is also suitable for polyamides, which are often used as matrix material for fiber reinforced polymeric composites. Unfavorable effects are higher investment costs and, in connection with the linear vibration welding, limited possibilities of shaping the components' geometry. It is more difficult to weld thin-walled components. Because of the lack of stiffness they might bend as a result of the joining force, and therefore it is not possible to convert the total vibration energy into frictional heat. The main problem is the abrasion of fibers within the joining area, which causes a decrease of seam strength [1].

HEATED-TOOL WELDING

In this process, the matrix of the components to be joined is melted in the joining area by having contact with one or more heated sheets called heating elements. Afterwards, the heating elements are removed and the components are pressed together. Heating temperature, pressure during joining, time of heating and time of cooling as well as cooling rate are the variable parameters of the process [2].

If thermoplastics without reinforcing structures are used, high values of seam strength can be achieved, which might even be equal with the strength of the material. Furthermore, compared with vibration welding, this requires only one-third of the investment costs. This process is suitable for components which are complex shaped in three dimensions and thin-walled, because heated-tool welding is a static process and therefore deformations hardly occur. Low-molecular and low-viscous thermoplastics tend to stick to the heating elements, which makes the above described process almost unsuitable for the welding of these materials. Other negative effects are high energy consumption and long cycle times.

HOT-AIR WELDING

This process is similar to the gas welding used for joining of metals. However, in this process a heated air stream is used instead of direct flame. This process is derived from fitting of zippers to rain resistant clothes made out of woven polymer fibers. Via a wedge-shaped guiding device, the joining areas of the components are moved together closely and heated up by a jet of hot air. Heating will be continued until the polymer in the welding zone is melted. Joining of components occurs by pressing together parts using rollers or a roller system. A variation of this process is to bring additional melted material into the joining area. Variable parameters of the process are air flow, joining pressure, chart speed of the jet as well as temperature of hot air [3].

With hot-air welding, continuous seams can be achieved. Our investigations showed that the seam strength is very high. The negative effects of this process are the limited geometry of components due to the wedge-shaped guiding, which is not very flexible. Furthermore, the use of this process for joining stiff components is inefficient.

INDUCTIVE WELDING

In this process, electrically conductive materials, so-called welding aids, are placed in the joining area of the components and then exposed to an electromagnetic field. The heating of the welding aids results from the hysteresis and eddy current losses. Via heat conduction, the matrix material is melted in the joining area. Depending on the conductivity of the welding aid material, different electromagnetic frequencies between 20 kHz to 10 MHz are necessary. The aim is to induce a voltage high enough to melt the matrix but low enough to provide a distributive discharge. After the joining of the parts, which occurs under pressure, the welding aid remains in the component.

One of the remarkable advantages of the induction welding is the easy detachability. By using an electromagnetic field, the components can be separated in the same way as they have been joined. If carbon fiber reinforced polymers are used, the inherent heating of composites via the fibers themselves is possible, because they function as reinforcing structure as well as welding aids. The process will be done without oxygen. This makes it suitable for the joining of polyamides, which react sensitive to oxygen. As it was confirmed in numerous investigations, it is possible to weld various types of polymers. Therefore, combinations of different materials can be joined with this procedure. In addition, this very flexible and mobile procedure requires low investment costs [4].

If metallic material is used as welding aids (e.g. wire) stresses arise in the seam. This fact and also the relatively long cycle times are the disadvantages of induction welding.

RESISTANCE WELDING

The principle of resistance welding is similar to the one of induction welding, except that heating of the welding aids does not occur indirectly by inducing a voltage, but directly via conduction of electricity. The variable parameters are intensity of current, joining pressure and time of heating as well as time of repressing. With resistance welding, it is also possible to produce disconnectable joints, and carbon fibers as welding aids can be used. Problems of this process are the relatively long cycle times, and the electrodes on the surfaces that easily corrode [5].

CLINCHING

Clinching is mainly used for the assembly of metallic parts. With clinching, metallic components are joined by cutting and pushing them into each other. From this process, a kind of joining spot results but with a geometrical connection. Applied to polymers or fiber reinforced composites, and by application with higher temperature, this clinching process will be an interesting joining technique for those materials.

Clinching of polymers and composites is also possible without a direct cutting step which positively effects the joining of fiber reinforced composites. Here fibers are not cut but pushed into the spot. Joining pressure, joining speed, joining temperature and the geometry of the punch can be varied accordingly.

Joinings produced with clinching possess a high strength. The only requirement for the joining of different materials is the existence of a

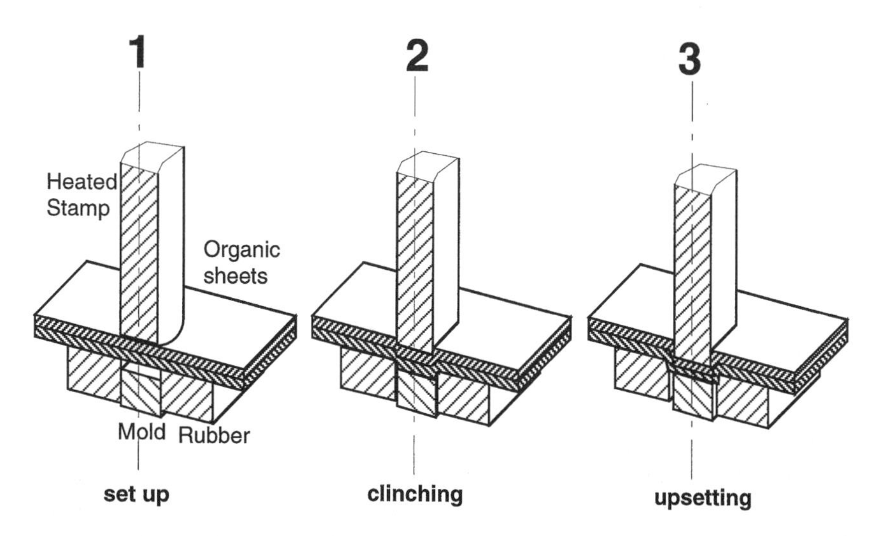

Figure 3 Clinching process steps for thermoplastic composites.

Table 1 Evaluation of relevant joining methods for fiber reinforced composites

Aspects	*Joining Method*					
	Vibration	*Heated tool*	*Hot gas*	*Inductive*	*Resistance*	*Clinching*
Joining strength	+	+	+	0	0	+
Available materials	+	0	+	+	+	++
Materials combinations	+	0	0	+	+	++
Cycle time	++	−	−	0	−	+
Investment costs	−	+	+	++	++	++
Automation	++	++	++	++	++	++
Component complexity	−	+	−	++	++	+
Detachability	−	−	−	++	++	−

Beneficial or positive (+), marginal or no (0) and adverse or negative (−) effects.

similar deep drawing behavior. The realization of this procedure is possible at low investment costs and furthermore this process can be highly automated. However, with clinching continuous seams cannot be produced. Basic processing steps of clinching are shown in Figure 3.

EVALUATION

All the advantages and disadvantages of the discussed procedures are summarized in Table 1. Since reliable data for composites do not exist up to now, the assessment of the different procedures is mainly based on data on joining of thermoplastics without reinforcing structures. This was extended by our preliminary results achieved on reinforced thermoplastics.

POTENTIAL OF JOINING TECHNIQUES

At the moment, fiber reinforced polymeric composites are exclusively joined by gluing, by screws or by rivets. But these techniques do not exploit the potential given by fiber reinforced polymers. The outcome is below the expectations, and thus fiber reinforced thermoplastics do not penetrate new applications.

In fact, various areas of application for fiber reinforced composites exist, e.g. in the automotive area they could be used for crash elements, dashboards, fenders or leaf springs. On the one hand, some of these components are already produced from fiber reinforced polymeric composites but, on the other hand, they usually just replace the original metal designs. So the peculiarities of fiber reinforced composites are not considered or only partially exploited (such as specific stiffness and strength as well as the high capacity of energy absorption). In order to make the best use of the properties of fiber reinforced composites, the design of the components already has to be suited to the material. Since these components are mostly complex structures composed of simple shaped parts, the requirements of joining have to be taken into consideration. If all requirements concerning the design of materials and components, processing and joining techniques are fulfilled, this might be a great achievement in the area of lightweight design and as such for the saving of resources. Whereas the areas of design, materials design and processing have already been well advanced, the area of joining techniques is still in its infancy.

REFERENCES

1. Corish, P.J. (ed.) *et al.* (1992) *Concise Encyclopedia of Polymer Processing and Applications*, Pergamon Press, Oxford, pp. 699–700.
2. Gehde, M., Giese, M. and Ehrenstein, G.W. (1997) Welding of thermoplastic reinforced with random glass mat. *Polymer Engng Sci.*, **37**, 702–714.
3. Watson, M.N. (1982) Hot gas welding of thermoplastics. *The Welding Institute Bulletin*, May/June, 77–80.
4. Hughes, C. (1995) The sealing and welding of thermoplastics materials, in *Plastics: Surface and Finish*, 2nd edn, (ed. Simpson G.W.), The Royal Society of Chemistry, Cambridge, pp. 71–89.
5. Neitzel, M. and Breuer, U. (1997) Die Verarbeitungstechnik der Faser-Kunststoff-Verbunde; Chapter 4.2.2, Hanser, Munich, pp. 161–170.

Keywords: joining, composites, welding, thermoplastic composites, vibration welding, heated tool welding, hot air welding, inductive welding, resistance welding, clinching.

Lamella dimension and distribution

Tomasz Sterzynski

The lamellae are structural elements produced by growth of a three-dimensional order, during cooling of melt of semicrystalline polymers. The fold-chain fibrils are formed by radial growth of crystal systems with in several cases less or more small-angle branching of fibrils. The lamellar habit is typical for linear macromolecules crystallized via chain folding at relatively slow rate to achieve sufficient crystal perfection [1]. Three reasons have been proposed [1] to explain the arrangement of the chain folding: the first based on the lattice dynamics of the crystal, the second based on structural considerations and the third based on kinetic considerations.

The lattice dynamics considerations showed that the thermal longitudinal and torsional oscillations of the chain tend to smudge the lattice potential energy field and therefore increase the free energy density of the crystal, an effect depending on the length of the crystallized polymer chain segment. On the contrary, the fold surface contributes a constant positive free energy contribution to the overall free energy of the crystal. Association of both effects may lead to a minimum surface energy at a finite chain length.

Structural considerations assumed that chain polymer molecules may pack together in a regular arrangement (crystalline) extended in the direction of the chain axis. This arrangement may become unstable when interchain and intrachain forces no longer fit to this or any other type of structure. The instability of the crystals is linked to the introduction of small strains due to mismatch of the dense packing and lowest energy

Polypropylene: An A–Z Reference
Edited by J. Karger-Kocsis
Published in 1999 by Kluwer Publishers, Dordrecht. ISBN 0 412 80200 7

chain conformation. A result may be a formation of amorphous regions. This effect may be a contributing factor to chain folding but it cannot clarify the supercooling dependence of the chain length.

The basic idea in a kinetics consideration of nucleation and crystallization of linear macromolecules from an already-polymerized mobile random state is that in crystallization a larger segment of the chain must be added to the crystal growth face via intermediate steps involving a maximum in free energy. In this case, the existence of a saddle point in the surface of free energy is assumed. The next step is the thickening process coupled with the decrease in free energy. The rate of thickening must be dependent on the crystal perfection; larger rates are expected immediately after crystallization. To achieve high molecular weight equilibrium crystals with chain folding, one must sufficiently intensify the thickening process, or crystallize during polymerization.

Such crystal structure elements, frequently creating bigger units like spherulites (see also the chapter entitled Spherulitic crystallization and structure), may be found in semicrystalline polymers, especially in polymers with lower crystal growth rates such as isotactic polypropylene (iPP) (20 μm/min) or polyethylene oxide (PEO) (200 μm/min), but also in polyethylene (PE) (growth rate $>$ 1000 μm/min). The macromolecular chains are placed in a perpendicular position to the growth direction, i.e. perpendicular to the spherulite radius. The thickness of the lamellae is about 10 nm. Depending on the crystallization temperature, various numbers of single lamellae may be overlaid forming the lamellar packs, usually having thickness of about 100 nm. If the thickness exceeds 1 μm, it becomes possible to observe the lamellae directly under the optical microscope.

OPTICAL ANISOTROPY

For both crystallographic forms (α and β) of the iPP, the monomeric units in chains has a threefold helix configuration, where the methyl groups determine the right- or left-handed conformation of the helix. Optical anisotropy of a single iPP helix indicates a positive birefringence for the polymer chain. The birefringence Δn is called positive when the refractive index of light polarized parallel to the long chain axis of the object (n_p) is larger than the index for the light polarized perpendicular to the long chain axis (n_q)

$$\Delta n = n_p - n_q \text{ (for an elongated object)}$$

Consequently, in the case of spherulite, the birefringence is called positive when the refractive index for the light polarized parallel to the radius of the spherulite (n_r), is larger then for the light polarized tangential (n_t):

$$\Delta n = n_r - n_t \text{ (spherulite)}$$

The same positive birefringence, as for the PP helix, is also observed for the stretched PP fibers. Comparing with the morphology of other polymers, such as PE, one could expect a tangential orientation of the chain axis in the spherulites, i.e. a negative birefringence of PP spherulites, which is the case but only for the β-spherulites.

RADIAL AND TANGENTIAL LAMELLAE IN iPP

Generally, depending on the polymer chemical structure and on its mobility, within one structure unit, two directions of fold-chain fibril growth may be observed: radial [R] and tangential [T]. In some cases, only radial [R] fibrils are visible, i.e. the growth of all lamellae starts in one point (germination centre) and follows only in a radial direction. By growth of the spherulite from very low concentrated solution in isothermal conditions, such growth leads to the creation of almost circular units. Another type of growth may be observed when, beside radial [R], the high angle branched tangential [T] fold-chain lamellae appear simultaneously. In this case, changes in the growth rate as well as other consequences, such as modifications of the static and dynamic mechanical properties [2], are usually observed. These changes in the growth rate are strongly dependent on the temperature, i.e. below a certain temperature the growth rate of the radial lamellae leads, but, above this temperature, the creation of both radial and tangential lamellae is more pronounced.

For several semicrystalline polymers, such as PE, PP, and polyamide (PA), the morphology is characterized by radial type of growth of fold-chain macromolecules. Such growth direction may be observed in the form of spherulites which usually are well visible in polarized light microscopy (PLM). The spherulites may be treated as a special kind of dendrite, which are always formed when crystals grow in the direction of a strong temperature gradient in supercooled melts. The liquid phase occurring between the dendrites often solidifies in a microcrystalline form.

LAMELLAR THICKNESS DEPENDENCE ON TIME AND TEMPERATURE

A sufficiently regular, flexible linear macromolecule, crystallized from the mobile random state, will always crystallize first in a chain-folded macroconformation.

If the degree of polymerization is sufficiently large, the lamellar height (also called the fold height [3]) at a given crystallization temperature is practically independent of the molar mass. On the other hand, the lamellar thickness for all semicrystalline polymers shows a considerable dependence on the crystallization conditions, especially on the crystallization temperature and the time of annealing. An increase of the crystallization temperature may increase the lamellar thickness even ten times; the same effect occurs for the annealing time, where also a significant increase of the lamellar thickness was noted. In some cases, a linear relationship between the lamellar thickness and logarithm of annealing time may be observed [1].

NATURE OF THE FOLD-CHAIN SURFACE

The crystallization by cooling leads to spherulites which are formed from the melt and show a fibrillar structure [3]. This kind of internal structure created by spherulitic growth results from the fractionation crystallization of semicrystalline polymers. More strongly branched and/or lower molecular mass polymer fractions have lower melting temperature and thus require a stronger undercooling in order to crystallize. Subsequently, the more poorly crystallizing fractions are excluded from the zones of crystal growth and accumulate between these zones. They impede crystallization between the zones and lead to preferential growth of the crystals' growth zones, giving a fibrillar structure.

Little may be said about the nature and structure of the surface of the folded chains. Usually, a relatively thick layer of an amorphous noncrystallizable polymer exists on the fold surface, where the main difficulty is to describe this layer with commonly used measurement methods. The lamellae are slightly curved and contain turned-back molecules on both surfaces, as well as the defects in the regular macromolecular structure. Therefore, both lamellar surfaces are covered with amorphous layers with certain amount of loose loops, chain ends, branching sites, etc. Overlaid on this rough surface of regular folds are a number of loose cilia making up the major portion of irregular material. In multilayer crystals, the tie molecules transferring the amorphous layer may also connect two crystalline layers, being responsible for the shape memory. The relative ratio of all these surface structures may vary with crystallization conditions and with polymer-nucleating agent composition.

NUCLEATION-INDUCED MODIFICATION OF THE LAMELLAR STRUCTURE

The selective formation of α or β spherulites may be controlled by application of specific additives, called nucleating agents (NA) [4]. Well-known α form NAs are derivatives of sorbitols, sodium benzoate and talc which is less expensive but also an efficient NA [5]. The *trans*-quinacriodone (known as Red Pigment E3B) is one of the most popular β nucleators (see the chapter entitled Beta modification of isotactic polypropylene). Recently, a mixture of equal quantities of pimelic acid and calcium stearate was also patented as a β nucleator. A Gauss-type curve of the relationship between the β-phase content (*k*-value) and the quantity of E3B was noted in the nucleated iPP [6].

Two various effects may be mentioned regarding the influence of nucleation on the iPP crystal structure:

1. an increase of the germination density leading to an increase of the crystallization temperature; and
2. a selective formation of various crystallographic phases, such as monoclinic α or hexagonal β, characterized by existence of only radial (β-phase) or both radial and tangential (α-phase) lamellae, with various dimensions (Figure 1).

By use of microscopic observation it was found that [6] the spherulites' dimensions are modified significantly by increasing quantity of E3B. The spherulites' dimensions decrease dramatically with the increasing quantity of the NAs, and for the highest concentration of E3B, instead of spherulites, only single lamellae may be observed.

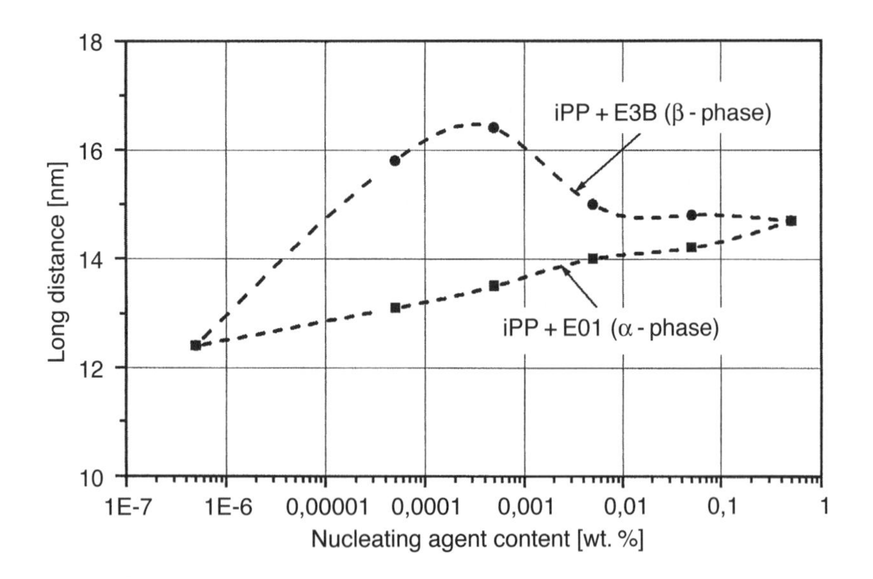

Figure 1 SAXS-determined lamella thickness distribution for iPP nucleated with E01 (α-phase) and E3B (β-phase) as a function of the nucleating agent content.

RELATIONSHIP BETWEEN LAMELLA THICKNESS AND MELTING TEMPERATURE

The melting temperature, defined as temperature at which the crystalline layer is in thermodynamic equilibrium with the melt, depends on the lamellar thickness before the onset of the melting process. The enthalpy of fusion is a sum of contribution of the monomeric unit contributions (ΔH_m) lowered be an amount equal to the surface enthalpy (ΔH_f) of both sides of the lamellae. Therefore, if a lamella is composed of N_u monomeric units, the enthalpy of fusion:

$$\Delta H_m = N_u(\Delta H_m)_u - 2\ \Delta H_f$$

and the melting temperature:

$$T_m = \frac{\Delta H_m}{\Delta S_m}$$

The thermodynamic melting temperature T°_m may be found by extrapolation of a plot of the melting temperature T_m against the number of monomeric units per lamella, i.e. by the lamellar height. The relationship between melting and crystallization temperature may be written as [3]:

$$T_m = T^\circ_m\ (1 - \gamma^{-1}) + \gamma^{-1}\ T_{cryst}$$

For most polymers the value of γ was found to be about 2. Consequently, the plot of melting temperature against their crystallization temperature is a straight line. To find the thermodynamic melting temperature T°_m, an intersection of this line with the line $T_m = T_{cryst}$ has to be drawn.

SECONDARY CRYSTALLIZATION

As secondary crystallization, one may understand the increase of the lamellar thickness by simultaneous decrease of the amorphous layer thickness. Generally speaking, this effect may be observed just after processing, up to some days or even weeks, as long the temperature of the polymer $T > T_g$. The increase of the density during secondary crystallization is rather low and the increase of the lamellar thickness depends strongly on the temperature. Two explanations for this effect are possible: (1) this may be a slow crystallization of some difficult crystallizable components, which were not crystallized during the first stage, or (2) there is an increase of perfection of previously crystallized fibrils.

This phenomenon is usually related to annealing, and in view of the fact that it tends to occur in polymers with mobile chain segments rather than in materials with stiff molecular components (such as polyesters), the interference is more likely to be associated with a reorganization of previously crystallized material. Other features, such as the increase in melting point of the sample during the second crystallization stage, tend to support this idea.

DETECTION OF THE LAMELLAE

Several techniques are used to detect the existence of the lamellar morphology. Among the microscopic techniques, the most popular are TEM (transmission electron microscopy), SEM (scanning electron microscopy), and PLM (polarized light microscopy) [4]. All these techniques differ as to the sample preparation, observation methods and consequently on the resolution which may be achieved.

Usually a cut of thin films or etching technique has to be applied in order to prepare the samples of the investigated polymers. The etching technique, by properly adjusted reaction conditions, make it possible to reveal details of the structure of a crystalline sample by progressive removal of non-crystalline or poorly crystallized material. The residual crystalline material can usually be related to the original structure. In the case of iPP, the permanganic etching may be realized applying a mixture of potassium permanganate ($KMnO_4$) with concentrated sulfuric acid (H_2SO_4) and orthophosphoric acid (H_3PO_4) [7].

Electron microscopy (see the related chapter in this book) examination of replicas of etched thin films allows distinction of the α and β type lamellae of iPP. Scanning electron microscopy (SEM) is a powerful investigation method, providing the possibility of detecting the type and the average dimensions of lamellae, by application of reflection of electron beam on the etched surface of polymer covered with a thin layer of gold. Depending on the apparatus resolution and sample preparation structure units of about 10^{-1} μm may be detected by SEM.

Optical microscopy in polarized light is done for samples of thin film (maximum thickness of the film should be less than a few micrometers in order to get a sharp picture of the structure). This optical transmission method allows the detection of the type of birefringence (negative or positive) but is insufficient to assign the lamellar structure. Satisfactory results are achieved for structure units with dimensions of some micrometers.

The above-mentioned microscopic methods, operating in transmission or reflection mode, are especially useful in cases where qualitative answer of existence or nonexistence of lamellar structure is awaited.

The average lamellae thickness distribution may be described quantitatively by small angle X-ray diffraction (SAXS), or indirectly by differential scanning calorimetry (DSC). According to the Bragg relation, the SAXS measurements permit the determination of the structure units with dimension of some hundreds angstrom ($1\,nm = 10^{-9}\,m = 10\,Å$). The

SAXS diffractograms, by applying the Lorentz correction, give the adjusted graphic presentation in a form of Is^2 as a function of scattering vector, $s = 2 \sin \theta / \lambda = 1/d$ [Å^{-1}], where I is the measured diffraction intensity, θ is the diffraction Bragg angle, λ is X-ray wavelength and d the interplanar distance. The maximum on the curve $Is^2 = f(s)$ allows the evaluation of the average value of the long period, i.e. the distance between two crystalline layers in a multilayer lamellar model. The long period L_p for α and β nucleated iPP changes as a function of the NA concentration (Figure 1). Almost a linear increase of the L_p for the α-phase iPP is probably related to the increase of the crystallization temperature due to the higher content of the E01. For iPP nucleated with E3B (β phase) a higher value of the long period was noted, where the maximum on the L_p curve corresponds to the maximum of the k-value [6].

DSC is another tool giving access to the lamellar thickness distribution. Based on the form of the DSC melting curve, the average distribution of the lamellae thickness may be determined. This procedure is based on the assumption that the melting temperature T_m is related to the thickness L of a crystalline lamella, by the Thomson–Gibbs relationship [8]:

$$L = \frac{2\sigma_e}{\Delta H_{f,th} \rho_c} \left(\frac{T_0}{T_0 - T} \right)$$

where T_0 is the melting temperature of a lamella with an infinite thickness, σ_e is the lamella surface energy, $\Delta H_{f,th}$ is the melting enthalpy of a perfect crystal and ρ_c is the density of the crystalline phase. Such calculations give satisfactory results for polyethylene and polypropylene samples, although the resolution is lower than that of the SAXS measurements, especially in the case of polypropylene where the gradual melting of β phase, followed by recrystallization of α phase and final melting of one or two various α phases, tends to obscure the shape of the DSC curve.

REFERENCES

1. Wunderlich, B. (1973) *Macromolecular Physics*, Academic Press, New York.
2. Karger-Kocsis, J. (1996) *Polymer Engng Sci.*, **36**, 203.
3. Elias, H.-G. (1984) *Macromolecules 1, Structure and Physics*, Plenum Press, New York.
4. Varga, J. (1995) Crystallization, melting and supermolecular structure of isotactic polypropylene, in *Polypropylene: Structure and Morphology*, Vol. 1. (ed. J. Karger-Kocsis), Chapman & Hall, London.
5. Sterzynski, T., Lambla, M., Crozier, H. and Thomas, M. (1994) *Adv. Polymer Technol.*, **13**, 25.
6. Sterzynski, T., Lambla, M., Calo, P. and Thomas, M. (1997) *Polymer Engng Sci.*, **37**, 1917.

7. Olley, R.H. and Bassett, D.C. (1989) *Polymer*, **30**, 399.
8. Huang, Y.L. and Brown, N. (1991) *J. Polymer Sci.*, **29**, 129.

Keywords: chain fold crystal growth, lamella, lamellar thickness distribution, nucleation, secondary crystallization, etching, lamellae detection, small angle X-ray scattering (SAXS), differential scanning calorimetry (DSC), optical microscopy, spherulite, morphology, electron microscopy.

'Living' or plastic hinges

I. Naundorf and P. Eyerer

INTRODUCTION

'Living', plastic or molded-in hinges are integrated and flexible connections between two pieces of a molded polymer part. Due to the hinges, a motion of the two pieces relative to each other is possible. The main advantage of plastic hinges is the simultaneous production of the polymer part as well as the hinge. Because of this single production step, a manufacturer will save time and money.

There exist multiple applications with different requirements: In some cases, only a few bending cycles with a high bending angle are needed; in other cases, many cycles with a low angle are required (Table 1). This results in different hinge geometries. A simple geometry is a thin film of rectangular shape between the two pieces (Figure 1). In the following, the two pieces are called 'plates'. On the other hand, hinges with a complex geometry will be made of two triangles, for example. These triangles are connected to each other by one of the tips and to the polymer part (e.g. a cap or a side of a box) by the baseline. Because of an additional stress while bending, there exists two rest positions (open and closed).

During material selection, the requirements of the whole polymer part are considered and not only those of the hinge. But hinges made of brittle polymers, such as polystyrol (PS) or polycarbonate (PC), are not useful and will break after a few bending cycles. Often-used polymers are polypropylene (PP), polyethylene (PE) or polyoxymethylene (POM). Sometimes, also, nylon (PA) is used. When the material and the geometry

Polypropylene: An A–Z Reference
Edited by J. Karger-Kocsis
Published in 1999 by Kluwer Publishers, Dordrecht. ISBN 0 412 80200 7

Table 1 An overview on the application fields of plastic hinges [1].

Field of application	*Number of cycles, N*	*Bending angle, α*	*Frequency, ν*	*Geometry*	*Example*
Housing edges	1	90°	–	Simple	Coffee machine
Connection cap–reservoir	approx. 10	±90°	–	Simple	Cable tube
Connection of two sides	>1000	90°	<1 Hz	Simple	Foldable box
Fixing element	>10 000	45°	>1 Hz	Simple	Accelerator pedal
Snap-in cap	>1000	±90°	–	Complex	Shampoo bottle
Element with a technical function	>10 000	<90°	–	Complex	Lock of a glove compartment
Element with a technical function	>100 000	<10°	>100 Hz	Complex	Swinging plate of an orbital sander
Element with a technical function	>100 000	<45°	>100 Hz	Complex	Cardan joint

of the hinge are defined, its quality and reliability can be improved by selecting suitable processing parameters (e.g. melt temperature etc.).

Although plastic hinges are often used there exist only few investigations [2], and a mathematical treatment of the mechanical behavior of a plastic hinge is difficult. One of the best models and calculation instructions is given by Tres [3].

THE HINGE MORPHOLOGY

PP is an excellent material for plastic hinges. It is a semicrystalline thermoplastic polymer (like PE, POM or PA) and normally shows a spherulitic morphology in a light microscope under crossed polars. A hinge reduces the cross-section between the polymer parts (showing a thickness of some millimeters) since the thickness of the hinge is in 0.1 mm range. The velocity of the melt will increase and this results in three layers inside of the hinge: two highly oriented layers and one almost isotropic layer in between (Figure 1). Here, the highly oriented layers appear dark, the layer in the core and the plates appear bright.

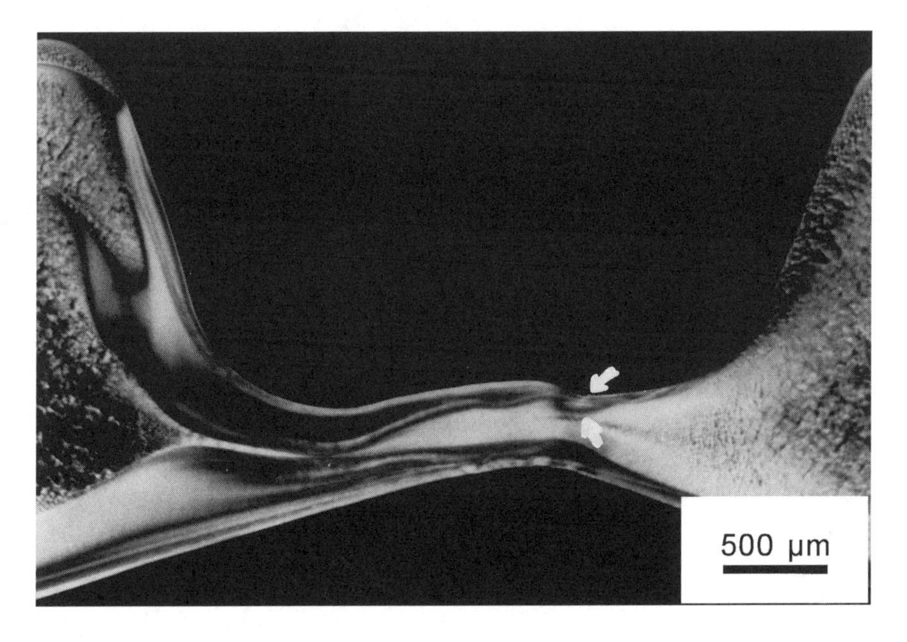

Figure 1 Optical micrograph of a U-shaped hinge after 100 bending cycles, using crossed polars. The highly oriented layers appear dark, the isotropic central layer is bright. One can see the small deformation area not in the middle of the hinge but near to the gate (arrows indicate). The flow direction of the polymer melt is from the right to the left side.

While the isotropic layer in the core shows a small-sized spherulitic morphology (less than 1 μm in diameter), the oriented layers show no spherulites. These layers are not amorphous but also consist of a crystalline structure. The fact is that the crystallinity – determined by differential thermal analysis (DSC) – inside of the hinge is even slightly higher than in the plates: Usually, inside the plate, we will find a crystallinity of about 50%; the crystallinity inside the hinge is 1–3% higher.

Spherulites consist of crystal lamellae of about 10–20 nm in thickness and a lateral range of some micrometers. In PP, the spherulites show a 'cross-hatched' lamellae structure [4] with radial lamellae as well as tangential lamellae, which grow up on the radial lamellae ('homo-epitacticity').

Investigations by transmission electron microscopy (TEM) exhibit the known cross-hatched structure in the central layer but inside the oriented layers there are stacked lamellae. Their orientation is perpendicular to the flow direction of the polymer melt (Figure 2, upper side). The lateral range of these lamellae are shorter than the range of the radial lamellae

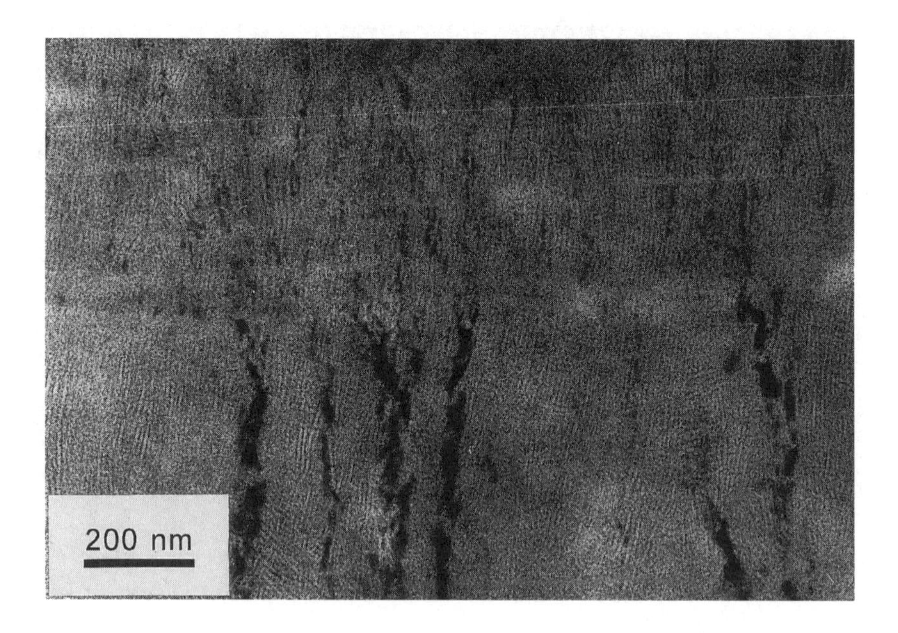

Figure 2 Transmission electron micrograph of the different deformation behavior of the highly oriented layer (top) with stacked lamellae and the typical cross-hatched lamellae structure of PP (bottom). The lamellae appear bright, the amorphous phase dark and the damaged structures black. The melt flow direction inside of this micrograph is horizontal.

of spherulites. Often, they are stopped by white streaks which are oriented parallel to the flow direction.

This high molecular orientation and special lamellae structure is similar to that of the so-called 'hard-elastic morphology' (see related chapter in this book) [5]. It was found in uniaxially drawn fibers and films. This structure leads to a very good mechanical behavior: A high stiffness (the Young's Modulus is about 2–5 times higher than that of isotropic PP) and a good elastic recovery. Sprague [6] found a strain recovery from 90% to nearly 100% after a tensile strain of 50% and 100%, respectively. Also Göritz and Müller [7] found that a leaf spring model (in contrast to a coil spring) will explain most of the features of hard elasticity.

Variation of the process parameter does not much affect the morphology of a hinge. Due to a lower melt or mold temperature, the thickness of the oriented layers increase only by some micrometers. A similar effect will be caused by the reduction of the melt velocity. A lower mold temperature and velocity results in a more turbulent flow of the polymer melt, causing a variation of the layer thicknesses along the hinge. But, in all cases, birefringence measurements show that the grade of molecular orientation inside the oriented layers is increased markedly. This means that in spite of a perfect alignment of the stacked lamellae inside the oriented layers (Figure 2, upper side) differences in the molecular orientation may still be present.

THE MECHANICAL BEHAVIOR OF PLASTIC HINGES

The hinge geometry affects the mechanical behavior of the hinges. It is evident that a notch hinge has a poorer quality than a U-shape hinge. But, up to now, it is not clear whether a hinge shaped like a semicircle, a hinge with a U-shape or a long U-formed hinge (Figure 3) has the best properties.

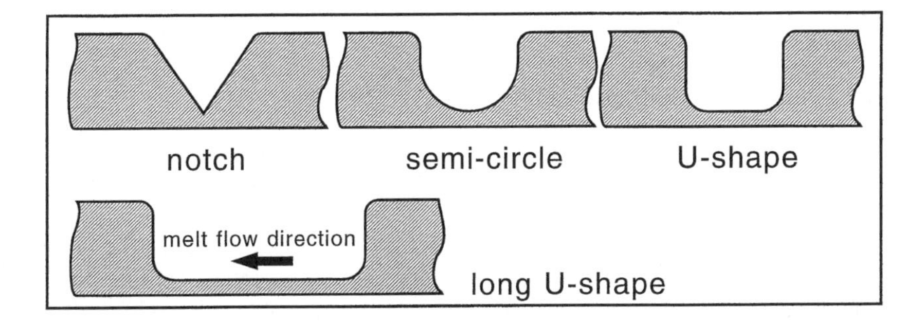

Figure 3 Schematic view on the different hinge geometries.

The main difference between the U-shaped and the semicircle shaped hinge is that in the first case the smallest cross-section of the hinge is defined but not in the second. So a U-formed hinge will bend at the weakest point. Figure 1 shows that the deformation area is not in the middle of the hinge but near to the gate where the thickness of the oriented layers is small.

The main difference between the U-formed and the long U-formed hinge is the possibility to define the deformation area for the latter. By applying an external load, the axis of rotation can be determined. When the axis runs at the 'end of the hinge', i.e. far from the gate, the bending properties will be improved because in this area the oriented layers are thicker (Figure 1). However, if the axis of rotation is close to the gate, there is no difference between these two U-types.

In general, the thicker the oriented layers the better are the hinge properties. A higher grade of molecular orientation will also improve the hinge quality. To characterize the quality of a hinge, we can determine the remaining or residual yield strength of a hinge after a defined number of bending cycles. In contrast to the yield point, the break point is less precisely defined because the hinges often fail statistically.

The remaining strength of most of the U-shaped hinges remains constant over some 1000 bending cycles and then reduces. After about 100 000 cycles, the strength reaches values of 75–90% of the starting value. In contrast, the slope of the strength of long hinges keeps constant over 100 000 bending cycles if the deformation area is far from the gate.

Lower melt and – less pronounced – lower mold temperatures result in more resistant hinges. The starting value of the residual yield strength differs between 30 and 50 MPa (melt temperature 300 and 200°C, respectively; mold temperature 60°C) and between 40 and 47 MPa (mold temperature 60 and 20°C, respectively; melt temperature 230°C).

It is an interesting result that after about 100 000 bending cycles hinges formed by a notch show the same remaining strength, like the other hinge types (excluded the long one). This could be explained by the deformation mechanism explained below.

THE DEFORMATION MECHANISM

Bending causes a tensile force on one side of the hinge (upper side in Figure 1) and a compressive force on the other. The deformation area is localized to a small region inside the hinge (some 0.1 mm in diameter, Figure 1). Investigations by light microscopy show that up to 1000 bending cycles the cross-section of the hinge only decreases slightly and that shear bands arise on both sides of this area [8]. After 10 000 bending cycles, the oriented layer of the 'upper' side become fibrillar and, after more than 100 000 bending cycles, both oriented layers are completely destroyed. Only the core layer remains. This explains why a hinge formed by a notch has the same remaining strength as the other types.

TEM showed that the deformation behavior of the oriented layers with the stacked lamellae differs from that in the central layer with the cross-hatched morphology (Figure 2): In contrast to the cross-hatched lamellae morphology, no crazing is observed in the oriented layers. Here the deformation develops in the following steps. First, the amorphous phases between the lamellae are broadening (Figure 2, upper side). This broadening partly disappears when the external load is removed. Second, the lamellae tilt partly into shear bands.

In the third step, the lamellae break partly. Microvoids arise between broken lamellae and inside of the amorphous areas. These microvoids grow to cavities of some micrometers, (Figure 4). The origin of the microvoids and cavities is always some 10 μm beneath the surface. Microvoids and cavities form hole bands and grow from inside of the hinge to the surface. Their position is defined by the location of the former shear bands. In the last step, the hole bands tear open both to the

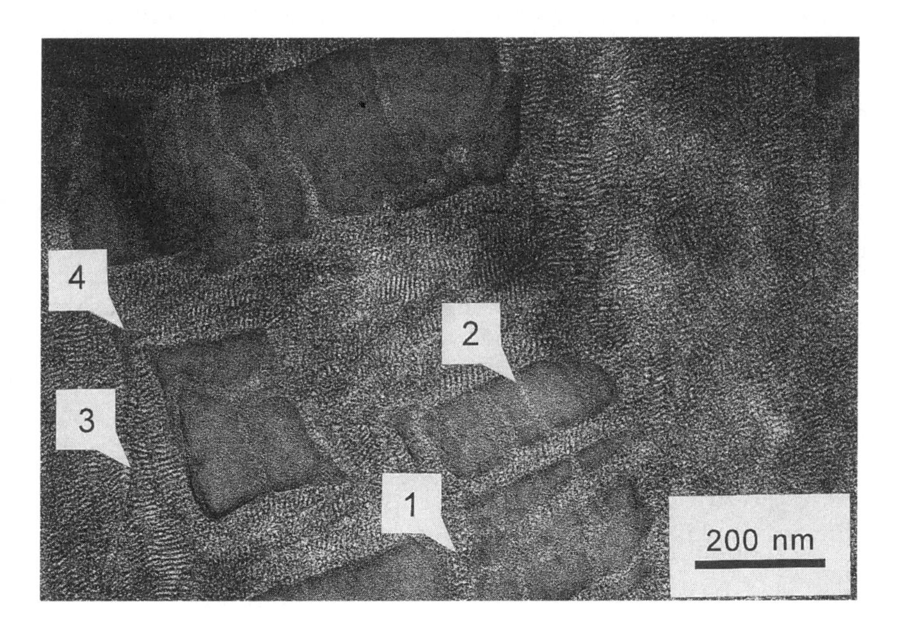

Figure 4 TEM micrograph of microvoids and cavities of a hole band. The rectangular cavities are filled by an epoxy resin and appear light grey. Around the cavities lamellae tilting (up to 90°) is visible. Between the edges of the cavities broad PP-connections with undamaged lamellae (1) but also fibrillian structures (2) can be found. Perpendicular to the stacked lamellae black lines arise (3), which represents the first step of the formation of microvoids and cavities (4).

surface and to the core layer which becomes visible on the macroscopic scale.

Using a PP block copolymer with ethylene (PP/PE), in general the hinge has less strength. But the interface between the 'hard' PP matrix and the 'soft' PE inclusion has a stopping function for shear bands. Small shear bands – composed of tilted lamellae – are not able to induce shear bands inside of the PE inclusion. Because of further bending, small shear bands grow up into large bands. And, in contrast to the small bands, large shear bands can induce shear bands, inside of the PE phase. But this growth takes time and causes a retardation of the damage process. After many bending cycles (more than some 1000 cycles), a second deformation mechanism appears. The PE inclusion peels off from the PP matrix releasing the stopping mechanism. After that the 'normal' deformation process works again.

CONCLUSION

Essentially a hinge is composed of three layers caused by a high velocity of the polymer melt: two highly oriented layers and one isotropic core layer. The mechanical behavior of the oriented layers is similar to that of 'hard elastic fibers' which show a high stiffness and a high strain recovery. The higher the molecular orientation and the layer thickness, the better the hinge properties are. The layer thickness can be increased by low temperatures (melt and/or mold).

Because of the increasing thickness of the oriented layers along a plastic hinge, the axis of rotation should be put to 'the end', i.e. far from the gate. This is easier when a long hinge is applied. When short hinges are used, the semicircle formed hinges have a good quality because the deformation area is well defined. Sometimes an unsymmetric hinge geometry (close to the gate formed like a semicircle, far from the gate U-shaped) will deliver the best quality: The smallest cross-section is now located at the 'end of the hinge'.

Additionally, the use of a PP/PE copolymer can retard the damage process because of the stopping mechanism at the interface between the PP matrix and the PE inclusions.

REFERENCES

1. Hoechst, A.G. (1990) *Filmgelenke aus Technischen Kunststoffen.* Firmenschrift B.3.5.
2. Clark, E.S. and Spruiell, J.E. (1976) Unlimited flex-life in moulded-in hinges in PP. *Polymer Engng Sci.*, **16**, 176.
3. Tres, P.A. (1995) Living Hinges. In: *Designing Plastic Parts for Assembly,* (ed. P.A. Tres), Hanser, Münich, p. 140.

4. Khoury, F. (1966) The spherulitic crystallization of iPP from solution: On the evolution of monoclinic spherulites from dendritic chain-folded crystal precursors. *J. Res. NBS*, **70A**(1), 29.
5. Cayrol, B. and Petermann, J. (1975) Elektronenmikroskopische Beobachtungen an elastischen Hartfasern aus Polyäthylen. *Colloid Polymer Sci.*, **253**, 840.
6. Sprague, B.S. (1973) Relationship of structure and morphology to properties of hard elastic fibers and films. *J. Macromol Sci.*, **B8**(1–2), 157.
7. Göritz, D. and Müller, F.H. (1975) Morphologie und mechanisches Verhalten von elastischen Hart-Fasern. *Collold Polymer Sci.*, **253**, 844.
8. Naundorf, I., Osterloh, S., Kech, A. and Fischer, G. (1995) Struktur und Eigenschaften von Filmscharnieren aus PP, Teil 2: Biegebelastetes Scharnier. *Plastverarbeiter*, **45**(12), 18.

Keywords: homopolymer, block copolymer, hinge, living hinge, plastic hinge, hinge geometry, process parameter, morphology, oriented layers, stacked lamellae, molecular orientation, hard elasticity, deformation mechanism, transmission electron microscopy (TEM).

Long term properties and lifetime prediction for polypropylene

M. Gahleitner and J. Fiebig

PHYSICAL CHANGES IN THE MATERIAL

Even if a part from a polymer is normally considered finished and solid after the processing step, its properties continue to change for a very long time. Most of the work on this process, normally called 'physical aging', of polymers has been focused on amorphous materials, as can be seen in the main textbook on this field by Struik [1]. According to Struik [1], aging is a thermoreversible process that occurs in all glassy polymers and affects their properties primarily by changing their relaxation times; the most important region of occurrence is between the glass transition temperature (T_g) and the first secondary relaxation (T_α). In the linear (small strain) region, a behavior like:

$$J(t) = J_0 e^{(t/t_0)^m} \tag{1}$$

with $J = \gamma/\sigma$ (in shear) or $J = \varepsilon/\sigma$ (in extension) being the compliance and $m \approx 1/3$ (meaning log–linear behavior) is observed.

Reorganization processes, however, are in principle more developed in semicrystalline polymers, where the application temperature T_{app} lies above T_g, which for polypropylene (PP) homopolymers is at about 0–+4°C (in random copolymers with ethylene, T_g can be lowered to −10°C). Struik [1] considers crystalline polymers as inhomogeneous systems with reduced mobility. Below the glass transition T_g, the same behavior as for amorphous ones is expected (in the amorphous regions); above T_g, the main effect should be due to a 'widened glass transition'; Struik does not accept any contribution of the crystalline regions.

Polypropylene: An A–Z Reference
Edited by J. Karger-Kocsis
Published in 1999 by Kluwer Publishers, Dordrecht. ISBN 0 412 80200 7

This is in contrast to the general opinion, according to which post-processing changes are separated into changes regarding crystallinity, which are termed 'postcrystallization' and changes in the amorphous regions, which are termed 'physical aging'. Solid PP consists of three phases:

- crystalline (α (monoclinic) or β (hexagonal) modifications);
- mesomorphic (also termed 'smectic' or 'paracrystalline'; consisting of small α-crystallites without superstructure); and
- amorphous (between and within crystalline regions).

Generally, the mechanical properties of semicrystalline polymers, such as PP, are significantly influenced by their crystalline structure. Apart from the overall crystallinity also the morphology, e.g. the dimension of spherulites or layered structures developed by shear-induced crystallization, play an important role here. These structures are in turn influenced by the molecular structure of the material, on one hand, and the solidification (or processing) conditions on the other hand. As the degree and organization of crystallinity also determines the amount of material 'fixed' with respect to reorganization processes, it should also have significant influence on the aging phenomena.

One has to distinguish [2] between processes occurring at normal ambient temperature (i.e. 0–+30°C) or during annealing at elevated temperatures up to the melting temperature T_m, which for PP-homopolymers is at about +165°C (in random copolymers with ethylene or terpolymers with ethylene and another α-olefin, T_m can be lowered to +140°C). These two regions are separated by the crystalline mobility transition, T_α, which is clearly detectable in dynamic-mechanical analysis (DMTA). A more complete scheme has been put together in Table 1.

Studies of conventional aging [3] show no change in the crystallinity as investigated in wide-angle X-ray scattering (WAXS), but an increasing density (with approximately log–linear behavior with time) and an approach in density between crystalline and amorphous regions, as seen from small-angle X-ray scattering (SAXS). All this points towards an increasing degree of order in the amorphous phase as the dominant factor for aging. From gas diffusion experiments (Ar, Ne, CH_2Cl_2), the dimensional scale of structural changes can be assessed. Two time regions are discernible: at times up to 10 h after cooling from the melt, most of the change occurs in the crystalline fraction (also visible in helical content from infrared (IR) measurements); for longer times, changes are slower and mainly in the amorphous region. The latter effect can be fully erased by thermal treatment.

The situation is even more complex in case of impact (heterophasic) copolymers with ethylene, which consist of a PP-matrix and ethylene-propylene rubber/polyethylene (EPR/PE) inclusions. Here, structural

Table 1 Mobility regions for physical aging of polypropylene (temperature values refer to homopolymers produced with conventional catalysts)

Region	T_{min} (°C)	T_{max} (°C)	Changes affect
Glass	–	0–4°C (T_g, T_β)	Density, mechanics (slightly)
Amorphous mobility	0–4°C (T_g, T_β)	50–100°C (T_α)	Density, mechanics, order in amorphous phase (strongly)
Crystalline mobility	50–100°C (T_α)	145–160°C (T_r)	Density, mechanics, crystallinity in mesomorphic phase
Recrystallization	145–160°C (T_r)	162–167°C (T_m)	Melting point, crystallinity (transition $\alpha_1 \rightarrow \alpha_2$)
Melt relaxation	162–167°C (T_m)	–	Phase morphology in heterophasic systems, overall structure

T_r, recrystallization temperature.

changes in the elastomeric phase with T_g at −60 to −30°C can take place at lower temperatures to a great extent already. As indicated in Table 1, phase relaxation takes place here during actual melting.

CHEMICAL CHANGES IN THE MATERIAL

Apart from purely physical changes, aging effects involving the chemical structure of the polymer have also to be considered. In contrast to other polyolefins, such as PE or most olefin-elastomers (EPR, ethylene-propylene-diene rubber (EPDM)), radical reactions in PP cause mainly a degradation effect, reducing the average chain length of the polymer and especially affecting the high molar weight fraction. As these are of primary importance for the mechanics of the system – through their activity as inherent nucleants as well as their function as 'tie molecules' between different crystalline sections – a significant reduction of mechanical properties can also be expected. The normal consequence is embrittlement, a massive decrease in toughness.

The main possible sources of harmful radicals are:

- combination of oxygen and heat;
- ozone;
- UV radiation; and

- radioactive radiation (relevant because of application of γ-irradiation for sterilization in, for example, medical applications of PP).

If the degradation process is set off by one of these effects, even storage at normal (ambient) temperature leads to a continuation; also, processes follow an Arrhenius behavior and are accelerated by higher temperatures. The process may be catalyzed by metal ions present in the polymer (through impurities, fillers etc.); especially critical are copper and iron. Apart from the aforementioned effect on the molar weight distribution (MWD), the phase structure of impact copolymers may also be changed.

Stabilizers, or more precisely antioxidants, are used to capture radicals in the system and thereby inhibit the degradation process. Mostly O-radicals are captured, but there are also some developments towards capturing C-radicals. Primary types (e.g. phosphites) mainly show activity in processing (melt temperature), secondary types (e.g. sterically hindered phenolics) in the normal application temperature range (long term stabilizers). Ultraviolet stabilizers are an additional means of preserving material properties by inhibition of the primary radical creation through radiation.

In their activity, stabilizers normally are consumed; other losses may occur through migration into elastomer inclusions or to the surface as well as through fixation to filler/reinforcement surfaces (adsorption). Under low loading levels, chemical aging will definitely determine the lifetime of a PP part. Not only does it create discolorations and aesthetic failure of the surface, but also cracks are induced which increase the stress sensitivity dramatically.

EFFECT OF STRUCTURAL CHANGES ON MATERIAL PROPERTIES

Physical aging of a PP material can only be monitored indirectly via the changes in its bulk properties, such as density, crystallinity as assessed via DSC or relaxation processes as seen in DMTA. If better organization in amorphous phase is accepted as the dominant effect, this can well explain the increasing stress transfer between crystalline regions (sometimes called 'pseudonetwork effect') having a positive effect on stiffness.

The primary mechanical profile of material can be summarized in two groups of parameters: (1) stiffness (flexural or tensile modulus, yield stress, heat deflection temperature); and (2) toughness (impact strength, breaking stress).

Changes induced by physical aging or 'postcrystallization' go on for a very long time on a log–linear scale (Figure 1). The value of 'standardized' measurements (ISO standard: $>96\,h$) for comparison seems therefore doubtful. Increase of stiffness in ambient-temperature aging always

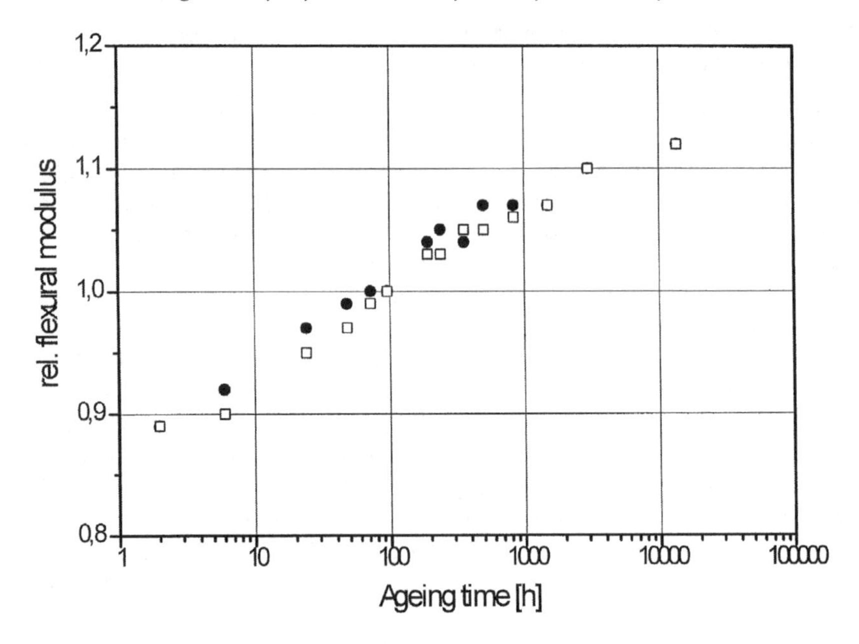

Figure 1 Influence of storage time on the relative flexural modulus of injection-molded parts from 2 PP-homopolymer grades (•) – Daplen KS 10, standard grade, FM (96 h) = 1300 MPa; □ – Daplen KS 44 N, nucleated grade, FM (96 h) = 1900 MPa).

goes along with reduction of the toughness. The latter effect is even more developed in parts of initially low crystallinity, such as films, which are often quenched in processing [2].

Apart from mechanics, crystallinity also influences secondary properties, such as transparency or light scattering, as well as stress whitening, which is an especially unpleasant effect in PP. All of these are also negatively influenced by aging.

INVESTIGATION METHODS UNDER LOADING AND PREDICTION PROCEDURES

In technical applications of PP, the aging under load has to be considered which is quite different from normal aging. The problem of predicting actual long term behavior under load [4] lies in a superposition of creep and aging phenomena with significant interaction. Normally, creep is mainly reduced by physical aging phenomena. In a tensile creep-experiment, the deformation can be defined either by:

$$\varepsilon_H = \ln\left(\frac{L}{L_0}\right) \text{ (Hencky strain) or} \tag{2}$$

$$\varepsilon_R = \frac{L - L_0}{L} \text{ (relative extension)} \tag{3}$$

with L_0 being the original length. Generally, ε will depend on time (t), stress (σ; either actual stress or nominal stress as related to the original dimension) and temperature (T). In the most simple case of logarithmic aging and an exponential temperature dependence, the equation:

$$\varepsilon = f(t, \sigma, T) = a_0 t^n \sigma^m e^{-E/RT} \tag{4}$$

holds. For a more detailed modelling, a generalized Voigt–Kelvin model can be applied, for example.

Technically, creep tests are carried out under tensile or flexural load [5]. One has to distinguish between creep and purely viscous flowing (at higher stress levels), which may also occur in actual applications and leads to effects such as 'de-aging', reducing the mechanical strength of the material. For actual lifetime predictions from such tests, one problem has to be faced: even if creep can be predicted correctly, the problem of finding correct failure criteria remains. Actual failures (crazes, cracks) are mostly connected with flaws either already present in the original material or generated in the application, possibly by mechanical or chemical damage. Creep tests are not suitable for determination of failure behavior, which can better be detected in dynamical tests at higher stresses (fracture mechanics) and simulating the actual application in temperature, exposure etc.

To reduce the actual time demand for testing, extrapolation from shorter tests can be used. In Figure 2, approaches to the problem using rather simple approaches are presented showing the underestimation of creep compliance at long time [4]). The applied equations are:

$$J(t) = J_0 + \Delta J(1 - e^{(-t/t_0)^m}) \text{ (Williams–Watts)} \tag{5}$$

$$J(t) = J_0 + \Delta J(t/\tau)^m \text{ (power law).} \tag{6}$$

From general experience, extrapolations of creep curves should be limited to one time decade to retain accuracy.

Other ways to predict long term behavior from short term tests are the 'accelerated' methods, mostly using the generally accepted principle of time–temperature superposition or through working at higher stresses. However, the applicability of such methods in crystalline systems is much worse than in amorphous ones, not at least through the interaction between creep and aging.

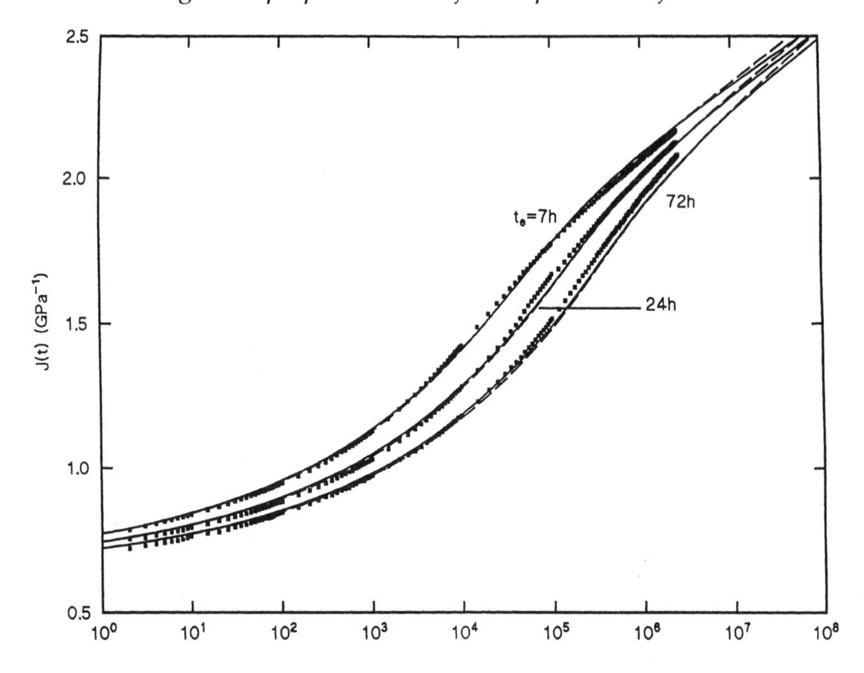

Figure 2 Comparison of the fits to long-term data ($\sigma = 3$ MPa) of a PP-creep experiment using equation (5) (——) and equation (6) (••••); from [4] (Crown copyright 1996; reproduced by permission of the controller HMSO, NPL).

REFERENCES

1. Struik, L.C.E. (1978) *Physical Aging in Amorphous Polymers and Other Materials*, Elsevier, Amsterdam.
2. Greco, R. and Ragosta, G. (1988) Isotactic polypropylenes of different molecular characteristics: Influence of crystallization conditions and annealing on the fracture behaviour, *J. Mat. Sci.*, **23**, 4171–4180.
3. Agarwal, M.K. and Schultz, J.M. (1981) The physical aging of isotactic polypropylene, *Polymer Engng. Sci.*, **21**, 776–781.
4. Tomlins, P.E. and Read, B.E. (1996) *Creep and Physical Ageing of Polypropylene: A Comparison of Models*, NPL-report CMMT(A)23, National Physics Laboratory, Teddington, Middlesex TW11 0LW, UK.
5. ISO 899–1 (1993) *Plastics – Determination of Creep Behaviour, Part 1: Tensile Creep, Part 2: Flexural Creep by Three-point-bending.*

Keywords: physical aging, crystallization, mechanics, stabilizers, glass transition, morphology, creep, lifetime prediction, postcrystallization, Voigt–Kelvin model, compliance, molecular mobility, Hencky strain, relaxation.

Mathematical modelling of propylene polymerization

João B.P. Soares and Archie E. Hamielec

INTRODUCTION

The complexity of molecular weight distributions (MWDs), chemical composition distributions (CCDs) and isotacticity distributions (IDs) of homo- and copolymers of propylene made with Ziegler–Natta catalysts constitutes a challenging problem for polymer quality control. These distributions affect the final mechanical and rheological properties of polypropylene (PP) and ultimately determine its applications. This issue becomes very complex with PP and copolymers of propylene and α-olefins made with heterogeneous Ziegler–Natta catalysts because polymers with broad and sometimes multimodal MWDs, CCDs and IDs can be produced. The major objective of propylene polymerization models is to predict these distributions and ultimately correlate them to mechanical and rheological properties [1].

MATHEMATICAL MODELS FOR HETEROGENEOUS CATALYSTS

There is vast experimental and theoretical evidence supporting the existence of multiple site types on heterogeneous Ziegler–Natta catalysts. From a mathematical modelling point of view, each site type has distinct polymerization kinetic constants and consequently produces polymer chains with different average properties.

The mechanism of propylene polymerization is reasonably well established. A generic kinetics scheme involves steps of chain propagation, chain transfer to small molecules such as hydrogen, monomer and

Polypropylene: An A–Z Reference
Edited by J. Karger-Kocsis
Published in 1999 by Kluwer Publishers, Dordrecht. ISBN 0 412 80200 7

cocatalyst and chain transfer by β-hydride elimination. β-methyl elimination has also been proposed for some metallocene catalysts. The mechanism of site formation by reaction of catalyst and cocatalyst is generally modelled as a first order reaction in both catalyst and cocatalyst concentrations. A general polymerization mechanism is given by the following chemical equations:

$M_j^* + A$	$\rightarrow$	$P(0)_j$	Site formation
$P(0)_j + C_3H_6$	$\rightarrow$	$P(1)_j$	Initiation
$P(r)_j + C_3H_6$	$\rightarrow$	$P(r+1)_j$	Propagation
$P(r)_j$	$\rightarrow$	$D(r) + P(0)_j$	β-hydride or β-methyl elimination
$P(r)_j + H_2$	$\rightarrow$	$D(r) + P(0)_j$	Transfer to hydrogen
$P(r)_j + C_3H_6$	$\rightarrow$	$D(r) + P(1)_j$	Transfer to propylene
$P(r)_j + A$	$\rightarrow$	$D(r) + P(1)_j$	Transfer to cocatalyst

where M_j^* is the catalyst, A is the cocatalyst, $P(r)_j$ is a living polymer of chain length r, and $D(r)$ is a dead polymer of chain length r. Each site type, symbolized by j, is assumed to have a different set of polymerization kinetics constants. This generic kinetics model has been modified to include steps of catalyst deactivation, poisoning and site transformation as needed for different polymerization systems.

One common modelling approach is to develop population balances for each site type and formulate moment equations to calculate molecular weight averages for the formed polymer chains. The major disadvantage of the method of moments is that it is not possible for complex distributions (distributions with shoulders or multimodal) to predict the whole MWD only from the knowledge of the molecular weight averages. Polymer populations with same number, weight and z-average molecular weights can have distinct MWDs and consequently different rheological and mechanical properties.

A more general and computationally efficient approach uses distributions that have a theoretical basis to predict detailed molecular properties of polymers in general, and of PP in particular. Flory's most probable distribution can be used to model the instantaneous MWD of PP made on each active site type of heterogeneous catalysts [2]:

$$w(r) = \tau^2 r \exp(-\tau r) \tag{1}$$

where $w(r)$ is the weight chain length distribution (MWD), r is chain length and τ is the ratio of chain transfer to propagation rates, given by:

$$\tau = \frac{R_\beta}{R_p} + \frac{R_M}{R_p} + \frac{R_H}{R_p} + \frac{R_A}{R_p} = \frac{k_\beta}{k_p[M]} + \frac{k_M}{k_p} + \frac{k_H[H_2]}{k_p[M]} + \frac{k_A[A]}{k_p[M]} \tag{2}$$

where R_p, R_β, R_M, R_H, R_A are the rates of propagation, β-hydride or β-methyl elimination, transfer to monomer, to hydrogen, and to cocatalyst, respectively, k_p, k_β, k_M, k_H, k_A are their equivalent rate constants, [M] is monomer concentration, $[H_2]$ is concentration of hydrogen, and [A] is cocatalyst concentration. Equation (1) applies for copolymerization as well as for homopolymerization.

The instantaneous MWD for the polymer made on all site types can be calculated as the average of the MWDs for each site type, weighted by the mass of polymer made on each site type:

$$W(r) = \sum_{j=1}^{n} m_j w_j(r) \tag{3}$$

where $W(r)$ is the overall chain length distribution of PP made on n different site types and m_j is the mass fraction of polymer made on site type j.

For a continuous stirred-tank reactor (CSTR) at steady state, the instantaneous MWDs and those of the accumulated polymer are the same, and therefore equation (3) can also be used to calculate the MWD of PP in a CSTR at steady state. For non-steady-state reactor operation (for instance, during grade transitions and reactor start-up or shut-down), the instantaneous MWD will vary in time.

For copolymerization of propylene and α-olefins, the terminal model is generally assumed, i.e., only the monomer bound to the active site and the monomer adding at the active site are considered to affect the rate of polymerization:

$$\begin{array}{llll} P(r,i)_j + M_i & \rightarrow & P(r+1,i)_j & k_{p\,i,i} \\ P(r,k)_j + M_i & \rightarrow & P(r+1,i)_j & k_{p\,k,i} \\ P(r,i)_j + M_k & \rightarrow & P(r+1,k)_j & k_{p\,i,k} \\ P(r,k)_j + M_k & \rightarrow & P(r+1,k)_j & k_{p\,k,k} \end{array}$$

thus requiring the knowledge of four propagation rate constants, $k_{p\,i,i}$, $k_{p\,k,i}$, $k_{p\,i,k}$, $k_{p\,k,k}$, and an appropriate number of transfer, initiation and deactivation rate constants [3]. Average copolymer chemical compositions are generally estimated from a simple molar balance for all comonomers, but the state of the accumulated polymer can be more completely defined from a knowledge of its bivariate distribution of composition and chain length.

A similar approach can be used to predict the bivariate distribution of MWD and CCD for binary copolymers of propylene and α-olefins [1]. Stockmayer's bivariate distribution can be conveniently used to calculate the instantaneous MWD and CCD of binary, linear copolymers, such as the ones formed by copolymerizing propylene and α-olefins:

$$w(r,y)\mathrm{d}r\mathrm{d}y = r\,\tau^2 \exp(-r\,\tau)\mathrm{d}r \frac{1}{\sqrt{2\pi\,\beta/r}} \exp\left(-\frac{y^2 r}{2\beta}\right)\mathrm{d}y \tag{4}$$

where y is the deviation from the average molar fraction, F_1, of propylene in the copolymer and:

$$\beta = F_1(1 - F_1)K \tag{5}$$

$$K = [1 + 4F_1(1 - F_1)(r_1 r_2 - 1)]^{0.5} \tag{6}$$

where r_1 and r_2 are reactivity ratios for copolymerization.

Similarly, the overall CCD and MWD, $W(r,y)$, can be obtained by the weighted average of Stockmayer's distributions for each site type:

$$W(r,y) = \sum_i m_i w_i(r,y) \tag{7}$$

Unfortunately, instantaneous CCDs and MWDs are not available for terpolymers or higher copolymers. In this case, one must rely on average copolymer composition values.

The distribution of stereoregularity has been less studied, from a mathematical modelling approach, than CCDs and MWDs. It can, in principle, be simulated in a similar way to the one for the copolymerization of propylene and α-olefins. Different orientations for monomer insertion can be assigned different polymerization kinetic constants per site type and therefore the problem of modelling ID becomes identical to the one of modelling CCD.

Mass and heat transfer resistances within the polymer particle have also been used to explain the broad MWD and CCD of PP and copolymers of propylene and α-olefins, although it is generally accepted that, under most polymerization conditions, the effect of multiple site types on polymer properties is far more important than that of mass and heat transfer resistances. During polymerization, the growing polymer chains break up the catalyst particle, forming an expanding particle of polymer and catalyst fragments. If diffusion resistances in the polymer particle are significant, catalyst fragments in different radial positions are exposed to different concentrations of monomer and chain transfer agent and consequently produce polymer with chain length averages that differ spatially inside the polymer particle. For copolymerization, comonomer with different ratios of propagation to diffusion rates will have distinct radial concentration profiles, leading to spatial compositional heterogeneity in the polymer particle. In addition, if there is appreciable heat transfer resistance, hot spots can occur inside the polymer particle, altering reaction rates and further broadening MWD and CCD (Figure 1). Evidently these transport models fail to explain the simultaneous production of isotactic and atactic PP on the same catalyst particle [3].

The effects of transfer resistances and of multiple site types on MWD and CCD broadening are evidently not exclusive. The heterogeneity caused by the presence of multiple site types can be increased by mass and heat transfer resistances during polymerization.

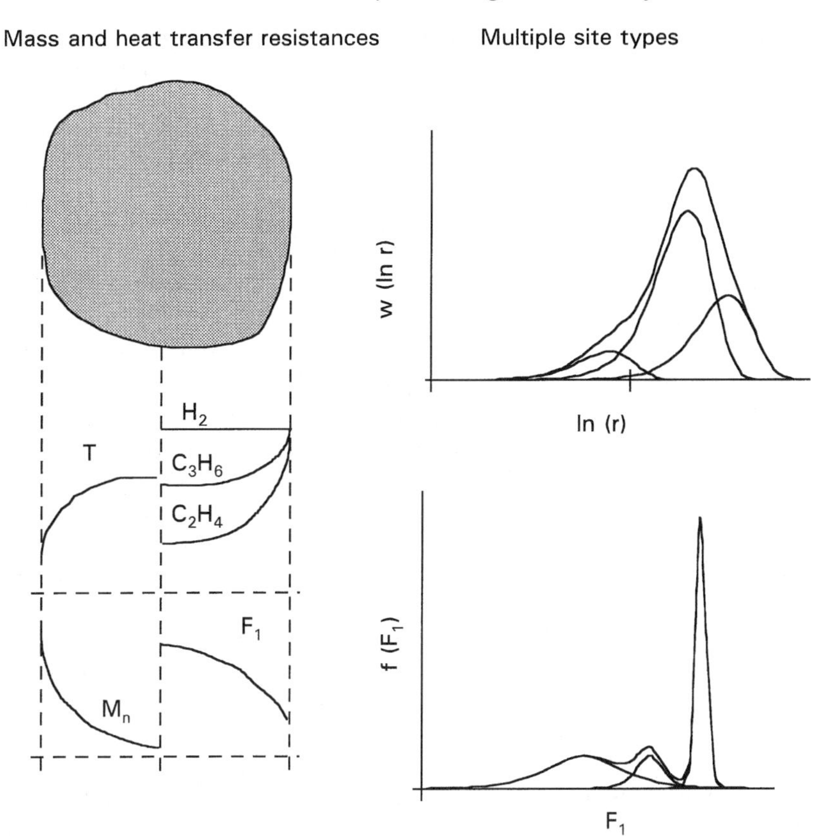

Figure 1 Effect of multiple site types and mass and heat transfer resistances on the microstructure of polypropylene made with heterogeneous Ziegler–Natta and metallocene catalysts. The overall MWD and CCD are assumed to result from the superposition of individual MWDs and CCDs for three site types (T = temperature, M_n = number average molecular weight, H_2 = hydrogen, C_3H_6 = propylene, C_2H_4 = ethylene, F_1 = molar fraction of propylene in copolymer, $f(F_1)$ = copolymer composition distribution, r = chain length, $w(r)$ = weight chain length distribution).

The models that best describe intraparticle mass and heat transfer resistances are the so-called *expansion models*. Expansion models consider the fragmentation of the catalyst particle and the formation of an expanding particle of polymer and catalyst fragments. Modelling equations for intraparticle radial profiles of monomer concentration and temperature for all expansion models are very similar. The radial profile of monomer concentration and temperature within the polymer particle is described by the well-known diffusion-reaction equations in spherical coordinates. Equations for the moments of MWD and copolymer composition averages have to be solved simultaneously with the intraparticle mass and heat transfer equations for each radial position in order to calculate the molecular weight and composition averages in the polymeric particle, generating a system of partial differential equations with moving boundaries.

Growing polymer chains and catalyst fragments are treated as a continuum in the *polymeric flow model*. It has been shown that if the polymerization is diffusion controlled, the radial profiles of monomer and chain transfer agent in the particle may cause MWD and CCD broadening. The *multigrain model* postulates two levels of mass and heat transfer resistances. The polymer particle (secondary particle) is considered to be formed by an agglomerate of primary particles. Each primary particle consists of a fragment of the original catalyst particle, with all active sites on its surface, surrounded by dead and living polymer. Monomer diffuses through the pores of the secondary particle and through the layer of polymer surrounding the catalyst fragment in the primary particles. Polymerization occurs on the surface of the catalyst fragments [4]. The existence of yet another level of mass transfer is considered in the *double grain model*, in which primary particles agglomerate to form subparticles. The polymer particle results from the agglomeration of these subparticles. Growing polymer rapidly fills the pores of the subparticles and therefore the monomer diffusion in the subparticle is of the order of magnitude of diffusion in pure polymer. This model evidently predicts higher diffusion resistances than the predicted by the multigrain model, since the effective diffusion in the larger subparticles is of the same magnitude as in the primary particles. In the *polymeric multigrain model*, catalyst fragments are assumed to be in a continuum of polymer with only one level of diffusional resistance for the monomer. In the polymeric multilayer model, the polymer particle is divided into concentric spherical layers but microparticles are not explicitly considered. The polymeric multilayer model estimates MWDs and CCDs in each model layer and for the whole polymer using Stockmayer's bivariate distribution. The expansion models previously discussed can only estimate averages of molecular weight and composition [3].

The conclusions of all expansion models are similar and can be summarized as follows: (1) mass transfer resistances can reduce the polymerization rate, decrease molecular weight averages, increase the polydispersity index of the MWD, and affect copolymer composition for large, highly active catalysts; (2) particularly important for supported catalysts, the concentration of high-activity catalyst sites can increase mass transfer resistance effects and have undesirable results in catalyst performance and product quality; (3) mass transfer resistances may also be a source of copolymer composition heterogeneity for highly active and large catalyst particles, if the comonomers have reactivities that differ significantly; (4) in most cases, intraparticle temperature gradients are negligible, with the remarkable exception of gas phase reactors with very active and large catalyst particles during the initial stages of polymerization.

The conclusions obtained with single site-type expansion models are especially important for the technology of supported metallocene catalysts, where single site, highly active catalytic species may be subjected to significant mass and heat transfer resistances.

SOLUBLE CATALYSTS

Soluble Ziegler–Natta catalysts, including metallocene catalysts, are generally of single site type and can therefore make PP and copolymers of propylene and α-olefins with narrow MWD and CCD.

In most processes using homogeneous Ziegler–Natta catalysts in general, and metallocenes in particular, the polymer is not soluble in the reaction medium and precipitates after a critical chain length is achieved. If after chain termination the active site returns to the solution, then interparticle mass and heat transfer should not influence the polymerization. However, if the active sites are trapped inside polymer particles, interparticle mass and heat transfer resistances could become significant. A mathematical model for particle growth during polymerizations catalyzed with metallocenes has been proposed recently [5]. It is possible that, as the polymer particles grow by agglomeration, some active sites will remain inside the polymer particles. This is especially more likely in processes where the active sites are exposed to high concentrations of monomer, as in liquid monomer polymerizations. In this case, due to the high activity of metallocene catalysts, mass and heat transfer resistances may play a role in broadening MWD and CCD.

Very few mathematical models have been proposed to describe polymerization of propylene using homogeneous Ziegler–Natta catalysts. Soares and Hamielec [3] proposed that the polymeric multilayer model could be used to simulate polymerization with homogeneous Ziegler–Natta catalysts if one neglects mass and heat transfer resistances in the polymeric particle. This would be rigorously valid only if active sites are not trapped inside polymer particles during polymerization. The underlining assumptions for MWD represented by Flory's distribution and CCD by Stockmayer's distribution should evidently be applicable to soluble metallocene and Ziegler–Natta catalysts as well.

Some metallocene catalysts, specially monocyclopentadienyl complexes, can be used to synthesize polyolefins with narrow MWD and significant degrees of long-chain branching. Some recent experimental evidence seems to indicate that long chain branches can occur as well in PP synthesized with certain metallocene catalysts. An analytical solution for the instantaneous MWD, CCD, and degree of long-chain branching has been recently proposed [1].

REFERENCES

1. Dubé, M., Soares, J.B.P., Penlidis, A. and Hamielec, A.E. (1997) Mathematical modelling of multicomponent chain-growth polymerizations in batch, semi-batch and continuous reactors: A review. *Ind. Engng Chem. Res.*, **36**, 966–1015.
2. Soares, J.B.P. and Hamielec, A.E. (1995) Deconvolution of chain-length distributions of linear polymers made by multiple-site-type catalysts. *Polymer*, **36**, 2257–2263.
3. Soares, J.B.P. and Hamielec, A.E. (1995) General dynamic mathematical modelling of heterogeneous Ziegler-Natta and metallocene catalyzed copolymerization with multiple active sites and mass and heat transfer. *Polymer React. Engng*, **3**, 261–324.
4. Ray, W.H. (1988) Practical benefits from modelling olefin polymerization reactors, in *Transition Metal Catalyzed Polymerization*, (ed. R.P. Quirk), Harwood, New York, pp. 563–590.
5. Herrmann, H.F. and Böhm, L.L. (1991) Particle forming process in slurry polymerization of ethylene with homogeneous catalysts. *Polymer Commun.*, **32**, 58–61.

Keywords: polymerization kinetics, polymerization reactors, mathematical modelling, molecular weight distribution (MWD), chemical composition distribution (CCD), Ziegler–Natta catalysts, metallocenes, microstructure, isotacticity distribution, mass transfer resistances, heat transfer resistances, effects of multiple site types.

Mechanical and thermal properties of long glass fiber reinforced polypropylene

J.L. Thomason

INTRODUCTION

In the past few years, the growth in structural composite usage has resulted in the need for higher output manufacturing processes than have been used previously. This has provided the impetus for the development of techniques to produce long fiber reinforced thermoplastic matrix composites which possess both high performance and mass processability. In particular, the long (but discontinuous) fiber reinforced materials, such as random in-plane fiber reinforced polypropylene (PP) sheets (better known as glass mat thermoplastic or GMT) and long fiber reinforced injection molding pellets prepared by wire coating, cross head extrusion, or thermoplastic pultrusion techniques, have recently received much attention. High performance levels can only be obtained from a composite part with high fiber concentrations and if the reinforcing fibers in the final product have a sufficiently high aspect ratio (length/diameter). It is, therefore, essential to be able to set appropriate targets for the desired fiber concentration and fiber aspect ratio in a composite part. In this chapter, we review some of the available equations for the prediction of GMT properties from the constituent properties and compare these predictions with experimental data. We also review some of the pitfalls in matching theory and experiment.

Polypropylene: An A–Z Reference
Edited by J. Karger-Kocsis
Published in 1999 by Kluwer Publishers, Dordrecht. ISBN 0 412 80200 7

MODELLING OF THE TENSILE MODULUS

The most commonly used theory used to model the stiffness of this type of composite was developed by Cox [1] and further improved by Krenchel [2, 3]. Cox's 'shear lag' model was developed for aligned discontinuous elastic fibers in an elastic matrix. The applied load is transferred from the matrix to the fiber via interfacial shear stresses, with the maximum shear at the fiber ends decreasing to zero at the centre. Thus, the tensile stress in the fiber is zero at the ends and maximum in the middle. Thus, although the efficiency of stress transfer increases with fiber length, it can never reach 100%. In order to accommodate this dependence of reinforcement efficiency on fiber length, Cox introduced a fiber length efficiency factor η_1 into the 'rule-of-mixtures' equation for the composite modulus E_c:

$$E_c = \eta_l V_f E_f + (1 - V_f) E_m \tag{1}$$

where E_f, E_m, V_f, are the fiber and matrix stiffness and the fiber volume fraction respectively and:

$$\eta_l = \left[1 - \frac{\tanh(\beta L/2)}{\beta L/2}\right] \tag{2}$$

where

$$\beta = \frac{1}{r}\left[\frac{2\ G_m}{E_f \ln(\sqrt{r/R})}\right]^{1/2} \tag{3}$$

where L is the fiber length, G_m is the shear modulus of the matrix and r is the fiber radius and $2R$ is the mean spacing of the fibers. The r/R factor can be related to the fiber volume fraction by:

$$\ln(\sqrt{r/R}) = \ln(\sqrt{\pi / X_i V_f}) \tag{4}$$

where X_i depends on the geometrical packing arrangement of the fibers [2]. Krenchel extended this theory to take fiber orientation into account by adding a fiber orientation factor η_0 into the 'rule-of-mixtures' equation.

$$E_c = \eta_0 \eta_l V_f E_f + (1 - V_f) E_m \tag{5}$$

For the random in-plane orientation of the fibers, it can be shown that $\eta_0 = 0.375$ [2, 3]. Equation (5) has been verified experimentally for glass-fiber polypropylene GMT over a wide range of fiber length and concentration and over the temperature range −50 to 150°C when the appropriate temperature-dependent parameter values are used [4].

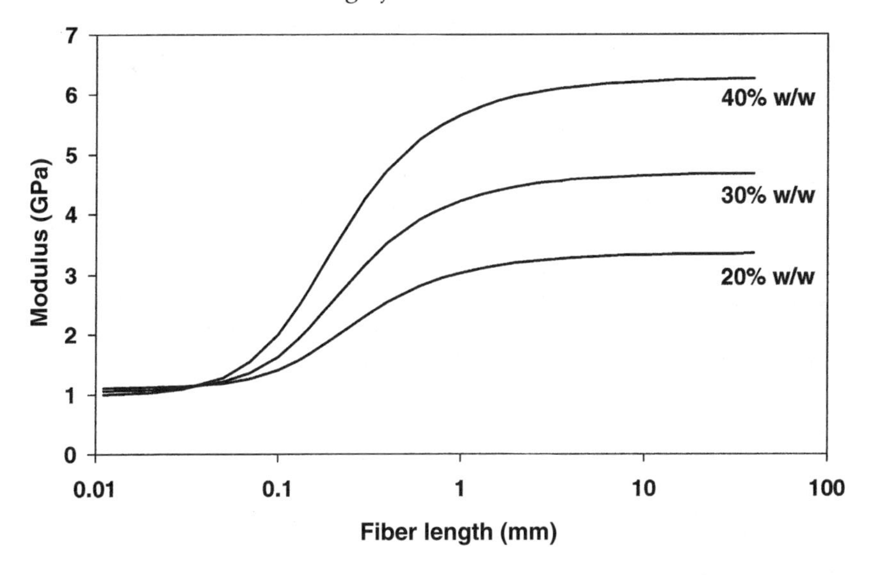

Figure 1 GMT modulus versus fiber length (from equation (5)).

As an example of the use of equation (5), Figure 1 shows the predicted stiffness of polypropylene GMT containing 20–40% w/w of 15 μm diameter fibers as a function of the fiber length. It can be seen that little improvement in stiffness is obtained above a fiber length of 2 mm.

Experimental deviation from equation (5) can be obtained for a number of reasons. In general, it can be stated that the stiffness of GMT laminates is not directly sensitive to the molecular weight of the PP matrix or to the level of fiber–matrix adhesion. However, it should be noted that the matrix viscosity plays an important role in the level of fiber wet-out, dispersion and breakage during processing. The level of interfacial interaction will also affect wetting. Equation (5) assumes perfect wetting of the fibers by the PP matrix. However, imperfect wetting, poor fiber dispersion and unexpected fiber length reduction can all lead to a lower stiffness than the predictions of equation (5).

At higher fiber contents, we must also consider the relationship between fiber aspect ratio and the theoretical maximum achievable volume fraction. As fibers are laid down randomly in a plane, the probability of fibers crossing each other and thus being oriented out of plane increases as the fiber length and concentration are increased. This will result in a lower stiffness in the *X–Y* plane and a subsequent increase in stiffness in the *Z* direction. For instance, it has been shown [2] that, in the fiber aspect ratio range of 300–1100, the maximum achievable volume fraction in a two-dimensional random in-plane laminate is 18–20% v/v. If the laminate volume fraction is greater than this, then there are a number of possible scenarios. One may obtain a higher fiber fraction by poor dispersion of the fibers leading to the presence of fiber bundles which results in a reinforcement with a much lower aspect ratio and effective stiffness than well-dispersed fibers and consequently gives a lower than predicted laminate stiffness. If the GMT has undergone aggressive processing then the fiber length (and aspect ratio) may be decreased to allow the fibers to pack into the available space. If the GMT has experienced relatively mild preparation conditions then it is not the aspect ratio that is reduced but the void content is increased when the maximum fiber volume fraction is exceeded, i.e. there is insufficient matrix available to fill the gaps between the fibers. This lowers the actual fiber volume fraction of the sample and the relative stiffness of the matrix.

THERMAL PROPERTIES

During their service lifetime, many of the products made with these materials will experience extremes of temperature and a large number of temperature cycles. It is therefore important for designers to have a clear understanding of the influence of temperature on composite properties. Furthermore, the intrinsic inhomogeneous, anisotropic nature of these materials results in a complex thermal expansion behavior and can also result in the presence of large internal stresses in composite products. In terms of thermal expansion behavior, most of the theoretical and experimental studies of fiber reinforced polymer composites have focused on continuous unidirectional reinforced composites, for the main part with thermosetting matrices. The dependence of the composite linear coefficients of thermal expansion (LCTEs) of unidirectional fiber composites containing isotropic fibers and matrix are given [4] by:

$$\alpha_c^L = \frac{V_f E_f \alpha_f + V_m E_m \alpha_m}{V_f E_f + V_m E_m} \tag{6}$$

$$\alpha_z^T = (1 + \nu_f) V_f \alpha_f + (1 + \nu_m) V_m \alpha_m - \nu_c \alpha_c^L \tag{7}$$

where α_c^L and α_c^T are the composite LCTEs in the fiber longitudinal and transverse directions, ν is the Poisson ratio where $\nu_c = V_f \nu_f + V_m \nu_m$. Due to the low strain rates present during most thermal expansion measurements, it can be assumed that we are operating in the elastic region. Thus by using a similar approach to Cox and Krenchel we obtain the following equation for the in-plane LCTE α_{xy}:

$$\alpha_{xy} = [\eta_0 \eta_l V_f E_f \alpha_f + V_m E_m \alpha_m]/E_c \tag{8}$$

where E_c is given by equation 5, η_l is given by equation (2), and $\eta_0 = 0.375$ for random in-plane fiber orientation. Equation (8) has been shown to give good correlation with experimental values from glass fiber PP GMT over a wide range of temperature, fiber length and fiber concentration [4].

However equation (7) is not a good predictor of α_z for GMT. In random in-plane laminates, the matrix expansion is restricted in the two dimensional *X-Y* plane in which the fibers are randomly oriented. Consequently, the relaxation of the resultant compression of the matrix is concentrated in the *Z* direction transverse to the axis of the fibers. This can result in the higher values of α_z made up of a contribution due to the normal transverse expansion of the components and a contribution due to the compressive stresses developed in the matrix in the *X-Y* plane due to the restriction of the matrix expansion by the fibers. Thus we obtain:

$$\alpha_z = V_f\alpha_f + V_m\alpha_m + 4\nu_m V_m(\alpha_m - \alpha_{xy}) \quad (9)$$

Since α_m, E_m and E_c are all temperature sensitive α_{xy} and α_z can exhibit complex temperature dependence. Figure 2 shows experimental values for α_m, E_m and E_c and the resulting calculated vales of α_{xy} and α_z for a PP randomly reinforced with 30% w/w 12 mm long fibers. Equation (9) has also been well verified experimentally [4].

It should be noted that some GMT samples can undergo a significant degree of expansion in the out-of-plane direction when heated close to or above the polypropylene melting point. This phenomenon is called lofting and is well known in the GMT world. It is the result of

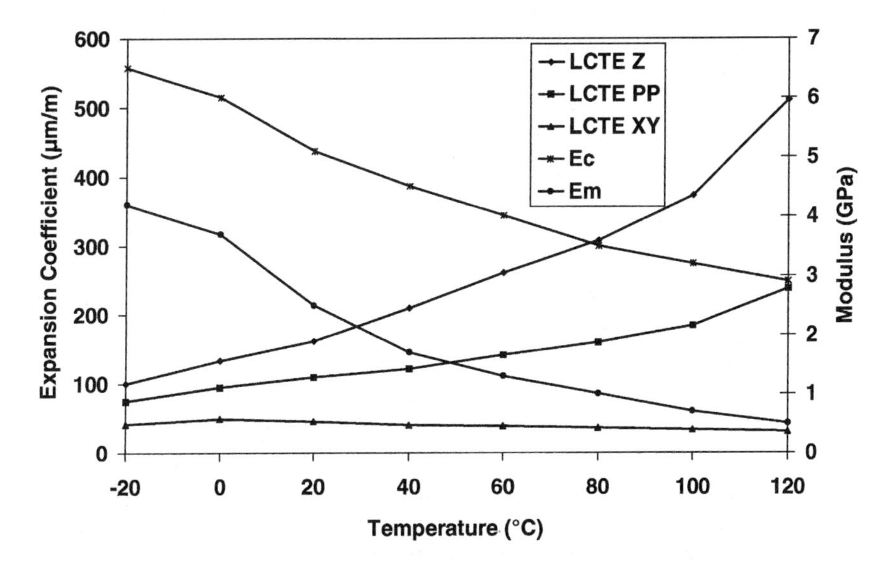

Figure 2 Experimental values for GMT and PP expansion coefficient and modulus.

compression of the fiber mat during production of the laminate; out-of-plane fibers are bent and held in place by the polymer matrix. On heating the polymer above its melting temperature, these fibers may straighten out and the mat returns to its original thickness. The degree of loft is a complex function of the temperature, the fiber length and concentration, combined with the original processing parameters of the GMT sample. Lofting will lead to large experimental deviations from the prediction of equations (8) and (9), particularly in the values of α_z.

STRENGTH MODELLING

The Kelly–Tyson model for the prediction of the strength (σ_{uc}) of a polymer composite reinforced with discrete aligned fibers [5,6] is given by:

$$\sigma_{uc} = \Sigma_i \left[\frac{\tau L_i V_i}{D}\right] + \Sigma_j \left[\sigma_{fj} V_j \left(1 - \frac{L_c}{2 L_j}\right)\right] + (1 - V_f)\,\sigma_{um} \qquad (10)$$

where τ is the interfacial strength, $V_{i,j}$ the volume fraction of fibers of length $L_{i,j}$, σ_{fj} is the fiber strength, σ_{um} is the matrix strength at the fiber failure strain equal to $E_m \sigma_f / E_f$, and L_c is a critical fiber length defined by $L_c = \sigma_f D / 2\tau$. The two summation terms arise from the contributions of fibers of subcritical and supercritical length. The Kelly–Tyson model assumes that all the fibers are aligned in the loading direction which, in practice, rarely occurs and is difficult to achieve experimentally. In GMT laminates, there is an approximately random fiber orientation distribution and equation (10) cannot be integrated to give a simple numerical orientation factor to take account of the random fiber orientation as is the case in modelling of laminate stiffness. Despite this problem, the Kelly–Tyson model is often referred to in relation to the analysis of the strength of discontinuous fiber composites and in particular the calculation of L_c is often made and quoted as an important parameter. However, it is possible to fit experimental data using a simple numerical orientation factor approach. For glass fiber PP GMT an excellent fit to experimental data was obtained [6] when the fiber contribution in equation (10) was modified by a factor $\eta_0 = 0.2$.

Figure 3 shows the predicted strength of PP GMT containing 20–40% w/w of 15μm diameter fibers as a function of the fiber length. We have used a value of $\tau = 4$ MPa which is fairly typical for a glass fiber adhesion to unmodified polypropylene [6, 7] and a fiber strength $\sigma_f = 2000$ MPa, which results in $L_c = 3.75$ mm. It is important to note that at this value the fiber contribution to the laminate strength has only reached 50% of its maximum value. To attain greater than 90% of the maximum attainable strength, we actually need to use fibers with length $> 5L_c$. This is a much

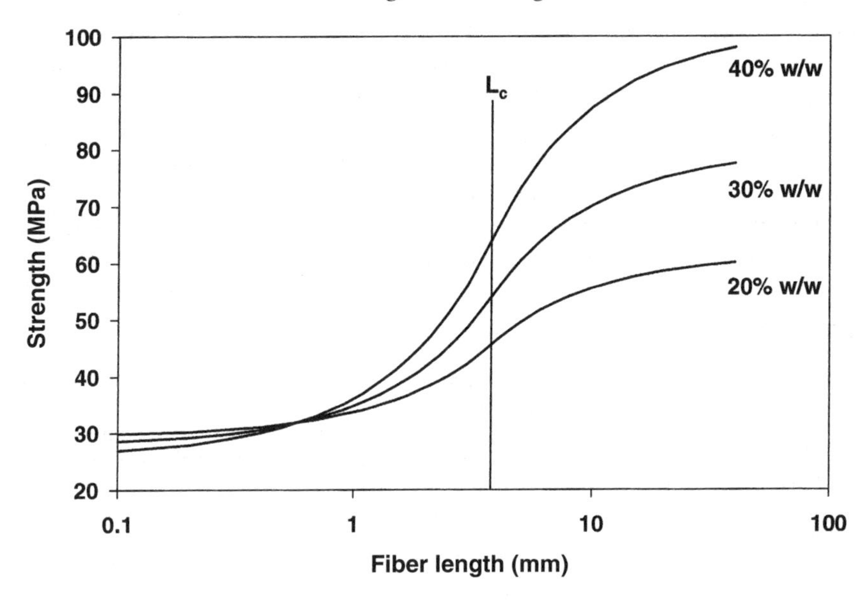

Figure 3 GMT strength versus fiber length (from equation (10)).

greater value than the value of 2 mm observed previously as necessary to exceed the 90% level of maximum laminate stiffness.

Most of the points of discussion raised in the section on the modulus (i.e. imperfect wetting, poor fiber dispersion, unexpected fiber length reduction and voids) are also relevant to the comparison of experimental and calculated GMT strength. Two other keys points that should be mentioned relate to the sensitivity of equation (10) to the values of σ_f and τ. It has been shown that σ_f in real composites is nearly always much lower than the pristine filament strength values quoted for glass fibers (e.g. see the chapter on 'Properties of glass fibers for PP reinforcement' in this book). Glass fiber strength is strongly affected by processing damage and wherever possible should be measured directly for the composites to which equation (10) is be applied. Furthermore, σ_f is a function of gauge length and so one should make these measurements as close to the average fiber length in the composite as possible. The reader should also give due consideration to the value of τ which is to be used. Currently, there is no universally accepted method for determining τ, so one needs to consider which of the various techniques is best suited for the system under study. We have found that the fiber pullout test gives [6, 7] acceptable results for glass fiber and PP, although other test methods are also possible [8].

As a final comment, the effect of dispersity in the factors (V_f, σ_f, D, L, τ, etc.) used in the above equations has not been fully explored, either experimentally or theoretically. However, GMT is generally an inhomogeneous material [9] and there will be variation in all of these parameters to some degree and this will affect the experimentally measured data. It is, for instance, a fairly simple exercise to show that using an average length value in equation (10) may give a very different result from using the full distribution of lengths. Notwithstanding these numerous pitfalls, when applied intelligently, the equations shown here can prove to be invaluable tools to developers and users of GMT.

REFERENCES

1. Cox, H.L. (1952) The elasticity and strength of paper and other fibrous materials. *Br. J. Appl. Phys.*, **3**, 72–79.
2. Thomason, J.L. and Vlug, M.A. (1996) The influence of fibre length and concentration on the properties of glass fibre reinforced polypropylene: 1. Tensile and flexural modulus. *Composites*, **27A**, 477–484.
3. Folkes, M.J. (1985) *Short Fibre Reinforced Thermoplastics*, Chap. 2, Research Studies Press, Chichester, p. 16.
4. Thomason, J.L. and Groenewoud, W.M. (1996) The influence of fibre length and concentration on the properties of glass fibre reinforced polypropylene: 2 Thermal properties. *Composites*, **27A**, 555–565.
5. Kelly, A. and Tyson, W.R. (1965) Tensile properties of fibre-reinforced metals. *J. Mech. Phys. Solids*, **13**, 329–350.
6. Thomason, J.L. Vlug, M.A., Schipper, G. and Krikor, H.G.L.T. (1996) The influence of fibre length and concentration on the properties of glass fibre reinforced polypropylene: 3. Impact properties and overall conclusions. *Composites*, **27A**, 1075–1084.
7. Thomason, J.L. and Schoolenberg, G.E. (1994) An investigation of glass fibre/polypropylene interface strength and its effect on composite properties. *Composites*, **25**, 197–203.
8. Herrera-Franco, P.J. and Drzal, L.T. (1992) Comparison of methods for the measurement of fibre/matrix adhesion in composites. *Composites*, **23**, 2–27.
9. Berglund, L.A. and Ericson, M.L. (1995) Glass mat reinforced polypropylene, in *Polypropylene: Structure, Blends and Composites*, Vol. 3, Ch. 5 pp. 202–227, (ed. J. Karger-Kocsis), Chapman & Hall, London.

Keywords: long glass fiber, glass mat thermoplastic (GMT), shear lag model, rule of mixture, tensile modules, Cox–Krenchel model, aspect ratio, thermal properties, linear thermal expansion coefficient, lofting, tensile strength, Kelly–Tyson model, adhesion, wetting, glass mat, random fiber orientation.

Melt blowing technology

Michael Wehmann and W. John G. McCulloch

INTRODUCTION

The most commonly accepted and current definition for the melt-blown process is: 'a one-step process in which high-velocity air blows molten thermoplastic resin from an extruder die tip onto a conveyor or takeup screen to form a fine fibered self-bonded web'. Melt-blown microfibers generally have an average fiber diameter range of 2–4 μm, although they may be as low as 0.3–0.6 μm and as high as 15–20 μm. Even higher fiber size webs are attainable with the process but, to date, have generated limited commercial interest. The authors believe the melt blowing process is one of the most, if not the most, cost-effective and versatile processes commercially available to produce microfibrous products which are essential to separate the 'good from the bad' inherent in today's air and liquids.

HISTORICAL BACKGROUND

The basic technology to produce these extremely fine fibers was first developed under USA government sponsorship in the early 1950s. The Naval Research Laboratory initiated this work to produce microfilters for the collection of radioactive particles in the upper atmosphere [1]. The significance of this work was recognized by Exxon Chemical, who developed the process further and was the first company to demonstrate the process as a cost effective process to produce microfiber nonwoven webs of PP. The Exxon Chemical melt blowing technology (Figure 1) was

Polypropylene: An A–Z Reference
Edited by J. Karger-Kocsis
Published in 1999 by Kluwer Publishers, Dordrecht. ISBN 0 412 80200 7

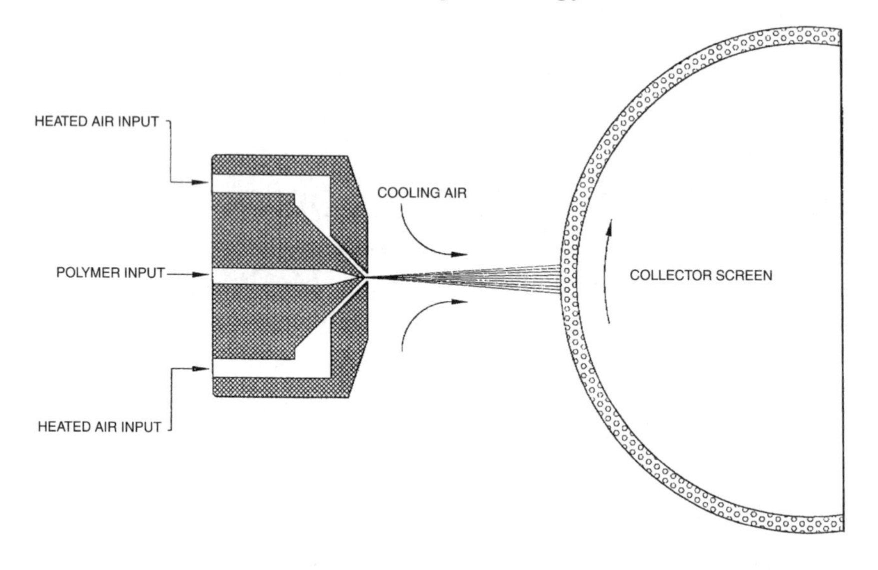

Figure 1 Exxon Chemical melt blowing technology.

licensed to selected equipment manufacturers, including Accurate Products Co. (Hillside, NJ, USA) followed by Reifenhäuser GmbH (Troisdorf, Germany).

J and M Laboratories, Inc. (Dawsonville, GA, USA) developed and patented their own polymer melt blowing equipment based on their expertise with hot melt equipment and became one of the main suppliers of innovative melt blowing equipment [2] for both polymer and adhesive melt blowing.

POLYMER MELT BLOWING EQUIPMENT

System description

Melt blowing technology involves the controlled melting of thermoplastic polymers and subsequent transfer and metered distribution to a multi-orifice nose-piece. Polymer exiting from these orifices form continuous or discontinuous fibers, which are blown onto a conveyor screen or substrate. They are selfentangling and form a self-bonded web, which can be transported to a winder.

Polymer or resin feed

The melt blowing process has the inherent ability to convert almost any material into microfibrous webs provided the material is fibrous formable and the viscosity is sufficiently low at the die tip to permit attenuation by high-velocity hot air. Accordingly, the type of feed equipment is dependent on the material to be fed. Specific additives (colorants, wetting agents, stabilizers, electrostatic agents, etc.) can also be readily incorporated into the feed to tailor the web properties to meet customer needs. Polypropylene (PP) is, by far, the most widely used polymer for melt blowing.

Fiber forming

Metering pumps, resin filter systems and means for intermittent operation are often included between the extruder and the die to control or pattern the fiber laydown and to ensure a consistent flow of clean polymer or resin melt to a distribution system to provide uniform flow to each hole in the die tip. The die tip is designed in such a way that the holes are in a straight line with high-velocity air impinging from each side. A typical die will have 0.38 mm diameter holes spaced at 1200 or 1600 holes per meter. The impinging hot, high-velocity air attenuates the filaments and forms the desired microfibers. Typical air conditions range from 200°C to 400°C at velocities of 0.5–0.8 the speed of sound.

Fiber quenching

Immediately below the die, a large amount of ambient air is drawn into the hot air stream containing the microfibers. This air cools the hot gas and solidifies the fibers. The distance between the die and the forming belt is relatively short and can readily be adjusted from a few centimeters to 50 cm in a properly designed melt blowing line. This is an important process parameter.

Secondary quenching

J and M Laboratories has designed equipment to obtain additional quenching by three different methods. These methods provide substantially improved productivity and modified web formation to obtain improved filtration.

Web formation

The filaments are deposited on a cicular drum equipped with a porous cover or on a horizontal moving forming belt. Vacuum is applied to the inside of the drum or under the belt to collect the primary hot air and the entrained cool air. The combination of fiber entanglement and fiber-to-fiber bonding generally produces enough web cohesion so that the web can be readily handled on a winder without bonding.

Bonding

When additional bonding is required, a number of different techniques are used. Calendering will reduce the thickness and will increase web density. Other bonding methods include surface or area bonding, point or emboss bonding, ultrasonic bonding etc. Recently oxidation of the outer surface of the filaments by hotter than normal air is claimed to provide improved bonding.

Slitting/winding

During winding, the edges are trimmed off to prepare uniform final rolls. The winder may be equipped with a number of slitting knifes to cut narrower rolls from a larger parent roll.

Special treatments/additives

During processing, multiple treatments can alter the nature of the final product. Additives can be directed into the hot air attenuating stream. Various surface modifications can be sprayed onto the web before winding to impart desired characteristics. Substrates can be placed under the melt-blown web to create composite structures. Electrostatic charging units can be integrated to provide improved filtration efficiency [3].

PROCESS VARIABLES

The melt blowing process responds to process variables more readily than most of the other nonwoven processes to produce a wide variety of end products. The actual process parameters that determine final web properties involve the selection of polymer or resin, conditions in the extruder or melt tank, geometry and condition at the die tip and fiber distribution and separation at laydown. Modern melt-blown lines are designed with highly automated machine and process control equipment, which allow for quick selection of particular parameters and easy machine adjustments.

MACHINE DESIGN

Machines are designed in modular systems, which allow an easy arrangement of the components depending on individual requirements. For the production of melt-blown rollgoods, the components are typically arranged on a mezzanine, which carries extruder, die, etc. In case of combinations with other processes, such as nonwoven production (Spunbond, Carded, etc) or paper production and film production, the components can be integrated into the other existing machine structures.

PRODUCTS AND MARKETS

Because of the early development, the North American market is the largest market in the world. Business activities are accelerating in Europe and areas in Asia. World growth rates are currently estimated at 6–8% per year [4]. Based on the melt-blown consumption in Europe, the following segments are important.

Sorbents and wipes

Due to the hydrophobic character of PP and the fine porous web structure, a meltblown web can absorb up to 15–20 times of oil compared to its own weight. The sorbents classify into oil spill sorbents, which are used in sea and land oil spills and in industrial sorbents. These are used for maintenance in industrial plants, walkways, etc. Additives are used to provide hydrophilic sorbents and wipes.

Filtration

In air and liquid filtration, the air segment is more important today. Government regulations and the capability to form composite fabrics with melt-blown layers (if required, with different fiber diameters) are driving the development of air filtration. Composite Media for Heating, Ventilation and Air Conditioning (HVAC) are used in filter classes EU 7 to EU 9 (DIN 24 185).

Hygiene

Melt-blown fibers have considerable use in disposable diapers for babies and adults, sanitary napkins, feminine hygiene, etc., since eliminating plastic backsheets, integrating wicking layers and standing leg cuffs requires the use of meltblown fabrics.

Apparel

Melt-blown microfibers are used for insulation in durable apparel for cold weather. Composite fabrics, e.g. spunbond/melt-blown/spunbond (S/M/S) or polyethylene-spunbond/polyethylene-melt-blown, combined with a variety of special PE films have been used in disposable industrial products.

Hot-melt adhesives

Specialty adhesives such as EVA-copolymers, polyurethanes, etc., are formed by the melt-blown technology into webs for the assembly of components in diapers, automotive, furniture, carpets, bags and other industrial products.

Medical

The microfiber structure of the melt-blown web provides excellent barrier and breathability properties, which are required in this market segment. It is dominated by composite fabrics (mainly S/M/S), because, for example, surgical gowns need to be strong, breathable and barrier resistent.

Miscellaneous

These include elastomeric fabrics, high temperature webs, such as those based on ethylene/chlorotrifluoroethylene copolymers (E-CTFE), water soluble webs based on polyvinylalcohol (PVOH) and others [5].

The melt blowing process by itself, or by integration with other nonwoven or film and paper processes, has tremendous flexibility to produce new and specialty products and to improve existing products, provided that new product conceptualizers and equipment manufacturers effectively interact to meet existing and new market needs.

REFERENCES

1. US Dept of Commerce Office of Technical Services, *Manufacture of Superfine Organic Fibers*.
2. Allen, M.A. and Fetcko, J.T. (1992) US Patent 5 145 689.
3. Tsai, P.T. and Wadsworth, L.C. (1995) US Patent 5 401 446.
4. Exxon Corporation (1994) *Worldwide Meltblown Markets*, November 1994.
5. Fagan, J.P. and Wadsworth, L.C. (1991) *Meltblowing Process and Characteristics of Fluoropolymers*, Textiles and Nonwoven Development Center, University of Tennessee, February 1991.

Keywords: melt blowing, web, microfibers, microfilters, hot melt equipment, thermoplastic polymers, PP, extruder, filaments, air quenching, fiber entanglement, substrates, composite structures, nonwoven, sorbents, filtration, hot-melt adhesives, melt-blown textiles, microfibers, melt tank, die.

'Melt fracture' or extrudate distortions

R.J. Koopmans

INTRODUCTION

Molten polymers, when forced or extruded through dies at high rates, will show extrudate distortions. Such phenomena have been described since the advent of thermoplastic polymeric materials. Over the years, a very large number of publications has been devoted to this subject in efforts to define the nature and origin of these extrudate distortions, commonly referred to as 'melt fracture' [1]. Many observations have been made regarding the varying degrees in which extrudate distortions can manifest themselves [2]. Visual observations range from smooth-glossy extrudates (no distortions), over extrudates showing degrees of surface irregularities defined as matt, loss of gloss, 'orange peel', 'shark skin', wavy, screw thread, to extrudates which are volumetrically distorted as a consequence of their oscillating emergence from the die (often referred to as 'spurt') or their very irregular, 'chaotic' shape. For most polymers, such a succession of extrudate distortions can be observed with increasing flow rates. Certainly, the common terminology, melt fracture, has a too narrow connotation to capture all observed phenomena as described above. It only suggests that the melt emerging from the die fractures. Therefore, the term 'extrudate distortion' seems more appropriate to refer to all these phenomena. The long-standing academic and industrial interest in this subject may be associated with several driving forces. One, which poses a continuing intellectual challenge, is related to the elusiveness of finding the origin of extrudate distortions in relation to the

Polypropylene: An A–Z Reference
Edited by J. Karger-Kocsis
Published in 1999 by Kluwer Publishers, Dordrecht. ISBN 0 412 80200 7

molecular composition of the polymer. The complexity of the issue is related to the many melt processing variables (e.g. temperature, die geometry and alloy) and polymer-type variables (e.g. molecular weight, molecular weight distribution and branching) which need to be considered. The effect of each variable on extrudate distortion is difficult to determine experimentally. Often contradictory findings are reported. Another driving force is the evolution to faster melt processing equipment and a shift to higher flow rates. Accordingly, the practical barrier to extrudate distortion, which is considered to limit industrial melt processing operations, needs to be shifted. Existing solutions to alleviate or reduce extrudate distortion such as the use of extra additives or changing die geometry all tend to have a direct (negative) impact on the economics of the industrial practice of melt processing.

THE NATURE OF EXTRUDATE DISTORTIONS

Most studies on extrudate distortions have focused on polyethylene (PE). Relatively little attention has been given to the study of these phenomena for polypropylene (PP). Furthermore, most research is done using capillary rheometers to keep the experimental set up simple.

In general, when visually examining capillary extrudates for distortions, it is possible to distinguish between:

- *Surface distortions*: the extrudate emerges straight down, out of the die and shows some perturbations located at the surface of the strand. Various degrees of intensity in the perturbations can be observed. The perturbations may range from regular periodic to highly irregular aperiodic with varying amplitude and often having the shape of a screw thread.
- *Volume distortions*: the extrudate emerges out of the die in a helicoidal fashion or at higher flow rates in a totally irregular, 'unorganized' fashion. The distortion affects the total volume of the extrudate. The volume defects can be accompanied by surface distortions although smooth surface helicoidal extrudates have been observed (e.g. low density PE (LDPE)). For many linear polymers (e.g. high density PE (HDPE), linear low density PE (LLDPE), polybutadiene (PB)), an oscillating volume distortion exists in which a strand emerges straight out of the die but in periodic bursts. It is possible to observe a regular succession of zones showing no (smooth), surface and volume distortions.

The observable extrudate distortions are a consequence of a melt flow instability which is initiated somewhere along the flow channel. Surface distortions are generally considered to originate due to melt flow instabilities near the die exit, while volume distortions are associated with vortex formation at the die entrance.

Usually, capillary rheology is performed in a constant flow rate mode and the steady-state pressure (or apparent-shear stress) is graphically represented versus each set flow rate (or apparent-shear rate) as shown in Figure 1(a). Such a flow curve, showing a discontinuity, suggests there may be two critical apparent-shear stress values. These critical apparent shear stress values σ_{min} and σ_{max} are generally associated with the onset of surface defects and volume defects respectively. For PP (as with, for example, LDPE and polystyrene (PS)), this type of curve will not show a discontinuity and typically looks as shown in Figure 1(b). Typically it is possible to identify a critical apparent shear stress value of 0.10–0.13 MPa related to the onset of screw thread-like, surface distortions quickly giving rise to helicoidal and eventually highly irregular volume distortions at the highest flow rates. The critical apparent shear stress values are found to be independent of temperature [3]. Figure 2 shows pictures of the typical PP extrudate defects at increasing flow rates.

To capture the onset of extrudate distortions which can be associated with melt flow instabilities in the die, several modelling approaches have been followed [4]. Two common hypotheses are forwarded and centre around the so-called constitutive and slip instability issues. The constitutive approach starts with the premise that, on the basis of some viscoelastic theory, the shear stress becomes a many-valued function of shear rate. As a consequence of this nonmonotone function, a melt flow instability and the associated distorted extrudate will develop. For many commercial polymers, the nonmonotone function could be considered as the sum total of many nonmonotone functions, each associated with a specific molecular weight fraction. The associated experimental apparent shear stress–apparent-shear rate curve could then become monotone, i.e. as in Figure 1(b), as is the case for PP. It should be noted that for viscoelastic materials, no direct linear relation exists between a constitutive shear stress–shear rate function and the experimental pressure (apparent shear stress)–flow rate (apparent shear rate) curve.

The slip approach defines the onset of surface defects as related to a slippage of the polymer at the die wall. Accordingly, the zero-wall velocity concept, as used in the constitutive approach, does not apply. Many attempts have been made to experimentally measure the velocity at the wall, however, with mixed success. Still, theories have been developed which support a shear-dependent slippage at a polymer/solid interface. Slippage may be understood as the buildup of a layer of polymer molecules bonded to the die surface, which undergo a coil-stretch transition under the shear stress. In light of this understanding, the microscopic slip layer development may be in line with the possible formation of a macroscopic surface layer as a consequence of a many

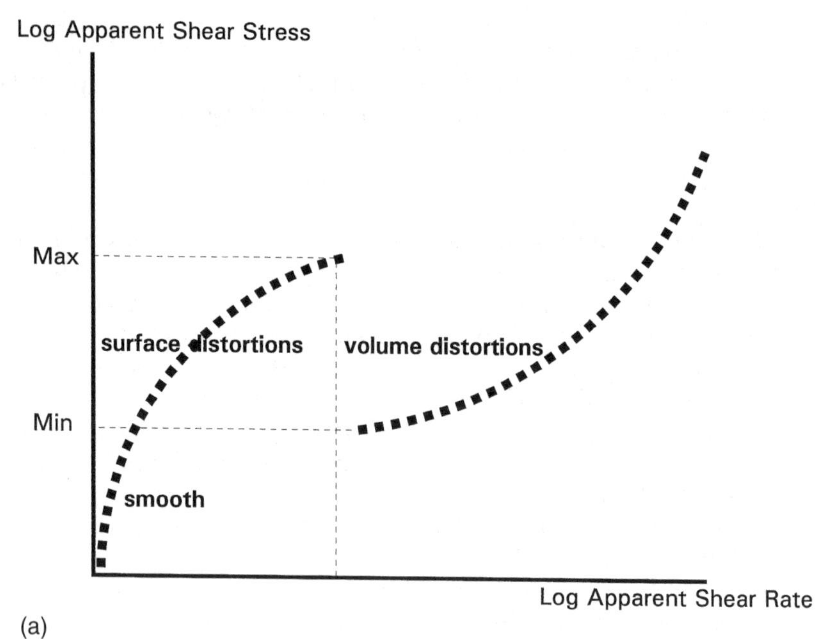
Log Apparent Shear Stress
Max
Min
surface distortions
volume distortions
smooth
Log Apparent Shear Rate

(a)

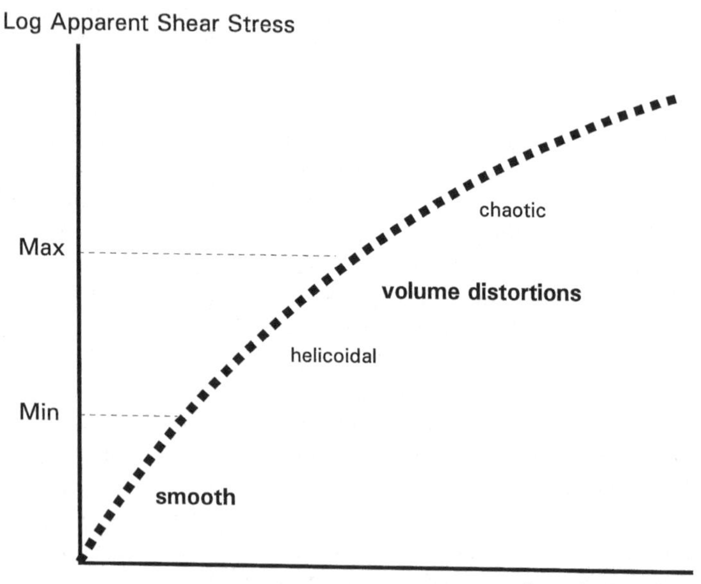
Log Apparent Shear Stress
Max
Min
chaotic
volume distortions
helicoidal
smooth
Log Apparent Shear Rate

(b)

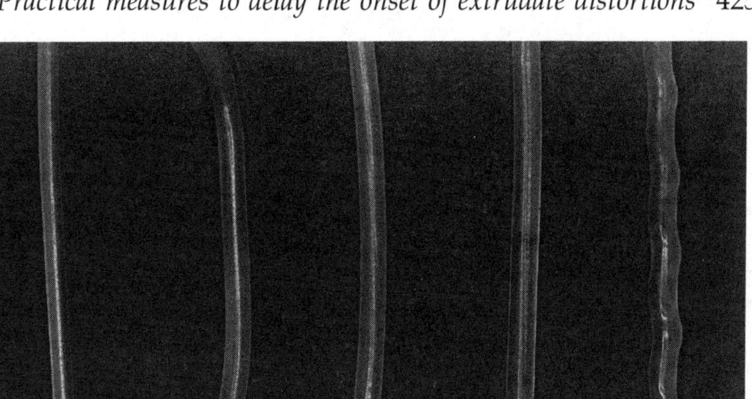

Figure 2 Different extrudate distortions for PP as function of flow rate. The experiment was performed on a Göttfert 2003 capillary rheometer using a die of dimensions $L{:}D = 20{:}1$ at 180°C. The extrudates are shown in order of increasing distortion and are obtained at apparent wall shear rates of 144, 360, 540, 720 and $12800\,s^{-1}$ (from left to right).

valued shear stress–shear rate function. Accordingly, the two theories could be considered as equally valid for describing melt flow instabilities and the associated extrudate distorions.

PRACTICAL MEASURES TO DELAY THE ONSET OF EXTRUDATE DISTORTIONS

From a pragmatic point of view, the variety of extrudate distortions has little relevance for the industrial practice. For most applications, especially those where free surfaces are formed (e.g. film, blow molding), the first onset of a visually observable extrudate distortion is sufficient to take corrective measures. The possible variables to tinker with are either

---

Figure 1 (opposite) Schematic representation of an experimental nonmonotone (a) and monotone (b) flow curve with indications of the sequence of characteristic extrudate distortions, typical for linear polymers, e.g. HDPE and PP, respectively. The apparent shear stress and shear rate are directly related to the measured pressure and set flow rate in a capillary experiment and presented on logarithmic axes.

related to the polymer composition or the melt processing conditions and/or geometries.

The PP molecular composition can be modified by adding additional components which act as lubricants. For PP, some lubricating effect may come already from the level of atactic PP, manufacturing process residuals, e.g. catalyst residues, and special additives, e.g. antioxidants or ultraviolet-stabilizers. In some cases, vis-breaking improves the lubrication action via the formation of relatively small molecules. The latter effect may also be induced by the thermomechanical action of the extrusion machines. As lower flow rates are generally not an option in industrial operations, increasing the melt processing temperature tends to reduce the appearance of extrudate defects. Such action may yield a positive effect by way of lower stress for the same flow rate and an increase of the thermomechanical degradation of PP. The critical challenge is to process a melt at the highest possible flow rate and the lowest stress. Increasing the die gaps and coating the die lips can be possible solutions. As some of the extrudate distortions can have a very local origin as reflected in, for example, die lines it is essential to keep the dies clean and undamaged.

It is interesting to note that extrudate distortions are only considered as a problem when visually observed. In contrast, extrudate distortions are induced on purpose in some applications. In those cases, all the above measures do not apply.

REFERENCES

1. Larson, R.G. (1992) Instabilities in viscoelastic flows. *Rheologica Acta*, **31**, 213–63.
2. Agassant, J-F., Avenas, P., Sergent, J-P., Vergnes, B. and Vincent, M. (1996) *La Mise en Forme des Matières Plastiques*, 3rd edn. Techniques and Documentation, Paris.
3. Kazatchkov, I.B., Hatzikiriakos, S.G. and Stewart, C.W. (1995) Extrudate distortion in the capillary/slit extrusion of molten polypropylene. *Polymer Engng Sci.*, **35**, 1864–71.
4. Dealy, J.M. and Wissbrun, K.F. (1990) *Melt Rheology and Its Role in Plastics Processing*. Van Nostrand Reinhold, New York.

Keywords: melt fracture, shark skin, lubrication instabilities, extrudate distortion, slip, processing.

Melt spinning of polypropylene

J.E. Spruiell and Eric Bond

INTRODUCTION

Melt spinning of polypropylene is done, almost exclusively, with the semicrystalline isotactic form. Melt spun isotactic polypropylene (iPP) filaments and fibers have many textile applications, including apparel, home funishings, wall coverings and carpets. One of the largest markets for iPP filaments and fibers is the nonwovens industry, in both staple fiber nonwovens and in 'spunbonded' and 'melt-blown' nonwoven materials. Staple fiber nonwovens and spunbonded nonwovens compete for such applications as diaper coverstock, feminine hygiene products, surgical gowns and numerous other applications. Melt blowing is a special form of melt spinning process which produces very fine but weak fibers which are useful for filters and other applications where high surface area is important and strength is a secondary factor. In this article, we will deal primarily with classical melt spinning of yarns, but the information will apply directly to the formation of filaments during spunbonding. In general, we will not cover here the secondary processes that generally follow the filament formation step, such as drawing or texturing of the as-spun yarns or bonding of the filaments into a spunbonded nonwoven web.

THE MELT SPINNING PROCESS

A basic design of a melt spinning process is illustrated in Figure 1. Polymer, in the form of pellets or granules, is fed into an extruder, where it is melted and pumped via a positive displacement pump to the melt spin pack. The spin pack consists of filters and channels that supply

Polypropylene: An A–Z Reference
Edited by J. Karger-Kocsis
Published in 1999 by Kluwer Publishers, Dordrecht. ISBN 0 412 80200 7

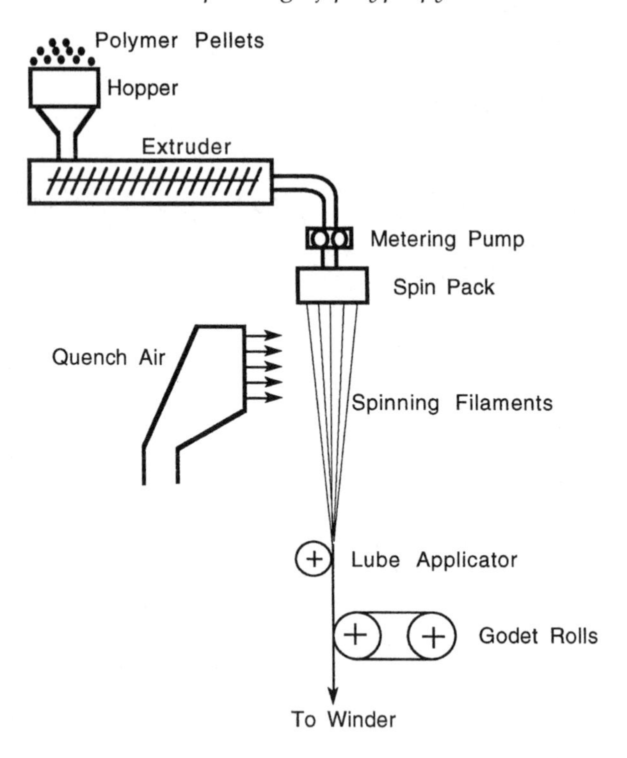

Figure 1 Schematic of the melt spinning process.

molten polymer to a multifilament spinneret. The molten filaments that exit the spinneret enter a cooling zone which cools and solidifies the filaments. The cooling zone can be as simple as a region in which quenching air is blown across the filaments, or it may be an elaborate chamber constructed so the cooling environment can be strictly controlled. In addition to being cooled and solidified, the molten filaments exiting the spinneret are simultaneously drawn down to a smaller cross-sectional area by the application of a drawdown force. This force is applied by action of a godet which controls the final linear velocity, or take-up velocity, of the spinning filaments. This force can also be supplied by air drag in special cases such as spunbonding. The take-up process is largely dependent upon the end use of the polymer filaments. In the case of yarn production, the multifilament yarn is usually wound onto a bobbin or sent directly to a drawing and/or texturing process.

The key variables of the melt spinning process are (1) take-up velocity, (2) length of the spinline, (3) cooling conditions along the spinline, (4) extrusion temperature, (5) mass throughput per spinneret orifice, and (6) the size and shape of the spinneret holes. These process variables interact with the resin characteristics to control the processability and the structure and properties of the as-spun filaments. The important resin variables are those that affect the rheological and crystallization behavior of the polymer. For iPP, these include (1) molecular weight and its distribution, (2) isotacticity, (3) comonomer type (ethylene is common) and content, (4) content, if any, of a nucleating agent, and (5) presence and amounts of antioxidants and stabilizers. The latter are necessary to control thermooxidative degradation during the spinning process. In this article, we will consider the influence of both process and resin variables on the spinnability and the development of the structure and properties of the melt spun filaments.

STRESS IN THE SPINLINE

Because of the many process and materials variables, melt spinning is a complicated process requiring, in general, detailed mathematical simulation for its understanding. One important concept that has emerged from our attempts to understand the melt spinning process is that stress developed in the spinline largely controls both the spinnability and the resulting morphology of the spun filaments [1, 2]. Higher stresses can lead to instabilities that cause the spinline to break. They also generally lead to greater deformation rates which produce higher molecular orientation. The higher molecular orientation leads to faster crystallization kinetics (stress-enhanced crystallization) which sometimes leads to higher crystallinity.

The spinline stress, σ_{zz} is related to the viscoelastic properties of the polymer and to the processing conditions. As a first approximation we can write:

$$\sigma_{zz} = \eta(T,\dot{E})\,\frac{dV}{dz} \tag{1}$$

where dV/dz is the velocity gradient along the spinline (V is velocity of the running spinline at any given distance z from the spinneret), $\eta(T,\dot{E})$ is a temperature and deformation rate dependent elongational viscosity of the resin. According to equation (1), the spinline stress is expected to increase, for given process conditions, as $\eta(T,\dot{E})$ increases. This suggests that resin characteristics that affect the viscoelastic properties of the polymer such as molecular weight and polydispersity will have important effects on the spinning process.

Neglecting radial variations in the filaments, the spinline stress at a given distance z from the spinneret is also given by:

$$\sigma_{zz} = \frac{F_{rheo}}{\pi D^2/4} \tag{2}$$

where D is the equivalent filament diameter and F_{rheo} is the rheological force acting in the filament. F_{rheo} is given by an overall force balance on a single filament in the spinline [3]:

$$F_{rheo} = F_o + F_{inert} + F_{drag} - F_{grav} + F_{surf} \tag{3}$$

where F_o is the rheological force at the exit of the spinneret, F_{inert} is the inertial force produced by the acceleration of the polymer mass along the spinline, F_{drag} is the drag force caused by the fiber moving through the cooling medium, F_{grav} is the gravitational force acting on the spinline and F_{surf} is the surface tension force at the fiber/cooling medium interface. Under normal spinning conditions, F_{inert} and F_{drag} are the major components of the rheological force. Clearly, equation 3 indicates that for a given polymer, the spinline stress increases rapidly as filament take-up velocity increases due to an increase in both the inertial and air drag terms.

INFLUENCE OF PROCESS VARIABLES

The processing variable effects are considered from the point of view that changes in them are made with all other conditions held constant.

Take-up velocity

The take-up velocity is generally the most important process variable. As shown in Figure 2, the speed at which the molten resin is drawn down significantly affects the resulting crystallinity, as measured by density, and overall molecular orientation, as measured by birefringence. It should be noted that the results shown in Figure 2 are for a constant mass throughput, so that the final filament diameters are decreasing with increasing take-up velocity and the cooling rate along the running spinline is increasing. As the spinning speed increases, the spinline stress increases and both crystallinity and molecular orientation increase in spite of the increase in cooling rate which should suppress the development of crystallinity. This is strong evidence that the stress and molecular orientation lead to stress-enhanced crystallization. Online measurements of temperature, birefringence and X-ray diffraction patterns [1, 5] show that crystallization begins at higher temperatures in the spinline than under quiescent conditions with similar cooling rate. This is further evidence of stress-enhanced crystallization.

The crystal structure of melt spun iPP is most often the α-monoclinic form [2]. However, if the fiber is quenched rapidly or the stress in the

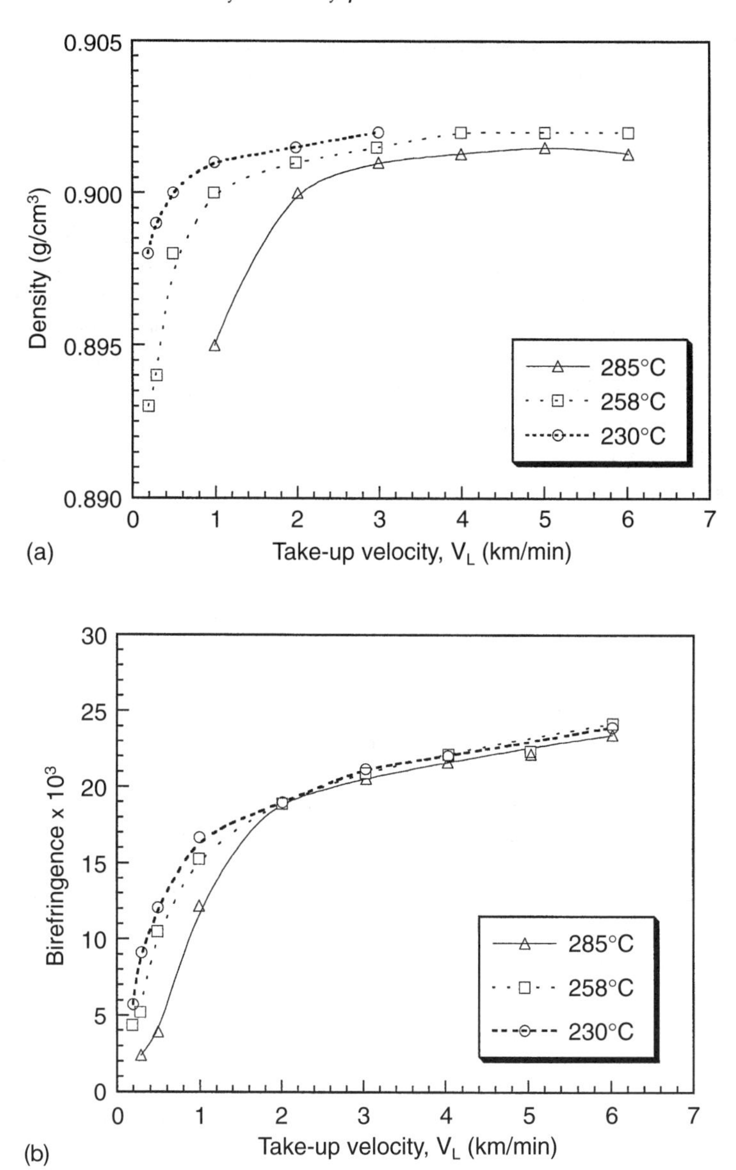

Figure 2 Influence of spinning speed and extrusion temperature on (a) the density, (b) the birefringence of melt spun iPP filaments (data of Shimizu *et al.* [4]).

spinline is low, as for the lower spinning speeds of Figure 2, a conformationally disordered (condis) or smectic structure is formed. The smectic structure is not thermally stable and transforms to the monoclinic form when annealed above 70°C.

Extrusion temperature

Figure 2 also shows the effect of changing the melt extrusion temperature of the polymer. As the extrusion temperature is increased, the elongational viscosity in the upper part of the spinline decreases, thereby lowering the stress in the spinline and reducing the molecular orientation and crystallinity developed at a given spinning speed. However, as the spinline stress is increased (by increasing the spinning speed), the density and birefringence tend to become independent of the extrusion temperature, indicating that the increased stress due to increased spinning speed eventually overrides the decreased stress due to higher extrusion temperature.

Mass throughput

Increasing mass throughput per spinneret hole, while keeping take-up speed constant, produces larger filament diameter (continuity condition) and lowers the stress in the spinline. This in turn leads to lower filament birefringence. Increasing mass throughput also reduces the cooling rate in the spinline due to the greater amount of material that must be cooled. One might expect that the slower cooling rate would increase crystallinity and density due to more available time for crystallization. However, the lower stress decreases molecular orientation and causes less stress-enhancement of the overall crystallization rate. The final crystallinity results from a balance between these two factors which also depends on the nature of the cooling process and the characteristics of the specific resin.

Cooling conditions

As with the take-up velocity, the cooling conditions have a significant influence on the crystallization behavior during the spinning process. As the cooling air temperature decreases or volume of air blown across the filament increases, the cooling rate of the spinning filaments increases. This moves the solidification point in the filament closer to the spinneret. As the freezing point moves closer to the spinneret, the deformation rate increases since the filament must reach its final diameter closer to the spinneret; this produces an increase in the spinline stress. This increase in stress closer to the spinneret results in an increase in molecular orientation which is more readily frozen-in due to the higher cooling rate. There is again a balance between the retardation of crystallinity development due to faster cooling and the enhancement of crystallinity development due to increased stress.

Length of spinline

The length of the spinline is important for two reasons. First, the spinline must be long enough to allow the polymer to freeze before being contacted by guides or godet rolls. Because of this, the spinline length is controlled by the efficiency of spinline cooling; longer spinlines are required for slower cooling. However, longer spinlines also increase the gravitational contribution to F_{rheo} which increases the spinline stress. In most cases, especially for fine filaments, spinline length will have relatively little effect on the structure and properties of the spun filaments due to the low value of F_{grav} relative to other contributions to F_{rheo}.

Filament shape

The filament shape and die orifice design affect the heat transfer rate between the filament and surroundings. The larger the surface area for a given volume of material, the faster the polymer will transfer heat away and crystallize. This case is similar to the effect of increasing the cooling rate by other means described above.

MATERIAL VARIABLES

Molecular weight

Both molecular weight and its distribution (MWD) strongly affect the spinnability of iPP. Generally, spinnability, as measured by the maximum take-up speed at which spinning can be carried out in the absence of instabilities, increases as molecular weight distribution is narrowed. There tends to be a maximum in the spinnability with increase in molecular weight. At very low molecular weight, the spinline exhibits instabilities due to low melt strength. For high molecular weight, various other instabilities, including melt fracture and spinline breaking, arise due to the high stresses in the spinneret capillaries and in the spinline.

Within the range of molecular weight and MWD where spinnability is relatively good, these characteristics also affect the resulting crystallinity and molecular orientation in the spun filaments [2] as illustrated in Figure 3. This figure shows density and birefringence for three iPP samples having molecular weights and polydispersities shown in Table 1. Samples E-035 and E-300 are narrow distribution iPPs with substantially different molecular weights. Sample E-012 has both higher weight average molecular weight, M_w, and much broader MWD.

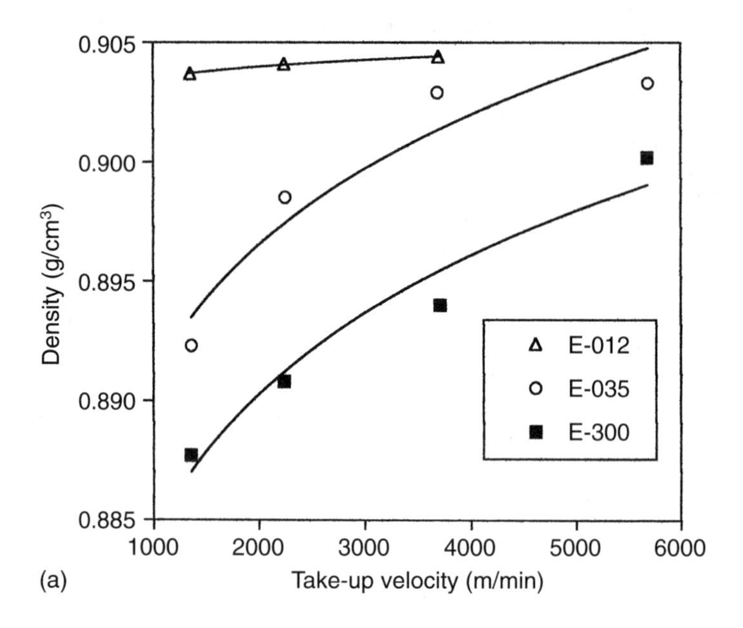

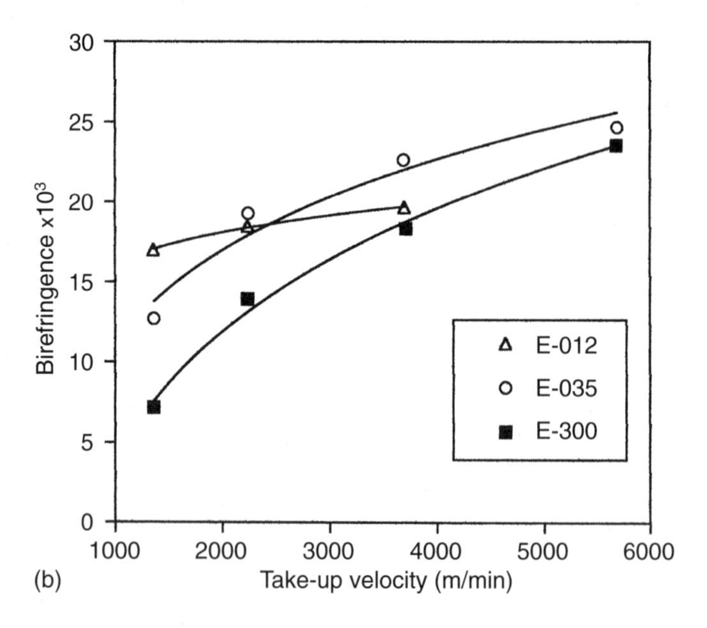

Figure 3 Influence of resin molecular weight and MWD on (a) the density, (b) the birefringence of as-spun iPP filaments (data from [2]). See Table 1 for characterization of the three resins.

Table 1 Characteristics of iPP samples used for data of Figure 3

Code name	*Melt flow rate (MFR, g/10 min)*	*Intrinsic viscosity (dl/g)*	M_w *(by GPC)*	*Polydispersity* (M_w/M_n)
E-012	12	1.51	238 000	4.76
E-035	35	1.21	170 400	2.16
E-300	300	0.93	124 400	2.85

Figure 3(a) shows that the density (and hence crystallinity) increases as M_w increases at a given take-up velocity. This can be understood on the basis that increasing M_w produces higher spinline stress and molecular orientation in the melt which, in turn, causes crystallization rates to increase and crystallization to occur closer to the spinneret and at higher temperatures. There is also an effect of MWD; broader distribution resins with the same MFR develop higher crystallinity at equivalent spinning speeds. This appears to be caused by broader MWD resins being more susceptible to stress-enhanced crystallization as a result of the high molecular weight tails in the distribution providing a source of 'row nuclei' which seed the stress-induced crystallization [6]. Note the high density of the broad distribution sample, E-012, at spinning speeds as low as 1000 m/min.

The higher molecular orientation in the melt state produced by higher molecular weight usually results in higher birefringence in the spun filaments, Figure 3(b). Note, however, that the broad distribution sample, E-012, does not develop as high a birefringence as the narrow distribution E-035 sample at spinning speeds above about 2000 m/min in spite of its higher M_w. The lower orientation in the broader distribution sample has been attributed to a bimodal orientation that occurs preferentially in the broader distribution resins [6]. This bimodal orientation has one component in which the polymer chains are parallel to the fiber axis and a second component in which the chains are approximately perpendicular to the fiber axis. This latter component lowers the overall molecular orientation with respect to the fiber axis and hence the fiber birefringence. Another possible contribution to the difference between the broad and narrow distribution samples is a difference in the noncrystalline orientation developed by a given spinning condition.

Percent isotacticity

As percent isotacticity decreases, both the crystallization rate and ultimate crystallinity of iPP generally decrease. During melt spinning this causes the crystallization to occur farther from the spinneret and at lower

Table 2 Characteristics of iPP samples used for the data of Figure 4

Sample Code	*MFR*	*% Ethylene (by FTIR)*	*Isotacticity (%)*	M_w	*Polydispersity* (M_w/M_n)
HT-PP	36	0	99	131 000	2.22
MT-PP	34.2	0	93	143 800	2.38
LT-PP	37	0	91	134 800	2.36
1.5% RCP	35	1.4	96	158 600	3.22
3.0% RCP	35.4	3.0	94	142 200	2.31
5.0% RCP	31	4.9	92	150 100	2.26
1.0% NA-PP	35	0			
3.0% RCP +1% NA	36	2.6	94	144 400	2.13

crystallization temperature [7]. The resulting as-spun filaments have lower density, but their overall molecular orientation and birefringence are mostly unaffected. This behavior is illustrated in Figure 4(a) and (b). Characteristics of the various samples are given in Table 2. All are narrow distribution resins with similar M_w. The first three resins are high, medium and low isotacticity homopolymers.

Propylene/ethylene copolymers

The addition of ethylene units to form a random copolymer (RCP) lowers the crystallinity that can be developed, by virtue of the lower stereoregularity of the chain. This is analogous to the effect of stereo defects when decreasing the isotacticity of the chain. The influence of ethylene copolymer content is also illustrated in Figure 4.

Nucleating agent additions

The addition of nucleating agents raises the crystallization temperature and the resulting crystallinity substantially for quiescently crystallized iPPs. These effects also occur during melt spinning of iPP fibers [7], but they are more pronounced under conditions of low spinline stress (i.e. low spinning speeds). At high spinning speeds and/or other conditions that produce high stresses in the spinline, the differences due to nucleating agents are smaller, but not eliminated, arising from the influence of stress in raising the rate of crystallization. The effects of a 1% nucleating agent addition on the behavior of an iPP homopolymer and a 3.0% RCP are also illustrated in Figure 4. Nucleating agents tend to lower the molecular orientation in the filaments. This is a result of crystallization

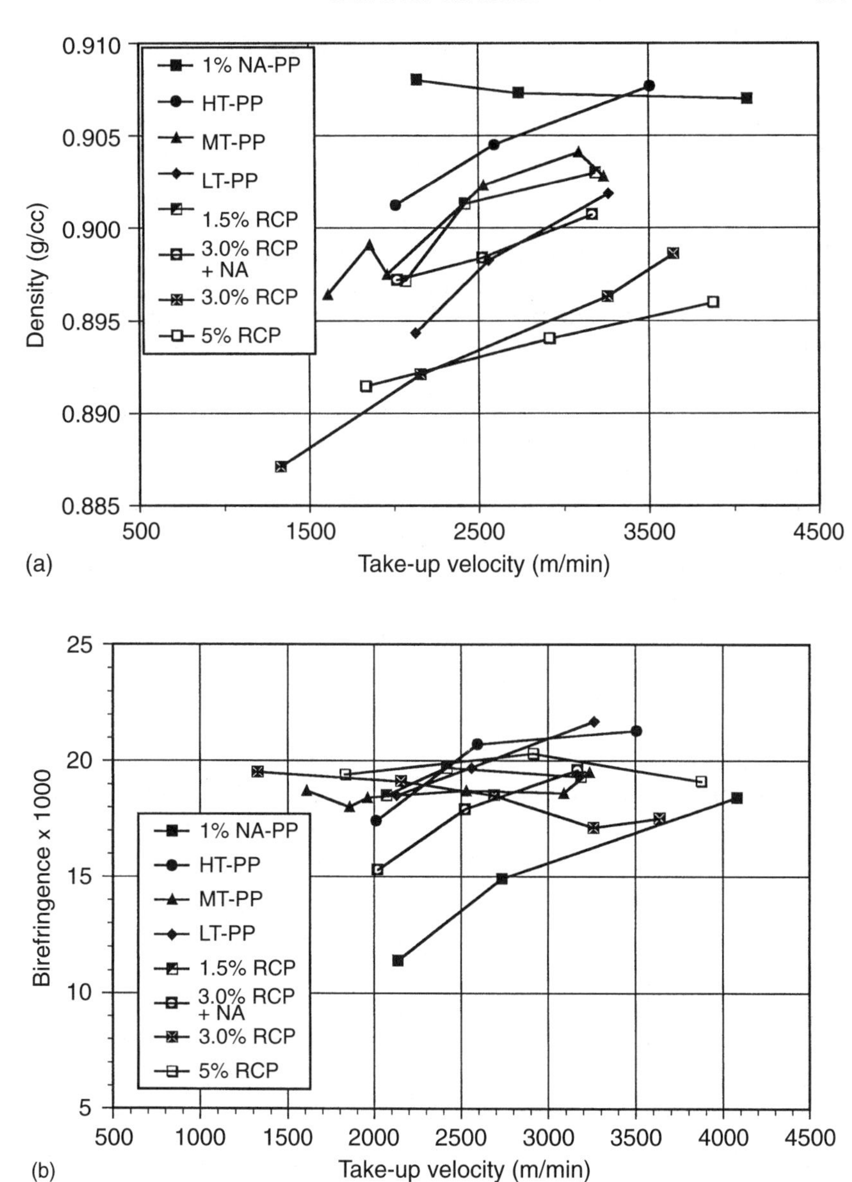

Figure 4 Influence of several other resin variables on (a) the density, (b) the birefringence of as-spun iPP filaments with similar molecular weight and polydispersity (data from [6]). See Table 2 for characterization of the various resins.

ocurring at higher temperatures and nearer the spinneret, where the molecular orientation in the melt is lower.

TENSILE PROPERTIES OF MELT SPUN iPP FILAMENTS

As noted above, increasing the stress in the spinline tends to increase the overall molecular orientation resulting from the spinning process. Recall that overall molecular orientation (commonly measured by birefringence) is composed of both crystalline and amorphous orientation. Research has shown that the amorphous region most strongly contributes to the tensile strength of iPP fibers. This arises from the fact that tie molecules connect the various crystalline regions, thereby allowing the transfer of stress and the load from one crystallite to another. The crystalline regions are important in providing a medium in which to anchor the various molecules that make up and run throughout the fiber, but the crystallinity alone is a poor indicator of mechanical properties. Under normal spinning conditions, a strong correlation of tensile strength with birefringence is observed as shown in Figure 5. This figure contains data from a wide range of iPPs and spinning conditions, including the samples of Table 2 and samples containing differing molecular weights and polydispersities. Figure 5 illustrates the fact that factors that increase the overall molecular orientation in the filaments also increase the tensile strength. The elongation-to-break obeys a somewhat similar, but inverse dependence on the molecular orientation in the spun filament; it decreases from several hundred percent (~ 1000%) at low birefringences to less than 200% at a birefingence of about 0.025.

Although the initial elastic modulus (Young's modulus) increases as molecular orientation increases, it is also a function of crystallinity. Since

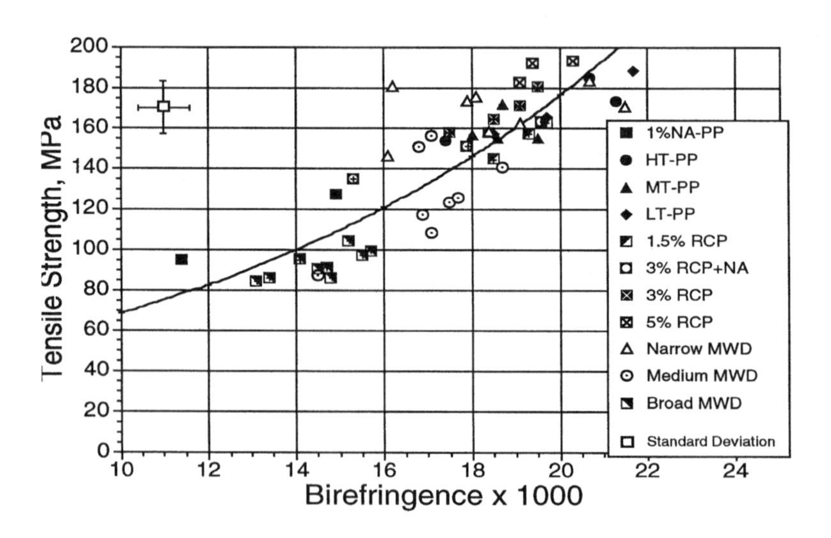

Figure 5 Correlation of as-spun iPP filament tensile strength with birefringence for a number of differing resins (data from [7]).

the higher crystalline modulus contributes roughly according to the rule of mixtures, the overall modulus of the filament increases as the crystallinity increases. For this reason, the modulus does not correlate well with birefringence alone, as do the tensile strength and elongation-to-break. In the case of modulus, both the level of molecular orientation and degree of crystallinity must be considered. For this reason, the modulus of ethylene copolymer melt spun filaments is substantially lower than for the homopolymers when prepared under similar spinning conditions, even though their tensile strengths are equal or slightly higher. On the other hand, nucleating agent additions may lead to lower modulus because of lower molecular orientation, in spite of a higher degree of crystallinity.

REFERENCES

1. Nadella, H.P., Henson, H.M., Spruiell, J.E. and White, J.L. (1977) *J. Appl. Polymer Sci.*, **21**, 3003.
2. Lu, F.-M. and Spruiell, J.E. (1987) *J. Appl. Polymer Sci.*, **34**, 1521.
3. Ziabicki, A. (1976) *Fundamentals of Fiber Formation*, Wiley, London.
4. Shimizu, J., Okui, N. and Imai, Y. (1979) *Sen-i Gakkaishi*, **35**, T-405.
5. Lu, F.-M. and Spruiell, J. E. (1987) *J. Appl. Polymer Sci.*, **34**, 1541.
6. Misra, S., Lu, F.-M., Spruiell, J.E. and Richeson, G.C. (1995) *J. Appl. Polymer Sci.*, **56**, 1761.
7. Spruiell, J.E., Lu, F.-M., Ding, Z. and Richeson, G.C. (1996) *J. Appl. Polymer Sci.*, **62**, 1965.

Keywords: crystal structure, process variables, molecular weight, molecular weight distribution (MWD), polydispersity, copolymers, nucleating agents, melt spinning, spinline, effects of molecular variables, effects of processing variables, tensile properties of filaments, birefringence, take-up velocity, conformationally disordered structure, amorphous orientation, crystalline orientation.

Melt spinning: technology

K. Schäfer

Polypropylene (PP) is becoming increasingly prominent within the family of polymer materials for fiber spinning. The results of the polymer and process developments over the past years have placed PP in an extremely competitive position compared to other man-made fibers. The disproportionately high rate of growth of PP fibers is not only due to economic reasons, but fiber properties, such as easy processability, excellent melt dyeability, good insulating rating, no moisture absorption and excellent wicking effect, are very desirable [1]. If we are looking to the market for PP fiber products, four main different product groups can be identified: (1) multifilaments; (2) staple fibers; (3) film tapes and fibrillated yarns; (4) spunbonded fabrics [2]. This overview will deal with the melt spinning technology and key components for the production of pre-oriented yarns (POY), fully drawn yarns (FDY) and bulked continuous filaments (BCF) (Figure 1). The development of the processes and their key components is generally aimed at reducing investment costs (through higher production speeds and/or more yarn ends per spinning position) as well as reducing energy, maintenance and labor costs. By optimizing the equipment and its operation, it is possible to increase the spinning performance and reduce the amount of waste, thereby increasing process efficiency. This led to the possibility of adding more and more yarn ends to each spinning position for different POY, FDY and BCF processes. Continuous development of the key components of spinning plants, such as extruder, spinning beams, packs, godets and winding equipment, allowed us to configure machines for the most

Polypropylene: An A–Z Reference
Edited by J. Karger-Kocsis
Published in 1999 by Kluwer Publishers, Dordrecht. ISBN 0 412 80200 7

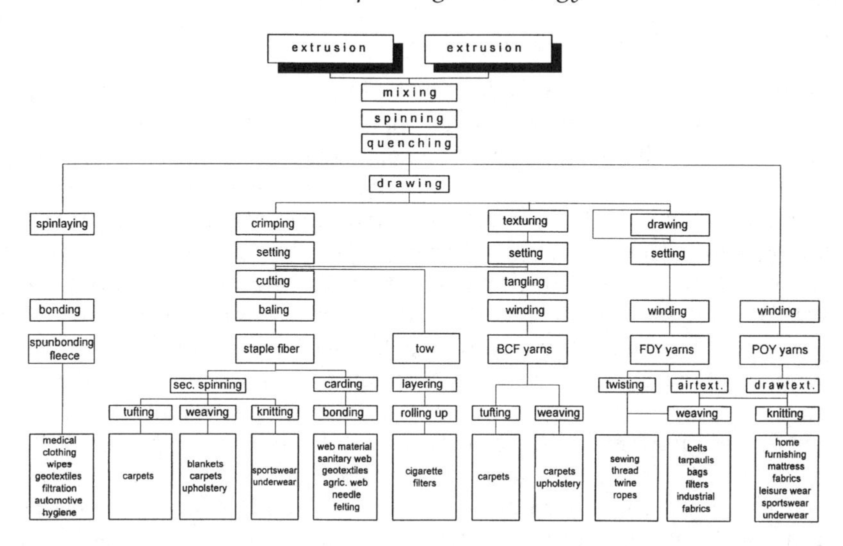

Figure 1 Process steps, downstream processes and main applications for PP fiber and filament production.

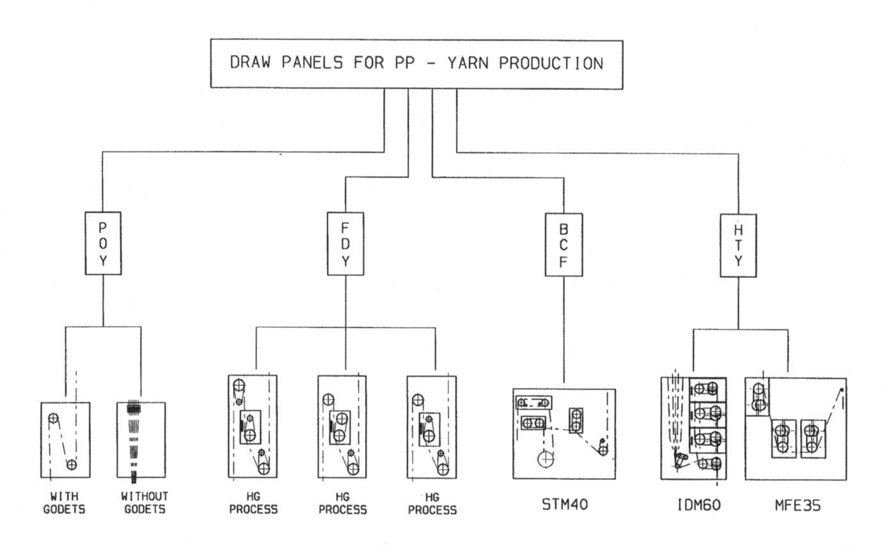

Figure 2 Draw panels for PP yarn production.

diverse processes and filament properties (Figure 2). Next, the main aspects of the developments are briefly described.

SINGLE SCREW EXTRUSION

Extrusion technology is aimed at achieving a high output capacity in connection with an optimization in melt quality, respectively product quality. Melt quality means a homogeneous melt stream at any time and any place. Homogeneity in time can be described through constant throughput, constant temperature, constant concentration of additives. Homogeneity in place means uniform temperature and uniform dispersion of additives over the melt cross-section. The most important component to fulfill these requirements is the screw design. Short residence times are the targets when designing the screw, in order to keep the thermal and mechanical loading of the melt to a minimum. Conventional and/or barrier screws can be adapted to customer requirements by varying the number of zones, their length, numbers of flights, flight depths, flight pitches and arrangement and geometry of mixing and shearing components. Separation of melt and solid parts, by means of additional barrier flights, proves to be advantageous in most cases. With the proper design, the barrier screw provides a higher melt performance compared with conventional screws, a more stable pressure/speed ratio and a bigger melt throughput range. In the production of fibers and filaments, additives are frequently used for ultraviolet stabilization and/or for melt dyeing. The demand for a high degree of color uniformity, and/or dispersion of a second polymer component in the PP matrix, makes the finest dispersion of the additives necessary. This condition can be achieved by using a special additive feeding/mixing system. The color concentrate and/or other additives are melted and homogenized in a side extruder and subsequently passed through a metering pump to a dynamic mixer. This mixer is a part of the screw of the main extruder. The advantages of such a mixing system are: (1) injection of the additives, even the smallest quantity, is exact and reproducible; (2) excellent mixing quality is achieved even with wide differences in the melt viscosity between polymer and additives; and (3) quick color changes to minimize the production waste.

SPINNING SYSTEMS

The basic requirement of a modern spinning system is a melt guidance system which eliminates differences in residence time and especially dead spots, guarantees uniform heating (to avoid temperature differences over the spinneret) and uniform quenching of the individual filaments. State-of-the-art technology consists of using bottom loading round self-sealing spin packs the spinning beam, in order to prevent undesired cooling as a result of the chimney effect [4]. The spin pack body is threaded for screw connection of the spin pack to the pump block. The spin beam is Dowtherm heated and provides an optimal energy transfer for spinnerets and melt pipes. The melt is metered from the gear pump to the spin pack via melt pipes which allow an optimal heat transfer, a low pressure drop, equal residence time due to identical tube length, and no dead spots and optimized bending radii. All of this leads to an extremely high temperature uniformity less than $\pm 1°$ at the spinneret. After metering and sizing, the filament will be cooled in a very uniform cross-flow quenching system.

GODETS

Heated godets are used in spinning machines for the purpose of imparting specific filament properties. To a considerable extent, these properties are determined by the temperature and speed accuracy of the godet. The temperature profile of the godet surface should be within $\pm 1°$ so that the multiple threadlines running over one godet receive identical heat treatment. Speed fluctuations of greater than 0.01% may lead to an unacceptable fiber quality in some processes.

WINDING HEADS

The most important step is the production of a package. The package shape and formation are important for good and stable transportation, storage and an excellent take-off performance at high speeds in subsequent processes. For the production of such packages automatic winders for speeds up to 8000 m/min and up to ten packages per chuck will be used. The yarn traversing system for higher speeds is the Birotor system, which works without reciprocating mass with excellent transfer reliability and offers gentle treatment of the yarn. To avoid yarn movement during transportation, the bobbin has to be ribbon free. To achieve this, ribbon-breaking systems are integrated, e.g. the Barmag Helicont system.

PROCESSES

The different processes of filament manufacture can be built on the basis of these key components. Pre-oriented yarns (POY) are normally spun or wound within a speed range from 2500 to 4000 m/min. The typical range is from 40 to 200 dtex (1 dtex equal 1 g/ 10000 m) for textile applications. Individual filaments can range from 0.5 to 4 dpf (dtex per filament). Sound research and development have bought an increase in speeds together with an increase in ends per position so that today, ten ends can be processed on one winder. In order to achieve an optimum yarn quality, a parallel arrangement of spin beam and winder chuck gives a yarn path which is virtually free of yarn friction due to minimized yarn deflection. This is a key factor in achieving yarn ends of uniform top quality with improved Uster values. Processes for fully drawn yarn (FDY) and other flat yarn manufacturing processes mainly utilize hot godets either as a two-step process (spin process separated from the draw process) or one-step (spin process integrated with draw process) process. The most modern FDY manufacturing processes are one-step. In this case, yarn is passed over heated godets with their associated idler rolls and/or second godet and gets drawn between the sets. Depending on the required yarn properties, one to three drawsteps are used. After drawing, the yarn has to be crystallized on a heated godet for the reduction of shrinkage. Only with low shrinkage values can a proper package formation be achieved. The crystallization of PP should occur with a controlled low yarn tension; this may require additional heated godets. Depending on the required shrinkage value, one or two relaxation zones are used. Winding speeds up to 5000 m/min with draw ratios of 3:1 up to 8:1 are normal. For the dtex range from 50 to 300 dtex with 0.5 to 4 dpf for textile applications and 500–2000 dtex for industrial applications (mainly high tenacity up to 8 g/dtex) flat yarn warp draw processes are also used. In this case up to 120 yarn ends are drawn simultaneously in a draw unit with speeds up to 300 m/min. The largest usage for PP filament is in bulked continuous filament (BCF) yarns. These yarns are used mainly for carpet and upholstery fabrics. The BCF yarn range is from 150 to 3600 dtex with 1–40 dpf. BCF yarns, today, are mainly produced in a one step-process from granules to finished package [3]. All required production steps are combined into a single process. After spinning, the filaments are drawn in a single step between duo and heated take-off godet. After wrapping around the duo, the threadlines pass through the texturing chamber. The heart of any BCF machine is its texturing jet and the three-dimensional crimp that it imparts on the yarn filaments.

In this jet, the yarn is pulled in the first stage, introduced to a high-temperature medium, compacted in a stuffer box and then cooled on a vented wheel. The magnitude of this crimp is determined by the following characteristics: yarn temperature at the first stage, interval between the injector jet and the plug formation, plug density, temperature and pressure of the medium, size and geometry of the stuffer box, total bundle denier, denier per filament, and production speed. As a result of the increase in production speeds to the currently higher operating speeds, and also due to the higher operating speeds of tufting machines and weaving looms, the number and uniformity of the tangle knots have become increasingly important factors. For this reason, tangling takes place between two separately driven godets for controlling and maintaining the optimum tension for tangling. Following the draw texturing process, the yarn is taken up without waste on a fully automatic winder.

THE FUTURE

PP will grow in importance in relation to polyester and PAs especially in the field of nonwoven (staple fiber and spunbonded) and carpet yarns. This is due to the fact that PP is a readily available and inexpensive raw material. Growth in most of the multifilament applications, such as industrial, furnishings and apparel will suffer due to PP's poor fiber dyeability. However PP will expand and penetrate into numerous special market niches.

REFERENCES

1. Moore, E.P., Jr. (ed) (1996) *Polypropylene Handbook*, Hanser, Munich.
2. Lines for an efficient PP-Carpet, -Fibers, -Filaments and Tape Production, Proceedings of the Conference 5th Annual World Congress on Polypropylene '96, Zurich, Switzerland.
3. Schäfer, K. and Mayer, M. Modern production methods for BCF yarns, 32th International Fiber Conference, Dornbirn, Austria.
4. Fourné, F. (ed) (1995) *Synthetic Fibers*, Hanser, Munich.

Keywords: melt spinning, extrusion, filaments, pre-oriented yarn (POY), fully-drawn yarn (FDY), bulked continuous filament (BCF), filament winding, drawing, shrinkage, spinning head, godet, texturing, crimping, linear density (titre).

Metallocene catalyzed polymerization: industrial technology

Archie E. Hamielec and João B.P. Soares

INTRODUCTION

Metallocene catalysts are organometallic coordination compounds in which one or two cyclopentadienyl rings or substituted cyclopentadienyl rings are π-bonded to a central transition metal atom (Figure 1). The most remarkable feature of these catalysts is that their molecular structure can be designed to create active centre types to produce polymers with entirely novel properties.

Currently, there are six review articles published in the literature on metallocene catalysts, covering different aspects of metallocene catalyst synthesis, nature of active sites, polymerization mechanisms, metallocene catalyst patents, and polymerization reaction engineering [1–6]. The present review will highlight some of the particular features of metallocene catalysts that make them attractive for the industrial production of polypropylene (PP).

ZIEGLER–NATTA VERSUS METALLOCENE CATALYSTS

Most polyolefin manufacturing processes today utilize conventional heterogeneous Ziegler–Natta catalysts. Several types of these Ziegler–Natta catalysts are stereospecific, i.e. the insertion of asymmetric monomers into the growing polymer chain in a given orientation is favored over all other possible orientations, leading to the production of isotactic

Polypropylene: An A–Z Reference
Edited by J. Karger-Kocsis
Published in 1999 by Kluwer Publishers, Dordrecht. ISBN 0 412 80200 7

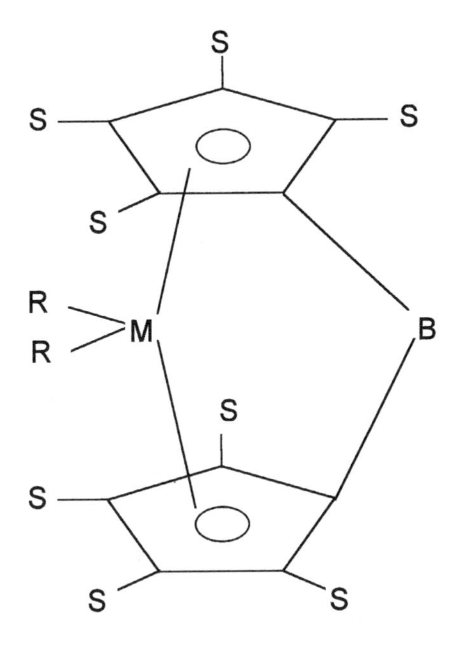

Figure 1 Generic structure of a metallocene catalyst.

and syndiotactic PP. Because these catalysts have more than one type of active site, they produce PP with broad molecular weight distribution (MWD) and nonuniform stereoregularity. In contrast, metallocenes can be synthesized as single site-type catalysts to produce polymers with narrow MWD (with the theoretically predicted polydispersity index of two) and uniform stereoregularity.

More importantly, while it is very difficult to control the nature of the site types on conventional heterogeneous Ziegler–Natta catalysts, metallocene catalysts can be designed to synthesize PP with different chain microstructures. PP chains with atactic, isotactic, isotactic-stereoblock, atactic-stereoblock and hemiisotactic configurations can be produced with metallocene catalysts (Figure 2). It is also possible to synthesize PP chains that have optical activity by using only one of the enantiomeric forms of the catalyst. Additionally, copolymers of propylene, ethylene and higher α-olefins made with metallocene catalysts have random (or near random) comonomer incorporation and narrow chemical composition distributions (CCD).

METALLOCENE CATALYTIC SYSTEMS

Metallocene catalyst systems can be conveniently divided into two categories. In the first category, an aluminoxane or an alkylaluminum, or

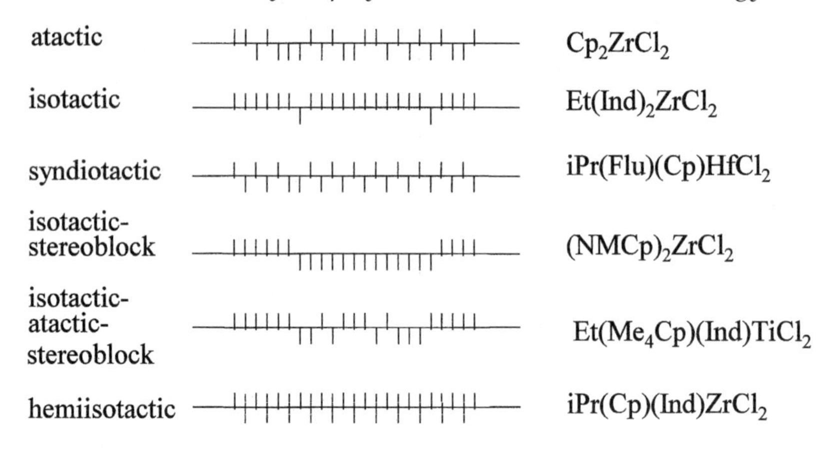

Figure 2 Types of polypropylene chains produced with metallocene catalysts.

a combination of aluminoxanes and alkylaluminums, are used to activate the metallocene catalyst. In the second category, an ion exchange compound, comprising a cation and a noncoordinating anion, is combined with the metallocene catalyst. The cation reacts irreversibly with at least one of the metallocene ligands. The anion must be capable of stabilizing the transition metal cation complex and must be labile enough to be displaced by the monomer. The metallocene complex is therefore a cation associated with a stable anion. There is now enough experimental evidence to support the hypothesis that all active center types operative with metallocenes are cationic.

Ethylene was the first olefin to be polymerized using metallocene/ aluminoxane catalysts. Ethylene polymerization with metallocenes has been extensively reviewed elsewhere [1–6]. The metallocenes commonly used for ethylene polymerization, such as nonchiral cyclopentadienyl derivatives, are also able to polymerize propylene with high productivities, but only atactic chains are produced.

For the synthesis of stereospecific metallocene catalysts for propylene polymerization, C_2 symmetric precursors are necessary to obtain a catalyst for isospecific polymerization, and C_5 symmetric precursors to produce a catalyst for syndiospecific polymerization. Asymmetric precursors can be used to synthesize metallocene catalysts that produce hemiisotactic and isotactic-stereoblock PP. Farina *et al.* [7] have proposed an useful classification of metallocene catalysts based on their symmetry (Figure 3).

Chiral metallocenes, such as $Et(H_4Ind)_2ZrCl_2$ and $Et(Ind)_2ZrCl_2$, can make PP with a high degree of stereoregularity. These zirconocenes are obtained as racemic mixtures of the D- and L-forms. The spatial arrange-

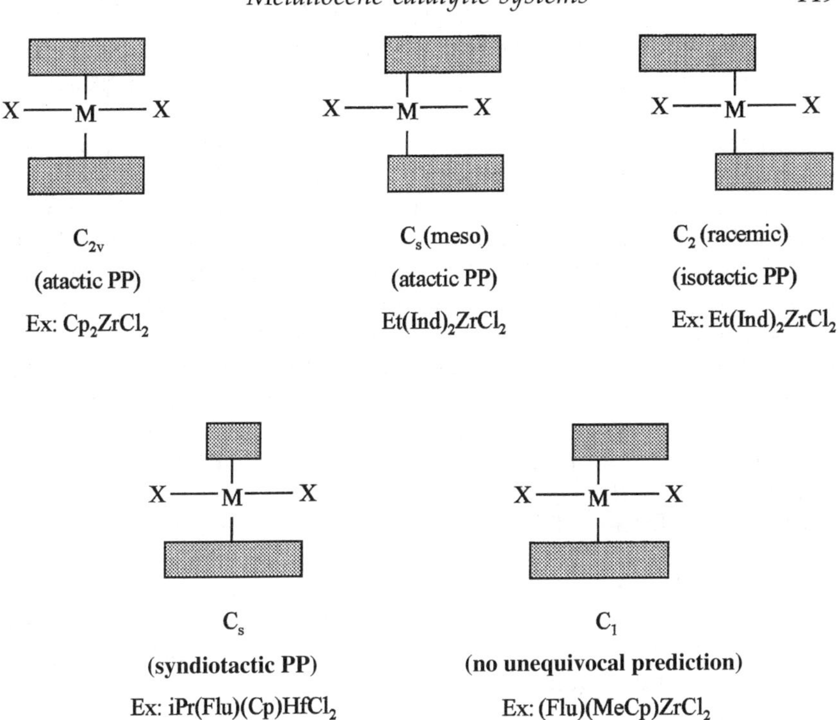

Figure 3 Symmetry requirements for polypropylene synthesis.

ment of the chiral racemic isomer favors the coordination of propylene molecules in such a way as to produce mainly isotactic chains. For the meso form, both monomer orientations are equally favored and therefore only atactic chains are formed. Unfortunately, these catalysts produce polymer with low molecular weight averages and although the dominating propagation mechanism is 1,2-insertion, a small amount of 2,1- and 1,3-insertion takes place. Consequently the polymer melting temperature is lower than that of isotactic PP obtained with heterogeneous Ziegler–Natta catalysts.

Molecular weights and stereoregularity of PP made with metallocene catalysts are very sensitive to polymerization temperature because of the flexibility of the ligands; more atactic polymer is formed and the rate of β-hydride elimination increases at higher polymerization temperatures. This somewhat disadvantageous behavior can, however, provide a convenient route for the production of PP waxes.

Syndiotactic PP with high molecular weight, narrow MWD and high productivity can be synthesized with C_5 symmetric metallocenes such as iPr(Flu)(Cp)$HfCl_2$/MAO. Since syndiotactic PP is more resistant to ultraviolet radiation than isotactic PP, it can be used for medical applications which require sterilization.

Stereoblock PP, with sequence lengths between two and seven and narrow MWD, can be produced with $(NMCp)_2ZrCl_2$/MAO.

PP having the properties of a thermoplastic elastomer can be synthesized with asymmetric metallocenes such as Et(Me_4Cp)(Ind)$TiCl_2$/MAO and Et(Me_4Cp)(Ind)$TiEt_2$/MAO. This is the first example of a thermoplastic elastomer consisting of only one monomer type. To account for the properties of this new PP, it is supposed that the active sites can exist in two different states, one stereospecific and the other nonstereospecific. Since the site types can change states during the lifetime of a polymer molecule, the chain has alternating blocks of atactic and isotactic PP. The isotactic domains act as physical cross-links and give the polymer its elastomeric properties.

Chiral metallocene catalysts are also effective for the copolymerization of ethylene, propylene and higher α-olefins. $Et(Ind)_2ZrCl_2$/MAO can be used to copolymerize ethylene and propylene, producing a copolymer richer in propylene than the one produced with the achiral Cp_2TiCl_2/MAO or Cp_2ZrCl_2/MAO under the same polymerization conditions. One of the great advantages of metallocene catalysts for the copolymerization of propylene with ethylene and other α-olefins is the formation of random copolymer with narrow CCD.

SUPPORTED METALLOCENE CATALYSTS FOR USE IN EXISTING INDUSTRIAL PLANTS

Since most conventional Ziegler–Natta polyolefin manufacturing plants are designed to use heterogeneous catalysts (with exception of plants producing ethylene-propylene-diene (EPDM) rubbers that use soluble vanadium-based catalysts), the commercial application of soluble metallocene catalysts would require the design of new plants or the adaptation of existing ones to operate with soluble catalysts. One way of overcoming this problem is by supporting the metallocene catalyst on an adequate carrier.

Metallocene catalysts have been effectively supported on several inorganic compounds, such as SiO_2, $MgCl_2$, Al_2O_3, MgF_2, and CaF_2. Synthetic and natural polymers have also been used to support metallocenes.

The type of support as well as the technique used for supporting the metallocene and MAO have a crucial influence on catalyst behavior. By the appropriate choice of supporting conditions, stereo- and regioselectivity can be improved and transfer reactions can be minimized with consequent production of polymers with improved stereoregularity and higher molecular weights.

Supporting can also affect the stereochemical control of the metallocene. Depending on the supporting technique employed, catalysts that are syndiospecific when soluble might produce mainly isotactic PP when supported on SiO_2. This inversion of stereoselectivity has also been observed when the type of cocatalyst is changed.

Additionally, supported metallocenes usually require smaller aluminum to transition metal ratios than the equivalent soluble systems and some can be activated in the absence of aluminoxanes by common alkylaluminums. This reduced dependence on the presence of aluminoxanes and on high ratios of aluminum to transition metal has been related to a reduction in catalyst deactivation by bimolecular processes due to the immobility of the active sites on the surface of the support.

Additionally, it is possible to support different metallocene types onto the same carrier to produce a designed multiple site-type catalysts for applications where broad MWD and CCD are required.

Unfortunately, the catalytic activity of supported metallocenes is usually inferior to that of the equivalent soluble catalysts, probably due to deactivation of catalytic sites during the supporting process. Broadening of the MWD and CCD for supported catalysts can also occur under certain supporting conditions.

ADAPTATION OF METALLOCENE CATALYSIS TO EXISTING OLEFIN POLYMERIZATION PROCESSES

Although metallocene-produced polyolefins can compete with commodity polyolefins synthesized with conventional Ziegler–Natta catalysts, they will probably not be restricted to the polymer commodity market. Because of the better polymer microstructure control obtained with metallocene catalysts, it will be possible to produce specialty polyolefins to compete with nonolefinic polymers, thus opening an entire new market for polyolefin applications.

GAS PHASE PROCESSES

Metallocene catalysts need to be supported to be used in gas phase reactors, such as Union Carbide's fluidized-bed Unipol process, or BASF's stirred-bed Novolen process. For these processes, it is necessary to have a free flowing catalyst powder which will form polymer particles with adequate size distribution, avoiding the formation of fine powder or particle agglomerates. In other words, good replication of the catalyst particle size distribution is essential for the efficient performance of these reactors.

Langhauser *et al.* [8] reviewed the industrial production of PP (homopolymer, random copolymer, and impact copolymer) using Me_2Si(2-

MeBeInd)$_2$ZrCl$_2$-supported catalyst and the Novolen-BASF process. This catalyst can produce PP with high molecular weight even at elevated temperatures. The polymer particles replicate well the size distribution of the catalyst particles. This catalyst can produce PP with new properties, such as low extractables for food wrapping and medical applications, which is a consequence of the uniform stereoregularity of polymers produced with a single site-type catalyst.

Impact copolymers can also be produced with this catalyst. Impact copolymers made with conventional heterogeneous Ziegler–Natta catalysts show some crystalline domains in the amorphous elastomeric phase, while the elastomeric phase of the metallocene-produced copolymer is entirely amorphous. This new microstructure will likely enable the production of copolymers with enhanced impact properties.

According to Langhauser *et al.* [8], this catalytic system can be adapted to their existing gas phase polymerization process without any significant technical change.

LIQUID-BULK OR DILUENT SLURRY PROCESSES

Supported metallocenes and conventional heterogeneous Ziegler–Natta catalyst will have a similar behavior regarding macroscopic phenomena in the reactor, provided that there is no desorption of the active sites during the polymerization. It is expected that most existing slurry polymerization reactor processes, either in continuous stirred-tank or loop reactors, can be easily adapted to use supported metallocene catalysts. For these processes, metallocenes can actually be an improvement over conventional heterogeneous Ziegler–Natta catalysts, since the lower extractables content of the produced polymer will imply lower slurry viscosity or permit higher comonomer levels for copolymer products.

For homogeneous catalysts, the process of polymer particle formation generally leads to porous, low-density polymer particles, which can cause significant increase in slurry viscosity and reactor fouling, leading to inadequate reactor temperature control. Additionally, polymer particles with poor powder properties are undesirable for postreactor polymer processing. These problems must be addressed before using soluble metallocene catalysts for industrial production of polyolefins.

REFERENCES

1. Gupta, V.K., Satish, S. and Bhardwaj, I.S. (1994) Metallocene complexes of group 4 elements in the polymerization of monoolefins. *J.M.S.–Rev. Macromol. Chem. Phys.*, **C34**(3), 439–514.

2. Huang, J. and Rempel, G.L. (1995) Ziegler–Natta catalysts for olefin polymerization: Mechanistic insights from metallocene systems. *Prog. Polymer Sci.*, **20**, 459–526.
3. Reddy, S.S. and Sivaram, S. (1995) Homogeneous metallocene-methylaluminoxane catalyst systems for ethylene polymerization. *Prog. Polymer Sci.*, **20**, 309–367.
4. Soares, J.B.P. and Hamielec, A.E. (1995) Metallocene/aluminoxane catalysts for olefin polymerization. A review. *Polymer React. Engng*, **3**(2), 131–200.
5. Hamielec, A.E. and Soares, J.B.P. (1996) Polymerization reaction engineering – Metallocene catalysts. *Prog. Polymer Sci.*, **21**, 651–706.
6. Kaminsky, W. (1996) New polymers by metallocene catalysis. *Macromol. Chem. Phys.*, **197**, 3907–3945.
7. Farina, M., Di Silvestro, G. and Terragni, A. (1995) A stereochemical and statistical analysis of metallocene-promoted polymerization. *Macromol. Chem. Phys.*, **196**, 353–367.
8. Langhauser, F., Kerth, J., Kersting, M., Kölle, P., Lilge, D. and Müller, P. (1994) Propylene polymerization with metallocene catalysts in industrial processes. *Angew. Makromol. Chem.*, **223**, 155–164.

NOMENCLATURE

CCD	Chemical composition distribution
Cp_2TiCl_2	Biscyclopentadienyl titanium dichloride
Cp_2ZrCl_2	Biscyclopentadienyl zirconium dichloride
$Et(Ind)_2ZrCl_2$	Ethylenebis(indenyl)zirconium dichloride
$Et(H_4Ind)_2ZrCl_2$	Ethylenebis(tetrahydroindenyl)zirconium dichloride
$Et(Me_4Cp)(Ind)TiCl_2$	Ethylidene(1-tetramethylcyclopentadienyl)(indenyl) titanium dichloride
$Et(Me_4Cp)(Ind)TiEt_2$	Ethylidene(1-tetramethylcyclopentadienyl)(indenyl) titanium diethyl
$iPr(Flu)(Cp)HfCl_2$	Isopropyl(cyclopentadienyl-1-fluorenyl)hafnium dichloride
MAO	Methylaluminoxane
$Me_2Si(2\text{-}MeBeInd)ZrCl_2$	Dimethylsilanediyl-bis(2-methylbenzo[e]indenyl) zirconium dichloride
MWD	Molecular weight distribution
$(NMCp)_2ZrCl_2$	Bis(neomenthylcyclopentadienyl)zirconium dichloride

Keywords: metallocene catalyst, Ziegler–Natta catalyst, olefin polymerization, polyolefins, homogeneous catalysts, supported catalysts, stereoregularity, molecular weight distribution (MWD), chemical composition distribution, Unipol®, Novolen®, stereoselectivity, single site catalyst, multiple site catalyst, gas phase process, slurry process, homopolymerization, copolymerization.

Metallocene catalysts and tailor-made polyolefins

R. Mülhaupt

INTRODUCTION

Since the early 1980s innovations in metallocene catalyzed olefin and styrene polymerization have stimulated progress in polymer synthesis and development of new polymeric materials derived from well-known and readily available petrochemical feedstocks, such as ethylene, propylene, higher 1-olefins, cycloolefins, and styrene. Recent advances and new insights into reaction mechanisms and basic correlations between metallocene structure and polymer properties were reviewed by Brintzinger *et al.* [1], Kaminsky [2–4], Hamielec and Soares [5, 6], Gupta *et al.* [7], Reddy and Sivaram [8], Huang and Rempel [9], Tian and Huang [10], Mülhaupt [11–13], Aulbach and Kübor [14], Coates and Waymouth [15], and Albizzati *et al.* [16]. The state of the art in catalyst development during the mid-1990s is reflected by contributions compiled in proceedings [17, 18] of conferences held in honor of Karl Ziegler and Giulio Natta.

During the pioneering days of Ziegler–Natta catalysis, metallocenes, such as Cp_2TiCl_2, were activated with conventional aluminum alkyls and used in the laboratories of Natta *et al.* [19] and Newburg and Breslow [20] to polymerize ethylene. Due to rather poor catalyst activity with respect to highly active supported catalysts and the inability to catalyze propylene polymerization, early metallocenes served mainly as model systems to understand elementary reactions of Ziegler–Natta catalysis. The breakthrough in metallocene technology stems from long-standing effort in basic sciences combined with serendipity. During the 1970s it

Polypropylene: An A–Z Reference
Edited by J. Karger-Kocsis
Published in 1999 by Kluwer Publishers, Dordrecht. ISBN 0 412 80200 7

was discovered accidentally by Reichert and Meyer [21] that traces of water, which represents a well-known catalyst poison, can enhance catalyst activity of $Cp_2TiEtCl/EtAlCl_2$ when using Al/H_2O ratios >3. This unexpected effect is more pronounced in the case of $AlMe_3$ activators. This discovery led Kaminsky and Sinn [22–24] to the preparation of methylaluminoxanes, referred to as MAO and produced by controlled hydrolysis of $AlMe_3$. MAO proved to be a highly effective activator for metallocene catalysts in ethylene as well as for the first time propylene polymerization. Preferably linear or slightly branched MAO corresponding to $Me_2AlO\text{-}(MeAlO)_x\text{-}OAlMe_2$ with high molecular weight and $5<x<20$ were used. Sinn and Kaminsky filed a patent jointly with BASF AG on MAO-activated propylene polymerization which initiated exciting developments in metallocene catalysis during the 1980s [25]. The first catalysts were completely nonstereoselective, producing ideally atactic polypropylene (aPP; see also the chapter entitled Amorphous or atactic polypropylene) with random distribution of the two configurations of the stereogenic carbon atoms of the propylene repeat units along the PP chain. At that time, such aPP was only available by epimerization of isotactic PP [26] because conventional catalysts produce rather ill-defined stereoirregular PP fractions as byproducts in stereoselective polymerization. As reported below in more detail, Wild *et al.* [27] introduced in 1982 novel chiral ansa-metallocenes, such as racemic ethylenebisindenylzirconiumdichloride. Upon activation with MAO, chiral metallocenes gave isotactic PP, which was reported by Ewen [28], and Kaminsky *et al.* [29]. This pioneering research marked the beginning of an outstanding development of great industrial significance.

POLYMERIZATION MECHANISMS AND STERIC CONTROL

Today it is well established that MAO activators alkylate metallocene halides to form the corresponding alkyl halides. The Lewis acid MAO can abstract chloride to produce the cationic 14-electron alkyl $[Cp_2MR]^+$ and the sterically hindered counterion. Such cationic metallocene complexes were prepared by Turner *et al.* [30, 31], Jordan [32] and others, using borane counterions such as $B(C_6F_5)_4^-$. The development of such MAO-free cationic metallocenes, which are activator-free and therefore are not considered to be Ziegler catalysts, confirmed the mechanistic scheme displayed in Figure 1 and led to the development of highly active catalyst systems, thus eliminating the need for addition of expensive MAO. Most conventional MAO-activated metallocenes require a large excess of MAO with [Al]/[trans.met.] molar ratios >100 in order to shift the equilibrium towards formation of active cationic metallocenes at the expense of inactive neutral complexes. Higher concentration of

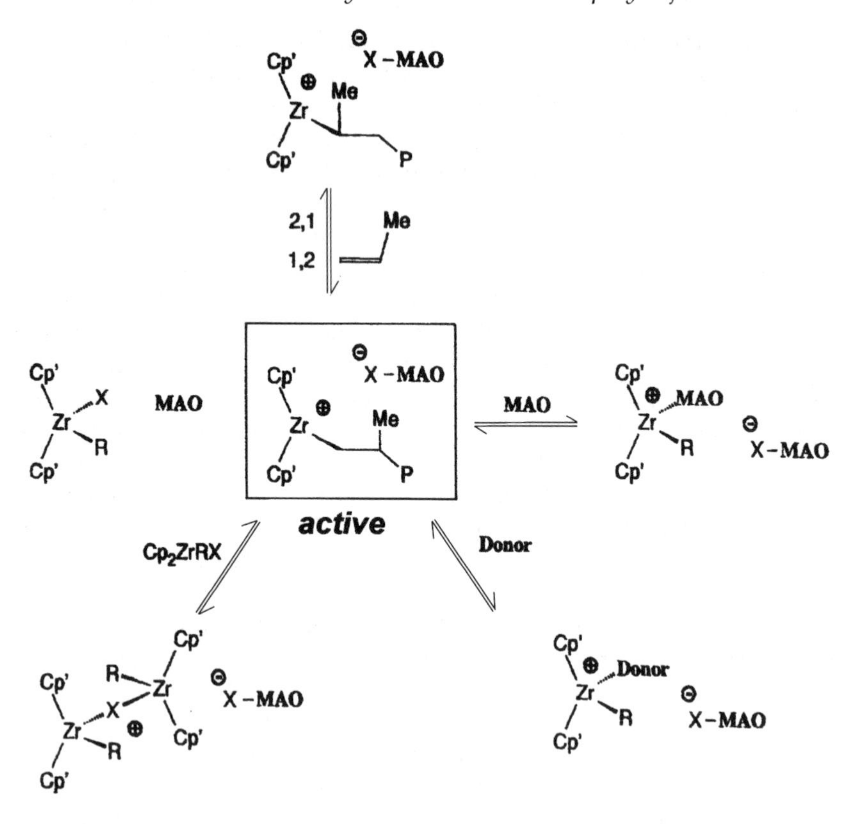

Figure 1 Cationic metallocene alkyls are catalytically active and can form temporarily inactive ('dormant') sites.

cationic metallocene alkyls is also promoted by increasing propylene pressure and polymerization performed in liquid propylene.

The steric control of metallocene-based catalysts depends primarily upon the structure and symmetry of the metallocene catalyst. As is apparent from Figure 2, the changes of symmetry account for drastic variation of PP microstructure. The C_s symmetric isopropylidene-bridged zirconocene with cyclopentadienyl and fluorenyl ligands produces exclusively syndiotactic PP (sPP). When bulkiness of the ligand framework is increased by attaching a tert.butyl group in the 3-position of the cyclopentadienyl ring system, the corresponding C_1 symmetric zirconocene, similar to isospecific C_2 symmetric isospecific metallocenes, produces exclusively isotactic PP. Interestingly, in the case of methyl substitution, the resulting PP is hemiisotactic, where stereoselective and nonstereoselective insertions are alternating. Although hemiisotactic PP did not find industrial applications, research by Fink *et al.* [33] and others

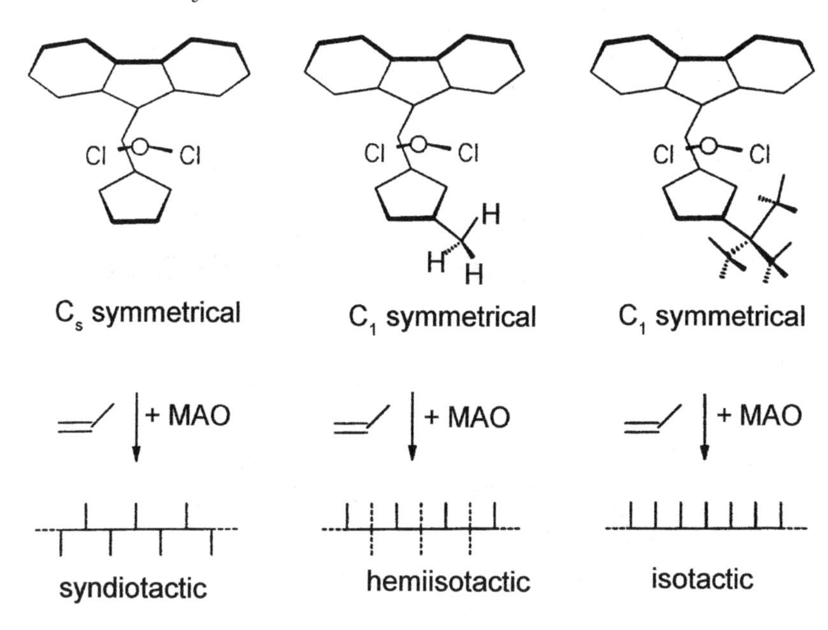

Figure 2 The role of zirconocene ligand substitution pattern on stereoselectivity in propylene polymerization.

demonstrates that the microstructure of hemiisotactic PP represents a fingerprint of the catalysts and is the key to better understanding of polymerization mechanisms.

Remarkable features of metallocenes are: (1) very uniform homo- and copolymers with narrow molecular weight distribution with $M_w/M_n = 2$, typical for the presence of only one type of catalytically active site ('single site' catalysts); (2) excellent control of molecular weight, end groups, stereochemistry, long- and short-chain branching; (3) molecular weight-independent comonomer incorporation including less reactive comonomers, such as long-chain 1-olefins, and styrenes; (4) polymerization of cycloolefins; (5) morphology control and potential for production of reactor blends using metallocene blends; (6) tailor-making of molecular weight distributions in cascade reactors or hybrid-metallocene-based processes; (7) new catalysts can be introduced in old plants to make new polymers ('drop-in' technology). The outstanding precision affords clean polymers without byproduct formation.

While some metallocene catalysts of the early 1980s gave rather low molecular weight PP and low stereoselectivity, new generations of the 1990s give much higher performance. This is apparent in Figure 3 and Table 1 where different isoselective metallocenes are listed. The groups of Spaleck [34, 35] and Brintzinger [36, 37] demonstrated the role of the

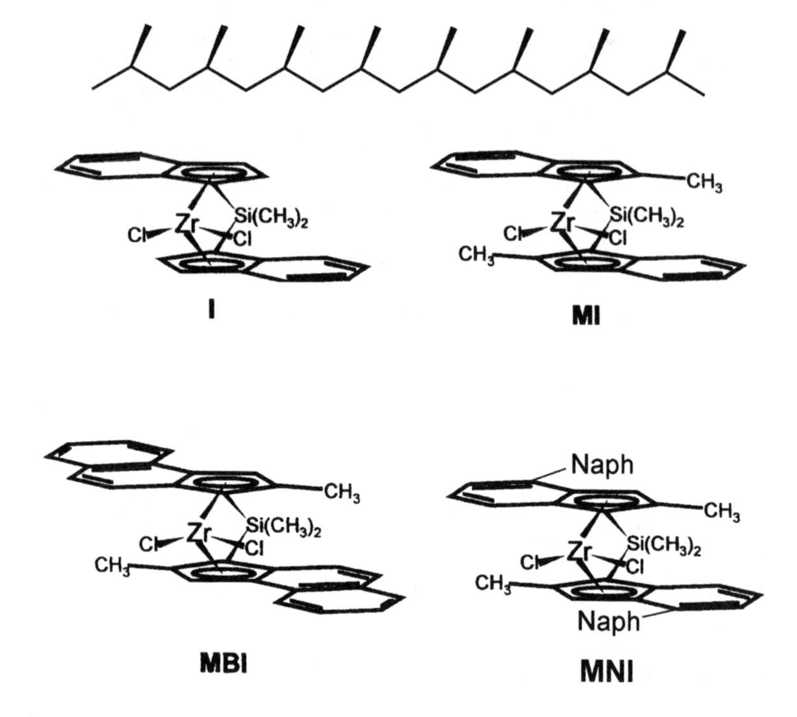

Figure 3 Isotactic PP and evolution of isoselective metallocene catalysts (cf. Table 1).

Table 1 The influence of the bisindenyl-ligand substitution pattern of isoselective metallocene catalysts*

Metallocene	*Activity*†	*Molecular weight* M_w (10^3 g/mol)	*PP melting temperature* (°C)
I	190	36	137
MI	99	195	145
MBI	248	330	146
MNI	875	920	161
Ng/Ti/Al‡	20	100–900	162

* source: Aulbach, M. and Küber F. (Hoechst AG), *Chemie in unserer Zeit*, **28**(4), 203 (1995); † kg PP/mmol Zr×h; ‡ conventional $MgCl_2$-supported catalyst.

ligand substitution pattern. When a methyl group is introduced in the 2 position, much higher molecular weights are obtained because chain termination via β-hydride transfer to propylene is eliminated in favor of hydride transfer to the transition metal. Substitution in the 4-position and benzannelation of the the indenyl ligand affords higher catalyst activity, most likely because formation of dimers is hindered due to unfavorable steric interactions. When monomer is exclusively involved in chain propagation, molecular weight, which reflects the ratio of propagation rate to chain termination rate, increases with increasing propylene pressure [38]. The presence of a large number of dynamic equilibria, involving catalytically active as well as dormant sites, was proposed by Fischer and Mülhaupt [39]. Recently, this hypothesis was confirmed by kinetic studies of Ko *et al.* [40], who concluded that higher active center concentration and larger chain propagation rate constants (k_p) are found with increasing Al/Zr molar ratio above 2000. Also increasing propylene concentration can enhance activity significantly due to drastic increase of k_p.

ISOTACTIC POLYPROPYLENE

On one hand, metallocene catalysts offer the opportunity of improving stereoselectivity, thus increasing chain rigidity and stiffness of PP. Recently, Kaminsky *et al.* [41] reported that isotacticity and melting temperature of PP produced with the C_2-symmetrical zirconocene rac.-Et(2,4,7-Me_3Ind)$_2$ZrCl$_2$, activated with $Ph_3C^+B(C_6F_5)_4^+$, increased significantly with decreasing temperatures. At −78°C the PP melting temperature was found to be 168.9°C, whereas PP produced with the C_1 symmetrical zirconocene Me_2C(3-tBu-Cp)(3-t.Bu-Ind)ZrCl$_2$ was unaffected under the same condition. This discovery is stimulating efforts to improve properties of highly isotactic PP. On the other hand, metallocene catalysts can be tailored to introduce random stereo- and regioirregularities into the stereoregular polymer chain. In addition to typical stereoirregularities of the (mrrm)-type, regioirregularities can produce conventional head-to-head units resulting from '2–1'-type insertion (see the chapter on Ziegler–Natta catalysis in this book) as well as ethylene units according to '1–3' insertion, which results from β-hydride elimination and reinsertion shown in Figure 4. This '1–3' insertion [42, 43] produces propylene/ethylene copolymers without using ethylene as comonomer! Most catalyst systems, however, producing such *in-situ* copolymers exhibit low activity and produce fairly low molecular weights [44, 45], probably due to the competing chain termination via β-hydride elimination without reinsertion. The mechanism for controlling stereoselectivity was associated with chirality of the transition metal or the configuration of the last propylene unit of polymer chain (see the chapter on Ziegler–Natta catalysis in this book). Recently, Fink *et al.* [46]

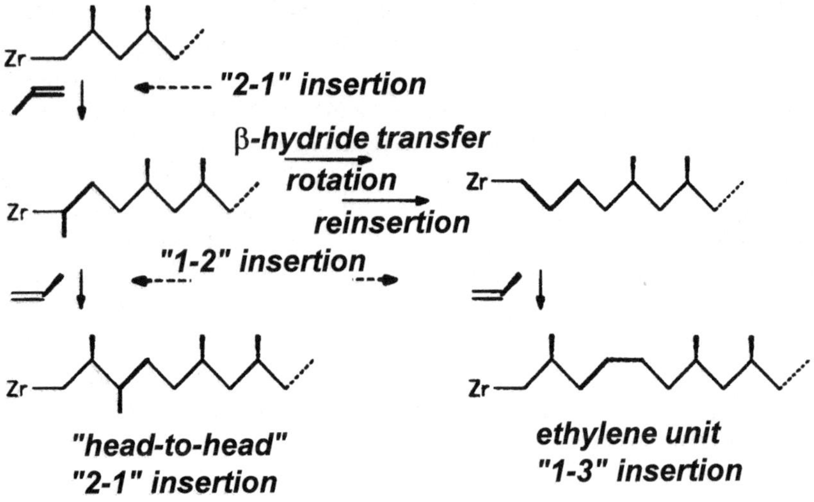

Figure 4 Formation of regioirregularities in 1-olefin polymerization.

have proposed an intriguing alternative mechanism taking into account the four lowest energy conformers of the metallocene species, which coordinates the prochiral propylene, and the preferred position of the polymer chain.

PP containing randomly distributed irregularities is shown in Figure 5. Most conventional catalysts were only able to prepare highly isotactic PP accompanied by formation of low molecular weight PP with rather ill-defined molecular architectures. Excellent control of PP microstructure is the key to controlled formation of superstructures. According to Fischer and Mülhaupt [47], there exists a linear correlation between the average isotactic segment length n_{iso} and the formation of the γ-modification of PP. Moreover, with decreasing n_{iso}, the melting temperature decreases and varies between 120 and 162°C. Such low melting PPs are attractive materials for application in packaging with improved heat sealability. Atomic force microscopic (AFM) studies by Thomann *et al.* [48] revealed that large spherulitic structures, typical for α modification, are absent when pure γ-PP is crystallized. Due to nanostructure formation during crystallization, the resulting PP exhibits markedly improved optical clarity.

Properties of metallocene-based PP are compared with properties of conventional PP in Table 2. Metallocene-PP exhibits higher stiffness and improved optical properties. Improved orientation of metallocene-based PP offers opportunities for production of biaxially oriented PP film (Table 3) and PP fibers (Table 4). In fact, metallocene-PP can be spun at

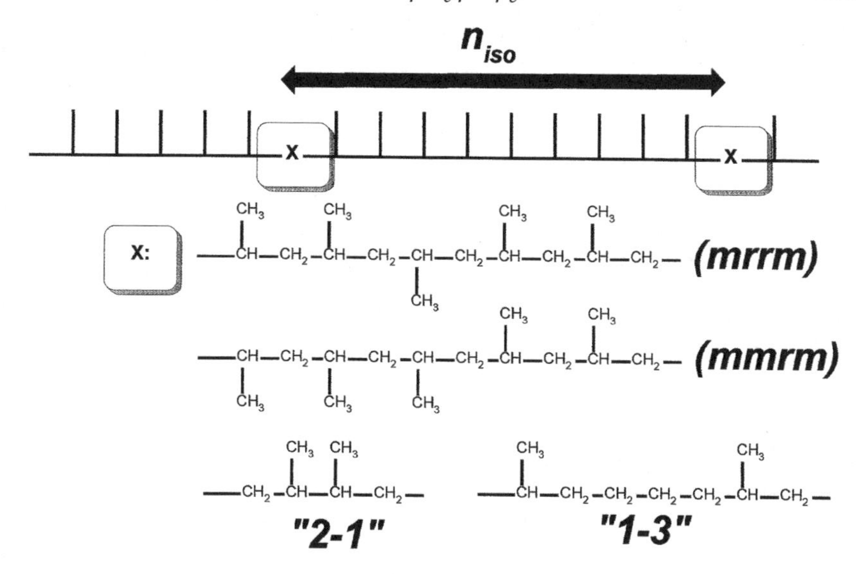

Figure 5 Variation of the isotactic segment length between two steric irregularities.

Table 2 Comparison of conventional and metallocene-based propylene homo- and copolymers (data from Hoechst AG)

Properties	*Metallocene-based copolymer*	*Metallocene-based copolymer*	*Metallocene-based homopolymer*	*Conventional copolymer*
MFI* (g/10 min)	4	4	4	4
Melting temp. (°C)	130	140	140	140
Ethylene cont. (wt.%)	3.0	1.8	0	4.8
Young's modulus (MPa)	730	950	1100	640
Hardness† (MPa)	43	54	63	40
Notched Izod (mJ/mm²)	12.5	9.5	6.5	13.5
Clarity 1 mm sheet (%)	75	65	60	58
Extractables‡ (wt.%)	1.2	1.2	0.8	8.5

* 230 °C, 2, 16 kg; † ball hardness; ‡ in n-hexane at 69°C.

much higher speeds, similar to those of the polyamide spinning processes. Improved performance of PP films may allow the reduction of film thickness and save weight in packaging applications. At lower melting temperatures, there exist more 'tie' molecules between PP crystallites, thus enhancing tensile strength and elongation at break. High gloss and excellent optical properties of metallocene-based PP are particularly attractive in injection molding.

Table 3 Metallocene-based biaxially oriented PP films (data from BASF AG)

Properties	*Metallocene-PP*	*Conventional PP*
MFI* (g/10 min)	8	2.5
Film thickness (μm)	12	18
Processing temp (°C)	140–150	165–170
Tensile strength (MPa) MD/TD	134/362	120/260
Elongation at bread (%) MD/TD	210/37	200–240/40–50
Young's modulus (MPa) MD/TD	1700/3110	2000/2800
haze (%)	0.1	2.5
gloss	116	85–90

* 230 °C, 2, 16 kg.

Table 4 Metallocene-PP fibers (data from BASF AG)

Spinning speed (m/min)	*Fiber diameter* (μm)	*Conventional PP tensile strength* (cN/dtex)	*Conventional PP elongation at break* (%)	*Metallocene PP tensile strength* (cN/dtex)	*Metallocene PP elongation at break* (%)
2500	26	2.23	265	3.63	154
3000	26	2.20	255	3.69	142
3500	26	n.s.*	n.s.*	3.65	145
4000†	26	n.s.*	n.s.*	3.64	137

* n.s.: not processable, MFI (230°C/2, 16 kg) = 18 g/10 min; † range typical for polyamide spinning processes.

SYNDIOTACTIC POLYMERS

Production of sPP, prepared with catalysts, such as $VCl_4/AlEt_2Cl$ at −78°C, was not practical on an industrial scale and gave rather poor regio- and stereoselectivities [49]. In 1988 Ewen *et al.* [50] developed novel metallocene-based catalyst systems, such as MAO-activated Me_2C-(Flu)(Cp)$ZrCl_2$, which form sPP in very high yields according to catalytic-site control mechanism.

The reaction mechanism of syndioselective propylene polymerization using $Me_2C(Cp)(Flu)ZrCl_2/MAO$ has been elucidated by Ewen and coworkers [51, 52]. Upon each insertion of propylene, the position of the growing chain changes and changes chirality of the transition metal center which switches back and forth between the two possible configurations. Unless the chain does not flip back to the other side, the next propylene approaches the metal center from the opposite side. As a consequence, the configuration of the stereogenic carbon atom of the repeat unit alternates to form syndiotactic PP. This mechanism is shown in Figure 6.

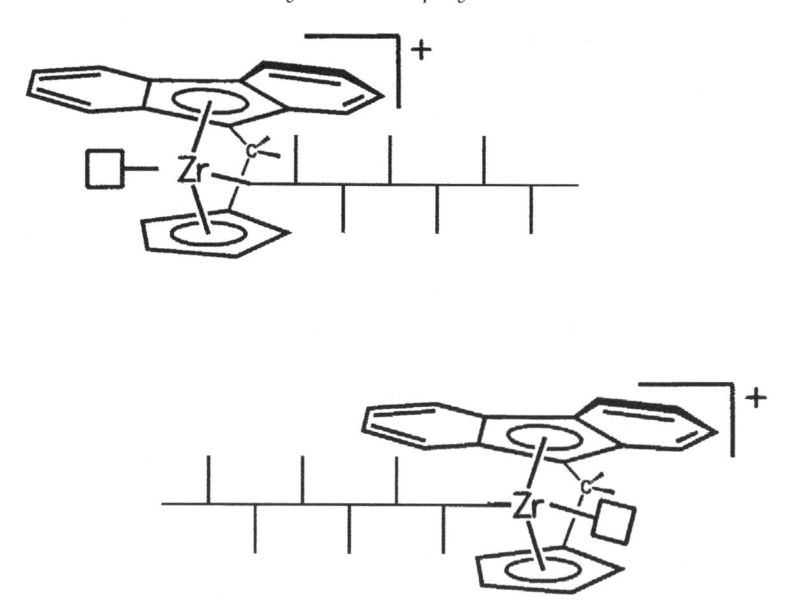

Figure 6 Mechanism of syndioselective propylene polymerization using a C_s-symmetrical metallocene catalyst. The configurations of active centers in subsequent insertion steps are shown.

Today, typical stereoregularities of syndiotactic PP are in range of 93% racemic (rrrr) pentads and corresponding melting temperatures are somewhat lower than those known for conventional isotactic PP. In comparison to conventional isotactic PP, toughness and optical clarity of such sPP are markedly higher at the expense of stiffness. Most early research on crystallization of sPP required major revisions due to the much higher precision of syndioselective polymerization using metallocenes (see the chapter on syndiotactic PP in this book) [53–56]. Structure and morphology correlations were reviewed by Lotz *et al.* [57]. According to Thomann *et al.* [58] sPP, containing 91% (rrrr) pentads, isothermally crystallized from the melt forms needle-like structures of crystalline entities at crystallization temperatures of 115–150°C. Large bundles of lamellae and rectangular entities with single crystal character were the main morphological structures (see the chapter on syndiotactic PP in this book). In the absence of spherulitic structures, optical clarity is much higher with respect to conventional isotactic PP. Rheological properties of sPP differ from those of isotactic PP. Eckstein *et al.* [59] used rheological methods to determine the entanglement molecular weights M_e which is approximately 2000 g/mol for sPP and approximately 7000 g/mol for both isotactic and atactic PP. This was attributed to differences in conformations. In contrast to iPP, where 3_1 helical conformations are preferred, sPP forms 2_1 helices.

Since the pioneering advance of Ishihara and coworkers in 1985, it is well recognized that metallocene catalysts, such as $(Me_5Cp)TiCl_3/MAO$, produce syndiotactic polystyrene (sPS). Synthesis and reaction mechanisms were reviewed by Po and Cardi [60] and Kuramoto [61]. The highly syndiotactic PS [62] represents a new engineering thermoplastic with melting temperature of 270°C and much higher crystallization rate with respect to isotactic PS, which melts at 225°C. Syndiotactic PS exhibits high heat distortion temperatures, stiffness, excellent water/steam resistance, attractive optical properties, and dielectric properties required by electrical insulation applications. At present, development of catalyst and process technology as well as compounding is being aimed at overcoming the inherent brittleness typical for the first generation sPS.

OLIGOMERS AND FUNCTIONALIZED POLYMERS

Metallocene catalysts can be applied to prepare olefin oligomers with well-defined end groups. Vinylidene-terminated propylene oligomers have been prepared by Mülhaupt and coworkers [63, 64] who converted the olefinic end group into a variety of polar end groups by means of double bond conversion (Figure 7). Examples of functional end groups are as hydroxy, bromo, epoxy, ester, carboxylic acid, amine, thiol, silane, borane and anhydride. This includes novel PP macromonomers bearing a polymerizable end group, such as methacrylate [65] and oxazoline [66]. When oxazolines are polymerized or grafted onto poly(ethylene-co-methacrylic acid) novel polymers with pendant PP side chains are obtained. Tailor-made succinic-anhydride-terminated oligopropylene was applied to elucidate the basic correlation between compatibilizer molecular architecture and performance of PP/PA6 blends [67]. By a

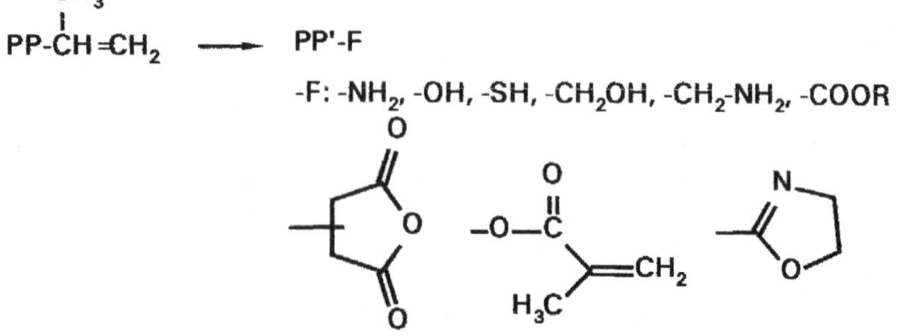

$PP\text{-}C(CH_3)H{=}CH_2 \longrightarrow PP'\text{-}F$

$\text{-F: -}NH_2\text{, -OH, -SH, -}CH_2OH\text{, -}CH_2\text{-}NH_2\text{, -COOR}$

Figure 7 PP with functional end group.

similar route, Shiono and Soga obtained olefinic and aluminum alkyl end groups [68]. Waymouth *et al.* describe aluminum-alkyl-terminated poly(1,5-hexadiene) cyclopoymers which are converted into the mono hydroxy-terminated derivatives and used to prepare poly(caprolactone) block copolymers [69, 70].

HIGH MOLECULAR WEIGHT ATACTIC POLYPROPYLENE AND STEREOBLOCK POLYPROPYLENE

A low stereoregular PP may contain isotactic segments distributed in a flexible atactic chain. In typical stereoblock polymers, atactic and isotactic blocks alternate. Conventional supported catalysts produce highly isotactic PP containing small amounts of low molecular weight low stereoregular polymers as impurities. Low molecular weight aPPs were only useful as components of adhesive formulations (see related chapter in this book). In the early days of Ziegler catalysis such byproducts had to removed by solvent extraction. It was Wisseroth [71] at BASF AG who pointed out in 1977 that high molecular weight low stereoregular polymers give good mechanical strength. His special, 'glove test' demonstrated that low stereoregular PP did not exhibit 'peanut-butter-like consistency' as he referred to reports by Natta's group, where only low molecular weight stereoisomers were obtained as byproduct during propylene polymerization. Due to very low molecular weights, no mechanical strength was achieved. As shown in Figure 8, several catalyst families have emerged to produce elastomeric PP (ELPP) based upon stereoblock PP containing flexible atactic and crystallizable isotactic segments.

During the late 1980s workers at Du Pont [72, 73] discovered ZrR_4/Al_2O_3 catalysts with $R{=}CH_2C_6H_5$ or $CH_2C(CH_3)_2C_6H_5$, which produce simultaneously highly crystalline isotactic polypropylene and very high molecular low stereoregular polypropylene. The resulting ELPP (see also the chapter entitled Elastomeric polypropylene homopolyers using metallocene catalysts) was blended together with crystalline isotactic PP to

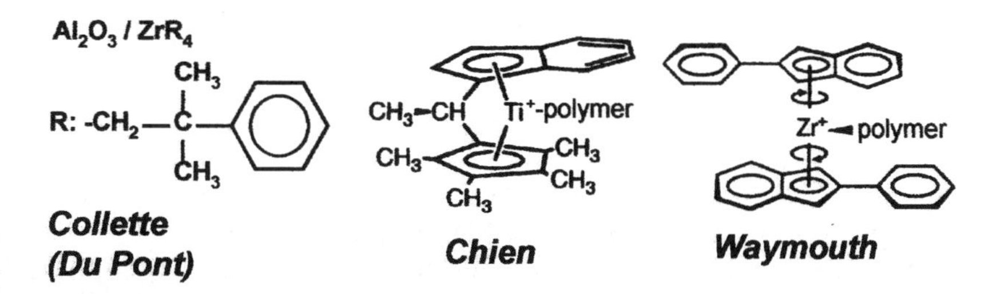

Figure 8 Catalysts for producing elastomeric PP (ELPP).

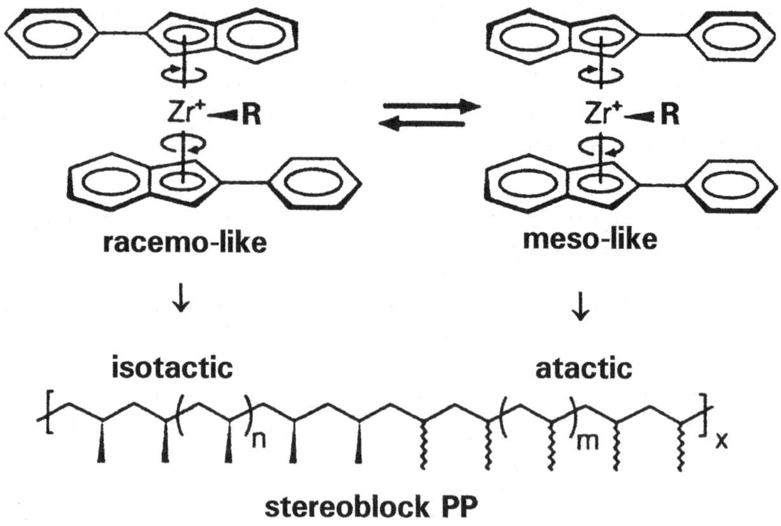

Figure 9 Oscillating metallocene catalysts.

tailor elastomeric properties such as tensile set, elongation at break, resilience. During the late 1980s, Chien *et al.* [74] discovered special metallocene catalysts which produce high molecular weight stereoblock polypropylenes consisting of alternating isotactic and atactic segments. Mechanisms were investigated by Gauthier and Collins [75]. Similar to Du Pont's ELPP, Chien and Collins describe formation of thermoplastic elastomers by means of blend formation. Blends of stereoblock ELPP and isotactic PP were characterized by Canevarolo and Decandia [76–78].

In 1996, Waymouth *et al.* [79, 80] developed a new class of oscillating metallocenes. As schematically represented in Figure 9, during polymerization there exists a very rapid exchange by means of rotation between nonstereoselective meso-like and stereoselective racemic-like rotamers where racemic-like rotamers produce rigid crystalline isotactic and meso-like conformers give flexible atactic segments. The segment distribution can be influenced by the polymerization temperature.

It was Resconi *et al.* [81, 82] at Montell who realized that also nonstereoselective metallocenes, such as $Me_2Si(Flu)_2ZrCl_w$/MAO, form high molecular weight aPP with molecular weights exceeding 500 000 g/mol exhibiting elastomeric properties. Similar to ELPP, the prime requirement for formation of thermoplastic elastomers is the presence of flexible atactic polypropylene chains containing at least two isotactic segments which can cocrystallize. With decreasing stereoregularity, the molar mass must increase drastically to afford cocrystallization. Drawback of ELPP is its relatively high glass temperature in range of −5°C. This ELPP glass

catalyst system (Chien):
$L_2ZrCl_2 + AliBu_3 + Ph_3C^+ B(C_6F_5)_4^-$

catalytically active sites (C*):
$L_2Zr^+R\ B(C_6F_5)_4^-$

$AliBu_3$-mediated chain transfer between sites:

$$C^*_{syndio}\text{-sPP} + C^*_{iso}\text{-iPP}$$

$$+ AliBu_3 \longrightarrow C^*_{syndio}\text{-iPP} + C^*_{iso}\text{-sPP}$$

$$\downarrow \qquad\qquad \downarrow$$

$$C^*_{syndio}\text{-sPP-block-iPP} \quad C^*_{iso}\text{-iPP-block-sPP}$$

Figure 10 Stereoblock polypropylenes via AlR_3-mediated transfer of growing PP chains between catalytically active sites with different stereoselectivity.

transition temperature can be reduced by incorporating comonomers or by adding plasticizers. ELPP is currently being considered for industrial use as PP blend component and material to produce elastomeric PP fibers.

Novel metallocene-based catalysts have been reported by Chien [83], who claims that aluminium-alkyl mediated transfer of growing PP chains between different catalytically active sites can yield stereoblock polymers such as PP containing isotactic and atactic or syndiotactic segments, respectively. The reaction pathway leading to formation of such block copolymers is schematically presented in Figure 10. This research is expected to stimulate development of novel thermoplastic elastomers.

NOVEL ETHYLENE COPOLYMERS

One of the key features of metallocene catalysts is their ability to copolymerize a wide variety of olefins as well as styrene without sacrificing the extraordinary uniformity with respect to narrow molecular weight distribution and especially molecular weight independent comonomer incorporation. Most conventional catalysts are multisite catalysts and contain catalytically active centers with greatly different reactivity towards insertion of ethylene and other olefins. Frequently,

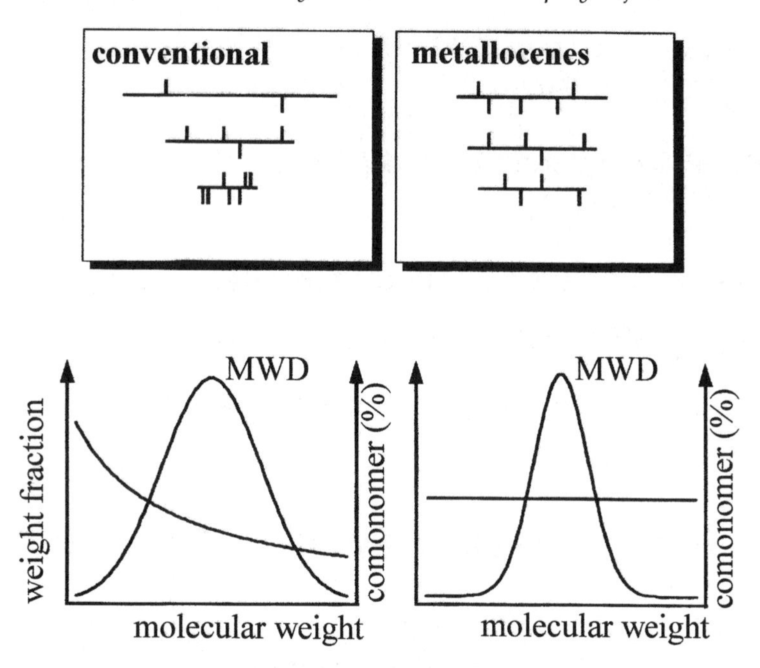

Figure 11 Comparison of conventional multisite catalyst and modern single site metallocene catalysts in copolymerization.

most of the higher 1-olefin comonomers are found in the low molecular weight fraction of ethylene copolymers (Figure 11). Typical molecular weight distributions were found in range of $3 < M_w/M_n < 50$. Therefore, in the past there existed a large technological barrier between the production of LLDPE containg less than 5% 1-olefin comonomer and that of EPM rubber containing >20 wt.% propylene. When adding styrene together with ethylene to conventional catalysts, less than 1 wt.% styrene is found in a copolymer and large fractions of polystyrene homopolymer are formed as an undesirable byproduct.

Today metallocene catalyst technology is overcoming this technological barrier (see the chapter entitled Copolymerization in this book). As with the homopolymers, copolymers exhibit very narrow molecular weight distributions ($M_w/M_n = 2$) and uniform comonomer incorporation covering the entire feasible composition range. Materials cover the range from high and low density polyethylene to plastomers, rubbers, impact-modified PP and PP. Comonomer incorporation and molecular weight distribution are independent of copolymer molecular weight. The versatility of metallocene catalysts has been demonstrated for ethylene copolymers with 1-olefins, including the less reactive higher 1-olefins [84–97]. Schneider *et al.* [98, 99] have investigated the influence of indenyl ligand substitution pattern on ethylene copolymerization, where benzannelation promoted both comonomer incorporation and high activity, whereas 2-methyl-substitution gave higher molecular weight.

For the first time ethylene/styrene with large amount of styrene incorporation [100, 101], ethylene/1-olefin/styrene terpolymers [102], and cycloolefin copolymers (see next section) are now at hand. A novel process uses cyclocopolymerization of ethylene with 1,5-hexadiene to incorporate cyclopentane units into the PE backbone [103]. The typical copolymer structures are displayed in Figure 12. Halfsandwich metallocenes, also referred to by Dow Chemical as constrained geometry catalysts (see the chapter on copolymerization in this book), are very effective in copolymerization. When vinyl end groups of polyethylene are copolymerized, the resulting polyethylenes and LLDPE contain long-chain branches. Long-chain branched LDPEs, such as Dow's Affinity® products, give improved processability as evidenced by substantially lower melt viscosity upon applying high shear during extrusion [104]. The random placement of 1-olefin comonomers in the polyethylene backbone gives excellent control on crystallization. At high comonomer

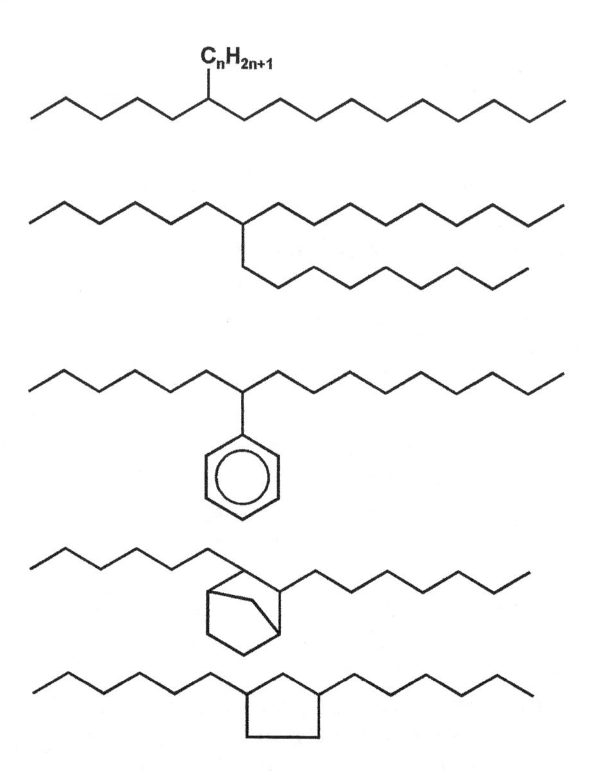

Figure 12 Molecular architectures of metallocene-based copolymers.

content, PE chain folding is prevented and fringed-micelle-type nanostructures [105] are formed. Similar to PP, as reported above, such nanostructures account for improved optical clarity of such materials. Random placement of comonomers into sPP backbone produces single crystals at low supercooling [106]. Metallocene-based ethylene/propylene and ethylene/propylene/diene copolymers are attractive new polyolefin elastomers. [107] As will reported in the next section, cycloolefin copolymers, such as poly(ethylene-co-norbornene), are expected to expand the polyolefin frontiers into applications which are currently being dominated by conventional engineering thermoplastics. Prime advantage of all metallocene-based polymers is the very high purity as reflected by their low content of extractable low molecular weight fractions. The aspects of metallocene catalyzed olefin copolymerization are discussed in more detail in the chapter on copolymerization in this book.

CYCLOALIPHATIC POLYMERS

Two metallocene-based synthetic routes lead to cycloaliphatic polyolefins: cyclopolymerization of nonconjugated 1,5-hexadiene, first described by Miller and Waymouth [108], and metallocene-catalyzed homo- and copolymerization of cycloolefins, discovered by Kaminsky [109]. While traditional heterogeneous Ti-based Ziegler catalysts catalyze ring-opening metathesis polymerization (ROMP), homogeneous metallocene catalysts do not open the ring during polymerization. Polyinsertion of cycloolefins gives highly rigid polymers with very high heat distortion temperature, well above decomposition temperature. Crystalline polycycloolefins exhibit melting temperatures above 380°C [110, 111]. Recently ribbon-like cycloaliphatic polymers were prepared via [2 + 2] cycloaddition of norbornadiene [112]. In the case of cyclopentene, first polymerized by Kaminsky *et al.* [113], Kelly and Collins [114] found that isomerization during polymerization accounts for formation of poly(1,3-cyclopentene). In order to render highly infusible and insoluble polycycloolefins processable, cycloolefins were copolymerized with ethylene to produce a new family of thermoplastics [115–117]. As a function of metallocene type, completely amorphous random copolymers, alternating copolymers or semicrystalline copolymers with melting temperature of 270°C are formed. As shown in Figure 13, incorporation of cyclic structures into the polymer backbone increases glass transition temperatures dramatically. Cycloolefin copolymers (COCs) were introduced commercially by Hoechst AG under the trade name of Topas®. Optically transparent COCs exhibit very low water uptake and are being evaluated as materials for manufacturing of compact discs and medical

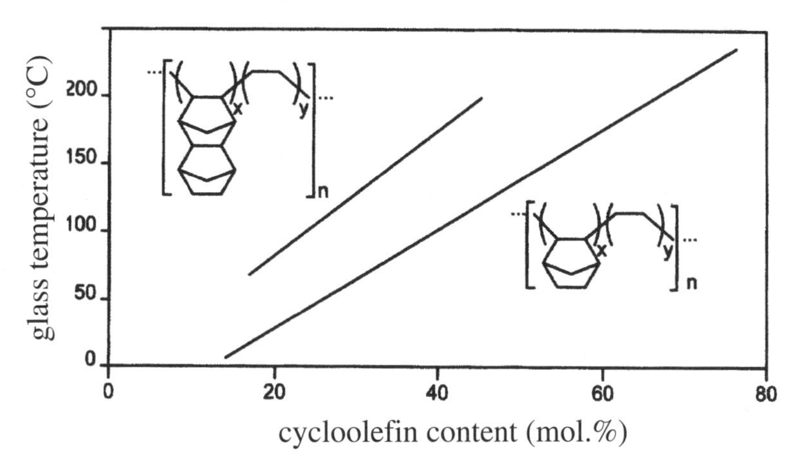

Figure 13 Glass transition temperature of cycloolefin copolymers as a function of cycloolefin content.

packaging. COCs based upon propylene copolymerization with cyclopentene using $Me_2Si(Ind)_2ZrCl_2$/MAO have been reported by Arnold *et al.* [118].

In Waymouth's route to cycloaliphatic polymers [119–121], the stereoselective cyclopolymerization of 1,5-hexadiene produces a variety of cycloaliphatic stereoisomers, including optically active polymers when using optically active metallocene catalysts [122].

OUTLOOK

Since the discovery of the first stereoselective metallocene catalysts during the early 1980s, a wide array of catalyst generations are emerging, which give unprecedented precision in polyolefin synthesis and produce a variety of novel polymeric materials ranging from waxes to commodity resins, engineering thermoplastics, rubbers, thermoplastic elastomers, and optically active polyolefins. For the first time, there exist clear correlations between catalyst structure, polymer microstructure and polymer properties. Metallocene catalysts and processes are designed to meet highly diversified demands of different applications. With metallocene-based model polymers it is possible to improve understanding and modelling of basic structure/property relationships of modern polymeric materials. At the beginning of the 21st century, metallocene catalysts are excellent tools to tailor-make polyolefins and to improve the utilization of readily available cheap petrochemical feedstocks.

REFERENCES

1. Brintzinger, H.H., Fischer, D., Mülhaupt, R., Rieger, B. and Waymouth, R.M. (1995) *Angew. Chem. Int. Ed. Engl.*, **34**, 1143.
2. Kaminsky, W. (1997) *Macromol. Chem Phys.*, **197**, 3907.
3. Kaminsky, W. (1994) *Angew. Makromol. Chem.*, **223**, 101.
4. Kaminsky, W. (1994) *Catalysis Today*, **20**, 257.
5. Hamielec, A.E. and Soares, J.B.P. (1996) *Prog. Polymer Sci.*, **21**, 651.
6. Soares, J.P.B. and Hamielec, A.E. (1995) *Polymer React. Engng*, **3**(2), 131.
7. Gupta, V.K., Satish, S. and Bhardwaj, J.S. (1994) *J. Macromol. Sci. Macromol. Chem. Phys.*, **C34**(3), 439.
8. Reddy, S. and Sivaram, S. (1995) *Prog. Polymer Sci.*, **20**, 309.
9. Huang, J., Rempel, G.L. (1995) *Progr. Polymer Sci.*, **20**, 459.
10. Tian, J. and Huang, B. (1996), in *Polymeric Materials Encyclopedia*, Vol. 6, (ed. J.C. Salamone), CRC Press, Boca Raton, FL, p. 4740
11. Mülhaupt, R. (1995), in *Ziegler Catalysts*, (eds G. Fink, R. Mülhaupt and H.H. Brintzinger), Springer-Verlag, Berlin.
12. Mülhaupt, R. and Rieger, B. (1996) *Chimia*, **50**, 10.
13. Mülhaupt, R. (1996) *Gummi Fasern Kunststoffe*, **49**, 394.
14. Aulbach, A., and Küber, F. (1994) *Chemie in unserer Zeit*, **28**(4), 197.
15. Coates, G.W. and Waymouth, R.M. (1995), in *Comprehensive Organometallic Chemistry II*, Chap. 12.1, (eds E.W. Abel, F.G.A. Stone and G. Wilkinson), Pergamon, Elsevier Science, Oxford p. 1193.
16. Albizzati, E., Giannini, U., Collina, G., Noristi, L. and Resconi, L. (1996), in *Polypropylene Handbook*, (ed. E.P. Moore, Jr.), Hanser, Munich.
17. Tritto I. and Giannini, U. (eds) (1995) *Macromol. Symp.*, **89**.
18. Fink, G., Mülhaupt, R. and Brintzinger H.H. (eds.) (1995) *Ziegler Catalysts*, Springer-Verlag, Berlin.
19. Natta, G., Pino, P., Mazzanti, G. and Giannini, U. (1957) *J. Am. Chem. Soc.*, **79**, 2975.
20. Breslow, D.S. and Newburg, N.R. (1957) *J. Am. Chem. Soc.*, **79**, 5072.
21. Reichert, K.H. and Meyer, K.R. (1973) *Makromol. Chem.*, **169**, 163.
22. Andresen, A., Cordes, H.-G., Herwig, J., Kaminsky, W., Merck, A., Mottweiler, R., Pein, J., Sinn, H. and Vollmer, H.-J. (1976) *Angew. Chem.* **88**, 688.
23. Kaminsky, W. (1986), in *History of Polyolefins*, (eds R.B. Seymour and T. Cheng), Reidel, Dordrecht, p. 257.
24. Sinn, H. and Kaminsky, W. (1980) *Adv. Organomet. Chem.*, **18**, 99.
25. Sinn H., Kaminsky, W., Vollmer, H. and Woldt, E. (1984) US Patent 4 404 344 (Feb., 27, 1984), assigned to BASF AG.
26. Suter, U.W. and Neuenschwander, P. (1981) *Macromolecules*, **14**, 528.
27. Wild, F.R.W.P., Zsolnai, L., Huttner, G. and Brintzinger, H.H. (1982) *J. Organomet. Chem.*, **232**, 233.
28. Ewen, J.A. (1984) *J. Am. Chem. Soc.*, **106**, 6355.
29. Kaminsky, W., Külper, K., Brintzinger, H.H. and Wild F.R.P.W. (1985) *Angew. Chem. Int. Ed. Engl.*, **24**, 507.
30. Hlatky, G.G., Eckmann, R.R. and Turner, H.W. (1992) *Organometallics*, **11**, 1413.
31. Turner, H.W. and Schrock, R.R. (1983) *J. Am. Chem. Soc.*, **105**, 4942.
32. Jordan, R.F. (1991) *Adv. Organomet. Chem.*, **32**, 325.
33. Van der Leck, Y., Angermund, K., Reffke, M., Kleinschmidt, R., Goretzki, R. and Fink, G. (1987) *Chem. Eur. J.*, **3**(4), 585.
34. Antberg, M., Dolle, V., Klein, R., Rohrmann, J., Spaleck, W. and Winter, A. (1990) *Stud. Surf. Sci. Catal.*, **56**, 502.

35. Spaleck, W., Antberg, M., Aulbach, M., Bachmann, B., Dolle, V., Haftka, S., Küber, F., Rohrmann, J. and Winter, A. (1983), in *Ziegler Catalysts*, (eds G. Fink, R. Mülhaupt and H.H. Brintzinger), Springer-Verlag, Berlin, p. 83.
36. Burger, P., Hortmann, K. and Brintzinger, H.H. (1993) *Macromol. Symp.*, **66**, 127.
37. Stehling, U., Diebold, J., Kirsten, R., Röll, W., Brintzinger, H.H., Jüngling, S., Mülhaupt, R. and Langhauser, F. (1994) *Organometallics*, **13**, 964.
38. Jüngling, S., Mülhaupt, R., Stehling, U., Brintzinger, H.H., Fischer, D. and Langhauser, F. (1995) *J. Polymer Sci., Part A, Polymer Chem.*, **33**, 1305.
39. Fischer, D. and Mülhaupt, R. (1991) *J. Organomet. Chem.*, **417(1–2)**, C7–C11.
40. Ko, Y.S., Park, Y.W. and Woo, S.I. (1996) *Macromolecules*, **29**, 7305.
41. Deng, H., Winkelbach, H., Taeji, K., Kaminsky, W. and Soga, K. (1996) *Macromolecules*, **29**, 6371.
42. Rieger, B. and Chien, J.C.W. (1989) *Polymer Bull.*, **21**, 159.
43. Grassi, A., Zambelli, A., Resconi, L., Albizzati, E. and Mazzocchi, P. (1988) *Macromolecules*, **21**, 617.
44. Spaleck, W., Küber, F., Winter, A., Rohrmann, J., Bachmann, B., Antberg, M., Dolle, V. and Paulus, E.F. (1994) *Organometallics*, **13**, 954.
45. Stehling, U., Diebold, J., Kirsten, R., Röll, W., Brintzinger, H.H., Jüngling, S. and Mülhaupt, R. (1994) *Organometallics*, **13**, 964.
46. Vanderleck, Y., Angermund K., Reffke, M., Kleinschmidt, R., Goretzki, R. and Fink, G. (1997) *Chem. Eur. J.*, **3**(4), 585.
47. Fischer, D. and Mülhaupt, R. (1994) *Makromol. Chem. Phys.*, **195**, 1433.
48. Thomann, R., Wang. C., Kressler, J. and Mülhaupt, R. (1996) *Macromolecules*, **29**, 8425.
49. Zambelli, A., Bajo, E. and Rigamonti, E. (1978) *Makromol. Chem.*, **179**, 1249.
50. Ewen, J.A., Jones, R.L., Razavi, A. and Ferrara, J.D. (1988) *J. Am. Chem. Soc.*, **110**, 339.
51. Ewen, J.A., Elder, M.J., Jones, R.L., Curtis, S. and Cheng, H.N. (1990) *Stud. Surf. Sci. Catalysis*, **56**, 439.
52. Ewen, J.A., Elder, M.J., Jones, R.L., Haspeslagh, L., Atwood, J.L., Bott, S.G. and Robinson, K. (1991) *Macromol. Symp.*, **48/49**, 253.
53. Lovinger, A.J., Lotz, B., Davis, D.D. and Schumacher, M. (1994) *Macromolecules*, **27**, 6603.
54. Stocker, W., Schumacher, M., Graff, S., Lang, J., Wittmann, J.C., Lovinger, A.J. and Lotz, B. (1994) *Macromolecules*, **27**, 6948.
55. Schumacher, M., Lovinger, A.J., Agarwal, P., Wittmann, J.C. and Lotz, B. (1994) *Macromolecules*, **27**, 6956.
56. Petermann, J. (1996) *Polymer*, **37**, 4417.
57. Lotz, B., Wittmann, J.C. and Lovinger A.J. (1996) *Polymer*, **37**, 4979.
58. Thomann, R., Wang, C., Kressler, J., Jüngling, S. and Mülhaupt, R. (1995) *Polymer*, **36**, 3795.
59. Eckstein A., Suhm. J., Friedrich C., Maier, R.-D., Sassmannshausen, J., Bochmann, M. and Mülhaupt, R., manuscript in preparation
60. Po, R. and Cardi, N. (1996) *Prog. Polymer Sci.*, **21**, 47–88.
61. M. Kuramoto (1996), in *Polymeric Materials Handbook*, Vol. 9, (ed. J.C. Salamone), CRC Press, Boca Raton, FL, p. 6828.
62. Ishihara, N. (1995) *Macromol. Symp.*, **89**, 553.
63. Mülhaupt, R., Uschek, T., Fischer, D. and Setz, S. (1993) *Polymer Adv. Technol.*, **4**, 439.
64. Mülhaupt, R., Duschek, T. and Rieger, B. (1991) *Makromol. Chem., Macromol. Symp.*, **48–49**, 317.

65. Duschek, T. and Mülhaupt, R. (1992) *Polymer Preprints*, **33**(1), 170.
66. Wörner, C., Rösch, J., Höhn, A. and Mülhaupt, R. (1996) *Polymer Bull.*, **36**, 303.
67. Mülhaupt, R., Duschek, T. and Rösch, J. (1993) *Polymer Adv. Technol.*, **4**, 465.
68. Shiono, T. and Soga, K. (1992) *Macromolecules*, **25**, 3356.
69. Mogstad, A. and Waymouth, R.M. (1992) *Macromolecules*, **25**, 2282.
70. Mogstad, A., Kresti, M.R., Coates, G.W. and Waymouth, R.M. (1993) *Polymer Preprints*, **34**, 211.
71. Wisseroth, K. (1977) *Chemiker Zeitung*, **101**, 271.
72. Shih, C.K., and Su A.C.L. (1987), in *Thermoplastic Elastomers*, (eds N.R. Legge, G. Holden and H.E. Schroeder), Hanser, Munich, p. 91.
73. Collette, J.W., Tullock, C.W., MacDonald, R.N., Buck, W.H., Su, A.C.L., Harrell, J.R., Mülhaupt, R. and Anderson, B.C. (1989) *Macromolecules*, **22**, 3851.
74. Llinas, G.H., Dong, S.H., Mallin, D.T., Rausch, M.D., Lin, Y.G., Winter, H.H. and Chien, J.W.C. (1992) *Macromolecules*, **25**, 1242.
75. Gauthier, W.J. and Collins, S. (1995) *Macromol. Symp.*, **98**, 223.
76. Canevarolo, S. and Decandia, F. (1995) *J. Appl. Polymer Sci.*, **57**, 533.
77. Canevarolo, S.V., Decandia, F. and Russo, R. (1995) *J. Appl. Polymer Sci.*, **55**, 387.
78. Canevarolo, S and Decandia, F. (1994) *J. Appl. Polymer Sci.*, **54**, 2013.
79. Coates, G.W. and Waymouth, R.M. (1995) *Science*, **267**, 217.
80. Hauptmann, E., Waymouth, R.M. and Ziller, J.M. (1995) *J. Am. Chem. Soc.*, **117**, 11586.
81. Resconi, L. Jones, R.L., Rheingold, A.L. and Yap, G.P.A. (1996) *Organometallics*, **15**, 998.
82 Resconi, L. and Silvestri, R., in *Polymeric Materials Encyclopedia*, (ed. J.C. Salamone), CRC Press, Boca Raton, FL, p. 6609.
83. Chien, J.C.W., manuscript in preparation
84. Chien, J.C.W. and Nozaki T. (1993) *J. Polymer Sci., Part A, Polymer Chem.*, **29**, 227.
85. Herfert, N. and Fink, G. (1992) *Polymer Mat. Sci. Engng*, **67**, 31.
86. Chien, J.C.W. and He, D. (1991) *J. Polymer Sci., Part A, Polymer Chem.*, **29**, 1585.
87. Chien, J.C.W. and He, D. (1991) *J. Polymer Sci., Part A, Polymer Chem.*, **29**, 1609.
88. Chien, J.C.W. and He, D. (1991) *J. Polymer Sci., Part A, Polymer Chem.*, **29**, 1595.
89. Chien, J.C.W. and He, D. (1991) *J. Polymer Sci., Part A, Polymer Chem.*, **29**, 1603.
90. Zambelli, A., Grassi, A., Galimberti, M., Mazzocchi, R. and Piemontesi, F. (1983) *Macromol. Rapid Commun.*, **4**, 417.
91. Ewen, J.A. (1986) *Stud. Surf. Sci. Catalysis*, **25**, 271.
92. Kaminsky, W., Miri, M., Sinn, H. and Woldt, R. (1983) *Makromol. Chem. Rapid Commun.*, **4**(6), 417.
93. Busico, V., Mevo, L., Palumbo, G., Zambelli, A. and Tancredi, T. (1983) *Macromol. Chem. Phys.*, **184**, 2193.
94. Kaminsky, W., Külper, W. and Niedoba, S. (1986) *Macromol. Symp.*, **3**, 377.
95. Kaminsky, W. and Miri, M. (1985) *J. Polymer Sci., Part A, Polymer Chem.*, **23**, 2151.
96. Uozomi, T. and Soga, K. (1992) *Macromol Chem. Phys.*, **193**, 8223.

97. Soga, K., Shiono, T. and Doi Y. (1992) *Polymer Bull.*, **10**, 168.
98. Schneider, M.J. and Mülhaupt, R. (1997) *Macromol. Chem. Phys.*, **198**, 1121.
99. Schneider, M.J., Suhm, J., Mülhaupt, R., Prosenc, M.-H. and Brintzinger, H.H. (1997) *Macromolecules*, **30**, 3164.
100. Cheung, Y.W. and Guest, M.J. (1996) *ANTEC 96*, 1634.
101. Sernetz, F.G., Mülhaupt, R., and Waymouth, R.M. (1996) *Macromol. Chem. Phys.*, **197**, 1071.
102. Sernetz, F.G. and Mülhaupt, R. (1997) *Polymer.*
103. Sernetz, F.G., Mülhaupt, R. and Waymouth, R.M. (1997) *Polymer Bull.*, **38**(2), 141.
104. Batistini, A. (1995) *Macromol. Symp.*, **100**, 137.
105. Minick, F., Moet, A., Hiltner, A., Baer, E. and Chum, S.P. (1995) *J. Appl. Polymer Sci.*, **58**, 1371.
106. Thomann, R., Kressler, J. and Mülhaupt, R. (1997) *Macromol. Chem. Phys.*, **198**, 1271.
107. Galimberti, M., Martini, E., Piemontesi, F., Sartori, F., Camurati, I., Resconi, L. and Albizzati, E. (1995) *Macromol. Symp.*, **89**, 259.
108. Miller, S.A. and Waymouth, R.M. (1995) in *Ziegler Catalysts*, (eds G. Fink, R. Mülhaupt and H.H. Brintzinger), Springer-Verlag, Berlin, p. 440.
109. Kaminsky, W. and Noll, A. (1995) in *Ziegler Catalysts*, (eds G. Fink, R. Mülhaupt and H.H. Brintzinger), Springer-Verlag, Berlin, p. 151.
110. Kaminsky, W. and Spiehl, R. (1989) *Macromol. Chem. Phys.*, **190**, 515.
111. Kaminsky, W. and Noll, A. (1993) *Polymer Bull.*, **31**, 175.
112. Huang, D.J. and Cheng, C.H. (1995) *J. Organomet. Chem.*, **490**, C1.
113. Kaminsky, W., Bark, R., Spiehl, R., Möller-Lindenhof, N. and Niedoba, S. (1988), in *Transition Metals and Organometallics as Catalyst for Olefin Polymerization*, (eds W. Kaminsky and H. Sinn), Springer-Verlag, Berlin, p. 291.
114. Collins, S., and Kelly, W.M. (1992) *Macromolecules*, **25**, 233.
115. Chedron, H., Brekner, M.-J. and Osan, F. (1994) *Angew. Makromol. Chem.*, **223**, 121.
116. Herfert, N. and Fink, G. (1993) *Macromol Symp.*, **66**, 157.
117. Benedikt, G.M., Goodall, B.L., Marchant, M.S. and Rhodes, L.F. (1994) *New J. Chem.*, **18**, 105.
118. Arnold, M., Henschke, O. and Koller, F. (1996) *J. Macromol. Sci., Pure Appl. Chem.*, **A33** (Suppl. 3–4), 219.
119. Resconi, L. and Waymouth, R.M. (1990) *J. Am. Chem. Soc.*, **112**, 4953.
120. Cavallo, L., Guerra, G., Corradini P., Resconi, L. and Waymouth, R.M. (1993) *Macromolecules*, **26**, 260.
121. Coates, G.W. and Waymouth, R.M. (1992) *J. Mol. Catalysis*, **76**, 189.
122. Coates, G.W. and Waymouth, R.M. (1993) *J. Am. Chem. Soc.*, **115**, 91.

Keywords: metallocene catalysis, polymerization mechanisms, steric control, syndiotactic PP, atactic PP, tacticity, stereoregularity, regioregularity, constrained geometry catalyst, single site catalyst, cycloolefin copolymer, (COC), 'oscillating' metallocene catalyst, Ziegler–Natta catalysts, chain branching, properties of metallocene PP, morphology, conformation, functionalized polymers, block copolymers, copolymerization, elastomeric PP (ELPP), blending, polyethylene, cycloaliphatic polymers, ethylene/propylene rubber (EPM).

Microporous polypropylene films and fibers

Yukio Mizutani

INTRODUCTION

Recently, microporous polypropylene (PP) materials (sheets, films and hollow fibers) have gained acceptance in various fields of separation technology. Usually, these microporous materials are prepared by using uniaxial or biaxial stretching of adequate 'preforms'. Some examples of processing routes are:

1. Preparation of a PP film having a row lamellar morphology by uniaxial stretching to develop microporous structure [1];
2. Preparation of a PP film containing an organic substance, such as dioctylphthalate, and fine powdery SiO_2 by extraction of the organic substance and SiO_2 and stretching in order to modify the microporous structure;
3. Extrusion of a mineral oil extended PP on a chilling roll to produce a gelled film by stretching to develop microporous structure;
4. Preparation of a PP film containing a suitable filler by biaxial stretching to develop microporous structure [2].

Also, microporous PP hollow fibers and fibers are prepared by uniaxially stretching PP microtubes and fibers containing various fillers, respectively [3–5].

Polypropylene: An A–Z Reference
Edited by J. Karger-Kocsis
Published in 1999 by Kluwer Publishers, Dordrecht. ISBN 0 412 80200 7

MICROPOROUS PP FILMS CONTAINING FILLERS [2]

Microporous PP films were first prepared by biaxial stretching of PP films containing fillers in 1973. Now, such films are commercially available.

Preparation of preforms

PP powder, filler (e.g. $CaCO_3$ or SiO_2), and additives (antioxidant and plasticizer) are well mixed and extruded to pellets, from which a thin base film is extruded.

The filler content of the base film is usually between 40–70 wt.%. Afterwards, this base film is stretched successively in the machine direction (MD) and the transverse direction (TD) in order to develop microporous structure. The main advantages of this process are: (1) easy control of pore size and porosity, which significantly affect the gas permeability; (2) tailoring of the mechanical properties (tensile strength at yield and elongation to break) in both MD and TD, which is of great practical importance; (3) uniform stretchability of the base film, which is one of the most important issues in manufacturing of microporous PP films.

Microporous structure

It is straightforward to show how the microporous structure is formed. Figure 1 shows scanning electron micrographs (SEMs) taken on the surface and cross-section of the base film, in which well dispersed $CaCO_3$ filler particles can be resolved. Figure 2 demonstrates the effect of the stretching in the MD. One can see many elliptical pores with various sizes in the surface, aligned parallel to the MD. The SEM picture from the cross-section in the MD is very different from that in the TD. The PP matrix is stretched in the MD at first and transformed to an interlocking structure, in which filler particles can be observed. On the other hand, SEM taken from the cross-section in TD shows no layered structure and the PP phase exhibits a cellular structure packed with filler particles. Figure 3 shows SEM pictures from the surface and the cross-section of the biaxially stretched film, which was produced by stretching the film shown in Figure 2 in the TD. In the surface, the shape of the pores changes from elliptical to circular and the continuous PP phase develops fibrous characteristics. On the other hand,the cross-section consists of a layered fibrous structure. Consequently, the mechanism of pore formation can be explained as follows. The continuous PP phase is split at the poles of filler particles by stretching in the MD, and thus voids are created. Subsequent stretching in the TD changes the voids to round pores. The resultant pores become larger with increasing stretching ratio and coalesce. This results in an interpenetrating structure since both

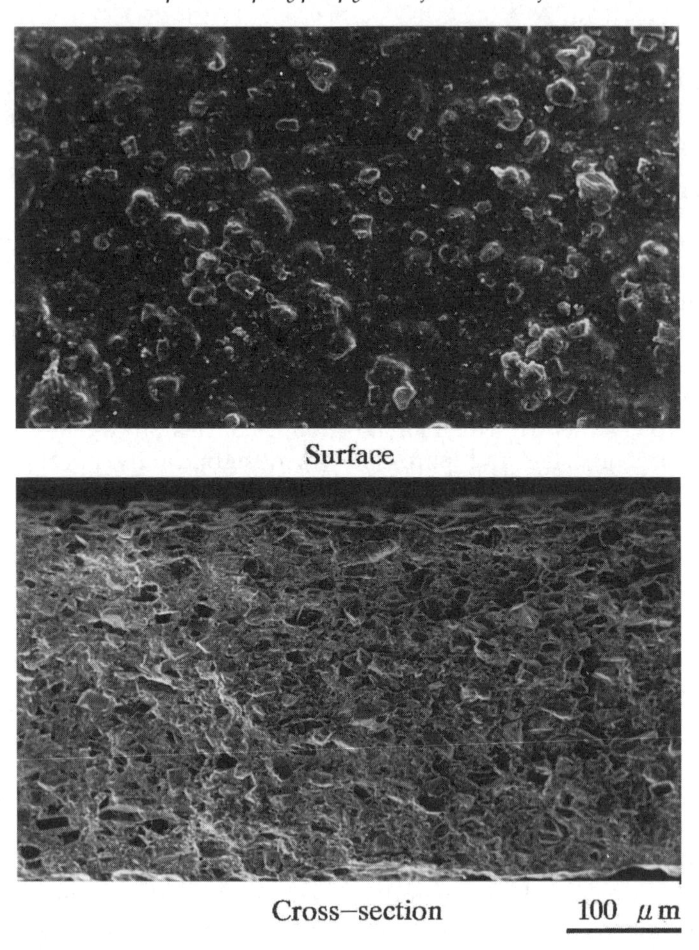

Figure 1 SEM pictures on a base film containing $CaCO_3$ filler, (68 wt.% with a mean particle size of 3 μm).

phases, i.e. pores (or air) and PP, become continuous. In addition, since the base film is biaxally stretched in both surface directions, a beneficial layered structure is generated. The resultant microporous structure is considerably complex, because the tortuosity factor of the pores is rather large i.e. 20–40. Recall that the tortuosity factor is unity when the shape of the pore is cylindrical and oriented perpendicular to surface.

Properties

Microporous films are soft and dry to the touch and also required to have well-balanced properties as follows: (1) flatness and uniformity of the

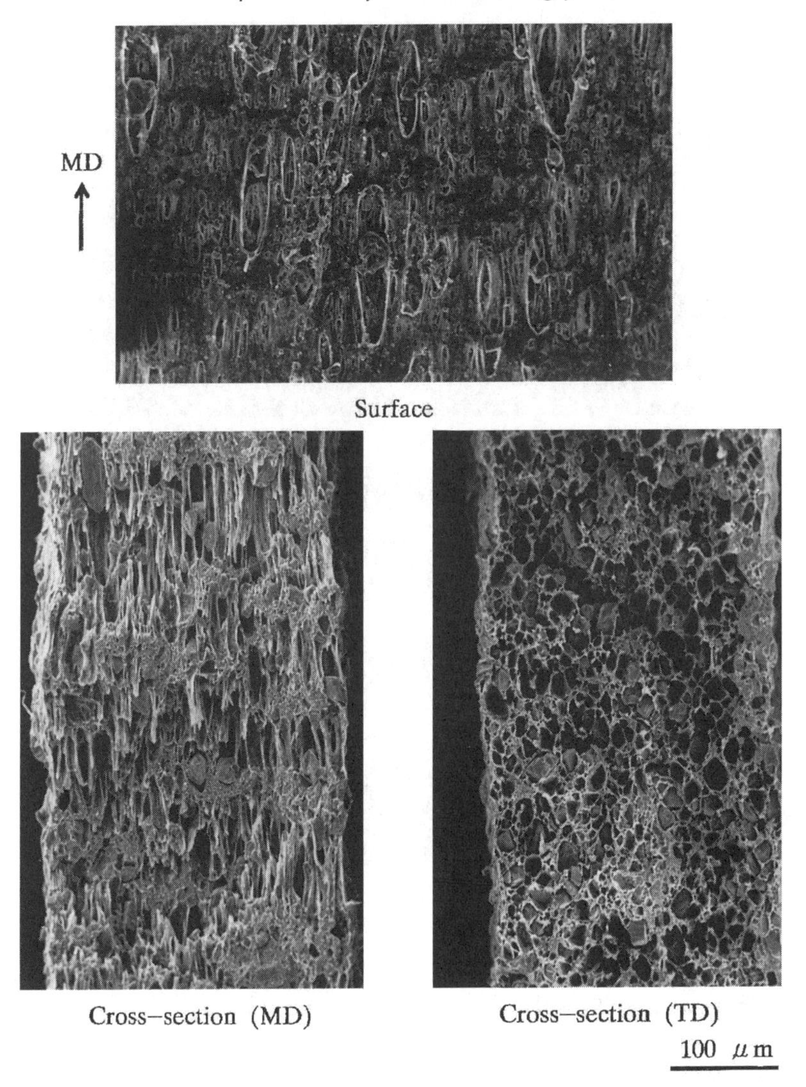

Figure 2 SEM picture on the PP film in Figure 1 stretched in MD (stretching ratio, 2).

properties over wide film area; (2) appropriate porosity and pore size; (3) well-balanced mechanical properties; (4) reasonable cost. Their properties can be controlled by adjusting the filler content, particle size of filler, stretching ratio, etc. The main properties of microporous films used to control gas permeation are listed in Table 1. It should be mentioned that

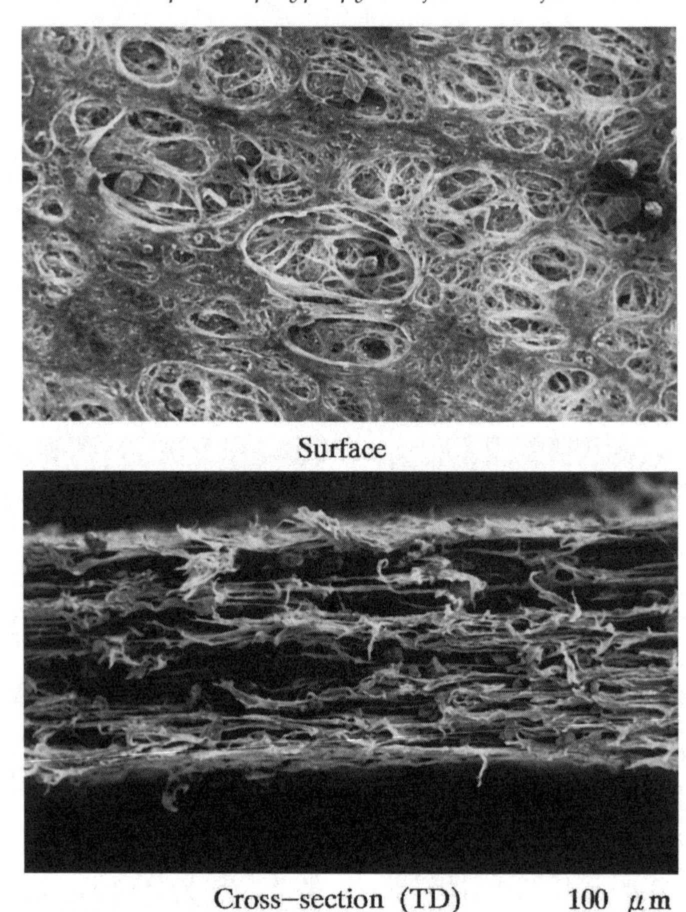

Figure 3 SEM picture on the PP film in Figure 2 after subsequent TD stretching (stretching ratio in TD is 5).

water vapor permeates through the films but not liquid water. The hydrophobicity of PP can be modified by using surface active agents.

Application

Figure 4 shows the position of microporous PP films among various separators. The microporous films are positioned between microfilters and nonwoven fabrics with respect to the pore size. There is some overlapping with microfilters, but the cost of microporous PP films is similar to that of nonwoven fabrics. Currently, microporous films are used in various ways, e.g. wrapping materials for moisture and oxygen

Table 1 Properties of microporous PP sheet

Property	*Standard*	*Grade*			*Celgard*
		120	S140	190	2500
Thickness (μm)		120	140	190	25
Apparent density (g/km^3)	JIS K-6767	0.55	0.45	0.58	0.47
Tensile strength (MD/TD) (kg/15 mm)	JIS K-7127	3.1/1.4	3.7/1.7	3.8/1.8	4.4/0.5
Gurley air permeability (s/100 ml)	JIS Z-0208	130	120	50	200
Water vapor permeability* (g/m^2 24 h)		4400	4500	4600	3900
Average pore size,† (μm)		1.2	0.6	1.6	0.06
Maximum pore size (μm)	ASTM F-316	3.7	1.9	4.6	0.27
Water-pressure resistance (mmH_2O)	JIS L-1092	4800	5300	4200	>20 000

* Determined under relative humidity of 90% at 40°C. †Value at 50 vol.% of the pore size distribution curve.

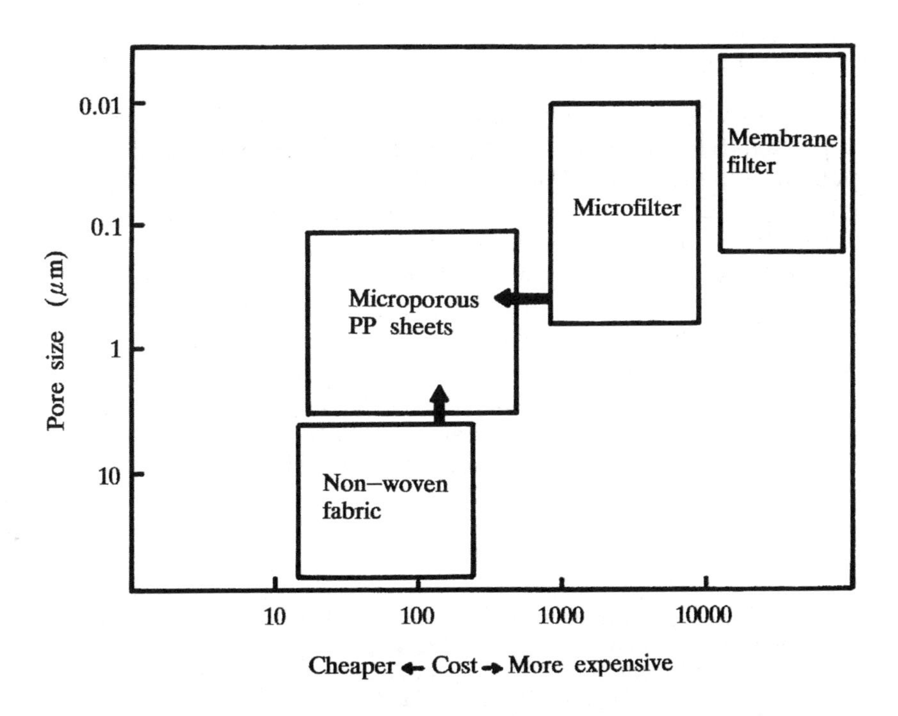

Figure 4 Position of microporous PP films among other separation materials.

absorption, which keep food fresh, etc. Furthermore, microporous PP films can be laminated with cloths, nonwoven fabrics, etc, and the resultant materials exhibit good mechanical strength and toughness. Accordingly, the application of microporous PP films can be further diversified.

MICROPOROUS PP HOLLOW FIBERS [3, 4]

Microporous PP hollow fibers are prepared as follows: blending PP powder, filler and additives, followed by extrusion to prepare pellet, extrusion to prepare microtube, and stretching. The properties of hollow fibers can be controlled by the filler content (40–60 wt.%), particle size of filler (0.1–3 μm), and the stretching ratio (4–6), respectively. Porosity and pore size of the resultant hollow fiber are in the range of 0.44–0.52 and 0.1–0.8 μm, respectively. In these hollow fibers, there are many slits (parallel to the fiber axis) in the surface whereas the inner section consists of fine interconnected PP fibrils. Also, microporous hollow fibers with double layers can be prepared by using coextrusion technique for preparation of the microtubes. These hollow fibers are used as a membrane filters (e.g. for water purification).

MICROPOROUS PP FIBERS [5]

Microporous PP fibers are prepared by the following route: blending of PP powder, filler and additives, followed by extrusion to prepare pellet, spinning and stretching. Their properties can be tailored by the filler content, particle size of filler and stretching ratio. Here, it is preferable to use a filler with relatively smaller particle size, and there is an upper limit of the filler content so as not to clog the spinneret, of which the diameter is usually small. In the cases of $CaCO_3$ and poly(methylsilsesquioxane) fillers (particle size, 0.09 and 0.3 μm in diameter, respectively), the upper limits are 25 and 35 wt.%, respectively, when the diameter of the spinneret is 0.7 mm. Porosity, pore size and specific surface area are in the range of 0.1–0.2, 0.01–0.05 μm, and 50–140 m^2/g, respectively. Their practical application is an object of research and development activity.

REFERENCES

1. Bierenbaum, H.S., Isacson, R.B., Druin M.L. and Plovan, S.G. (1974) Microporous polymeric film. *Ind. Engng Chem.*, **13**(1) 2–9.
2. Mizutani, Y., Nakamura, S., Kaneko, S. and Okamura, K. (1993) Microporous PP sheets. *Ind. Engng Chem. Res.*, **32**(1) 221–227.
3. Nago, S. and Mizutani, Y. (1994) Microporous PP hollow fibers containing poly(methylsilsesquioxane) filler. *J. Appl. Polymer Sci.*, **53**(12), 1579–1587.

4. Nago, S. and Mizutani, Y. (1995) Microporous PP hollow fibers with double layers, *J. Appl. Polymer Sci.*, **56**(2), 253–261.
5. Nago, S. and Mizutani, Y. (1996) Micoroporous PP fibers containing $CaCO_3$ fillers. *J. Appl. Polymer Sci.*, **62**(1), 81–86.

Keywords: microporous film, microporous hollow fiber, microporous fiber, filler, uniaxial stretching, biaxial stretching, permeation, water vapor permeation, air permeation, tortuosity factor.

Miscibility and phase separation in polypropylene blends

David J. Lohse

Blends of polypropylene (PP) with other polymers are used widely and are of great commercial importance. Many of these are immiscible at all compositions and temperatures of interest; for example, the blends of PP and polyamides (PAs) that are used in fiber applications are highly incompatible. However, recent work has shown that PP is miscible with a number of other polyolefins, and the strength of the interactions between these polymers has been measured. Here the data that have been obtained on PP blend miscibility are compiled and work on the kinetics of phase separation for such blends is described.

POLYPROPYLENE BLEND MISCIBILITY

It is clear that blends of most polymers with PP are highly incompatible, due to the large phase sizes exhibited by such materials. However, on the basis of chemical similarity, one might suspect that PP would be miscible with other saturated hydrocarbon polymers, such as the other polyolefins, at least under certain conditions. This has been an area of much investigation in the last fifteen years, and this is summarized in this section.

The investigation of polymer blend miscibility has been greatly enhanced in recent years by the use of small angle neutron scattering (SANS). This is a very powerful tool in examining the miscibility of blends since it can be used to directly measure the interactions between polymers. The most commonly used expression for the free energy of

Polypropylene: An A–Z Reference
Edited by J. Karger-Kocsis
Published in 1999 by Kluwer Publishers, Dordrecht. ISBN 0 412 80200 7

mixing two polymers (designated by the subscripts 1 and 2) per unit volume, ΔG_m is the Flory–Huggins–Staverman (FHS) form:

$$\frac{\Delta G_m}{kT} = \frac{\phi_1}{N_1 v_1} \ln\phi_1 + \frac{\phi_2}{N_2 v_2} \ln\phi_2 + \frac{\chi_{12}}{v} \phi_1\phi_2 \tag{1}$$

where ϕ_i is the volume fraction of polymer i, N_i its degree of polymerization, v_i its monomer volume, χ_{12} the interaction parameter, v a reference volume (usually $v = (v_1 v_2)^{1/2}$), k Boltzmann's constant, and T the temperature. Since the definitions of N and χ_{12} depend on the choice of reference volume, when comparing the results for a wide range of polymers it is more convenient to express the free energy of mixing in terms of directly measurable quantities. Let M_i and ρ_i be the molecular weight and density of polymer i, and also define X_{12} as the interaction energy density (or interaction strength); in the literature the symbols B and Λ are also used for this quantity. This is given by:

$$X_{12} = \frac{\chi_{12}}{v} kT. \tag{2}$$

Using these, the free energy of mixing becomes:

$$\Delta G_m = RT\left\{\frac{\phi_1\rho_1}{M_1} \ln\phi_1 + \frac{\phi_2\rho_2}{M_2} \ln\phi_2\right\} + X_{12}\phi_1\phi_2 \tag{3}$$

where R is the gas constant. This is all written in terms of measurable quantities; the interaction energy density can be directly obtained from SANS for a miscible polymer blend.

The first attempt to get some direct, quantitative information of PP miscibility was by Wignall *et al.* [1]. They examined blends of PP with polyethylene (PE) by SANS, and saw that they were highly immiscible, so they could not measure X_{12} for this system. They were able to show that the phase domains in these blends were quite large. Lohse [2] also used SANS to study blends of PP with ethylene–propylene random copolymers (EP) of various compositions. He showed that these were immiscible when the ethylene content of the EP was more than about 12% for molecular weights of 100 000 g/mole.

The first work which directly measured the interaction strengths for PP blends involved the use of model PPs made by the saturation of polydienes [3]. Two PP models can be produced in this way, one with a head-to-tail structure, as in commercial PP (htPP), and the other with a head-to-head conformation (hhPP); these two polymers have strikingly different phase behavior with other polymers. Table 1 shows some examples of the values of the interaction strength that have been

Table 1 Interaction energy density of selected polypropylene blends

Blend	$X_{12}(T)$ (MPa)				
	27°C	*51°C*	*82°C*	*121°C*	*167°C*
htPP/EB78		0.0643	0.0457	0.0300	0.0247
htPP/PB	0.148	0.118	0.0881	0.0605	
hhPP/EB66	0.0619	0.0398	0.0214	0.00811	0.00467
hhPP/EB78	0.118	0.103	0.0901	0.0815	0.0702
hhPP/PEP	0.0285	0.0187	0.0110	0.0134	0.0229
hhPP/htPP	0.0883	0.0780	0.0690	0.0634	0.0562
htPP/PEP	0.106	0.123	0.148	0.165	0.177
DhhPP/PIB	−0.327	−0.248	−0.138		0.0434

htPP, head-to-tail polypropylene; hhPP, head-to-head polypropylene; DhhPP, a partially deuterated version of hhPP; EB*x*, random ethylene-butene copolymer with *x* weight percent butene; PB, poly(1-butene); PEP, alternating ethylene–propylene copolymer; PIB, polyisobutylene.

measured for these polypropylenes. It should be noted that many kinds of phase diagrams have been seen for the PP blends in this work: ones showing an upper critical solution temperature (UCST; X decreasing with temperature, e.g. hhPP with htPP), others with a lower critical solution temperature (LCST; X increasing with temperature, such as htPP with PEP, the alternating ethylene–propylene copolymer, and hhPP with polyisobutylene), and even blends showing both an LCST and a UCST (hhPP with PEP). Note also how different the values are for blends of the two different versions of PP with the same polymer (e.g. hhPP/PEP versus htPP/PEP), and that even the hhPP/htPP blend displays a fairly large interaction energy. This is a much greater variety than would at first be expected for systems where only dispersive van der Waals forces are involved, so the behavior of PP blends is quite surprising.

Several of these values have been confirmed by other work as well. For blends of htPP and PEP a very similar value of X_{12} was found by the determination of the phase diagrams for both blends and block copolymers [4]: 0.0981 MPa at 25°C versus a value of 0.104 MPa from the SANS work. Microscopic and thermal analysis of blends of PP with polybutene showed that they had a UCST behavior consistent with the value shown in the table [5]. On the other hand, work on block copolymers of PP and PE [6] showed that X_{12} for these blends must be at least 1.93 MPa, which is far greater than the values derived from the SANS studies of the blends [3].

The miscibility of PP with other hydrocarbon polymers has also been studied recently. Yamaguchi *et al.* [7] examined blends of PP with ethylene–hexene and ethylene–butene copolymers by microscopy and thermal analysis. The copolymers that had large amounts of the comonomer were seen to be miscible with the PP, which is consistent with the values of X_{12} in Table 1 for the model ethylene–butene copolymers. Also, Cimmino *et al.* [8] showed that hydrogenated oligomers of cyclopentadiene were miscible with PP. These molecules were not well characterized and so no quantitative determination of the interactions could be made, but these blends were clearly seen to be single phase. Putting together all of these studies, there is now a good deal of data on PP miscibility, so that a reasonable understanding of the origins of such miscibility may now be possible [3].

PHASE SEPARATION

Much less work has been done on the kinetics of phase separation for PP blends. This is in general quite difficult to study for highly incompatible mixtures, since one cannot get to a well-mixed starting state. Careful studies of the phase separation can only be done for those blends that are just slightly incompatible, which means that again this has to be done on blends of PP with other polyolefins.

An early study of this was done by Inaba *et al.* [9] on blends of PP with EP copolymer. They prepared the blends in solution so that, by rapidly quenching the mixture as it was precipitated, they could preserve a miscible state of the solution. They then investigated the blend by light scattering and microscopy to see how quickly the domains grew. They were able to show that the domains grew over time, t, with a $t^{1/3}$ power law. The 50/50 blends also maintained a bicontinuous domain morphology while separating, consistent with a spinodal decomposition mechanism; this was also seen by Lohse [2].

The phase separation of blends of PP and EP copolymer was also studied by Mirabella [10]. He followed the morphology of blends with EP as the minor component ($\sim 17\%$) by electron microscopy and also saw the same $t^{1/3}$ power law measured by Inaba *et al.* [9]. He showed that this was in good agreement with the predictions of an Ostwald ripening mechanism. The magnitude of the domain growth rate was consistent with the value of X_{12} that can be inferred from the SANS work.

CONCLUSION

Our understanding of the miscibility of polypropylene blends has been greatly enhanced in the last few years. Basic data on the strength of the interactions of PP with other polymers and the rates of the phase separation in such systems have now been obtained. This information should prove quite valuable in developing improved materials from such blends.

REFERENCES

1. Wignall, G.D., Child, H.R. and Samuels, R.J. (1982) Structural characterization of semicrystalline polymer blends by small-angle neutron scattering. *Polymer*, **23**, 957–964.
2. Lohse, D.J. (1986) The melt compatibility of blends of polypropylene and ethylene-propylene copolymers. *Polymer Engng Sci.*, **26**, 1500–1509.
3. Graessley, W.W., Krishnamoorti, R., Reichart, G.C., Balsara, N.P., Fetters, L.J. and Lohse, D.J. (1995) Regular and irregular mixing in blends of saturated hydrocarbon polymers. *Macromolecule*, **28**, 1260–1270.
4. Lohse, D.J., Fetters, L.J., Doyle, M.J., Wang, H.-C. and Kow, C. (1993) Miscibility in blends of model polyolefins and corresponding diblock copolymers: Thermal analysis studies. *Macromolecules*, **26**, 3444–3447.
5. Cham, P.M., Lee, T.H. and Marand, H. (1994) On the state of miscibility of isotactic poly(propylene)/isotactic poly(1-butene) blends: Competitive liquid-liquid demixing and crystallization processes. *Macromolecules*, **27**, 4263–4273.
6. Sakurai, K., MacKnight, W.J., Lohse, D.J., Schulz, D.N., Sissano, J.A., Wedler, W. and Winter, H.H. (1996) Dynamic viscoelastic properties of poly(ethylene-propylene) diblock copolymers in the melt state and solutions. *Polymer*, **37**, 5159–5163.
7. Yamaguchi, M., Miyata, H. and Nitta, K.-H. (1996) Compatibility of binary blends of polypropylene with ethylene-α-olefin copolymer. *J. Appl. Polymer Sci.*, **62**, 87–97.
8. Cimmino, S., DiPace, E., Karasz, F.E., Martuscelli, E. and Silvestre, C. (1993) Isotactic polypropylene/hydrogenated oligo(cyclopentadiene) blends: Phase diagram and dynamic-mechanical behavior of extruded isotropic films. *Polymer*, **34**, 972–976.
9. Inaba, N., Sato, K., Suzuki, S. and Hashimoto, T. (1986) Morphology control of binary polymer mixtures by spinodal decomposition and crystallization. 1. Principle of method and preliminary results on PP/EPR. *Macromolecules*, **19**, 1690–1695.
10. Mirabella, F.M. (1994) Phase separation and the kinetics of phase coarsening in commercial impact polypropylene copolymers. *J. Polymer Sci. Polymer Phys.*, **32**, 1205–1216.

Keywords: blend miscibility, interaction energy, phase diagram, phase separation kinetics, small angle neutron scattering (SANS), Flory–Huggins–Staverman equation, free energy of mixing, upper critical solution temperature (UCST), lower critical solution temperature (LCST).

Modelling and analysis of composites thermoforming

D. Bhattacharyya and G.R. Christie

The recent advent of thermoplastic sheets reinforced with continuous fibers, mats and woven fabrics has greatly interested the manufacturers of engineering components. However, the understanding of the deformation mechanisms for these materials is much more demanding due to a number of factors including: (a) the time-dependent visco-elastic behavior of the matrix at high temperatures; (b) the possibility of hyper-anisotropic material behavior due to fiber collimation; (c) fiber movement during forming; and (d) the complicated fiber/matrix and tool/workpiece interactions. It is also acknowledged that when producing a three-dimensional doubly curved component, especially using unidirectional preimpregnated sheet material, all the possible flow mechanisms involving resin percolation, transverse flow, interply slip and intraply slip will take place [1]. A complete analysis of any thermoforming process is quite involved and may become quite difficult to achieve due to the difficulties mentioned above. However, reasonably successful attempts have been made for developing kinematic, finite-element and analytical models that can help in understanding the various aspects of a thermoforming process, including the influences of different process parameters on the ultimate result. The present section briefly describes some of the techniques available to date for modelling and analyzing the thermoforming of composite sheets.

Polypropylene: An A–Z Reference
Edited by J. Karger-Kocsis
Published in 1999 by Kluwer Publishers, Dordrecht. ISBN 0 412 80200 7

KINEMATIC APPROACH

The main advantage of a kinematic approach is the absence of any need to have the constitutive behavior of the deforming material known *a priori* for macro-understanding of the deformation pattern. However, with the additional information on the constitutive characteristics, it may also be possible to generate more knowledge about the stress magnitudes from the strain distribution.

Although in order to make rational decisions on how to improve sheet forming operations, it is often necessary for the manufacturers to visualize and understand the deformations taking place, this knowledge can seldom be gained by merely observing the shape of the deformed part. Thickness variations are measurable and interesting, but alone do not describe the overall deformation since appreciable in-plane strains may take place without changing the sheet thickness, as is the case in pure drawing. In forming fiber-reinforced materials, the final fiber orientations give some clue to the deformations that have taken place. This is especially true with certain woven fabric composites for which such information is sufficient for determining overall forming strains. However, in the less constrained case of laminates made from unidirectional plies, the fiber orientations are much less useful since they give no measure of in-plane shear and transverse flow.

A greater insight into sheet forming processes can be gained by the so-called grid strain analysis (GSA) techniques [2]. These involve printing or etching a regular grid pattern onto the undeformed blank, forming it into its final shape, and then measuring the deformed grid positions. The undeformed and deformed grid points then describe the kinematics of the deformation, allowing forming strains to be determined. An obvious requirement is that the grid density be sufficient to describe the deformation of interest, without losing too much detail. Grid strain analysis is principally used to determine in-plane strains on the surface of the sheets. However, applying the condition of incompressibility, the thickness strain may also be determined. Similarly, knowledge of the curvature can also be used to determine the variation of strain through the thickness of the sheet in cases where through-thickness shear is negligible.

Grid strain analysis provides an understanding of the deformation by the surface only but can give an idea of how other plies influence its behavior. Recently GSA has been extensively used by several researchers [3–5] to analyze the deformation pattern while forming composite sheets and important information has been obtained regarding thickening and thinning of material, fiber buckling, blank shape optimization and the influence of process parameters. Martin *et al.* [3] have given a comprehensive account of GSA's application in composite sheet forming, where the strain analysis is based on a continuum approach, recently developed by Christie [5]. Unlike the previous methods where the deformed shape is approximated as a polyhedral surface or splines are fitted to the deformed geometry, this new approach uses the grid data to fit continuous parametric representations of the undeformed and deformed surfaces. Strains are then calculated at any point on the surface by finding the displacement gradient mapping the undeformed geometry to the deformed. The use of higher-order surface patches improves the accuracy of deformed surface representation. Strains can be displayed by arrow diagrams (orthogonal pairs of arrows in principal directions) or contour maps. These diagrams can be used to highlight various strain patterns, such as biaxial stretching strains, drawing strains (tensile/compressive combination) and regions of plane strain, which, in turn, indicate the areas of possible problems. A large stretching strain under biaxial or plane strain condition leads to material thinning which might eventually lead to sheet/tow splitting. On the other hand, fiber buckling is shown to be associated with large compressive strain gradient areas [3].

In continuous-fiber reinforced thermoplastic (CFRT) materials, the fibers can be treated as inextensible as the material deforms by in-plane shearing. This type of deformation is normally referred to as *trellis* action due to its geometric similarity with the garden trellis. The kinematics of this deformation suggests that a tensile strain in one direction is accompanied by a compressive strain in the perpendicular direction. It is also known [5] that for initially orthogonal fibers, the principal compressive strain always exceeds the principal tensile strain, suggesting sheet thickening. Figure 1 shows a good correlation of strains calculated by trellis kinematics with those determined by GSA on a Plytron® (35% glass fiber/PP composite) dome. The results support the notion that bidirectional laminates deform like networks of inextensible cords. They also show that, unlike metallic materials, thermoplastic sheets can undergo a fairly large magnitude of compressive strain (and accompanying thickness increase) without buckling. Martin *et al.* [3] have shown some interesting GSA results and their implications for various parts with different fiber architecture.

Another variation of the kinematic approach may be achieved by applying the fabric draping analogy [3] with the assumptions of fiber inextensibility and the hinging of fibers at their intersections. Furthermore, the location of an intersecting warp and a weft yarn is found by the intersection point between the surface and two spheres whose radii are equal to the fiber arc length. A further kinematic constraint is achieved by completely specifying the paths of one warp yarn and one weft yarn on the die surface. This analysis gives the postdeformation positions of the originally orthogonal fibers on the draped surface from

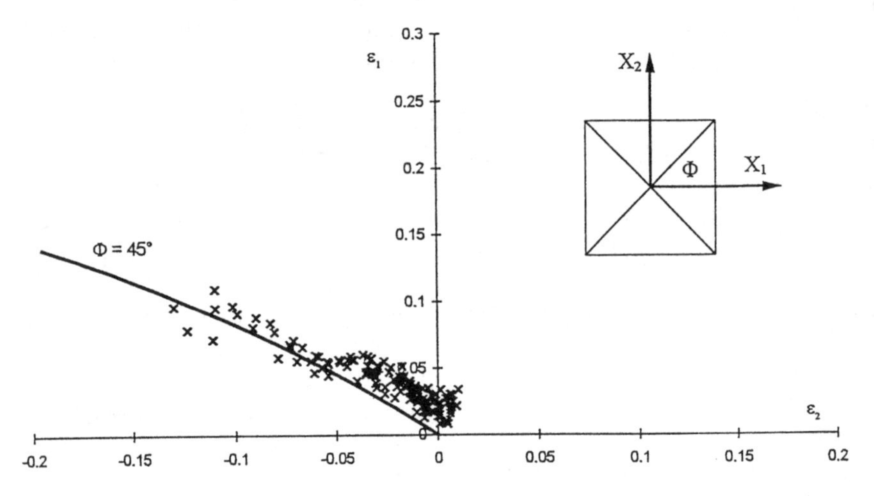

Figure 1 Strain distribution on a $[0°, 90°]_{2s}$ Plytron® (35% vol. fraction glass/polypropylene composite) dome. (After Martin *et al.* [3].)

which the nodal coordinates can be determined. If required, strain magnitudes may also be determined from the deformed coordinates, which compare very well with GSA results.

FINITE-ELEMENT APPROACH

Numerical solution techniques, such as the finite-element method, provide an accepted and systematic approach to applying theories of material behavior over continua of arbitrary geometry. As such, a numerical model of a molten composite could be developed simply by applying an appropriate constitutive equation along with suitable elements. However, when laminates are produced from a stack of discrete plies containing continuous (or long discontinuous) fibers, matrix-rich interlayers remain in the laminate. During thermoforming, these layers provide low viscosity zones over which interply slip can occur and create discontinuities that have to be accounted for during modelling. CFRT prepregs are also crossed by numerous zones of weakness and low volume fraction stemming either from interbundle spaces or the random fiber distribution. Therefore, along with through-thickness intraply shear, in-plane longitudinal intraply shear may also occur [6]. However, in contrast to interply slip where slip movements may be large, intraply slip occurs between parallel fibers, making it possible to macroscopically model the processes as continuous, using some average shear properties. It is then appropriate to treat a ply as the sensible unit of continuous deformation in a model of molten CFRT laminate.

Smiley [7] has developed a decoupled model of pressure forming of axisymmetric parts using superplastic aluminium alloy diaphragms. Decoupling has been achieved by assuming forming geometry and loads depend only on the behavior of relatively stiff diaphragms. Between the diaphragms, transverse flow in each ply is independently computed on a series of two-dimensional planes, using a Newtonian fluid model, while a kinematic scheme is used to update fiber orientations. This model has several major drawbacks and lacks sufficient flexibility to be applied to more general forming situations. In particular, the decoupling between the diaphragms and laminate, and between the plies themselves is not physically realistic.

Scherer [6] has carried out a large number of experiments and finite-element simulations to characterize the interply slip phenomenon. Ply-pull-out tests have been performed and smooth curves relating the applied force to the slipped distance and velocity have been generated. Then, attempts have been made to reproduce ply-pull-out results with a finite-element model, first by placing thin elements of fluid between elastic plies, and eventually by adopting a viscous-type contact/friction relation over a negligible inter-layer thickness. A further series of two-ply, two-dimensional forming simulations have investigated the influence of various slip conditions in practical forming situations. Since the study has focused on the interply slip phenomenon, only a simple, orthotropic elastic ply model has been employed. Despite this, Scherer's use of a friction model for interply slip is crucial in reducing the size and complexity of the problem.

O'Brádaigh and Pipes originally studied the plane stress flows of ideal fiber reinforced fluids (IFRF) and have used a penalty method to impose the fiber inextensibility constraint with biquadratic velocity/bilinear discontinuous tension elements. A linear and quasistatic scheme has been used to calculate instantaneous velocities which are multiplied by the time step to displace the mesh. Such an approach involves the build-up of considerable error unless time steps are extraordinarily small. Recently, this model has been improved by developing a mesh updating scheme that incorporates finite incompressibility and inextensibility constraints [8]. The new model also uses large displacement contact/friction elements to model tool contact and interply slip between layers of IFRF, in plane strain.

More recently Christie [5] has developed a finite-element simulation package, *SimForm*, using transversely anisotropic, hyperelastic discrete ply deformation as a first step to simulate the viscoelastic shear response discussed earlier in this section. The program employs three-dimensional continuum elements to describe deformations in each ply, while a contact procedure couples the motion of neighbouring plies to simulate the interply slip process. Interply slip is given a linear velocity-dependent shear stress response, corresponding to shear of thin interlayers of Newtonian fluid, which reproduces the main features of ply-pull-out tests. An added justification for this elastic ply model comes from the fact that through-thickness longitudinal intraply shear appears negligible in CFRT sheet forming. The use of C_1-continuous bicubic Hermite interpolation in the plane of the ply enables a fairly high anisotropy to be handled quite adequately. The simulations predict the deformation phenomena such as the onset of fiber instability and fiber reorientation, in two- and three-dimensional situations very well and the results are supported by experiments. However, the contact procedure may suffer slow convergence and stability problems when applied between highly anisotropic layers. Of the major deformation processes involved in forming CFRT laminates, only in-plane intraply shear and transverse flow are not well described by the *SimForm* model.

A group at Engineering Systems International, France, has adapted the company's PAMCRASH metal stamping and crash simulation code to the problem of thermoforming [9]. Using this explicit, dynamic code, the group has been able to carry out some very large thermoforming simulations of multiply laminates. Shell elements are used across each ply, with a reinforced fluid model for intraply shear and a viscous contact/friction relation between the plies. Thermal effects have also been considered. The use of an explicit model involves the approximation of mass lumping and other matrix diagonalization, and its conditional stability requires the use of very small time steps. Nevertheless, this approach has produced some very impressive results and has enormous potential for industrial application.

ANALYTICAL APPROACH

Tam and Gutowski [10] have developed an elegant, semi-analytical molten laminate model consisting of a number of elastic plies separated by thin layers of Newtonian fluid to simulate the matrix-rich interlayers. By assuming a strain field for the part, forming loads have been predicted for several bending examples. One of the model's assumptions is that the elastic layers have only extensional stiffness and no resistance to bending, making it most applicable to laminates containing very thin plies. Despite handling the coupling between laminae introduced by the viscous interlayers, the model seems only suited to the simple, two-dimensional flows studied in the paper.

REFERENCES

1. Cogswell, F.N. (1992) *Thermoplastic Aromatic Polymer Composites*, Butterworth–Heinemann, Oxford.

2. Zhang, Z.T. and Duncan, J.L. (1990) Developments in nodal strain analysis of sheet forming, *Int. J. Mech. Sci.*, **32**, 717–727.
3. Martin, T.A., Christie, G.R. and Bhattacharyya, D. (1997) Grid strain analysis and its application in composite sheet forming, in *Composite Sheet Forming*, Vol. 11, (ed. D. Bhattacharyya), Composite Materials Series, Elsevier, Amsterdam, pp. 217–246.
4. Krebs, J., Bhattacharyya, D. and Friedrich, K. (1997), Production and evaluation of secondary composite aircraft components – a comprehensive study, *Composites: Part A*, **28A**, 481–489.
5. Christie, G.R. (1997) Numerical modelling of fibre-reinforced thermoplastic sheet forming, University of Auckland, PhD dissertation.
6. Scherer, R. (1995) Thermoforming of unidirectional continuous fibre-reinforced polypropylene laminates and their modeling, in *Polypropylene: Structure, Blends and Composites*, Vol. 3, Chap. 8, (ed. J. Karger-Kocsis), Chapman & Hall, London, pp. 293–315.
7. Smiley, A.J. (1988) Diaphragm forming of carbon fibre reinforced thermoplastic composite materials, University of Delaware, PhD dissertation.
8. O'Brádaigh, McGuiness, G.M. and McEntee, S.P. (1997) Implicit finite element modelling of sheet forming processes, in *Composite Sheet Forming*, Vol. 11, (ed. D. Bhattacharyya), Composite Materials Series, Elsevier, Amsterdam, pp. 247–322.
9. Pickett, A.K., Queckbörner, T., de Luca, P. and Haug, E. (1995) An explicit finite element solution for the forming prediction of continuous fibre-reinforced thermoplastic sheets, *Composites Manufact.*, **6**(3/4), 237–244.
10. Tam, A.S. and Gutowski, T.G. (1989) Ply-slip during the forming of thermoplastic composite parts, *J. Composite Mat.*, **23**, 587–605.

Keywords: grid strain analysis, Plytron®, trellis effect, kinematic approach, finite-element approach (FEM), analytical approach, anisotropy, hyperelastic, interply slip, intraply slip, thermoforming, modelling sheet forming, composite laminates, continuous fiber-reinforced composites.

Modelling of the compression behavior of polypropylene foams

R. Denzer, F. Möller and M. Maier

INTRODUCTION

Due to the good energy absorbing properties under impact loading and deformability, polypropylene (PP) foams possess great potential for technical applications. Their possible usages cover different fields of packaging and safety systems (such as bumper cores, side impact paddings, knee bolsters, inserts of helmets, etc.). There are also structural applications (as core material in sandwich panels) with PP foams envisaged.

CELLULAR STRUCTURE AND DEFORMATION MECHANISMS

Figure 1 shows the cellular structure of a typical PP foam, which was manufactured by molding PP foam beads with superheated steam. Each bead consists of closed polyhedral cells, with no visible accumulation of material in the corners. The PP foam beads themselves are melted together at their surfaces and form a secondary cellular structure.

The compression behavior of PP foams is influenced by the cellular structure and by the mechanical properties of the PP matrix polymer. Figure 2(a) shows the compressive stress–strain curve of a PP foam schematically. The curve displays linear elastic behavior at low strains followed by a long collapse plateau, truncated by a regime of densification in which the stress rises steeply [1, 3].

Polypropylene: An A–Z Reference
Edited by J. Karger-Kocsis
Published in 1999 by Kluwer Publishers, Dordrecht. ISBN 0 412 80200 7

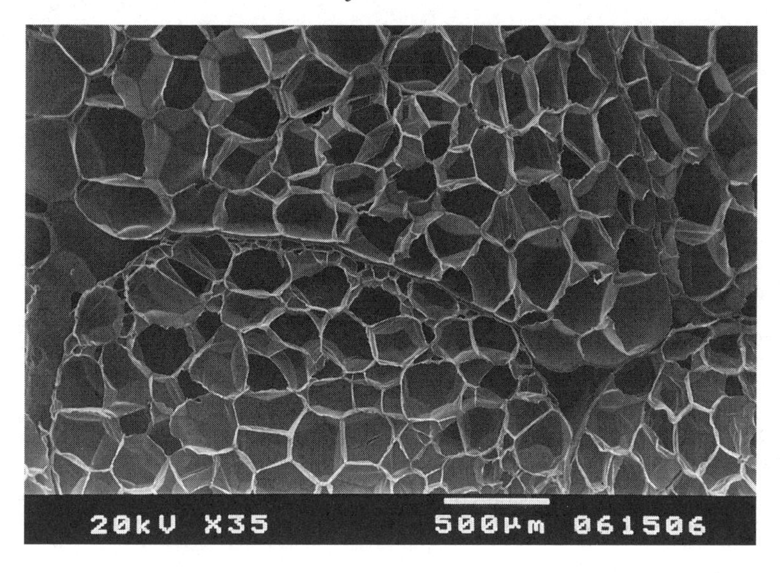

Figure 1 SEM of the cellular structure of a PP foam.

The linear elastic regime is caused by bending and stretching of the cell walls or faces. At higher strains, the cell walls begin to collapse by elastic or elastic–plastic buckling which results in the plateau. At even higher strains, the neighboring cell walls interact. This leads to the regime of densification, where the solid polymer is partly compressed itself. That is the reason for the steep stress increase at high strain values.

The compression behavior of the enclosed gas in the cells is also involved in the deformation mechanism of the cells. So, the compression behavior of a closed cell PP foam can be considered as a parallel connection of the compression behavior of the cells (the so-called skeleton of the foam) and the enclosed gas.

ANALYTICAL MODELS

In the literature [1, 2, 3], some analytical models have been developed to describe the compression behavior of closed-cell foams. Christensen [2] introduced a model to derive the Youngs's modulus E^*, the Poisson's ratio ν^* and the compressive strength σ^*_{el}, which denotes the beginning of the collapse plateau, of a foam from the corresponding material properties E_s, ν_s of the solid material. In this model, it is assumed that the material of the foam is distributed randomly in a three-dimensional network as thin membranes under plane stress conditions. By the use of an averaging integral over all orientations of the membranes the final formulations for the elastic properties are:

$$\frac{E^*}{E_s} = \frac{2(7-5\nu_s)}{3(1-\nu_s)(9+5\nu_s)}\phi \tag{1}$$

$$\nu^* = \frac{1+5\nu_s}{9+5\nu_s} \tag{2}$$

where ϕ is the volume fraction of the solid material.

The compressive strength is defined by the elastic instability of the thin cell wall membranes. The closed-cell foam is presumed to have a cell geometry represented by a pentagonal dodecahedron. Based on the buckling load for a simple supported thin circular plate which has the same area as the pentagonal faces, the final result for the beginning of the collapse plateau is:

$$\sigma^*_{el} = 0.235\,\phi^3 \frac{E_s}{1-\nu_s^2} \tag{3}$$

Another model was presented by Gibson and Ashby [1], based on a cubic cell model for a closed-cell foam, which takes into account the enclosed gas. As shown in Figure 1, the thickness of the edges and the faces of PP foam cell are approximately equal, which means there is no accumulation of material in the corners. Therefore, the main deformation mechanisms are the stretching of the cell walls and the compression of the enclosed gas. As a result, the elastic properties of the closed-cell foam are described by:

$$\frac{E^*}{E_s} \approx \frac{\rho^*}{\rho_s} + \frac{p_0(1-2\nu^*)}{E_s(1-\rho^*/\rho_s)} \tag{4}$$

$$\nu^* \approx \frac{1}{3} \tag{5}$$

where ρ^* is the density of the foam, ρ_s is the density of solid polymer and p_0 is the initial gas pressure. The compressive strength, or elastic collapse stress σ^*_{el}, is derived from the behavior of an open-cell foam where the struts of the cells reach the load of elastic buckling. This leads to:

$$\frac{\sigma^*_{el}}{E_s} \approx 0.05\left(\frac{\rho^*}{\rho_s}\right)^2 \tag{6}$$

All models mentioned above imply an isotropic material behavior of the closed-cell foam. For foams with orthotropic behavior further refined models are given in [1, 2, 3].

NUMERICAL MODELS

Although the analytical models describe important mechanical properties of PP foams, they are limited to small strains, typically 5% or less. When using PP foam in typical applications, such as crash padding, the compressive strain reaches values of 50% and higher. For that case numerical models were developed by Neilson *et al.* [4, 5] and were implemented in simulation techniques, such as finite-element (FE) analysis. Although these models were developed for low-density closed-cell polyurethane foams, they were adopted for the compression behavior of PP foams.

The constitutive theory is based on a decomposition of the foam in two parts (Figure 2(b)): a skeleton, i.e. the cell walls, and a nonlinear elastic continuum. The skeleton accounts for the foam behavior in the linear elastic and collapse plateau regimes and is approached by a coupled plasticity and continuum damage theory. The diffuse nonlinear elastic continuum describes the densification of the foam caused by the compression of the enclosed gas and the interaction between the cell walls.

By using Neilson's theory [4, 5], a mathematical model for the description of PP foam was developed. In the model, the skeleton is assumed to occupy the same space as the diffuse continuum. The continuum consists of a gas representing the enclosed gas in the foam cells, and polymer particles, which represent the damaged cell wall in the densification regime. Hence, the initial response of the diffuse continuum to compression loading is a slight increase of the compression stress, caused by the compressed gas. With increasing compression, the volume of the continuum reaches the volume of the polymer particles and the densification starts (accompanied by the steep stress rise). An expression for the hydrostatic pressure p_c of the continuum was obtained by assuming that the gas is in series with the polymer particles, the gas is ideal and the foam compression is an isothermal process. The equation of state for the gas is:

$$p^{gas}\, V^{gas} = p_0^{gas}\, V_0^{gas} \tag{7}$$

where p^{gas} and p_0^{gas} are current and the initial gas pressure, V^{gas} and V_0^{gas} are the current and initial gas volume. The initial gas volume is related to the foam volume V_0^* by:

$$V_0^{gas} = V_0^* \,(1 - \phi) \tag{8}$$

where ϕ is the volume fraction of the polymer in the undeformed foam. The change of the internal gas pressure due to the loading $p^* = p^{gas} - p_0^{gas}$ is related to the change of the gas volume:

$$\Delta V^{gas} = \frac{-p^* V_0^*(1-\phi)}{p_0^{gas} + p} \tag{9}$$

The relation between the volume of the polymer particles, representing the cell walls and the bulk modulus of the solid polymer K_s is given by:

$$\Delta V_s = \frac{-p^* V_0^* \phi}{K_s} \tag{10}$$

With the volumetric strain, ε_{vol}, the internal hydrostatic pressure of the diffuse continuum is given by:

$$p_c = A - \sqrt{\left(A^2 - \frac{\varepsilon_{vol} K_s p_0^{gas}}{\phi}\right)}, \text{ where } A = \frac{p_0^{gas}\phi + K_s(1 - \phi + \varepsilon_{vol})}{2\phi} \tag{11}$$

Therefore, the bulk modulus of the diffuse continuum is calculated by:

$$K_c = \frac{p_c}{\varepsilon_{vol}} \tag{12}$$

The stress–strain behavior of the skeleton is superimposed to the pressure–volumetric strain behavior of the diffuse continuum. It is assumed that the skeleton behaves nonlinearly elastic–plastic with a strain-controlled damage parameter and that no coupling between the cell walls exists. This leads to a Poisson's ratio of zero. Here the damage parameter means that after the load has reached a certain value, the stiffness of the material decreases and the deformation, caused by further loading, is partly plastic and partly elastic (weighted by a user defined parameter). The loading and unloading behavior of the skeleton of the foam model is shown in Figure 2(c).

The stress–strain relation of the skeleton can be described as:

$$\begin{bmatrix} \sigma_1^{sk} \\ \sigma_2^{sk} \\ \sigma_3^{sk} \\ \sigma_4^{sk} \\ \sigma_5^{sk} \\ \sigma_6^{sk} \end{bmatrix} = \begin{bmatrix} E_{11}(\varepsilon_1^{sk}) & 0 & 0 & 0 & 0 & 0 \\ 0 & E_{22}(\varepsilon_2^{sk}) & 0 & 0 & 0 & 0 \\ 0 & 0 & E_{33}(\varepsilon_3^{sk}) & 0 & 0 & 0 \\ 0 & 0 & 0 & E_{44}(\varepsilon_4^{sk}) & 0 & 0 \\ 0 & 0 & 0 & 0 & E_{55}(\varepsilon_5^{sk}) & 0 \\ 0 & 0 & 0 & 0 & 0 & E_{66}(\varepsilon_6^{sk}) \end{bmatrix} \cdot \begin{bmatrix} \varepsilon_1^{sk} \\ \varepsilon_2^{sk} \\ \varepsilon_3^{sk} \\ \varepsilon_4^{sk} \\ \varepsilon_5^{sk} \\ \varepsilon_6^{sk} \end{bmatrix} \tag{13}$$

where σ_i^{sk} and ε_j^{sk} designate the components of the stress and strain tensor and E_{ij} is the user defined material matrix of the skeleton.

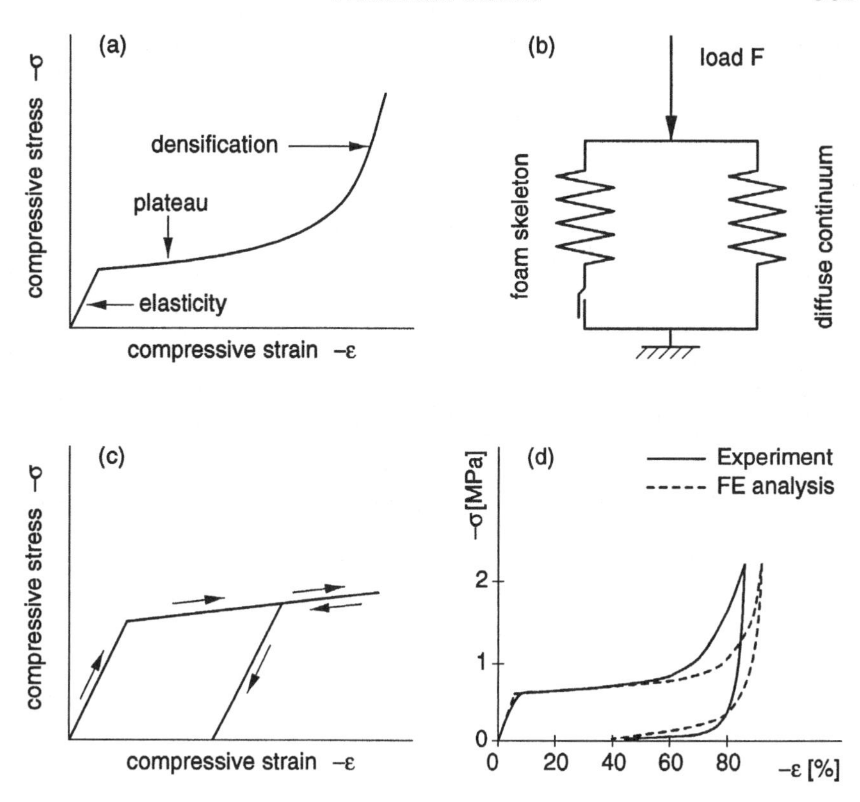

Figure 2 (a) Schematic compressive stress-strain curves for PP foam. (b) Schematic description of the numerical foam model. (c) Loading and unloading behavior of the foam skeleton. (d) Comparison of experimental data and finite-element analysis result.

Because no coupling between the neighboring cell walls is assumed, only the main diagonal of the material matrix has nonzero values and, due to the nonlinear elastic–plastic structural behavior with damage of the cell walls, their components are functions of the corresponding strain components. As mentioned above, the overall stress–strain relation of the foam is obtained by adding both contributions of the skeleton and the diffuse continuum:

$$\sigma_{ij}^{*} = \sigma_{ij}^{sk} + p_c \delta_{ij}, \tag{14}$$

where δ_{ij} denotes the Kronecker symbol.

Figure 2(d) shows the comparison of a compression test of a PP foam specimen with a density $\rho = 60\,\text{kg/m}^3$ and the FE results. The predicted behavior is qualitatively correct. However, the densification of the PP foam in the experiment begins at lower compressive strain values than predicted by the FE analysis. One reason for this discrepancy is that neither the diffuse continuum nor the skeleton model describe the interaction of the neighboring cell walls adequately.

REFERENCES

1. Gibson, L.J. and Ashby, M.F. (1988) *Cellular Solids*, Pergamon, Oxford.
2. Christensen, R.M. (1986) Mechanics of low density materials. *J. Phys. Solids*, **34**, 563–578.
3. Krynik, M. and Warren, W.E. (1994) The elastic behavior of low-density cellular plastics, in *Low Density Cellular Plastics* (eds N.C. Hilyard and A. Cunningham), Chapman & Hall, London, pp. 187–225.
4. Neilson, M.K., Morgan, H.S. and Krieg, R.D. (1987) A phenomenological constitutive model for low density polyurethane foams. *Sandia Report SAND86–2927*, Sandia National Laboratories, Albuquerque, NM.
5. Neilson, M.K., Krieg, R.D. and Schreyer, H.L. (1995) A constitutive theory for rigid polyurethane foam. *Polymer Engng Sci.*, **35**, 387–394.

Keywords: foam, compression behavior, constitutive models, numerical simulation, energy absorption, finite-element modelling, compressive stress–strain behavior, damage, modelling.

Molecular structure: characterization and related properties of homo- and copolymers

J-L. Costa

INTRODUCTION

It is well known that the heterogeneous nature of Ziegler–Natta catalysts, combined with the use of cocatalysts and electron donors as activating agents will lead to the presence of multiple types of active sites on the catalyst. Each type of site will have its own steric environment and reactivity toward (co)monomer and hydrogen. The consequence is that the polyolefin will be a mixture of macromolecular chains with various molecular weights and structure (isotacticity and/or comonomer incorporation).

Isotactic polypropylene (PP) is a partially crystalline thermoplastic, therefore, its properties are strongly dependent on both molecular weight and stereo/regio defects' distributions. They affect both rheological and crystallization behavior of the PP product, imparting its processability and end-use properties, as well.

'Tailoring' the properties of a PP, even for general-purpose applications, requires a precise control of its molecular architecture.

HOMOPOLYMERS

Most end-use properties of PP homopolymers, such as stiffness, hardness and high temperature mechanical properties, are positively influenced

Polypropylene: An A–Z Reference
Edited by J. Karger-Kocsis
Published in 1999 by Kluwer Publishers, Dordrecht. ISBN 0 412 80200 7

by their overall crystallinity. This is related to both molecular weight and stereoregularity of the PP. The impact strength and elongational properties are generally negatively influenced by crystallinity.

Molecular weight

PP can be considered as a linear polymer; consequently, different methods of determining the molar mass from dilute solutions can be used: they can be absolute and related to the weight average molecular weight (M_w), such as light scattering, or relative and dependent on the hydrodynamic volume (itself linked to the molecular weight), such as size exclusion chromatography (SEC) and viscosimetry. The most interesting information comes from SEC measurements which give values for M_w, number average molecular weight (M_n) and molecular weight distribution (MWD) [1].

Despite the useful solution characterizations techniques evidencing the dispersity of catalytic active sites, effects of cocatalytic composition, etc., a correlation with the properties in the molten state is of paramount interest as far as the processing of the material is concerned.

The traditional melt index test (230°C/2.16 kg for PP) is widely used in the plastics industry to estimate the polymer melt processability by a melt flow index (MFI) which is inversely related to the viscosity and the average molecular weight of the material. The MFI has to be considered very cautiously because polymers are usually processed under very different shear rates and this fairly simple index is far from giving a complete picture of the material's flow behavior.

Therefore, it is more suitable to measure the shear sensitivity of polypropylene. Capillary rheometers or oscillatory shear flow rheometers are widely used for that purpose. Moreover, an investigation of elongational flow properties of molten PP can be used to check, for example, the presence of long-chain branching in some speciality grades of PP (to study the strain-hardening effect).

The effect of molecular weight on flexural modulus is rather low, at least in the usual MFI range, while broadening of MWD allows us to produce more oriented and crystalline products.

Isotacticity

Isotacticity level has a stronger effect on crystallinity and related properties of polypropylene. PP grades with an isotactic index (II) of 0.96–0.98 are obtainable which can offer a 20% increase in flexural modulus in comparison to standard grades (II = 0.93–0.95). This enhanced stereoregularity has very interesting consequences on the properties of biorientated PP films (BOPP). A 25% increase in flexural stiffness and 25%

decrease of both oxygen and water vapor transmission rates can be obtained.

Carbon-13 nuclear magnetic resonance (NMR) spectroscopy represents the only direct method to analyze the polymer regio- and stereostructure (isotacticity level). The structural information derived from this technique is not limited to diads (sequence of two monomer units) but generally includes triads or pentads and, in some cases, nonad or undecad levels have been reached, as well. Nevertheless, it is important to keep in mind that these values are only mean values referring to the whole polymer sample and, in some cases, they are not sufficient to discriminate samples with a different distribution of defects (see later).

Infrared spectroscopy (IR), differential scanning calorimetry (DSC) and soluble fractions are indirect methods to estimate the stereoregularity of the sample as they are related to its crystallinity and, therefore, are dependent on its thermal history.

Fractionation of PP

The most simple methods of fractionating PP are based on extraction in Soxhlet or Kumagawa type devices with boiling solvents [2], usually hexane (C_6) and heptane (C_7). The difficulty of these methods is that equilibrium is often impossible to reach and other parameters, such as the specific surface of the sample, are of importance. Moreover, the separation is controlled by diffusion, therefore the molecular weight will have a considerable effect on the amount of extracted material, even for products with the same isotacticity! (Figure 1; the figures are obtained for

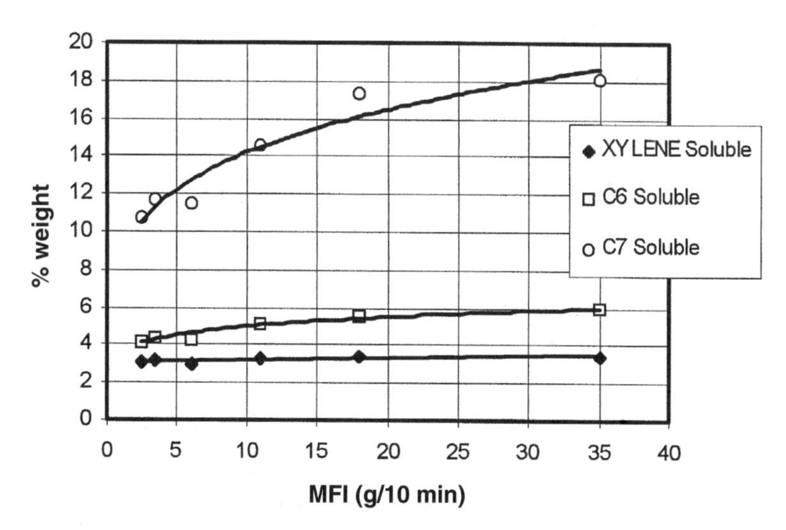

Figure 1 MFI effect on soluble fraction percent.

reprecipitated samples which all have a mean isotacticity index of 0.96, measured by ^{13}C NMR).

Other methods, based on slow recrystallization of a polymer solution (usually in xylene or decaline) give results better correlated to the isotacticity of the sample and are relatively independent of its molecular weight (Figure 1).

The analysis of the soluble fractions indicates that the nonrecrystallizable material obtained in xylene is essentially atactic material (II = 0.33 by ^{13}C NMR), whereas C_6 and particularly C_7 soluble fractions are mixtures of lower molecular weight atactic and isotactic chains giving consequently a bad estimation of the mean isotacticity level of the sample and, in addition, this result depends on the molecular weight [3].

Temperature rising elution fractionation (TREF)

A more accurate characterization of PP can be obtained by fractionations based on solvent/non-solvent mixtures or by a progressive dissolution of carefully crystallized samples (temperature rising elution fractionation, TREF).

The microtacticity of the isotactic parts of PP is known to vary subtly with the catalytic system. It is possible to discriminate samples by TREF (compare the weight distribution of high elution fractions for B and C homopolymer in Figure 2). The presence of more stereodefects in the B

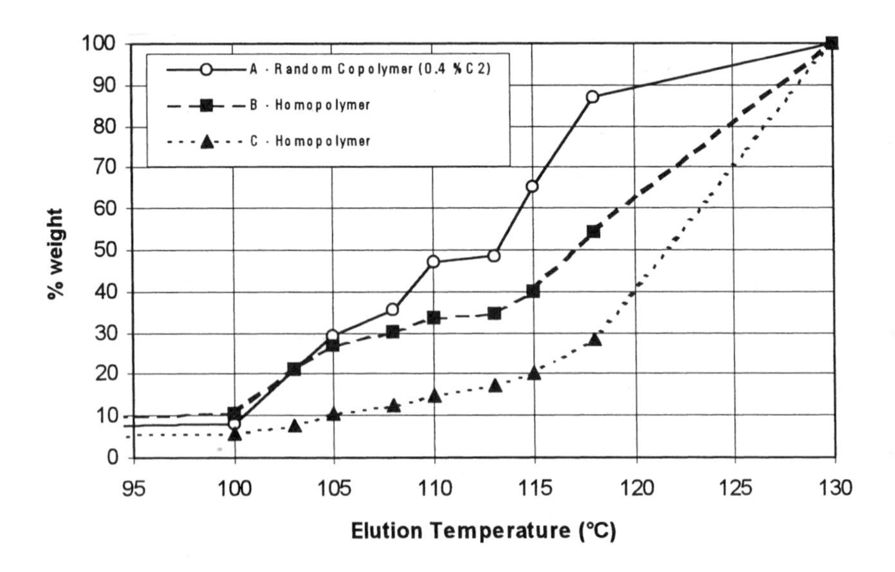

Figure 2 TREF of three PP with different distributions of defects.

sample has the same effect on crystallization as the presence of a small amount of comonomer (A sample)

The balance between these characteristics and the molecular weight can explain some important differences in flow-induced crystallization during transformation processes for these samples (skin-core morphology, biorientability, etc).

RANDOM COPOLYMERS

The introduction of small amounts of comonomer units acting as defects in the macromolecular chain of polypropylene will decrease the length of isotactic segments leading to different thermal, mechanical and processing properties compared with the homopolymer (i.e. lowering of melting and sealing temperatures, flexural modulus, broadening of the melting range). Another very important effect is the improvement of the optical properties by the reduction of the apparent densities of the crystalline zones.

These effects depend on the nature of the comonomer and on its distribution within the chain. The most efficient way is the introduction of ethylene (C2) fragments into an isotactic chain. On the other hand, the introduction of comonomer in poorly isotactic chains will only affect the amount of amorphous fraction but will not affect the melting properties of the copolymer.

An illustration of this nonhomogeneous distribution of ethylene comonomer can be given by a coupled SEC/FTIR analysis of a random copolymer where the C2 content is defined as the infrared absorbance ratio: A $733\,cm^{-1}/(A\ 733\,cm^{-1} + A\ 973\,cm^{-1})$ (Figure 3).

This feature is due to the difference in reactivity ratios between comonomers which depend on the type of catalytic active site. The lower isospecific centres of traditional Ziegler–Natta catalysts are much more active toward ethylene than toward propylene or butene-1. Moreover, these less isotactic sites being the more reactive toward hydrogen, the atactic fractions will have lower mean molecular weight than isotactic ones.

It was found that butene-based copolymers, in comparison to ethylene-based ones, contribute less to the overall amorphous character of the polymer. Therefore, butene-1 co- or terpolymers, have higher melting points, lower extractable content, higher stiffness than random ethylene copolymer with the same molar content of comonomer (exploited in BOPP packaging films).

The nature of the (co)catalytic system will also influence the randomness of comonomer incorporation (see later); ^{13}C NMR or IR are widely used to illuminate these characteristics.

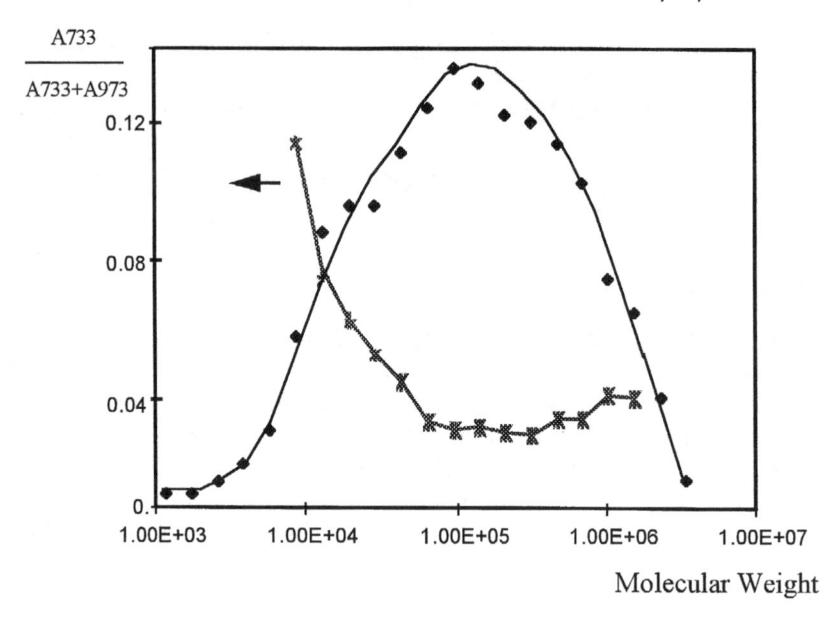

Figure 3 SEC/FTIR analysis of a random copolymer.

IMPACT COPOLYMERS ('BLOCK COPOLYMERS')

It is important to note that these 'block copolymers' are, in fact, mixtures of isotactic homopolymer and ethylene–propylene random copolymer (EPR) with a high content of ethylene (45–55 wt.%) produced by sequential polymerization in at least two reactors.

The specific properties of these 'reactor blends' and especially the stiffness/ impact toughness balance will strongly depend on the amount of EPR (acting as energy dissipating species) in addition to the molecular composition of both continuous (PP) and dispersed (EPR) phases.

Viscosity ratio

Therefore, the ratio between the viscosity of continuous and dispersed phases at a given shear rate strongly affects the final morphology of the blend (particle diameter and interparticular distance) both in the core of the material (with a strong effect on the impact resistance) and near to the surface (with a series of consequences for the optical properties and the paintability of the block copolymer) [4].

Comonomer distribution

The ethylene content of EPR will strongly affect its glass transition temperature and its adhesion with PP matrix which, in turn, will modify low-temperature impact properties of the copolymer.

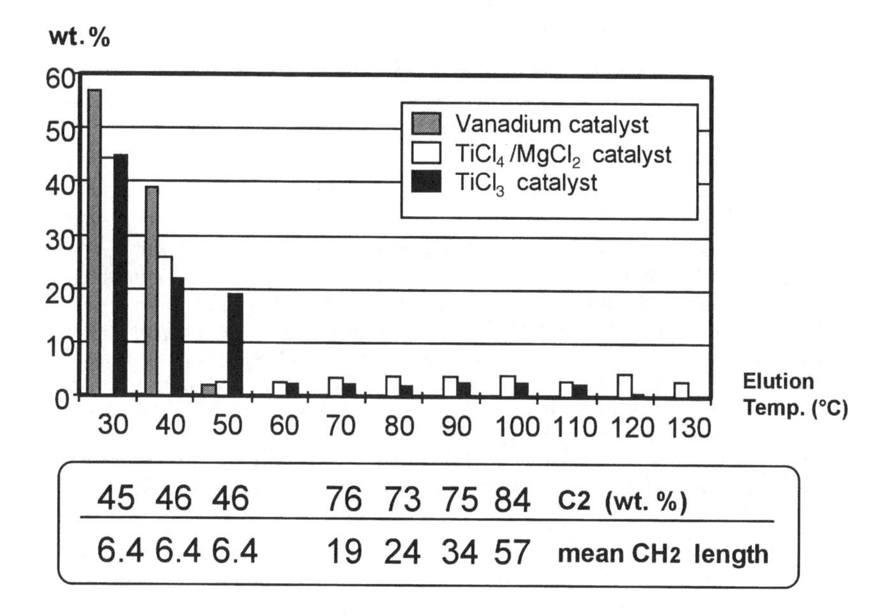

Figure 4 TREF of three EPR (total C2 = 50 wt.%) produced by different catalytic systems.

The comonomer distribution along the macromolecular chain will also affect the tendency of EPR to contain crystallized polyethylenic structures that will modify some properties of the blend (i.e stress whitening due to impact).

Figure 4 illustrates the influence of the catalyst nature on the statistical distribution of ethylene units and thus on the amount of crystallizable polyethylenic fractions (eluted at high temperatures in the TREF experiment) coming from three EPR with the same total amount of ethylene (50 wt.%). The ethylene content of each fraction of the second EPR is given, as an example, on the bottom line Figure 4.

^{13}C NMR analysis gives complementary information on the mean CH_2 length of a sample, at a given total C2 amount, related to the randomness of C2 incorporation which is, in turn, characteristic of the catalytic system used.

CONCLUSION

The use of more sophisticated analyses and fractionation methods together with a better knowledge of the action of the whole catalytic system helps to understand the structure–properties relationships and allow us to produce homo- and copolymers with the desired end-use properties.

A key step, in that context, will be the use of metallocene catalysts. They have well-defined single catalytic sites allowing control of the MWD and the comonomer and stereo- or regioerror distributions of polypropylene homo- and copolymers more precisely.

REFERENCES

1. Monasse, B. and Haudin, J-M. (1995) Molecular structure of polypropylene homo- and copolymers in *Polypropylene: Structure, Blends and Composites*, Vol. 1, (ed. J. Karger-Kocsis), Chapman & Hall, London, pp 4–30.
2. Ser van der Ven (1990) *Polypropylene and Other Polyolefins: Polymerization and Characterization*, Vol. 7, Studies in Polymer Science, Elsevier, London.
3. Paukkeri, R. (1994) Molecular structure, crystallization and melting behavior of fractionated polypropylenes. *Acta Polytechnica Scandinavia, Chemical Technology and Metallurgy Series*, **216**, 1–37.
4. Phillips, R. and Wolkowicz, M. (1996) Structure and morphology, in *Polypropylene Handbook*, (ed E. Moore, Jr), Hanser, Munich, pp 113–176.

Keywords: molecular weight, isotacticity, isotacticity index, fractionation, solubility, temperature rising elution fractionation (TREF), size exclusion chromatography (SEC), Fourier-transform infrared spectroscopy (FTIR), 13C NMR, homopolymer, random copolymer, block copolymer, structure–property relation.

Morphology and nanostructure of polypropylenes by atomic force microscopy

G.J. Vancso

During the last ten years, atomic force microscopy (AFM) has played an increasingly important role in polymer science because of its unique ability (1) to characterize surface structure and morphology (without extensive sample preparation), and (2) to study surface properties from the angstrom to the millimeter scales. Since the principle of AFM was first described in 1986 [1], a large number of spin-off techniques have been invented which fall into the general category of scanning force microscopy (SFM). The main common feature of all these instruments is that the measurements are performed with a very sharp tip probe (with a radius of curvature at the apex in the 20–50 nm range). The tip is attached to the end of a soft cantilever beam which typically has a length of 100 μm and a spring constant of 1 N/m. In contact mode AFM, the sharp probe is brought down to within close proximity of the surface of the sample. The sample surface is then raster scanned by moving either the sample or the tip probe. High-precision piezoelectric scanners are used for positioning. Molecular (sometimes even atomic) lattice resolution can be achieved for surfaces with two-dimensional translational symmetry. For example, one can resolve methyl groups on flat areas of crystalline polymer samples (Figure 1(a) (b)). It must be emphasized that no coating or tedious sample preparation is required and that images can be obtained in air (or in liquids) directly on the surface of the specimen.

Polypropylene: An A–Z Reference
Edited by J. Karger-Kocsis
Published in 1999 by Kluwer Publishers, Dordrecht. ISBN 0 412 80200 7

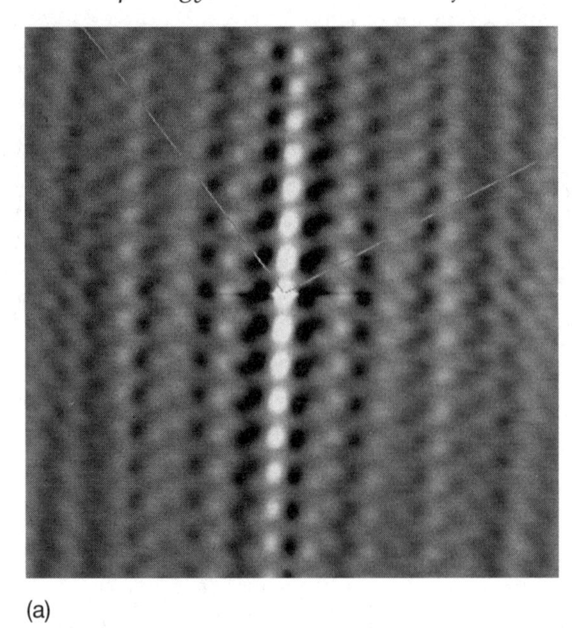

(a)

Figure 1 (a) A 10 nm × 10 nm nanograph (autocorrelation pattern) obtained on the surface of uniaxially oriented iPP showing PP chains with predominantly vertical orientation. Brighter spots correspond to images of methyl groups exposed. (Unfiltered image is published in [3].) (Reprinted from [3], Copyright 1994, with kind permission from Elsevier Science.) (b) Fourier-filtered AFM image of the bc contact face of sPP epitaxially crystallized on *p*-terphenyl. The zig-zag pattern of CH_3—CH_2—CH_3 units along the chain is also shown (computer generated picture, on a different length scale). (Reproduced with permission from [4], granted by the American Chemical Society. Courtesy of Prof. B. Lotz.)

Minute deflections of the cantilever (in the order of less than 0.1 nm) are usually measured with an optical lever technique. This technique functions by detecting the positional changes of laser light reflected off the back of the cantilever. During height imaging of morphological features at the sample surface, a feedback loop of the instrument ensures that the distance between the probe and the sample is kept constant to within atomic dimensions. Thus the tip 'follows' the sample surface. True three-dimensional images are created by capturing the x, y and the $z(x, y)$ coordinates of structural features (Figure 2). The ability of SFM to produce these quantitative three-dimensional images is unique among all microscopy techniques. In another imaging mode, the deflection of the cantilever beam is measured. This deflection is related to detecting the variation of forces between the probing tip and the sample as a

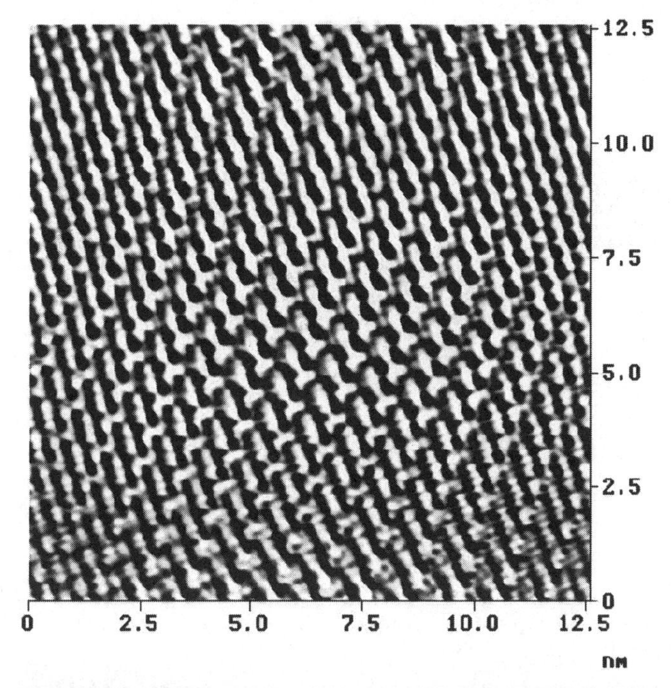

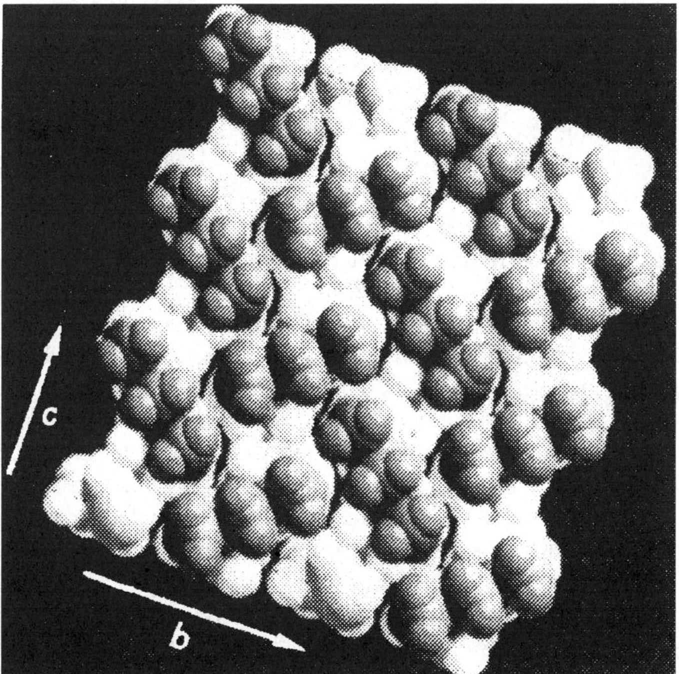

Figure 1 (b)

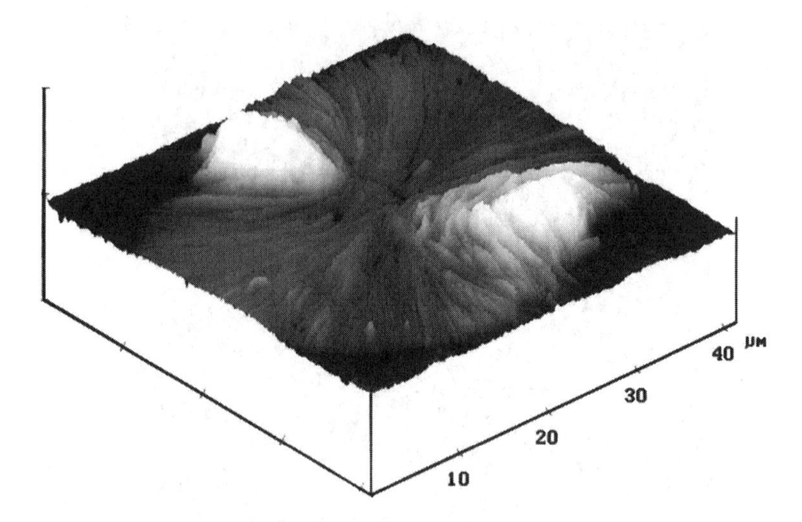

Figure 2 AFM surface plot of a hedritic structure grown in β-iPP (vertical scale: 3 μm/div.). The lamellae are vertical near the diagonal connecting the top and bottom corners. (Trifonova, D., Varga, J., Ehrenstein, G.W., Vancso, G.J. (1996) *ACS Polymer Prepr.* **37**(2), 563.)

function of tip position (x, y). Such 'deflection' mode scans are very useful when recorded simultaneously with height images, as they show height changes with sharp contrast.

Other imaging modes include lateral force microscopy (LFM) and various dynamic techniques, such as tapping mode (TM). In LFM, the torsion (twist) of the cantilever beam is measured. This allows one to obtain friction maps which are chemically specific and can be used to study friction (nanotribology), and to identify chemical components in heterogeneous materials. In dynamic modes, e.g. TM, the cantilever beam is oscillated at its resonance frequency and 'feels' the sample surface only very briefly when it passes the minimum of its vibration. As the scanning frequency (1–10 Hz) is much slower than the resonance frequency of the vibrating tip (e.g. between 300–400 kHz), the shear forces between sample and tip are virtually eliminated during TM imaging (minimum surface damage). In addition, valuable information can be obtained about local stiffness variations (viscoelasticity) by studying the relationship between the amplitude change (or phase shift) and the tip position.

SFM techniques are developing with an astonishing speed. Every year significant improvements are made by instrument manufacturers. Thus it is to be expected that this brief description of imaging techniques will be out of date within a few years. Due to the 'tender' age of force microscopy, the results on polypropylenes discussed below cannot be considered as conclusive, completed work.

Isotactic polypropylene (iPP) imaging with molecular resolution was performed on epitaxially crystallized [2] and on uniaxially oriented [3] samples. Specimens obtained by epitaxial crystallization on benzoic acid showed methyl group resolution within the captured (010) planes [2]. The observed pattern was consistent with imaging 3/1 helices and with the 'four face' type in which one methyl group per helix turn is exposed. AFM scans of cleaved samples of mechanically oriented iPP with uniaxial texture unveiled microfibrils with 100–150 nm diameter [3]. Chain orientation and twisting was observed in the (110) crystal facet (Figure 1(a)) In this crystal face, alternating left- and right-handed helices were captured and identified in one scan. The helical hand of individual chains was identified by comparing the symmetry of the imaged methyl groups with computer generated patterns of molecular packing in the (110) crystal plane.

Syndiotactic polypropylene (sPP) was also imaged by SFM with collective (lattice) methylene group resolution (Figure 1(b)). Samples were obtained by epitaxial crystallization during which the *bc* crystallographic facet grew in contact with the substrate. SFM scans show the *bc* planes viewed from the crystallographic *a* direction. The molecular helix of sPP in this view is expected to show patterns of linear blocks containing CH_3, CH_2 and CH_3 units. The array of these rod-like units of groups of atoms makes an angle of 45° with respect to the chain direction in either (1) clockwise direction for right-handed, or (2) anticlockwise direction for left-handed helices, respectively. For a single chain one would expect a 'fishbone-like' pattern of the blocks containing CH_3, CH_2 and CH_3 units. High-resolution scans [4] show alternating packing of left- and right-handed helices in the *bc* facet of sPP which enabled the authors to identify the helical hand for individual sPP chains.

The crystal structure of the β form of iPP was only recently described despite the fact that this phase has been known to exist for 40 years. The structure consists of sections of three isochiral helices packed in a trigonal cell. One of the three helices is embedded in a crystallographic environment which is different from the other two, i.e. the structure is 'frustrated' [5]. AFM studies of the (110) plane confirmed this frustrated character. In addition, it seems that in epitaxy the chains in the contact plane of polymer crystals (grown on dicyclohexylterephthalamide) expose rows of 2, 2 and 1 methyl groups.

The studies discussed above, which yielded images of various known crystal facets of polypropylenes with molecular resolution, established the credibility of AFM for studying crystal structure at surfaces in combination with other techniques (X-ray and electron diffraction, computer simulation, etc.) Features of the various planes of the crystalline lattice were reproduced accurately to within 5% by imaging either in air or in liquids (e.g. water).

Thin films of iPP were crystallized *in situ* using a microscope hot stage and were first investigated with AFM by Schönherr *et al.* [6]. The crystallization conditions were selected to produce different types of spherulites including α-I, α-II, mixed, and β-III types. AFM images unveiled morphological details such as mother/daughter lamellae, cross hatching, thickness, and the thickness distribution and orientation of lamellae (Figure 2).

The morphology of high-modulus carbon-fiber (HM-CF) reinforced iPP was investigated by AFM using chemically etched specimens [7]. The images exhibited typical features of α-transcrystalline morphology for samples which were crystallized from quiescent melts, and nucleated on HM-CF. In melts sheared by fiber pulling, αβ-cylindritic columnar

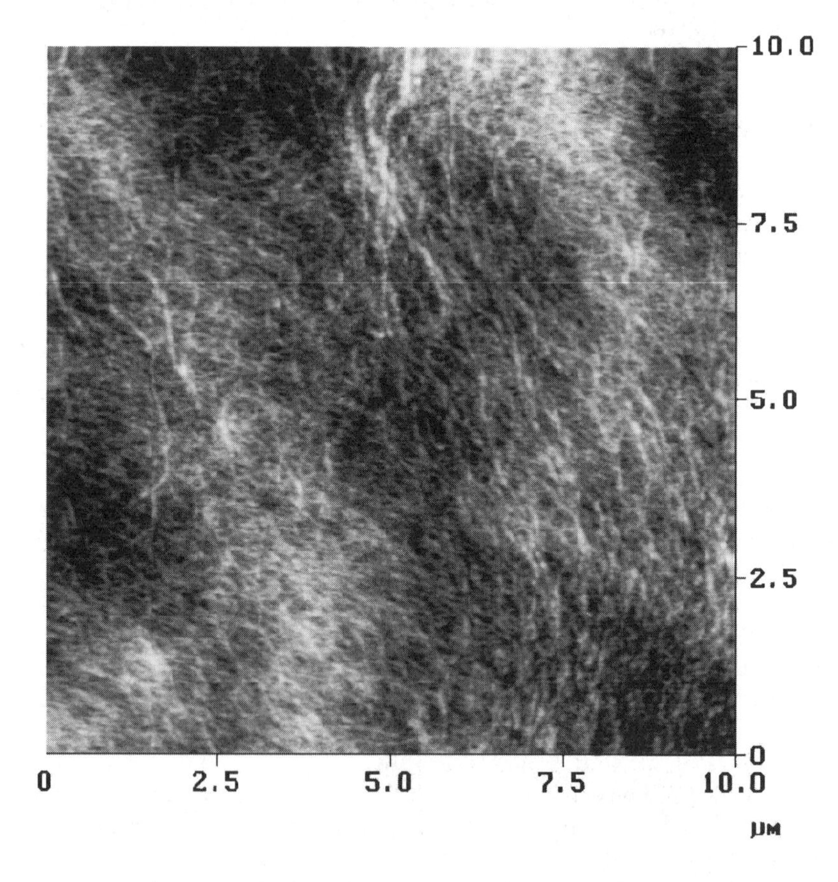

Figure 3 Surface morphology (TM image) of a biaxially oriented iPP film. (P. Smith, G.J. Vancso, unpublished data.)

morphology was observed. In the interfacial region of the β-cylindrites fine morphological details of α-row nuclei were resolved.

AFM, especially when used in noncontact modes, is an ideal tool to image delicate morphological details of surfaces of polymer films [8a]. When these films are coated with gold for scanning electron microscopy (SEM), features with typical dimensions of less than ≈ 50 nm become buried and may disappear from SEM images. In AFM imaging this is not a problem as no conductive coating is required.

Images of technical PP films obtained by cast-film extrusion and subsequent biaxial orientation consist of branched fibrillar features. These features form during the orientation when the lamellae become torn apart and subsequently form fibrils (Figure 3). The orientation of the fibrillar axes is distributed according to the mechanical stress. Striae can occasionally be seen in particular surface areas of these films. However, striated surface structures are predominant for uniaxially oriented films [8b]. Corona treatment results in the formation of 400–500 nm large 'droplet-like' features at the film surface [8]. The surface coverage of these droplets depends on the corona dose, while the diameter and height (i.e. surface roughness) increase with increasing dose [8b]. In addition to morphology imaging, surface roughness and fractal analysis provide quantitative measures which can be obtained from AFM image analyses. For metallized PP films manufactured for barrier applications, the quality of the metal layer (e.g. grain boundaries, grain size) can be characterized by AFM.

In conclusion, force microscopy techniques, in spite of the fact that they are not fully developed, have significantly contributed to structural analyses of polypropylenes in areas where classical morphology characterization methods have serious inherent limitations.

REFERENCES

1. For reviews, see Miles, M.J. (1994) New techniques in microscopy, in *Characterization of Solid Polymers*, (ed. S.J. Spells), Chapman & Hall, London, Chapter 2; Magonov, S.N. and Whangbo, M.H. (1996) *Surface Analysis with STM and AFM*, VCH, Weinheim. For a collection of recent work on SFM of polymers see articles in *ACS Polymer Preprints* (1996), **37**(2), pp. 546–625.
2. Stocker, W., Magonov, S.N., Cantow, H.J., Wittman, J.C. and Lotz, B. (1993) *Macromolecules*, **26**, 5915.
3. Snétivy, D. and Vancso, G.J (1994) *Polymer*, **35**, 461.
4. Stocker, W., Schumacher, M., Graff, S., Lang, J., Wittmann, J.C., Lovinger, A.J. and Lotz, B. (1994) *Macromolecules*, **27**, 6948.
5. Stocker, W., Lovinger, A.J., Schumacher, M., Graff, S., Wittmann, J.C. and Lotz, B. (1996) *ACS Polymer Preprints*, **37**(2), 548.
6. Schönherr, H., Snétivy, D. and Vancso, G.J. (1993) *Polymer Bull.*, **30**, 567; Castelein, G., Coulon, G., Aboulfaraj, M., G'Sell, C. and Lepleux, E. (1995) *J. Phys. III France*, **5**, 547.

7. Vancso, G.J., Liu, G., Karger-Kocsis, J. and Varga, J. (1997) *Colloid Polymer Sci.*, **275**, 181.
8. (a) Vancso, G.J., Allston, T.D., Chun, I., Johansson, L.-S., Liu, G. and Smith, P.F. (1996) *Int. J. Polymer Anal. Charact.* **3**, 89. (b) Overney, L.M. *et al.* (1993) *Appl. Surf. Sci.*, **64**, 197.

Keywords: atomic force microscopy (AFM), structure, morphology, spherulite, hedrite, lamella, conformation, chain packing, films, scanning force microscopy (SFM), tapping mode, syndiotactic PP, β-PP, transcrystallinity, corona treatment, biaxial orientation.

Morphology–mechanical property relationships in injection molding

Mitsuyoshi Fujiyama

INTRODUCTION

Polymeric materials are used as products and 'shaped' in processing. Accordingly, when determining the quality of a polymer, it is necessary to consider its processability (processing properties) and the quality of the final part (product properties). The ranking of product properties is made with the end-use of the product in view.

The processing properties are controlled by the primary structures of the raw material and by processing conditions, while the product properties depend on both the primary and higher-order structures. The higher-order structures formed during processing depend on both the primary structures and processing conditions. Therefore, it is necessary to clarify, by analyzing the higher-order structures, how the material and processing conditions affect the product properties.

The higher-order structures (morphologies) of injection-molded polypropylene (PP), which should be considered, are crystalline form, lamellar thickness, spherulite size, crystallinity, molecular orientation, crystal orientation, dispersion state of a blended polymer, length and orientation of reinforcing fibers, crystalline texture (skin-core structure), etc.

Meanwhile, the mechanical properties to be considered are also manyfold (tensile, flexural, compression, shear, twist, hardness, creep, stress relaxation, impact, fatigue, friction, abrasion, tear properties, etc.).

In this chapter, relationships between the morphologies and some mechanical properties in injection-molded PPs are described.

Polypropylene: An A–Z Reference
Edited by J. Karger-Kocsis
Published in 1999 by Kluwer Publishers, Dordrecht. ISBN 0 412 80200 7

SKIN-CORE STRUCTURE

When a cross-section of an injection-molded PP is observed under a polarizing microscope, a skin-core or skin-shear-core structure is seen as shown in Figure 1. A cross-section cut perpendicular to the flow direction (MD) shows a skin-core structure, while a cross-section cut parallel to the MD shows a skin-shear-core structure. The skin layer is composed of a kind of fiber structure oriented in the MD, the shear layer is composed of a row structure oriented in the MD, and the core layer is composed of almost unoriented spherulites. There are also papers claiming that, in injection-molded PP, more than five different layers can be resolved through the thickness.

The skin thickness increases with increasing molecular weight and its distribution. The skin layer increases by particulate filling and decreases by copolymerization with ethylene and by glass fiber (GF) filling. In respect to the molding conditions, the skin layer becomes thicker when the thickness of the molded item is reduced and by decreasing the cylinder and mold temperatures, and by reducing the injection speed. The injection pressure scarcely affects its thickness [1].

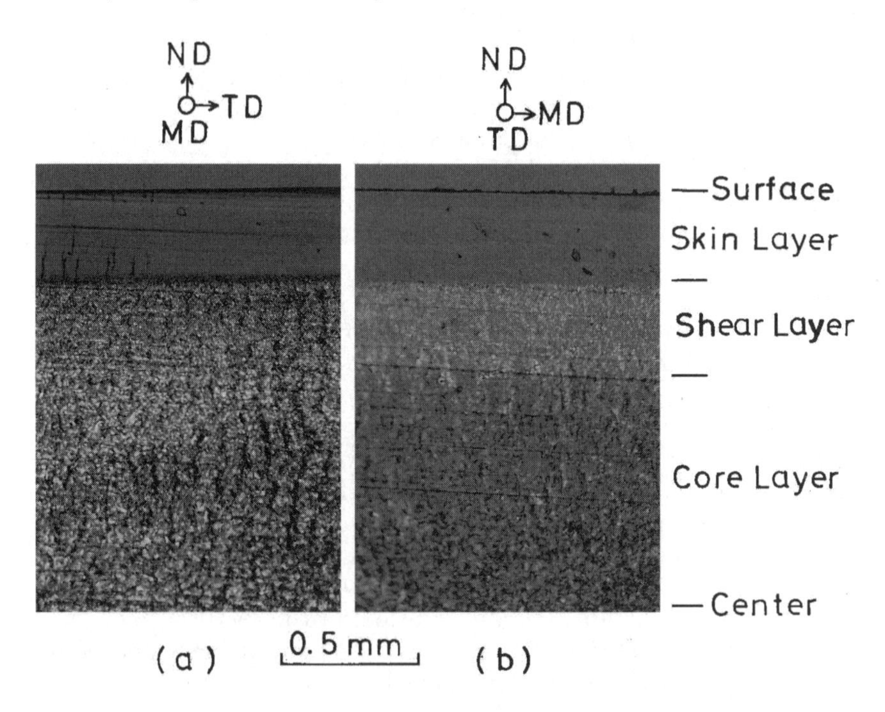

Figure 1 Polarizing micrographs of thin sections cut: (a) perpendicular; (b) parallel to the flow direction (MD). (TD: transverse direction, ND: thickness direction).

Because the molecular chains in the skin or shear layer are highly oriented, it can be presumed that the mechanical and thermal properties are better for the article with thicker skin or shear layer.

Next, the mechanical properties in the MD will be briefly treated. Mechanical properties, such as flexural modulus and strength, tensile modulus, yield strength, necking stress, and hardness, increase almost linearly with increasing the skin thickness or skin and shear layer fraction [1]. The Izod impact strength increases and the maximum load at high-speed impact increases linearly with the skin thickness. The elongation at break and the ductile coefficient H/F (where H and F are the half-value width and the maximum load of load–displacement curve in the high-speed impact test, respectively) decrease linearly with increasing skin thickness. Accordingly, items of high mechanical performance can be molded, in which thicker skin layers are generated. Such relationships between mechanical properties and skin thickness hold not only for homo-PPs but also for copolymers with ethylene, for particulate and GF filled PPs.

In the case of injection moldings of PPs filled with GF, a skin-core structure with different states of fiber orientation develops: the fibers align parallel to the MD in the skin layer, while they adopt a perpendicular orientation to the MD in the core layer [2]. Accordingly, the tensile modulus and strength increase linearly with increasing skin thickness. In the case of injection moldings of PPs filled with two-dimensional particulates, such as glass flakes, a skin-core structure with different states of flake orientation appears: the flakes are oriented parallel to the surface of the molding in the skin layers and perpendicular to the surface and MD in the core layer. Accordingly, the molding with thicker skin layer shows higher tensile modulus.

Due to molecular orientation in the skin and shear layers, the injection-molded part shows considerable mechanical anisotropy. Figure 2 displays the relationships between the tensile yield strength measured in various directions, and the skin thickness. The yield strength is in the order of MD > TD > 45° and its value in each direction increases linearly with increasing skin thickness [1]. As far as the relationships between the flexural modulus and strength in various directions and the skin thickness are concerned, these flexural properties in the MD and TD increase while those in the 45° direction decrease linearly with increasing skin thickness. The impact strength in the MD increases with increasing skin thickness as shown before, while that in the TD decreases with it. Under cyclic fatigue loading of an injection-molded PP, the crack always initiates at the shear layer and propagates toward the core. The fatigue lifetime is longer when loading occurs in the MD than in the TD. Furthermore, the fatigue performance is better when the skin layer is thinner and the crystalline texture in the core is more uniform [2].

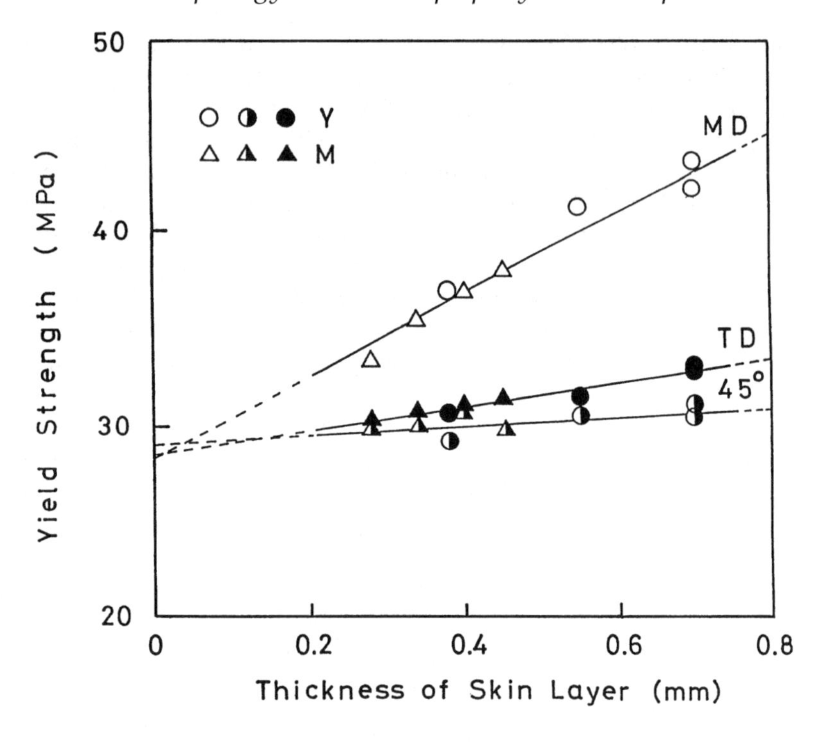

Figure 2 Relationships between tensile yield strengths in various directions and skin thickness. The overall thickness of the molding was 1.6 mm.

MOLECULAR ORIENTATION AND CRYSTAL ORIENTATION

The orientation functions and birefringence are widespreadly used as measures of molecular orientation of injection-molded PPs. In addition, the crystalline orientation fraction, skin thickness, and annealing shrinkage are determined in the practice.

The tensile yield strength in the MD increases linearly with the average molecular orientation expressed by the birefringence. The compression strength, on the other hand, exhibits a downward-curved (acceleration-type) increase with increasing the average orientation. The correlation between birefringence and elongation across the thickness direction showed that the elongation is the lowest at the highest birefringence.

The flexural modulus, flexural strength and Izod impact strength in the MD increase linearly or in a downward-curved manner with increasing crystalline *c*-axis orientation function (f_c) [1]. The ratio of the tensile impact strength in the MD to that in the TD increases linearly with f_c.

The flexural modulus, flexural strength and Izod impact strength in the MD show positive linear relationships with the crystalline orientation fraction (OF). Such relationships hold not only for homo-PPs but also for copolymers with ethylene, nucleated PPs, particulate and GF filled PPs. The linear regression of the mechanical properties versus OF is usually clearer than versus f_c.

When the PP is subjected to shear-controlled orientation injection molding (SCORIM), the molecular orientation may be enhanced, resulting in a 'self-reinforced' molding. For example, by the SCORIM method, moldings of improved stiffness (by 75%) can be obtained [3] – see the related chapter in this book.

In the injection molding of PP containing even small amount (0.5 wt.%) of talc, the *b*-axes of PP crystals orient to the thickness direction [4]. This is associated with improved surface hardness.

SPHERULITE SIZE AND LAMELLAR THICKNESS

Only a few studies are available on the relationships between spherulite size or lamellar thickness and mechanical properties in injection-molded PPs. A possible reason for this is that, when the spherulite size is large, the lamellar thickness and crystallinity also usually have high values, which makes the separation of these factors difficult.

The spherulite size tends to increase toward the interior of the molding where the cooling rate is low. However, it is also reported that since the crystallization rate goes through a maximum as a function of cooling rate (due to a balance between the rates of nucleus initiation and molecular diffusion), the spherulite size has a minimum between the quenched surface and the slowly cooled center [5]. The spherulite size usually increases with increasing the cylinder and mold temperatures. Its size is decreased by adding nucleating agents to PP.

With increasing spherulite size, the tensile modulus increases, however at costs of the tensile yield strength, elongation, and Izod impact strength. The deterioration in the latter properties with increasing spherulite size is due to an increased amount of material less prone for crystallization, which is deposited at the spherulite boundaries. This results in weak spherulite boundaries and promotes brittle fracture.

There are even fewer reports available on the effects of lamellar thickness. It is claimed, however, that the hardness slightly increases with increasing lamellar thickness.

CRYSTALLINITY

The crystallinity tends to rise with increasing the tacticity and melt flow index (MFI) of PP. The opposite occurs by copolymerization with ethylene, particularly by random copolymerization. Incorporation of nucleants yields higher crystallinity. Concerning the effects of the molding conditions, higher mold temperature leads to higher crystallinity due to the slower cooling. The effect of cylinder temperature depends on the thickness of the mold on a balance between effects of orientation crystallization and cooling rate. In the case of thin moldings, the orientation crystallization dominates and the crystallinity is higher when the cylinder temperature is lowered. In the case of thick moldings, the effect of slow cooling dominates and the crystallinity increases when the cylinder temperature is raised. The crystallinity is increased by annealing.

The tensile modulus, flexural modulus, flexural strength, compression strength and hardness increase linearly with increasing the crystallinity. The impact strength tends to decline with increasing crystallinity.

Because mechanical properties are affected by both the molecular orientation and crystallinity, the relationship between the mechanical properties and molecular orientation is 'smeared' by crystallinity effects. For example, the relationship between the flexural modulus and the skin thickness or crystalline *c*-axis orientation function (f_c) are greatly improved when corrected by the crystallinity influence [6].

CRYSTALLINE FORM

PP exists in several crystalline forms: α-form (monoclinic), β-form (hexagonal), γ-form (triclinic), and smectic form (pseudohexagonal) are known. The α-form is the usual one, the γ-form appears in an injection-molded PP copolymer with disturbed molecular structure and due to the SCORIM molding [3]. The smectic form can be detected at the quenched surface layer (in 10–50 μm thickness) of a molding. The β-form is mostly present in the shear layers. The formation of the β-form is suppressed by copolymerization with ethylene and by the addition of α-crystal nucleating agents. Its development is promoted by the addition of β-crystal nucleating agents.

When a β-nucleated PP is molded, the β-crystal content is lower in the skin and higher in the core layer of the molded item. It is possible to obtain moldings with a β-crystal content higher than 90%. Low cylinder temperature is favored for the formation of the β-crystals. There is an optimum amount of the β-crystal nucleating agents, e.g. the maximum β-crystal content was obtained at a γ-quinacridone content of about 10 ppm. With increasing γ-quinacridone content, the flexural modulus and strength decrease, pass minima at 3–10 ppm, and increase after that. The heat distortion temperature rises uniformly with increasing γ-quinacridone content. The Izod impact strength shows a maximum at a content of about 100 ppm whereby a threefold impact strength can be achieved [7]. Although there is almost no difference between the fracture toughness (K_c) of a β-PP injection molding and that of usual PP molding, the fracture energy (G_c) of the former is about twice that of the latter. The superior toughness of the β-PP injection molding is a combined effect of the following: morphology (skin-core structure, crystal structure, spherulite size, tie molecules); mechanical damping; and phase transformation (β-to-α transition) [8]. The higher impact strength and thermal resistance of β-PP injection moldings are achieved at cost of the stiffness.

FIBER LENGTH AND ORIENTATION

The modulus and strength of injection moldings of PP containing fibers, such as glass fibers (GF) and carbon fibers, are affected not only by fiber content but also by fiber length and orientation. The fibers in the resin are broken by the plasticating action of the screw in the cylinder and by the flow in the sprue, runner, gate and cavity. As a consequence, their length is reduced in respect to the initial one. In addition, they are oriented by the action of stresses developed during the flow.

The tensile modulus of a composite, E_c, is given by the following equation [9]:

$$E_c = \eta_0 \eta_L E_F V_F + E_m(1 - V_F)$$

where E_c, E_F, and E_m are the tensile moduli of the composite, fiber and matrix, respectively, η_0 is the Krenchel's orientation efficiency, η_L is the Cox's fiber length efficiency factor, and V_F is the volume fraction of fibers. With increasing V_F, the breakdown of fibers becomes more notable, leading to lower η_L and, at the same time, the fiber orientation decreases, resulting in lower η_0. Accordingly, the increase in E_c with increasing V_F is smaller than the predicted value according to the simple volumetric additivity ('rule of mixtures').

When the logarithms of the flexural modulus, tensile strength, flexural strength and Izod impact strength of injection moldings of PPs filled with GF are plotted against the logarithm of average fiber length, the above properties increase linearly with increasing GF length (in the range of fiber lengths shorter than 10 mm) [10].

REFERENCES

1. Fujiyama, M. (1995) Higher order structure of injection-molded polypropylene, in *Polypropylene: Structure, Blends and Composites, Vol. 1, Structure and Morphology.* (ed. J. Karger-Kocsis), Chapman & Hall, London, pp. 167–204.
2. Karger-Kocsis, J. (1991) Microstructural aspect of the fatigue crack growth in polypropylene and its chopped glass fiber reinforced composites. *J. Polymer Engng*, **10**, 97–121.

3. Kalay, G., Allan, P. and Bevis, M.J. (1995) Microstructure and physical property control of injection molded polypropylene. *Plast. Rubber Composite Process. Appl.*, **23**, 71–85.
4. Fujiyama, M. and Wakino, T. (1991) Crystal orientation in injection molding of talc-filled polypropylene. *J. Appl. Polymer Sci.*, **42**, 9–20.
5. Fitchmun, D.R. and Mencik, Z. (1973) Morphology of injection-molded polypropylene. *J. Appl. Polymer Sci.*, **11**, 951–71.
6. Phillips, R., Herbert, G., News, J. and Wolkowicz, M. (1994) High modulus polypropylene: Effect of polymer and processing variables on morphology and properties. *Polymer Engng Sci.*, **34**, 1731–43.
7. Fujiyama, M. (1995) Structures and properties of injection molding of β-crystal nucleator-added polypropylenes. Part 1 Effect of β-crystal nucleator content. *Int. Polymer Process.*, **10**, 172–8.
8. Karger-Kocsis, J., Varga, J. and Ehrenstein, G.W. (1997) Comparison of the fracture and failure behavior of injection-molded α- and β-polypropylene in high-speed three-point bending tests. *J. Appl. Polymer Sci.*, **64**, 2057–66.
9. Darlington, M.W., Gladwell, B.K. and Smith, G.R. (1977) Structure and mechanical properties in injection molded discs of glass fiber reinforced polypropylene. *Polymer*, **18**, 1269–74.
10. Grove, D.A. and Miller, D.E. (1994) The effects of fiber length on the mechanical performance of injection moldable long glass reinforced polypropylene. *SPE Tech. Pap., 52nd. ANTEC*, **40**, 2306–10.

Keywords: injection molding, morphology, mechanical properties, skin-core structure, skin layer, shear layer, core layer, molecular orientation, orientation function, orientation fraction, crystal orientation, spherulite size, lamellar thickness, crystallinity, crystalline form, β-form, fiber length, fiber orientation, tensile modulus, tensile yield strength, flexural modulus, flexural strength, compression strength, hardness, impact strength, fatigue properties, shear controlled orientation injection molding (SCORIM).

Natural fiber/polypropylene composites

Klaus-Peter Mieck

There are ecological and economical reasons in favor of the use of natural fibers as reinforcing fibers for plastics. A lower price, better recyclability, hygienic harmlessness and a carbon dioxide neutral combustion are the advantages of natural fibers compared to glass fibers (GF).

PROPERTIES OF NATURAL FIBERS

All bast (stem) fibers (flax, kenaf, ramie, nettle, hemp, jute) as well as hard fibers (caroá, sisal) are suitable as for reinforcing fibers for natural fiber reinforced polymer composites, if they have a high tensile modulus and sufficient tensile strength. In addition to cultivation site, type and harvest, the properties of natural fibers depend significantly on the fiber extraction method. An extraction to technical fiber grades, i.e. production of bundles with different number of single fibers, is generally sufficient for use in plastics composites. The properties of such extracted fibers may be described as follows:

- the fineness covers a range from 3.1 to 20.8 tex, that corresponds, supposing a round cross-section, to a diameter range of 51–140 μm;
- the tensile strength of technical natural fibers is between 500–780 N/mm^2 (Figure 1). The natural fibers do not reach the strength parameters of glass fibers, which is beyond of 3500 N/mm^2;
- the Young's modulus can be found between 18–90 kN/mm^2 (Figure 1). Sisal is at the lower and nettle at the upper threshold. The Young's

Polypropylene: An A–Z Reference
Edited by J. Karger-Kocsis
Published in 1999 by Kluwer Publishers, Dordrecht. ISBN 0 412 80200 7

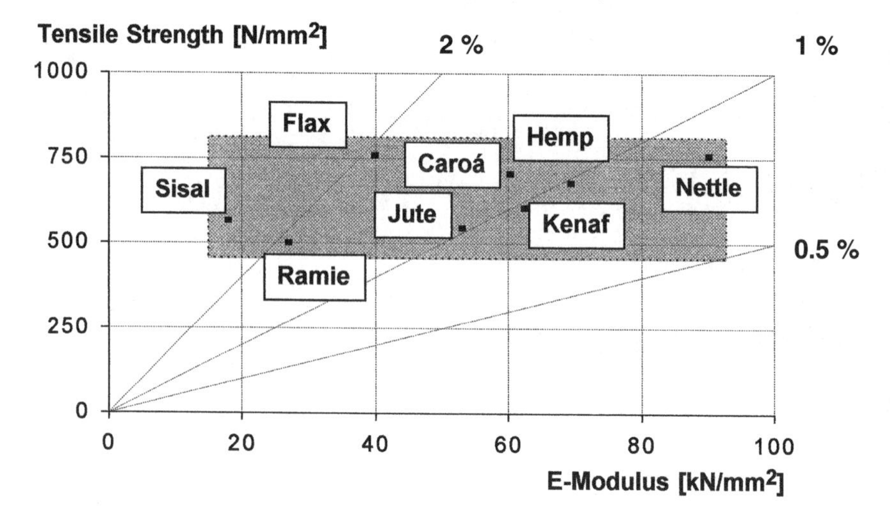

Figure 1 Tensile strength and Young's modulus of technical grade natural fibers.

modulus of some natural fibers corresponds to the values of glass (≈ 73 kN/mm^2) and aramide fibers (between 60 and 124 kN/mm^2);
- the mean of the moisture content at normal climate (20°C, 65% relative humidity) is at 11.6%;
- the density is in a range between 1.41–1.56 g/cm^3;
- considering the thermal stability of natural fibers, the processing temperatures should not exceed 200°C in order to avoid deterioration in the tensile strength.

An extraction up to individual fibers may result in higher tensile strength for flax, ramie or similar natural fibers. The related additional expenditures and the poorer wetting out by the matrix are against an extraction to individual fibers.

METHODS OF OBTAINING THE FIBER

The traditional extraction of bast fibers, mainly used for textile applications, is done by retting and scutching. In case of dew flax retting, the flax remains for 30 days on the field after pulling. Thereby the microorganisms, formed in the flax stem, will extract the fibers from the compound. The fibers can be cleared mechanically from accompanying substances by scutching (breaking/batting). The retting time can be reduced by chemical, enzymatic or with combined steam explosion processes.

Technical bast fibers are obtained by mechanical extraction [1]. These methods are based generally on fibers initially decorticated on the field. Next they are extracted stationary by converting the decorticated aggregates to technical fiber grades. The hard (leaf) fibers are extracted by traditional methods. After harvesting, the leaves are depulped, the fibers are washed, dryed and combed by rotating brushes.

The technical grade natural fiber will get a form appropriate to use in plastics by forming fiber bands or nonwovens. Parallelizing of the fibers on a card and subsequent compressing and a light twisting will lead to formation of a fiber band. Short cut fibers are obtained from the fiber band by a cutting machine.

Nonwovens can be formed by carding on a needle punching machine [2] or aerodynamically by using a machine for random laid nonwovens. In case of 100% natural fiber nonwovens, coarse fibers, such as hard (leaf) fibers, are usually mixed to the finer technical natural fibers in order to improve their wetting out by the matrix.

PRODUCTION OF TEXTILE COMPOSITE PREFORMS

A usual method is the blending of natural and polypropylene (PP) fibers before formation of bands or nonwovens [3]. So preforms are obtained in form of ribbons or nonwovens ready for thermoforming. The grade of the matrix-forming PP has less effect on the tensile and flexural properties of the composites. Therefore, there are no special requirements for the blended, matrix-forming PP fibers. PP fibers of a fineness between 1.7–6.7 dtex with lengths of 40–60 mm, even PP obtained from recycling processes, may also be used. There are no differences in a composites' mechanical performance when homo- or copolymers are used.

In principle, woven, knitted or braided structures could also be prepared as natural fiber/PP-composites. But the higher manufacturing costs do not justify the use of these fabrics. The main reason for this is that natural fiber containing composites are 'low-tech' and low-cost products nowadays.

MANUFACTURING, PROCESSING, APPLICATION

Appropriate manufacturing and processing methods for natural fiber/PP-composites are:

- extrusion coating and impregnation for manufacturing of prepregs (tapes, sheets and other profiles);
- press molding for door-, side-, seat-, A-/B-/C-pillar casings, trunk walls, boards or other large area parts (automotive sector);

- injection molding for appliances, dashboards, automotive A-/B-/C-pillar casings, compact parts; and
- filament and tape winding processes.

For the manufacturing of injection-moldable natural fiber/PP-compounds, the band of the fiber is impregnated by the PP-melt in a continuously running extruder line. The natural fiber reinforced compound is granulated at the extruder exit. Melt compounding of short-cut fibers has not proved viable because of metering problems with the natural fibers.

In case of molding according to the EXPRESS-method [4], 100% natural nonwovens and PP-melt films are laid on one another by using a transportable extruder prior to pressing. In the case of PP-containing (hybride) nonwovens, the usual press molding can be adopted. In order to reduce the cycle time, a preliminary heating of the preform is recommended. In addition, the cooling after shaping may occur in a separate (cold) press.

PROPERTIES OF NATURAL FIBER/PP-COMPOSITES

In Figure 2 the tensile strength and Young's modulus of natural fiber reinforced PP-composites are shown and compared to those achieved by GF-reinforcement at the same reinforcement content (30 wt.%).

Because of the low tensile strength of the natural fibers, these composites reach only of about 60% of the tensile strength of the GF/PP-composites. But the Young's moduli are commonly higher than those of GF/PP-composites. Similar relations were observed for the flexural stress, as well.

For the improvement of tensile and flexural strength parameters a 'compatibilization' by silanes or functionalized PP becomes necessary. A better adhesion toward natural fibers can be obtained when the fibers are treated by vinyl or methacryl functionalized silanes. Adding an adhesion promoter, such as maleic anhydride (MA) modified PP, into the PP matrix is more simple and thus widely used. The anhydride groups undergo esterification reactions with cellulosic hydroxylic groups resulting in good bonding. The grafting degree and the average molecular weight of the modified PP controls the efficiency of compatibilization [5]. Commercial compatibilizers are Hercoprime HG 210 (Himont), Hostaprime HC 5 (Hoechst), Exxelor PO 10/5 (Exxon Chemical), Polybond 3002 (BP Chemical), added generally in low wt.% to the matrix. In this way, an $>100\%$ increase in the interlaminar shear strength (ILSS) and $>25\%$ in the tensile strength were registered for flax-fiber reinforced PP-composites.

In comparison to GF-reinforcement, the natural fiber/PP-composites show a different crash behavior. In case of impact, they do not split but

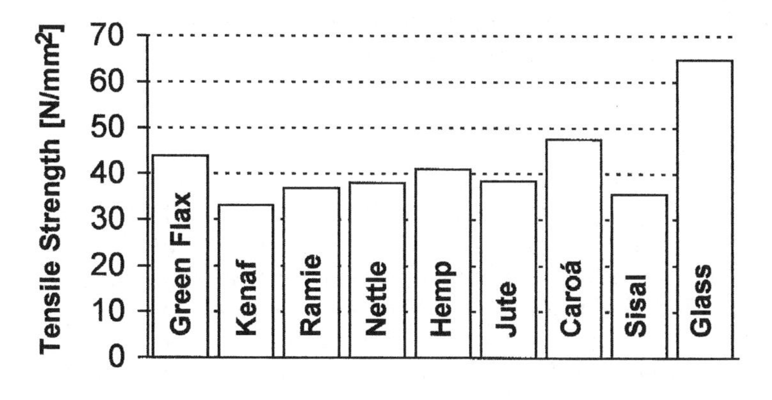

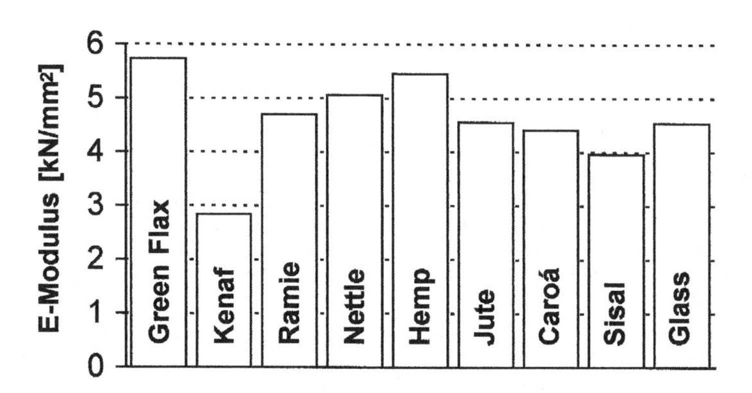

Figure 2 Tensile strength and Young's modulus of natural fiber and glass fiber reinforced PP (fiber content mass = 30 wt.%).

tear. The impact resistance in flexural impact and the perforation energy in falling-weight-impact tests are approximately half of the related GF-PP composites at the same reinforcement content.

The fatigue strength is in the range of glass mat reinforced PP (GMT-PP) with the same fiber mass content. A higher mechanical damping of the natural fiber/PP-composites is typical. This is caused by the structure of natural fibers being themselves composite structures.

Natural fiber/PP-composites absorb moisture from the environment. This moisture absorption does not lead to a change in width or length but to an alteration in thickness (about 1% at 1% water absorption). The tensile and flexural properties depend also on the moisture content. So the effects of water uptake are only partly reversible. The relative reduction of the tensile strength after the boiling test is about 4.5% and so is comparable to a glass fiber reinforcement.

RECYCLING

After use, the natural fiber/PP-composites can be re-utilized without problems. Cutting the parts in a suitable mill results in ground stocks that can be fed to a twin-screw extruder by appropriate dosage devices in order to produce regranulates. Regranulates with different fiber contents and decor contents can be manufactured by blending with original PP. Without further addition of compatibilizers, the ground content can be raised to 70 wt.%. By addition of compatibilizers, the ground stock content can be increased to 80 wt.%. The regranulates serve as feed-stock for injection- or extrusion-molding technologies. The obtainable tensile and flexural strengths will exceed that of PP filled with sawdust.

REFERENCES

1. Mieck, K.P., Nechwatal, A., Knobelsdorf, C. (1994) Potential applications of natural fibres in composite materials, *Mellianal Textilberichte* **75**, 892–98.
2. Mieck, K.P., Reußmann, T. (1995) Flax versus glass, *Kunststoffe* **95**, 366–70.
3. Mieck, K.P., Lützkendorf, R., Reußmann, T. (1966) Needle-punched hybrid nonwovens of flax and PP fibres, *Polymer Composites* **17**, 873–78.
4. Schlößer, T., Fölster, T. (1995) Automobile construction and ecology, *Kunststoffe* **85**, 319–21.
5. Mieck, K.P., Nechwatal, A., Knobelsdorf, C., Faser Matrix Haftung in Kunststoffverbunden aus thermoplasticher Matrix und Flacks, 2, (1995), *Die Angewandte Makromolekulare Chemie*, **225**, 37–49.

Keywords: compatibilizer, flax, natural fibers, natural fiber composites, processing, recycling, moisture content.

Nuclear magnetic resonance spectroscopy of polypropylene copolymers

Eddy W. Hansen and Keith Redford

During the last two decades, nuclear magnetic resonance (NMR) spectroscopy has become a valuable and indispensable analytical tool among polymer chemists. For example, the determination of n-ad distribution numbers is frequently a pre-requisite in the characterization of copolymers.

To illustrate the information obtainable from PP copolymers with the aid of NMR, this article will use copolymers of propylene and ethylene (PP/PE copolymer) as an example. The techniques described are, however, equally applicable to other copolymers. The initial and important phase of such an analysis involves peak assignment, and this is rather tedious. However, the availability of a broad spectrum of dedicated one-dimensional and two-dimensional NMR techniques and improved instrumental design makes this activity easier and more attractive. Figure 1 shows a ^{13}C NMR spectrum of a PP/PE copolymer of low-ethylene content (about 4%). A copolymer with low ethylene content has been selected as it enables information regarding regio-irregularities and tacticity in the PP chain to be extracted. A higher concentration of ethylene results in severe overlap of peaks and renders this type of information somewhat more difficult [1].

As can be seen in Figure 1, ethylene units along the main PP chain can be identified as structures B and X where the former represents regio-irregular incorporation of propylene [2]. Peaks marked with an asterisk

Polypropylene: An A–Z Reference
Edited by J. Karger-Kocsis
Published in 1999 by Kluwer Publishers, Dordrecht. ISBN 0 412 80200 7

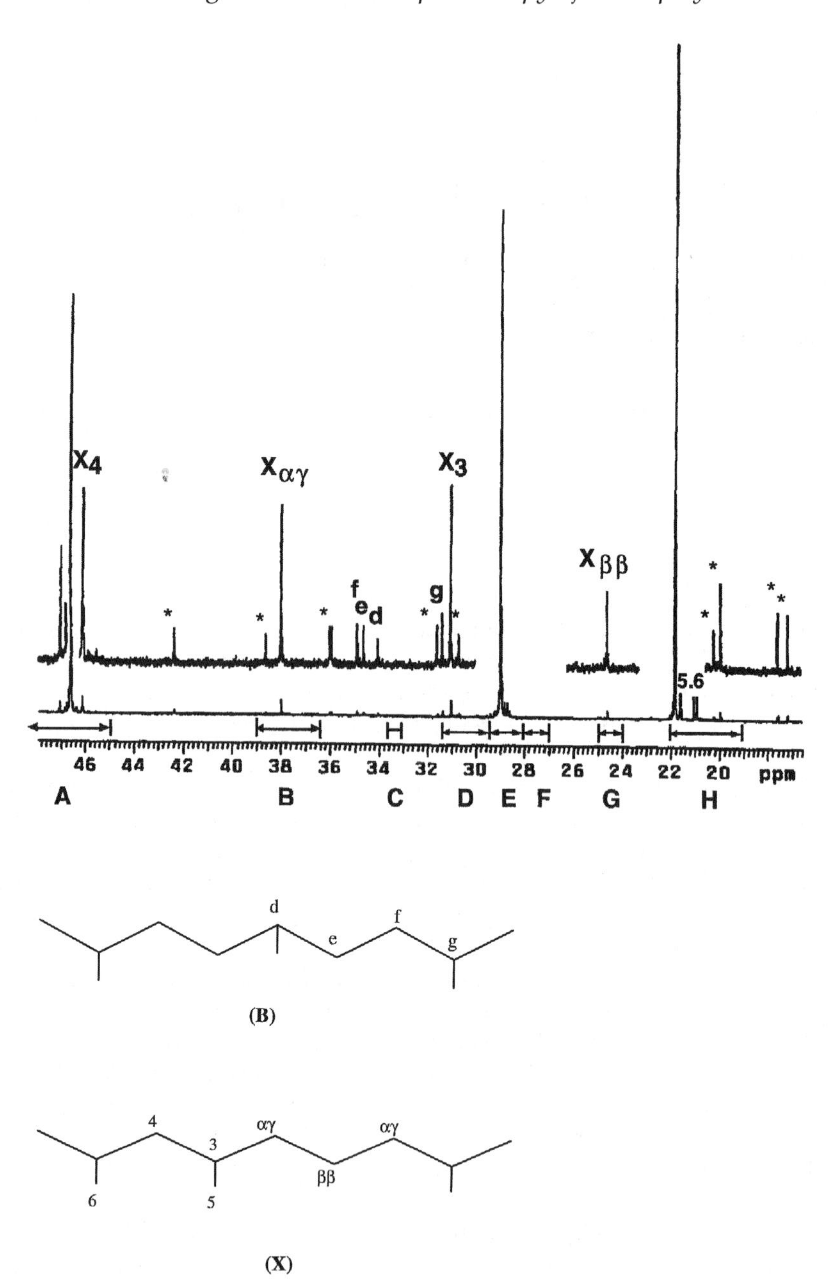

Figure 1 ^{13}C-(^{1}H) NMR spectrum of PP containing approximately 4% ethylene. Refer to structures B and X for peak assignments. See text for further details (SINTEF Chemistry 1996, unpublished work).

represent resonance lines from irregular 2.1 insertion of propylene in meso placement along the chain. Information regarding the type and amount of regio-irregularities is important from a catalyst viewpoint.

The tacticity is determined by first correcting (subtracting) the resonance peaks of structures X and corresponding peaks originating from regio-irregular structure (B) in the methyl chemical shift region. For larger content of ethylene, overlap of peaks originating from triad concentrations and different tacticity must be considered [3].

Concerning the derivation of triad distributions from ^{13}C NMR spectra, these can be derived from the eight specific integral regions A–H indicated on the spectrum (Figure 1), and result in the following eight linear equations [4].

$$T_A = k\left\{[PPP] + \tfrac{1}{2}[PPE + EPP]\right\} \tag{1a}$$

$$T_B = k\left\{[PEP + \tfrac{1}{2}[PEE + EEP] + [EPE] + \tfrac{1}{2}[PPE + EPP]\right\} \tag{1b}$$

$$T_C = k\ \{[EPE]\} \tag{1c}$$

$$T_D = k\left\{2[EEE] + [PPE + EEP] + \tfrac{1}{2}[PEE + EEP]\right\} \tag{1d}$$

$$T_E = k\ \{[PPP]\} \tag{1e}$$

$$T_F = k\ \{[PEE + EEP]\} \tag{1f}$$

$$T_G = k\ \{[PEP]\} \tag{1g}$$

$$T_H = k\ \{[PPP] + [PPE + EEP] + [EPE]\} \tag{1h}$$

where k is a normalization constant, T_X is the area (integral) of region X (= A–H) and the symbols in brackets represent the respective triad concentrations. This over-determined set of equations can be solved analytically as described in a recent publication by Hansen *et al* [1]. For PP/PE-copolymers containing only small amounts of ethylene some of the distribution numbers derived by solving equation (1) could become negative. On these grounds, another strategy – involving the use of a simplex algorithm with suitable constraints – should be considered [5].

The corresponding dyads and monads are derived from the so called 'necessary relationships' [4]:

$$[EE] = [EEE] + \tfrac{1}{2}[EEP + PPE] \tag{2a}$$
$$[EP + PE] = [EEP + PEE] + 2[PEP] = [EPP + PPE] + [EPE] \tag{2b}$$
$$[PP] = [PPP] + \tfrac{1}{2}[EPP + PPE] \tag{2c}$$

and

$$[E] = [EEE] + [EEP + PEE] + [PEP] \tag{3a}$$

$$[P] = [PPP] + [EPP + PPE] + [EPE] \tag{3b}$$

Also, the average sequence lengths of ethylene, $n(E)$, and propylene, $n(P)$, can be calculated from the derived triad concentrations:

$$n(E) = \frac{[EEE] + [PEE + EEP] + [PEP]}{[PEP] + \frac{1}{2}[PEE + EEP]} \tag{4a}$$

$$n(P) = \frac{[PPP] + [PPE + EPP] + [EPE]}{[EPE] + \frac{1}{2}[EPP + PPE]} \tag{4b}$$

Likewise, the sequence number, S, which reflects the number of times there is a switch from ethylene to propylene or propylene to ethylene per 100 repeat units can be derived from either the E-centered ($S(E)$), or P-centered ($S(P)$) triads [4]:

$$S(E) = 100\left\{[PEP] + \tfrac{1}{2}[PEE + EEP]\right\} \tag{5a}$$

$$S(P) = 100\left\{[EPE] + \tfrac{1}{2}[EPP + PPE]\right\} \tag{5b}$$

Since the sequence number can be determined independently by two different equations, equation (5) permits an internal consistency check of the quantitative NMR data. It is also possible to implement these equations, i.e. $S(E) = S(P)$, as an additional constraint when solving equation (1) by a simplex algorithm.

DISPERSION

There is other useful information that can be determined from an n-ad distribution, for example how well a secondary comonomer is dispersed among the primary repeat units. Two approaches are commonly used to indicate the degree of 'clustering' of ethylene units in an ethylene-1-olefin copolymer [4]. One involves a direct empirical measurement of clustering, the monomer dispersity (MD), and the other utilizes the Bernouillian distribution as a reference point for establishing the degree of clustering and is denoted the cluster index (CI). These numbers are given by the following equations, which are valid for ethylene units dispersed in an ethylene-propylene copolymer:

$$MD = \frac{S(E)}{[E]} = 100 \frac{[PEP] + \frac{1}{2}[PEE + EEP]}{[E]} \tag{6}$$

$$CI = 10 \frac{[E] - [PEP]}{2[E]^2 - [E]^3} \tag{7}$$

If all the 1-olefin repeat units are incorporated as [*PEP*], the MD will be 100, which is the case for alternating copolymers and medium density ethylene-1-olefin copolymers where the 1-olefin content is less than approximately 3 mol.%. The largest errors in monomer dispersity will occur for values between 0 and 1, which occur for average sequence lengths of 100 and higher. Likewise, the cluster index has a useful range of zero to approximately 20.

The accuracy and precision of the parameters derived from NMR measurements are solely controlled by how well the integrals of the different regions in the NMR spectrum can be determined [1, 4].

^1H NMR

The potential use of ^1H NMR to estimate the amount of ethylene in a PE/PP-copolymer is rather exciting because the time needed to acquire a ^1H NMR spectrum is of the order of minutes compared to C^{13} NMR which needs hours. Figure 2 shows a typical ^1H NMR spectrum of a PP/PE copolymer and suggests that four regions (A–D) can be clearly separated.

The different protons contributing to the four chemical shift regions have been identified by two-dimensional NMR [5]. Due to the severe overlap of resonance peaks, a direct determination of the copolymer composition is not possible. However, applying different probability models (Markovian or Bernouillian) gives not only information on composition, but also on the comonomer reaction probabilities.

In the case of a random distribution of ethylene the Bernouillian probability model can be applied resulting in the following simplified equations:

$$I_A = k\,[2\alpha^2 - \alpha^3] \tag{8a}$$

$$I_B = k\,[4 - 3\alpha - \alpha^2 + \alpha^3] \tag{8b}$$

$$I_C = k\,[2\alpha - 2\alpha^2] \tag{8c}$$

$$I_D = k\,[3\alpha + \alpha^2] \tag{8d}$$

where $1 - \alpha = P_{22} = P_{32}$ and $P_{23} = P_{33} = \alpha$ and P_{ij} is the probability of a chain terminating in unit i and adding to monomer j, 2 and 3 correspond

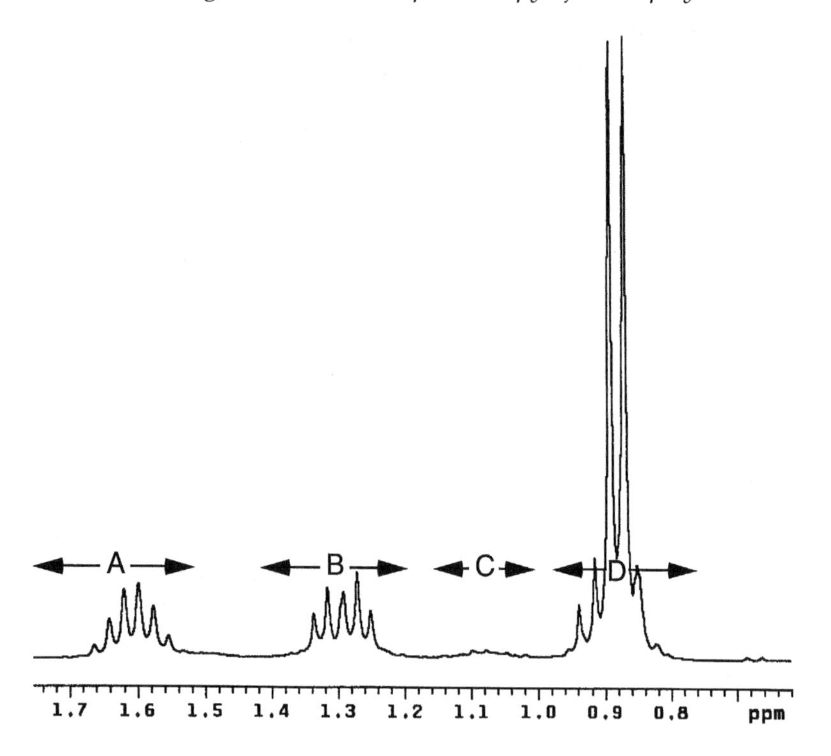

Figure 2 ^1H NMR spectrum of a PP sample containing approximately 4% ethylene. See text for further details (SINTEF Chemistry 1996, unpublished work).

to ethylene and propylene, respectively. Here k is a normalization equal to $1/(4+2\alpha)$ and the molar composition is given by $C_3 + k\alpha$ and $C_2 = k(1-\alpha)$.

REFERENCES

1. Hansen, E.W., Redford, K. and Øysæd, H. (1996) Improvements in the determination of triad distributions in ethylene-propylene copolymers by ^{13}C NMR. *Polymer*, **37**, 19–24.
2. Mizuono, A., Tsutsui, T. and Kashiwa, N. (1992) Stereostructure of regio-irregular unit of PP obtained with rac-ethylene-bis(1-indenyl)zirconium dichloride/methylaluminoxane catalyst system studied by ^{13}C-^1H shift correlation two-dimensional NMR. *Polymer*, **33**, 254–258.
3. Kanezaki, T., Kume, K., Sato, K. and Asakura, T. (1993) ^{13}C NMR determination of the isotacticity of propylene homopolymer part in ethylene-propylene block copolymer. *Polymer*, **34**, 3129–3131.

4. Randall, C.J. (1989) A review of high resolution liquid ^{13}C NMR characterization of ethylene-based polymers. *Macromol. Sci. Rev. Macromol. Chem. Phys. (C)*, **29**, 201–317.
5. Cheng, H.N. (1989) Analytical and synthetic approaches for the NMR characterization of polymers. *J. Appl. Pol. Sci.: Appl. Pol. Symp.*, **43**, 129–163.

Keywords: cluster index, comonomer distribution, monomer dispersity, NMR, ^{1}H NMR, ^{13}C NMR, tacticity, sequence length, copolymer.

Nuclear magnetic resonance spectroscopy of polypropylene homopolymers

Eddy W. Hansen and Keith Redford

The main advantage of NMR in the characterization of polymers is the unique molecular structural information which can be obtained. For instance, the relative configurational distribution (tacticity) and regio-irregularities (defects) in homopolymers of polypropylene (PP) can be mapped out and represent a unique microfingerprint of the polymer. In particular, this information contains indirect and vital information about the stereospecificity of the catalyst, which in turn enables one to improve the design of the catalyst.

This leaves us with the question of what kind of information can be obtained from NMR experiments and how to extract it. To answer this question, a specific example will be discussed which illustrates the general procedure involved. Figure 1 shows a typical ^{13}C NMR spectrum of a highly isotactic polypropylene homopolymer which can be inferred from the intense peak at $\delta = 21.84$ ppm, corresponding to the mmmm-pentad configuration [1].

If no peaks corresponding to regio-irregularities are observable in the spectrum, the n-ad distribution of the configurational distribution (tacticity) can be established from different chemical shift domains in the spectrum (methine region, methylene region or methyl region). Often, the methyl region is chosen for tacticity analysis due to a better chemical shift resolution. Figure 1 illustrates the ten chemical shift regions corresponding to the ten pentad configurations [1]. Increasing the magnetic

Polypropylene: An A–Z Reference
Edited by J. Karger-Kocsis
Published in 1999 by Kluwer Publishers, Dordrecht. ISBN 0 412 80200 7

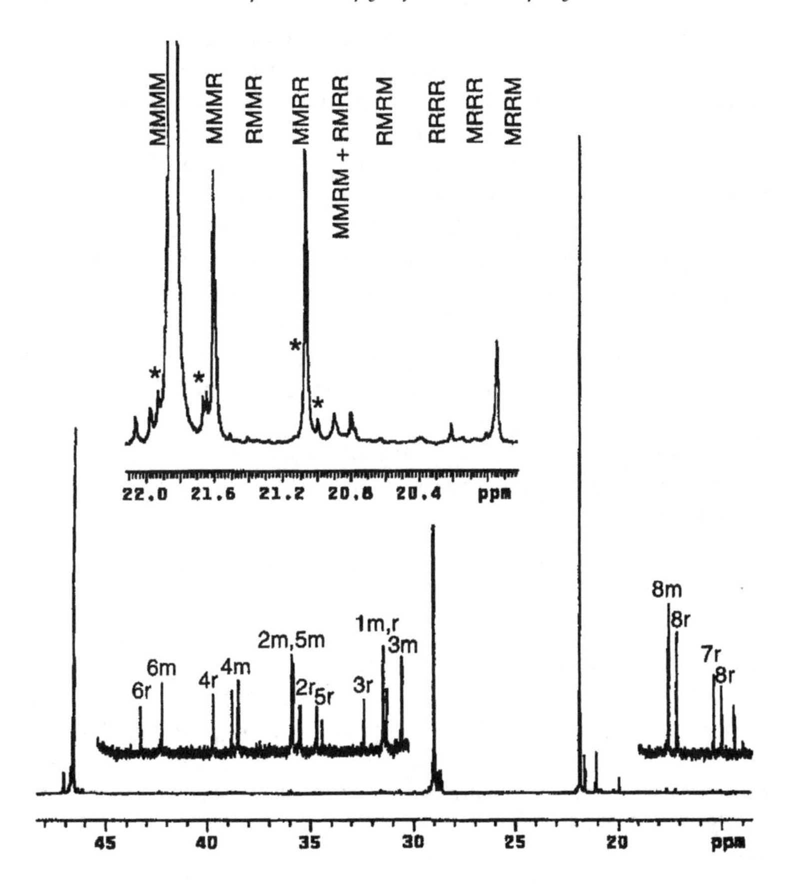

Figure 1 ^{13}C-(^{1}H) NMR spectrum of PP. Refer to structures A_m and A_r for peak assignments (SINTEF Chemistry, 1996, unpublished work).

field strength improves the resolution and enables heptads and probably higher degrees of tacticity to be identified. Combining all three shift regions (methine, methylene and methyl regions) in the analysis of tacticity is more cumbersome, but represents a good test of internal consistency regarding assignment and quantification.

Spectral resolution is also limited by sample characteristics as mirrored in the line width of the resonance peaks which is dependent upon polymer concentration and acquisition temperature. Dissolving the polymer at relatively high temperature followed by temperature cycling improves the resolution, i.e. reduces the line width. Experience suggests that a polymer concentration of approximately 10–15 wt.% gives the narrower lines with an acceptable signal-to-noise ratio.

Another subject of concern relates to the intrinsic NMR characteristics of the polymer, i.e. T_1 and nOe (nuclear Overhause enhancement factor). The latter parameter is tentatively assumed to be the same for all carbons and seems to be indirectly supported by a large number of published data. However, to our knowledge, this has not been unambiguously and experimentally verified. If the nOe of the methyl carbons differs with relative configuration this will certainly affect the quantitative interpretation when using proton decoupling experiments. The spin–lattice relaxation time (T_1) versus temperature of the carbons is well known and determines the necessary time between successive rf-pulses to be used. For quantitative purposes, the minimum interpulse time should be three times the longest T_1 of the carbons. The longest T_1 for PP is of the range 3–5 s.

When comparing the 'pentad' integral areas of the methyl resonances two 'necessary relations' (equations (1) and (2)) should always be considered [2].

$$(\text{mmmr}) + 2(\text{rmmr}) = (\text{mmrm}) + (\text{mmrr}) \qquad (1)$$

$$(\text{rrrm}) + 2(\text{mrrm}) = (\text{rmrr}) + (\text{mmrr}) \qquad (2)$$

As can be seen from the ^{13}C NMR spectrum of Figure 1, these relations are not fulfilled. This observation is not caused by errors in assignment but relates to the existence of other 'species' in the polymer. These other species represent regio-irregularities along the polymer chain and are of considerable importance with regard to the catalyst properties and need to be identified. The many small peaks in the chemical range $\delta = 30$–45 ppm enables these regio-irregularities to be assigned. In this particular case, the irregularities shown in Figure 2 are identified.

Structure A_m represents a 2,1-insertion of propylene, i.e. meso placement and the other, A_r, represents the racemic 2,1-replacement of propylene. Of main importance is that the large peak in the methyl chemical shift region of $\delta = 24.06$ ppm does not belong to a mmrr-configuration but to the methyl resonance '10m' of structure A_m. The presence of A_r suggests that the 2,1-insertion of propylene unit has taken place stereospecifically which is of great interest from the viewpoint of catalyst performance. In general, other types of regio-irregularities can appear, depending on the catalyst system used. The important thing is that these irregularities can be identified in the NMR spectrum and quantified.

Concerning the calculation of relative configurational distribution, the peak intensities originating from stereo-irregularities (marked with an asterisk in Figure 1) are subtracted from the respective areas before determining the actual tacticity distribution (pentads) from the methyl chemical shift region and enables equations (1) and (2) to be applied (Figure 1).

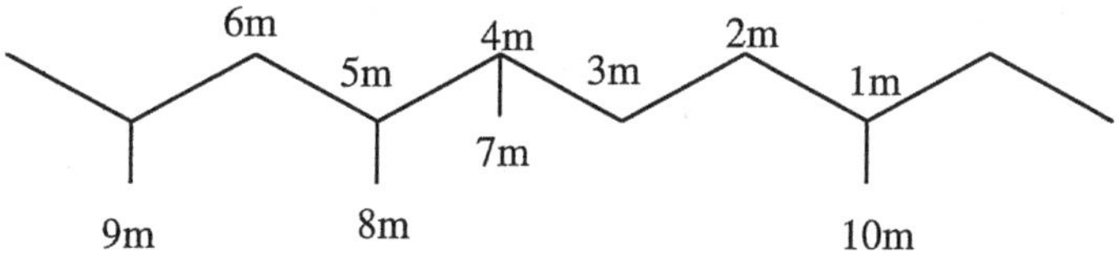

6m 4m 2m
5m 3m 1m
7m
9m 8m 10m

(A_m)

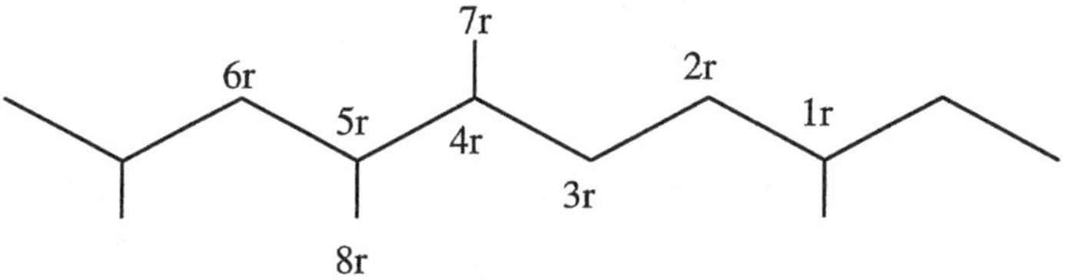

7r
6r 2r
5r 1r
4r
3r
8r

(A_r)

Figure 2

SOLID-STATE NMR

Solid-state NMR spectroscopy provides a unique method of characterizing molecular structure. The problems of solubility and high boiling point solvents encountered in solution NMR are avoided, but at a price, since sensitivity is lost. In return, information is obtained relating to the morphology and molecular dynamics. The same sequence distribution data as seen in the high-resolution spectra contribute to the spectra but more importantly information relating to the packing of the monomer units can be explored. The temperature dependence of molecular motions can also be probed.

The potential use of solid-state CPMAS ^{13}C NMR to detail the molecular structure of PP is demonstrated by many authors and is nicely reviewed in a recent paper by Tonelli [3]. The key parameter in such an analysis is the γ-gauche effect which enables the ^{13}C NMR chemical shifts of vinyl polymers to be assigned and provides an opportunity to test or derive rotational isomeric state (RIS) model descriptions of their conformational characteristics.

The isotropic chemical shifts of γ-trans/γ-trans (tt) and γ-gauche/γ-gauche (gg) conformations known from sPP and of γ-trans/γ-gauche (tg) from the 3_1 helix of iPP give the basis for the interpretation of the spectra of aPP. At 20° above the glass transition the exchange of conformations is rapid.

Three crystalline polymorphs produced by alteration of crystallization conditions of iPP have been identified. In each of these polymorphs (α, β and γ), the individual iPP chains adopt the same 3_1-helical conformation, but they are characterized by different modes of crystalline chain packing: α-monoclinic; β-hexagonal and γ-triclinic crystal forms. As a consequence of its unusual bilayer structure, the γ-polymorph, where alternating bilayers of parallel iPP chains are arranged in a nearly perpendicular fashion, is by NMR seen to be unique. This confirms the X-ray diffraction interpretation [3].

In quenched iPP, there is also observed a phase described as mesomorphic or smectic. This phase consists of the 3_1-helical conformation of the other phases but in a more open structure. NMR has been used to show the conversion of the mesomorphic phase in a quenched iPP sample to the α-phase on annealing [3].

Sozzani and coworkers [4] have presented an experimental scheme which makes it possible to convert between the stable 4_2-helical conformation of sPP (ttggttgg) to a stretched planar, zigzag conformation (ttttttt) and subsequently to a solvent-induced ttggtttt conformation and provides a particularly elegant example of utilizing the conformationally sensitive γ-gauche-dependent ^{13}C chemical shift effect.

Zeigler has measured the fraction of CH in mobile chains as a function of temperature for PP and compared this with the observation from differential scanning calorimetry (DSC). At 100°C, NMR shows that 70% of the CHs are in mobile propylene units. DSC however indicates that the melting endotherm is only just beginning. NMR shows that at 100°C most of the chains are reorientating rapidly on an NMR time scale. This is much below the 160°C where the endothermic peak from the DSC indicates that there is translational chain motion occurring [5].

REFERENCES

1. Cheng, H.N. (1986) Statistical modelling and NMR analysis of polyolefins. *Int. Symp. on Transition-Metal Catalyzed Polymerization, 2: Ziegler–Natta and Metathesis Polymerization*, (ed. R.P. Quirk), pp. 599–623.
2. Randall, C.J. (1989) A review of high resolution liquid ^{13}C NMR characterization of ethylene-based polymers. *Macromol. Sci. Rev. Macromol. Chem. Phys. (C)*, **29**, 201–317.
3. Tonelli, A.E. (1995) High resolution NMR as a local probe of structure, conformation, and mobility in solid polymers. *J. Mol. Struct.*, **355**, 105–119.
4. Sozzani, P., Simonutti, R. and Comotti, A. (1994) Phase structure and polymorphism of highly syndiotactic polypropylene. *Magn. Res. Chem.*, **32**, S45–S52.
5. Zeigler, R.C. (1994) Dynamics of propylene and propylene-ethylene copolymers at temperatures above ambient. *Macromol. Symp.*, **86**, 213–227.

Keywords: NMR-high resolution, NMR-solid state, ^{13}C NMR, tacticity, regio-irregularity, spin–lattice relaxation, necessary relationship, homopolymer, chain mobility, chain packing, molecular conformation.

Nucleation

A. Galeski

Primary nucleation is the precursor of crystal growth and of overall crystallization processes. Crystallization in polypropylene (PP) starts at seeds which pre-exist or are formed in molten PP. Seeds give rise to crystal lamellae which grow then radially to form discs or spheres, forming crystalline aggregates called spherulites. At later stages, growing spherulites impinge and form a spherulitic structure. The classical concept of crystal nucleation is based on the assumption that fluctuations in the supercooled phase can overcome the energy barrier at the surface of the crystal (see the general survey of nucleation for macromolecules by Price [1]). The rate of nucleation, I^*, has been derived using the absolute rate theory:

$$I^* = (NkT/h)\exp[-(\Delta G^* + \Delta G_\eta)/kT]$$

where N is related to the number of crystallizable elements, ΔG^* is the energy of formation of a nucleus of critical size and ΔG_η is the activation energy for chain transport, k is the Boltzmann constant and h is the Planck constant. Generally, in polymers as the temperature is lowered from the melting temperature a rapid decrease in ΔG^* and a slow increase in ΔG_η occur causing I^* to increase. As the temperature is lowered even further, the decrease in ΔG^* becomes moderate but the increase in ΔG_η is more significant resulting in a decrease in I^*. Therefore, a maximum in I^* exists which is related to the ease with which crystallizable elements can cross the phase boundary. In PP, the maximum is rarely observed because the primary nucleation is so intense and the growth of spherulites is so fast that, except for very special cases, the

Polypropylene: An A–Z Reference
Edited by J. Karger-Kocsis
Published in 1999 by Kluwer Publishers, Dordrecht. ISBN 0 412 80200 7

Table 1 Spherulite nucleation density of isotactic polypropylene of M_w=51 200, heptane extracted, melt annealed before crystallization at 180°C for 15 min. (von Falkai, B. and Stuart, H.A. (1959) *Kolloid Zeitschrift*, **162**, 138–140)

Crystallization temperature (°C)	*Nucleation density* ($10^6 \times cm^{-3}$)
122.0	2.36
125.0	1.56
127.5	1.02
130.0	0.85
132.5	0.73
135.0	0.65
138.0	0.58
140.0	0.53
145.0	0.47

material crystallizes completely before reaching the respective temperature.

The microscopic data on primary nucleation of isotactic polypropylene (iPP) for melt samples annealed at 180°C prior to crystallization and then crystallized isothermally in the temperature range from 122 to 145°C are presented in Table 1. The nucleation in iPP in these experiments was found to be heterogeneous and instantaneous.

The course of primary nucleation in iPP down to 70°C was first demonstrated employing the droplet technique i.e. PP was dispersed into small droplets in hot oil and crystallized as the temperature of the oil bath decreased. The dispersion of polypropylene was fine enough to obtain more droplets than the number of pre-existing nuclei in the polymer. The nuclei are not activated at a particular temperature but they appear at increasing supercooling which is connected with their increasing value of ΔG^*. One can recognize four distinct regions in the nucleation of iPP melt:

1. Immediately below the melting point (165–167°C) determined by differential scanning calorimetry (DSC), there is a gap where crystal nucleation and growth hardly take place. Neither the present heterogeneities nor an introduced nucleating agent can accelerate the nucleation.
2. Most published nucleation data concern the region of temperature below 150°C but above 115°C, where regular spherulites are nucleated. This region is the extended region of activity of heterogeneous nuclei. The number of heterogeneous nuclei is limited during the crystallization.

3. Some of the heterogeneous nuclei become active at even a lower temperature which follows from their smaller size or lower perfection. These nuclei are also limited in number.
4. Finally at about 80–85°C and below is a region of homogeneous nucleation. The number of nuclei in this region increases rapidly with the decrease of the temperature.

Figure 1 illustrates the nucleation activity of isotactic polypropylene in thin films determined in isothermal experiments. The data are differentiated to represent the contribution of new nuclei activated by the temperature decrease by 1°C. Samples in the form of thin films (20–30 μm) were crystallized isothermally on a microscopic hot stage. It is seen that the number of nuclei increases initially as the temperature of crystallization decreases. The number of new nuclei at the temperature below 115°C levels off. Apparently all heterogeneous nuclei present in

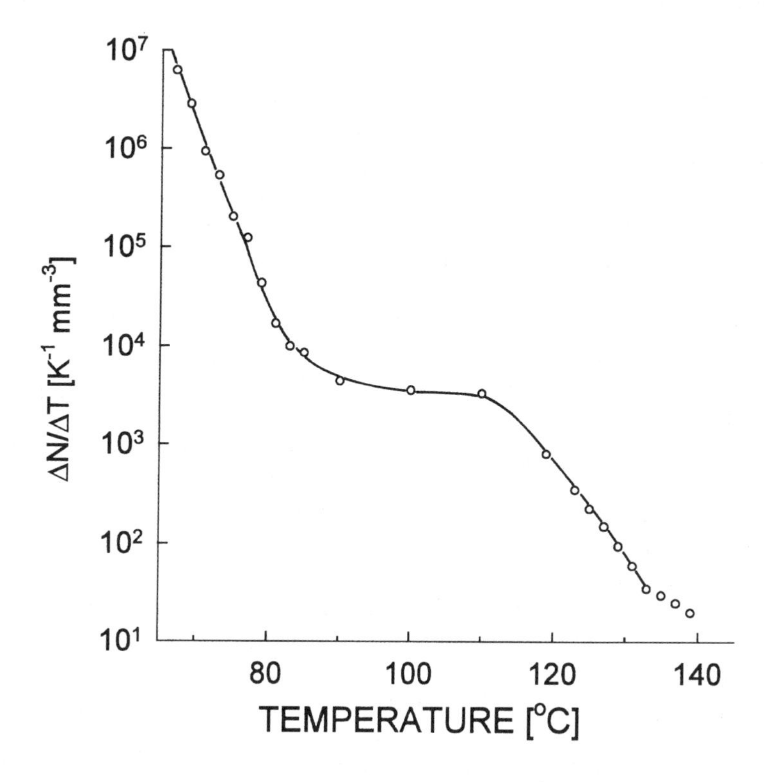

Figure 1 Nucleation activity in iPP (RAPRA, iPP1, $M_w = 3.07 \times 10^5$, $M_n = 1.56 \times 10^4$ g/mole, density = 0.906 g/cm^3, melt flow index = 3.9 g/10 min). The number of nuclei are differentiated to represent the contribution of new nuclei activated by a temperature decrease of 1°C.

the sample were able to show up within the time of crystallization slightly below 115°C.

At temperatures below 85°C, a new intense process of homogeneous nucleation takes place. A rapid increase in the number of formed nuclei with the decrease of crystallization temperature is observed (Figure 1). Droplet experiments during isothermal crystallization also showed that the nucleation in droplets of iPP is thermally activated below 85°C and the droplets crystallize sporadically in time. Investigations of nucleation performed during continuous cooling could, however, resolve only a singular large peak of nucleation at one particular supercooling.

Annealing of PP melt prior to crystallization decreases the active fraction of primary nuclei. The crucial factor is the temperature of melt annealing below or above the equilibrium melting temperature, T_m^0. At 190°C, a vast number of the thermally sensitive nuclei remain untouched while even a short exposure to a temperature around 220°C decreases the number of active nuclei by orders of magnitude. The nuclei disappearing due to annealing are thermally activated heterogeneous nuclei, the so-called 'self-nuclei'. Self-nucleation is a general term describing nucleation of a melt or solution by its own crystals grown previously. Self-nucleation in polymers is particularly strong because of the large temperature range where crystals do not melt entirely. There are suggestions that most of previous observations of the behavior of primary nucleation concerns the self-seeding. The equilibrium melting temperature must be exceeded on melt annealing in order to remove entirely the self-seeded nuclei from the melt.

A technique of self-nucleation was developed based on the observation that the critical nucleus size decreases with decreasing temperature. After melting, the temperature is reduced below the melting point, preferably 10–15°C higher than the required crystallization temperature, and kept for a period of time to produce embryos. Then, the temperature is lowered to the crystallization temperature, when most embryos reach the critical size and become stable nuclei. Using this technique, one can increase the number of nuclei by a few orders of magnitude and significantly reduce the crystallization time.

The other source of thermally sensitive nuclei are polymer crystal fragments in small cracks and cavities on a foreign surface which survive melting because of an increase of their melting temperature due to stiffening effect. At moderate supercooling, they originate the growth of spherulites.

The crystallization of iPP from melt could be enhanced in the temperature region where heterogeneous nucleation is observed by adding some extra heterogeneous nuclei. The interest in such procedure was stimulated by industrial efforts to decrease the size of spherulites for improvement of transparency and mechanical properties. By adding finely subdivided foreign material, it was shown that solids, liquids, and even gas bubbles, are able to nucleate PP spherulites. The list of most active nuclei for isotactic polypropylene, in decreasing activity order, contains: sodium tertiary butylbenzoate, monohydroxyl aluminum p-tertiary butyl benzoate, sodium p-methylbenzoate, sodium benzoate, colloidal silver, colloidal gold, hydrazones, aluminum salts of: aromatic and cyclo aliphatic acids, aromatic phosphonic acids, phosphoric acid, phosphorous acid, several salts of Ca^{+2}, Ba^{+2}, Cu^{+2}, Co^{+2}, Ga^{+3}, In^{+3}, Ti^{+4} and V^{+4}, with the above-mentioned acids, indigo, talc, also air bubbles. From the above list, sodium benzoate, giving the best transparency, and talc, are most frequently used on an industrial scale. Good nucleating agents are insoluble in PP, solid or solidified before crystallization of PP. The important feature of a good nucleating agent appears to be the existence of a crystal lattice matching for epitaxial growth of PP. Earlier, the alternating rows of polar and nonpolar groups at the seed surface was considered to be the prerequisite for a strong nucleating activity. The best cleavage planes of seed crystals do not necessarily expose crystallographic planes matching the growth faces of PP or planes with alternating polar–nonpolar rows.

The extensive data on the nucleating agent of crystallization of PP do not completely satisfactorily match the theory of heterogeneous nucleation which is based on the surface free-energy consideration. The accommodation coefficient was introduced to account for the reduction of interfacial free energy by increased epitaxy through lattice matching and other effects. The existence of alternating rows of polar and nonpolar groups on nucleating surfaces may be the cause of a large value of the accommodation coefficient.

The presence of a large quantity of filler in polypropylene changes drastically the crystallization conditions. Some active fillers, such as talc, aluminum oxide, etc., act as strong nucleating agents while inert fillers, such as chalk, show little nucleating activity. Because of strong nucleating activity of active fillers, the crystallization of filled PP occurs at higher temperatures during quenching. However,the presence of a filler usually increases the thermal diffusivity of the composition which enables inert fillers to reach lower crystallization temperatures during quenching.

Primary nucleation in polymer blends is a special case because of high concentration of the second component. Other phenomena, such as migration of nuclei during mixing, rejection of the second component, occlusions, etc., play an important role. The primary nucleation of PP in blends differ from those for plain PP. The behavior of three basic modes of primary nucleation, homogeneous, heterogeneous and self-seeding, was investigated in these materials. However, the changes of the overall nucleation density observed in blends can be attributed mainly to the changes in the heterogeneous nucleation mode. There is a variety of phenomena causing the changes of heterogeneous nucleation in blends and copolymers as compared to plain crystallizing homopolymer. The most important are miscibility, presence of the interfaces in the system and their properties, migration of the heterogeneities from one component to another during blend mixing and crystallization ability of the second blend component. Two groups of factors have decisive influence on the primary nucleation behavior in blends: (1) the properties of the second component including those dependent on the major crystallizing component, such as miscibility or interfacial energies; (2) the method of preparation of the blend and the parameters of the resulting phase structure of blends, such as the size of inclusions and the total area of the interfaces in the case of phase separated systems. In copolymers, the chemical structure of the copolymer and the phase separation phenomena seem to be decisive for the nucleation behavior.

A special case of interest is reinforced polypropylene with various fibers. Often transcrystallinity in polypropylene occurs which is due to dense heterogeneous nucleation by a substrate. The occurrence of transcrystallinity depends on the type of fiber and the temperature. In contrast to transcrystallinity in quiescent crystallization, the application of stress at the interface between a fiber and a PP melt results in the crystallization of polypropylene on a row-nuclei around a fiber. This effect is caused by strain-induced nucleation via some self-nucleation mechanism and is independent of the type of fiber and less dependent on the temperature of crystallization [5, 6]. Axial stress arises also during cooling of two materials with a large difference in thermal expansion coefficients. As such, the stress-induced nucleation in reinforced PP depends also on the cooling rate, fiber length, position along the fiber and viscoelastic properties of the PP melt [5].

Knowledge of nucleation data is essential for controlling physical properties of a polymer which depend to a great extent on the spherulite average size, size distribution and the size of so-called 'weak spots' – defects of spherulitic structure including cavities and frozen stresses which resulted from volume contraction during crystallization, all determined by the primary nucleation process.

For some applications, it is sufficient to determine only the total number of nuclei activated during the crystallization. The simplest way of determining this value is from the average spherulite size for samples which are filled with spherulites. The average spherulite size can be obtained from the first moment of the size distribution or of the distribution of chord intercepts with spherulite boundaries. The higher moments of spherulite size distribution can be determined on the basis of direct characterization of spherulite patterns, on the basis of the small angle light scattering (SALS) or the light depolarization technique.

If the time dependence of activation of nuclei is required, the data are usually obtained by direct microscopic observation of a crystallizing sample. The difficulty of this method is the necessity of using thin samples and the conditions of crystallization allowing for counting the spherulite centers. These limitations can be overcome by applying a method of reconstructing the sequence of nucleation events from shapes of spherulite boundaries in already crystallized films and in thin sections for bulk samples. The time lag between the nucleation of two neighboring spherulites can be found from the curvature of their common boundary and this procedure repeated for the chain of neighboring spherulites delivers the information on the time distribution of the activation of nuclei. The time distribution of activation of nuclei should be expressed in number of activated nuclei per volume unit of untransformed fraction of the sample. Calibration of the time axis in that method is made by measurements of the spherulite growth rate.

A given brand of PP is characterized by an average number of primary nuclei per volume unit or an average spherulite size in bulk at certain crystallization conditions. However, for sample thickness below the average spherulite size in bulk, the thinner the sample, the larger are the spherulites. The apparent increase in spherulite sizes in thin films results from the constant average number of nuclei per volume unit. The factor complicating this simple relation is the nucleating ability of sample surfaces.

It is usually considered that nuclei are spread randomly over the sample, except for nuclei formed on outer surfaces, on surfaces allowing for transcrystallinity and on surfaces of the second dispersed component, e.g. short fibers. However, if the nucleation events occur not only at the very beginning of crystallization but also during crystallization, then the volume occupied by already advanced spherulites is excluded from further nucleation. Such an exclusion always produces a kind of a distance correlation, if the nucleation process is prolonged in time. The close vicinity of an arbitrarily chosen nucleus is then poorer in other nuclei than more distant regions.

The nucleus is built of more or less regular lamellae showing a multilayer arrangement. Early object develops by branching, usually at rather large angles, and splaying apart the dominant lamellae. A sheaf-like center is usually seen for the crystallization temperature above 155°C or a cross-hatched structure when viewed in plan for the crystallization temperature 155°C and below. The multilayer arrangement of the very center of a spherulite extends as far as several lamellae thicknesses, i.e. 50–100 nm.

The nucleation of the β-crystallographic form occurs significantly less often in bulk samples than the predominant α-form. The β-form generally occurs at a level of only few percent. Nucleation of β-spherulites is preferred in a temperature gradient, in the presence of shearing forces, in certain brands of PP, especially those having high molecular weight or containing a low amount of ethylene groups incorporated within polyproplynene chains. Certain heterogeneous nuclei preferentially nucleate β-spherulites. A quinacridone dye known as permanent red E3B is very effective in generating spherulites of β-form for iPP crystallized below 130°C. However, its effectiveness depends on the nucleant concentration, dispersion and cooling rate. A series of other crystalline substances were found to nucleate β-form in isotactic polypropylene: 2-mercaptobenzimidazole, phenothiazin, triphenodithiazine, anthracene and phenanthrene and pimelic acid (see the related chapter in this book).

After completion of a layer on the surface of the crystal, a new surface nucleus must be created for the formation of a new layer. This is called a secondary nucleation process. The growth of the crystal is then a series of secondary nucleation events and completions of new layers. The most widely accepted expression for the secondary nucleation rate is given by

$$I = A\beta_g \exp[-CT_m^0/(T\Delta T)]$$

where T_m^0 is the equilibrium melting temperature, ΔT is the supercooling, C is a parameter depending on energetical conditions of the secondary nucleus formation, the β_g factor is the retardation factor because at a large supercooling polypropylene becomes viscous and the chain motion is retarded. The temperature dependence of the secondary nucleation rate at high supercoolings is determined by the retardation factor while at low and moderately high supercoolings is determined by $T_m^0/(T\Delta T)$. As the temperature is lowered from the melting temperature, I increases. When the temperature is lowered even further, the increase of $T_m^0/(T\Delta T)$ becomes large with a maximum at around 80°C. The decrease of the β_g factor is less significant in this temperature range. Therefore, the secondary nucleation rate reaches very high values at a similar temperature range, as does the primary nucleation.

REFERENCES

1. Price, F.P. (1969) Nucleation in polymer crystallization, in *Nucleation*, (ed. A.C. Zettlemoyer), Dekker, New York.
2. Hoffman, J.D., Davies, G.T. and Lauritzen Jr., J.I. (1976) The rate of crystallization of linear polymers with chain folding, in *Treatise on Solid State Chemistry*, Vol. 3, Chap. 7, (ed. N.B. Hannay), Plenum Press, New York, pp. 497–614.
3. Galeski, A. (1995) Nucleation of polypropylene, in *Polypropylene: Structure, Blends and Composites*, Vol. 1, (ed. J. Karger-Kocsis), Chapman & Hall, London, pp. 116–139.

4. Galeski, A., Bartczak, Z. and Martuscelli, E. (1996) Nucleation processes, in *Toughened Plastics*, (eds E. Martuscelli, P. Musto and G. Ragosta), Elsevier, Amsterdam, pp. 157–241.
5. Thomason, J.L. and van Rooyen, A.A. (1992) Transcrystallized interphase in thermoplastic composites. Part II. Influence of interfacial stress, cooling rate, fibre properties and polymer molecular weight, *J.Mat. Sci.*, **27**, 897–907.
6. Varga, J. and J. Karger-Kocsis (1996) Rules of supermolecular structure formation in sheared isotactic polypropylene melts, *J. Polymer Sci., Part B, Polymer Phys.*, **34**, 657–670.

Keywords: nucleation, primary nucleation, nucleation rate, heterogeneous nucleation, supercooling, homogeneous nucleation, annealing, self-nucleation, nucleating agents, epitaxy, nucleation in blends, nucleation in copolymers, spherulite, lamellae, β-crystal form, secondary nucleation, nucleation rate.

Optical clarity of polypropylene products

Béla Pukánszky

INTRODUCTION

The number of products where transparency and optical clarity are basic requirements increases continuously. Articles used in medicare represent very good examples of such applications, where optical clarity is not only an aesthetical demand, but a necessary condition for the quick identification of the product. Earlier amorphous polymers, e.g. polyvinyl chloride (PVC) and polystyrene (PS), were used in such areas, since, in semicrystalline polymers, light scattered on the different units of their structure made the products opaque. In polyolefins, the incident light is scattered on the crystallites, spherulites and also on the interface between the amorphous and crystalline phases having different refractive indices [1]. Often the size of the crystalline units is large enough to interfere with visible light and this interference results in considerable haze. The advantageous price/performance relations of polypropylene (PP) and development both in processing as well as additive technology lead to a breakthrough for its use also in fields where optical clarity is needed.

DEFINITION AND MEASUREMENT OF CLARITY

When an object is viewed through an imperfectly transparent specimen, there are two ways in which the visibility may be reduced, thus two criteria exist for transparency. The first is the degree to which the specimen reduces the apparent contrast of the object; this aspect of transparency is commonly referred to as milkiness or haze. The second

Polypropylene: An A–Z Reference
Edited by J. Karger-Kocsis
Published in 1999 by Kluwer Publishers, Dordrecht. ISBN 0 412 80200 7

measure of transparency describes the degree to which fine detail is resolved in the object; the ability of a specimen to transmit fine detail is known as clarity. For the complete characterization of transparency, both must be determined, since they can change independently from each other, i.e. clarity improves while haziness deteriorates [2]. Clarity is determined by the part of the incident light scattered at low angles, while haze by the part scattered at larger angles. Overall transparency of a sample is determined by the ratio of the total transmitted and scattered light.

As was mentioned in the introduction, optical properties of a polymer are determined by the complex combination of its basic scattering, transmission and reflection characteristics, but they depend also on the technique and conditions of illumination and detection. The number of determining factors and the complexity of the relations are in strong contradiction with the natural demand in practice to characterize optical properties with a simple number, such as transparency, haze etc. In spite of this controversy and difficulties in the interpretation of the dominating factors, optical properties of a polymer can be determined, reproduced and compared by the strict application of relevant standards.

Haze is used the most often for the characterization of the transparency of a plastic product. Haze is the total flux of light scattered within the angular range from 2.5 to 90° and normalized to the total transmitted flux. The determination of haze is described by national and international standards (ASTM D 1003, BS 2782), which define the light source, the size of the incident light, detection and other conditions of the measurement. Commercial instruments are available for the determination of haze according to the above-mentioned standards. Less unambiguous is the determination of clarity, i.e. the ability of a specimen to transmit fine details of an object. Several quantities are used for its characterization and, unlike haze, no standardized, universally accepted technique exists for its determination [2].

CLARITY OF PP PRODUCTS

Lately considerable effort is concentrated on the improvement of the transparency of PP products. In spite of this fact, the number of papers reporting a systematic study of the factors influencing transparency is surprisingly low. Bheda and Spruiell [3] carried out extensive experiments to determine the correlations between the type of the polymer, the amount of nucleating agent, processing conditions and the optical properties of blown PP films. They changed the melt flow index (MFI) of the polymer, its molecular weight distribution, the amount of sodium benzoate nucleating agent, the blowing and drawing ratios, as well as the position of the frost line. Beside transparency they measured the crystallinity, morphology, surface quality and light scattering of the films. Their most important conclusions can be summarized in the following points:

1. Transparency is influenced both by material properties and processing conditions. The effects of these factors are interrelated.
2. Increased orientation and rate of crystallization improve light transmission. As a consequence, increasing drawing and blowing ratios improve clarity.
3. Transparency is primarily determined by the quality (smoothness) of the surface; internal morphology has secondary importance.
4. Increasing cooling rate, narrower molecular weight distribution, lower MFI and introduction of a nucleating agent, all result in the decrease of surface roughness, thus increase transparency. Surprisingly, these changes slightly decrease transparency in the inner part of the film.
5. Changes in the smoothness of the surface can be traced back to changes in the crystallization process and they are not caused by the variations in the processing conditions.

The importance of the type of polymer used and the quality of the surface is emphasized also by the results of White and Donn [4] who studied the transparency of stretch blow-molded PP bottles. They observed that bottles prepared from a polymer with a wider molecular weight distribution are more opaque than the ones produced from a polymer with a narrower distribution. The increased opacity is the result of surface cracks.

These studies emphasize the decisive role of surface quality in the determination of transparency. In Figure 1, haze of a PP random copolymer containing an efficient commercial nucleating agent in two concentrations is plotted against the thickness of the specimen. A continuous increase of haze is observed as a function of thickness at both nucleating agent contents, indicating that the structure of the internal part of the product also contributes to its transparency. In the case of thin films, surface quality dominates optical properties.

IMPROVEMENT OF TRANSPARENCY

The results presented above and Figure 1 indicate that transparency is closely related to the crystallization and the crystalline structure of the polymer. Any effect which influences these factors will change transparency as well. Several publications indicate that the introduction of heterogeneous nucleating agents or clarifiers into PP homo- and copolymers considerably improves the transparency of the product.

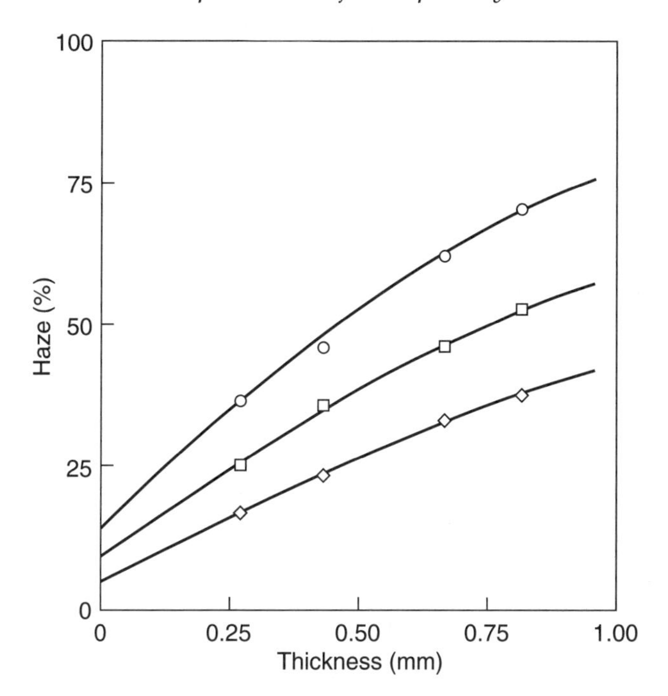

Figure 1 Effect of specimen thickness on the haze of a PP random copolymer containing a commercial nucleating agent at (○) 0, (□) 0.1 and (◇) 0.3 wt.% concentrations.

In the last 10–15 years, very powerful organic nucleating agents have been developed which decrease the average size of the spherulites below the wavelength of visible light. This development was especially important in PP, which usually has quite large spherulites. This development led to PP products with increased optical clarity, thus PP containing the proper additive package and homogenized appropriately can compete in areas where earlier only styrene polymers were used. At the moment, sorbitols are considered to be the most effective clarifiers.

Sorbitols were originally used for the gelation of organic solvents, but found use as nucleating agents in the 1970s. These materials have from moderate to strong nucleating activity, but most publications or patents mention them as clarifiers. Their importance has increased continuously, as shown by the large number of patents related to them. Because of their relative novelty, less information is available concerning their effect or mechanism.

However, beside sorbitols other traditional nucleating agents were also reported to improve transparency of PP products [3]. Since transparency is evidently related to the crystalline structure of the polymer and

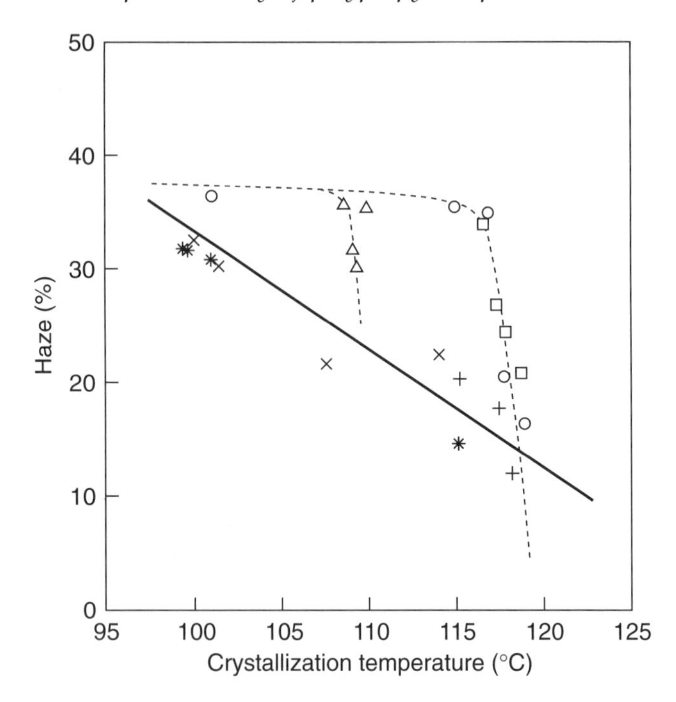

Figure 2 Correlation of haze and nucleating efficiency for selected clarifiers and nucleating agents in a PP random copolymer; (○, □) commercial nucleating agent, (△) talc, (×, +, *) clarifier of sorbitol type.

structure is determined by the conditions of crystallization, parameters characterizing crystallization must be also connected with the optical properties of a PP product. The peak temperature of crystallization (T_c) is one of the quantities often used for the characterization of the crystallization process and the efficiency of nucleating agents. With increased crystallization temperature, the thickness of the lamellae increases as well [5]. Larger efficiency and concentration of nucleating agent lead to an increase of T_c and decrease of the size of spherulites [6]. In Figure 2, haze of a PP random copolymer containing various clarifiers and nucleating agents is plotted against the peak temperature of crystallization. Two types of correlations can be observed, depending on the type of the additive used. Haze decreases more or less linearly with the increasing efficiency of clarifiers, which are different sorbitols in this case. Heterogeneous nucleating agents, on the other hand, seem to have a limiting concentration above which they improve haze. Up to a certain efficiency (T_c), there is hardly any change in transparency, while above it clarity increases very fast. The same symbols belong to various concentrations

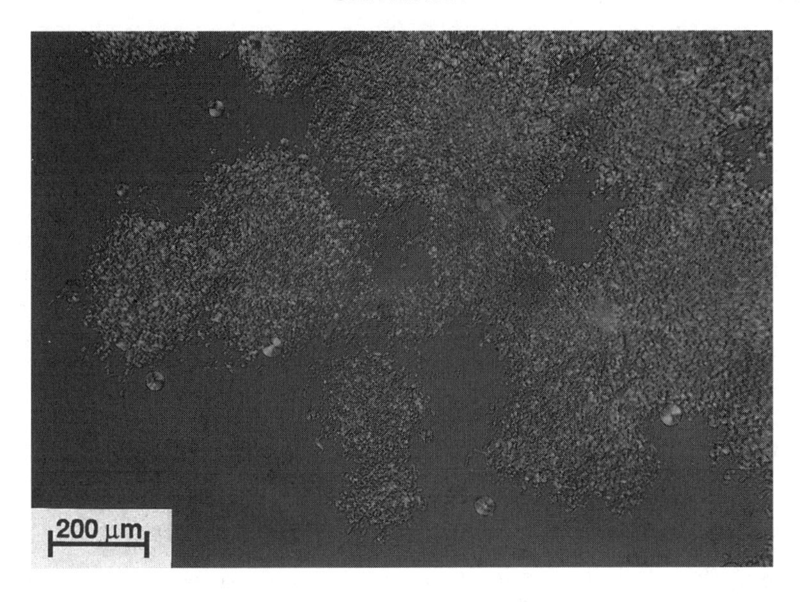

Figure 3 Crystalline structure of a PP homopolymer containing an efficient clarifier at 0.1 wt.%; polarization optical micrograph.

of the additives, i.e. transparency, can be regulated also by the amount of nucleating agent introduced into the polymer.

The two correlations indicate some differences in the mechanism of clarifiers and nucleating agents. In the presence of the latter, the well-known spherulitic morphology of PP develops in the polymer. Increased efficiency or concentration lead to an increase in the number of nuclei, which results in a considerable decrease of spherulite size. When spherulite size reaches a critical value, i.e. becomes smaller than the wavelength of the incident light, transparency improves considerably. However, clarifiers, i.e. sorbitols, induce a different morphology. They form a three-dimensional network structure on which PP crystals grow epitaxially. Spherulitic structure usually cannot be observed in the presence of clarifiers, the developed structure is demonstrated in Figure 3. Since spherulites do not form in the products containing clarifiers, transparency increases linearly with increasing efficiency of the additive.

CONCLUSIONS

Transparency of PP products can be efficiently improved both with traditional nucleating agents and clarifiers, which are various sorbitol derivatives. The improvement in the optical properties depends on the efficiency in destroying the spherulitic structure of PP or at least decreasing the size of the spherulites below a critical value. Strong nucleation efficiency is usually accompanied by an improvement of clarity as well. The effect is similar in PP homo- and random copolymer, but in block copolymers transparency cannot always be improved due to the inherent heterogeneity of this polymer. If the particle size of the additive exceeds a certain value or if it is not properly dispersed, transparency may decrease in spite of an increased nucleation effect. In thin-walled products, surface quality plays an important role in the determination of transparency; also, transparency can be improved with the introduction of efficient nucleating agents and clarifiers.

REFERENCES

1. Meeten, G.H. (1986) *Optical Properties of Polymers*, Elsevier, New York.
2. Willmouth, F.M. (1986) Transparency, translucency and gloss, in *Optical Properties of Polymers*, (ed. G.H. Meeten), Elsevier, New York, p. 265.
3. Bheda, J.H. and Spruiell, J.E. (1986) The effect of process and polymer variables on the light transmission properties of polypropylene tubular blown film, *Polymer Engng Sci.*, **26**, 736.
4. White, S.A. and Donn, S.K. (1992) The development of surface texture in injection stretch-blow molded polypropylene, *Polymer Engng Sci.*, **32**, 1426.
5. Bassett, D.C. (1981) *Principles of Polymer Morphology*, Cambridge University Press, Cambridge.
6. Sterzynski, T., Lambla, M., Crozier, H. and Thomas, M. (1994) Structure and properties of nucleated random and block copolymers of polypropylene, *Adv. Polymer Technol.*, **13**, 25.

Keywords: refractive index, optical clarity, haze, transparency, nucleation, opacity, clarifers, sorbitol, crystallization.

Orientation characterization in polypropylene

A. Ajji and K.C. Cole

INTRODUCTION AND DEFINITIONS

Orientation of polymers enhances many of their properties, particularly mechanical, impact, barrier and optical properties. Orientation processes can generally be classified into three categories: fibers, films and parts (sheets, bottles, rods, etc.). Polypropylene (PP) is widely used for all three applications. In the case of fibers, the fiber spinning process and fiber/textile applications of PP have a share of about 25% of total production [1]. Films, on the other hand, constitute a major application for PP (about 35% share of the total production) [1] and it is the one that involves biaxial orientation. Biaxial orientation has the added advantage of allowing property enhancement in two directions (i.e. avoiding any weakness in the transverse direction). The most widely used biaxial orientation processes for films are tubular film blowing and cast film biaxial orientation (or tentering). Finally, other processes, such as blow molding, compression and injection molding, and thermoforming produce oriented articles of PP for diverse applications. The applications that take most advantage of these oriented forms of PP are the textile and packaging industries.

Thus, a knowledge of the polymer orientation produced by the different processes mentioned above is critical for establishing the process conditions and the final properties of the oriented PP. There is a long history of investigations of orientation in polymers [2]. The simplest

Polypropylene: An A–Z Reference
Edited by J. Karger-Kocsis
Published in 1999 by Kluwer Publishers, Dordrecht. ISBN 0 412 80200 7

representation of the orientation is that corresponding to uniaxial symmetry, which simply defines the orientation factor (f) in terms of the angle (ϕ) that the polymer chains make with respect to the draw direction, as:

$$f = \frac{3\langle \cos^2 \phi \rangle - 1}{2}$$

where the brackets indicate an average over all the angles. For crystalline materials, there are three crystallographic axes and the orientation of the material is defined by the orientation of these axes. For the uniaxial orientation case, the orientation of the j crystallographic axis can thus be written as:

$$f_j = \frac{3\langle \cos^2 \phi_j \rangle - 1}{2}$$

Two of the three factors are independent since the sum of the $\langle \cos^2 \phi_j \rangle$ is equal to 1. In the case of biaxial orientation, more than one orientation factor is needed to specify the orientation states of the polymer. White and Spruiell [3] defined the following biaxial orientation factors for a polymer chain, where the indices 1 and 2 refer to the orientation directions (machine and transverse):

$$f_1^{\mathrm{B}} = 2\langle \cos^2 \phi_1 \rangle + \langle \cos^2 \phi_2 \rangle - 1$$
$$f_2^{\mathrm{B}} = 2\langle \cos^2 \phi_2 \rangle + \langle \cos^2 \phi_1 \rangle - 1$$

For a crystalline material, these factors can be written as:

$$f_{1j}^{\mathrm{B}} = 2\langle \cos^2 \phi_{1j} \rangle + \langle \cos^2 \phi_{2j} \rangle - 1$$
$$f_{2j}^{\mathrm{B}} = 2\langle \cos^2 \phi_{2j} \rangle + \langle \cos^2 \phi_{1j} \rangle - 1$$

There are in total four independent crystalline biaxial orientation factors because the sums of the $\cos^2\phi_{1j}$ and the $\cos^2\phi_{2j}$ are each equal to one. These factors assume machine and transverse symmetry as is the case in most processes; if this symmetry is not present, these factors should be modified [2]. The orientation factors defined above are generally called the second moment of the orientation distribution function. In fact, the orientation distribution has an infinity of moments (Legendre polynomials of $\langle \cos^2\phi \rangle$), but for a moderately oriented polymer (orientation factor f in the range of -0.4 to 0.8), these moments are negligible. However, for a highly oriented polymer, the second moment f may saturate between 0.9 and 1 and remain constant, thus not allowing the full characterization of the polymer. In this case, higher moments of the orientation function have to be determined. The complete definition of these moments as well as a more rigorous representation of general orientation factors can be found in references [2] and [4].

Many techniques can be used for the characterization of the orientation states of polymers. One can mention birefringence, infrared spectroscopy, wide angle X-ray diffraction (WAXS), small angle X-ray scattering (SAXS), small angle light scattering (SALS), small angle neutron scattering (SANS), fluorescence spectroscopy, nuclear magnetic resonance (NMR) and ultrasonic measurements. Each of these techniques has its advantages and disadvantages. It will not be possible to describe them all in detail here; instead, we describe in more detail in the next section three of the most widely used techniques, namely birefringence, infrared dichroism, and X-ray diffraction. More information on the other techniques can be found elsewhere [5].

ORIENTATION CHARACTERIZATION TECHNIQUES

Details on PP structure and crystalline forms can be found elsewhere in this handbook.

Birefringence

Birefringence is the anisotropy of refractive index resulting from a variation in the polarizability along the different directions. It can be measured using various techniques [5, 6]. Birefringence can be determined in two ways. The first amounts to determining the refractive index in the three directions with the aid of a refractometer, and then calculating the birefringence as the difference between the different refractive indices; the orientation function can also be calculated, with some assumptions, directly from the refractive indices [6]. In this case, a contact liquid (generally α-bromonaphthalene) is used to avoid reflection at the interface between the prism and the sample. The second method measures the birefringence directly using optical methods with polarized light (from the retardation in polarized light on going though an oriented sample). Different optical set-ups may be used. The most complete is the one using a multiwavelength source and detector (array detectors), which allow the determination of the absolute birefringence without a limit of its order, as obtained with monochromatic methods. Birefringence is a global measurement and does not discriminate between the different phases present (amorphous and crystalline). For a semicrystalline polymer it can be expressed as:

$$\Delta n = X\Delta n_{\mathrm{c}} + (1 - X)\Delta n_{\mathrm{a}} + \Delta n_{\mathrm{form}}$$

where Δn_{form} is the form birefringence (generally negligible) and X is the crystalline fraction. The intrinsic birefringence can be written similarly (by neglecting the form contribution) as:

$$\Delta^{\mathrm{o}} = \mathrm{X}\Delta^{\mathrm{o}}_{\mathrm{c}} + (1 - \mathrm{X})\Delta^{\mathrm{o}}_{\mathrm{a}}$$

where Δ_c^o and Δ_a^o are the intrinsic birefringences of the crystalline and amorphous phases respectively. The most cited and widely used values for these are $\Delta_c^o = 0.0291$ and $\Delta_a^o = 0.060$ [5, 6]. Values of $\Delta_c^o = 0.0416$ and $\Delta_a^o = 0.0379$ also have been reported in the literature [6].

The orientation is calculated from birefringence simply by the ratio:

$$f_{xz} = \frac{\Delta n_{xz}}{\Delta^o} = \frac{3\langle\cos^2 \phi_{xz}\rangle - 1}{2}$$

where Δn_{xz} is the birefringence in the direction xz (measured in the xz plane), Δ^o the absolute birefringence, ϕ_{xz} the angle between the chain and stretching direction in the xz plane, and f_{xz} the second moment of the orientation function. Let M designate the stretching (machine) direction, T the transverse direction, and N the normal direction. In the case of uniaxial orientation, there is isotropy in the plane normal to the stretch direction, thus $\Delta n_{MT} = \Delta n_{MN}$ and the orientation function is unique. For the case of biaxial orientation, two independent orientation factors (described in the Introduction) can be calculated:

$$f_1 = \frac{\Delta n_{MN}}{\Delta^o} \text{ and } f_2 = \frac{\Delta n_{TN}}{\Delta^o}$$

Infrared spectroscopy

Infrared dichroism is based on the interaction between linearly polarized infrared radiation and the oriented material. The atoms of a polymer molecule vibrate in characteristic normal modes, each of which produces a change in dipole moment (the transition moment) that has a specific direction. Each mode absorbs infrared energy at a characteristic frequency, giving rise to peaks in the infrared spectrum. The peak intensity (i.e. the absorbance) depends on the angle between the transition moment and the electric field vector of the radiation, and it is this that provides information on the molecular orientation. The orientation is defined in terms of the second moment of the orientation function $P_2(\cos \theta)$, where:

$$\langle P_2(\cos \theta)\rangle = \frac{D-1}{D+2} \Big/ P_2(\cos \theta_m)$$

The quantity $P_2(\cos \theta)$ is the second-order Legendre polynomial:

$$P_2(\cos \theta) = \frac{3\langle\cos^2 \theta\rangle - 1}{2}$$

The angle θ is that between the polymer chain axis and the draw direction, θ_m denotes the angle between the transition moment of a particular vibrational mode and the chain axis, and D is the dichroic ratio given by $D = A(\parallel)/A(\perp)$, i.e. the ratio of the absorbance when the electric field vector is parallel to the draw direction to that when the electric field vector is perpendicular. The angle brackets on the left-hand-side of $\langle P_2(\cos\theta)\rangle$ equation denotes an ensemble average.

One advantage of infrared dichroism is the fact that the vibrational modes are generally highly localized, i.e. associated with specific groups of atoms. Hence the spectrum is sensitive to molecular conformation and local structure. For example, in a semicrystalline polymer, some infrared bands are associated with the crystalline phase, others with the amorphous phase, and yet others with both. This makes it possible to obtain information on the degree of crystallinity as well as the orientation in both the crystalline and amorphous phases. Table 1 is a summary (from different sources) of the main infrared bands that can be used to characterize orientation in PP.

The approach used depends on the sample. For example, if transmission measurements are made on a thick sample, most of the bands will be totally absorbing, and it is necessary to use weakly absorbing bands like those at 1256, 1220, 528, and 400 cm^{-1}. For thinner samples, the bands of medium intensity can be used, the most common being those at 1167, 998, 972, 841, and 809 cm^{-1}. The bands at 1458 and 1376 cm^{-1} are

Table 1 Bands suitable for orientation measurement by IR in PP

Frequency (cm^{-1})	*Relative intensity*	*Phase(s)*	*Dichroism* (θ_m)
1458	Strong	Both	Weak
1376	Strong	Both	⊥ (70°)
1304	Weak	Both	∥
1256	Weak	Both	∥ (38°)
1220	Weak	Crystalline	⊥ (72°)
1167	Medium	Crystalline	∥ (0°)
1103	Weak	Crystalline	⊥
1045	Weak	Crystalline	∥
998	Medium	Crystalline	∥ (18°)
972	Medium	Both	∥ (18°)
941	Weak	Crystalline	⊥
899	Weak	Crystalline	⊥
841	Medium	Crystalline	∥
809	Weak	Crystalline	⊥
528	Weak	Both	⊥ (90°)
400	Weak	Crystalline	∥ (20°)

usually too strongly absorbing to allow accurate measurements, although the former is useful as an internal reference because of its low dichroism. Another approach, especially useful for thicker samples, is to use attenuated total reflection (ATR), which analyzes the surface of the sample to a depth of the order of a few micrometres. In this case, the medium intensity bands are most useful. For biaxially oriented material, the orientation can be determined either in transmission, by including measurements on tilted films, or in ATR, by rotating the sample. In all cases, orientation functions can be calculated from the dichroic ratios if θ_m is known or assumed. Generally speaking, the orientation function calculated for the crystalline phase peaks agrees well with that obtained from X-ray diffraction, while the 'overall' function calculated for the peaks arising from both phases agrees with that obtained by birefringence. Only infrared techniques provide both together. In addition, the ratio of the bands at 998 and 972 cm^{-1} can be used to quantify the helical content, an essential factor for crystallization.

X-ray diffraction

X-ray diffraction can yield all the moments of the orientation distribution of the crystalline phase of PP. Using the same notations for the drawing directions as indicated previously and taking O as the origin of the frame, the general description of the orientation of the crystallographic planes can be performed by measuring the diffracted intensity using the pole figure accessory. Thus two independent angles, ν and μ, serve to define the orientation of the normals to the (*hkl*) crystallographic plane; ν is the angle between the normal of the crystallographic plane (*hkl*) and the direction ON, and μ is that between the projection of the (*hkl*) plane normal on the OMT plane and the OT direction. If $I(\nu,\mu)$ is the intensity representing the relative amount of plane normals oriented in the ν,μ direction, the different $\langle\cos^2\theta\rangle$ can be determined as:

$$\langle\cos^2\theta_{hkl,N}\rangle = \frac{\int_0^{2\pi}\int_0^{\pi/2} I(\nu,\mu)\cos^2(\nu)\sin(\nu)\,d\nu\,d\mu}{\int_0^{2\pi}\int_0^{\pi/2} I(\nu,\mu)\sin(\nu)\,d\nu\,d\mu}$$

$$\langle\cos^2\theta_{hkl,T}\rangle = \frac{\int_0^{2\pi}\int_0^{\pi/2} I(\nu,\mu)\sin^3(\nu)\cos^2(\mu)\,d\nu\,d\mu}{\int_0^{2\pi}\int_0^{\pi/2} I(\nu,\mu)\sin(\nu)\,d\nu\,d\mu}$$

$$\langle \cos^2 \theta_{hkl,M} \rangle = \frac{\int_0^{2\pi}\int_0^{\pi/2} I(\nu,\mu)\sin^3(\nu)\sin^2(\mu)\mathrm{d}\nu\mathrm{d}\mu}{\int_0^{2\pi}\int_0^{\pi/2} I(\nu,\mu)\sin(\nu)\mathrm{d}\nu\mathrm{d}\mu}$$

Using the equations given above, the second moment of the orientation function can thus be determined. For higher moments, see [2–4]. The strongest of the WAXS diffraction planes for isotactic PP crystallized in the most common α-form correspond to the (031)/(041), (130), (110), (040), (111) and (150)/(060). For the β-form, the characteristic reflection planes are (300) and (301). For the α-form, the orientation of the *b*-axis is most readily determined using the 040 reflection. The *c*-axis (chain axis) orientation may most readily be obtained from a combination of the 110 and 040 reflections. Wilchinsky has shown [7] that:

$$\langle \cos^2\phi_{1c} \rangle = 1 - 0.901\langle \cos^2\phi_{040} \rangle - 1.099\langle \cos^2\phi_{110} \rangle$$

For sPP, depending on the crystalline structure, (200), (002), (020), (021) and (011) reflections can be observed.

SOME RESULTS ON CHARACTERIZATION OF ORIENTATION IN PP AND ITS EFFECTS ON PROPERTIES

There are numerous publications on the orientation of PP by various methods, such as stretching of films (uniaxial or biaxial), roll-drawing of sheets, die-drawing of rods or tubes, uniaxial compression (forging), etc. Many of them report on the mechanical performance as a function of the processing conditions, while others deal with characterization of the molecular orientation and structure, including crystallinity. However, there are very few which provide a direct quantitative correlation of mechanical properties with structure. Because of space limitations, it is impossible to give much specific data here. In any case, these depend on a number of factors, such as molecular weight, processing temperature, and strain rate, to mention only a few. We can only give examples and describe general trends.

In one study, X-ray diffraction, Fourier transform infrared (FTIR) spectroscopy and refractive index measurements were utilized to characterize the state of molecular orientation in one-way and two-way (biaxially) drawn isotactic polypropylene (iPP) films [6]. It was shown that the use of all three techniques leads to much greater confidence in the orientation averages deduced than can be obtained by using any two of the techniques. It was observed that, with one-way drawing, the chain axes of both crystalline and amorphous regions orient towards the direction of drawing. The crystalline chains are more highly oriented than the amorphous chains and tend to orient towards the plane of the film, whereas the amorphous chains tend to be more uniaxially oriented towards the draw direction. In balanced, simultaneously two-way drawn films, the crystalline chains are more highly oriented towards the plane of the film than the noncrystalline chains. For a sequentially, equi-biaxially drawn iPP film, the orientations of the chain axes of both the crystalline and amorphous regions were found to be higher in the second draw (i.e. transverse) direction than in the first draw direction. The orientation of the crystalline chains was very close to the plane of the film, whereas the amorphous chains were almost uniaxially oriented with respect to the second draw direction. In all the films, there is a strong tendency for the *b*-axes of the crystallites to align normal to the plane of the film.

The mechanical as well as the barrier and optical properties of oriented PP improve significantly with the degree of orientation. In order to give a general idea of the enhancement obtained, undrawn PP typically has a modulus around 1.3 GPa and a strength around 30–40 MPa. Drawing to a ratio of 5 increases these to about 2.4 and 190 respectively (in the draw direction). Typical orientation factors for such material are 0.4 for the amorphous phase and 0.8 for the crystalline phase. In cases where higher draw ratios can be achieved, the gains are even greater. The orientation factor for both phases can exceed 0.9 and tensile moduli and strengths in excess of 14 GPa and 400 MPa have been reported. Significant gains have also been reported for impact strength and barrier properties with respect to gases (e.g. water vapor, oxygen and chemical warfare agents).

REFERENCES

1. Beer, G. (1996), Polypropylene, *Kunststoffe Plast. Europe*, **86**, 14–16.
2. White, J.L. and Cakmak, M. (1988) Orientation, crystallization and haze development in tubular film extrusion, *Adv. Polymer Technol.*, **8**, 27–61, and (1990) Orientation, in *Encyclopedia of Polymer Science and Technology*, (eds F.M. Herman, N.G. Gaylord, N.M. Bikales), Wiley Interscience, New York.
3. White, J.L. and Spruiell, J.E. (1981) Specification of biaxial orientation in amorphous and crystalline polymers, *Polymer Engng Sci.*, **21**, 859–868.
4. Nomura, S. (1989) Oriented polymers, in *Comprehensive Polymer Science*, Vol. 2, (eds G. Allen and J.C. Bevington), Pergamon Press, Toronto, 459–489.
5. Samuels, R.J. (1974) *Structured Polymer Properties*, Wiley, Toronto.
6. Karacan, I., Taraiya, A.K., Bower, D.I. and Ward, I.M (1993) Characterization of orientation of one-way and two-way drawn isotactic polypropylene films, *Polymer*, **34**, 2691–2701.
7. Whilchinsky, Z.W. (1962) in *Advances in X-ray Analysis*, Vol. 6, (eds W.M. Mueller and M. Fay), Plenum, New York.

Keywords: orientation function, birefringence, infrared spectroscopy of oriented PP, X-ray scattering of oriented PP, infrared dichroism, ATR-IR, FTIR orientation factor, biaxial orientation factor, amorphous orientation, crystalline orientation, mechanical properties.

P–V–T data and their uses

Witold Brostow

INTRODUCTION

Pressure–volume–temperature (P–V–T) investigations represent an underappreciated area of polymer science and engineering. Such data can be obtained for both polymer solids and melts, sometimes in a single experiment (see the next section). Needless to say, such experiments can be conducted also for materials other than polymeric. The main applications of P–V–T results in the field of polymers are:

1. prediction of polymer + polymer miscibility;
2. prediction of service performance and service life of polymeric materials and components on the basis of free volume concepts;
3. evaluation of start and progress of chemical reactions in polymer melts in cases when volume effects accompany the reaction;
4. optimization of processing parameters, instead of establishing such parameters by trial and error;
5. calculation of the surface tension γ of polymer melts.

We shall briefly characterize below the experimental procedure for P–V–T determination, then parameters derivable directly from the experiments including the equations of state, the areas listed above, and finally the information on polypropylene (PP).

EXPERIMENTAL DETERMINATION OF P–V–T CHARACTERISTICS

An apparatus called Gnomix was constructed in 1976 by Zoller *et al.* [1]. The sample sits under mercury, which enables experiments in both solid and liquid phases. One can keep the temperature constant and vary the

Polypropylene: An A–Z Reference
Edited by J. Karger-Kocsis
Published in 1999 by Kluwer Publishers, Dordrecht. ISBN 0 412 80200 7

pressure by increments, or the other way around. Thus one obtains results which can be represented in the three-dimensional *P–V–T* space. Results for several polymers have been reported, largely by Zoller *et al.* [2], including also some polymer liquid crystals (PLCs) [3].

ISOBARIC EXPANSIVITY, ISOTHERMAL COMPRESSIBILITY AND THE EQUATIONS OF STATE

One works typically with twin quantities: the isobaric expansivity (also called the coefficient of thermal expansion, CTE):

$$\alpha = V^{-1}(\partial V/\partial T)_P \tag{1}$$

and the isothermal compressibility:

$$\kappa_T = -V^{-1}(\partial V/\partial P)_T \tag{2}$$

If κ_T does not change with *P*, then its reciprocal is the bulk modulus.

The full *P–V–T* results can be represented by an equation of state (EOS). There is a variety of such equations, but the Hartmann equation [4] renders good results for both polymer solids and melts [5]:

$$\tilde{P}\tilde{v} = \tilde{T}^{3/2} - \ln \tilde{v} \tag{3}$$

We have here reduced quantities defined as:

$$\tilde{P} = P/P^* \qquad \tilde{v} = v/v^* \qquad \tilde{T} = T/T^* \qquad \tilde{\gamma} = \gamma/\gamma^* \tag{4}$$

with the starred parameters characteristic for a given material. The first three of these parameters can be obtained from the experimental *P–V–T* data. The free volume is defined as:

$$v^f = v - v^* \tag{5}$$

All quantities in equation (5) are specific per unit volume, hence typically in $cm^3\,g^{-1}$, although, in principle, molar (or other well-defined) volumes could be used also.

POLYMER + POLYMER MISCIBILITY

The miscibility – or lack of it – is reflected by the values of the isobaric expansivity α of the components; see [3] and references therein. But there is also another option, namely using the interfacial tension γ_{ij} for the miscibility prediction; see below.

PREDICTIONS OF SERVICE PERFORMANCE AND RELIABILITY

Such calculations can be based on the free volume defined by equation (5). Reliable predictions include the ductile–brittle impact transition temperature, rapid crack propagation, slow crack propagation, and also a variety of properties obtained using the time–temperature correspondence principle [5]. To use the principle, one needs an equation for the temperature shift factor, a_T. A general formula is [5]:

$$\ln a_T = A + B/(\tilde{v} - 1) \qquad (6)$$

The reduced volume $\tilde{v}$ can be obtained from the Hartmann equation (3).

PROGRESS OF CHEMICAL REACTIONS

Reactions which take place in the melt phase can be followed; this of course, if there are volumetric ΔV effects accompanying the reaction. Since such reactions do not occur in PP, we do not provide further details here. ΔV changes which take place, for instance, during epoxy curing are reported elsewhere by the present author.

OPTIMIZATION OF PROCESSING

A knowledge of both α and κ_T is necessary here. There exist computer programs which produce optimized parameters, for instance for injection molding. However, such programs use on input hypothetical P–V–T data since the amount of real data is limited. Therefore, the results produced by such programs have to be treated with caution – unless real EOS data for a given material have been used as input.

SURFACE AND INTERFACIAL TENSION

The electronics industry and also industries using adhesives and/or fiber reinforced heterogeneous composites (HCs) are increasingly interested in the surface tension γ (with respect to air) of polymer melts and also in the interfacial tension γ_{ij} between two condensed phases, such as two polymers or between a fiber and a polymer matrix. In turn, γ_{ij} depends on γ. While methods of experimental determinations of both γ_{ij} and γ exist, they are either time consuming or require fairly elaborate equipment to determine. However, there are theoretical approaches which enable the computation of γ from P–V–T data.

The γ-EOS connection is based on the theory of corresponding states formulated by Johannes van der Waals in 1881. His theory was extended to chain molecules by Ilya Prigogine and coworkers in 1950 and then further developed by Donald Patterson; see [3] and references therein. According to the Prigogine–Patterson theory, the characteristic surface tension γ^* may be calculated as:

$$\gamma^* = (kc)^{1/3}P^{*2/3}T^{*1/3} \qquad (7)$$

where k is the Boltzmann constant. The parameter $c \leq 1$ takes into account the fact that a segment in a polymer chain has $3c$ (rather than 3) degrees of freedom due to the connectedness. However, since this would result in an extra parameter, we can assume c equal to unity. The reduced surface tension is:

$$\tilde{\gamma}\tilde{v}^{5/3} = m - (1 - \tilde{v}^{1/3})\ln[(\tilde{v}^{1/3} - b)/(\tilde{v}^{1/3} - 1)] \qquad (8)$$

To minimize the degree of arbitrariness, one can take $b = 0.5$, so that m is the only parameter to be adjusted for a given polymer or class of polymers. Equations (7) and (8) have been applied with success to a number of polymers.

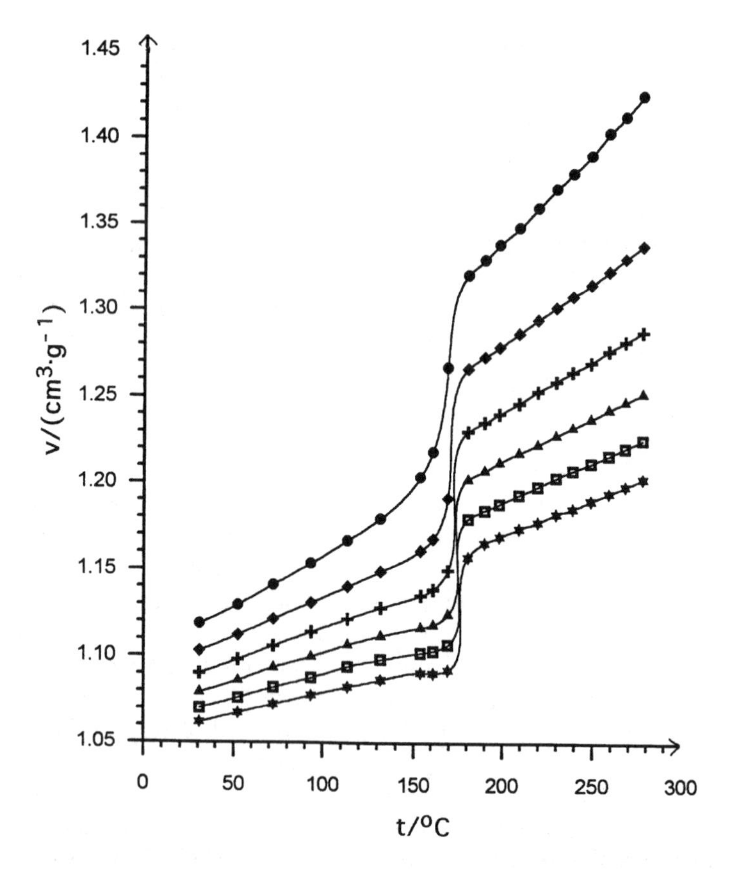

Figure 1 Specific volume of iPP as a function of temperature: ● $P = 0$; ◆ $P = 390$ bar; + $P = 780$ bar; ▲ $P = 1180$ bar; □ $P = 1570$ bar; ★ $P = 1960$ bar (data from [2]). 1 bar $= 10^5$ Pa $= 0.1$ J cm^{-3}.

P–V–T DATA FOR POLYPROPYLENE

The *P–V–T* results for several PPs have been collected by Zoller and Walsh [2]. Results for an isotactic PP (iPP) with the specfic volume at room temperature $v = 1.115\,\mathrm{cm^3\,g^{-1}}$ are shown in Figure 1. The glass transition temperature of iPP is ≈ -15°C, hence below the experimental range. If that range were included, one would see a change of slope in the v versus T curve. The melting point of iPP is ≈ 167°C [6] at atmospheric pressure, and is clearly visible in Figure 1. One sees how much the pressure change affects the specific volume - an important effect in polymer processing. Variations of pressure affect also the melting point, but to a distinctly lesser extent than v is affected.

REFERENCES

1. Zoller, P., Bolli, P., Pahud, V. and Ackermann, H. (1976) *Rev. Sci. Instrum.*, **47**, 948.
2. Zoller, P. and Walsh, D.J. (1995) *Standard Pressure–Volume–Temperature Data for Polymers*, Technomic, Lancaster, PA, and Basel, and references therein.
3. Berry, J.M., Brostow, W., Hess, M. and Jacobs, E.G. (1998) *Polymer*, **39**, 4081.
4. Hartmann, B. (1983) *Proc. Can. High Polymer Forum*, **22**, 20; Hartmann, B. and Haque, M.A. (1985) *J. Appl. Phys.*, **58**, 2831; Hartmann, B. and Haque, M.A. (1985) *J. Appl. Polymer Sci.*, **30**, 1553.
5. Brostow, W. (ed.) (1992) *Failure of Plastics*, Hanser, Munich; Brostow, W., Kubát, J. and Kubát, M.J. (1996), Mechanical properties, in *Physical Properties of Polymers Handbook*, Chap. 23, (ed. J.E. Mark), *Am. Inst. Phys. Press*, Woodbury, NY.
6. Brostow, W., Sterzynski, T. and Triouleyre, S. (1996) *Polymer*, **37**, 1561.

Keywords: equation of state, free volume, bulk modulus, isothermal compressibility, isobaric expansivity, surface tension, pressure–volume–temperature relationship (*P–V–T*), polymer miscibility, injection molding.

Particulate filled polypropylene composites

Béla Pukánszky

INTRODUCTION

Particulate filled polypropylene (PP) is widely used in many fields of application. The major advantage of the introduction of fillers into PP is the higher stiffness of the composites, which is especially important in applications at higher temperatures. Increased stiffness is usually accompanied by decreased impact resistance, which is not always acceptable. The selection of an appropriate PP copolymer or the incorporation of additional elastomer may lead to the often desired simultaneous enhancement of stiffness and toughness. Incorporation of a filler into the polymer changes all properties at the same time, therefore an optimization of properties is required during development. The major benefits of modified PP are versatility and an exceptional price/volume/performance ratio.

PREPARATION

The PP and the filler must be homogenized prior to the manufacture of a product. The thorough homogenization of the components is the minimum condition for the easy processing and application of these materials. Insufficient homogenization leads to aggregation. Aggregates cause surface blemish and deteriorate all properties especially impact strength, because they initiate cracks under external loading.

Numerous devices are mentioned in connection with the preparation of particulate filled PP composites, i.e. two-roll mills, internal mixers,

Polypropylene: An A–Z Reference
Edited by J. Karger-Kocsis
Published in 1999 by Kluwer Publishers, Dordrecht. ISBN 0 412 80200 7

single- and twin-screw extruders [1]. Two-roll mills are mostly used in the rubber industry; particulate filled PP is not produced with them on an industrial scale. High-intensity internal mixers can be used effectively for the homogenization of PP and various fillers. The disadvantages of batch operation can be overcome by attaching a discharge extruder to the mixer, or by continuous internal mixers, such as the Farrel extruder [2]. Single-screw extruders are not very efficient at homogenization even when special mixing elements are designed in the melt pumping zone of the screw. The co-kneader with its special operation principle, which includes the rotation and reciprocation of the screw, is a very efficient compounder especially for heat-sensitive materials. Its major disadvantage is the capital cost of installation.

However, most of the compounding is done by twin-screw extruders and their further development is expected in the future. The major advantage of this equipment is the continuous operation, high productivity and flexibility, which makes possible the change of machine configuration and processing conditions according to the properties of the processed material. Both the screw and the barrel of these machines is constructed from segments and can be assembled to produce the necessary conditions for efficient homogenization. High shear must be used for the homogenization of particulate filled PP to avoid aggregation, but low shear and intensive mixing is preferred in the case of shear-sensitive fillers, such as (starch) and glass fibers (GF). The attrition of fibers leads to the loss of their reinforcing effect and generally to the deterioration of composite properties.

FACTORS INFLUENCING THE PROPERTIES OF FILLED PP

The properties of particulate filled polymers are determined by the characteristics of the components, composition, interaction and structure. All four factors can be varied in a wide range thus making possible the design of 'tailor-made' composites.

Component properties

The most important properties of the matrix polymer are its chemical composition and melt viscosity. Copolymers have lower stiffness and higher impact strength than the homopolymer, which are transferred also to the composites. Products with higher modulus are usually prepared from homopolymers, while those subjected to dynamic loads during application are made from copolymers. Also the sequence distribution of the ethylene and propylene units is of importance, as random and block copolymers yield composites with significantly different properties.

The most important filler characteristics determining the properties of PP composites are particle size, particle size distribution, specific surface area and shape. None of these influence stiffness very much; the reinforcing effect is a result of the orientation of the anisotropic particles. All other properties are considerably affected by these filler characteristics. Yield stress and strength usually increase with decreasing particle size and increasing surface area, while deformability and impact resistance change in the opposite direction.

Composition

The reinforcing effect of the filler increases and the price of the composite usually decreases with increasing filler content. Thus, significant effort is concentrated on the introduction of the highest possible filler loadings into the polymer. Since other properties deteriorate at the same time, a composition must be found where the combination of the properties is optimal.

Stiffness increases with increasing filler content. Although occasionally a linear correlation is mentioned, theory predicts a nonlinear relationship [3]. The overall belief is that tensile yield stress and tensile strength decrease with increasing filler content when fillers with nearly spherical shape are used. The composition dependence of these properties are strongly influenced also by particle characteristics. As Figure 1 shows true reinforcing can be achieved with fillers of sufficiently low particle size [4]. The effect of fillers with anisotropic particle geometry depends on orientation. Deformability, yield strain and elongation-at-break always decrease with increasing filler content. Impact resistance often exhibits a maximum at a certain filler content, which depends on the characteristics of the filler, but also on interfacial interactions [5]. The maximum is a consequence of several competitive effects. Debonding of the filler during deformation leads to energy consumption, while increasing filler loadings lead to increased stiffness and decreased impact resistance. Most other properties (heat conductivity, heat deflection temperature (HDT), hardness) are more or less proportional to the filler content of the composite.

Interfacial interaction

Interfacial interaction strongly influences the properties of polymer composites. In PP, only van der Waals forces act between the polymer

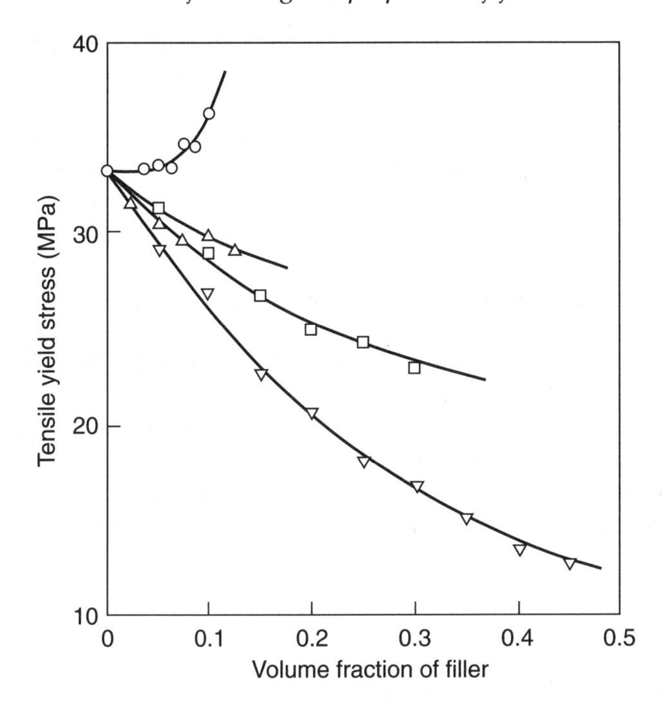

Figure 1 Effect of particle size on the tensile yield stress of PP composites. Particle diameter: (○) 0.01 μm, (△) 0.08 μm, (□) 3.3 μm, (▽) 58 μm.

and the filler. They are, however, sufficient to attach the polymer to the filler. Adsorption of the polymer on the filler surface leads to the formation of an interphase in which molecules have decreased mobility. The existence of the interphase is proved by the increase of composite viscosity with increasing specific surface area of the filler and by the similarly changing yield stress and strength. The effect of interaction depends on the size of the contacting surfaces (specific surface area of the filler) and the strength of the interaction. The effect of the former is clearly demonstrated by Figure 1. Smaller particles have larger surface area, leading to increased yield stress. Differences in the strength of the interaction are less significant than that of the specific surface area.

The importance of interfacial interactions is shown by the numerous attempts to modify them. Modification of interactions can be classified into four categories:

(1) *Nonreactive treatment* leads to the decrease of particle/particle and matrix filler interaction. The latter leads to decreased strength and increased deformability, occasionally to increased impact resistance,

while the former to improved processability, homogeneity and appearance.

(2) *Reactive treatment*, which assumes covalent coupling of both components, is difficult in PP due to its inert, apolar structure. Occasionally, reactive coupling can be achieved with certain aminosilane coupling agents, which polymerize on the filler surface and react chemically with the functional groups of the polymer formed during processing.

(3) Introduction of *functionalized PP* into the composite was shown to improve adhesion and increase yield stress and tensile strength in the case of various fillers, from $CaCO_3$, through wood flour and magnesium hydroxide to glass fiber.

(4) In multicomponent PP systems containing a filler and two polymers, a special morphology may form in which one of the polymer components *encapsulate the filler*. Naturally, such a structure is accompanied by significantly modified interaction and properties [4].

Structure

Particulate filled polymers are believed to have a simple structure consisting of uniformly dispersed particles in a homogeneous matrix. However, structure-related phenomena are at least as important in composites as in blends. Two important structural phenomena must be always kept in mind when filled polymers are prepared and applied. Small particles tend to form aggregates, which deteriorate properties. Above $5\,g/m^2$ specific surface area, the aggregation tendency of spherical, nontreated fillers becomes significant. Orientation of anisotropic particles determines their reinforcing effect, which can be really judged only if their orientation distribution or at least average orientation is known.

APPLICATIONS

The main fields of application of particulate filled polymers is shown in Figure 2. As was mentioned earlier, the major benefit of the application of fillers and reinforcement is increased stiffness and HDT. Specific applications which must be mentioned here are garden furniture ($CaCO_3$), air-filter covers, heater boxes (talc) for cars, washing machine soap dispensers (talc), etc. Bumpers are usually made from multicomponent PP systems containing a PP homo- or copolymer, an elastomer and a filler. In order to demonstrate the usual property range of particulate filled polymers, Table 1 presents the most important characteristics of several commercial grade PP composites.

Table 1 Properties of selected commercial grade PP composites; homopolymers and copolymers were both used as matrix polymer

Filler		ρ	*MFI*[a]	E^b	$\sigma_y{}^c$	σ^d	ε^e	$a_n{}^f$	*HDT*[g]
Type	*Content* (wt%)	(g/cm^3)	(g/10 min)	(GPa)	(MPa)	(MPa)	(%)	(kJ/m^2)	(°C)
Talc	20	1.04	2.6	2.15	23	19	130	20.0	57
Talc	30	1.13	1.6	2.50	29	22	85	14.0	62
Talc	40	1.25	1.7	3.65	35	25	18	3.8	74
$CaCO_3$	30	1.14	25.0	1.80	26	18	126	6.1	51
$CaCO_3$	40	1.24	4.7	1.90	23	21	330	28.6	56
Glass fiber	10	0.96	3.0	2.10	–	30	10	12.7	109
Glass fiber	30	1.14	3.5	5.10	–	85	8	5.3	144
$BaSO_4$	25	1.13	25.0	1.85	26	18	640	4.4	53
$BaSO_4$	50	1.51	20.0	2.10	22	19	56	3.7	58

[a] 230°C/2.16 kg; [b] Young's modulus; [c] tensile yield stress; [d] tensile strength; [e] elongation-at-break; [f] notched Charpy impact strength at 23°C; [g] Method A.

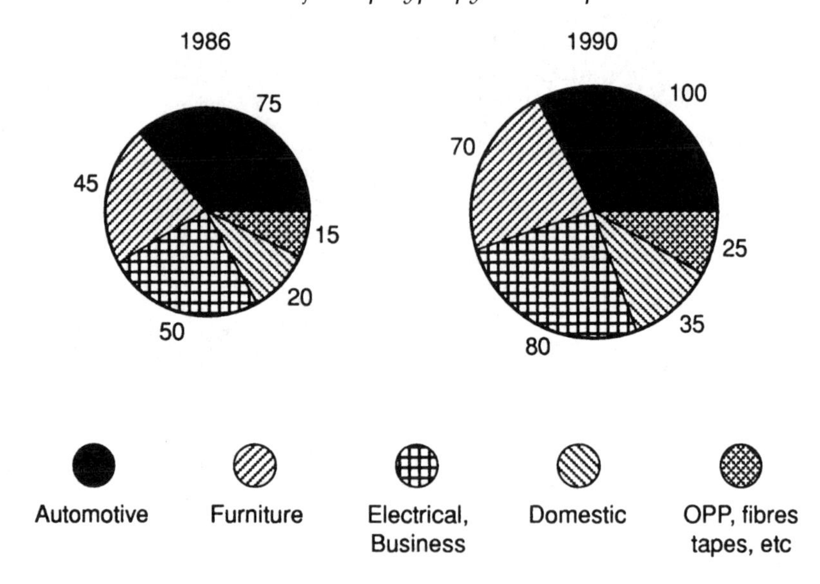

Figure 2 The most important fields of application for particulate filled PP (values in 1000 ton).

REFERENCES

1. Rothon, R. (1995) *Particulate-Filled Polymer Composites*, Longman, Harlow.
2. Manas-Zloczower, I. and Tadmor, Z. (1994) *Mixing and Compounding of Polymers; Theory and Practice*, Hanser, Munich.
3. Nielsen, L.E. (1974) *Mechanical Properties of Polymers and Composites*, Dekker, New York.
4. Pukánszky, B. (1995) in Particulate filled polypropylene: structure and properties, *Polypropylene: Structure, Blends and Composites*, Vol. 3, (ed. J. Karger-Kocsis), Chapman & Hall, London, p. 1.
5. Pukánszky, B. and Maurer, F.H.J (1995) Composition dependence of the fracture toughness of heterogenous polymer systems, *Polymer*, **36**, 1617.

Keywords: particulate filled composites, filler, aggregate, homogenization, mixing, internal mixer, single-screw extruder, twin-screw extruder, mechanical properties, tensile yield stress, tensile strength, stiffness, impact resistance, structure–property relationships, interface, interphase, reactive treatment, nonreactive treatment, surfactant, encapsulation, functionalized PP, coupling, specific surface area, application.

Photostabilizers

S. Al-Malaika

INTRODUCTION

The outdoor performance of polymers depends largely on their photo-oxidative (combined action of ultraviolet (UV)-light and oxygen) stability, although generally the overall effect is that of a combined photo- and thermal-oxidation of the polymer. Commercial polypropylene (PP) is highly sensitive to UV light and its photostabilization is essential for outdoor and indoor end-use applications. The shorter wavelength of the sun's radiation does not reach the surface of the earth and, in most locations, there is little incident radiation below 285 nm. It is the incident UV-part of the sun's radiation, in the region 290–400 nm, which is mainly responsible for the light-induced polymer oxidation (photooxidation). Saturated hydrocarbon polymers (e.g. 'pure' PP) are intrinsically transparent to the incident sun spectrum and should not be affected by the solar radiation. Commercial PP, however, absorbs weakly at wavelengths above 285 nm and its sensitivity to sunlight is a consequence of light-absorbing impurities, particularly oxygen-containing species, and trace levels of metals and other species present arising from production processes, i.e. polymer manufacture, melt processing and fabrication. These impurities sensitize and accelerate PP photooxidative degradation under service conditions involving exposure to light (outdoors and indoors) resulting in discoloration, gradual loss of mechanical properties and ultimate embrittlement of polymer artifact with adverse economic and, possibly, health consequences. Fortunately, the vulnerability of PP to the deleterious effects of their outdoor service environment can be

Polypropylene: An A–Z Reference
Edited by J. Karger-Kocsis
Published in 1999 by Kluwer Publishers, Dordrecht. ISBN 0 412 80200 7

greatly controlled, and the outdoor performance can be markedly improved, by appropriate choice of photostabilizers used either separately or in synergistic combinations. Indeed, the largest market for photostabilizers is in polyolefins and in particular in PP.

PHOTOOXIDATION OF POLYPROPYLENE

The basic free radical chain mechanism involved in light-initiated oxidation of PP is essentially the same as that involved in its thermally-initiated oxidation, with the main difference lying in the rate of chain initiation which is much higher in photooxidation resulting in a shorter kinetic chain.

In view of the important role of oxygen-containing species as photoinitiators, the prior thermal-oxidative history of PP determines, to a large extent, its photooxidative behavior in service. Hydroperoxides formed during melt processing (and to a lesser extent during manufacturing and storage) are the most important photoinitiators. They play a major initiating role during the early stages of PP photooxidation, while the derived carbonyl-containing products exert deleterious effects during later stages of photooxidation (e.g. by Norrish types I and II reactions), see Figure 1. The initiating species, e.g. hydroperoxides and their decomposition products, are responsible for the changes in molecular structure and overall molar mass of PP which are manifested in practice by the loss of mechanical properties (e.g. impact, flexural, tensile strengths, elongation) and by changes in the physical properties of the polymer surface (e.g. loss of gloss, reduced transparency, cracking, chalking, yellowing). Photooxidation of PP occurs preferentially in the amorphous phase and, fortunately, this phase retains the maximum concentration of stabilizer, especially during the crystallization of the polymer melt.

Photostabilizers for polypropylene and factors affecting their performance

Photostabilizers are incorporated in PP during its melt processing and are generally used at higher concentration levels than that of thermal antioxidants (up to 1 wt.%). The effectiveness of photostabilizers is assessed by first subjecting the stabilized polymer to accelerated weathering and/or outdoor exposure conditions. Accelerated weathering devices have different types of light sources and configuration and include, for example, a combination of fluorescent and sun lamps, xenon arc or carbon arc and operate in the presence or absence of a combination of other factors, e.g. humidity, temperature, dark and light cycles. The 'weathered' polymers are tested in a manner similar to that discussed in the chapter 'Thermal antioxidants' in this book, by examining physical

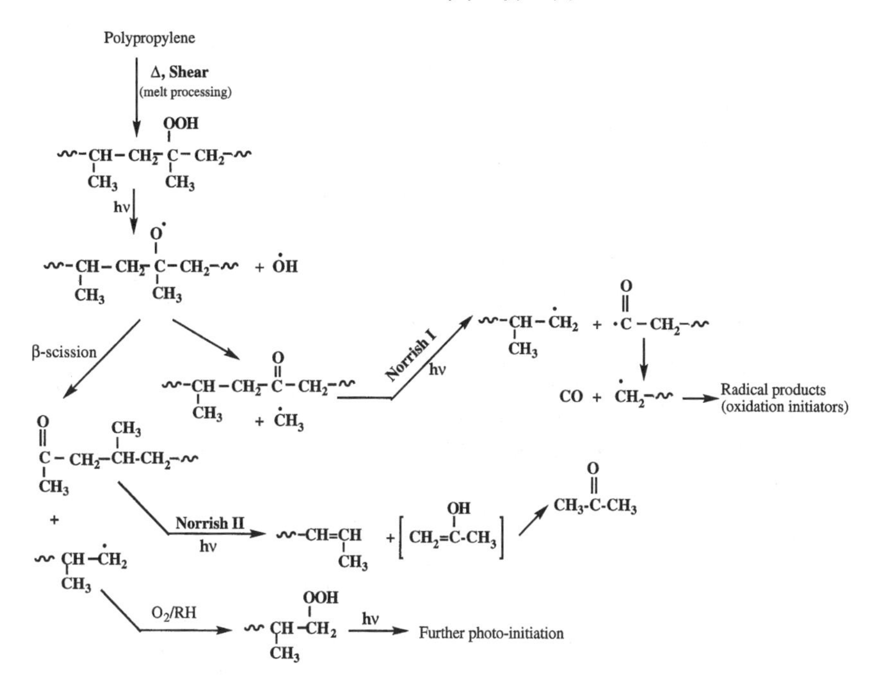

Figure 1 Polymer hydroperoxidation during processing and further photo-initiation by the hydroperoxides and the derived carbonyl compounds.

and chemical changes in the presence and absence of antioxidants. These include changes in physical properties (e.g. molar mass, mechanical properties, time to embrittlement), chemical microstructure (e.g. buildup of carbonyl functions during photooxidative degradation), as well as visual characteristics (e.g. gloss, color, chalking).

To be effective photostabilizers, these compounds must satisfy not only the basic chemical (i.e. intrinsically active) and physical (i.e. solublility, low diffusion and volatility, nonextractability) requirements, but must also be stable to UV light and must withstand continuous periods of exposure to UV light without being destroyed or effectively transformed into sensitizing products. Other factors that can affect the ultimate photostability of the polymer are sample thickness, polymer crystallinity and presence of other additives, e.g. pigments, fillers [1, 2].

Three different classes of compounds form the major and most important commercial categories of photostabilizers for PP. These are based on nickel complexes (those containing sulfur ligands function primarily as peroxide decomposers, PD), UV-absorbers (UVA), e.g. based on 2-hydroxybenzophenone and 2-hydroxyphenylbenztriazole, and sterically hindered amine light stabilizers (HALS).

The chain breaking donor (CB-D) antioxidants, such as hindered phenols, which have been shown to be very effective thermal stabilizers for PP (see the chapter 'Thermal antioxidants' in this book), are relatively ineffective under photooxidative conditions both because they are unstable to UV light and because some of their oxidation transformation products (formed during their antioxidant action) are photosensitizing. Similarly, many simple sulfur-containing antioxidants, such as the thiodipropionate esters (see in the chapter 'Thermal antioxidants', Figure 6, for the structure of Irganox PS800, as an example) which are very effective peroxide decomposers (PD) and thermal antioxidants are not effective photoantioxidants due to the dissociation of their primary oxidation products to initiating radicals under the effect of UV-light [1]. However, both the hindered phenols and these sulfide antioxidants can synergize with UV stabilizers and become much more effective photostabilizers.

In contrast to simple sulfides, many metal dithiolates (e.g. MDRC, MDRP; see the structure in Figure 2) are very effective photoantioxidants due to their much higher UV stability. For example, nickel complexes of dithioic acids are excellent photo- and peroxidolytic (PD-C) antioxidants. The metal ion in these compounds plays a crucial part in their overall effectiveness as UV stabilizers. Transition metal complexes containing Ni, Co and Cu are more photostable than group II metal complexes, e.g. Zn, leading to better overall photostabilizing effectiveness (Table 1). Like most other metal dithiolates, the primary oxidation product formed during the antioxidant function of NiDRP is the corresponding disulfide which undergoes further oxidation to give different sulfur acids (the real catalysts for peroxide decomposition) (Figure 2) [3].

The UV stabilizing action of nickel and iron complexes (e.g. dithiocarbamates, NiDRC and FeDRC) is strongly concentration dependent; nickel complexes are much more stable to UV light than iron complexes (Table 1). Further, the behavior of the iron complexes (e.g. FeDRC) changes with concentrations; at low concentration (below 0.05%, in polyolefins), the iron complexes photooxidize rapidly, eliminating the thiocarbamoyl ligand and freeing the metal ion which is a very powerful sensitizer of photooxidation through its reaction with light and hydroperoxides (Figure 3 reactions (e), (f)). Conversely, at high concentration, FeDRC shows a photo-induction period during which the iron complex behaves as a photoantioxidant due to the oxidation of the sulfur ligand to low molar mass sulfur acids (by a mechanism similar to that of the dithiophosphate discussed above) (Figures 2 and 3, reactions (a) and (b), Table 1). This dual function of the iron dithiocarbamate (i.e. being a weak UV-stabiliser at high concentrations and a photosensitizer at lower concentrations), together with its excellent melt stabilization at all concentrations, has been exploited commercially for the precise time-

Table 1 Ultraviolet-embrittlement times (EMT) of PP films containing different concentrations of antioxidants (processed in an internal mixer at 190°C and exposed to ultraviolet light in an accelerated sunlamp-blacklamp ultraviolet-aging cabinet). For chemical structures, see Figures 2, 4 and 5

Antioxidant	*Concentration* (mol/100 g × 10^4)	*EMT* (h)
Controls (no antioxidant)	0	90
ZnDEC	2.5	175
NiDEC	0.25	140
NiDEC	2.5	740
NiDEC	3	790
NiDEC	10	840
FeDMC	0.25	85
FeDMC	2.5	150
FeDMC	5	336
CuOctX	3	250
FeOctX	3	330
NiOctX	3	700
CoOctX	3	1600
Cyasorb UV 531	3	245
Tinuvin 770	3	950
NiDBP + UV 531	6	2650
Tin 770 + UV 531	4	1750

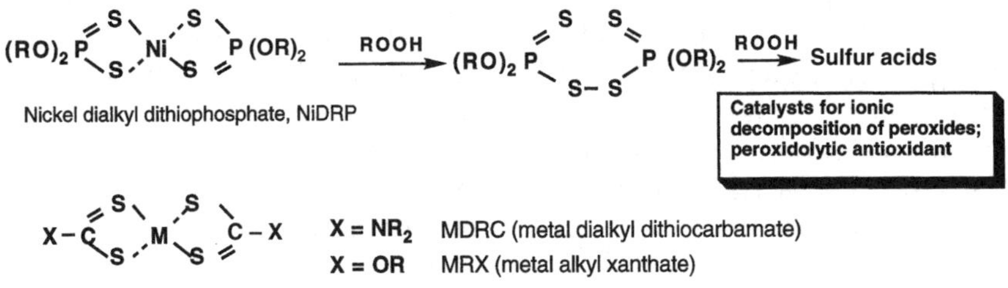

$(RO)_2P(S)S$–Ni–$SP(S)(OR)_2$ $\xrightarrow{ROOH}$ $(RO)_2P(S)$–S–S–$P(S)(OR)_2$ $\xrightarrow{ROOH}$ Sulfur acids

Nickel dialkyl dithiophosphate, NiDRP

Catalysts for ionic decomposition of peroxides; peroxidolytic antioxidant

X = NR_2 MDRC (metal dialkyl dithiocarbamate)
X = OR MRX (metal alkyl xanthate)

Figure 2 Simplified antioxidant mechanism of metal dithiolates (dithiophosphates, dithiocarbamates, xanthates) exemplified with nickel dialkyl dithiophosphate.

controlled stabilization required for agricultural applications, e.g. in mulching films [4].

The photostabilizing role (in PP and other polymers) and the antioxidant mechanisms of UV absorbers based on hydroxybenzophenones

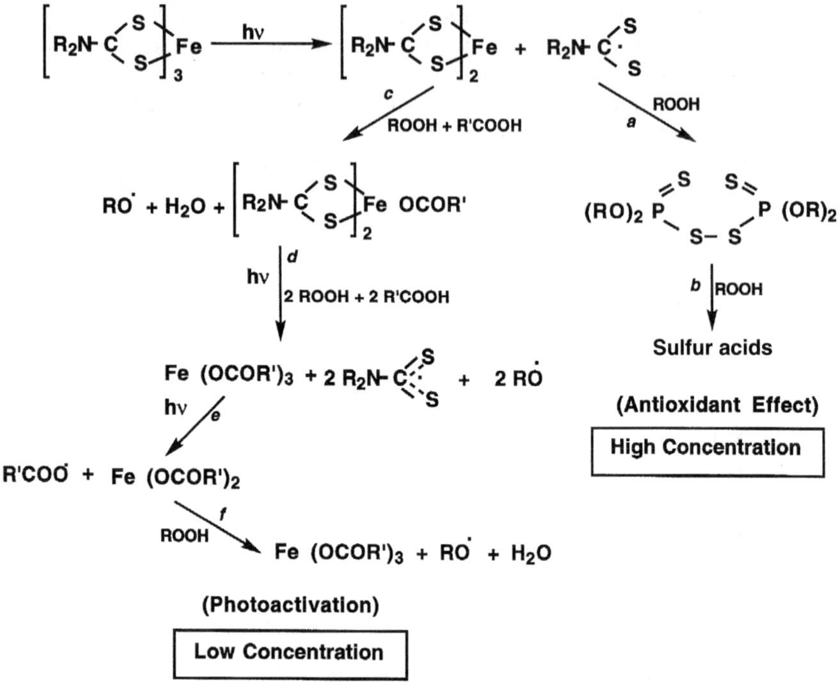

Figure 3 The antioxidant mechanism of iron dithiocarbamate.

and hydroxyphenylbenzotriazoles (see Figure 4 for structures) have been studied extensively. These compounds are stable to UV light and have high extinction coefficients in the region 330–360 nm. They operate primarily by absorbing UV light and dissipating it harmlessly as thermal energy, e.g. via an excited state keto-enol tautomerism (Figure 4). It is important to point out that most known UV stabilizers do not act by a single mechanism, e.g. Cyasorb UV531 functions not only as a UV screen but also as a sacrificial antioxidant, removing chain initiating radicals (e.g. alkoxyl radicals) via a weak chain breaking-donor (CB-D) mechanism. The limited effectiveness of Cyasorb UV531 as a UV stabilizer (Table 1) is not due to its photolability but because of its instability towards hydroperoxides and carbonyl compounds which are formed under photooxidative conditions. Thus, UV531 is a much more effective photostabilizer for mildly processed saturated polyolefins than for severely oxidized polyolefins. UV absorbers, such as Cyasorb UV531, synergize effectively with peroxide decomposers, e.g. nickel dithiophosphate (Table 1).

The latest class of photostabilizers developed commercially is based on hindered amines (known as HALS, hindered amine light stabilizers)

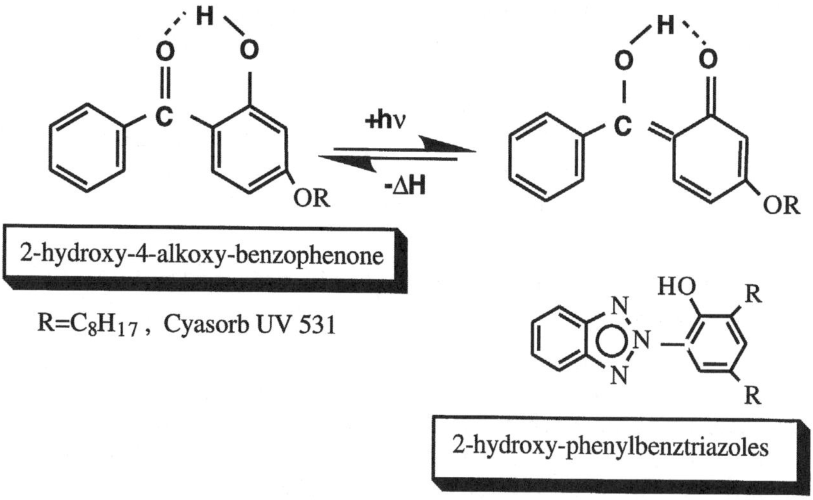

2-hydroxy-4-alkoxy-benzophenone

$R=C_8H_{17}$, Cyasorb UV 531

2-hydroxy-phenylbenztriazoles

Figure 4 The main photostabilizing reaction of ultraviolet-absorbers examplified with substituted hydrobenzophenomes. The structure of hydroxyphenylbenztriazole (UVA) is also shown.

which have proved to be amongst the most effective photostabilizers for a large number of commercial polymers. The first commercially developed HALS, Tinuvin 770, proved to offer much higher UV-stability to PP than any other conventional UV-stabilizer available at the time, such as UV-absorbers, nickel compounds and benzoates (Table 1). HALS is a unique class of photostabilizers which do not subscribe to the general mechanisms of photostabilization; they are not UV absorbers, do not quench singlet oxygen or triplet carbonyls, and do not catalyse hydroperoxide decomposition. Their effectiveness, however, is due to their transformation product, the corresponding nitroxyl radicals, which is the real stabilizing species, (Figure 5(a)). Hindered nitroxyl radicals are effective chain

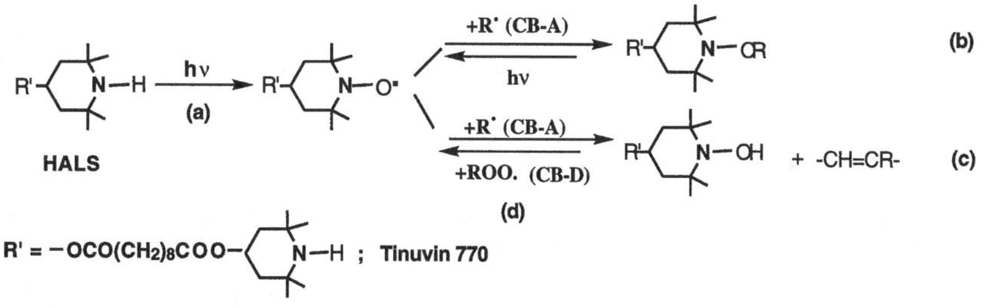

HALS

R' = $-OCO(CH_2)_8COO-$; Tinuvin 770

Figure 5 Simplified antioxidant mechanism of hindered amine light stabilizers (HALS).

breaking-acceptor (CB-A) antioxidants which act by trapping the macro-alkyl radicals to give hydroxylamines (Figure 5(b)) or/and alkylhydroxylamines (Figure 5(c)); the former regenerates the nitroxyl radical via a CB-D process (Figure 5(d)). The overall high efficiency of HALS as photostabilizers in polymers is attributed to the regeneration of nitroxyl radical and the complementarity of the CB-A/CB-D antioxidant mechanisms involved (for further details, see reference [5]).

OTHER DEVELOPMENTS TOWARDS EFFICIENT AND SAFE STABILIZATION

Since the developments of hindered amine photostabilizers (in the mid-1970s), no other new class of antioxidant has emerged. A number of new stabilizers, however, have since been commercialized but these are, by and large, based on modifications of existing antioxidant structures. This apparent state of stagnation has more to do with the passing of more stringent safety regulations in many countries, on health hazards, and the very expensive toxicity tests required for the clearance of new antioxidants, especially in applications destined for contact with the human environment. Research over the last two decades has been devoted, however, to manipulating the known antioxidant structures in order to achieve the efficiency, safety and environmental requirements consequent on the use of antioxidants, especially in food, children's toys, pharmaceutical and medicinal applications.

The efficiency of antioxidants depends not only on their inherent chemical activity but also on their permanence in polymers. Physical loss of antioxidants and stabilizers leads not only to premature failure of the polymer article but also to problems associated with health hazards and toxicological effects (if used in applications requiring contact with the human environment). Overall, a number of approaches have been developed with the aim of enhancing the efficiency, permanence (in order to withstand the ever more demanding applications of polymers) and safety of antioxidants and stabilizers used in polymers. These include the use of large molar mass antioxidants, copolymerizing antioxidant functions in polymers (during polymer synthesis) and *in-situ* grafting of antioxidant functions onto polymer backbones during the fabrication and manufacture of polymer articles using reactive processing methods.

A number of oligomeric photostabilizers (e.g. based on HALS structure) have been developed and commercialized. For example, in the case of PP fibers, where volatility, extractability and migration of photo-antioxidants become important, the performance of low molar mass hindered amine photostabilizers would be considerably reduced, and the use of polymeric HALS has been shown to be very advantageous leading to much improved performance. Furthermore, often a combination of low molar mass HALS and high molar mass HALS is synergistic [5b]. The approach of copolymerization of photoantioxidant functions onto PP (during the polymer synthesis) has not received much commercial attention since this method, which is less suitable for crystalline polymers, is expensive because new specialty polymers must be produced for each end-use product.

Chemical attachment of photoantioxidant functions on the polymer backbone during processing (utilization of the processing machine as a chemical reactor for the grafting reaction) leads to the ultimate permanence of antioxidants since they cannot be detached from the polymer matrix except through severance of chemical bonds. Reactive antioxidants containing one of the basic photoantioxidant functions (e.g. HALS, UVA) as well as a polymer reactive function can be made to chemically react with PP during melt processing, normally in the presence of a very small concentration of a free radical initiator (and sometimes also in the presence of a coagent). Both chemical factors and processing parameters can be manipulated to achieve high levels of grafting of the functional antioxidant on PP.

Examination of Table 2 clearly shows the advantages, especially under extractive conditions, of using such reactive antioxidants (photostabilizers) which are highly grafted on PP, compared to traditional (non-graftable) analogues. The grafted HALS photostabilizer AOTP, for example, shows a superior photostabilizing performance under extractive conditions compared to the commercial HALS, Tinuvin 770, tested under the same conditions. These reactive photostabilizers are much less readily lost from the polymer during processing and under aggressive

Table 2 Ultraviolet embrittlement times (EMT) of PP films containing different grafted reactive photostabilizers, based on HALS and UVA functions, when used alone and in combination with other stabilizers (extraction in dichloromethane)

Antioxidant combinations containing a grafted (g) photoantioxidant	*Concentration in PP film (%)*	*EMT* (h)	
		Before extraction	*After extraction*
AOTP-g	0.4	–	1200
AATP-g	0.4	1850	–
AOTP-g + (UV 531)	0.6	3000	–
AATP-g + (UV 531)	0.5	2050	–
HAEB-g + (DBBA-g)	0.4	1160	1130
Tinuvin 770	0.4	1500	130
Cyasorb UV 531	0.4	330	110
PP no additive (control)		90	–

service conditions. Furthermore, both grafted HALS antioxidants, AOTP and AATP, synergize effectively when used in combination with other classes of photostabilizers, e.g. UVA, such as Cyasorb UV531. A graftable UVA based on a hydroxybenzophenone structure, HAEB, is also shown to give effective synergism in combination with a highly grafted hindered phenol, DBBA [6]. It is important to mention that there are various approaches to the grafting of antioxidants on PP and these have widely different degrees of success and practicality. However, in general reactive antioxidants grafted on polymer melts behave in similar ways to 'traditional' low molar mass antioxidants and they offer more 'permanently' stabilized polymer compositions under the more demanding conditions of PP applications.

REFERENCES

1. Al-Malaika, S. and Scott, G. (1983) Photostabilisation of polyolefins, *Degradation and Stabilisation of Polyolefins*, (ed. N.S. Allen), Applied Science, London, pp. 283–336
2. Billingham, N. (1990) Physical phenomena in the oxidation and stabilisation of polymers, in *Oxidation Inhibition of Organic Materials*, Vol. II, (eds. P. Klemchuk and J. Pospisil), CRC Press, Boca Raton, FL, pp. 249–298.
3. Al-Malaika, S. (1990) Antioxidant mechanisms of derivatives of dithiophosphoric acid, in *Mechanisms of Polymer Degradation and Stabilisation*, (ed. G. Scott), Elsevier, New York, pp. 61–108.
4. Scott, G. (1995) Photo-biodegradable plastics, in *Degradable Polymers: Principles and Applications*, (eds. G. Scott and D. Gilead), Chapman & Hall, London, pp. 169–185.
5. (a) Sedlar, J. (1990), Hindered amines as photostabilisers, in *Oxidation Inhibition of Organic Materials*, Vol. II, (eds. P. Klemchuk and J. Pospisil), CRC Press, Boca Raton, FL, pp. 1–28. (b) Gugumus, F., Photoxidation of polymers and its inhibition, *ibid*, pp. 29–162.
6. Al-Malaika, S., Scott, G. and Wirjosentono, B. (1993) Mechanisms of antioxidant action: Polymer-bound hindered amines by reactive processing, part III – Effect of reactive antioxidant structure, *Polymer Degrad. Stab*, **40**, 233–238; Al-Malaika, S., Ibrahim, A.Q. and Al-Malaika, S. (1988) Mechanisms of antioxidant action: Photoantioxidant activity of polymer-bound hindered amines I: Bis maleate esters, *Polymer Degrad. Stab.*, **22**, 233–239 ; Al-Malaika, S. and Scott, G. (1995) US Patent 5 382 633.

Keywords: photooxidation, UV absorbers (UVA), outdoor performance, weathering, accelerated weathering, kinetic chain length, photoinitiator, Norrish reactions, peroxide decomposer, hindered amine light stabilizers (HALS), photosensitizer, transition metal complex, UV stabilizer, time-controlled stabilization, reactive antioxidants, polymeric antioxidants.

Pigmentation of polypropylene

B. Pietsch

Pigmentation is the most frequently used method of modifying the optical appearance of polypropylene (PP) for design, styling and functional purposes. The process of incorporating pigments is called mass-coloration and can be combined with other techniques such as printing and painting. The coloration of PP by use of pigments, which are per definition insoluble in the medium in which they are applied, has the advantage that there is no or only little migration of the colorant out of the polymer.

PIGMENTS

Pigments can be classified into inorganic and organic pigments, by their color or functional properties or by their chemical constitution and crystal morphology. The color index (CI) [1] lists pigments by their generic names and constitution numbers, together with an indication of main use. The color and fastness properties of pigments with identical chemical composition, however, can vary significantly, depending on their crystal morphology, particle size distribution or additionally applied finishing techniques such as surface treatments. Basically suitable for PP are inorganic and organic pigments which show sufficient heat resistance to withstand processing temperatures of 200–300°C. Pigments of the following categories are widely used for PP.

Inorganic pigments

Most inorganic pigments show low to moderate color strength combined with good to excellent heat resistance, lightfastness and weather resistance [2]. They are applied in concentrations up to 1% in PP and keep

Polypropylene: An A–Z Reference
Edited by J. Karger-Kocsis
Published in 1999 by Kluwer Publishers, Dordrecht. ISBN 0 412 80200 7

their fastness properties at very low concentrations of use. Inorganic pigments are normally easily dispersible but are often hard and abrasive.

White pigments and carbon blacks

Titanium dioxide white pigment based on the rutile modification is mainly used for PP due to its high refractive index of 2.8. ZnS and ZnS/$BaSO_4$-based lithopone pigments are also used due to their low hardness and abrasivity, e.g. in glass reinforced PP types.

Channel and furnace type carbon blacks, distinguished by their manufacturing process, are of major importance for the coloration of PP. The status of the various carbon blacks for food applications [3] differ according to type and purity.

Lead chromate yellow and molybdate orange pigments

These brilliant, inorganic yellow and orange pigments are widely used for PP. Specially treated types show adequate heat resistance for PP processing also at higher temperatures. Both pigment types are under legislative pressure in several countries, due to their lead and chromium VI content [4].

Cadmium yellow and red pigments

Cadmium sulfide and sulfoselenide pigments have been widely used for their attractive color, high heat resistance and excellent fastness properties. Due to environmental concerns the use of cadmium-based pigments is increasingly restricted.

Iron oxides

Special types of iron oxide yellows and zinc iron brown can have a moderate heat resistance up to 260°C. Iron oxide reds show limited coloristic properties but excellent heat resistance and fastness properties and are widely used for PP.

Other metal oxides and mixed metal oxide pigments

Outstanding heat resistance and fastness properties are the characteristic of nickel and chromium titananate and bismuth-vanadate yellows, chromium oxide green and cobalt green and blue. Further ultramarine blue and violet pigments are used.

Special effect pigments

Metal effect pigments are mainly based on small metallic flakes of aluminium or bronze. In another class are mica–metal oxide multilayer interference pigments for perlescent effects.

Organic pigments

Organic pigments [5] provide high color strength and chroma and are used in concentrations of 0.01% up to 0.5%. The fastness properties and heat resistance of organic pigments vary significantly for the various chemical types, and are dependent on their concentration of application and on the presence of TiO_2 (white reductions). A selection of organic pigments having heat resistance of at least 240°C in polypropylene is given below.

Organic yellow pigments

These include the yellow monoazo- and monoazo-like pigments CI pigment yellow (P.Y.), 120, 151, 181 and P.Y. 62, 168, 183, 191, 191:1, disazo- and azo condensation yellow P.Y. 93, 95, 155, 180, the heterocyclic P.Y. 182, isoindoline yellow P.Y. 139, isoindolinone yellow P.Y. 109, 110, 173, the quinophthalone P.Y. 138, the Ni-complex P.Y. 150 and the recently developed anthraquinone yellow P.Y. 199. Diarylide pigments such as P.Y. 13, 17, 83 are not recommended for PP, as processing above 200°C can lead to the formation of hazardous decomposition products, which do not necessarily become visible.

Organic orange pigments

These include benzimidazolones pigment orange (P.O.) 72 and 64, iso-indolinone P.O. 61, perinone P.O. 43, the Ni-complex P.O. 68 and the new diketo-pyrrolo-pyrrole (DPP) P.O. 71.

Organic red and violet pigments

These include monoazo pigments and monoazo salts P.O. 38, pigment red (P.R.) 170, 175, 176, 187 and 253, P.R. 53:1, 247:1 and 247, azo condensation reds P.R. 144, 166, 214, 220, 242, diketo-pyrrolo-pyrrole pigments P.R. 254, 264 and 272, the anthraquinone P.R. 177, pryenes P.R. 149, 178, 179, quinacridone pigments P.R. 122, 202, 209 and P.V. 19. Further are used thioindigo violet P.R. 88 and the dioxazine violets pigment violet (P.V.) 23 and 37. The monoazo BONA pigment lakes P.R. 48:2, 48:3 and 57:1 are restricted to low processing temperatures.

Organic blue and green pigments

Of importance are indanthrone blue pigment blue (P.B.) 60, alpha-stabilized and beta-phthalocyanine blues P.B. 15:1, 15:2 and 15:3 and the halogenated phthalocyanin green pigments (P.G.) P.G. 7 and P.G. 36. These pigments are all high heat resistance.

DETERMINATION OF PIGMENT PERFORMANCE

The properties of a pigment depend on the medium in which it is applied. Quality and performance determinations of a pigment are therefore ideally carried out after having applied it in PP or other representative polymers. Furthermore, different requirements exist for fibers and plastics applications. Standardized test methods for plastics applications, defined by organizations such as ASTM, ISO and DIN, are available for the assessment of many pigment properties in PP.

Color properties

For the visual assessment of a color, a reference sample is needed. Systems of physical color samples are available for reference purposes, such as the Munsell system, DIN color chart, Natural Color System and the more customized RAL and Pantone color charts. The ISO grey scale is defined for the visual assessment of color differences and used for the specification of color tolerances for quality control purposes.

Colorimetry

Beside visual assessment, the color of an object can be characterized more precisely in terms of mathematical models, which are based on the remission and transmission spectra in the visible light area [6]. The most common CIELAB system, defines a color space and allows one to describe a specific color in terms of L (lightness) and the color coordinates $\boldsymbol{a}$ and $\boldsymbol{b}$, and color differences or tolerances by related ΔL, $\Delta \boldsymbol{a}$ $\Delta \boldsymbol{b}$ and a combined ΔE value. $\boldsymbol{L}$, $\boldsymbol{a}$, $\boldsymbol{b}$, values can be transformed into the probably more convenient terms of $\boldsymbol{L}$ (lightness), $\boldsymbol{h}$ (hue) and $\boldsymbol{c}$ (chroma). Today, colorimetry software is used for color matching and allows very precise calculations of pigment formulations.

Tintorical strength

The tintorical strength of a pigment in PP is normally expressed as the amount of pigment needed to achieve a defined standard depth of color, e.g. according to DIN 53235. It depends to a certain extent on the method of dispersion and sample preparation.

Hiding power and transparency

Hiding power and transparency are standardized for paint applications in ISO 6504–1, ASTM D 2805 and DIN 55987, transparency in DIN 55988. The hiding power of an pigmented sample depends on the opacity contribution and the concentration of the applied pigment. A color difference of CIELAB $\Delta E = 1$, measured of a sample over a white and a black background is used as hiding criterion. Transparent pigments show high ΔE values, when measured over a black and a white background at a defined concentration in a standardized PP test pattern.

Heat resistance

The heat resistance of a pigment applied in PP is expressed as the maximum processing temperature at which any change of color does not exceed a defined limit, e.g. CIELAB $\Delta E = 3$ according DIN 6174, after a dwelling time of 5 min in the polymer melt. Fiber filament extrusion or injection molding of test specimens at different processing temperatures is used for the determination of the heat resistance. The heat resistance of organic pigments in PP depends on their concentration and the dwell time of the polymer melt during processing.

Migration resistance

The low glass transition temperature of PP enables partly dissolved pigment particles to migrate to the surface of a pigmented sample (blooming) or to migrate into a contact layer (bleeding). Standardized methods are defined in DIN 53775/54002.

Lightfastness

The lightfastness of a pigmented PP sample is mainly dependent on the pigment and its concentration, the processing temperature (heat history), the presence of TiO_2, but also depends on the substrate, the thickness of the layer and the presence of ultraviolet (UV) stabilizers. Test methods for the determination of lightfastness, expressed in ratings of 1–8 of the Blue Wool Scale, define exposure to glass filtered xenon light at a black panel temperature of 50°C (ISO 105–B02).

Hot-light fastness

To simulate the conditions in a car, methods for the determination of the lightfastness at elevated temperatures of, for example 83°C, have been defined by the automotive industry. Standard methods are the FAKRA test (DIN 75202), SAE J-1885 and DIN 75202.

Weather resistance

The weather resistance of pigmented PP, as defined by color consistency after exposure, is a function of the pigment fastness and the substrate. Fading or darkening of the pigments or the formation of microcracking (chalking) at the surface of thick-walled PP samples are the common effects after exposure. An appropriate UV-stabilization can achieve color stability of pigmented PP over several years.

Outdoor exposure

The weather resistance of a pigmented PP sample depends on the intensity and duration of the sunlight irradiation (measured in kLangley), humidity and also air pollution, such as salts and acidic gases at the exposure location. Exposures of two years in Florida (USA) or Bandol (France), for instance, have reached the status of industry references for the determination of weather resistance for highly demanding applications.

Artificial weathering

Accelerated weathering test conditions (DIN 53387, ASTM G26–70, ISO 4892) defining cycles of exposure to light, humidity and water spraying, have been developed. Xenon light but also carbon arc is still used as light source in weatherometers.

Chemical resistance of pigments

Unexpected color changes of pigmented PP can be caused by interactions of pigments with additives in the polymer. Alkaline nucleating agents or antistatica can react with alkaline sensitive pigments such as isoindoline yellow P.Y. 139 and provoke a dramatic drop of the heat resistance. Ion exchange reactions of monoazo lake pigments, such as the Sr-salt P.R. 48:3 with Mg- or Zn-stearate, can cause color shifts of PP compounds during processing. Color shifts of, for example, P.R. 247 due to interaction with modified masterbatch carriers, have also been observed. Especially, polymorphous pigments can undergo crystal modifications under certain conditions, resulting in color shifts and/or change of fastness properties.

INCORPORATION OF PIGMENTS

Pigment powders consist primarily of particles, aggregates and agglomerates. Primary particles, which are true single crystals with a specific X-ray pattern, and aggregates consisting of primary particles that have grown together at their common surface, determine the particle size distribution after dispersion of a specific pigment and therefore its coloristic properties.

Aggregates

Aggregates formed by primary particles and agglomerates have to be broken down by a dispersion process. An adequate dispersion of pigments is a basic requirement of any incorporation process [7].

Powder pigments

The direct incorporation of powder pigments during the final conversion process of PP or during PP compounding is practically restricted to easily dispersible inorganic pigments. The pigment powder or powder-blends can be metered directly to the screw of a compounder. For smaller batch processing, PP powder can be preblended with inorganic pigments. PP granules have to be wetted with mineral oil before blending with pigments, so that the pigment can stick to surface of the granules.

Pigment elaborations

Pigment elaborations are mixtures of organic pigments with dispersion aids, such as stearates, which enable the use of the more easily dispersible organic pigments on machinery equipped for direct metering of powders. Organic pigments which are sensitive to compaction can form hard, practically indispersible aggregates when exposed to pressure during the blending with polymer granules. Organic powder pigments blended with chalk or other fillers are also used to avoid compaction.

Pigment monoconcentrates, masterbatches

Pigment monoconcentrates or masterbatches are granules of 0.5–4 mm diameter, which contain the pigments in a predispersed form, usually in a polyolefinic carrier such as low density polyethylene (LDPE). The pigment loading is normally in the range of 20–50% but can reach 80% in case of inorganic pigments. Masterbatches can be metered directly on a compounder or extruder or preblended with PP granules. Custom color masterbatches, applied at 0.5–4%, contain all pigments in a definite ratio needed to achieve a specific shade. Pigment incorporation via masterbatch is today the most important process in the coloration of PP for all applications.

Liquid color

Liquid colors are high loaded concentrates based on a liquid carrier, e.g. soya-bean oil. They can be dosed directly into the machine hopper or directly into the PP melt, using special dosing pumps and equipment, and homogenize easily into the polymer.

APPLICATIONS

For a specific PP-application a selection of suitable pigments is required. Besides the economic aspect, the technical requirements can be categorized under processing requirements and end-use requirements.

Processing requirements

Pigment-inherent properties, such as heat resistance and chemical inertness to other additives present in PP, have to be considered to achieve color consistency during the whole conversion process. High quality of pigment dispersion is required for film and especially for continuous filament fiber extrusion, to avoid breakage of filaments and to ensure process stability.

Although overlapping, in practice there are processes which are often run at rather moderate temperatures below 240°C, e.g. film extrusion, film coextrusion, blown film and the recently developed PP calendering. For these applications, most of the above-listed pigments are suitable due to their heat resistance.

Processes often run at temperatures above 240°C are pipe and sheet extrusion. Blow molding and especially injection molding are run at process temperatures up to 300°C. Fiber filament extrusion, processed at 240–300°C, requires pigments which are thermally stable at these temperature even for extended periods.

End-use requirements

Specific requirements, which are relevant in making the choice of an appropriate pigment, originate from the various applications PP is used for.

PP fibers

A major outlet for PP fibers is in carpet, upholstery and other textile applications. Pigments for these applications have to show specific textile fastness properties. These are, for example, washing fastness for which specific laundering tests are standardized (ISO 105–C03–1989), color fastness to peroxide containing washing agents, to bleaching with hypochlorite, to perspiration, to organic solvents, to dry cleaning and to dry and wet rubbing. Automotive interior applications require hot-light fastness additionally to common textile fastness properties. Different levels of hot-light fastness are usually specified for upper parts and lower parts in a car.

Outdoor applications

Artificial lawn, stadium seating, garden furniture and automotive exterior parts, such as side panels and bumpers, are typical PP outdoor applications. Inorganic pigments and selected high weather resistant organic pigments, such as isoindolinone, quinacridone, perylene, diketopyrollo-pyrrole, indanthrone and phthalocyanine pigments, are used in combination with an appropriate light-stabilization packages, mainly combinations of HALS and a UV absorber.

Food packaging, toys and consumer goods

In many countries, different legislative regulations exist for food packaging and other consumer applications. These regulations specify purity requirements, weather for the pigment itself or the colored end application. Positive lists such as the French positive list, the JOHSPA list in Japan, 'The Code of Federal Regulations' of the FDA in the USA, are pigments which are approved for food packaging applications. In other countries, limits for migration and extraction of defined heavy metals and organic substances are defined for pigments or pigmented plastics. Other applications, such as consumer goods or toys, can be covered by such regulations or are separately regulated as, for example, in the European Norm EN 71–3–1994; Colourants for Toys.

Transparent PP

Transparent random copolymer PP types and nucleated transparent PP are gaining increasing importance in various application areas. For the coloration of such PP types, highly transparent pigment grades have been developed, which maintain the transparency of the polymer and provide all advantages of pigments in the coloration.

REFERENCES

1. *Color Index*, The Society of Dyers and Colourists, Bradford, England.
2. Endriss, H. (1997) *Aktuelle anorganische Bunt-Pigmente*, Curt R. Vincentz Verlag, Hannover.
3. Buxbaum, G. (1993) *Industrial Inorganic Pigments*, VCH, Weinheim.
4. *Safe Handling of Pigments*, Verband der Mineralfarbenindustrie (VdMi), Frankfurt.

5. Herbst, W. and Hunger, K. (1993) *Industrial Organic Pigments*, VCH, Weinheim.
6. Völz, H.G. (1990) *Industrielle Farbpüfung*, VCH, Weinheim.
7. McKay, R.B. (ed.) (1994) *Technological Applications of Dispersion*, Dekker, New York.

Keywords: pigments, pigmentation, carbon black, color, color index (CI), inorganic pigment, organic pigment, tintoric strength, hiding power, transparency, migration, lightfastness, weather resistance, masterbatch, regulations, chalking.

Polymer blends: fundamentals

L.A. Utracki

INTRODUCTION

A polymer blend is a mixture of at least two macromolecular substances, polymers or copolymers, in which the concentration of each polymeric ingredient is above 2 wt.%. Blends are either miscible or immiscible [1]. The miscible polymer blends are homogeneous down to the molecular level. Their free energy and heat of mixing, $\Delta G_m \cong \Delta H_m \leq 0$ and $\partial^2 \Delta G_m / \partial \phi^2 > 0$, where ϕ is the volume fraction. In these blends, the domain size is comparable to the dimension of the macromolecular statistical segment, $d < 20$ nm. The immiscible polymer blends are those where $\Delta G_m > 0$, and/or $\partial^2 \Delta G_m / \partial \phi^2 < 0$. Most polymer blends are immiscible.

A polymer alloy is an immiscible, compatibilized blend with modified interface and tailored morphology. Compatibilization modifies the interfacial properties by either a chemical or physical means. The interphase is the third phase in binary blends, formed in between domains of the two polymeric components; its thickness is $\Delta l = 2\text{–}60$ nm.

THERMODYNAMICS: MISCIBILITY AND PHASE SEPARATION

It is customary to discuss blend miscibility starting with the Huggins–Flory relation [1]:

$$\Delta G_m = \Delta H_m - T\Delta S_m; \; \Delta H_m = RTV\chi_{12}\phi_1\phi_2; \; \Delta S_m = RTV\sum_{i=1}^{2}(\phi_i \ln\phi_i)/V_i \quad (1)$$

where ΔS_m is the configurational entropy of mixing, R is the gas constant, T is temperature, V_i is molar volume, $V = \Sigma V_i$, and χ_{ij} is the binary

Polypropylene: An A–Z Reference
Edited by J. Karger-Kocsis
Published in 1999 by Kluwer Publishers, Dordrecht. ISBN 0 412 80200 7

interaction parameter, usually a complex function of ϕ, T and pressure, P. In polymer blends $\Delta S_m \to 0$, thus miscibility originates mainly from the enthalpic effects, $\Delta H_m < 0$, viz. specific interactions (hydrogen bonding, ionic, or dipole interactions). It may also be caused by the interactions between two macromolecules that reduce the internal steric stresses in at least one polymeric chain.

Owing to the presence of specific interactions, most blends have a phase separation diagram with a lower critical solution temperature, LCST, i.e. phase separation occurs upon heating. Two separation mechanisms are known: spinodal decomposition (SD), and nucleation-and-growth (NG). The morphology generated in NG is dispersed, whereas that in SD is co-continuous. Cahn and Hilliard's theory describes well the SD kinetics [1, 2].

INTERPHASE AND COMPATIBILIZATION

The interphase is a separate phase with its own characteristics and two interfacial tension coefficients, $\nu_1 + \nu_2 = \nu_{12}$, with ν_{12} being the experimental quantity. The lattice theories predict that in binary blends: (1) there is a reciprocity between ν_{12} and the interphase thickness, $\nu_{12}\Delta l$ = constant; (2) the surface energy is proportional to $\chi_{12}^{1/2}$; (3) polymer chain ends concentrate at the interface; and (4) any low molecular weight component migrates to the interface. In consequence, the interphase is characterized by low entanglement density and viscosity, often evidenced by the interlayer slip [3].

Compatibilization must: (1) reduce ν_{12} and parallel with it the size of the dispersed phase; (2) stabilize the morphology against changes during the subsequent processing steps; and (3) ensure good interaction between domains in the solid state. Compatibilization involves either addition of a compatibilizer (e.g. a block or graft copolymer), co-solvent or reaction between the principal blend components during the compounding or processing. The latter method is more advantageous and economic, thus it dominates commercial blending.

MORPHOLOGY

Performance of immiscible blends depends on the composition, interphase and morphology. At equilibrium, for $\phi < \phi_c \cong 0.16$ droplets are expected, while at higher concentrations fibers and lamellae are found [4]. Further increase of concentration leads to phase inversion at $\phi = \phi_{iI}$:

$$\lambda = [(1 - \phi_c - \phi_{2I})/(\phi_{2I} - \phi_c)]^{[\eta]\phi_m}; \quad \phi_{1I} + \phi_{2I} = 1 \tag{2}$$

where λ is the viscosity ratio, $\phi_m = 1 - \phi_c$ is the maximum packing volume fraction, and $[\eta]$ is the intrinsic viscosity.

MICRORHEOLOGY AND FLOW

In steady-state uniform shear flow at low concentrations and stresses, the drop deformation can be expressed using the three dimensionless parameters: the capillarity number, κ, the viscosity ratio, λ, and the reduced deformation time, t* [5]:

$$\kappa = \sigma_{ij} d / \nu_{12}; \quad \lambda = \eta_{\text{dispersed}} / \eta_{\text{matrix}}; \quad t^* = t\dot{\gamma}/\kappa \tag{3}$$

where σ_{ij} is the local stress (shear: $ij = 12$; elongation: $ij = 11$), d is the initial drop diameter, t is the deformation time, and $\dot{\gamma}$ is the deformation rate. In shear flow, for $\lambda > 3.8$, drops may deform but not break, while in extensional flow this limitation does not exist [5, 6]. Drop breakup occurs for $\kappa \geq \kappa_{\text{crit}} \cong 1$. However, the equilibrium drop deformation or break may occur only when t^* exceeds the required time: $t^* \geq t^*_{\text{d}} \cong 25$, or $t^* \geq t^*_{\text{b}} \cong 160$, respectively. When $\kappa > 2$, drops deform affinely with matrix into fibers that disintegrate by the capillary instability mechanism into minidrops when $\kappa < 2$. The diameter of these minidrops is about twice as large as the diameter of the disintegrated fiber.

Microrheology considers only individual drops in an infinite sea of the matrix fluid. At concentrations with $\phi \geq 0.005$, the coalescence effects must be taken into account. Coalescence can be driven either by the thermodynamics (i.e. minimization of the interfacial energy), or by flow (shear coalescence). During compounding the latter type dominates. It has been shown that the dynamic coalescence increases with $\dot{\gamma}^2$ and $\phi^{8/3}$, thus at equilibrium between dispersion and coalescence the drop diameter can be expressed as [7]:

$$d_{\text{equilibrium}} = d^{\phi=0}_{\text{equilibrium}} + (6C\kappa_{\text{crit}} t^*_{\text{break}} \phi^{8/3})^{1/2} \tag{4}$$

where C is an experimentally determined constant.

It is convenient to separate effects flow has on morphology into those that affect individual particles (local) and the global blend structure. For example, the local morphology includes variation of the degree and type of dispersion, including fibrillation, orientation, shear coalescence, etc. The 'global' morphology pertains to the whole extrudate or injection-molded piece. It comprises the interlayer slip, encapsulation, gradient composition and skin-core structures. The effects depend on the imposed strains and stresses.

The concentration dependence of blends' viscosity (at constant T and σ) can be classified in reference to the log-additivity rule, $\ln \eta = \Sigma\phi_i \ln \eta_i$, as showing a positive deviation, PDB, negative, NDB, or mixed, PNDB or NPDB. Treating blends as emulsions of viscoelastic liquids, leads to prediction of PDB (found in 60% of blends). The mechanism that explains NDB is the interlayer slip, caused by the thermodynamically driven low entanglement density in the interphase [4, 5]. In non-compatibilized, immiscible blends, the interphase viscosity can be calculated as being several orders of magnitude smaller than viscosities of component polymers. For example, in a polystyrene/polymethylmethacrylate blend, PS/PMMA, the interphase viscosity was determined as $\zeta_{interph} = 90$ Pa s, while viscosities of the two polymers were three orders of magnitude larger [8].

In polymer blends, both the morphology and flow behavior depend on the deformation field. Under different flow conditions the blend may adopt different structures, hence behave as different materials. Note that in multiphase systems, the relationships between the steady state, dynamic and elongational viscosities (known for simple fluids) are not observed. Similarly, the time–temperature (t–T) superposition principle that has been a cornerstone of viscoelastometry is not valid.

PERFORMANCE

The performance of polymer blends is controlled by morphology. For example, the optimum toughening of a brittle resin is usually obtained by dispersing $\phi \leq 0.1$ of an elastomer. The optimum drop size derived from the cavitation mechanism, $d_{opt} = 0.3\ \mu m$ for acrylonitrile–butadiene–styrene copolymer, ABS, $>1\ \mu m$ for PS, $< 0.1\ \mu m$ for polyvinylchloride, PVC, and $0.5\ \mu m$ for propylene blend with ethylene–propylene–diene terpolymer, PP/EPDM. One of the three compatibilization tasks is to ascertain the required d_{opt} value. The most efficient method of dispersing high-viscosity elastomeric particles in a thermoplastic matrix is in extensional flow field. It has been shown that an extensional flow mixer attached to a single-screw extruder can outperform a twin-screw machine. The new mixing device was designed to force a polymer blend through a series of convergences and divergences [6].

The phase co-continuity provides the best balance of properties, e.g. maximum ductility, high rigidity and large elongation. Two basic methods are used to generate this structure: (1) SD by control of T, P, and solvent evaporation rate; and (2) rheologically controlled phase inversion. Compatibilization is used to optimize the degree of dispersion [1, 9].

In blends developed to improve the barrier properties against permeability by gases, vapors needed, or liquids, the lamellar morphology is needed. To create this structure, biaxial extensional flow is required in molding or extrusion. Best results are obtained preblending a low permeability resin, e.g. $d \simeq 50\ \mu m$ drops of polyamide, PA, or polyvinyl alcohol, PVAl, in an olefinic matrix, then processing it in a biaxial flow field, e.g. by film blowing or blow molding.

REFERENCES

1. Utracki, L.A. (1989) *Polymer Alloys and Blends*, Hanser, Munich.
2. Utracki, L.A. (1994) Thermodynamics and kinetics of phase separation, in *Interpenetrating Polymer Networks*, (eds D. Klempner, L.H. Sperling, and L.A. Utracki), American Chemical Society, Washington, DC, pp. 77–123.
3. Luciani, A., Champagne, M.F., and Utracki, L.A. (1996) Interfacial tension in polymer blends, *Polymer Networks Blends*, **6**, 41–50, 51–62.
4. Utracki, L.A. (1991) On the viscosity–concentration dependence of immiscible polymer blends, *J. Rheol.*, **35**, 1615–1637.
5. Utracki, L.A. (1995) The rheology of multiphase systems, in *Rheological Fundamentals*, (eds J.A. Covas, J.F. Agassant, A.C. Diogo, J. Vlachopoulos and K. Walters), Kluwer, Dordrecht, pp. 113–137.
6. Utracki, L.A., and Luciani, A. (1996) Mixing in extensional flow fields. *Int Plast. Engng Technol.*, **2**, 37–54.
7. Huneault, M.A., Shi, Z.-H., and Utracki, L.A. (1995) Development of polymer blend morphology during compounding in a twin screw extruder. Part IV: A new computational model with coalescence. *Polymer Engng Sci.*, **35**, 115–127.
8. Valenza, A., Lyngaae-Jørgensen, J., Utracki, L.A., and Sammut, P. (1991) Rheological characterization of polystyrene/polymethylmethacrylate blends *Polymer Networks Blends*, **1**, 79–92.
9. Ajji, A., and Utracki, L.A. (1996) Interphase and compatibilization of polymer blends, *Polymer Engng Sci.*, **36**, 1574–85.

Keywords: blends, alloys, miscibility, interphase, phase inversion, compatibilization, phase diagram, spinodal decomposition, nucleation and growth, coalescence, rheology, extensional flow, mixing, compounding, blending, microrheology, morphology, performance.

Note: Polymer abbreviations are listed on page 626.

Polymorphism in crystalline polypropylene

S. Brückner and S.V. Meille

ISOTACTIC POLYPROPYLENE [1,2]

The isolated chain

Configurational and conformational properties of the isolated chain of isotactic polypropylene (iPP) are essential for explaining the polymorphic behavior of crystalline polymer. The two main chain bonds astride a tertiary carbon atom can be distinguished from a configurational point of view, and we shall refer to them as (+) or (−) bonds. A (+) bond is always followed by a (−) bond or vice versa, consequently two distinct directions in moving from one end to the other of the chain [1] exist (Scheme 1).

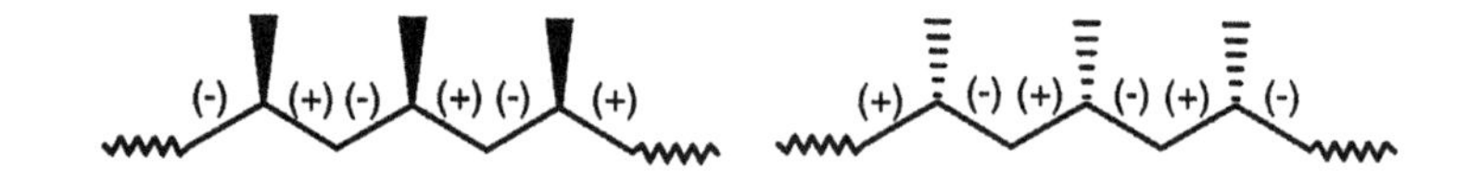

Scheme 1

T (180° trans) and G^+(60° gauche) conformations are available to a (+) bond whereas T and G^-(−60°) are available to a (−) bond. Considering a

Polypropylene: An A–Z Reference
Edited by J. Karger-Kocsis
Published in 1999 by Kluwer Publishers, Dordrecht. ISBN 0 412 80200 7

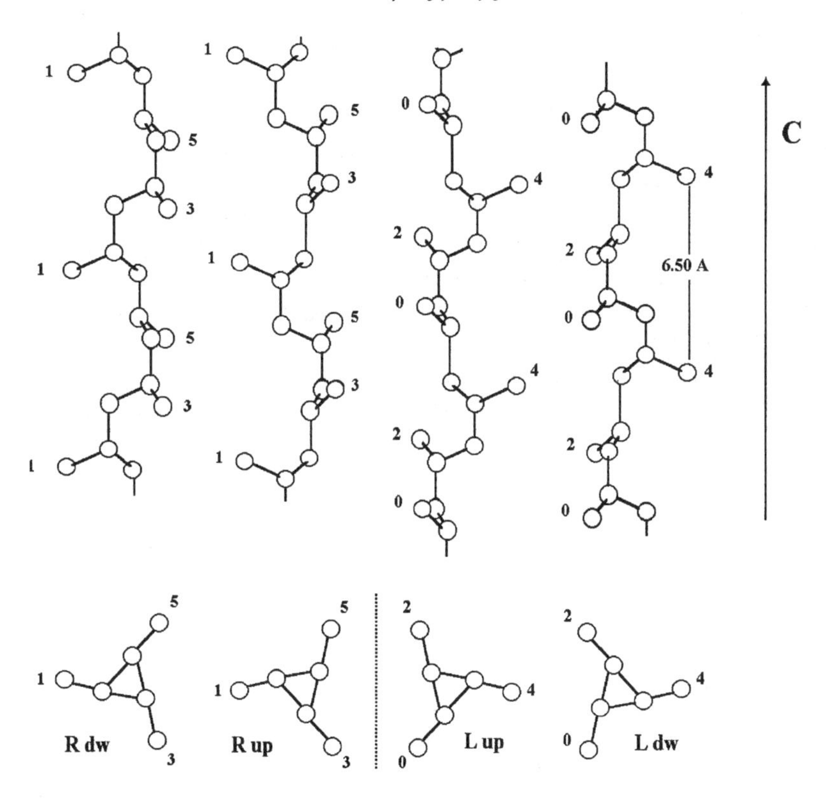

Figure 1 The four possible insertions of an iPP chain in the crystal lattice.

pair of consecutive bonds, e.g. (+)(−), two equi-energetic and symmetric conformations are available: TG^- and G^+T. Applying the equivalence principle of monomeric units, these conformers give rise to a right-handed (R) 3_1 helix and to a left-handed (L) 3_1 helix respectively. If we arbitrarily choose, as in Figure 1, a positive direction on the helix axis then a $[G+T]_n$ chain gives rise to a L_{up} helix, where *up* indicates that CH_3—CH bond has a positive component along helix axis. The same chain in $[TG^-]_n$ conformation gives rise to a R_{dw} helix, i.e. a helix of opposite chirality and negative component (*dw* = down) of CH_3—CH bond on the helix axis. Reversing the directionality of the chain does not change chirality of helices but simply changes *up* into *dw* or vice versa, so L_{dw} and R_{up} helices are generated respectively. The four ways of insertion of iPP chain into a crystalline lattice are shown in Figure 1. Finally we point out that isochiral *up* and *dw* helices are 'isosteric', i.e. their side methyl groups, determining the main intermolecular contacts, are superimposable.

α-form

This is the most common crystalline form of iPP. It is observed for both melt- and solution-crystallized samples prepared at atmospheric pressure. Accurately annealed samples can melt at temperatures as high as 180°C and their density is 0.94 g/cm^3. Some disorder is always present in the crystal structure of α-iPP, in fact an *up* chain can substitute a *dw* chain at almost the same packing energy expense. The first model, by Natta and Corradini [3], assumes 50% statistical substitution of *up* and *dw* chains while L and R helices occupy well-defined positions. This is the α_1-structure (monoclinic, C2/c, Figure 2). A second structure (α_2) was later proposed by Mencik, who showed that, upon accurate annealing, new diffraction peaks appear revealing a monoclinic $P2_1/c$ symmetry (Figure 2).

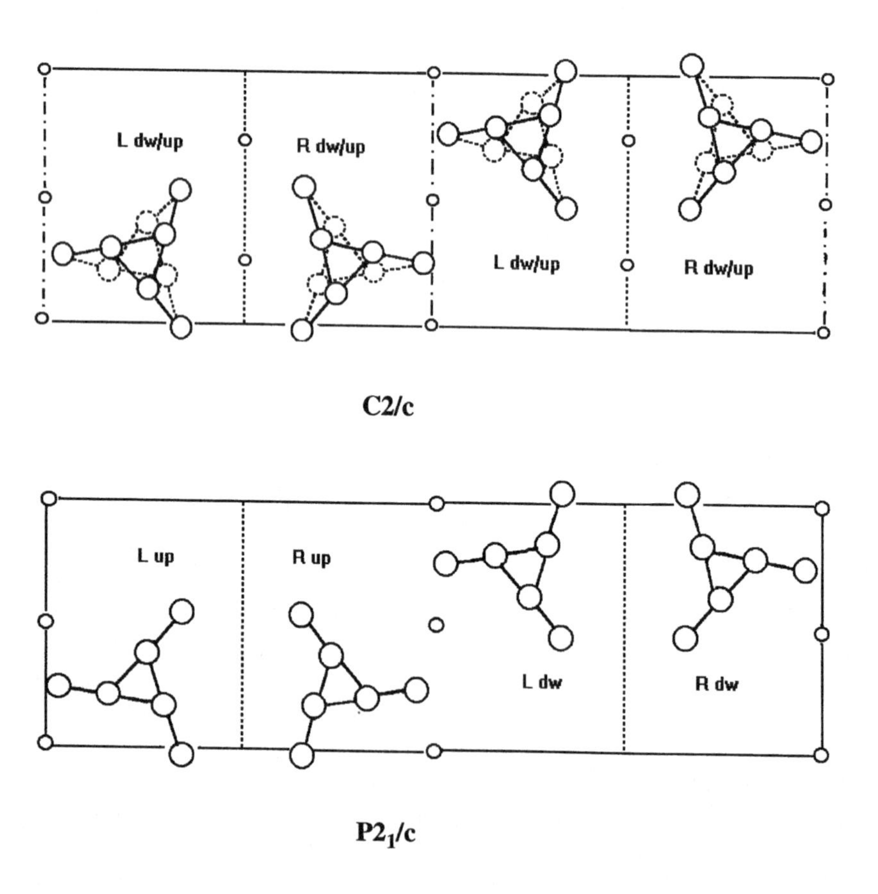

Figure 2 The α_1-form (C2/c) and the α_2-form ($P2_1/c$) of α-iPP. Unit cell constants are $a = 0.666$ nm, $b = 2.078$ nm, $c = 0.6495$ nm, $\beta = 99.6°$.

This is due to a reduction of the local *'up–dw'* chain disorder, a process which is never complete, making α_2 only a limit situation.

β-form

This is normally referred to as 'hexagonal iPP' and was identified in 1959 by Keith and coworkers. β-phase spherulites, characterized by strong negative birefringence, could be sporadically obtained when iPP was crystallized in the 128–132°C temperature range. Pure β-phase can be obtained with the aid of crystalline nucleating agents (see the chapter on 'Beta-modification of isotactic PP' in this book). Crystallization in a temperature gradient is also an efficient route to produce oriented iPP samples with predominant β-crystallinity. The growth rate of β-spherulites is up to 70% faster than that of α-spherulites. β-iPP is metastable relative to α-iPP ($T_m \approx 155$°C versus 180°C), it has lower density (0.92 g/cm^{-3}) and is unstable upon stretching, which produces a transition to α-iPP or to the 'smectic' form (*vide infra*) depending on whether the sample is processed above or below 60°C. The unit cell of β-iPP is trigonal and it contains three isochiral helices with up-down statistics (Figure 3). Diffraction patterns display, however, a symmetry higher than trigonal (hexagonal) and are consistent with averaging effects produced by disorder or twinning. Transition of the β-form to α-iPP occurs via a melting–recrystallization process.

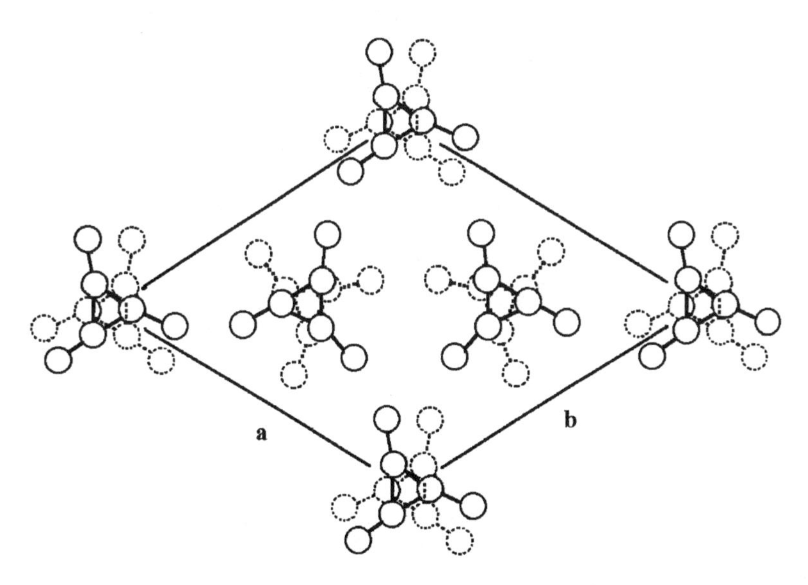

Figure 3 The unit cell and structure of β-iPP. Unit cell constants are $a = b =$ 1.101 nm, $c = 0.65$ nm, trigonal.

γ-form

This is the most peculiar iPP polymorph and was identified in 1961 by Addink and Beitema who immediately related it to the presence of short chains. Further studies revealed that also small amounts (4–10%) of comonomers (ethylene, 1-butene, 1-hexene) could induce crystallization of γ-iPP. The main route, however, is crystallization under pressure and at about 5000 bar also high molecular mass, highly stereoregular iPP crystallizes in practically pure γ-form. This modification is not stable upon stretching and the reasons of this behavior are now clear. The structure of γ-iPP, proposed by Brückner and Meille [4] (Figure 4), is unique in the field of synthetic polymers since nonparallel polymer chains coexist in the same crystal lattice. In fact, macromolecules belonging to different layers have helix axes that make an angle of 81°. γ-iPP

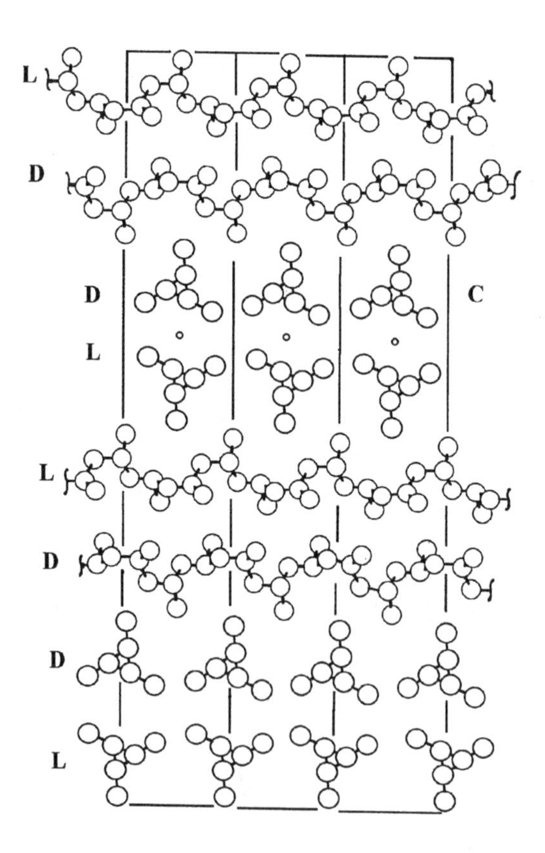

Figure 4 The structure of γ-iPP viewed along the chain axis of macromolecules belonging to one bilayer. Unit cell constants are $a = 0.854$ nm, $b = 0.993$ nm, $c = 4.241$ nm, orthorhombic.

has a density equal to that of α-iPP and packing energy calculations showed that it is at least as stable as α-iPP. Predominance of γ-iPP in samples crystallized under pressure has been interpreted as suggesting that the crystalline–amorphous interface plays an important role stabilizing γ-iPP.

Mesomorphic form

This form is often referred to as 'smectic' and was first mentioned in 1958 by Slichter and Mandell who observed a peculiar wide angle X-ray diffraction (WAXD) pattern in a sample melted and then rapidly quenched with dry ice. It is characterized by an order intermediate between those found in crystalline and in amorphous phases and is metastable since annealing at temperatures higher than 70°C leads to the crystallization of α-iPP. While density is low ($0.88\,g/cm^3$), infrared (IR) spectra indicate that iPP chains adopt the usual 3_1 helix conformation. Solid state ^{13}C nuclear magnetic resonance (NMR) shows a closer resemblance to β-iPP while WAXD patterns are in favor of a predominance of very local (pairs of chains) arrangements similar to those found in α-iPP.

SYNDIOTACTIC POLYPROPYLENE [2, 5]

The isolated chain

Main chain bonds in syndiotactic polypropylene (sPP) can be classified like in iPP as enantiomorphic (+) and (−) bonds. The syndiotactic chain, however, has no directionality (Scheme 2).

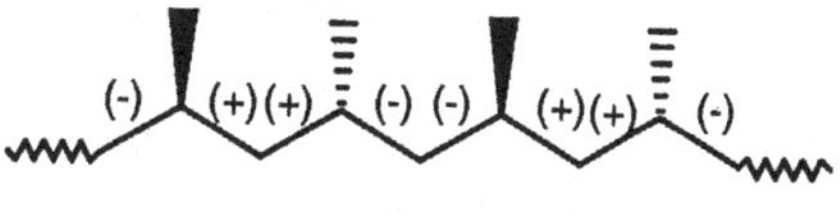

(-) (+)(+) (-) (-) (+)(+) (-)

Scheme 2

Conformational rules similar to those stated for iPP hold, i.e. a (−) bond can be T or G^- and a (+) bond can be T or G^+. A pair of bonds astride the tertiary carbon atom, e.g. a (+)(−) pair, can be TG^-, G^+T or TT. The equivalence principle allows for the sequences $...T|G^-G^-|TT|G^-G^-|T..$, $...G^+|TT|G^+G^+|TT|G^+..$ or $...T|TT|TT|TT|T...$ The first two conformations correspond to the left-handed and the right-handed twofold helices respectively and the third one is the planar zigzag conformation which is forbidden to iPP. Low energy conformations available to syndiotactic chains may be obtained generalizing the twofold —$(T_2G_2)_2$— helical conformation to ...$T_{2(2n+1)}G_2T_{2(2n'+1)})G_2T_{2(2n''+1)})G_2$... The resulting helices are not necessarily crystallizable. We finally note that right- and left-handed $(G_2T_2)_2$ helices have almost identical outside envelopes and this may produce substitutional disorder even more extensive than in iPP structures.

High-temperature orthorhombic form

This is the most stable crystal form with equilibrium melting temperature about 155–165°C and a crystallographic density of 0.93 g/cm^3. The chain conformation is —T_2G_2— with R and L helices. However, this structure does not seem to be unique and different packing arrangements have been proposed with the same unit cell dimensions, or multiples of the same subunit. Natta's group first proposed, from oriented fibers WAXD data, structure I reported in Figure 5. More recently, Lotz *et al.* proposed structure II from single crystal electron diffraction data and, ultimately, structure III from studies on crystals obtained at very low supercooling. The structures differ for the chirality of neighboring chains: in structure I, they are isochiral, while structures II and III are respectively characterized by alternation of enantiomorphic helices along *a* and along both *a* and *b*. Since R and L helices can substitute each other having nearly identical steric requirements, structures I, II and III may represent coexisting limit local situations. Intermediate arrangements may result due to averaging over entire crystal domains and even over different crystals.

Low-temperature orthorhombic form

This form is obtained by cold-drawing the polymer quenched from the melt. IR spectroscopy and WAXD techniques indicate that in this modification sPP adopts the all-trans conformation and the crystallographic density is 0.945 g/cm^3. Upon annealing fibers of this form at about 100°C for a few hours, the more stable helical form results, without losing the preferred chain orientation along the stretching direction.

Triclinic form

This form has been characterized by Chatani *et al.* from WAXD data from fibers originally in the planar zigzag polymorph and then exposed to benzene, toluene, or p-xylene vapours below 50°C for several days. A triclinic cell was determined containing six monomer units and with a crystallographic density of 0.939 g/cm^3. The chain conformation is —$(T_6G_2T_2G_2)$— which may be considered as intermediate between the original —(TT)— conformation and the helical conformation

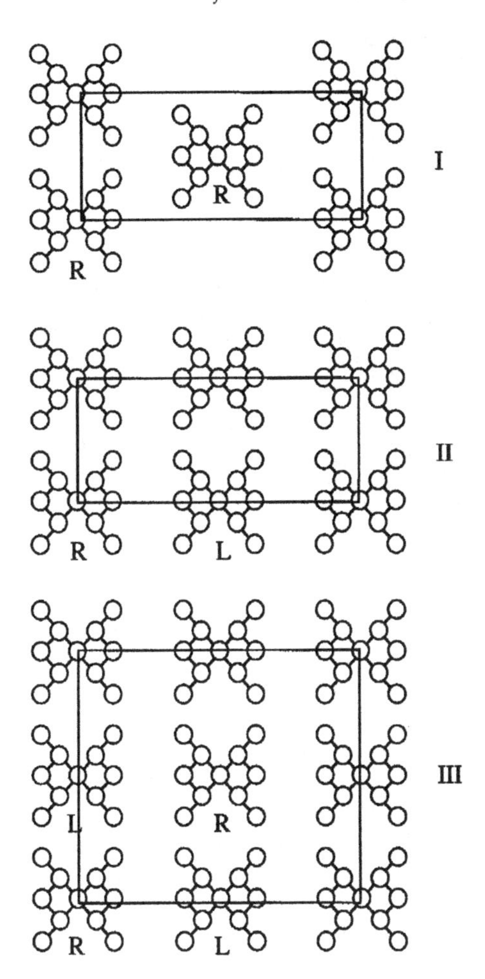

Figure 5 The proposed structures for the most stable form of crystalline sPP. Unit cell dimensions of structures I and II are $a = 1.45$ nm, $b = 0.58$ nm, $c = 0.74$ nm, structure III has a doubled b axis.

—$(G_2T_2G_2T_2)$— which is attained upon annealing at temperatures higher than 50°C.

REFERENCES

1. Brückner, S., Meille, S.V., Petraccone, V. and Pirozzi, B. (1991) Polymorphism in isotactic polypropylene. *Prog. Polymer Sci.*, **16**, 361–403.
2. Lotz, B., Wittmann, J.C. and Lovinger, A.J. (1996) Structure and morphology of poly(propylenes): A molecular analysis. *Polymer*, **37**(22), 4979–4992.

3. Natta, G. and Corradini, P. (1960) Structure and properties of isotactic polypropylene, *Nuovo Cimento Suppl.*, **15**, 40–51.
4. Brückner, S. and Meille, S.V. (1989) Non-parallel chains in crystalline γ-isotactic polypropylene, *Nature*, **340**, 455–457.
5. Rodriguez-Arnold, J., Zhengzheng, B. and Cheng, S.Z.D. (1995) Crystal structure, morphology, and phase transitions in syndiotactic polypropylene. *JMS Rev. Macromol. Chem. Phys.*, **C35(1)**, 117–154.

Keywords: conformation, packing, polymorphism, disorder, crystallographic density, isotactic, α-form, β-form, γ-form, mesophase, WAXS, WAXD, syndiotactic, orthorhombic form (high temperature, low temperature), triclinic form.

Polypropylene blends with commodity resins

L.A. Utracki

INTRODUCTION TO PP BLENDS

Isotactic polypropylene (iPP or PP) has several disadvantages that are corrected by blending, viz. it is brittle at $T \leq T_g \simeq 0°C$, difficult to process, and it has poor weatherability [1]. The search for methods to improve the low-temperature impact strength of iPP started with its discovery in 1954. Thus, PP was blended with either low density polyethylene (LDPE), and/or elastomers, either ethylene–propylene rubber (EPR), chlorosulfonated polyethylene rubber (CSR), or ethylene–propylene–diene copolymer (EPDM). PP blends with polyisobutene (PIB) were patented in 1958, those with polyethylene (PE), ethylene-vinylacetate (EVAc), polycarbonate of bisphenol-A (PC) soon followed. In the 1960s, PP was blended with its homologues: 3–50 wt.% syndiotactic polypropylene (sPP). This resulted in materials having excellent low temperature impact strength and freedom from surface crazing upon repeated flexing. For improvement of impact strength at low temperature, PP was also blended with 3–30 wt.% of amorphous polypropylene (aPP).

Blending also addressed two aspects of PP processability: high viscosity and poor melt strength. Several 'external lubricants' were incorporated into the resin, viz. ≤5 wt.% of either polydimethylsiloxane (PDMS), fluorinated polymers or elastomers. Usually, these blends also showed improved impact strength and elasticity. Another modification involved either copolymerization or grafting with acidic monomers, thus forming a transient crosslinking by the ion clusters. The presence of

Polypropylene: An A–Z Reference
Edited by J. Karger-Kocsis
Published in 1999 by Kluwer Publishers, Dordrecht. ISBN 0 412 80200 7

acidic groups also created opportunities for PP blending with engineering, polycondensation resins, viz. polyamides (PA) and polyesters (PEST).

Modification of the PP chain was also used as a preliminary step for improving the weatherability by blending with resins that have good resistance to ultraviolet (UV) irradiation, viz. acrylics, vinyls and/or vinylidenes [2, 3].

Almost all PP blends are immiscible; the only miscible ones are those with aPP, some sPP, EPR containing $w \leq 10\%$ of ethylene, and atactic polybutene-1 (PB-1) [4]. However, the isotactic PB-1 was reported immiscible with PP [5]. In immiscible blends, the interfacial adhesion in both molten and solid states is usually poor, thus compatibilization is advisable. However, as for any immiscible polymer pair, there are three cases where acceptable product is obtained without compatibilization: (1) at low concentration of either component, i.e., at $\phi = 1 - \phi \leq 0.1$; (2) blends with phase co-continuity, e.g., $\phi = \phi_{inversion}$; and (3) for applications not requiring high strength.

Compatibilization of PP blends is based on either copolymerization or grafting [3, 4]. In both cases, one part of the resulting compatibilizer should be miscible with the PP phase, the other with the second component. For example, EPR is used in PP/PE blends; the propylene part of EPR provides miscibility with PP and the ethylene part does so with PE. For the purpose of compatibilization, the block-type EPR is more desirable than random. Acid or anhydride modified PP is used for blends with PA, PC or PEST. Here, the PP part provides miscibility with the PP phase, while the acidic or anhydride groups usually react with chain-end groups of the other polymer to form a compatibilizing block copolymer.

The blend properties depend on composition, morphology, as well as the compounding and processing methods. The morphology is defined by the size and shape of the two phases, their distribution and orientation. In blends of semicrystalline polymers, blending affects crystallinity, often referred to as micromorphology. Thus, blending of PP must take into account the effects that the presence of other ingredients and interfaces have on the nucleation of PP and the kinetics of crystal growth.

Three types of blend crystallization are distinguished: (1) simultaneous of both components; (2) in the presence of molten second component; and (3) in the presence of solid second component. The case (1) takes place when ranges of the crystallization temperatures, T_c, of the two polymers overlap and their crystallization rates are similar; their mixed spherulites usually have additive physical properties. For the case (2), Paul and Barlow predicted five crystallization patterns: (i) no effect; (ii) retardation; (iii) delay; (iv) acceleration; and (v) crystallization of a polymer unable to crystallize by itself. Crystallization of PP with EPR, PB, SEBS, PE, and PB-1 or PCL followed pattern: (i), (ii), (iii), (iv), and (v), respectively. The case (3) can be treated as a special case of heterogeneous nucleation. Since nucleation is usually the rate controlling step, here the crystallization is expected to be faster than that of neat resin. The enhanced nucleation also tends to increase T_c. PP blends with PA or polyvinylidenefluoride (PVDF) follow this pattern [6].

PP/PE BLENDS

Blends of PP with polyethylenes (PEs) started with the intention of improving the low-temperature impact strength of PP [1]. Marked differences in the properties of PP blends containing different polyethylenes have been observed. Most of the PP/PE blends contain the traditional, low, linear low, or high density polyethylenes, LDPE, LLDPE or HDPE, respectively, but blends with ultra-low density PE (ULDPE, $\rho = 890\,kg/m^3$), and with ultra-high molecular weight (MW) HDPE (UHMW-PE, MW $\geq$ 1000 kg/mol), are also known. Miscibility of PP with PE depends on the molecular structure and MW of the ingredients. Most patents claim 5–50 wt.% of PE content, suggesting toughening by the presence of dispersed PE-phase. However, owing to the absence of antagonistic interactions, particularly interesting performance can be obtained from the blends having co-continuity of phases.

Compatibilization of PP/PE blends can be accomplished by addition of a copolymer, reactive blending (usually accomplished by addition of a peroxide), or by post-blending treatment, e.g. using chemical cross-linking, electron beam or γ-radiation. Blends of PP with PE have been compatibilized by addition of 5–15 wt.% of either EPR, EPR-block copolymer, PIB, EPDM, or PB-1. In the late 1960s, introduction of reactor blending was a major step in the evolution of PP/PE blend technology. The directly formed PE/PP/EPR blends became known as reactor-grade thermoplastic polyolefin (R-TPO), commercialized in 1979 (some companies market these materials as PPCP–PP copolymer). With time, the PE/PP blends are becoming more complex, e.g. comprising PP, LLDPE, 10–45 wt.% LDPE, and an EPR-block copolymer. Newest blends contain the metallocene resins, e.g. PP/LLDPE compatibilized by addition of a low MW plastomer (metallocene-catalyzed ethylene–butene copolymer).

Most PE/PP blends show a two-phase structure that may be detrimental to blend performance at large strains. Blending enhances the crystallinity and reduces the spherulite size, which may improve performance especially at small deformations, viz. increased modulus, where the effect of immiscibility is small. However, at large strains poor adhesion between the two phases results in a low value of the stress and strain at break. Either improved compatibilization or orientation can enhance this performance. For example, uniaxially drawn PP blends with 15–20 wt.% of HDPE showed increased Young's modulus as well as tensile strength at break.

PP/PIB BLENDS

These blends (introduced in 1958; ≤ 40 wt.% PIB) were developed to enhance the low-temperature mechanical performance (especially flexibility and impact strength). Since PIB is miscible with both, PP and LLDPE, soon the three-component blends were developed. They also showed good water vapor properties, and high UV resistance [4].

PP/PS BLENDS

Uncompatibilized blends of PP with 3–30 wt.% of either polystyrene (PS) or high-impact-PS (HIPS) were developed in 1971 for soda-straw tubes with pearly luster. Later, for improved mechanical performance, these blends were compatibilized by addition of either a tri-block styrene–butadiene–styrene copolymer (SBS), a multiblock $(SB)_n$ copolymer. Addition of hydrogenated-SBS, i.e. styrene–ethylene/butylene–styrene block copolymer (SEBS) was found to improve impact and flexural strength, whereas incorporation of a linear three-block copolymer: S1-D-S2 (where S1 ≥ S2 are PS-blocks and D = polydiene block) gave PP alloys with good crack and impact stress resistance [4].

Frequently, it is impossible for a single compatibilizer to accomplish all three tasks: control the degree of dispersion, stabilization of the morphology, and good adhesion between phases in the solid state. Thus, with growing frequency, combination of two or three compatibilizers is being used. For example, a combination of EPR and SBS provides a good example. Blends of PP with PS, upon addition of about 20 wt.% of these two copolymers, showed good synergistic enhancement of the impact performance [7]. Later, the company started production of reactor-made PP alloys. Using the proprietary 'Spheripol' technology, styrene, methyl methacrylate or styrene and maleic anhydride were polymerized inside the PP particles. These materials are commercialized as Hivalloy™.

Recently, the single-site metallocene catalyst was used to prepare random and block copolymers of syndiotactic-PP (sPP) with syndiotactic-PS (sPS). These were used as compatibilizers for these syndiotactic resins.

PP/EVAc BLENDS

Since 1960, PP was blended with a copolymer from ethylene and vinyl acetate (EVAc) and/or its hydrolyzed homologue, ethylene-vinylalcohol

(EVAl) to improve dyeability, flexibility, toughness, antistatic and/or barrier properties. Other ethylene copolymers (e.g. ethylene-ethylacrylate, EEA) were also used to improved impact strength, elongation, low brittleness temperature, radiation-resistance, and heat-sealability. Many of these blends also comprise other polymers, e.g. polyoxymethylene (acetal, POM), polymethylmethacrylate (PMMA), styrene-methylmethacrylate (SMM), poly(vinylchloride-co-vinylacetate) (PVCAc), maleated-LDPE, PC, PS, HDPE, etc. These alloys show good low-temperature impact resistance, melt strength, rigidity, and electrostatic charge dissipation.

PP blends with EVAl, and maleated polypropylene (PP-MA), can be formed by biaxial stretching into lamellar films or sheets. Their overlapping layer morphology result is significantly reduced permeability by gases or liquids [8].

PP/PVC BLENDS

PP has been blended with polyvinylchloride (PVC) for extrusion of pipes or electrical insulation [4]. The blends are compatibilized by addition of either methylmethacrylate-butadiene–styrene copolymer (MBS), a graft copolymer of acrylonitrile–butadiene–styrene–methylmethacrylate (ABSM), ethylene–vinylchloride copolymer (VCE), ethylene–vinylacetate–vinylchloride copolymer (EVAc-VC), EVAc, etc. The blends also comprised other resin, viz. PE, PS or SBR, PMMA or PC, etc.

OTHER PP BLENDS

PP blends with elastomers and those with engineering resins are discussed in separate chapters.

REFERENCES

1. Utracki, L.A. and Dumoulin, M.M. (1994) Polypropylene alloys and blends with thermoplastics, in *Polypropylene: Structure, Blends and Composites*, Vol. 2, (ed. J. Karger-Kocsis), Chapman and Hall, London, pp. 50–94.
2. Utracki, L.A. (1989) *Polymer Alloys and Blends*, Hanser Verlag, Munich.
3. Utracki, L.A. (ed.) (1994) *Encyclopaedic Dictionary of Commercial Polymer Blends*, ChemTec, Toronto.
4. Utracki, L.A. (1998) *Commercial Polymer Blends*, Chapman & Hall, London.
5. Cham, P.M., Lee, T.H. and Marand, H. (1994) On the state of miscibility of isotactic polypropylene/isotactic poly(1-butene) blends, *Macromolecules*, **27**, 4263–4273.
6. Nadkarni, V.N. and Jog, J.P. (1991) Crystallization behavior in polymer blends, in *Two-Phase Polymer Systems*, (ed. Utracki, L.A.), Hanser, Munich.
7. Okamoto, K.T., DeNicola, A.J., Huang, M.C.T. and Cleuvenbergen, M. (1993) A synergistic impact modifier combination for polypropylene/polystyrene alloys, *SPE Techn. Paper*, **39**, 2103–06.

8. Kamal, M.R., Garmabi, H., Hozhabr, S. and Arghyris, L (1995) The development of laminar morphology during extrusion of polymer blends, *Polymer Engng Sci.*, **35**, 41–51.

Keywords: blends, alloys, miscibility, compatibilization, crystallization, nucleation, polyethylenes (PE, LDPE, LLDPE, HDPE, UHMWPE), EPR, EPDM, ethylenevinylacetates (EVAc), EVAl, EVAc-VC, PB, PIB, styrenics (SBS, SEBS), PVC, PDMS, TPO, R-TPO, metallocene grades.
Note: Polymer abbreviations are listed on page 626.

Polypropylene blends with elastomers

L.A. Utracki

INTRODUCTION

Discovery of isotactic polypropylene (PP) in 1954 started a search for methods to improve its low-temperature impact strength [1, 2]. These activities commenced in the late 1950s by blending PP with polyethylenes (PE) or elastomers, then by copolymerizing it with ethylene into ethylene-propylene rubber (EPR) and ethylene–propylene–diene (EPDM).

PP BLENDS WITH ELASTOMERIC POLYOLEFINS

PP blends with polyisobutene, PIB, $—CH_2—C(CH_3)_2—$

The first PP blends with polyisobutylene, 5–20 wt.% PIB, were patented in 1958. The same year, three-component blends of PP with PIB and low density polyethylene (LDPE) were also disclosed. Owing to miscibility with PP and linear low density polyethylene (LLDPE), the low molecular weight (MW) PIB, having a narrow molecular weight distribution (MWD) was particularly useful as a low-temperature impact modifier and compatibilizer. Thus, PP blends with PIB (viz. Lupolen^{TM} O 250, or Pax-Plus^{TM}) gave alloys with good low-temperature flexibility, improved transparency, and strength. They also showed improved resistance to stress cracking, water-barrier properties and ultraviolet (UV) degradation [3].

Polypropylene: An A–Z Reference
Edited by J. Karger-Kocsis
Published in 1999 by Kluwer Publishers, Dordrecht. ISBN 0 412 80200 7

The PP/PIB blends are used for general-purpose films, boil-in bags, medical over-pouches, industrial sacks and liners. The blends compounded with carbon black served as over-wraps for the manufacture of tarpaulins, photographic films and papers, as well as in a variety of outdoor applications (weatherability of clear grades is poor).

PP blends with polybutene-1, PB-1, —CH_2—$CH(C_2H_5)$—

The commercial, isotactic polypropylene (iPP or PP) was reported miscible with PB-1, lowering the glass transition temperature, T_g, thus 'plasticizing' PP. PB-1 was used as a compatibilizer in blends of PP with either LLDPE or high density polyethylene (HDPE). These alloys were reported to have good balance of properties. Blends of this type were announced in 1993–95. These new compositions comprised PP, PB-1, and low MW, poly(α-olefin-co-ethylene) metallocene resin. The latter resin formed small pockets of amorphous material within the crystalline matrix of PP, thus improving spinnability, processability, impact strength and optical properties. Since PP/PB-1 blends showed low sensitivity to ionizing radiation, they have been targeted for the pharmaceutical and medical applications [3].

PP/ELASTOMER BLENDS

PP/Rubber blends

The butadiene–styrene rubber (SBR), or butadiene–acrylonitrile rubber (NBR) and elastomeric graft copolymers were found particularly valuable for impact modification of PP. The PP alloys (with 5–20 wt.% of an elastomer) were reported to have advantageous properties for blow molding of bottles free from brittleness and stress cracking. Blends with natural rubber (NR) require sulfur-curing [4]. Blending with NBR dramatically increased the modulus, but the material was brittle [5].

In 1958, PP blends with chlorinated butyl rubber (CBR) were prepared by Esso using dynamic vulcanization. The new process resulted in materials useful for high tensile strength applications. In 1976, the method was reapplied by Monsanto to several PP blends comprising, e.g. butyl rubber, ethylene-vinyl acetate copolymer (EVAc), chlorosulfonated polyethylene rubber (CSM) and EPDM. These covulcanized blends showed excellent toughness, elongation, impact strength, a wide range of Shore hardness and dimensional stability [6]. Later, a compatibilizing block copolymer was also added, which nearly doubled the blend tensile strength and increased its elongation at break by a factor of four.

PP/EPR or EPDM blends

Blends of PP with either EPR or EPDM provide the means of modifying the crystallinity of the rigid component, thus adjusting the properties, as well as the economy. In most blends, the elastomer forms a dispersed phase. The PP/EPR or EPDM blends show excellent processability. Additives, such as carbon black or sterically hindered amines (HALS), are recommended for protection against UV irradiation. The blends are characterized by high stiffness, high softening temperature, excellent low-temperature modulus and impact strength, dimensional stability, low shrinkage, good mechanical properties in a wide range of temperatures, from −40°C to 150°C, ozone resistance, dynamic fatigue and abrasion resistance, as well as high tear strength.

Numerous commercial PP/EPR blends are on the market. Reinforced and filled grades are also available [6, 7]. They have been used in automotive industry, in appliances, hardware and plumbing, medical tubing, shoe industry, sports equipment, toys, cases, etc.

The first disclosure describing these blends dates from the 1960s. Compositions containing PP with 0.1–60 wt.% EPR (containing 2–25% ethylene) have been disclosed. These blends can be treated as polymeric emulsions – addition of EPR increases viscosity of PP well above the log-additivity line [3].

Starting in the early 1970s, three-component blends, PP/LLDPE with either EPR or EPDM were introduced. These materials showed unexpectedly high tensile strength. The formulations could also be masticated and partially crosslinked with peroxides or sulfur (dynamic vulcanization), that could be shaped into articles with good properties without further vulcanization.

Another modification of the PP/EPR blend technology dates from 1981, when Montedison announced PP blends with an amorphous and a crystalline EPR. The alloys showed good balance of properties and improved impact strength. During the following years, many patents were deposited where advantages of addition of two or more elastomers to PP have been explored: PP blends with butyl rubber (BR) or chlorinated butyl rubber (CBR), and EPDM; PP with either EPDM or EPR and polydimethylsiloxane (PDMS); PP with EPDM and an ionomer ethylene-methacrylate-zinc, glycidyl methacrylate-acrylate; PP with EPR and metallocene plastomer; PP with EPR and either EPDM or LLDPE; PP with partially crosslinked rubber selected from between: natural rubber (NR), polyisoprene (IR), EPDM, EPM, BR, CBR, copolymers of iso-monoolefin and para-alkylstyrene, etc. To these blends a low molecular weight plasticizer could also be added.

In the 1980s metallocene catalysis created a revolution in polyolefin technology. New types of PP/EPR/LLDPE reactor blends were obtained in two stages, using a chiral metallocene catalyst and an aluminoxane. The blends showed good flow properties and a very good low-temperature impact strength. The first EPR blends with syndiotactic polypropylene (sPP) were disclosed in 1991. The blends were transparent and impact resistant. Metallocene homopolymer PP was blended with either EPR, EPDM or their mixture and a metallocene-type very-low density polyethylene (VLDPE). The blend showed excellent moldability, surface appearance and hardness, as well as good impact resistance. The blends were used for production of films having excellent low-temperature heat sealability and blocking resistance.

PP/Chlorinated elastomer blends

To this category belong PP blends with 4–6 wt.% of either chlorosulfonated polyethylene (CSR), or chlorinated polyethylene (CPE). These blends show good processability and mechanical properties. In 1990, PP blends with CSR and ultra-low density polyethylene (ULDPE) were first partially vulcanized then dispersed within a thermoplastic CSR. The alloys showed good processability, hot-weld strength, interplay adhesion and crack resistance.

PP/Styrenic elastomer blends

To this group of blends belong mixtures of PP with styrene–elastomer copolymers, styrene–diene blocks: styrene–butadiene–styrene (SBS), styrene–ethylene/butylene–styrene (SEBS), styrene–isoprene–styrene (SIS), with acrylonitrile–butadiene–styrene terpolymers (ABS), acrylonitrile–styrene–acrylate (ASA), or with EPR/EPDM grafted with styrene and acrylonitrile (AES or AXS). The first blends of this type date from the early 1960s. In these systems, PP is either the main component to be modified, or an additive to enhance performance of the styrenic matrix.

Blending PP with 20 wt.% SBS was found to greatly improve impact strength, without adversely affecting the other properties. Later, the reversed composition blends were announced. They had improved mechanical properties at lower cost. Similarly, on the one hand, PP was blended with 6–8 wt.% hydrogenated butadiene–styrene or isoprene–styrene block copolymer, SEBS, to give good transparency and high impact strength. On the other hand, addition of 2–30 phr PP to high impact polystyrene (HIPS) blend with SEBS, enhanced toughness and flexural strength, as well as retention of tensile yield stress. Recently, blends of PP with styrene–butadiene rubber (SBR) were compatibilized by addition of SBS, an acrylic elastomer, and PP grafted with styrene, acrylonitrile, methacrylates, etc. The alloys were used as stand-alone structural materials.

The earliest PP/ABS blends date from 1963; 0.5–15 wt.% PP was used to reduce the viscosity of ABS and improve its processability. Later, PP blends with 15–85 wt.% ABS, and chlorinated polyethylene (CPE) were reported to have good weatherability, flame resistance and impact behavior.

PP/Acrylic elastomer blends

These multicomponent PP blends have been developed during the last ten years. For example, they comprise: (1) either an acidified-PP, its mixture with PP, or a mixture of PP with carboxylic acid-modified EPR; (2) poly(methylmethacrylate-co-styrene-co-maleic anhydride); and (3) either ethylene-methylmethacrylate-glycidylmethacrylate, or ethylene-vinylacetate-glycidylmethacrylate. The compatibilization is obtained by chemical reactions between the acid and epoxy groups, as well as by forming ionic clusters capable of forming thermoreversible crosslinks. The blends were used to mold car bumpers and fenders. The products showed good stiffness and low-temperature impact resistance [8].

The blends exhibited high sag resistance without increase in melt viscosity, good stiffness, weatherability, and low-temperature impact resistance. These alloys can be molded (e.g. into car bumpers), extruded, or used for fiber spinning (that after stretching at $T = 60–160°C$ with the draw-down ratio, DR = 3–8, were formed into, e.g. strappings, netting, slit tape, twines, tarpaulins, sutures, nonwoven fabrics, etc.).

OTHER PP BLENDS

PP blends with commodity and those with engineering resins are discussed in separate chapters.

REFERENCES

1. Utracki, L.A. (1989) *Polymer Alloys and Blends*, Hanser, Munich.
2. Utracki, L.A. and Dumoulin, M.M. (1995) Polypropylene alloys and blends with thermoplastics, in *Polypropylene: Structure, Blends and Composites*, Vol. 2, (ed. J. Karger-Kocsis), Chapman and Hall, London, pp. 50–94.
3. Utracki, L.A. (1998) *Commercial Polymer Blends*, Chapman & Hall, London.
4. Choudhury, N.R. and Bhowmick, A.K. (1990) Micromechanism of failure of thermoplastic rubber, *J. Mat. Sci.*, **25**, 2985–89.
5. Choudhury, N.R. and Bhowmick, A.K. (1990) Micromechanism of failure of thermoplastic rubber, *J. Mat. Sci.*, **25**, 161–167.
6. Coran, A.Y. and Patel, R.P. (1995) Thermoplastic elastomers by blending and dynamic vulcanization, in *Polypropylene: Structure, Blends and Composites*, (ed. J. Karger-Kocsis), Chapman & Hall, London.
7. Utracki, L.A. (ed.) (1994) *Encyclopaedic Dictionary of Commercial Polymer Blends*, ChemTec, Toronto.

8. Abe, H., Fujii, T., Yamamoto, M. and Date, S. (1994), US Patent 5 278 233, to Sumitomo Chemical Company.

Keywords: ABS, acrylic-elastomers, ASA, AXS, blends, CBR, compatibilization, CPE, CSM, CSR, elastomer, EPDM, EPR, EVAc, EVAl, LLDPE, miscibility, NBR, PB, PIB, rubber, SBR, SBS, SEBS, SIS.

ABBREVIATIONS

aPP = amorphous polypropylene; ABS = thermoplastic terpolymer, an acrylonitrile-butadiene-styrene copolymer; ABSM = graft copolymer of acrylonitrile-butadiene-styrene-methylmethacrylate; ACM = acrylic elastomer, e.g. alkyl acrylate-2-chloroethyl vinyl ether copolymer; AES, AXS = terpolymer from acrylonitrile, ethylene-propylene elastomer, and styrene; ASA, AAS = thermoplastic copolymer from acrylonitrile, styrene and acrylates; BR = butadiene rubber; CBR = chlorinated butadiene rubber; CPI = *cis*-polyisoprene; CPE = chlorinated polyethylene; CR = chloroprene, or neoprene rubber; CSM, or CSR = chlorosulfonated polyethylene; EEA = elastomeric copolymer from ethylene and ethyl acrylate; EPDM = elastomeric terpolymer from ethylene, propylene and a non-conjugated diene; EPDM-MA = maleic anhydride-grafted ethylene-propylene-diene terpolymer; EPR, EPM = elastomeric copolymer of ethylene and propylene; EVAc = copolymer from ethylene and vinyl acetate; EVAc-VC = ethylene-vinyl acetate-vinyl chloride copolymer; EVAl, EVAL = copolymer of ethylene and vinyl alcohol; HALS = hindered amines (antioxidants); HDPE = high density polyethylene (ca. 960 kg/m^3); HIPS = high impact polystyrene; iPP = isotactic polypropylene (see PP); IR = synthetic *cis*-1,4-polyisoprene, synthetic isoprene rubber; LDPE = low density polyethylene (ca. 918 kg/m^3); LLDPE = linear low density polyethylene; MDPE = medium density polyethylene (ca. 930 to 940 kg/m^3); NBR = elastomeric copolymer from butadiene and acrylonitrile; nitrile rubber; NR = natural rubber; PA = polyamide; the first and second number indicate the number of methylene groups of aliphatic di-amines, and the number of carbon atoms of aliphatic di-carboyxlic acids, respectively; PB-l = poly-l-butene; PC = bisphenol-A polycarbonate; PCL = polycaprolactone; PDMS = polydimethylsiloxane; PE = polyethylene; PEST = thermoplastic polyesters, e.g. PBT, PET; see also TPES; PIB = polyisobutene; PMMA = polymethylmethacrylate; PO = polyolefin; POM = polyoxymethylene, polyformaldehyde, polyacetal or 'acetal resin'; PP = polypropylene; isotactic; PPCP = polypropylene copolymer (containing a small amount of ethylene); PP-MA = maleic anhydride-modified polypropylene; PS = polystyrene; PVAc = polyvinyl acetate; PVAl = polyvinyl alcohol; also PVOH; PVC = polyvinyl chloride; PVCAc = poly(vinyl chloride-co-vinyl acetate); PVDF = polyvinylidene fluoride; also PVF_2; sPP = syndiotactic polypropylene; sPS = syndiotactic polystyrene; $(SB)_n$ = multiblock of styrene-b-butadiene; SBR = thermoplastic copolymer from styrene and butadiene; also PASAB, S/B; SBS = styrene-butadiene-styrene triblock polymer; SEBS = styrene-ethylene/butylene-styrene triblock polymer; SIS = styrene-isoprene-styrene block polymer; SMM = copolymer from styrene and methyl methacrylate; TPO = thermoplastic olefinic elastomer; UHMW-PE = ultrahigh molecular weight polyethylene (over 3 Mg/mol); ULDPE = ultra low density polyethylene (ca. 900 to 915 kg/m^3); VLDPE = very low density polyethylene (ca. 885 kg/m^3).

Polypropylene blends with engineering and specialty resins

M.M. Dumoulin and L.A. Utracki

INTRODUCTION

Engineering thermoplastic resins (ETP) are those whose set of properties (mechanical, thermal, chemical) allows them to be used in engineering applications. They are more expensive than commodity thermoplastics and generally include polyamides (PA), polycarbonate (PC), linear polyesters such as polyethylene terephthalate (PET) or polybutylene terephthalate (PBT), polyphenylene ether (PPE) and polyoxymethylene (POM). Specialty resins show more specialized performance, often in terms of a continuous service temperature of 200°C or more and are significantly more expensive than engineering resins. This family include fluoropolymers, liquid crystal polymers (LCP), polyphenylene sulfide (PPS), aromatic polyamides (PARA), polysulfones (PSO), polyimides and polyetherimides.

Polypropylene (PP) alloys with ETP are in an early stage of development. Blends with specialty resins are even less advanced. The main source of difficulties is the 'antagonistic' immiscibility of PP with these polymers. In many cases, addition of a compatibilizer to PP/ETP blends reduces the crystallinity, which in turn lowers the performance. Optimization of composition and blending methods is not easy. The most commonly used compatibilizing agents contain maleic anhydride functionality, e.g. maleated-PP (PP-MA). Their reaction with such ETP as PA or PET depends on the reaction conditions: time, temperature and stress. The compatibilizing copolymer content increases upon reprocessing.

Polypropylene: An A–Z Reference
Edited by J. Karger-Kocsis
Published in 1999 by Kluwer Publishers, Dordrecht. ISBN 0 412 80200 7

There is intensive search for better compatibilizers and/or compatibilization methods.

Among the PP/ETP blends, only these with PA achieved a commercial success [1, 2]. They show good processability, low water absorption, dimensional stability, low density, low liquid and vapor permeability, moderate impact strength, as well as good resistance to alcohols and glycols. Applications include shoes, gaskets, medical bags and tubings, coatings, automotive interior and under-the-hood parts, cable ties, fasteners, pump housings, impellers, gears and bearing retainers, automotive parts, electromechanical parts. Blends of PP with PC are being introduced as high-performance 'plastic paper'. More recently, recycling automotive scrap showed that blends of PP with up to 10% POM have properties superior to virgin PP, and can be reused for molding small automotive parts.

PP blends with ETP will be systematically discussed in the following text.

BLENDS WITH POLYAMIDES

PP has been blended mainly with PA-6, but also with PA-66, PA-12, and PARA. The main commercial blends are these with PA-66 and PA-6 (viz., Orgalloy™, Elf Atochem) [3, 4]. The aim is to bridge the property gap between the two resins, viz. reduce PA water absorption and cost, as well as improve PP processability and impact strength. The blend has better impact resistance than either component. The good melt flow of PP and high crystallization rate of PA makes the blend suitable for thin-wall molding. Blends with PA-66 show higher mold shrinkage, modulus, hardness and heat deflection temperature than those with PA-6. It should be noted that the inherent moisture absorbency of the PA macromolecules is not affected by blending. Water absorption and permeability to moisture are reduced by virtue of the lesser amount of PA in the system. Because of the lower permeability, PA/PP blends require longer drying than virgin PA.

Uncompatibilized PA-6/PP blends are coarse mixtures with large particles ($d \leq 50\ \mu m$), easily deformable into fibrils. The morphology is sensitive to concentration, stress and temperature, particularly near the phase inversion composition. Compatibilization reduces the particle size to $d \approx 1\ \mu m$. In commercial alloys, PP forms the dispersed phase even at $w(PP) = 40$–60 wt.%. However, processing of these well-dispersed blends can lead to flow segregation, coarsening of morphology, and fibrillation. In extruded sheets, the originally spherical PP particles are deformed into large, extended and oriented PP domains. The degree of PP phase deformation and fibrillation depends on the local stress: larger near the surface, smaller in the core.

PA alloying is a two- or three-step reactive process: (1) modification of PP; (2) reaction with PA; and (3) mixing with PA and PP resins. Maleation of PP (in the presence of peroxides) is the most common approach, but other anhydrides, acids or epoxies have been used as well. Compatibilization of PA/PP blends requires a relatively large quantity (up to 4 wt.%) of the first stage (acidic) compatibilizing agent. Since this compatibilizer reacts with PA, the resulting concentration of PA-PP copolymer found in the blends varies from about 10 to 20 wt.%. Addition to PP/PA-6 blends of an ionomer can significantly improve the degree of dispersion and performance. The efficiency of compatibilization provided by copolymer addition and *in situ* reactive compatibilization was compared for a PP/PA-66 blend [5]. The final morphology was similar, although the latter method resulted in finer dispersion. The methods show similar efficiency.

Blending usually affects the crystallinity. For example, addition of PP to PA has decreased the melting temperature (T_m) by either 4°C (PP/PA-6 blends) or 6°C (PP/PA-6 blends with PP-MA). In the latter blends, increasing PA-6 content reduced T_m(PA-6), but it increased that of PP; the total crystallinity decreased with an increase of the compatibilizer content and the compounding time. For the PP/PA-6/PP-MA blend, the tensile strength and modulus increased monotonically with PA-6 content while the tensile strain at break showed a large negative deviation.

For uncompatibilized PP/PA blends, the concentration dependence of η at constant shear stress, σ_{12}, usually shows a negative deviation behavior, NDB, indicating the predominant effect of interlayer slip; its magnitude increases with σ_{12} [2]. Microscopic examination indicated that NDB was largest when the interfacial area was the largest. In compatibilized blends, rigidification of the interphase engendered a suspension-like behavior of the alloy leading to a positive deviation behavior (PDB). It should be noted that, as a rule, compatibilization increases the resistance to deformation, viz. shear viscosity. However, the rheological changes depend on the compatibilizer, compatibilization method and its extent, as well as on the intensity and type of the flow field.

The extrudate swell ratio, *B*, plotted as a function of composition, was found to go through a local maximum within the phase inversion region. As a rule, *B* in blends is large and shows PDB. The effect is not related to deformation of the macromolecular coil (as in homopolymer melts), but rather to the form recovery of the dispersed phase.

Addition of PA-6 to PP severely lowered the melt strength, probably due to the low viscosity of the interphase. Compatibilization of the system by addition of PP-MA enhanced drawability and fiber performance.

BLENDS WITH POLYCARBONATE

Blending and compatibilization of PP with PC is not as developed as with PA [1]. Addition of 5 wt.% PP to PC reduced melt viscosity, thus improved processability. It also increased the impact strength (especially in thick sections) and modulus, and reduced notch sensitivity. On the other hand, addition of about 10 wt.% of PC to PP improved its processability and product appearance. Furthermore, it increased hardness and impact strength, as well as reducing mold shrinkage. The key for achieving these advantages is control of PC concentration and morphology. As expected, the strain at break versus composition for these two component blends showed strong NDB.

There are several patents describing PC compositions with $w(\text{PP}) \geq 10$ wt.%. For example, addition of an acrylic copolymer enabled an increase in PP content, which led to better gasoline resistance. For packaging applications, addition of styrene–ethylene–butadiene–styrene copolymer (SEBS) made it possible to increase the PC content to 40 wt.%. PP/PC blends have also been compatibilized using: maleated-PP; a mixture of ethylene-vinyl alcohol (EVOH) and grafted-SEBS (with either MA or fumaric acid); styrene-acrylonitrile grafted either PP or EPDM; and PS containing oxazoline moieties.

For noncompatibilized PP/PC blends, the compositional dependence of the heat deflection temperature (HDT), modulus, yield and fracture strength showed a negative deviation from additivity. Addition of a compatibilizer improved the blend performance. Excellent mechanical properties, impact strength, HDT, moldability, and solvent and chemical resistance have been claimed for compatibilized blends containing 5–95 wt.% of either PP or PC. The best overall effects were observed using poly(propylene-g-styrene/acrylonitrile).

The degree of dispersion in PP/PC blends depends strongly on the viscosity ratio, λ, particularly when the high viscosity polymer forms the matrix. Both the particle size and its polydispersity increase with the minor phase content. The phase inversion region is affected by λ as well as by the relative elasticity of the components.

Addition of PC to PP leads to enhancement of PP crystallinity and a small increase in the crystallization temperature (T_c). The effects are more pronounced upon addition of a compatibilizer (note that compatibilization is limited to the amorphous phase).

BLENDS WITH LINEAR POLYESTERS

Ester linkages in the backbone chain are common to PC and polyesters, and there are also similarities in the chain-end groups. Thus, the blending philosophy for these polymers is common. For example, addition of a small amount of either PP to PET or vice versa is beneficial, even without compatibilization. Excepting the region of phase inversion, blends comprising more than 10 wt.% of either ingredient should be compatibilized. Blends of PP with PCTG (amorphous copolyester containing cyclohexane and terephthalate units) showed good impact properties at *w*(PP) = 30 and 50 wt.%.

At present, most PP/polyester blends are formulated within the low dispersed phase concentration region, where the mutually beneficial effects are evident [1]. Small amounts of PP have been used for toughening polyesters. However, in spite of several decades of activity, there is no commercial PP/polyester blend on the market. As discussed before for PP/PC blends, there is a problem to find good compatibilizer/impact modifier, capable of enhancing performance of the mid-range compositions of PP/polyester mixtures. Such additives as SEBS, PP-MA or maleated-SEBS, did not engender the desired performance at a reasonable cost. It was reported that PO/polyester blends with SEBS had good balance of properties, making them suitable for manufacturing tubing, hoses, wire and cable jacketing, etc.

In extrusion of PP/PET blends, although PET had lower η than the blends, the blends extruded at higher volumetric rates, indicating that the output was controlled either by melting or the interlayer slip. It was shown that even if the sheath-and-core monofilaments of PP/PET blends could not be oriented (because of poor adhesion between the two polymers), melt blending still produced useful oriented product. The maximum draw ratio for PET was 7, whereas for PP/PET blends it was 11. The dynamic mechanical testing of blend monofilaments, containing *w*(PET) = 50 and 70 wt.%, indicated that T_g(PET) remained constant and the dynamic moduli were approximately additive.

With time, the blends are getting progressively more complex. For example, multicomponent blends comprised polyester (e.g., PET or PBT with sodium dimethyl 5-sulfo-isophthalate), an olefinic copolymer (e.g., PP-MA) and PA-6 or PA-66. These had fine dispersion with $d = 0.05$–$1\ \mu m$, high tensile and impact strength. To improve performance of PBT/PC blends, PE, PP, or EPR grafted with GMA, and/or MA were added. The alloys showed excellent processability, toughness, rigidity, strength, dimensional stability and flexural modulus.

BLENDS WITH POLYPHENYLENE ETHER

PP and PPE are antagonistically immiscible, and since PPE alone is virtually nonprocessable, it must be 'plasticized' with another resin. Nevertheless, PP has been used in many PPE blends, to improve solvent resistance and processability. For example, PPE-MA was blended with PP-MA and a styrene-grafted elastomer. The alloys showed excellent solvent resistance, good moldability, impact and mechanical properties. A similar system comprised either PPE with PP, PE, PA-6, PA-66, PBT, PET, POM, PPS or PEEK, and a compatibilizer selected from ethylene–propylene rubber-MA (EPR-MA), ethylene-vinyl acetate-glycidyl methacrylate (EVA-GMA), monomeric MA or bis(4-phenyl isocyanate). Reactive blending in a high-speed twin-screw extruder resulted in alloys with PPE dispersed spheres having the diameter, $d = 0.01–10\,\mu m$. The blends could be formed into variety of parts for the automobile, electrical and electronic industries.

BLENDS WITH POLYOXYMETHYLENE

PP and POM are both semicrystalline and immiscible, thus at $w \geq 10$ wt.% they should be compatibilized and impact modified. Since POM can develop strong hydrogen bonding with acidic or epoxy groups, usually acidified or epoxidized PO has been used. For example, POM blends with PP were compatibilized by addition of muconic acid-grafted PP. Another system comprised POM, thermoplastic urethane (TPU), 10–30 wt.% of either PP, PA, or polyester, and ethylene–butyl acrylate–GMA copolymer, with (EBA-GMA) as compatibilizer.

OTHER PP BLENDS

To improve low-temperature impact performance of PP, it has been blended with elastomers, viz. TPU or nitrile–butadiene rubber (NBR). Addition of w(TPU) < 15 wt.% improved impact and other mechanical properties. Significant improvements of the tensile strength, as well as the tensile strain and strength at break, were obtained by adding to PP 2 wt% of di-methylol-*p*-octyl phenol, than blending it with NBR.

BLENDS WITH SPECIALTY RESINS

The high-temperature specialty resins have rigid macromolecules with strong intra- and inter-molecular interactions. By contrast, PP has simple molecular structure, and is usually processed at 220–250°C. In consequence, few blends of PP with specialty resins have been disclosed and none is commercial.

Two types of PP/LCP blends have been described: (1) with LCP forming either a lubricating or a protective skin; and (2) with LCP domains stretched into reinforcing fibrils or platelets. While exploitation of these systems depends on the advanced methods of processing, compatibilization has been found advantageous, especially for the second type of applications. For example, addition of LCP to PP followed by two-step orientation, increased the tensile modulus in the machine direction by one order of magnitude, and decreased its value in the transverse direction. Compatibilization by addition of PP-MA further improved these properties (see also the chapter '*In-Situ* reinforced PP blends' in this book).

LCPs show a sharp melting point above which their viscosity decreases significantly. Addition of these materials to PP can serve a dual purpose. At low concentration, they reduce pressure drop across the die, thus increasing throughput. When the concentration is high enough, they form a protective layer, which improves weatherability, solvent and chemical resistance. An apparent yield stress was observed at concentrations $w(\mathrm{LCP}) \approx 1$ wt.%. Plots of constant stress η versus composition showed NDB, indicating an interlayer slip.

Aromatic amorphous or semicrystalline polyamides, PARA, such as copolyamide from caprolactam, hexamethylenediamine and *iso*- and *tere*-phthalic acids, PA-6IT6, or copolyphthalamide, PPA, are frequent components of polymer blends. Their advantage is low permeability to liquids and gases, stiffness and high HDT. Several blends of PP with PARA have been described, e.g. with either PP-MA, PE-MA, or PP grafted either with acrylic acid or GMA. Usually, these blends must be heat stabilized with phosphite ester or hydrazine. The alloys are often reinforced by addition of glass fibers and talc.

Addition of low concentration of PP, $w(\mathrm{PP}) \leq 15$ wt.%, was found to be beneficial for processability and performance of several specialty polymers. For example, it improved the notched Izod impact strength of PEI as well as its ultimate flexural strength, but it decreased the unnotched impact strength, modulus and the tensile properties. Added to PAES, it improved processability as well as solvent resistance and impact strength.

REFERENCES

1. Utracki, L.A. and Dumoulin, M.M. (1995) Polypropylene alloys and blends with thermoplastics, in *Polypropylene: Structure, Blends and Composites*, Vol. 2, (ed. J. Karger-Kocsis), Chapman & Hall, London, pp. 50–94.
2. Utracki, L.A. (1989) *Polymer Alloys and Blends*, Hanser, Munich.
3. Utracki, L.A. (ed.) (1994) *Encyclopaedic Dictionary of Commercial Polymer Blends*, ChemTec, Toronto.
4. Utracki, L.A. (1997) *Commercial Polymer Blends*, Chapman & Hall, London.
5. Helmert, A., Champagne, M.F., Dumoulin, M.M. and Fritz, H.G. (1995) Compatibilization of polypropylene/polyamide-66 blends via reactive blending with maleated polypropylene, in *Proceedings of the Conference on Polyblends '95*, October 19–20, Boucherville, Canada, National Research Council Canada, Boucherville, Canada.

Keywords: blends, alloys, miscibility, compatibilization, crystallization, nucleation, polyamide (PA-6, PA-66), polycarbonate (PC), thermoplastic polyesters (PET, PBT), polyoxymethylene (POM), polyphenylene ether (PPE), ethylenevinylacetate (EVA), grafting with maleic anhydride (MA), grafting with glycidyl methacrylate (GMA), liquid crystal polymers (LCP), copolymer compatibilizer.

Polypropylene foams

M. Rätzsch, H. Bucka and U. Panzer

INTRODUCTION

Polyolefin foams are manufactured in a variety of processes and in low density (25–250 kg/m^3) or high density (250–700 kg/m^3) versions. Foamed polyolefins have a long tradition in research, production and application. The consumption in 1995/96 all over the world was more than 190 000 tonnes, including almost 18 000 tonnes of expandable polyolefin beads. Polyolefin foams are expected to continue growing at a faster rate than other plastic foams. A major part of the polyolefin foams is based on polyethlyene (PE) and its copolymers. Nearly 50% of the PE-based foam is partly cross-linked material.

The reason for the use of partly cross-linked PE resins is the broader foaming window in the melt on the one hand, and the improvement of rigidity and strength of the foamed material on the other hand. Figure 1 demonstrates the influence of crosslinking of PE on the optimum melt viscosity for the foaming process [1].

The optimum crosslinking is found to be at 30–40% gel content and the expansion ratio decreases with further increasing of the gel content as shown in Figure 2 [1]. The gel content is practically realized by the decomposition of peroxides in the melt in continuous melt mixing machines or in the solid state by irradiation or electron beam processing.

High viscosity of the polyolefin melt is necessary, especially in foaming equipment to produce sheets, tubes and shapes in batch and in continuous processes. On the contrary, polyolefin bead production is not so sensitive to melt viscosity.

Polypropylene: An A–Z Reference
Edited by J. Karger-Kocsis
Published in 1999 by Kluwer Publishers, Dordrecht. ISBN 0 412 80200 7

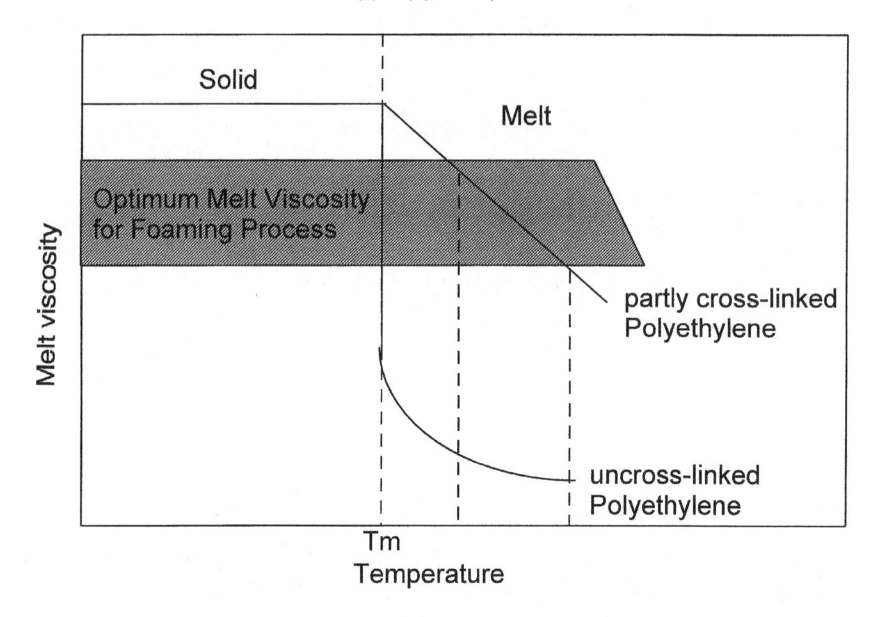

Figure 1 Effect of cross-linking on foaming window of polyolefin resin. (Courtesy of Chemical Economic Research Inst.)

The foaming of polyolefins is realized by incorporating either heat-decomposable chemical blowing agents, such as azodicarbonamide, or volatile physical foaming agents, such as low-boiling hydrocarbons, new chlorine–fluorine compounds (CFC) without the well-known ozone depletion potential, or gases such as N_2 or CO_2 prior to thermoprocessing. Endothermic chemical blowing agents, consisting of an organic acid, such as citric acid plus sodium carbonate, are most frequently utilized. A short review of the conventional chemical blowing agents has been published by Hurnjk and Facklam [2].

It is easy to understand that the solubility of volatile, or the dispersity of the nonsoluble, chemical blowing agents influences the homogeneity of the foamed material.

The cell structure of the polyolefin foam can be further influenced by adding cell nucleation agents into the polymer melt. Highly dispersed fillers and primary gas generators which build small voids (cavities) in the interface, or in the homogeneous melt, act as nucleation agents for further cell formation.

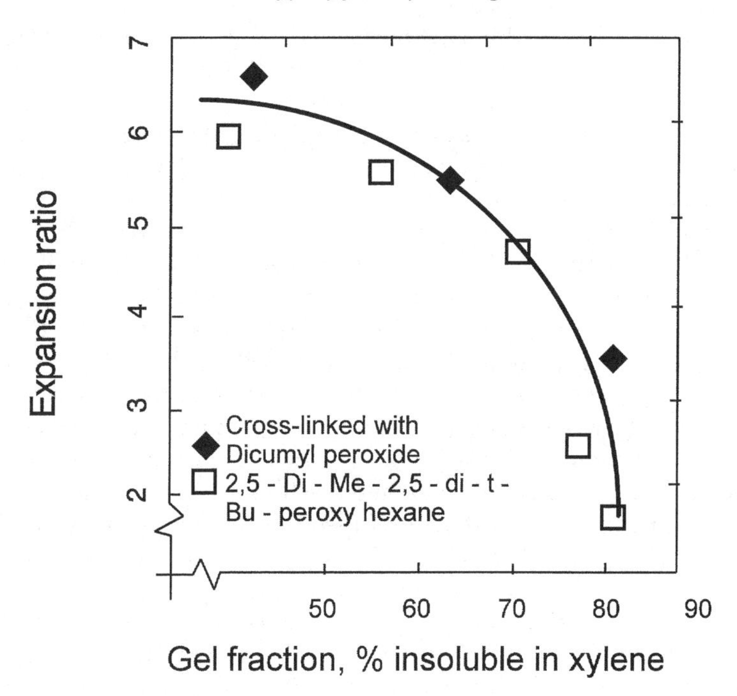

Figure 2 Expansion characteristics of peroxide-cross-linked polyethylene. (Courtesy of Society of Plastics Engineers, Inc.)

POLYPROPYLENE FOAMING

Structure and rheology of polypropylene

Polypropylene (PP) homo- and copolymers being linear chain molecules possess a narrow processing window in the melt.

As with the polyethylenes, the PP melts need a broader foaming window. Because of the molecular structure, crosslinking of PP by irradiation processes or with peroxides in the melt is not possible; with cross-linking predominant degradation occurs simultaneously, especially at high temperatures.

In the past, this was one of the reasons why PP foam formed only a small part of the overall polyolefin foam production. Consequently, the PP foam beads (particle foam) with the lowest demand on the viscosity of the melt had the highest share of PP foams.

However, different methods to cross-link polypropylene have been developed in the past 30 years [3]. Sulfazides were one of the first crosslinking agents for PP. The first technical process was the silane-assisted crosslinking of PP. That means radical grafting of the silane (for instance, vinyltrimethoxysilane) onto the PP-backbone initiated by irradiation or by peroxides. The crosslinking reaction occurs by hydrolyzation of the methoxy groups with water in a first step and the catalytic coupling of the Si-OH groups with special catalysts (for instance dibutyl-Sn-dilaurate) in the second step.

Later on, irradiation and peroxide-initiated crosslinking of PP could be reached by adding multifunctional monomers, such as divinylbenzene, bi-, tri- and tetra-acrylates (as well as trimethylolpropane-triacrylates) or di- and triallylcompounds (as well as triallylcyanurate), to PP. The grafting reaction of these monomers onto the PP backbone requires in the first step the splitting of hydrogen atoms from the tertiary carbon atoms by irradiation or free radicals given by the peroxide decomposition at moderate temperatures. This is followed by the addition of one reactive group of the multifunctional monomer used as cross-linking agent. High reaction temperatures must be excluded because of resin degradation resulting in embrittlement of the PP. In analogy of the polyethylenes, high crosslinking is disadvantageous for the PP foaming process.

The best foamability of PP was found with long-chain branched polypropylenes with only low gel contents in the melt. The first technical process to produce long-chain branched PP was developed by Montell [4].

The Montell process is based on electron-beam irradiation of PP powder taken directly from the synthesis process (without oxygen contact) at low temperatures (60–120°C). In this temperature range, the β-scission of the tertiary C-radicals is very slow so that, due to recombination reactions between backbone radicals and chain-end radicals, the formation of long-chain branched PP is preferred.

The significantly improved foamability of long-chain branched PP can be demonstrated by rheological measurements. For example, frequency sweeps of long-chain branched PP by plate-plate rheometers show even at shear rates as low as 10^{-3} s^{-1} an increase in viscosity (e.g. zero shear viscosity has not been reached yet), hinting at high molecular weight contributions due to branching. A much better evaluation of changes in melt strength and drawability due to long-chain branching can be done with the so-called Rheotens apparatus. In the experimental setup, at constant temperature, a melt strand of material is taken off from the nozzle of a die with increasing velocity until break. At the same time, strength and draw-down velocity (as measure of drawability) are determined by the take-off device.

Compared to (linear) standard PP, long-chain branched PP shows, in analogy to the long-chain branched low density PE (LDPE), a significant increase in melt strength and drawability (Figure 3).

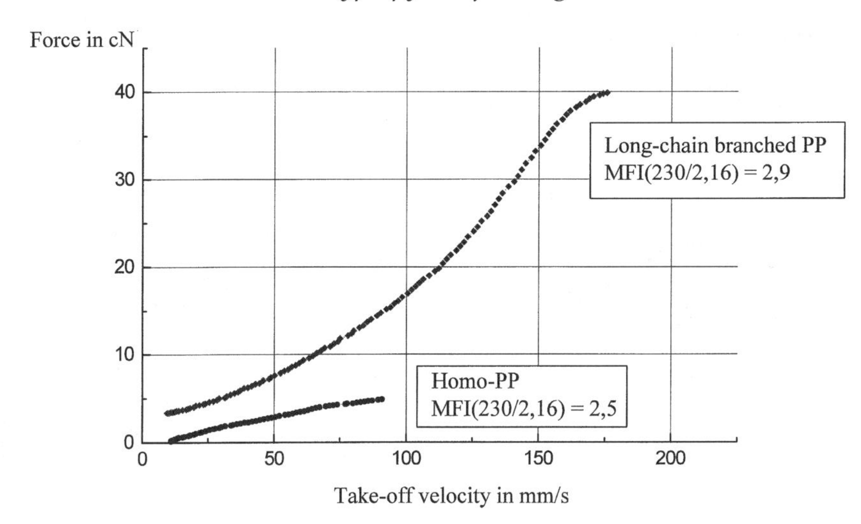

Figure 3 Strength and drawability of PP melts at 200°C (Rheotens apparatus fed by Rheograph 2001 (L/D = 30/2)).

Foaming technology

The foaming technology of PP generally corresponds to the processes known for PE. The choice of chemical or physical blowing agents depends on the foam density to be reached (e.g. the foam application) and influences the necessary foaming equipment and the costs of the foamed materials. The costs of chemical blowing agents are higher than those of physical agents, but they need no special processing technology. By extrusion with chemical blowing agents, PP is foamed into films and tapes in the density range of 400–700 kg/m^3. To foam PP into lower densities, low boiling volatile blowing agents have to be applied. Hereby, either special technologies or, more recently, specially adapted extruders and dies are required. For sheets and beads with high production rates (> 50 tonnes/month and unit) and low foam densities (50–200 kg/m^3), the continuous extrusion process with volatile blowing agents (direct gassing) yields the lowest processing costs. In addition, beads can also be made batchwise in a quite different, discontinuous process.

All current extrusion processes involve the following steps: melting, mixing with blowing agents, cooling of melt, expansion and degassing/aging. Figure 4 illustrates a flow diagram of a foam extrusion process. The steps in this process can be realized in different configurations of equipment, e.g. with long single-screw extruders, twin-screw extruders, or tandem extruder lines.

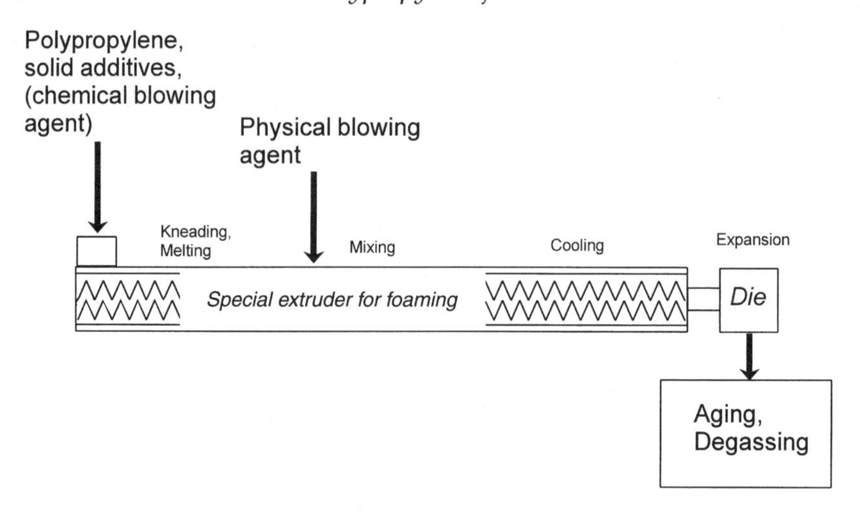

Figure 4 Typical process scheme for foam extrusion.

PP FOAM APPLICATIONS

The consumption of PP foam has increased steadily in the 1990s. The reason is the higher mechanical and thermal properties of PP in comparison to polystyrene (PS) and (PE), as well as the environmental acceptance in general (preferring 'all-PP solutions'). One of the main increasing markets is the food packaging sector with the demand for weight reduction, service temperatures up to 130°C, and high-impact strength at low temperatures (−40°C). Table 1 lists the different applications for PP foams.

As mentioned above, all PP materials have higher mechanical properties and especially abrasion resistance in comparison to PS and polyurethane (PU) foams, combined with higher modulus and heat distortion temperature (HDT) compared to polyethylene foams [5].

Elastomeric PPs are a new class of materials with a higher thermal stability than polyethylenes and a good elasticity without crosslinking. Elastomeric PPs have a low abrasion resistance and a high resilience at small elongations. Futhermore, new foamed materials based on thermoplastic olefinic elastomers (TPO) are important for the future. TPO consists of 50% of polypropylene and completes the range of PP foams for elastomeric foam applications.

Table 1 Applications and application potentials for polypropylene foams

Application	*Key properties*	*Examples*
Automotive uses	Light weight Energy absorption Resilience Higher service temperature	Interior: instrument panels, interior door panels, headliners, sun visors, trunk liners, gaskets, wheel house covers. Exterior: bumper inserts
Construction	Low thermal conductivity Energy absorption Compressibility High abrasion resistance Higher thermostability Sound absorption	Thermal pipe wraps, thermal insulations, eave and expansion joint fillers, sealant backer, floor underlay, roofing underlay
Food packaging	Light weight Higher service temperature Impact strength Thermoformability Low thermal conductivity	Food packaging trays for broad temperature range (from deep frozen to microwave heating). Cups, plates, etc.
Cushion packaging	Energy absorption Tensile and impact strength Compression-deflection behavior Resilience	Multiway packages and transport systems, pallets for internal transport, end-cap packages, packaging corners, wraps
Sport and leisure	Light weight Buoyancy Low water pick-up Energy absorption	Bodyboards, surfboards, fitness mats, turf underlay, helmets, life jackets, athletic protective paddings
Industrial uses	Compressibility Chemical and oil resistance Higher thermostability High abrasion resistance	Gaskets, vibration pads, buoys, oil booms, flotation collars
Miscellaneous uses	Cushioning behavior Thermal insulation	Prosthetic and orthopedic padding, shoe soles

REFERENCES

1. Park, C.P. (1991) Polyolefin Foam, in *Handbook of Polymeric Foams and Foam Technology*, Chap. 9, (eds D. Klempner and K.C. Frisch), Hanser, Munich, pp. 187–242.

2. Hurnjk, H. and Facklam, T. (1996) Chemische Treibmittel, *Kunststoffe*, **7**, 997–1001.
3. Nojiri, A., Sawazaki, T., Konishi, T., Kudo, S. and Onobori, S. (1982) *Furukawa Review*, **2**, 34–42
4. Schere, B.J. (1986) European Patent, EP 0190889B1.
5. Benning, C.J. (1969) *Chemistry and Physics of Foam Formation, Plastic Foams*, Vol. 1, Wiley Interscience, New York, pp. 191.

Keywords: foam, foam applications, foaming, structure and rheology, foaming technology, blowing agents, long-chain branching, cross-linking, electron beam irradiation.

Polypropylene in automotive applications

Just Jansz

AUTOMOTIVE MARKET TRENDS

Global light vehicle production, including cars and light commercial vehicles (LCVs), has grown at an average annual rate of 3% since 1994 to reach 51 million units in 1996. Further growth to 60 million units by 2001 is forecast, with the main contribution coming from emerging markets, although a negative impact of the Asia crisis may be anticipated. Consolidation is expected for the established markets in Western Europe, North America and Japan, reflecting the saturated state of these markets. Car production in Western Europe could still grow slightly from the current 13 million to 14 million in 2005, with LCVs remaining constant at 1.4 million annual units. Growth will come from the increasingly popular niche cars, e.g. multipurpose vehicles and small city cars, assuming no major changes in GDP development.

MATERIAL TRENDS IN THE AUTOMOTIVE INDUSTRY

The use of materials in automobiles has changed significantly over the past decades, with the total plastics content increasing from well below 10% in the early 1980s to levels of up to 15% today. The main driving forces causing this shift have been:

- weight reduction
- cost reduction
- design opportunities

Polypropylene: An A–Z Reference
Edited by J. Karger-Kocsis
Published in 1999 by Kluwer Publishers, Dordrecht. ISBN 0 412 80200 7

A new car built in Western Europe now contains an average of 100 kg of thermoplastic materials. In view of its unmatched versatility and cost–performance balance polypropylene (PP) is successfully substituting steel and other plastics such as ABS, PVC, polyamide and polyurethane. As a result, PP has advanced to become the single most important material, with some new car models already containing as much as 50 kg PP per vehicle. Global trends are similar, although design and manufacturing concepts may differ in detail.

POLYPROPYLENE CONSUMPTION

As an established and mature market, Western Europe can be used to illustrate the development of PP consumption in the automotive industry.

Since the introduction in the early 1970s, PP has penetrated an increasing number of applications. The steady growth in the use of PP, including a family of enhanced elastomer modified reactor thermoplastic polyolefins (TPOs) and compounded grades, has been driven by technological breakthroughs that have continuously extended the property range. This has provided the versatility needed to offer solutions for a number of challenges the auto industry is facing today.

For a wide range of interior and exterior functions, polyolefins offer favourable cost–performance versus other materials and allow innovative design concepts, e.g. for enhanced passenger safety and comfort.

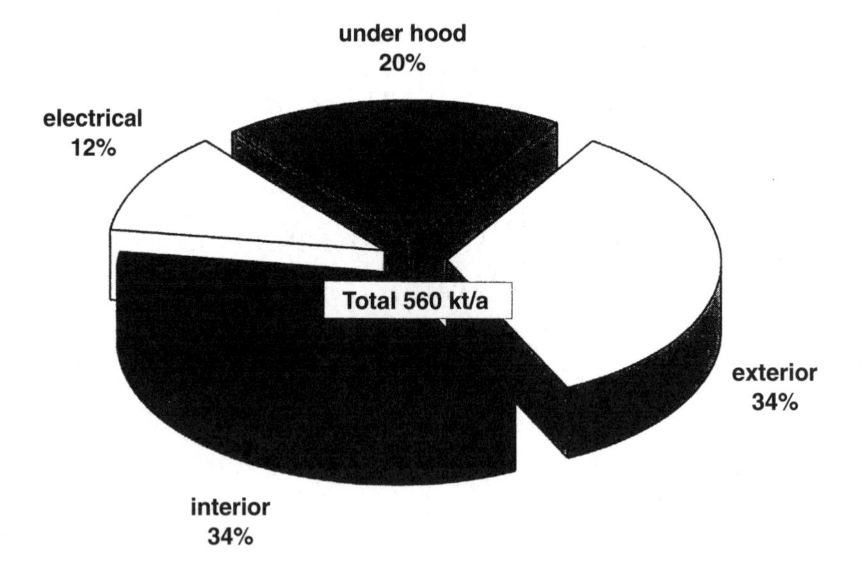

Figure 1 Automotive PP consumption in Western Europe, 1995, by application.

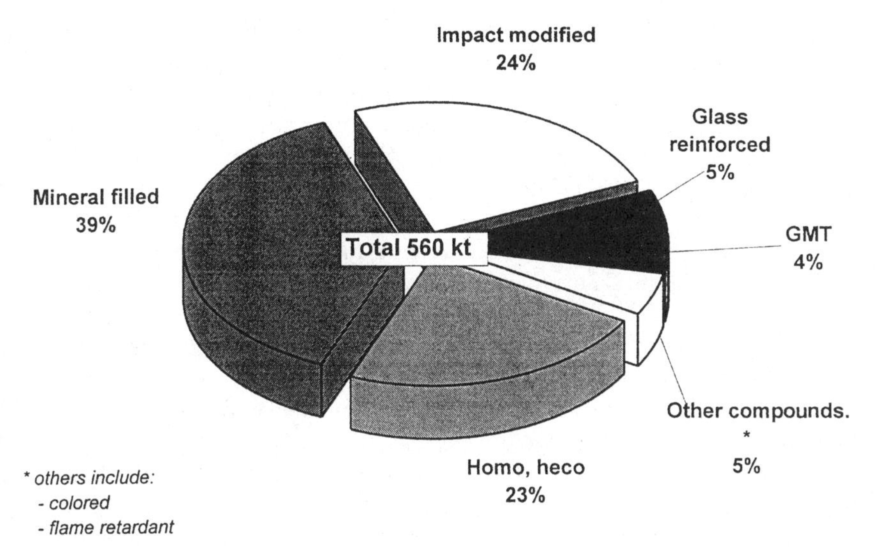

Figure 2 Automotive PP consumption in Western Europe, 1995, by material type.

The low specific density provides weight saving and helps to improve fuel economy, while also reducing cost. The intrinsically superior noise, vibration and harshness (NVH) properties contribute to improved comfort.

Figure 1 shows PP consumption by type of application, and Figure 2 shows consumption by PP material type in Western Europe. Total consumption in 1995 is estimated at 560 kt, of which close to 80% was compounded or modified. The transportation sector now accounts for approximately 12% of total PP consumption in Western Europe.

BUMPER SYSTEMS

Today, PP-based materials account for over 90% of bumper fascias in Europe, with the relative share still growing. Bumper designs have undergone rapid evolution in recent years, however. Depending on the concept, the application demands a very flexible or more rigid material. An excellent impact-rigidity balance is required to ensure the good low-temperature impact properties required to meet crash and safety criteria. In all cases, bumper systems are now an important design feature of the car as a whole, often closely integrated with the bodywork, more and more painted in body colors and incorporating additional functionality, such as headlights and grills.

Typical products are elastomer modified heterophasic copolymers (heco), as compounded or reactor TPOs. Talc, typically below 20%, is added for higher stiffness and lower coefficient of thermal expansion (LCTE). The latter is an important condition for zero-gap concepts, in which the bumper follows very closely the line of the body fascia. Now that bumpers painted in body colors have become a standard feature, paint adhesion and enhanced surface finish have become important criteria. These can be influenced by optimizing the polymer structure and rubber content. The recycling aspects are treated in the chapter on 'Bumper recycling technology' in this book.

Looking at recent developments, weight and cost reduction is being targeted with thin-wall concepts. This requires higher flexural modulus and significantly higher melt flow rate, which provides the additional benefit of lower molding cycle time.

Further improvement of cost–performance is being achieved by filling the bumper shell with an expanded PP foam (EPP) core, which gives a significantly higher energy absorption upon impact than conventional systems.

OTHER EXTERIOR PARTS

Side claddings and rocker panels are in some respect even more demanding than bumpers. Typically, a higher level of mineral fillers and elastomeric modification is used to ensure the required low thermal expansion as well as high impact, even at low temperature (see also the chapter 'Particulate filled polypropylene composites' in this book). Resistance to ultraviolet, heat and automotive fluids guarantee a long service life, whereas a high melt flow helps to mold the large, thin-walled components.

Another example of a demanding exterior application now made in PP is the cowl vent grill, which separates the windscreen from the upper edge of the bonnet. A one-shot dual-injection-molding process can be used to create a rigid structural section using a mineral filled copolymer in combination with a soft reactor TPO (Catalloy) seal.

INSTRUMENT PANEL SYSTEMS

PP materials have played a key role in the development of the instrument panel from a simple fascia to a complex multifunctional component. This application is very demanding in terms of safety standards (head impact) and aesthetics. Two different systems are used:

- *Rigid panels*. These are the most economic solution, using high-impact elastomer-modified and mineral filled compounds, which offer dimensional stability, scratch resistance as well as the ease of processing

needed to mold these large, complex structures. Moulded-in colors are common, but soft painting is also used to improve aesthetics.
- *Soft panels*. The standard system is based on a rigid carrier, made of high-impact elastomer modified PP with 20–30% mineral filler or glass fiber reinforcement, in combination with PU foam and PVC/ABS skin. A monomaterial solution using soft TPO-based foils and EPP foam could be envisaged in the near future.

INTERIOR TRIM

Substituting ABS on a large scale, PP has become the material of choice for a wide range of interior applications, meeting functional and aesthetic requirements as well as being low cost. Typical examples are pillar trim, seat backs, door cases and door panels. Interior trim components can be produced at various levels of complexity:

- *Simple trim components*, using mineral filled or unfilled copolymers, offer the lowest cost solution. Key properties are surface appearance, scratch resistance, impact strength, acoustic damping and low density.
- *Multilayer systems*, mostly consisting of a support made of filled copolymer with a good rigidity-impact balance and an outer layer of carpet or TPO foil. Special high melt flow grades have been developed to allow production in one shot by low-pressure injection molding. If safety regulations require a higher level of energy absorption, EPP foam can be used.

UNDER HOOD

Typical applications are air ducting, filter housings, heater and air conditioning units, headlamp housings, fan shrouds, battery trays and engine covers. High temperature and potential exposure to a range of automotive fluids are characteristic for the engine compartment, whereas the applications also call for a high level of rigidity and dimensional stability over a wide service temperature range. Depending on the mechanical requirements, mineral filled or glass fiber reinforced PP with special heat stabilization offer a very cost effective solution. Homopolymer is used as base if high rigidity is required, copolymer is used when impact properties are more critical. For demanding applications, pultruded long glass fiber pellets for injection molding could offer an interesting alternative, but need special tooling to retain fiber length. See the chapter 'Impregnation techniques for fiber bundles or tows' in this book. For less demanding applications, unfilled high crystallinity homopolymer allows weight and cost reduction.

Blow molded header tanks are now also commonly made in PP, using high-impact copolymer with a low melt flow and a high level of heat stabilization.

ELECTRICAL SYSTEMS

Battery cases are an important and established market for PP, in view of its excellent chemical resistance. Heterophasic copolymer with good stiffness–impact balance, high melt flow for short cycle time and good stress whitening properties is used. Flame-retardant compounds, based on soft and flexible reactor TPOs (Catalloy), are starting to replace PVC and cross-linked PE in automotive wire and cable systems.

SEMI-STRUCTURAL PARTS

Driven by weight saving and modular design concepts, long-fiber PP composites are being increasingly used to replace steel parts. In particular, PP-GMT (see the chapter entitled Glass mat reinforced thermoplastic, GMT) sheet composite, has found its way into front end modules (radiator end caps), seating structures, engine covers, bumper beams and most recently also a hatchback door.

The compression flow molding process, using preheated composite sheets, has proven to be very efficient and cost-effective in producing complex parts at the high output rates required by the automotive industry. The long glass fibers (GF), typically 50 mm, are critical in providing the unusually good dynamic toughness over a very broad temperature range.

Alternative processes have been developed that are also based on compression flow molding, but use plasticizing units for feed preparation instead of having ovens to preheat GMT sheets. This is done by mixing long (10–25 mm) glass fibers, PP and additives directly. Market share of these 'direct processes' is still low and properties do not seem to fully match those of GMT, but requirements for some applications could probably be met at lower overall costs.

The use of pultruded long fiber pellets for compression molding has been suggested but does not seem to offer any benefit.

OUTLOOK

Table 1 lists some typical property requirements for the main automotive applications. The penetration of PP in automotive applications has already reached a high level, but further growth is anticipated. Technical innovation, e.g. based on Montell's Catalloy technology for advanced reactor TPOs, is expected to further extend the property range and will open up new opportunities, while reducing the relative share of compounded products. A further shift from compounds to direct reactor polymer will come from the increasing use of color concentrate, to be added at the injection molding machine.

Weight reduction is envisaged to come from the successful introduction of a family of high-crystallinity materials and high-stiffness heterophasic copolymers for interior trim and under-hood applications. These products are expected to partly replace talc filled compounds.

Growth will also come from expandable PP foam with excellent energy absorption properties (see the chapter, 'Modelling of the compression behavior of polypropylene foams' in this book) which will allow new, innovative designs in bumper systems and monomaterial interior trim solutions, in combination with improved TPO foils for door trim and instrument panels.

Exciting opportunities are envisaged for PP in demanding exterior applications such as body panels and fenders. Key success factors will be further improvement in paint adhesion and the acceptance of off-line painting or lower in-line paint baking temperatures, that allow the use of PP-based materials.

The replacement of PVC and crosslinked PE in cable systems is anticipated to continue.

REFERENCES

1. Murphy, J. (1996) *Plastics and Elastomers in Automobiles,* Technical Industry Services, London.
2. *Global car and truck forecast, First quarter 1997,* LMC / J.D. Power LMC Automotive Services, London.
3. *Polypropylene Compounds in Western Europe* (1996), Applied Market Information, London.

Keywords: automotive, automotive application, car components, automotive requirements, substitution of metals and polymers, interior parts, exterior parts, electrical applications, filled PP, elastomer-modified PP, thermoplastic polyolefins (TPO), GMT-PP, long-fiber reinforced PP, composites, market trends, glass fiber (GF) reinforcement.

Table 1 Typical property requirements for selected automotive applications

Application type	*Example*	*Key functional properties*	*PP material family*
Exterior	Bumper fascia, soft	Low T impact	Elastomer modified
	Bumper fascia, rigid	CLTE, impact, rigidity, paint adhesion and durability (painted), UV resistance (non-painted)	Elastomer modified, mineral filled
	Spoiler	Low T impact	Elastomer modified
	Bumper beam	Energy absorption, rigidity	GMT
	Side cladding, strips	CLTE, impact, UV	Elastomer modified, mineral filled
	Rocker panels	CLTE, impact, paintability	Elastomer modified, mineral filled
	Wheel arch liners	Impact	Elastomer modified, mineral filled
Interior	Upper interior trim	Impact, scratch resistance, rigidity	Copolymer or elastomer modified, mineral filled
	Lower interior trim	Rigidity, impact	Copolymer; mineral filled copolymer
	Door panels, cases	Impact, rigidity	Copolymer or elastomer modified, mineral filled
	Seating structures	Rigidity, impact	PP-GF; GMT
	Seating trim	Rigidity, scratch resistance	Copolymer, mineral filled
	TPO foil	Softness, grain retention	Elastomer modified

	Rigid instrument panel	Impact, scratch resistance	Elastomer modified, mineral filled
	Centre console	Impact, scratch resistance, high T dimensional stability	Elastomer modified, mineral filled
	Instrument panel retainer	Impact, dimensional stability	PP-GF; mineral filled
Under hood	Filter housings	Rigidity, high T dimensional stability	HCPP; mineral filled homopolymer
	Fan shroud	Rigidity, dynamic toughness	PP-GF; GMT
	Fluid header tanks	Impact, temp. resistance	Low MFR copolymer
	Battery tray	Rigidity, chemical resistance, dynamic toughness	PP-GF; GMT
	Air ducts, HVAC ducting	Rigidity, high T dimensional stability	Mineral filled copolymer
	Engine covers	Impact, high-temperature rigidity	PP-GF; GMT
	Front end	Rigidity, dynamic toughness	GMT
Electrical	Battery cases	Chemical resistance, impact, low blush	Copolymer
	Cables	Flame retardancy, softness, temperature resistance	Elastomer modified, flame retardant

Polypropylene in cable applications

Conchita V. Tran

INTRODUCTION

Cables have two major functions: the transport of electrical energy through power cables, and the transfer of signals via telecommunication cables.

Some history

Polymers are used for the insulation and the jacketing of power and telecommunication cables. In the early 19th century, telegraph cables were the first to be produced and consisted of simple copper wires laid in hollow wooden blocks. With the advent of natural rubber (1850s), PVC (1920s) and polyethylene (PE) resin (1940s), the cable insulation and jacketing were made with these resins. Ziegler–Natta catalysts, hence polypropylene (PP), were developed only in the 1950s. The wire and cable market is a very conservative one and it is difficult to have a new resin accepted for an existing application. A minor change in a resin formulation requires approximately three years of testing before approval, and a new resin much longer. This cautiousness is explained by the fact that cables are guaranteed for 40 years and the burying of the cables is much more costly than the cables themselves. As PVC and PE were already well established in the cable insulation in the 1950s, PP has developed in special and new applications, e.g. where higher temperature resistance is required, such as for under-hood insulation.

Polypropylene: An A–Z Reference
Edited by J. Karger-Kocsis
Published in 1999 by Kluwer Publishers, Dordrecht. ISBN 0 412 80200 7

Power cables

Power cables are designed to transport high energies over long distances with a minimum of loss. Power cables are subdivided into two classes:

1. The function of the first class is to transport the energy from a power plant to a distribution centre. This is done at high voltage (20–175 kV) in order to reduce the resistivity losses. The buried high-voltage cables (HV) are insulated with peroxide-cross-linked PE (XLPE).
2. Medium voltage (MV) and low voltage (LV) cables have the function of distributing the electrical energy to the low voltage network and to the consumers respectively.

The higher the voltage, the more important the losses through heat, hence the more heat resistant the insulation has to be.

Telecommunication cables

Telecommunication cables establish a wired connection between users to allow data transmission and audiovisual communication. The main characteristics required from the polymer insulation are a low dielectric constant and low dissipation factor. As the best dielectric properties are given by air, the dielectric losses can be reduced drastically by expanding the insulation (30–50% vol.% inert gas); this is called cellular insulation, as opposed to solid insulation.

Optical cables, on the other hand, carry messages by light. The main function of the polymer is to give mechanical support to the brittle glass fibers.

Sheathing

The sheathing is usually made of the same resin for the power cables and the telecommunication cables. The main functions of the jacket are the mechanical and environmental protection of the cable; therefore, the cable jackets are commonly filled with carbon black, the best ultraviolet stabilizer known today.

USE OF PP IN CABLE APPLICATIONS

Over 60% of the polymer used in cable insulation and jacketing is plasticized PVC. This is flexible and has relatively good high-temperature resistance. It was also the first resin available commercially. PE comes second with approximately 35% of the world market. Cross-linked PE is the sole resin used for the insulation of HV cables and the most common also for the MV cables. PVC being polar, its insulation properties are not sufficient for HV cables. PP has a market share of 3–4%

in the USA, mainly in telecommunication cable insulation. Apart from the telecom insulation, PP is used only in 'niche' markets. It can be incorporated as additive or used in the form of elastomers, such as ethylene–propylene rubber (EPR) or ethylene–propylene–diene monomer (EPDM). Typically, EPR and EPDM are based on 30–50 mol.% propylene. Since the early 1980s, thermoplastic olefins (TPO) are being successful in applications where EPR and EPDM are used. TPOs consist of blends of EPDM with PP to combine the flexibility of elastomers with the processability of thermoplastics. Table 1 summarizes the main characteristics of the basic polymers commonly used in cable applications. The main advantages of PP over the other resins are its good resistance to high temperatures, its low thermal expansion and its excellent dielectric properties. The main limitations of PP are its premature embrittlement at low temperatures and its rigidity.

The polymers from Table 1 are often modified for their use in cables:

- PP is often used in the form of copolymer for better flexibility, which reduces its advantages at high temperatures over the other resins.
- TPO properties depend on the blend ratio of PP/EPDM. Higher EPDM amounts have a negative effect on dielectric properties.
- PVC properties depend on the ratio of PVC powder/plasticizer. Increasing plasticizer levels improve the flexibility at the detriment of the already weak dielectric properties. Plasticizer also reduces the flame retardancy of PVC.

PP in power cables

HV cable insulation is commonly made of peroxide-cross-linked PE (XLPE). The cross-linking enhances the resistance to high temperatures of the PE, increasing its maximum service temperature from 70°C to 90°C. Hence, PE is more attractive than PP which cannot be cross-linked with peroxide. The weakness of the XLPE is that it undergoes water-treeing (tree-like cracks initiated by the combination of a high electric field and the presence of moisture and metal ions). The addition of a small percentage of EPR, based on 35 mol.% propylene, is successfully used as water-tree retardant.

MV cables are, in some cases, insulated with EPR or EPDM. Compared to XLPE, EPR has lower mechanical properties, equivalent thermal resistance but much higher flexibility. EPR is also preferred to XLPE in radioactive environments.

EPR is making a breakthrough in the superconducting underground 500 kV power cables cooled to the temperatures of liquid helium (about −270°C). Despite its higher dielectric loss, EPR is preferred to PE because it retains a certain flexibility at these low temperatures, while PE insulation cracks.

Table 1 Properties of typical polymers for cable insulation and jacketing [1]

Property	*Polypropylene family*			*PVC*	*Polyethylene family*	
	PP homopolymer	*EPR/ EPDM*	*TPO*	*Rigid PVC*	*HDPE = High density PE*	*LDPE = Low density PE*
T_g (°C)	−7	−50	−50	80	−110	−90
Max. service temp. (°C)	100	80	80–120	65–85	70–80	60–75
Min. service temp. (°C)	−30	−55	−60	−5	−50	−50
T_m (°C)	176		150–165	185	135	115
Linear thermal expansion (°K^{-1})	0.5×10^{-4}			0.8×10^{-4}	2×10^{-4}	2.5×10^{-4}
Abrasion resistance	Good	Excellent	Excellent	Very good	Poor	Poor
Dielectric constant at 1 MHz	2.25	3.0	2.35	3.0–5.5	2.34	2.28
Dissipation factor at 1 MHz	5×10^{-5}	50×10^{-5}	20×10^{-5}	0.1	20×10^{-5}	8×10^{-5}
L.O.I. (%)[a]	18	18		47	17	17

[a] L.O.I. = amount of oxygen required for self-ignition of polymer, high LOI = good flame retardancy.

. TPOs combine the advantages of elastomers and the extrudability of thermoplastics. TPOs are used in special cases where requirements, such as heat aging resistance, ozone resistance, low-temperature flexibility, flame retardancy and generally good chemical resistance, are to be fulfilled. TPOs allow the combination of relatively good mechanical properties with high flame retardancy and low smoke emission; therefore, TPOs are preferred in critical environments, such as oil platforms, tunnels and underground. Their use is limited to LV cables.

PP in telecommunication cables

PP copolymers are used for the insulation of telephone singles for twisted pairs, in the form of solid or cellular insulation. Here again, PE is the usual resin for this application, except in the USA where PP is preferred for its higher mechanical properties and better thermal and flammability resistance. PP copolymers are preferred to homopolymers for their flexibility.

PP homopolymers are incorporated in some solid telecommunication PE insulation grades to improve their processability without affecting the dielectric properties of the insulation. The telecommunication insulation lines run at long distance and because of the dielectric properties requirements, the use of processing aids is limited.

PP is applied in expanded form to the coaxial cable insulation. PPs advantage over PE is its low dielectric losses.

In optical fiber cables [2], the PP is used to produce the slotted cores which hold the coated glass fibers together. The slotted core is the protective sleeve extruded directly onto the glass fibers. Radiation-crosslinked PP homopolymers are another possibility for this application. The PP plays the role of providing mechanical strength to the optical fiber cables.

PP in other cables

The main advantages of PP over PE in the automotive industry are its higher temperature resistance and better mechanical properties of non-halogen flame resistant formulations. Therefore, the use of PP in the 'under-hood' applications is common. The automotive industry wants to move away from PVC. Typically, the automobile ignition cables are made of EPDM or TPO.

PP in jacketing

Abrasion, flame and environmental resistance are the main requirements which can be fulfilled by TPOs or elastomers, and EPR or EPDM [3]. The former are easier to process, the latter are more flexible.

CONCLUSIONS

PP will continue to develop in niche markets where its advantages over PE and PVC are of importance, as well as in new applications where it is not confronted with existing specifications. TPOs are still in their growing period and their advantages over elastomers are not yet fully exploited. New improved PP are promised by different suppliers, based on catalyst modifications (metallocenes, bimodal polymers, etc.). The flexible PP copolymers are a possible candidate, especially in the telecommunication cables, provided they can be processed on standard PE and PVC equipment.

REFERENCES

1. Daniels, C.A. (1989) *Polymers: Structure and Properties*, Technomic, Basel.
2. Haugen, J. (1995) PP in fiber optic cables, Proceedings of the Conference on Plastics in Telecommunications, PIT VII, London, 4–6 Sept. 1995, p. 53–62.
3. Schwamborn, K. (1996) Draht- und Kabelummantelungen, *Kunststoffe*, **86**(7), 1034–1038.

Keywords: cable, sheathing, jacketing, insulation, telecommunication cable, power cable, electrical insulation, thermoplastic olefins, TPO, EPR, EPDM, dielectric properties, flammability, water-treeing.

Polypropylene/continuous glass fiber composite pipes: design principles

A. Cervenka

INTRODUCTION

If a thermoplastic composite pipe is pressurized under laboratory conditions (with the pipe ends freely moving) the stress/strain in the two principal, axial (ε_x) and hoop (ε_y), directions exhibit a behavior schematically illustrated in Figure 1. Three aspects of this response, namely (1) the initial stiffness of the pipe associated with the slopes S_x and S_y, (2) nonlinearity in the σ/ε response and (3) the ultimate failure σ_{ult} of the pipe corresponding to the burst pressure, are rationalized by means of micro- and macromechanical modelling. These types of modelling of the short term structural response are the initial steps in the overall design procedure.

BUILDING BLOCKS AND CONSTITUENTS

A thermoplastic composite pipe produced by the tape winding process consists of two building blocks: the mandrel (wall thickness t_m, internal radius r_i) and the load-bearing composite tape (thickness t_c) wound under a winding angle α with respect to the axial direction of the pipe and restraining the mandrel (Figure 2). The extruded mandrel is based on one constituent only, i.e. a viscoelastic polymer of the stress/strain response described by $\sigma = f(A, B, C, \varepsilon)$. The three parameters are relatable to the initial stiffness E_M and the coordinates of the yield point. The

Polypropylene: An A–Z Reference
Edited by J. Karger-Kocsis
Published in 1999 by Kluwer Publishers, Dordrecht. ISBN 0 412 80200 7

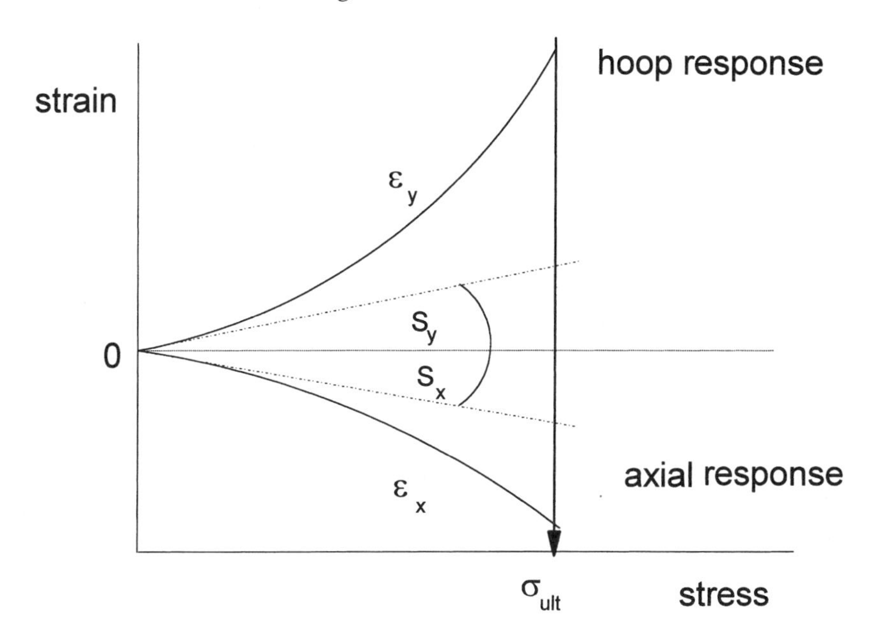

Figure 1 Schematic representation of the stress/strain response of a thermoplastic composite pipe.

restraining tape is a two-component material. In addition to the polymeric matrix identical in its nature to that in the mandrel, the tape also contains the volume fraction v_F of the continuous, unidirectional E-glass reinforcing fibers. Their alignment is achieved by means of the pultrusion process. Both constituents of the tape (lamina) are isotropic solids characterized by two elastic constants: the Young's moduli E_i and the Poisson ratios ν_i with the subscript i referring either to the matrix (M) or the fiber (F). Their strength characteristics are given by the ultimate tensile strength of the fiber σ_F and the yield stress of the matrix σ_M. Typically, $E_M = A = 1.7 \times 10^3$ MPa, $E_F = 7.2 \times 10^4$ MPa, $\nu_M = 0.35$, $\nu_F = 0.22$, $B = 27.2$, $C = 100$, $\sigma_M = 36$ MPa, $\sigma_F = 3.45 \times 10^3$ MPa.

In contrast to the isotropic constituents, the transverse isotropy of the lamina (pultruded tape) requires more stiffness and strength characteristics to be used. These are: the longitudinal and transverse moduli E_1 and E_2, the shear modulus and Poisson ratio in the lamina plane G_{12} and ν_{12}; the longitudinal and transverse strength in tension X and Y and the shear strength S. The stiffness characteristics can be either measured on unidirectional laminates or calculated micromechanically using a computer code [1]. The exact method of determining the strength characteristics is to measure them having manufactured testing sheets under the same conditions as used for the pipe (filament winding of unidirectional

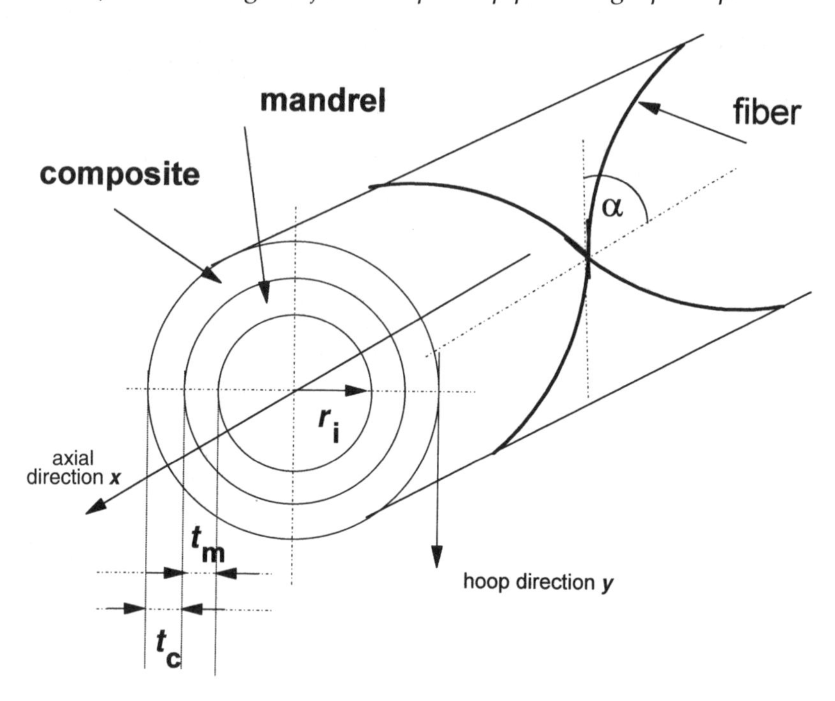

Figure 2 Definition of the characteristics defining the pipe construction.

race-track specimens). Alternatively, their upper bound can be roughly assessed by means of empirical rules digested from the behavior [2, 3] of polymeric composites: $X = v_F\sigma_F$, $Y = \sigma_M$, and $S = 0.8Y$. For the PP/glass lamina of the Plytron® type containing 28 vol.% fibers, we can thus assign: $E_1 = 21.73\,\text{GPa}$, $E_2 = 3.35\,\text{GPa}$, $G_{12} = 1.32\,\text{GPa}$, $\nu_{12} = 0.3$, $X = 966\,\text{MPa}$, $Y = 36$, and $S = 30\,\text{MPa}$.

ANALYSIS OF THE COMPOSITE PIPE

With both building blocks (the mandrel and the restraining composite tape-lamina) that form the monolithic structure of the pipe now described in the material terms, the pipe behavior can be analysed taking into account the manner of its construction. The construction is specified by the internal radius r_i, thicknesses t_m, t_c and the winding angle α (Figure 2). The analysis is easily accomplished by using one of several computer codes [4] based on the classical laminate theory. Constructing a symmetrical laminate, the half-stack consists of a single inner ply of thickness t_m, consisting of the isotropic PP matrix and two outer plies arranged in the $+\alpha/-\alpha$ angle-plied construction with the PP/glass laminae, each lamina $t_c/2$ thick, offset by $+\alpha$ and $-\alpha$ to the axial direction of the pipe. The construction details in combination with the material properties allow the analysis to be carried out. The output is all the stiffness characteristics of the pipe: the longitudinal (E_x), transverse (E_y) and shear (G_{xy}) moduli together with the Poisson ratio ν_{xy}. Defining either a loading spectrum or constraints, the principal stresses that develop in the pipe are also derived, namely the axial (σ_x), hoop (σ_y) and the interlaminar shear τ_{xy} stress. A failure criterion (Tsai–Hill [4]) is then used to decide whether the composite pipe remains intact or has been breached. The stress level relevant to the type of restrain is then readily related to the ultimate pressure sustainable by the structure. To illustrate, with the pipe pressurized under the conditions relevant to Figure 1, the burst-pressure p_{ult} is derived as the ratio $\sigma_y/\{(r_{\mathrm{i}} + t_{\mathrm{m}} + t_{\mathrm{c}})/(t_{\mathrm{m}} + t_{\mathrm{c}})\}$.

INITIAL PIPE STIFFNESS

Using the polypropylene (PP) glass composite pipe of internal radius 55 mm and based on 3.6 mm thick mandrel overwound by 0.4 mm thick laminae to form +57/−57 angle-plied construction as an example, the stiffnesses associated with this structure are: $E_x = 2.17$ GPa, $E_y = 4.34$ GPa, $G_{xy} = 2.26$ GPa and $\nu_{xy} = 0.37$. Furthermore, it can be shown that $S_y = 1/E_y$ and $S_x = -\nu_{xy}/E_x$.

NON-LINEAR STRESS/STRAIN BEHAVIOR

The stress/strain dependence shown in Figure 1 can be quantified considering (1) the strain softening of the matrix and (2) stress partitioning between the matrix and the fiber. The former is associated with the viscoelastic nature of the PP matrix and causes E_{M} to be a decreasing function of the strain $E_{\mathrm{M}}(\varepsilon) = \mathrm{d}f/\mathrm{d}\varepsilon$. Whilst the 'strain softening' is a simple function, the dependence of the matrix stiffness on the stress level (the independent variable in our case) is more complex. The problem is resolved by coupling E_{M} to the corresponding matrix stress σ_{M} by a double graphical interpolation working with $\sigma_{\mathrm{M}}(\varepsilon)$ and $\mathrm{d}\sigma_{\mathrm{M}}/\mathrm{d}\varepsilon$ plots. The procedure leads to pairing E_{M} to σ_{M}.

The need to partition the stress is associated with the fact that due to the presence of fibers, the stress exerted on the matrix, both in the mandrel and the composite tape, is lower than σ_{C} associated with the internal pressure in the composite pipe. The partition factor Ψ is proposed to be the simple ratio of the relevant stiffness, thus in our case $\Psi = \sigma_{\mathrm{C}}/\sigma_{\mathrm{M}} = E_y(E_{\mathrm{M}})/E_{\mathrm{M}}$ where $E_y\,(E_{\mathrm{M}})$ is the hoop stiffness of the pipe when the matrix modulus is E_{M}. This implies that the composite pipe during pressurization finds itself in different states $j = 1, 2 \ldots n$, each state governed by $E_{\mathrm{M}}(j)$ and associated with a specific set of the lamina characteristics, pipe stiffnesses and, consequently, stress partition factors

$\Psi(j)$. Decrementing E_M typically in steps of 100 MPa principal strains ε_x and ε_y are assigned to σ_C by means of a repetitive algorithm that generates the nonlinear behavior.

THE ULTIMATE STRESS OF THE PIPE

With the PP/glass composite pipe passing through different stages, the actual state at which the failure happens is not known. Thus, when decrementing the matrix modulus, we evaluate the pipe failure for all composite states $j = 1 \ldots n$ keeping the strength characteristics of the lamina and the mandrel constant. The repetitive algorithm yields a spectrum $\sigma_{ult}(j)$. When the counter j is small, lower failure stresses are predicted and the mechanism involves a simultaneous failure of both building blocks. As the matrix is forced to be more compliant, the ultimate failure increases and at a certain stage, i.e. when $j = 12$ or $E_M = 600$ MPa the model predicts 'a first-ply failure'. However, cascading twelve different composite states $1.7 > E_M > 0.6$ GPa, the ultimate failure stresses fall into a narrow band of 91–106 MPa. Thus the proposed analysis can be concluded to yield a narrow margin above the most conservative value of 91 MPa. This value can be translated as 89.3 bar internal pressure which is rather impressive considering that the overall thickness t_c of the restraining tape was only 0.8 mm!

REFERENCES

1. Whitney, J.M. and McCullough, R.L. (1990) Micromodels for composite materials – continuous fiber composites, in *Delaware Composites Design Encyclopaedia*, Vol. 2, (eds J.M. Whitney and R.L. McCullough), Technomic, Basel, p. 93.
2. Cogswell, F.N. (1992) *Thermoplastic Aromatic Polymer Composites*, Heinemann–Butterworth, Oxford.
3. Tsai, S.W. (1986) *Composite Design – 1986*, Think Composites, Dayton, OH.
4. *Laminate Analysis Program (LAP)* (1995) Imperial College of Science, Technology and Medicine & Anaglyph, London.

Keywords: composite pipe, design, modelling, tape winding, Plytron®, stress–strain behavior, failure criteria, laminate theory.

Processing of polypropylene blends

M.M. Dumoulin

INTRODUCTION

Only two classes of polypropylene (PP) blends have achieved commercial success: blends with polyolefins and with polyamides (PA). With PP/polyolefin blends, the goal is either to improve the impact resistance of the base resin (impact-modified PPs) or to produce thermoplastic elastomers (dynamically vulcanized blends). PA/PP blends aim at bridging the property gap between the two polymers. Therefore, significant information on processing is available only for these two families of blends.

BLENDS WITH POLYOLEFINS

Elastomer-modified PP [1]

Because of its glass transition temperature, around 0°C, PP shows brittle fracture in the subzero range. Therefore, most PP grades currently produced include an elastomeric phase in order to improve impact resistance, particularly at low temperature. The elastomer is usually of the polyolefin type. It can be an α-olefin random copolymer (e.g. Oleflex^{TM}, Showa Denko), ethylene–propylene–diene (EPDM) or ethylene–propylene (EPR) (Polytrope^{TM}, A. Schulman) copolymers. In recent years, the elastomer is being increasingly added at the reactor stage, by adding comonomers during polymerization. The elastomer-modified PPs (e.g. Hostalen^{TM}, Hoechst, Moplen^{TM} SP, Montell) show very high toughness,

Polypropylene: An A–Z Reference
Edited by J. Karger-Kocsis
Published in 1999 by Kluwer Publishers, Dordrecht. ISBN 0 412 80200 7

particularly at low temperature, excellent acoustic properties, good paint adhesion, wide viscosity range and good thermal oxidative stability. The material is injection molded into a wide variety of automotive exterior (bumpers, trim, spoilers, dashboards), interior (steering wheels) and engine compartment applications (battery casings, covers, ducts). It is also extruded or blow molded into pipes and ducts for domestic and automotive engine compartment uses. Nonautomotive applications include roofing membranes.

In general, recommended processing conditions are similar to those for the neat PP. For calendering, the melt temperature should be in the range 165–175°C. Fiber filled grades should be formed using a twin-screw extruder with screw length to diameter ratio (L/D) of 24–30, having the plasticizing zone equipped with kneading elements. However, sheet grades may be extruded using a single-screw extruder with $L/D = 24–30$, preferably with a two-stage screw.

Some PP/EPR systems show optimum impact strength for an elastomer particle size of approximately 0.5 μm. Such a degree of dispersion is difficult to achieve when the elastomer phase is compounded in. One method is to match viscosities of the two phases as closely as possible. The high viscosity of EPR requires the use of a high-viscosity PP, making the blend less suitable for injection molding, which is the principal forming method for these systems. Because the elastomer phase is neither grafted to the matrix nor crosslinked as is the case for high-impact polystyrene (HIPS) or acrylonitrile–butadiene–styrene copolymer (ABS), the blend morphology is complex and sensitive to flow conditions. Consequently, during injection molding, these blends are prone to show weak weld lines [2].

In PP/EPDM systems (e.g. Vestolen™ EM, Hüls), advantage has been taken of the combination of crystalline and amorphous polymer properties. The blends are formulated by selecting molecular weights of the components and concentration of the elastic segments in the EPDM. Some of these blends are prepared by dynamic vulcanization.

Dynamically vulcanized blends

Dynamically vulcanized blends ([3, 4]; see also the chapter on 'Thermoplastic dynamic vulcanizates' in this book) are produced by melt mixing an elastomer phase with a thermoplastic (the elastomer being the major component) and then curing the elastomer during melt mixing. The resulting blend comprises the thermoplastic resin filled with high concentration of rubber particles well bonded to the matrix. These blends are usually PP/EPDM (Santoprene™) or PP/polybutadiene (PB), but examples include PP/natural rubber, PP/styrene–butadiene rubber (SBR), PP/nitrile butadiene rubber (NBR) (Geolast™, Advanced Elastomer Systems), PVC/cross-linked ethylene copolymers (Alcryn™, Du Pont). The cured elastomeric phase provides high elastic character while the thermoplastic phase enables easy melt processing. Curing the elastomer phase provides improved melt strength and better morphology stability during processing while maintaining recyclability. When the degree of vulcanization is high, the EPDM particles may be broken into small elastomeric filler particles, physically bound to the matrix. Therefore, since the concentration of EPDM as well as the extent of its vulcanization can vary, a great variety of morphologies, and thus properties, can be obtained. A co-continuous morphology is observed for some PP/EPDM systems (e.g. Santoprene™, Advanced Elastomer Systems) and the melt is highly viscous.

Because of the ease of processing, dynamically vulcanized blends have displaced fully cured rubbers in many applications. Curing takes place during melt mixing, e.g. in a twin-screw extruder, and the blend pellets can then be used in typical thermoplastic forming processes. The blends can be extruded into window glazing, weather stripping or wire and cable jacketing. They can also be injection or blow molded into parts such as sleeves, air ducts, protective covers, in particular for automotive applications.

Despite their high viscosity, these PP/EPDM blends are easily extruded. The recommended extruder should have $L/D \geq 24$ and the screw a moderate compression ratio, e.g. 2.5–3.5, which is similar to those used for polyvinyl chloride (PVC) or PE extrusion. Temperature settings should be increased with the blend hardness. Owing to the two-phase nature of the blend, the material has relatively high extrudate swell which increases with shear stress, i.e. with resin hardness, at constant shear rate.

BLENDS WITH POLYAMIDES

Blends of PA-6 or PA-66 with PP (e.g. Orgalloy™, Elf Atochem, Ultramid™ KR, BASF) have flow properties similar to PP. This constitutes an improvement over the flow behavior of neat PA. The PA/PP blends are easier to extrude than PA because the melt viscosity is less sensitive to temperature variations. The processing conditions for PP/PA-66 blends are comparable to those normally used to process PA-66. However, it should be noted that drying of these blends is much slower than that for the neat PA. For single-screw extrusion, the barrel should have $L/D \geq 20$, fitted with a standard three-zone screw, with compression ratio of 2.5–3. Excessive shearing may cause destruction of morphology, and thus should be avoided. Gloss can be improved by raising tooling temperature. Purging with PP followed by low density polyethylene (LDPE) is recommended.

Owing to their low viscosity, PP/PA blends are well suited for processing in standard injection-molding machines. Three-zone PA-type screws are preferred, with $L/D = 15$ and compression ratio of 2–3. The clearance between screw and barrel should be low to ensure the best mold cycle repeatability. Due to high thermal stability, the blend can be processed over a wide range of temperatures, from 250 to 300°C, depending on the grade. It is generally advisable to use a high process temperature and a low injection pressure to achieve good mechanical properties, in particular, the impact strength.

Mold temperature strongly affects the product properties. Reduction of mold temperature increases orientation and internal stress, while increasing it augments the shrinkage, gloss, duration of dwell, cycle time and flow length. Usually, mold temperature should be 20–30°C for standard grades and 50–80°C for the glass fiber reinforced ones.

Dwell (or packing) pressure periods always demand close attention when processing alloys of semicrystalline polymers. An increase in degree of compaction causes increases in strength of weld lines, flashing, mass of item, risk of overfill and decreased shrinkage, sink marks and inclusions. It is interesting to note that shrinkage can be reduced by lowering the dwell pressure, mold temperature, material temperature and component thickness. In addition, it should be noted that in most moldings, PA/PP blends show anisotropic shrinkage, lower in machine direction (0.7–1.5%) than in transverse direction (1.2–2%).

Blends produced under the Eref™ (Solvay) trade name include both alloys rich in PP and alloys rich in PA where PA is PA-66 or PA-mXD6 (a semicrystalline PA, made by polycondensation of m-xylylenediamine and adipic acid). A compatibilizer is added in order to provide optimal adhesion between PP and PA. Injection-molding machines used for injecting PA are recommended, with peripheral screw speeds of 3–10 m/min. Cooling time can be estimated as $2.5 \times e$ s where e is the wall thickness in mm. Melt temperature for flame retarded grades must be well controlled and not exceed 270°C.

Bottles and containers can be blow molded using PA/PP blends, with either PA-6 or PA-66. Some grades are impact-modified through the addition of an elastomer. For extruding the parison, the extruder barrel should be at least 20*D* long, fitted with standard three-zone screw, with compression ratio of 2.5–3. Excessive shearing should be prevented. Purging with PP followed by LDPE is recommended. The blow-molding head may be either crosswise or in-line type. It was reported that the use of nitrogen instead of air injection resulted in improved weld strength.

PP/PA alloys can be spun into fibers. The large extensional strains induced by drawing can cause large changes in the morphology of blends. For example, during fiber spinning of PP/PA-6 blends (such as Orgalloy™ R6000, Elf Atochem), spherical particles with mean diameter $d \cong 0.9\,\mu m$ were found to elongate into 100 nm diameter fibrils [5]. The extent of particle elongation depends on the microrheological parameters, thus the interfacial forces, rheological characteristics of the polymers, hole geometry and the spinning parameters. For example, low interfacial tension and lower elongational viscosity of the dispersed phase than that of the matrix (at the extensional stresses occurring in the convergence) lead to the largest L/D ratios for the fibrils. Frequently a gradation of the fibrillar aspect ratio is observed along the radial direction. This originates from inhomogeneity of the stress field during the fibril formation as well as the presence of a thermal gradient during solidification.

REFERENCES

1. Martuscelli, E. (1995) Structure and properties of polypropylene-elastomer blends, in *Polypropylene: Structure, Blends and Composites*, Vol. 2, (ed. J. Karger-Kocsis), Chapman & Hall, London, pp. 95–140.
2. Thamm, R.C. (1977) *Rubber Chem. Technol.*, **50**, 24.
3. Coran, A.Y. and Patel R.P. (1980) *Rubber Chem. Technol.*, **53**, 141.
4. Coran, A.Y. and Patel R.P. (1995) Thermoplastic elastomers by blending and dynamic vulcanization, in *Polypropylene: Structure, Blends and Composites*, Vol. 2, (ed. J. Karger-Kocsis), Chapman & Hall, London, pp. 162–201.
5. Glotin, M. (1992) Proc. Colloque Génie de la transformation des polymères, Groupe Français des Polymères, Lyon, France, 23–25 November 1992.

Keywords: blends, processing, elastomer modification, dynamic vulcanization, morphology, injection molding, extrusion molding, thermoplastic elastomer, thermoplastic dynamic vulcanizates (TDV), application.

Processing-induced morphology

Mitsuyoshi Fujiyama

INTRODUCTION

The molecular structure of the polymer directly but partly affects the properties of processed articles. The properties of shaped, processed items mainly depend, on the other hand, on changes in the higher-order structures which are very sensitive to the processing conditions of semicrystalline polymers, such as polypropylene (PP).

The higher-order structures (morphologies) of processed PP which should be considered are crystalline form, lamellar thickness, spherulite size, crystallinity, molecular orientation, crystal orientation, amorphous orientation, dispersion state of blended polymer, length and orientation of filled fibers, crystalline texture (skin-core structure), etc. In this chapter, morphological changes induced by various processing methods are described.

PP exists in the following crystalline forms: α-form (monoclinic), β-form (hexagonal), γ-form (triclinic), and smectic form (pseudo-hexagonal). Their presence and content can be detected and measured by the wide angle X-ray diffraction (WAXS). The lamellar thickness is usually ten to several tens of nanometers and is measured with small angle X-ray scattering (SAXS). The spherulite size is usually several to several hundreds of micrometers and can be measured by a polarizing microscope, electron microscope or small angle light scattering. The crystallinity is determined by WAXS, density, infrared absorption or based on the heat of fusion. The molecular orientation is evaluated by X-ray diffraction, birefringence, or infrared dichroism. The crystal orientation is detected by WAXS and SAXS. The dispersion state of blended

Polypropylene: An A–Z Reference
Edited by J. Karger-Kocsis
Published in 1999 by Kluwer Publishers, Dordrecht. ISBN 0 412 80200 7

polymer can be observed in an optical or electron microscope. The length and orientation of fibers can be evaluated by optical microscopy or soft X-ray irradiation.

COMPRESSION MOLDING

In compression molding, although crystallization occurs under a high pressure and temperature gradient, the melt flow and solid-state deformation are slight, so that the molded part possesses a relatively uniform texture (composed of spherulites). Transcrystals form sometimes at the mold surface. Compression-molded PP items are composed of spherulites and exhibit a random crystal orientation. During compression molding of PP containing talc, the plate planes of the talc particles align parallel to the molding surface and the *b*-axes of PP crystals orientate in the thickness direction.

The lamellar thickness, spherulite size and crystallinity are the smaller, the larger the supercooling (defined as the difference between the melting point and crystallization temperature) is. High melting and cooling (mold) temperatures lead to a slow cooling and hence to a high crystallization temperature. This enhances both lamellar thickness, spherulite size and crystallinity. The copolymerization with ethylene results in an opposite tendency in this respect. The addition of nucleating agents increases the lamellar thickness and crystallinity and decreases the spherulite size. A high pressure has the same effect as the incorporation of nucleants [1].

While the tensile elongation decreases with increasing spherulite size, the yield strength and ultimate tensile strength scarcely depend on it. The tensile modulus, yield strength and hardness increase linearly with increasing crystallinity in contrast to the impact strength.

SHEET AND PIPE EXTRUSIONS

An extruded PP sheet shows a similar skin-core structure to that developed in injection molding. The skin layer is featureless and the core layer is composed of spherulites. The spherulite size increases toward the interior where the cooling is slower. A high draw ratio, creating high stresses, contributes to the initiation of crystal nuclei and thus to the formation of small spherulites [2]. The crystallinity is low at the sheet surface and increases toward the interior. The crystallinity increases when the sheet thickness and chill roll temperature are enhanced because of the slow cooling. A high draw ratio and short air gap, accompanied with a high melt orientation, initiates orientation crystallization and hence increases the crystallinity. The molecular orientation is high at the sheet surface and decreases toward the core. Its value is increased with increasing draw ratio or decreasing melt temperature. The molecular orientation at the sheet surface is higher when the die and chill roll temperatures are reduced. In a PP sheet containing talc particles, their plate planes orientate parallel to the sheet surface and the *b*-axes of PP crystals orient in the thickness direction. At present, simulation studies are in progress to estimate the average values of crystallinity and molecular orientation, covering also their changes through the thickness direction.

The tensile modulus increases with increasing crystallinity. A necessary condition for obtaining transparent sheets is to keep the spherulite size smaller than the wavelength of visible light. This can be achieved by the copolymerization method, by addition of nucleating agents or by quenching.

A similar morphology to that in sheet extrusion is formed in pipe extrusion, too. The spherulite size and crystallinity increase with increasing bath temperature. The molecular orientations in the axial, radial and tangential directions are high at the surfaces and decrease toward the interior. The orientation factors are higher when the melt flow index (MFI) of the PP is lowered or the extrusion speed is increased.

CAST FILM

A T-die cast film can be regarded as a thin sheet. While a sheet is composed mostly of the α-form crystals, in the case of the cast film, the cooling effect by the chill roll is so high that the α-crystals change gradually to the smectic form. This modification is triggered by decreasing chill roll temperature. A PP with higher melt flow index (MFI) crystallizes more easily in the smectic form than a lower MFI resin. Visbroken PP grades crystallize more easily in the smectic form than usual PPs at the same MFI under common casting conditions. The spherulite size increases with increasing film thickness and chill roll temperature and decreases with increasing draw ratio and die temperature [3]. The crystallinity is enhanced when the film thickness and chill roll temperature are raised. The molecular orientation increases with increasing draw ratio and take-up speed and with decreasing cylinder temperature. Crystallization during film casting is also the object of simulation work.

The tensile yield strength is higher and the ultimate elongation and strength are lower when the spherulite size becomes larger [3]. The tensile modulus increases linearly with increasing crystallinity. The heat-sealing strength decreases with increase in the crystallinity. A cast film rich in smectic form crystals, exhibits low rigidity but excellent transparency, impact strength and heat sealability.

INFLATION FILM

The processing variants of the inflation film technologies use water- or air-cooling methods. The air-cooled film is usually composed of α-form crystals, while in the water-cooled film the smectic form dominates. In an approximately uniaxial film containing α-form crystals, the *b*-axes of the crystals lie approximately perpendicular to the machine distribution (MD) and show an almost uniform distribution around the MD. With increasing blow-up ratio, the *b*-axes become concentrated in the thickness direction (ND). With increasing draw-down ratio, the *b*-axes show an increasing tendency toward symmetry about the MD [4]. The birefringences Δn_{12}, Δn_{13}, and Δn_{23} (where 1, 2, and 3 are the MD, TD, and ND, respectively) increase with increasing draw-down ratio and take-up speed. With increasing blow-up ratio, Δn_{23} increases, Δn_{12} decreases, and Δn_{13} scarcely changes [4]. The spherulite size increases with increasing MFI of the resin, temperature of cooling water, and height of the frost line. The crystallinity increases with increasing water temperature but is scarcely affected by the frost line height.

With increasing spherulite size, the tensile yield strength and tear strength increase and the elongation and puncture resistance deteriorate. In the case of films containing β-crystals, the dart impact strength increases with increasing β-crystal content.

FIBER SPINNING

In melt spinning, the diameter of a strand extruded from a spinneret gradually decreases and approaches a constant value at the point of crystallization initiation (where crystallization and molecular orientation start). Because the melt-spun PP fiber crystallizes under stresses, it is composed of a kind of shish-kebab structure with a mixed *c*- and *a**-axes orientation. With increasing spinning stress, the proportion of the *c*-axis-oriented component increases, while the *a**-axis-oriented component decreases after showing a maximum [5]. When the cooling conditions are severe, the smectic form appears. The formation of the smectic form is encouraged by using a low molecular weight resin and spinning it under a low draw ratio. The crystallinity increases with increasing take-up velocity and spinline stress. The molecular orientation characterized by the birefringence is higher near the fiber surface than in the bulk. Its value increases with increasing molecular weight, take-up velocity, draw ratio and spinline stress. With increasing take-up velocity or draw ratio, the crystalline *c*-axis orientation function (f_c) increases, while the *b*-axis orientation function (f_b) and *a**-axis orientation function (f_{a^*}) decrease.

These orientation functions, when plotted against the spinline stress, unify into three curves, respectively independent of the molecular weight of PP and spinning temperature used [5]. The crystallization and orientation during melt spinning are often simulated.

With increasing birefringence or f_c, the tensile modulus and yield strength increase, while the elongation decreases.

INJECTION MOLDING

The main aspects of the morphology-property relation are presented in another chapter (see 'Morphology–mechanical property relationships in

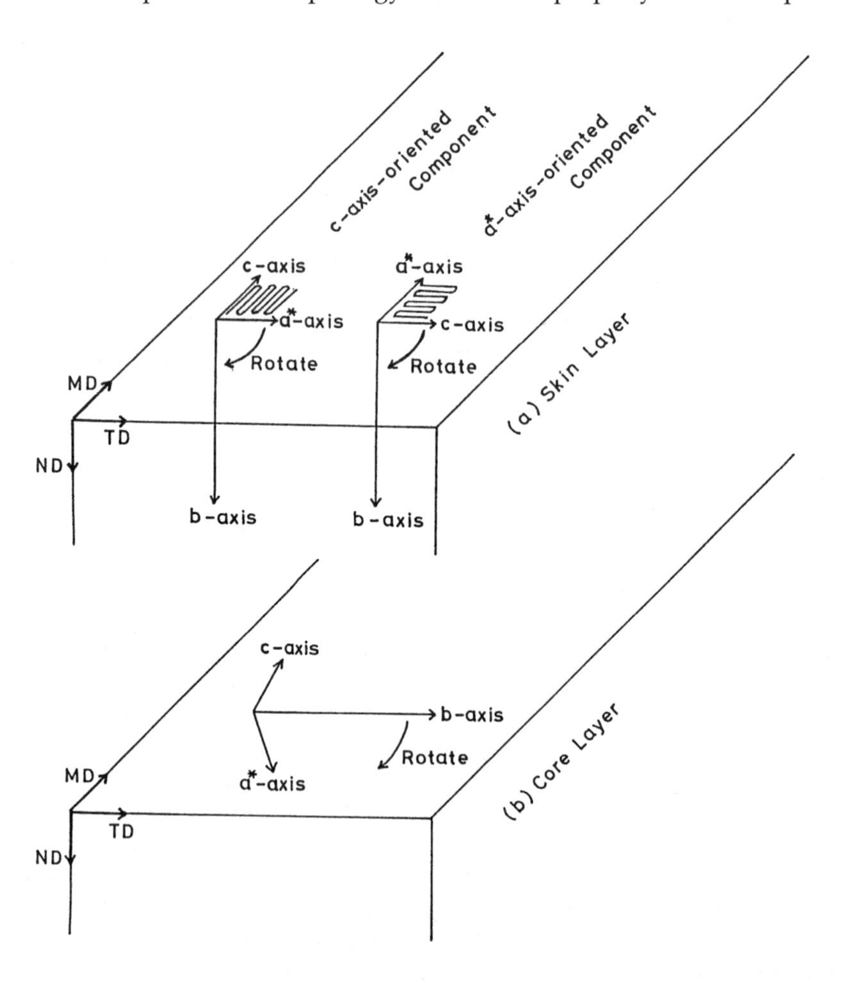

Figure 1 Crystal orientation states of the skin (a) and core layers (b) in an injection molding.

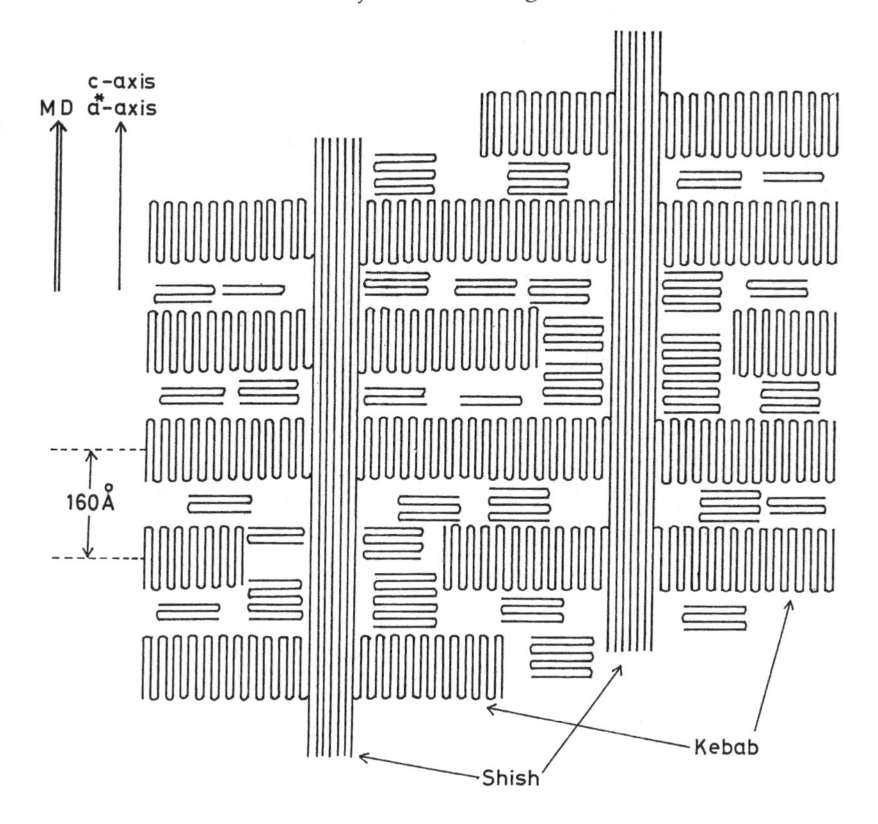

Figure 2 Crystalline structure of the skin layer developed in injection molding, schematically. Note: 10 Å = 1 nm.

injection molding'). Here, only some supplementary information is disclosed.

A cross-section of an injection-molded PP shows a clear skin-core structure under a polarizing microscope. Figure 1 shows the crystal orientation states of the skin and core layers [6]. In the skin layer, the *c*-axis-oriented and *a**-axis-oriented crystals are present. Their *b*-axis distribution is almost uniform around the MD. In the core layer, the *c*- and *a**-axes weakly orient in the MD and the *b*-axes uniformly distribute about it. Injection molding of talc filled PP shows a peculiar crystal orientation: the plate planes of talc particles orient parallel to the molding surface and the *b*-axes of PP crystals orient in the thickness direction [7].

Figure 2 shows the crystalline structure of the skin layer schematically [6]. It is composed of shish-kebab-like main skeleton structure, whose axis is parallel to the MD, piled epitaxially with a^*-axis-oriented imperfect lamellar substructure. This substructure works as a morphological plasticizer at bending. This is the reason for the good plastic hinge properties of injection-molded PP (see the chapter entitled Living or plastic hinges).

The skin thickness monotonously decreases or shows a maximum at a midway on going away from the gate in case of a molded plate. The crystallinity monotonously increases or goes through a maximum at midway toward the core. Its value decreases away from the gate. The molecular orientations assumed by the birefringence, crystalline orientation function, or amorphous orientation function monotonously decrease or decrease after a maximum on going away from the gate. The orientation monotonously decreases or decreases after passing a maximum just below the surface across the thickness.

During injection molding of a PP blended with another polymer, the dispersed polymer particles elongate parallel to the surface and MD in the skin layer. The degree of elongation becomes less pronounced toward the specimen bulk. The variation of the degree of elongation of the polymer particles in the thickness direction is very similar to that of the degree of molecular orientation.

During injection molding of a PP filled with glass or carbon fibers, the fibers orientate parallel to the MD at the surface region and orient perpendicular to the MD at the center region.

DRAWING

When a bulk PP composed of spherulites is drawn, a microfibrillar structure is formed by gradual chain tilting, slipping and breaking off blocks of folded chain lamellae. When the etched surface of the drawn article is observed under an electron microscope, a fibrillar structure elongated to the drawn direction is seen. With increasing draw ratio, the long period, crystallinity and the degrees of the average, crystalline and amorphous molecular orientations increase [8]. From around a draw ratio of 50%, f_c rapidly increases, approaching unity, while f_b and f_{a^*} decrease, approaching -0.5. From the fact that the decreases in f_b and f_{a^*} with increase in draw ratio are similar, it can be stated that the c-axes orientate in the draw direction and the b- and a^*-axes are randomly distributed around it. Drawing at a low temperature results in the smectic form which transforms to the α-form shish-kebab-like fiber structure upon annealing. With increasing molecular orientation, the tensile modulus and strength increase and the elongation decreases. The correlation with the tensile strength is better for the amorphous orientation than for the crystal or average orientation [8].

High-modulus and strength parts can be made by means of super drawings, such as high-temperature drawing, zone drawing, and gel drawing. Such parts have a crystalline structure resembling the extended chain crystals and possess high crystallinity and orientation. With increasing draw ratio, the crystalline molecular orientation increases and levels off at a relatively early stage, while the amorphous molecular orientation continues to increase. In addition, the number of tie molecules increases, imparting high modulus and strength.

When a PP is extruded and taken-up at a high draw ratio and crystallized under a high stress and annealed, a lamella-stacked structure is formed. In this structure, the lamellae are oriented perpendicular to the extrusion direction and connected by the tie molecules. When such an extrudate is stretched in the extrusion direction, the lamellae open elastically with the tie molecules working as fixing points. Therefore, after stress removal, the initial shape and structure are restored. Since such a part shows a high elastic recovery after a deformation and the elastic modulus is nearly the same as that of the usually processed article, it is called a 'hard-elastic' item. When the hard-elastic film or fiber is drawn beyond the yielding point, plastic deformation occurs, leading to void formation, and a microporous film or fiber can be obtained.

BIAXIALLY DRAWN FILM

In a successively biaxially drawn film (see also the chapter entitled Biaxially oriented polypropylene processes), the *b*-axes of the crystals orient in the thickness direction. The *c*-axes orient in the second drawing direction because the draw ratio of the second drawing direction is usually higher than that of the first step. In a simultaneously drawn film, the *b*-axes orient in the thickness direction and the *c*-axes are uniformly distributed in the film plane [9]. When the etched surface of a biaxially drawn film is viewed in an electron microscope, a fibrillar network structure can be resolved. With increasing planar draw ratio, the biaxial orientation factor increases, resulting in increased tensile modulus and strength and decreased elongation and dispensing force. In the successively biaxial drawing, when a cast sheet with the β-crystals is quickly heated to a temperature between the melting point of the β-crystals (about 145°C) and that of the α-crystals (about 165°C) at the first drawing and the β-crystal parts are 'melt-collapsed', a film with crater-like roughness, shown in Figure 3, is obtained.

HYDROSTATIC EXTRUSION AND ROLLING

In hydrostatic extrusion, the crystalline and amorphous orientations increase with increasing extrusion ratio. With increasing extrusion ratio,

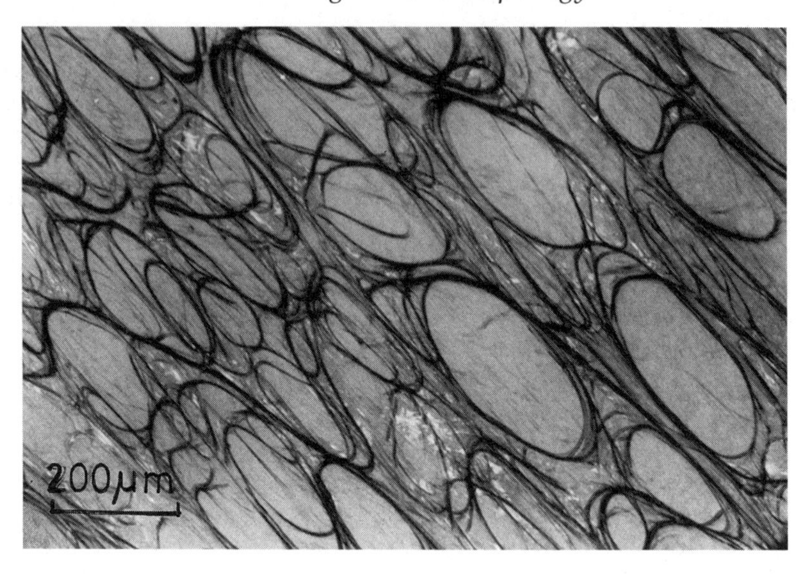

Figure 3 Photograph of biaxially drawn film with rough surface produced from β-PP.

f_c increases, approaching unity, while f_b and f_{a^*} decrease, approaching −0.5, which means that the *c*-axes orient in the extrusion direction and the *b*- and *a**-axes distribute uniformly about it. The tensile modulus increases linearly with the molecular orientation and the Vickers hardness increases linearly with the product of f_c and crystallinity [10].

In rolled (see the chapter entitled Rolltrusion processing of polypropylene for property improvement) sheet, the *c*- and *a**-axes are mixedly oriented in the rolling direction and the *b*-axes are random about it. With increasing compression ratio, the molecular orientation increases and the crystallinity decreases, resulting in an increase in the tensile modulus and a decrease in the elongation.

REFERENCES

1. Reinshagen, J.H. and Dunlap, R.W. (1975) The effects of melt history, crystallization pressure, and crystallization temperature on the crystallization and resultant structure of bulk isotactic polypropylene. *J. Appl. Polymer Sci.*, **19**, 1037–60.
2. Menges, G. and Winkel, E. (1982) Effect of the cooling rate on the morphology, density and Young's modulus of extruded films and sheets from polypropylene. *Kunststoffe*, **72**, 91–5.
3. Akay, M. and Barkley, D. (1983) On-line characterization in the quality control of extruded polypropylene film. *Proceedings of the 3rd International Conference on Polypropylene Fibers and Textiles III*, York, October 1983, Plastics and Rubber Institute, London, 14–1–11.

4. Shimomura, Y, Spruiell, J.E. and White, J.L. (1982) Orientation development in the tubular film extrusion of polypropylene. *J. Appl. Polymer Sci.*, **27**, 2663–74.
5. Nadella, H., Henson, H.M., Spruiell, J.E. and White, J.L. (1977) Melt spinning of isotactic polypropylene: structure development and relationship to mechanical properties. *J. Appl. Polymer Sci.*, **21**, 3003–22.
6. Fujiyama, M. (1995) Higher order structure of injection-molded polypropylene, in *Polypropylene: Structure, blends and composites*, Vol. 1 (ed. J. Karger-Kocsis), Chapman & Hall, London, pp. 167–204.
7. Fujiyama, M. and Wakino, T. (1991) Crystal orientation in injection molding of talc-filled polypropylene. *J. Appl. Polymer Sci.*, **42**, 9–20.
8. Samuels, R.J. (1970) Qualitative structural characterization of the mechanical properties of isotactic polypropylene. *J. Macromol. Sci. Phys.*, **B4**, 701–59.
9. Uejo, H. and Hoshino, S. (1970) Structure of biaxially oriented polypropylene film. *J. Appl. Polymer Sci.*, **14**, 317–23.
10. Nakayama, K., Kanetsuna, H. and Noda, E. (1979) Plastic deformation and molecular orientation of polypropylene – Hydrostatic extrusion of solid polymers IX. *J. Jpn Soc. Technol. Plast.*, **20**, 820–5.

Keywords: morphology, crystalline form, lamellar thickness, spherulite size, crystallinity, molecular orientation, crystal orientation, amorphous orientation, blend, fiber orientation, skin-core structure, molecular structures, processing conditions, properties, simulation, compression molding, sheet, pipe, cast film, inflation film, fiber spinning, injection molding, drawing, biaxially drawing, hydrostatic extrusion, rolling.

Properties of glass fibers for polypropylene reinforcement

J.L. Thomason and D.R. Hartman

INTRODUCTION

Continuous glass fibers were first manufactured in substantial quantities by Owens–Corning in the 1930s for high-temperature electrical applications. The raw materials are blended, in the proportions shown in Table 1, to form a glass batch which is melted in a furnace and refined during

Table 1 Composition ranges for glass fibers

Glass	*A*	*C*	*D*	*E*	*E-CR*	*AR*	*R*	*S-2®*
Oxide	%	%	%	%	%	%	%	%
SiO_2	63–72	64–68	72–75	52–56	54–62	55–75	55–65	64–66
Al_2O_3	0–6	3–5	0–1	12–16	9–15	0–5	15–30	24–25
B_2O_3	0–6	4–6	21–24	5–10		0–8		
CaO	6–10	11–15	0–1	16–25	17–25	1–10	9–25	0–0.1
MgO	0–4	2–4		0–5	0–4		3–8	9.5–10
ZnO					2–5			
BaO		0–1						
Li_2O						0–1.5		
Na_2O+K_2O	14–16	7–10	0–4	0–2	0–2	11–21	0–1	0–0.2
TiO_2	0–0.6			0–1.5	0–4	0–12		
ZrO_2						1–18		
Fe_2O_3	0–0.5	0–0.8	0–0.3	0–0.8	0–0.8	0–5		0–0.1
F_2	0–0.4			0–1		0–5	0–0.3	

Polypropylene: An A–Z Reference
Edited by J. Karger-Kocsis
Published in 1999 by Kluwer Publishers, Dordrecht. ISBN 0 412 80200 7

lateral flow to the forehearth. The molten glass flows to platinum/rhodium alloy bushings and then through individual bushing tips with orifices ranging from 0.75 to 2.0 mm and is rapidly quenched and attenuated in air (to prevent crystallization) into fine fibers ranging from 3 to 24 μm. Mechanical winders or choppers pull the fibers at lineal velocities up to 60 m/s over an applicator which coats the fibers with an appropriate chemical sizing to aid further processing and performance of the end product.

GLASS FIBER CHEMICAL COMPOSITIONS

Several types of silicate glass fibers are manufactured for the textile and composite industry. Their chemical compositions described below from ASTM C 162 were developed to provide combinations of fiber properties directed at specific end-use applications:

A-glass Soda lime silicate glasses used where the strength, durability, and good electrical resistivity of E-glass are not required;
C-glass Calcium borosilicate glasses used in corrosive acid environments;
D-glass Borosilicate glasses with low dielectric constant for electrical applications;
E-glass Alumina-calcium-borosilicate glasses used as general purpose fibers where strength and high electrical resistivity are required;
E-CR-glass Calcium aluminosilicate glasses used where strength, electrical resistivity and acid corrosion resistance are desired;
AR-glass Alkali resistant glasses composed of alkali zirconium silicates used in cement substrates and concrete;
R-glass Calcium aluminosilicate glasses used where added strength and acid corrosion resistance are required;
S-2® glass Magnesium aluminosilicate glasses for structural applications requiring high strength, modulus and stability in extreme temperature and corrosive environments.

Chemical composition variation within a glass type is from differences in the available glass batch raw materials, or in the melting and forming processes, or from different environmental constraints at the manufacturing site. These fluctuations do not significantly alter the physical or chemical properties of the glass type. Very tight control is maintained within a given production facility to achieve consistency in the glass composition for production capability and efficiency. Table 1 provides the oxide components and their weight ranges for the above glass fibers [1–4].

GLASS FIBER PROPERTIES

Glass fiber properties, such as tensile strength, Young's modulus and chemical durability, are measured on the fibers directly. Other properties, such as dielectric constant, dissipation factor, dielectric strength, volume/surface resistivities, and thermal expansion, are measured on glass that has been formed into a bulk sample and annealed (heat treated) to relieve forming stresses. Properties such as density and refractive index are measured on both fibers and bulk samples, in annealed or unannealed form. The properties presented in Tables 2 and 3 are representative of the compositional ranges in Table 1.

Physical properties

Density of glass fibers is measured and reported either as formed or as bulk annealed samples. ASTM C 693 is one of the test methods used for density determinations. The fiber density (in Table 3) is less than the bulk annealed value by approximately 0.04 g/cm^3 at room temperature. The glass fiber densities used in composites range from approximately 2.11 g/cm^3 for D-glass to 2.72 g/cm^3 for E-CR with E-glass having a value of 2.58 g/cm^3. Tensile strength of glass fibers is usually reported as the pristine single-filament or the multifilament strand measured in air at room temperature. The respective strand strengths are normally 20–30% lower than the values reported in Table 2 due to surface defects introduced during the strand-forming process. The various steps involved in

Table 2 Physical properties of glass fibers

Glass	*A*	*C*	*D*	*E*	*E-CR*	*AR*	*R*	*S-2*®
Density (gm/cm^3)	2.44	2.52	2.11	2.58	2.72	2.70	2.54	2.46
Refractive index	1.538	1.533	1.465	1.558	1.579	1.562	1.546	1.521
Softening point (°C)	705	750	771	846	882	773	952	1056
Annealing point (°C)		588	521	657				816
Strain point (°C)		522	477	615				766
Tensile strength (MPa)								
at −196°C		5380		5310	5310			8275
at 23°C	3310	3310	2415	3445	3445	3241	4135	4890
at 371°C				2620	2165		2930	4445
at 538°C				1725	1725		2140	2415
Young's modulus (GPa)								
at 23°C	68.9	68.9	51.7	72.3	72.3	73.1	85.5	86.9
at 538°C				81.3	81.3			88.9
Elongation % (23°C)	4.8	4.8	4.6	4.8	4.8	4.4	4.8	8.7

Table 3 Properties of glass fibers

Glass		*A*	*C*	*D*	*E*	*E-CR*	*AR*	*R*	*S-2®*
% weight loss in					**Chemical properties**				
H_2O :	24 h	1.8	1.1	0.7	0.7	0.6	0.7	0.4	0.5
	168 h	4.7	2.9	5.7	0.9	0.7	1.4	0.6	0.7
10%HCl :	24 h	1.4	4.1	21.6	42	5.4	2.5	9.5	3.8
	168 h		7.5	21.8	43	7.7	3.0	10.2	5.1
10% H_2SO_4 :	24 h	0.4	2.2	18.6	39	6.2	1.3	9.9	4.1
	168 h	2.3	4.9	19.5	42	10.4	5.4	10.9	5.7
10% Na_2CO_3 :	24 h		24	13.6	2.1		1.3	3.0	2.0
	168 h		31	36.3	2.1	1.8	1.5	1.5	2.1
					Electrical properties				
Dielectric constant									
1 MHz		6.2	6.9	3.8	6.6	6.9	8.1	6.4	5.3
10 GHz				4.0	6.1	7.0			5.2
Dissipation factor									
1 MHz			0.0085	0.0005	0.0025	0.0028		0.0034	0.0020
10 GHz				0.0026	0.0038	0.0031		0.0051	0.0068
Volume resistivity (ohm-cm × 10^{-12})		0.01			402	384		203	9.05
Surface resistivity (ohms × 10^{-14})					42	116		0.674	0.089
Dielectric strength (kV/mm)					10.3	9.8		10.8	13.0
					Thermal properties				
Specific heat (J/g°C)									
23°C		0.80	0.79	0.733	0.81				0.74
200°C			0.90		1.03	0.97			
Thermal expansion −30°C to 250°C (×10^{+6} K^{-1})		7.3	6.3	2.5	5.4	5.9	6.5	3.3	1.6

the processing of glass fiber with polypropylene (PP) into a composite part may also result in further reduction of the fiber tensile strength. Moisture has a detrimental effect on the pristine strength of glass as illustrated by the increase in the pristine single-filament strength measured at liquid nitrogen temperatures where the influence of moisture is minimized. The pristine strength of glass fibers decrease as the fibers are exposed to increasing temperature. E-glass and S-2 glass fibers have been found to retain approximately 50% of their pristine room-temperature strength at 538°C. The Young's modulus of elasticity of unannealed glass fibers ranges from about 52 GPa to 87 GPa. As a fiber is heated, its modulus gradually increases. E-glass fibers that have been annealed to compact their atomic structure will increase in Young's modulus from 72 GPa to 85 GPa. For most silicate glasses, Poisson's ratio falls between 0.15 and 0.26. The Poisson's ratio for E-glasses is 0.22 ± 0.02 and does not change significantly with temperature when measured up to 510°C.

Chemical resistance

The chemical resistance of glass fibers to the actions of acids, bases, and water is expressed as a percent weight loss. The lower this value, the more resistant the glass is to the corrosive solution. The test procedure involves subjecting a given weight of 10 μm diameter glass fibers, without binders or sizes, to a known volume of corrosive solution held at 96°C. The results reported are for 24 h (1 day) and 168 h (1 week) exposures. As Table 3 shows, the chemical resistance of glass fibers depends on the composition of the fiber, the corrosive solution, and the exposure time. It should be noted that glass corrosion in acidic environments is a complex time-dependent process. For a given glass composition, the corrosion rate may be influenced by the acid concentration, temperature, fiber diameter and the solution volume to glass mass ratio. In alkaline environments, weight loss measurements are more subjective as the alkali affects the network and reprecipitates the metal oxides. Tensile strength after exposure is a better indicator of the residual glass fiber properties.

Electrical properties

The electrical properties of annealed bulk glass samples are given in Table 3. The dielectric constant or relative permittivity is the ratio of the capacitance of a system with the specimen as the dielectric to the capacitance of the system with a vacuum as the dielectric. Permittivity and dissipation factor values are affected by test frequency, temperature, voltage, relative humidity and weathering. The dissipation factor of a dielectric is the ratio of the parallel reactance to the parallel resistance, or the tangent of the loss angle, which is usually called the loss tangent. In almost every electrical application, a low value for the dissipation factor is desired. This reduces the internal heating of the material and keeps signal distortion low. The dielectric breakdown voltage is the voltage at which electrical failure occurs under prescribed test conditions in an electrical insulating material that is placed between two electrodes. When the thickness of the insulating material between the electrodes can be accurately measured, the ratio of the dielectric breakdown voltage to the specimen thickness can be expressed as the dielectric strength in kV/cm.

Breakdown voltages are influenced by electrode geometry, specimen thickness, temperature, voltage application time, voltage wave form, frequency, surrounding medium, relative humidity, weathering and directionality in laminated and inhomogeneous plastics.

Thermal properties

As with most materials, the viscosity of a glass decreases as the temperature increases. Several reference viscosity points are defined by the glass industry as shown in Table 2. The softening point is the temperature at which a glass fiber of uniform diameter elongates at a specific rate under its own weight when measured by ASTM C 338; it occurs at a viscosity of approximately $10^{6.6}$ Pa s. The annealing point is the temperature corresponding to either a specific rate of elongation of a glass fiber when measured by ASTM C 336, or a specific rate of midpoint deflection of a glass beam when measured by ASTM C 598. At the annealing point of glass, internal stresses are substantially relieved in a matter of minutes. The viscosity at the annealing point is approximately 10^{12} Pa s. The strain point is measured following ASTM C 336 or C 598 as described above for annealing point. At the strain point of glass, internal stresses are substantially relieved in a matter of hours. The viscosity at the strain point is approximately $10^{13.5}$ Pa s. The mean coefficient of thermal expansion of annealed glass bars over the temperature range from −30 to 250°C is provided in Table 3. The specific heat data in Table 3 were determined using high-temperature differential scanning calorimetry techniques. In general, the average specific heat values can be represented as follows: 0.94 kJ/kg K at 200°C, 1.12 kJ/kg K just below the transition point, and 1.40 kJ/kg K in the liquid state above the transition. These values are accurate to about 5%. Above the transition temperature, no further increase in specific heat is observed. The transition temperature is nearly identical to the annealing temperature of bulk glass. The thermal conductivity of glasses drops steadily with temperature and reaches very low values near absolute zero. Thermal conductivity data for glass varies among investigators for materials which are normally identical. It is found that the approximate thermal conductivity of C-glass is 1.1 W/m K, E-glass is 1.3 W/m K, and S-2 is 1.45 W/m K near room temperature.

Optical properties

Refractive index is measured on either unannealed or annealed glass fibers. The standard oil immersion techniques are used with monochromatic sodium D light at 25°C. In general, the corresponding annealed glass will exhibit an index that will range from approximately 0.0030 to 0.0060 higher than the as-formed glass fibers given in Table 2.

GLASS FIBER SIZE TREATMENTS

The surface treatment chemistry of glass fiber is normally tailored to match the product function. Reinforcement size chemistries must be compatible with a multitude of processes and with the composite material end-use performance criteria. Processes such as injection molding require chopped fibers with compatibility for thermoplastic compounds. Filament winding and pultrusion require continuous fibers with utility in thermoset and thermoplastic compounds. Typically three basic components are used with glass size chemistries: a film former, lubricant, and coupling agent. Readers are directed to the patent literature or [1] and [2] for more specific details.

COMPOSITE PROPERTIES

The balance of properties obtained in a glass fiber reinforced PP composite will be strongly influenced by the various fiber properties discussed above. However, there are a number of other important parameters, such as PP properties, interfacial adhesion, interfacial morphology, voids, fiber content, fiber orientation, and residual fiber length. Furthermore, many of these factors are changed according to the processing history of the composite. A number of these points are illustrated in Table 4 which shows some typical property levels that can be obtained from three common forms of glass fiber reinforced PP. Processing history, fiber length and orientation, interfacial adhesion, and PP properties, all play a role in the differences observed Table 4. Here fiber length increases, and

Table 4 Properties of 40% w/w E-glass fiber reinforced polypropylene

Material	*PP*	*Tensile strength, MPa (D-638)*	*Flexural modulus, MPa (D-790)*	*Notched Izod impact J/m (D-256)*	*HDT °C at 1.8 MPa (D-648)*
Short fiber injection-molding compound	HP	73	6300	60	141
	CC	92	7700	50	149
Long fiber injection-molding compound	HP	76	6650	450	149
	CC	107	6720	370	152
Glass mat thermoplastic (GMT)	HP	98	5600	750	154

HP, homopolymer; CC, chemically coupled; HDT, heat distortion temperature.

the fiber orientation factor decreases, as we move from short fiber compound through long fiber compound to GMT. These factors are discussed in more detail in other chapters of this book.

REFERENCES

1. Hartman, D.R., Greenwood, M.E. and Miller, D.M. (1994) High strength glass fibers, *SAMPE Int.*, **39**, 521–533.
2. Loewenstein, K.L. (1973) *The Manufacturing Technology of Continuous Glass Fibers*, Elsevier, Amsterdam, pp. 28–30.
3. Miller, D.M. (1987) *Engineered Materials Handbook*, Vol. 1, *Composites*, ASM Int., Glass fibers in Metals Part, OH, pp. 45–48.
4. Watson, J.C. and Raghupathi, N. (1987) *Engineered Materials Handbook*, Vol. 1, *Composites*, ASM Int., Glass fibers in Metals Part, OH, pp. 107–111.

Keywords: glass fiber (GF), composition, physical properties, chemical resistance, electrical properties, optical properties, glass fiber production, glass fiber reinforced composites.

Pultrusion of glass fiber/ polypropylene composites

K. Friedrich and G. Bechtold

INTRODUCTION

Recent developments of impregnation technology for thermoplastic matrices have generated considerable interest in the possibility of thermoplastic pultrusion [1–3]. Successful work has, however, only been performed for simple cross-sections, and at pultrusion speeds not dramatically exceeding those known for commercial thermoset pultrusion (0.6–1.2 m/min). Major reasons for these deficits are the inherent difficulties associated with the thermoplastic matrices, such as high processing temperatures and high melt viscosities. An additional obstacle may be a lack of both fundamental understanding of the governing process mechanisms and adequate mathematical models for predicting the relationships between the various processing variables and the resulting structural/mechanical properties of the thermoplastic pultruded products [4].

Lee *et al.* [5] have presented a model which allows us to predict the effects of pulling speed and die geometry on the required temperatures, the degree of consolidation and the pulling force. Åström *et al.* [4, 6] pultruded rectangular beams through a heating and cooling die system. These authors were interested in a comparison of experimental pultrusion results with model predictions of the temperature and pressure distribution in the pultrusion die. Mechanical properties of their pultruded carbon fiber/polyetheretherketone (CF/PEEK) profiles were only slightly below those values presented in the literature for ideally

Polypropylene: An A–Z Reference
Edited by J. Karger-Kocsis
Published in 1999 by Kluwer Publishers, Dordrecht. ISBN 0 412 80200 7

compression-molded standard samples. Additional work was also carried out with glass fiber (GF)/polypropylene (PP) tape material.

Gibson and coworkers [2] emphasized that the preheating section of their thermoplastic pultrusion process was of crucial importance to a successful operation. Besides the results achieved with CF/PEEK tapes, these authors presented a wide range of interlaminar shear strength (ILSS) data for pultrudates from glass fiber (GF)/polyamide 12 (PA12) tapes.

An even wider set of tape materials, including CF/PEEK, CF/polyphenylenesulfide (PPS), GF/PPS, CF/PA12 and GF/PA12, was used by Taylor and Thomas [7] for demonstrating that the additional introduction of ultrasonic energy into the material during the consolidation process causes a better resin flow and therefore leads to much faster pultrusion rates (up to 6 m/min in the case of GF/PA12). As a quality criterion, the authors considered a void content in the final pultrudates of less than 2%. They also tested other types of preforms; for commingled yarns, they found that complete fiber wetting had first to be accomplished in the heated die. As this became the process rate controlling step, no further improvements in pultrusion speed could be achieved by additional ultrasonic energy during the subsequent consolidation step. The authors also succeeded in the pultrusion of tubular, hollow profiles of CF/PEEK and GF/PP.

Wilson *et al.* [3] showed that pultrusion is also applicable to other high-performance thermoplastics, in particular, amorphous polyetherimide (PEI). In this case, a wet-impregnation station for impregnating carbon fiber bundles with a PEI solution was put in line with a die system that was designed to consolidate multiple plies of prepreg tape by gradual reduction into rectangular profiles.

Polybutyleneterephthalate (PBT) powder-impregnated and sheath-surrounded glass fiber bundles have been pultruded by Kerbiriou [8] into rectangular profiles. It was found that the temperature conditions in both the preheating zone and the heated die had a remarkable influence on the parts' quality. In addition, the pressure conditions in the heated die (independently varied by the use of different taper angles and loads) and the pultrusion speed were of high importance. The results that were supported by a new model for predicting the remaining void content in the pultruded beams as a function of processing conditions led to the generation of an optimum processing window within which acceptable mechanical properties could be achieved. Additional results in Kerbiriou's previous works were also achieved with GF/PP-tapes, pultruded into circular rods and into hollow, rectangular profiles as well as with fabrics of carbon fiber reinforced polysulfone (PSU) pultruded into thin-walled tubes (Figure 1(a)).

(a)

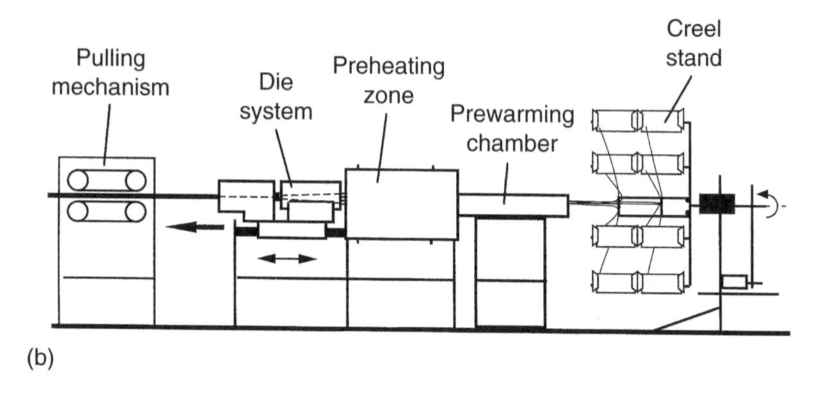

(b)

Figure 1 (a) Examples of profiles pultruded by V. Kerbiriou [8]. (b) Schematic of the pultrusion line used for the production of profiles shown in Figure 1(a).

Tomlinson and Holland [1] reported on the pultrusion and properties of GF/PP matrix composites. The properties of their beams, as pultruded from tapes at a relatively low line speed of 4 mm/min, were slightly better than their hot-pressed standard samples.

Additional studies on the pultrusion of unidirectional GF/PP-composites have been carried out by Michaeli and coworkers (e.g. [9]). By the use of various types of hybrid yarns, these authors demonstrated the effects of cooling conditions, pulling speed and pressure profile in the heated die on the consolidation quality and the surface properties of flat rectangular profiles. Their model for predicting the pulling force was in good agreement with the experimental data. Other pultrusion studies were focused on curvature phenomena of GF/PP-tubes due to non-uniform fiber distribution across the wall of the tubes.

In the following section, some new data on the pultrusion of glass fiber/PP fiber commingled yarn are presented and compared to results of previous investigators.

THE PROCESS

A schematic of a typical pultrusion line is shown in Figure 1(b) [8]. Fiber bundles are preheated in a hot air preheating zone and enter directly into the heated die. The cavity of the latter is tapered, and its angle can be varied without changing the final thickness of the beam. Two pressure sensors incorporated in the die continuously measure the internal pressure profile. A water-cooled die just behind the heated die is installed for further compaction and improvement of the pultrudate's surface quality. Heated and cooled die are fixed on a linear guide and are pulled against a load cell during the whole process, so that the pulling force is measured continuously. The beam is pulled by a pulling mechanism, which can realize speeds between 0.01 and 30 m/min.

The temperature of the fiber bundles just before entering the heated die is varied between 20°C and 163°C, the heated die temperature between 177°C and 225°C, and the angle α of the tapered section of the heated die between 5° and 8.5°. Speeds from 0.03 to 0.6 m/min are realized. The final cross section of the beams is $3.5 \times 10\,mm^2$, i.e. sufficient to be able to make mechanical test specimens.

PROCESSING WINDOW AND PROPERTIES

For the pultrusion experiments, a commingled continous GF/PP fiber yarn with a glass fiber volume content of 19% was used (Vetrotex, France). Due to a variation of processing conditions (preheating temperature T_{ph}, heated die temperature T_{hd} and pulling speed v_p) the mechanical properties of the pultruded beams, such as flexural modulus E_F, flexural strength σ_F and shear strength τ, reflected significant changes. Figure 2(a) shows qualitatively the influence of processing speed and processing time (i.e. time in the heated die) on the shear strength and the void content (as measured by the Archimedes' buoyancy method).

The maximum properties at four different preheating temperatures are presented in Table 1. Columns 3 and 4 represent the corresponding values of T_{hd} and v_p, at which individual maxima of E_F,σ_F and τ were

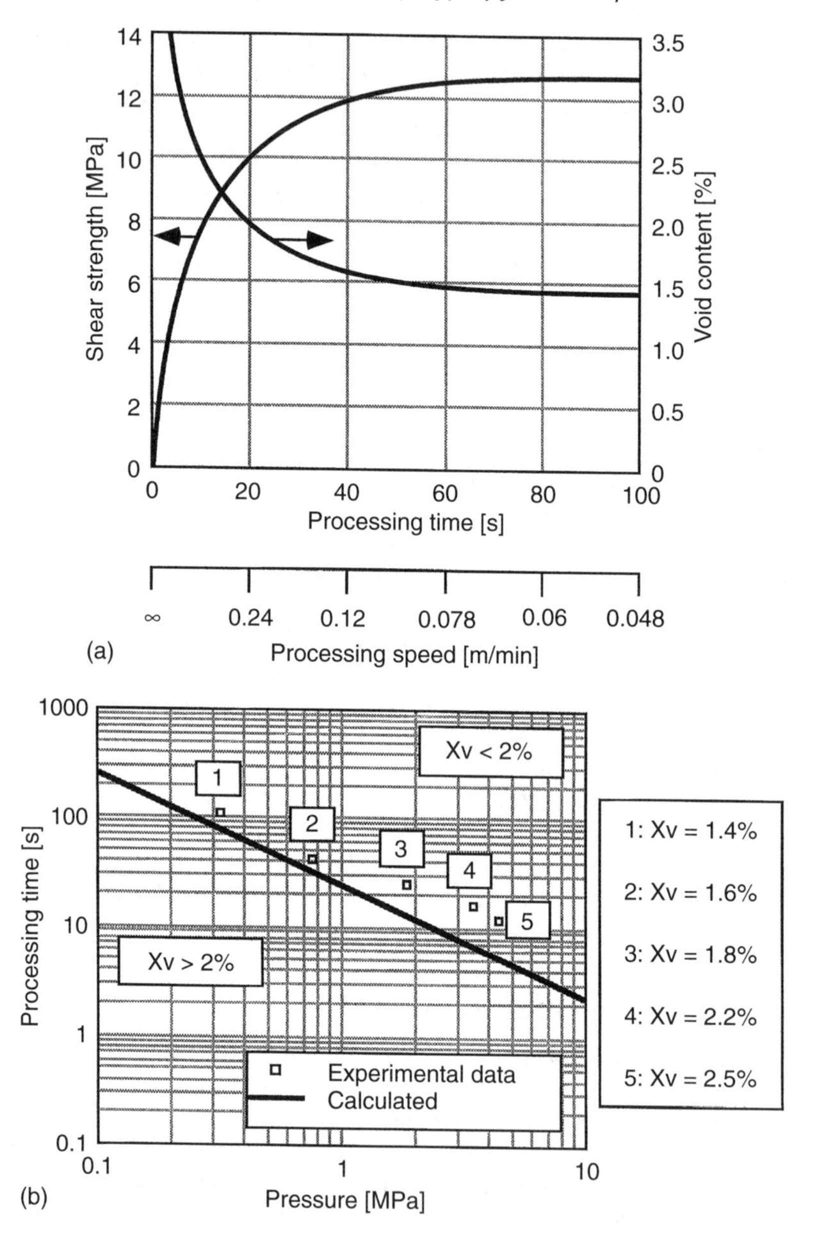

Figure 2 (a) Shear strength at different processing times ($T_{ph} = 143°C$). Further improvements are possible (i.e. higher strength, lower void content) when increasing (up to a certain degree) the pressure, the preheating temperature and/or the temperature of the heated die. (b) Required processing times at different pressures near die exit (almost maximum of the pressure profile), $T_{ph} = 143°C$, $T_{hd} = 203°C$. (Major support for these studies came from DFG FR 675/20–1.)

Table 1 Maximum properties at constant preheating temperatures

T_{ph} (°C)	*Properties*	T_{hd} (°C)	v_p (m/min)
20	E_F max = 12.5 GPa	215	0.225
	σ_F max = 339.8 MPa	203	0.068
	τ max = 13.03 MPa	203	0.068
114	E_F max = 12.5 GPa	177	0.080
	σ_F max = 339.8 MPa	185	0.077
	τ max = 14.11 MPa	185	0.050
163	E_F max = 12.3 GPa	177	0.146
	σ_F max = 349.5 MPa	177	0.060
	τ max = 12.79 MPa	225	0.273

achieved. The table reads as follows. At $T_{ph} = 20°C$, a maximum in E_F was achieved, when $T_{hd} = 215°C$ and $v_p = 0.225$ m/min, whereas a maximum of σ_F was achieved when $T_{hd} = 203°C$ and $v_p = 0.068$ m/min.

The results correspond quite well with those of Wilson and Buckley [10] and Tomlinson *et al.* [1], when compared on the basis of the same glass fiber volume fraction $V_f = 19\%$. Only a comparison of the shear strength data is limited due to different testing methods.

With the pultrusion line and materials used here, the following process parameters lead to at least 80% of the maximum shear strength ($\tau_{max} = 14$ MPa; as measured by the Lauke test [8]) at pulling speeds of at least 0.3 m/min:

Preheating temperature:	$T_{ph} = 143–163°C$
Heated die temperature:	$T_{hd} = 185–215°C$

It should be noted that the parameters partly influence each other or cannot be realized simultaneously.

A rectangular model based on Darcy's law for predicting the degree of impregnation (or the void content, respectively) during the pultrusion process was developed in [8]. It leads to the following equation for the void content X_v:

$$X_V = \frac{(h_0 - z)\ (1 - V_f)}{h_{1z}} \tag{1}$$

where h_0 and h_{1z} represent geometrical dimensions of the rectangular model arrangement between PP matrix fibers and the GF agglomeration to be impregnated. V_f is the global glass fiber volume fraction, and z the flow path of the polymer melt transverse to the glass fiber agglomeration. Note that longitudinal flow was not considered because the PP fibers in their initial state are already continously, longitudinally arranged, i.e. parallel to the glass fibers, so that most probably transverse flow will take place under pressure once the PP fibers exceed their melting temperature.

The impregnation time in order to reach a certain void content X_v can be calculated by inserting the corresponding value of z (according to equation (1)) into the following equation:

$$t = \frac{1}{n} \sum_{i=1}^{n} \left[\frac{z_{(x_i)}^2 \, 2 \, \eta \, k_{zz} \left(\frac{V_a}{V_{f(x_i)}} + 1 \right)}{r_{gf}^2 \left(\sqrt{\frac{V_a}{V_{f(x_i)}}} - 1 \right)^3 \frac{\Delta p}{\Delta L} \, x_i} \right] \tag{2}$$

where k_{zz} is the permeability constant of the fibers (Gutowski) = 0.044, V_a is the initial fiber volume fraction, r_f is the fiber radius = 7 μm, n is the number of sections in the heated die (in this case 80), z is the flow path through the agglomeration (calculated by equation (1)), $V_{f(x_i)}$ is the local fiber volume fraction (calculated), Δp is the pressure difference (measured by the sensors), ΔL is the length of the heated die = 80 mm, η is the viscosity of the matrix = 690 Pa s (temperature 185°C, shear rate 10 s^{-1}), and x_i is the position of the actual segment in the pulling direction.

Figure 2(b) contains the resulting information on required compaction time and pressure to reach a void content of less than 2%. Experimentally determined data enable a comparison to the calculated curves. The most significant sources of error in the latter are the viscosity and the permeability constant. The viscosity depends strongly on the temperature and the shear rate. Neither are constant at any point of the matrix. As the calculation estimated the processing time too short, it can be stated that both viscosity and the permeability constant have been chosen as too low. By a further development of the model, errors can be reduced, although the former model will then lose its advantage of simplicity.

CONCLUSIONS

Pultrusion of commingled GF/PP yarn leads to profiles with acceptable mechanical properties. The processing window shows that preheating temperatures marginally below the polymer melting point are ideal, whereas the heating die temperature should not exceed 215°C. Within the presented processing window, the pultrusion speed (as long as it stays in the range between 0.05 and 0.27 m/min) does not have a strong influence on the quality of the parts. A simple consolidation model can be applied to estimate the influence of processing speed and pressure on the void content of the final part. However, it is very sensitive to measurement errors of viscosity and permeability.

REFERENCES

1. Tomlinson, W.J. and Holland, J.R. (1994) Pultrusion and properties of unidirectional glass fiber-polypropylene matrix composites. *J. Mat. Sci. Letters*, **13**, 675–677.
2. Devlin, B.J., Williams, M.D., Quinn, J.A. and Gibson, A.G. (1991) Pultrusion of unidirectional composites with thermoplastic matrices. *Composites Manufact.*, **2**(3/4), 203–207.
3. Wilson, M.L., Backley, J.D., Dickerson, G.E., Johnson, G.S., Taylor, E.C. and Covington III, E.W. (1989) *Pultrusion process development of a graphite reinforced polyetherimide thermoplastic composite.* Proceedings of the 44th Annual Conference, Society of Plastics Industry, Reinforced Plastics / Composite Institute, Dallas, Texas, USA, February 6–10, Paper Session 8-D.
4. Åström, B.T. and Pipes, R.B. (1991) *Correlation between modelling and experiments for a thermoplastic pultrusion process.* Proceedings of the VIIIth International Conference on Composite Materials, ICCM-VIII (eds S.W. Tsai and G.S. Springer), Honolulu, Hawaii, USA, July 15–19, Section 13-A-1–13-A-10.
5. Lee, W.I., Springer, G.S. and Smith, F.N. (1991) Pultrusion of thermoplastics – A model. *J. Composite Mat.*, **25**, 1632–1652.
6. Åström, B.T., Larsson, P.H., Hepola, P.J. and Pipes, R.B. (1994) Flexural properties of pultruded carbon/PEEK composites as a function of processing history. *Composites*, **25**(8), 814–821.
7. Taylor, S.R. and Thomas, W.M. (1990) *High speed pultrusion of thermoplastic composites.* Proceedings of the 22nd International SAMPE Technical Conference, Anaheim, California, USA, November 6–8, Vol. 22, pp. 78–87.
8. Kerbiriou, V. (1996) Imprägnieren und Pultrudieren von thermoplastischen Verbundprofilen, Dissertation, Fachbereich Maschinenbau und Verfahrenstechnik, Universität Kaiserslautern, Germany.
9. Blaurock, J. and Michaeli, W. (1996) Pultrusion of endless fiber-reinforced profiles with a thermoplastic matrix system, *Engng Plast.*, **9**(4), 282–292.
10. Wilson, M.L. and Buckley, J.D. (1994) The potential for low cost thermoplastic pultrusion, *J. Reinforced Plast. Composites*, **13**, 927–941.

Keywords: pultrusion, mechanical properties, glass fiber, void content, commingled yarn, modelling, processing window, impregnation, Darcy's law, composite profiles, composite tubes.

Reactive compatibilization of polypropylene

M. Xanthos

INTRODUCTION

Polymers may be chemically modified in order to meet specific cost/performance/processability characteristics. Modification may involve single polymers or mixtures of two or more polymers. Reactive modification of single polymers may be accomplished through a variety of reagents or through radiation. The modified products (usually functionalized polymers), may be used as compatibilizers to enhance the properties of dispersed blends or as tie-layers in coextruded multilayered structures. Modification reactions can be carried out in solution, in bulk or on the surface of the plastic part or pellets. Recent advances in the technology and economics of polypropylene (PP) modification reactions on virgin polymers (particularly in the absence of solvents, as in reactive extrusion), suggest that this route of chemical conversion is also applicable to the recycling of polymer wastes. Note that polymer wastes may contain polymeric or other contaminants which may accordingly affect the rates and yield of modification reactions to different degrees.

COMPATIBILIZATION REACTIONS

Heterogeneous blends of technological importance are termed 'compatible' and they constitute the majority of the commercial blends introduced in the past 20 years. In such blends, satisfactory physical and mechanical properties are related to the presence of a finely dispersed phase and resistance to gross phase segregation. Blend composition,

Polypropylene: An A–Z Reference
Edited by J. Karger-Kocsis
Published in 1999 by Kluwer Publishers, Dordrecht. ISBN 0 412 80200 7

viscoelastic properties of the components and interfacial adhesion are among the parameters known to control the size and morphology of the dispersed phase and its stability to coalescence.

Polymer compatibility may be enhanced by various methods. Co-crystallization and co-cross-linking can often result in stable morphologies that are resistant to coalescence. Strong interactions, such as acid–base or ion–dipole, hydrogen bonding and transition metal complexation, have also been shown to enhance thermodynamic miscibility of suitably functionalized components and result in improved compatibility in a variety of systems of technological importance. More commonly, compatibility is promoted through copolymers (e.g. block, graft) with segments capable of specific interactions and/or chemical reactions with the blend components.

Modification reactions of PP aiming at producing suitably functionalized resins with enhanced affinity towards polar polymers are summarized in [1–3]. Important functionalization reactions involve the incorporation of reactive groups through, for example, halogenation, chlorosulfonation or peroxidation, reaction with azidosulfonyl benzoic acid and monomer grafting (e.g. maleic anhydride, acrylic acid) through the action of peroxides. Such modified materials are used to provide blends with enhanced properties.

In addition to copolymers, compatibility may also be enhanced through the addition of specific low molecular weight (MW) compounds which promote copolymer formation and/or crosslinking. The following is a summary of our present knowledge on the types and function of compatibilizers used in melt blended polymer mixtures with emphasis on PP blends.

In-situ formed copolymers

During melt blending of a pair of suitably functionalized polymers A and B, interchain block or graft copolymers may be formed at various concentrations through covalent or ionic bonding. The *in-situ* formed compatibilizers have segments that are chemically identical to those in the respective unreacted homopolymers and are thought to be located preferentially at the interface, thus lowering interfacial tension, and also promoting mechanical interlocking through interpenetration and entanglements. Examples of some important compatibilizing reactions that can take place easily across polymer phase boundaries involve functionalities, such as anhydride or carboxyl with amine, epoxy with anhydride or carboxyl, oxazolin with carboxyl, isocyanate with hydroxyl or carboxyl, and carbodiimide with carboxyl. Interchange reactions, macroradical recombination in polyolefins, and interchain ionic salt formation are other examples of compatibilizing reactions. A variety of stable polypropylene–polyamide (PP-PA) alloys can be produced through the reaction of carboxyl, anhydride or glycidyl functionalized PP resins with polyamides containing terminal amino groups.

Copolymers added separately

Nonreactive copolymers

Interfacially active graft or block copolymers of the type A-B or A-C may compatibilize the immiscible polymers A and B provided that C is also miscible or capable of strong interactions with B. Poly(ethylene-co-propylene) elastomer (EPR) or poly(ethylene-co-propylene-co-diene) (EPDM) is an example of such nonreactive compatibilizer for polyethylene/polypropylene (PE/PP) blends.

Reactive copolymers

Reactive copolymers or functionalized polymers of the type A-C (where C is a long reactive segment or a functional group attached to the main chain) may compatibilize a polymer pair A and B provided that C is capable of chemically reacting with B. The nonreactive segment of the polymeric compatibilizer often has different chemical and structural identity from component A, but is still capable of specific interactions leading to a certain degree of miscibility. The majority of PP blends in this category, employ polyamide as the constituent that may react with compatibilizers (e.g. PP, PE) containing anhydride or carboxyl functionalities to form amide or imide linkages. Other copolymers may contain the highly reactive oxazolin, epoxy, isocyanate, or carbodiimide groups discussed earlier.

Table 1 adapted from [2] summarizes recent literature data on the compatibilization of PP blends. In the majority of examples, the compatibilizer is a functionalized PP that has been either formed *in situ* or added separately. Rheological measurements through the use of on-line rheometers have been used to monitor viscosity increases due to chain extension/branching during the reactive compatibilization of PA/PP blends in a corotating twin-screw extruder [4]. The effectiveness of a given compatibilizer containing anhydride or epoxy groups and the choice of compounding conditions were parameters found to affect the on-line rheological data and correlate with the morphologies of the extrudates and the mechanical properties of injection-molded samples.

Low MW compatibilizing compounds

The addition of low MW compounds in a polymer blend may promote compatibility through the formation of copolymers (random, block,

Table 1 Compatibilization of polypropylene blends

Second polymer	*Compatibilizing agents/reactions*
ABS	PP-g-MA, (PAN-co-2-hydroxypropyl methacrylate-co-PS)
EPC	OH funct. EPC, PP-g-MA
EPDM	Dimethylol-phenolic curative, peroxide/coagents
EPR 1,3	(i-PP-g-SA)-g-(OH or NH_2 funct. EPR), hydrog. poly(dimethyl-butadiene-b-IPR)
EVA (saponified)	EDTA salt, Mg stearate
EVA/EPR	Terpene resin
EVOH	EA-MMA-PP graft copolymer, PP-g-VPD
NBR	NBR-b-PP
NBR (7%-COOH)	PP-g-2-isopropenyl-2-oxazoline
NR	Peroxide, bismaleimides, PP-g-MA/ENR, CPE, EPDM
PA	EGMA-co-VA, PE-co-MAA-co-IBA
PA,PA6	PP-g-MA
PA6	Modified polyolefin, PP-g-AA, organofunctional zirconate, peroxide, PP-g-MA, PP-g-AA, olefin copolymer grafted with reactive amine monomers and oligoamides, PBVE-g-PMOX
PA6, PA66	EPDM-g-MA/EPR, SEBS-g-MA, PP-g-MA
PA66	PP-g-MA,PP-g-AA, maleated SBR/trimellitic acid, modified hydrogenated butadiene copolymer
PA66/ADA-co-MXDA	PP-g-MA
PBT	Ionomer, PP-g-MA/2,2'-bis(oxazoline), PP-g-GMA, EAEGMA
PC	PET/EGMA, PBT/EGMA
PE	NR/NBR/IPR/vulcanizing agents, peroxide, EPR, SEBS, EAA precoated GRT, EAA, SEBS, BR, EPR, EPDM, PP-g-PE, 2,5-dimethyl-2,5-bis-(t-butyl-peroxy)hexyne
PE/PS	EVA, SEB, SEBS, PP-g-AA
PE/PVC/(PET,PA)	Ionomer
PET	PP-g-MA
PMMA	PP-g-MMA
PPE	(PP-g-MA + OH funct. PPE)
PPE/HIPS	Maleated PP-g-MA reacted with OH-terminated PS
PPS	PP-g-PS
PS	RPS/PP-g-AA, PP-g-PS, PS-b-PB, SEBS
PS-co-PMS	EPDM-g-(PS-co-PMS)
PVC	Bismaleimide (PDM, CP)
SMA	SBR

graft) or through the combined effects of copolymer formation and crosslinking. Low MW compounds are usually added at relatively low concentrations (typically 0.1–3 wt.%); thus, they may offer economic advantages versus polymeric compatibilizers that are usually effective at higher concentrations. Examples include compounds such as p-toluene sulfonic acid or phosphites that catalyze transamidation melt reactions in blends containing polyamides and polyesters, peroxides, (often in combination with coagents), that may promote compatibilization through the recombination of macroradicals in a variety of polyolefin-based systems, cross-linking additives in 'dynamically vulcanized' thermoplastic elastomers, common curatives in co-crosslinkable rubbers, etc. Examples of low MW compounds used with PP blends can be found in Table 1. In the case of PP/PE blends with mismatched rheological characteristics, peroxides may be used to promote the formation of equiviscous components having enhanced dispersive mixing characteristics [5]. Monomeric polar components, such as maleic anhydride and acrylic acid, may also fall in the present category when used in a sequential functionalization/compatibilization operation; in such a reactive twin-extruder operation, the modification of PP in the presence of monomer/peroxide is followed by the addition of either amine-end group containing polymers (e.g. polyamides) or aminosilane coated inorganic reinforcing fillers (1).

SUMMARY

Reactive processing of immiscible PP blends involves a variety of modification reactions aiming at the development of fine and stable dispersed morphologies; such morphologies are usually characteristic of 'compatibilized' blends and translate into improved physical and mechanical properties. Reactive compatibilization during compounding may involve, not only reactions leading to interchain block/graft copolymer formation, but also controlled degradation/cross-linking reactions that may enhance dispersive mixing of the components by modifying their rheological characteristics. Examples of functionalities that can react at phase interfaces within acceptable extrusion residence times are amine, anhydride, carboxyl, epoxy, isocyanate, carbodiimide, etc. Compatibilization can also be accomplished through free radical reactions, transamidation or transesterification and interchain salt formation. As in any reactive extrusion process, the type and kinetics of the reaction(s) involved and the type, form, characteristics and concentration of the reactants are controlling factors in the selection of process equipment and conditions, such as mode of addition, screw configuration, length of reaction zone, devolatilization, on-line quality control, etc.

REFERENCES

1. Brown, S.B. (1992) Review of reactive extrusion processes, in *Reactive Extrusion: Principles and Practice*, (ed. M. Xanthos), Hanser, Munich, pp. 75–199.

2. Xanthos, M. and Patel, S.H. (1994) Polymer modification/compatibilization, in *How to Manage Plastics Waste: Technology and Market Opportunities*, (eds A.L. Bisio and M. Xanthos), Hanser, Munich, pp. 145–165.
3. Fritz, H.G., Cai, Q. and Bölz, U. (1993) Modification of polypropylene by reactive blending, *Kunststoffe German Plast.*, **83**, 439.
4. Xanthos, M. (1995) Applications of on-line rheometry in reactive compounding, *Adv. Polymer Technol.*, **14**, 207.
5. Yu D-W, Xanthos, M. and Gogos, C.G. (1992) Peroxide modified polyolefin blends, Part II, Effects on LDPE/PP blends with polymer components of dissimilar initial viscosities, *Adv. Polymer Technol.*, **11**, 295.

ABBREVIATIONS OF MONOMERS/POLYMERS

AA	Acrylic acid
ABS	Poly(acrylonitrile-co-butadiene-co-styrene)
ADA	Adipic acid
BR	Butyl rubber
CP	Chlorinated polyolefin
CPE	Chlorinated polyethylene
EA	Ethyl acrylate
EAA	Poly(ethylene-co-acrylic acid)
EAEGMA	Ethylene acrylic ester glycidyl methacrylate terpolymer
EDTA	Ethylene diamine tetraacetic acid
EGMA	Ethylene glycidyl methacrylate copolymer
ENR	Epoxydized natural rubber
EPC	Ethylene-alpha-olefin copolymer
EPDM	Poly(ethylene-co-propylene-co-diene)
EPR	Poly(ethylene-co-propylene) elastomer
EVA	Poly(ethylene-co-vinyl acetate)
EVOH	Poly(ethylene vinyl alcohol)
GMA	Glycidyl methacrylate
GRT	Ground rubber tires
HIPS	High impact polystyrene
IBA	Isobutyl acrylate
IPR	Isoprene rubber
MA	Maleic anhydride
MAA	Methacrylic acid
MMA	Methyl methacrylate
MXDA	m-xylene diamine
NBR	Poly(acrylonitrile-co-butadiene) rubber
NR	Natural rubber
PA	Polyamide
PAN	Polyacrylonitrile
PB	Polybutadiene
PBT	Poly(butylene terephthalate)
PBVE	Poly(butyl vinyl ether)
PC	Polycarbonate
PDM	N,N'-m-phenylene bismaleimide
PE	Polyethylene
PET	Poly(ethylene terephthalate)
PMMA	Poly(methyl methacrylate)
PMS	Polymethyl styrene

PMOX	Poly(2-methyl oxazoline)
PP	Polypropylene
PPE	Poly(phenylene ether)
PPS	Polyphenylenesulfide
PS	Polystyrene
PVC	Poly(vinyl chloride)
RPS	Poly(styrene-co-vinyl oxazoline)
SA	Succinic anhydride
SBR	Poly(styrene-co-butadiene)
SEB	Styrene-ethylene/butylene diblock copolymer
SEBS	Styrene-ethylene/butylene-styrene triblock copolymer
SMA	Styrene-maleic anhydride copolymer
VA	Vinyl acetate
VPD	Vinyl pyrrolidone

Keywords: compatibilization, functionalization, reactive processing, blends, copolymer crosslinking, *in-situ* formed copolymers, low molecular weight compatibilizers, miscibility.

Recycling of polypropylene

Francesco Paolo La Mantia

INTRODUCTION

As for all the homogeneous polymers [1], the main problems in the recycling of PP arise from the easy degradability of this polymer both during its lifetime (mainly by photooxidation), and during processing and recycling operations. Heat, mechanical stress and ultraviolet radiation strongly modify the structure and the morphology and thus the characteristics of polypropylene (PP). Elongation at break and impact strength are the properties most influenced by the degradation phenomena but other effects (e.g. discoloration and other aesthetic damage) must be taken into account, as well.

Although the degradation behavior is common to all polymeric materials, the effects of photooxidative and thermomechanical degradation on PP are dramatic because of the tertiary carbon present in the chain of this polymer.

DEGRADATION AND RECYCLING

The degradation of PP occurs through the following series of reactions describing the oxidative degradation of the polyolefins:

$$\mathrm{P{-}H} \rightarrow \mathrm{P^{\bullet}} + \mathrm{H^{\bullet}}$$

$$\mathrm{P^{\bullet}} + \mathrm{O_2} \rightarrow \mathrm{P{-}O{-}O^{\bullet}}$$

$$\mathrm{P{-}O{-}O^{\bullet}} + \mathrm{P{-}H} \rightarrow \mathrm{POOH} + \mathrm{P^{\bullet}}$$

Although this scheme explains, at least qualitatively, the degradation of polyolefins whatever external forces (heat, mechanical stress, ultraviolet radiation, etc.) act on the polymer, some of the reactions are more

Polypropylene: An A–Z Reference
Edited by J. Karger-Kocsis
Published in 1999 by Kluwer Publishers, Dordrecht. ISBN 0 412 80200 7

specific in respect to the external attack. In particular, the photo-oxidative degradation shows some peculiar features and is mostly a surface reaction while the thermal and thermomechanical oxidation extend into the polymer buk. The degradation kinetics depends on many factors, in particular on the type and level of the external stresses and on some molecular and morphological characteristics of the polymer (molecular weight, crystallinity, etc.). Finally, oxygenated groups, formed during oxidative degradation, act as catalysts of the reactions, accelerating the degradation rate. The main effects of the degradation phenomena on the PP structure are decrease of the molecular weight, change of the molecular weight distribution (MWD) and formation of oxygenated functional groups. As a consequence of these structural changes, the morphology of PP can also be modified. The properties (rheological, mechanical, electric, etc.) strongly alter as a result of changes in the structure and morphology.

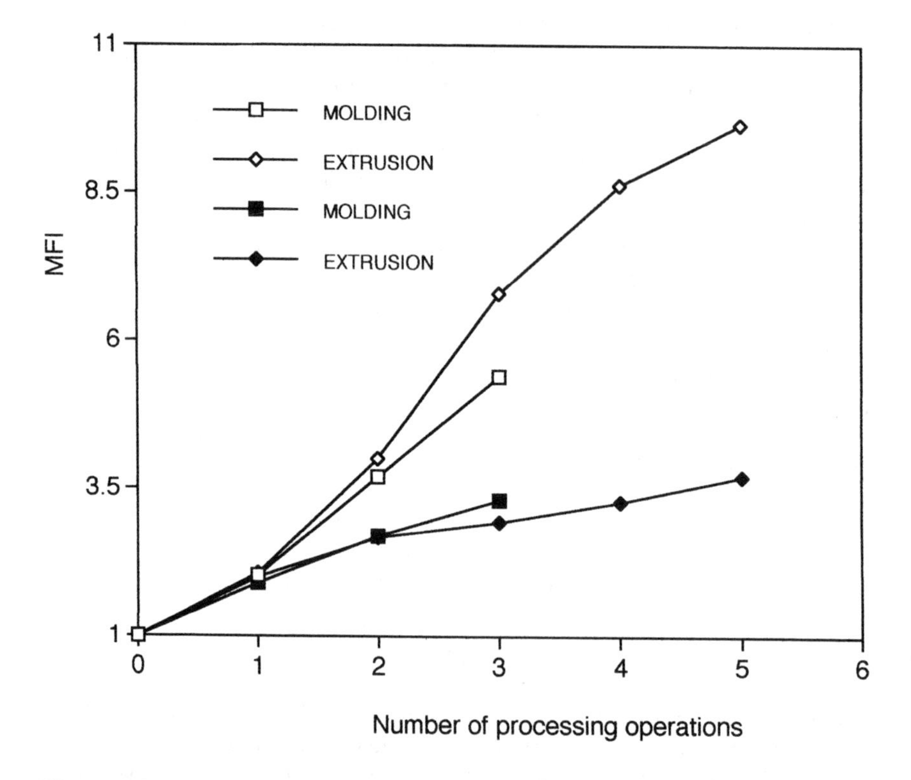

Figure 1 Dimensionless melt flow index as a function of the number of processing operations for unstabilized (open symbol) and stabilized (filled symbol; by adding 0.3 wt.% Irganox B900 prior to each reprocessing step) polypropylene (from [2] with permission).

In Figure 1, the dimensionless melt flow index (MFI) of an extrusion and of an injection-molding grade PP is shown as a function of the number of processing operations [2]. The dimensionless MFI has been calculated by dividing the value of the melt index after each operation by that of the virgin material. In the abscissa of Figure 1, 0 means the virgin polymer, 1 stands for the polymer after the first processing, 2 after the second processing operation, and so on. The MFI increases dramatically for both samples suggesting a drastic reduction of the molecular weight arising from chain scissions during processing. Although the behavior of the two samples is qualitatively similar, the increase of MFI of the high molecular weight sample (extrusion grade) is more pronounced. This can be attributed either to more severe processing conditions during extrusion or to larger mechanical stresses due to the higher viscosity of this sample. The reduction of the viscosity implies that in many cases the secondary material recyclate cannot be used in the same processing

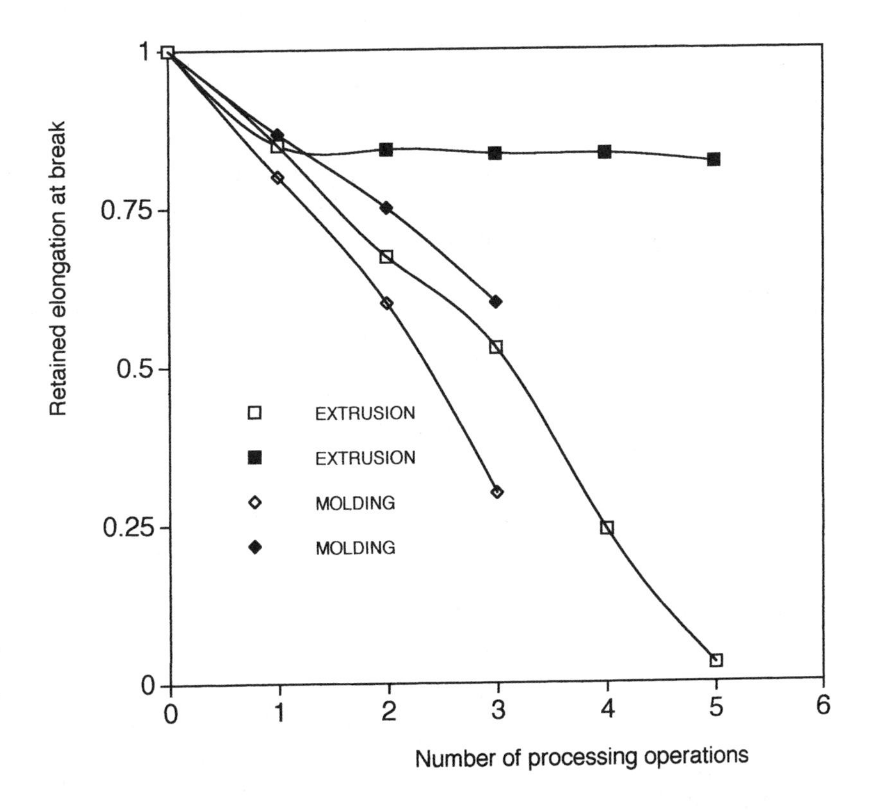

Figure 2 Elongation at break as a function of the number of processing operations for unstabilized and stabilized (by adding 0.3 wt.% Sandestab P-EPQ prior to each reprocessing step) polypropylene.

operation as the virgin polymer but must be reprocessed in a different one requiring a lower viscosity material. For example, in industrial practice recycled extrusion grade PP is used for injection molding.

In Figure 2, the retained elongation at break of the above samples is shown as a function of the number of processing operations [2]. The elongation at break drops dramatically even after the first recycling step and continues to decrease with increasing number of processing operations. For example, after four extrusions the ductile high molecular weight PP shows a very brittle behavior.

STABILIZERS AND FILLERS

As already mentioned, the dramatic reduction of the properties of the recycled PP is due to the easy degradation of this polymer not only during its lifetime, but mostly during the reprocessing operations. By adding suitable stabilizers (re- or post-stabilization) the degradation kinetics can be slowed down and the mechanical properties retained even after many recycling steps [3–5].

In Figures 1 and 2, the values of MFI and of elongation at break are reported also for samples stabilized by adding a phosphite compound (Irganox B900 of Ciba or Sandostab P-EPQ of Sandoz) in 0.3 wt.% before each recycling step. The dramatic changes of both MFI and elongation at break are drastically reduced and the recycled materials show properties similar to the virgin samples. The stabilization must be made before each processing step.

By adding inorganic fillers or polymeric impact modifiers, some mechanical properties of recycled PP can be improved. Inorganic fillers, such as mica, calcium carbonate, glass and wood fibers, etc., enhance the elastic modulus, the dimensional stability and the thermal resistance while at the same time a reduction of the elongation at break is observed. Because recycled PP is generally used for injection-molding items, the decrease of the elongation at break does not play any important role. Table 1 reports the values of elastic modulus and elongation at break of a recycled PP filled with different fillers.

Table 1 Elastic modulus and elongation at break of recycled PP containing 35 wt.% of different fillers

Filler	*Modulus* (GPa)	*Elongation at break* (%)
None	0.8	150
Glass fibers	2.6	11
Wollastonite	1.3	35
Calcium carbonate	1.1	40
Wood fibers	1.3	12

MARKETS AND APPLICATIONS

The main sources of PP to be recycled are packaging films (in particular biaxially oriented PP, but also multilayer films) and battery cases. PP is also found, commingled with other polymers, in postconsumer municipal plastic refuses. After separation of the higher density polymers such as polyethylene terephthalate (PET) and PVC, PP in heterogeneous streams is usually recycled as polyethylene-based blends in which PP is a minor component. During recycling of homogeneous PP (after separation, washing and milling), the material is melt processed (extrusion) into pellets. As already mentioned, the decrease of the molecular weight and thus of the viscosity, hampers the use of recycled PP in the same processes as the virgin material. Therefore the main market for recycled PP is injection molding. No definitive data on the amount of recycled PP are available; however, about 20 000 t/y of recycled PP coming from discarded batteries is used to manufacture new battery cases [6].

REFERENCES

1. La Mantia, F.P. (1996) Basic concepts on the recycling of homogeneous and heterogeneous plastics, in *Recycling of PVC and Mixed Plastics*, (ed. F.P. La Mantia), ChemTec, Toronto, pp. 63–76.
2. Marrone, M. and La Mantia, F.P. (1996) Re-stabilization of recycled polypropylenes, *Polymer Recycling*, **2**, 17–26.
3. La Mantia, F.P. (1998) Recycled plastics: additives and their effects on properties, in *Plastics Additives: An A-Z Reference*, (ed. G. Pritchard), Chapman & Hall, London, pp. 535–543.
4. Dietz, S. (1992) Phosphite stabilizers in postconsumer recycling, in *Emerging Technologies in Plastics Recycling*, (eds G.D. Andrews and P.M. Subramanian), ACS Symposium Series 513, Washington, DC, pp. 134–146.
5. Herbst, H., Hoffmann, K., Pfaendner, R. and Sitek, F. (1995) Stabilizers allow production of higher added-value post-use plastics, *Polymer Recycling*, **1**, 157–166.
6. Perlson, B.D. and Schababerle, C.C. (1992) Polyolefins, in *Plastic Recycling*, (ed. R.J. Ehrig), Hanser, Munich, pp. 73–108.

Keywords: recycling, stabilizer, filler, market, degradation, MFI, melt viscosity, residual properties, post- or re-stabilization.

Resistance to high-energy radiation

Tomasz Sterzynski

RADIATION-INDUCED STRUCTURE MODIFICATION

In many applications, ionizing radiation is used to improve specific properties of polymer materials, to sterilize medical devices, preserve food and to treat waste materials. Electrons and X-rays from particle accelerators or the gamma radiation from cobalt-60 are commonly used. In both cases, the transfer of the radiation energy to absorbing materials is achieved by means of secondary electrons, and the physical and biological effects observed for both types of radiation sources are very similar.

HIGH-POWER ELECTRON ACCELERATORS

Direct and indirect accelerators are used as radiation sources in industrial processes [1]. The direct accelerators, applied specially for low and medium energy applications, require the generation of a high electrical potential which is equal to the final energy of the electrons. In indirect acceleration, a high electron energy is produced by application of time-varying electromagnetic fields. These kinds of radiation sources are used when electron energy higher than 5 MeV is required.

A direct electron accelerator consists of a high-voltage generator connected to an evacuated system. The difference between the various devices is mainly the design of the high-voltage generator; they employ the same technique for electron emission, acceleration and dispersion. The following sources of the high voltage are frequently used:

Polypropylene: An A–Z Reference
Edited by J. Karger-Kocsis
Published in 1999 by Kluwer Publishers, Dordrecht. ISBN 0 412 80200 7

- *Electrostatic generators.* A high electric potential (up to 20 MV) is produced by mechanically moving a static charge between low and high voltage terminals. Because of a low competitivity, this kind of generators are seldom used today.
- *Resonant transformers.* High-voltage AC power is produced by exciting a large air-core winding at its natural resonant frequency and is subsequently applied directly to the electron guns without rectification to dc. The most powerful resonant transformers gives beam power up to 90 kW, with energies rating between 1.5 and 2.5 MeV.
- *Iron core transformers.* A potential of about 1 MV may be produced with this low-frequency transformer connected to rectifier circuits. The final energy obtained by an iron core transformer is usually between 0.3 and 1 MeV and the beam power may reach value of 100 kW.
- *Insulating core transformer.* A three-phase transformer with multiple secondary windings is energized serially by separated iron core segments. The energy ratings reach the values from 0.3 to 3 MeV and beam power capabilities up to 100 kW. A processing application is in the cross-linking of heat-shrinkable film, of tubing and electric wire.
- *Dynamitrons.* The high voltage DC power is generated by a cascade rectifier circuit that is energized by high-frequency (100 kHz) AC power. The energy from 0.5 to 5 MeV, and beam power up to 150 kW may be reached. A frequent use is in the cross-linking of polymeric materials. Units with energy above 3 MeV are used to sterilize medical product by electron treatment.

Indirect accelerators, frequently called linacs, are the machines producing high-energy electrons by injection of short pulses of low-energy electrons into a copper waveguide, containing intense microwave radiation. The final energy depends on the average strength of the alternating electromagnetic field, and on the length of the waveguide.

Other types of indirect accelerators are vhf resonant cavities and linear induction accelerators, such as travelling wave linacs, standing wave linacs, resonant cavity accelerators and linear induction accelerators [1].

REQUIREMENTS FOR RADIATION PROCESSING

The most significant parameters in the irradiation process are absorbed dose distribution, product size, shape, density and throughput rate. The dose absorbed by the irradiated material is defined as the energy absorbed per unit mass in a small volume, at a particular location. The SI unit of doses are gray (Gy) = 1 J/kg, and rad which is 100 times smaller than the gray.

The thermal effect of radiation absorption (for an absorbed dose of 10 kGy, corresponding to 2.4 calories per gram, the water temperature increases 2.4°C) or the radiation chemical effects (calibrated in uniform radiation fields), are used for dose measurement. The power requirement (emitted beam power, P, in kW) may be expressed as follows.

$$P = (D_a/3600)\ (M/t)\ (1/F_p),$$

where D_a is the average absorbed dose (kGy), M is mass of the absorber (kg), and t the exposure time (h). The ratio M/t denotes the throughput rate (kg/h) and F_p is the efficiency of power utilization (the fraction of the emitted beam power absorbed in the material).

The magnitude of the absorbed dose is proportional to the electrons fluency, i.e. the total number of electrons injected per unit area of material. The emitted beam current:

$$I = (D_0/K_0)\ (A/t)\ (1/F_i)$$

where D_0 is the dose measured at the surface of the material (kGy), A is the treated area (m^2) and t is the exposure time (min). The area throughput rate of the treatment process may be estimated as the ratio A/t. F_i is the efficiency factor representing the beam current utilization (typical values between 0.8 and 0.9). K_0 is the area processing coefficient and may be estimated as $K_0 = 6\ S\ J_0$; with S being the electron stopping power of the material (MeV cm^2/g); the factor 6 converts these units to more practical units (MeV cm^2/mA min), and K_0 is the value at the surface of the absorbing material.

IRRADIATION-INDUCED STRUCTURE CHANGES

Gamma radiation, frequently used for sterilization in diverse medical applications, and for crosslinking of polymers, may cause important modifications in both structure and properties of polypropylene (PP) and related copolymers. It was found that the radiation resistance of polymers strongly depends on their amorphous or crystalline structure.

Different effects of gamma radiation on polymers are known [2]: physical (absorption of energy and energy transfer); physicochemical (ionization of the polymers and excitation); chemical (reaction of the types: radical–molecule, ion–molecule, radical–radical and ion–ion); as well as changes of the morphology and of the materials' properties. The reason for property modification is usually the radiation induced changes of the molecular weight and macromolecular structure. Particularly, the degradation-induced decrease of the molecular weight, followed by crosslinking, may produce an increased rigidity of the polymers. Certain changes in crystalline and in amorphous domains of semicrystalline polymers may also be caused by radiation.

DECAY OF PEROXY RADICALS IN POLYPROPYLENE

The oxidative degradation of isotactic PP in the presence of air (as well as the postradiation oxidation) takes place frequently in industrial irradiation processes. The estimation of peroxy radicals, formed as a direct product of radiolytic oxidation reactions between oxygen and carbon-centred free radicals, is very useful in understanding the oxidation process. The application of electron spin resonance (ESR) is a practical method to follow the quantitative effect of this process [2].

The starting point for the decay reaction of the peroxy radicals is the introduction of air to vacuum irradiated samples. ESR spectra permit us to follow peroxy radicals trapped at 'mobile' and 'immobile' sites, where the decay of the radicals is mainly due to the decay of the mobile peroxy radicals. On the contrary, the immobile radicals are rather stable and do not participate in the decay reaction.

IRRADIATION-INDUCED DEGRADATION

The oxygen consumption and yields of oxidation products are frequently used indices of radiation-induced degradation by oxidation. The oxidation process depends on the irradiation conditions, i.e. the atmosphere, pressure, dose rate, temperature and sample dimensions. The oxygen uptake may be determined by measurements of the oxygen content before and after irradiation, where the oxygen products include carbon dioxide, carbon monoxide, water, hydroperoxide, alcohols, ketones and carboxylic acids [3].

Using the experimental equation [3] for the oxygen uptake and oxidative products, it may be seen that this process shows a clear dependence on the dose rate:

$$G = a + b\,[I]^{-n}$$

where a, b and n are constant and I is the dose rate. For most polymers, the experimentally obtained value of the constant is $n = 1/3$ to $1/2$.

OXIDATION MECHANISM

Arakawa *et al.* [4] proposed the following mechanism of oxidation caused by irradiation:

$$\mathrm{R{-}O{-}O^{\bullet}} + (3/2 + m)\,\mathrm{O_2} \rightarrow (\mathrm{{-}CH_2{-}CO{-}OH}) + \mathrm{H_2O} + \mathrm{CO_2(CO)}$$

In this most likely multistage reaction, the carboxylic acids and gaseous oxidative products are produce by peroxy radicals. Characteristic of the radiation-induced oxidation is the significant yield of carboxylic acids and low yield of hydroperoxides. The chain reaction involving hydroperoxides may be more important during the thermal oxidation.

The oxidation layer in the thick film can be determined from the gel fraction data since, in cross-linked polymers, the gel fraction increases upon irradiation under vacuum, except in the zone in which the oxidation takes place. The relationship between the gel fraction (Gel_{obs}) and the oxidation layer (La) is expressed as follow:

$$Gel_{obs} = Gel_{vac}\ (L - 2\ La)$$

where Gel_{vac} is the gel fraction by irradiation under vacuum, and L is the thickness of the sample [4].

The following thicknesses of the oxidation layer La [mm] have been reported for the copolymers of ethylene–propylene: $La = 0.1–0.2$ mm for 10^4 Gy/h; $La = 1–2$ mm for 10^2 Gy/h, and $La = 10–20$ mm for 10^0 Gy/h [1].

POLYMER DEGRADATION AND MECHANICAL PROPERTIES

The mechanical properties of polymers depend on the molecular structure (molecular weight or degree of crosslinking). For a lower dose rate, radiation-induced crosslinking predominates and an increase of the modulus may be observed. On the other hand, for a higher dose rate chain scission usually results in a dramatic decrease of the modulus. The same effects may be observed for the elongation at break. This is specially valid for thin films where a homogeneous action of irradiation in the whole volume of sample may be expected. In the case of thick irradiated samples, the depth of the oxidation layer is a function of the dose rate; the overall radiation resistance is proportional to the square root of dose rate $I^{1/2}$ [1]. The tensile test data shows the dose rate dependence which is nearly proportional to $I^{1/2}$ at higher doses.

STABILIZATION OF POLYPROPYLENE TO GAMMA RADIATION

To stabilize PP to high-energy radiation requires a solution to the problems of postirradiation embrittlement, discoloration and thermal instability. To control both autoxidation and heat-initiated oxidation, special additives, such as free radical scavengers, peroxide decomposers and other stabilizers are incorporated in the usual polymer formulation. As the radiation-induced radicals and peroxides formed in solid polymers are long lived, the stabilizer systems should also keep their activity for a long period.

COLOR FORMATION

Pure PP does not change color in an irradiation dose up to 100 kGy. On the other hand, most commercial-grade PPs shows a yellow discoloration as a result of irradiation. This effect may be attributed to the presence of phenolic additives, frequently used as inhibitors of irradiation and of high-temperature-induced oxidation. Upon oxidation, the phenolic antioxidants (usually trisubstituted phenols) may create phenolic radicals which can dimerize, disproportionate or react with other radicals, leading to formation of peroxycyclohexadienones and quinonoid compounds. These complex structures absorb light in a visible region of the spectrum and discolor the PP. To avoid color formation, the phenolics may be totally removed or used in a very low concentration. Another possibility is the use of additives with similar behavior which do not discolor the PP, such as dilaurylthiodipropionate (DLTDP) [1]. This is the case with medical PP products like syringes or bottles.

POLYMER MORPHOLOGY

Important for the stabilization of polymers are the transfer reactions and the chain termination by radiation-induced polymer/stabilizer radical reactions of the type:

$$R^{\bullet} + R'H \rightarrow RH + R'^{\bullet}$$

The addition of mobilizing additives, such as mineral oil, may effectively enhance the reaction of the chain termination, decreasing the number of radical sites, and significantly reducing polymer chain autoxidation. For example, the use of mineral oil as mobilizing additives [5] for two formulations of PP, with the same degree of crystallinity, leads to an increase of the chain termination by a factor of four. Consequently the use of a mobilizer additionally increases the radiation stability of PP, where the irradiation resistance depends on the level of mobilizer added.

The effect of mobilizing additives is directly reflected in a change of the mechanical properties. Even after receiving a high irradiation dose (which is the case for sterilization of syringes and other medical products), mobilization results in a higher impact resistance compared with sterilized products, but without mobilizing additives.

MOLECULAR WEIGHT DISTRIBUTION (MWD)

It is known that the bulk strength of semicrystalline polymers depends more significantly on the number of the tie molecules between the crystallites, and less on the average length of the molecular chains. Therefore, in certain cases, a remolded radiation-embrittled PP product may regain its original strength. The reason for this is the formation of new tie molecules upon remolding, which override the effect of the radiation-induced decrease in the molecular weight (Figure 1).

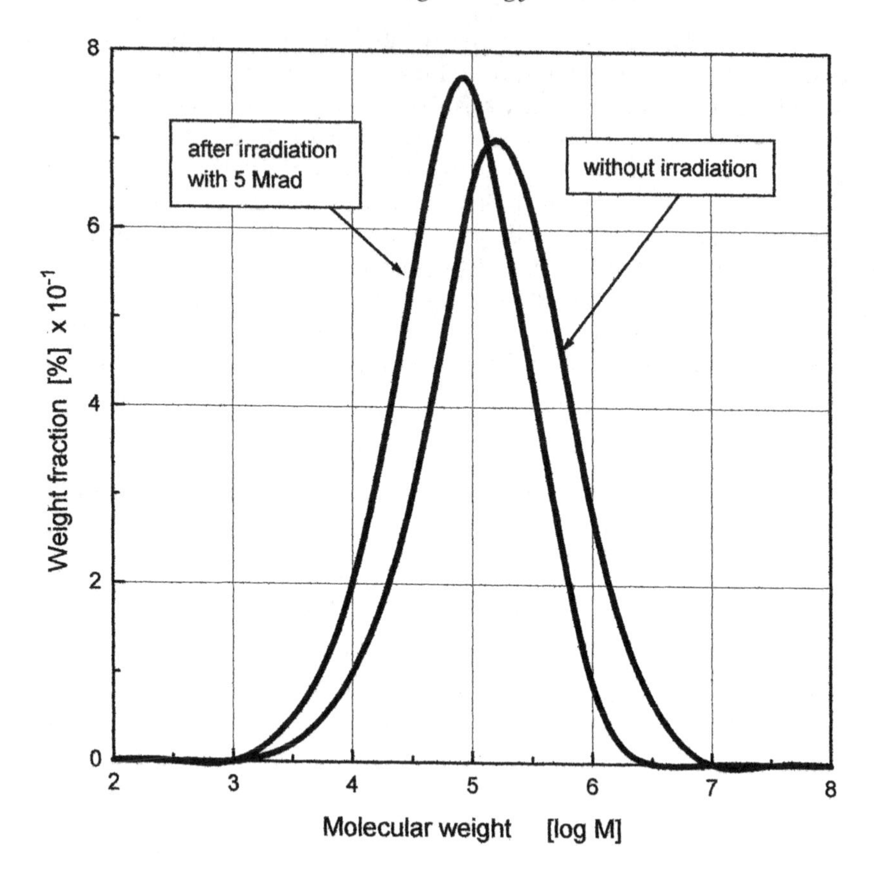

Figure 1 The molecular weight distribution of poly-(ethylene-b-propylene) without irradiation and after an irradiation dose of 5 Mrad.

As the irradiation causes the chain scission, with simultaneous creation of chain termination, induced by polymer/stabilizer radical reactions, a significant modification of the molecular weight $[M_n(5\,\text{Mrad})/M_n(0\,\text{Mrad}) = 0.5]$ may be observed. Consequently a decrease of the molten viscosity for irradiated PP with increasing irradiation dose rate is observed [6].

In PP with a narrow distribution molecular weight, more load-bearing chains or tie molecules are present in the amorphous domains, i.e. between crystallites, than in the polymer with a higher molecular weight distribution. The result is a better maintenance of mechanical properties after irradiation, for such a kind of PP.

The effect of the molecular weight distribution on the irradiation resistance should be considered jointly with the benefits by addition of the mobilizer additives. For medical applications it is found that the mobilizer enhances the radical scavenging activity, and yields PP stable to irradiation effects up to a dose of 70 kGy after aging for 12 months.

CRYSTALLINITY MODIFICATIONS BY IRRADIATION

About the dependence of crystallinity on irradiation, differing results have been published [6–8]. In some cases, a decrease, and in others an increase of crystallinity was found. The probable explanation is the presence or absence of superimposed thermal effects caused sometimes by irradiation. The 'heating' phenomenon, due to the high-temperature treatment by irradiation-processing, or to selfheating due to high radiation energy, significantly affects the formation of the crystalline order in PP.

For irradiation at ambient temperature, the wide angle X-ray scattering (WAXS) and differential scanning calorimetry (DSC) measurements confirmed a slight decrease in the melting temperature due to the molecular weight degradation, but almost no change of the crystallization temperature and of the degree of crystallinity (by WAXS and DSC). This is due to the limited influence of irradiation on the disappearance or formation of new tie molecules.

CONCLUSIONS

To achieve a high radiation resistance of PP, following conditions are found to assure the best results: a low crystallinity, narrow molecular weight distribution, as well as the inclusion of mobilizing additives, which make the chain termination more easy. A primary radical scavenger produces the minimum discoloration, by an increase of the thermal stability of PP, and by inhibition of the auto-oxidation. It was also shown that the gamma radiation, because of its outstanding penetration characteristic, is the best source for sterilization, compared with electron radiation sources.

REFERENCES

1. Singh, A. and Silverman, J. (eds.) (1991) *Radiation Processing of Polymers*, Hanser, New York.
2. O'Donnell, J.H. (1991) Chemistry of radiation degradation of polymers, in *Radiation Effects in Polymers*, (eds R.L. Clough and S.W. Shalaby), American Chemical Society, Washington, DC, pp. 402 ff.
3. Decker, C. and Mayo, F., (1973) *J. Polymer Sci. Polymer Chem.*, **11**, 2879.
4. Arakawa, K., Seguchi, T., Watanabe, Y. and Hayakawa, N. (1982) *J. Polymer Sci. Polymer Chem.*, **20**, 2681.
5. Dunn, T., Williams, E. and Williams, J. (1982) *Radiat. Phys. Chem.*, **19**, 287.
6. Sterzynski, T. and Thomas, M. (1995) *J. Macromol. Sci. Phys.*, B34(1, 2), 119.

7. Jayanna, H.S. and Subramanyam, S.V. (1993) *J. Mat. Sci.*, **28**, 2423.
8. Kostoski, D., Stojanovic, Z. and Kacarevic-Popovic, Z. (1988) in *Integration of Fundamental Polymer Science and Technology*, (eds P.J. Lemstra and L.A. Kleintjens), Elsevier, London, pp. 313 ff.

Keywords: radiation resistance, electron accelerators, gamma radiation, peroxy radicals, absorbed dose, irradiation-induced degradation, molecular weight distribution (MWD), crystallinity, stabilization to gamma radiation die molecules.

Rheology of polypropylene

L.A. Utracki and A. Luciani

INTRODUCTION

Amorphous polypropylene (aPP) of low molecular weight has been know for over 100 years as an oily or waxy substance of low viscosity. The isotactic polypropylene (iPP), discovered in 1954, has a regular structure, a high degree of order and crystallinity. Recently, two members were added to the industrial polypropylene family: isotactic PP with long-chain branching, and syndiotactic polypropylene (sPP), both produced using metallocene catalysts.

The molecular structure of PP results in (1) relatively high (in comparison to other polyolefins) glass transition temperature, $T_g = 0°C$, hence low-temperature brittleness, (2) low resistance to mechanochemical degradation, and (3) rheological behavior that depends on the molecular structure and tacticity. Furthermore, most commercial PP resins are 'modified' (prepared either inside a reactor or in a compounding extruder), blended or reinforced by addition of solid particles. There is great diversity of PP modifications aimed to improve processability, weatherability, rigidity or impact strength. These modification affect the resin flow behavior [1].

RHEOLOGY OF PP

As for other polymers, the viscosity of PP depends on temperature, deformation rate, additives, molecular weight (MW), molecular weight distribution (MWD), etc. For MW $> M_e = 7$ kg/mol (M_e is the entanglement MW) the zero-shear viscosity at $T = 190°C$ follows the dependence [2]:

Polypropylene: An A–Z Reference
Edited by J. Karger-Kocsis
Published in 1999 by Kluwer Publishers, Dordrecht. ISBN 0 412 80200 7

Table 1

No.	Polypropylene	T (°C)	$K \times 10^{17}$	a	b
1	Narrow MWD PP	190	5.32	3.60	0
2	Narrow MWD aPP	190	28.40	3.59	0
	$M_w/M_n \simeq 1$	75	5340.00	3.59	0
3	Polydisperse PP	190	5.32	3.60	1
	$M_w/M_n \simeq 4$	200	4.23	3.60	0.84

$$\eta_0(\text{Pas}) = KM_w^a(M_z/M_w)^b \tag{1}$$

with M_w and M_z being, respectively, the weight-average and z-average MW and exponents a and b are listed in Table 1..

According to Zeichner and Patel's observations, (Z-P), the crossover point of the storage and loss shear moduli: $G_x(\omega_x) \equiv G'(\omega) = G''(\omega)$, provides separate information on M_w (through the cross-point frequency, ω_x) and MWD (via the cross-point modulus, G_x). It has been shown that the Z-P method works quite well for peroxide-degraded PP, fairly well for reactor-grades, but poorly for the formulated resins. Furthermore, for low η_0 resins, the cross-point is not experimentally accessible. Thus, recently a new method was developed by Yoo [3]. The method is based on the 'separation modulus, Modsep', defined as a ratio of the frequencies, ω, at which $G' = G'' =$ const., e.g. 1 kPa: $\text{Modsep}_{G'=G''=1\,\text{kPa}} \equiv \omega'/\omega'' = 6.5354 - 4.3718\log_{10}\,[0.113355 + M_w/M_n]$. The relation is valid for $M_w = 170$–770 kg/mol; $M_w/M_n = 2.1$–6.3. Relationships between the relaxation spectra and MWD for PP have also been proposed.

The rate of shear dependence ($\dot{\gamma} = 0.1$–$2000\,\text{s}^{-1}$) of PP viscosity can be described as:

$$\eta = \eta_0[1 + (\eta_0\dot{\gamma}/\tau)^{m_1}]^{(n-1)/m_1}, \tag{2}$$

where $\tau = 1.58 \times 10^5$, $m_1 = 0.3982$, and $n = 0.0449$.

The temperature and pressure dependence of η (for $T = 180$–300°C; $P \leq 1$ kbar) is [4]:

$$\eta_0 = \eta_0^0 \exp\left\{\frac{a_0 + a_1P}{T - T_r} + a_2P\right\}, \tag{3}$$

where $a_0 = 4.034$ (K), $a_1 = 1.934 \times 10^{-5}$ (K/Pa), $a_2 = -2.150 \times 10^{-8}$ (Pa^{-1}), $T_r = 0$ (K).

The above relationships are valid for linear PP homopolymer, and should not be used either for branched PP, chemically crosslinked PP, or multiphase systems, e.g. for blends, composites, and lubricated formulations, viz. with siloxane or fluorinated additives. Furthermore, PP is sensitive to stress degradation (facilitated by peroxides, specially in the presence of metallic ions, e.g. Cu^{++}) that results in reduction of MW and MWD.

In elongational flows, the linear PP shows remarkable lack of strain hardening (only weak strain hardening was detected for high MW at high strain rates). In uniaxial elongation at small stresses, the Trouton rule holds: $\eta_e = 3\eta_0$. For PP with long-chain branching, strain hardening, similar to that of low density polyethylene, (LDPE) was observed [5].

RHEOLOGY OF PP BLENDS

The multiphase nature of polymeric systems (blends or composites) leads to complex rheological behavior [6]. Since flow depends on morphology, which in turn is determined by the flow field, the measurements in different flow fields usually show different behavior; the basic assumptions of the continuum mechanics are invalid for multiphase materials. Furthermore, since the deformation rate at the interface is discontinuous whereas stresses are continuous, the rheological dependencies should be analyzed as stress, not strain-rate dependent.

It is impossible to generalize the flow behavior of PP blends. The multitude of their types makes such an attempt impractical. However, the flow curves are expected to show a pseudoplastic behavior (equation (1)), with yield stress at higher loadings. When to a polymer-1, a polymer-2 is added, up to the percolation threshold volume fractions, $\phi_c \leq 0.16$, it forms a dispersed phase. When the concentration exceeds this limit, $\phi > \phi_c$, larger domains are formed. The co-continuity of phases increases with ϕ up to the phase inversion concentration, $\phi = \phi_I$, where the distinction between the matrix and the dispersed phase disappears [6].

Owing to the emulsion-like character of polymer blends, the system viscosity increases faster with ϕ than predicted by the log additivity rule: $\ln \eta = \Sigma\ \phi_i \ln \eta_i$. The maximum deviation from the additivity occurs at ϕ_I, hence this mechanism leads to a positive deviation, or PDB. At higher stresses and strains, a negative deviation (NDB) has often been observed. Its origin is in interlayer slip [7]: $1/\eta = \phi_1/\eta_1 + \phi_2/\eta_2 + k(\phi_1\phi_2)^{1/2}$, where k is a characteristic parameter. Note that the maximum negative deviation from log-additivity should occur at $\phi = 0.5$. Thus, the same type polymer blend at different stress levels can show five types of η–ϕ dependencies: additivity, PDB, NDB, PNDB, and NPDB.

Viscosities of PP blends with polyethylene (PE) are mostly additive, with slight either positive or negative deviations. In blends of PP with polystyrene (PS), even at low stresses the interlayer slip (evidenced by concentric 'tree-ring' layers in the extrudate) resulted in NDB. As a rule, the greater the immiscibility, the lower the viscosity of the interphase, the stronger the NDB. The corollary is that compatibilization tends to increase η and produce PDB [1].

During large strain elongational (e.g. through a convergence) or shear (e.g. through channels, pipes, or runners) flows, the dispersed domains elongate. Emerging from the confinement, under the influence of the interfacial energy, these elongated structures attempt to regain sphericity. The shrinkage of the elongated domains causes the stream to swell to a ratio *B*. In blends, the maximum swelling occurs near the phase inversion concentration, $\phi = \phi_I$. The B-value for blends is much larger than that for homopolymers. Since in blends, swelling is dominated by form recovery, its magnitude is unrelated to normal stresses (elasticity). Extrudate swell, *B*, should not be used to measure elasticity of a multiphase system [6].

The extensional stress growth functions in shear and in uniaxial extension were measured for neat PP and linear low density PE, LLDPE, as well as for their blends. Good agreement between the two types of deformation was obtained indicating that linear viscoelastic behavior was obtained; strain hardening was not observed [8].

Since viscosity of an elastomer is usually higher than that of PP, here the phase inversion is expected at high elastomer content. However, the morphology of these blends is far from equilibrium, mainly controlled by processing, not as much by composition. In these systems, mixing in a shear field is ineffective, and if a high degree of dispersion is needed, an elongational flow field should be used [9].

PP blends with either engineering or specialty resins are compatibilized, which usually causes the viscosity and the yield stress to increase. The compatibilizer may be viscous, causing rigidification of the interphase and, as a result, a suspension-like behavior of the alloy leading to PDB behavior. Evaluating the rheology of compatibilized blends, one must consider: (1) variability of the morphology with imposed stress; (2) the flow-dependent kinetics of phase ripening; and (3) effects of the third phase (interphase) with its own rheological characteristics.

For example, compatibilization of PP blends with either polyamide (PA), or polycarbonate (PC), requires 10–20 wt.% of copolymer. Owing to its relatively high molecular weight, the copolymer may significantly increase the system viscosity. The PP/PA blends showed large strain hardening effects in uniaxial extension. Large differences in flow behavior were recorded for flows in steady-state shear, dynamic shear, and uniaxial elongation. The directly measured elongational viscosity was one order of magnitude smaller than that calculated from the entrance pressure drop correction (Cogswell relation). The disparity is related to differences in blend morphology between the two types of flow [6].

RHEOLOGY OF FILLED AND REINFORCED PP

Many fillers and extenders are being incorporated into PP, e.g. $CaCO_3$, talc, mica, carbon black, TiO_2, glass fibers or beads, silica, etc. The rheological behavior of solid filled systems depends on the geometry of filler particles, total loading, dispersion of sizes and shapes of the solid additives, filler–matrix interactions, the rheological and thermal history, as well as on the type and the amount of the 'sizing' agent on the surface of the filler particles.

The flow of filled polymers is best described by the relationships between the relative viscosity and concentration, e.g. the relationship derived by Simha in 1952. Since the particles are suspended in viscoelastic PP melt:

$$\eta_r = \eta(\text{filled PP})/\eta(\text{neat PP})\big|_{\sigma_{12}, T} = 1 + [\eta]\tilde{\eta}\phi,$$

where (4)

$$\tilde{\eta} = 4(1 - Y^7)/[4(1 + Y^{10}) - 25Y^3(1 + Y^4) + 42Y^5];\ Y \equiv 1/[2(\phi_m/\phi)^{1/3} - 1].$$

Here, $[\eta]$ is the shape-dependent intrinsic viscosity, viz. $[\eta] = 2.5$ for spheres and ϕ_m is the maximum packing volume fraction. Owing to ϕ-dependent rotation and orientation of particles, the flow of suspensions with anisometric particles is more complex. Here also Simha's equation is applicable, but experimental values of the two parameters, $[\eta]$ and ϕ_m should be used.

The presence of a filler is expected to enhance viscosity. The behavior may be complicated by the presence of the yield stress (evident at low stress and high ϕ) and effects of the sizing agents. For example, diffusion of a low MW sizing agent to the pipe wall may lubricate the flow, hence the higher stress viscosities can be found independent of filler loading. Strong interactions between particles and PP cause the apparent ϕ and η increase. The η-measurements have been used to determine the thickness of immobilized PP layer on $CaCO_3$, glass fibers or talc. Owing to the yield stress, the presence of solid particles depresses the extrudate swell parameter, B.

In extensional flow, the morphological rearrangements lead to high extensional-to-shear viscosity ratios (both measured at the same ϕ and the deformation rate), and to yield stresses that are larger than in shear. The effects increase with concentration and anisometry of suspended particles [10].

The fabrication of polymer composites is complex. The nature and the shape of the materials used in the fabrication are numerous and the process may differ from one application to another. In the fiber composites materials, the fibers may be discontinuous or continuous. The former may be collimated; the latter may be used as preforms or prepregs with complex architecture and flow. Rheology of such materials depends on their structure and the process. In continuous or long fiber composites processes, the flow is dominated by the in-extensibility of the fiber and orientation present in the system. During preparation of prepregs, the polymer must flow through the fiber mat, where the elongational fields (converging and diverging) dominate.

REFERENCES

1. Utracki, L.A., and Dumoulin, M.M. (1994) Polypropylene alloys and blends with thermoplastics, in *Polypropylene: Structure, Blends and Composites*, Vol. 2, (ed. J. Karger-Kocsis), Chapman and Hall, London, pp. 50–96.
2. Wasserman, S.H., and Graessley, W.W. (1996) *Polymer Engng Sci.*, **36**, 852–61.
3. Yoo, H.J. (1994) MWD Determination of ultra high MFR polypropylene by melt rheology, *Adv. Polymer Technol.*, **13**, 201–205.
4. Kadijk, S.E., and van den Brule, B.H.A.A. (1994) *Polymer Engng Sci.*, **34**, 1535–46.
5. Hingmann, R., and Marczinke, B.L. (1994) Shear and elongational flow properties of polypropylene melts, *J. Rheol.*, **38**, 573–587.
6. Utracki, L.A. (1995) The rheology of multiphase systems, in *Rheological Fundamentals of Polymer Processing*, (eds J.A. Covas, J.F. Agassant, A.C. Diogo, J. Vlachopoulos, and K. Walters), Kluwer, Dordrecht, pp. 113–137.
7. Bousmina, M., Palierne, J.F., and Utracki, L.A. (1998) Modeling of immiscible polymer blends: flow in laminar shear field, *Polymer Engng Sci.*, (in press).
8. Dumoulin, M.M., Farha, C. and Utracki, L.A. (1984) Rheological and mechanical properties of LLDPE/PP/EP-block polymer blends, *Polymer Engng Sci.*, **24**, 1319–26.
9. Utracki, L.A., and Luciani, A. (1996) Mixing in extensional flow field. *Int. Plast. Engng Technol.*, **2**, 37–54.
10. Utracki, L.A., and Vu-Khanh, T. (1992) Filled polymers, in *Multicomponent Polymer Systems*, (eds I.S. Miles and S. Rostami), Longman, Harlow.

Keywords: blends, alloys, compatibilizer, rheology, flow, viscosity, extrudate swell, shear modulus, shear flow, elongational flow, orientation in flow, sizing, yield stress, fillers, talc, mica, glass fibers, reinforced system, interlayer slip, log-additivity rule, concentration dependence, Trouton rule.

Note: Polymer abbreviations are listed on page 626.

Roll forming of composite sheets

D. Bhattacharyya and T.A. Martin

This section provides a general discussion of the roll-forming process and highlights some aspects of this production process which need to be considered, in order to successfully apply it to continuous fiber reinforced thermoplastic (CFRT) materials. The text begins with a brief explanation of the roll forming process, followed by a description of CFRT material properties that make their formability unique with a focus on some important deformation parameters. It is intended that by carefully considering the various aspects of roll forming outlined in this section, a good understanding of the technology should be obtained.

ROLL FORMING

Roll forming is a continuous forming process, which utilizes consecutive roll stations to transform flat strips of material into profiled sections (Figure 1). The versatility and speed with which products can be manufactured make this forming operation particularly attractive for rapid manufacturing. The usual range of working speeds for this operation is 25–45 m/min in the case of sheet metals and the number of roll stations, each with a set of matching roll profiles, constitute the *roll schedule*. Different roll combinations can be utilized to produce an extensive range of cross-sections, from simple open sections to complex closed sections. Tubular sheet metal products are usually formed by butting the sheet edges together and welding them on-line with the forming operation. This auxiliary operation can also be carried out on thermoplastic composite materials with suitable modifications to the rolling mill. It should be noted that many apparently difficult sections

Polypropylene: An A–Z Reference
Edited by J. Karger-Kocsis
Published in 1999 by Kluwer Publishers, Dordrecht. ISBN 0 412 80200 7

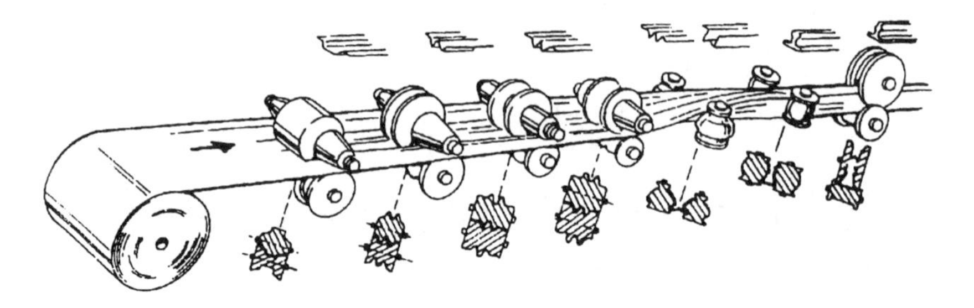

Figure 1 A schematic view of the roll-forming process.

can be broken down into simple sections that can be readily formed; the final product shape is only limited by the ingenuity of the designer and the formability of the material. In addition to the wide variety of shapes available, the roll-forming process offers the flexibility to change the fiber architecture of the laminate, the matrix polymer and the color of the material without seriously affecting the production cycle. This is particularly a major advantage when short runs of different products of the same sectional profile are required.

Roll forming is a relatively new forming technique, which has found widespread use in the sheet metal forming industry in the last few decades. Today, a very large percentage of total sheet metal products is roll formed, which find applications in automobile, building, construction and general manufacturing industries. However, a comparison of the amount of sheet metal being commercially roll formed with that of roll-formed CFRT material (practically none) makes two things apparent: (1) roll-forming technology can be applied to CFRTs only with some specialist modifications; and (2) the potential growth in this market is extremely high, given the vast demand for light-weight, corrosion resistant, high-strength structural components. Recently a good deal of international interest has been generated in using this technology with thermoplastic composite sheets, since it is one of the few rapid manufacturing methods likely to fully realize top quality CFRT products at economically competitive prices [1, 2].

CFRT CHARACTERISTICS

In their molten states, thermoplastic polymers behave like incompressible viscous fluids. By simply melting the matrix polymer and applying moderate forces, complex geometries can often be formed, even in the presence of fibers. Furthermore, it is interesting to note that polypropylene remains deformable over a wide temperature window while cooling from its melt temperature, before it recrystallizes. In this thermal region, it exhibits nonlinear viscoelastic fluid properties which are not easily quantified. In addition to the complicated response of the matrix, the fibers introduce a high degree of anisotropy as they severely limit any extensional flow along their paths. Normally, a hierarchy of four basic flow mechanisms are considered, namely resin percolation, transverse squeeze flow, interply shear and intraply shear [3]. These flow mechanisms are commonly present in combination depending on the forming operation and pertain to sheets with unidirectional plies. The first flow type, resin percolation, is typically associated with the consolidation process, because the matrix polymer flows through the fiber bed to wet the fibers. Transverse squeeze flow is essential in welding and compression-molding processes, as it enables the laps and gaps to be 'healed' between adjacent layers [4]. The third and fourth flow mechanisms, interply and intraply shear, are required in shaping processes which induce single and double curvature in a sheet. Consolidated laminates are often constructed from unidirectional prepreg materials with the fibers aligned in two or more directions, while other laminates are constructed from bidirectional cloths with the fibers woven together. Strain analysis results show that these particular types of CFRTs deform by a *trellis* mechanism [5]. Since the matrix polymer behaves like an incompressible fluid, the *trellis* effect causes the sheet to change its thickness during forming, as the fiber bundles move closer together or further apart.

SALIENT FEATURES OF ROLL FORMING

Roll forming is a plane-stress forming process, which causes a sheet to undergo three-dimensional bending and stretching in order to conform to the shape of the rolls. Furthermore, if the sheet thickness increases during forming, the fixed gap between rolls will might cause transverse squeeze flow and/or resin percolation. On the other hand, if the thickness decreases too much during forming, the material will not conform accurately to the roll profile. The tolerance on the roll gap, therefore, becomes particularly important while forming CFRT sheets.

In spite of its popularity as a process, there is a lot that remains unknown about the roll-forming process from a scientific perspective. In order to form a shaped section, a flat strip is passed through pairs of matching rolls of given profiles. This operation seems simple but the question remains as to how the strip is actually transformed from its initial to its final shape in a smooth and continuous manner. It is now known that the deformation takes place intermittently over a certain length under each roll station. This transition region has been termed as the *deformation length* [6]. Within this zone the sheet acquires a complex three-dimensional geometry that necessitates in-plane and out-of-plane deformations. Defects, such as fiber wrinkling, edge buckling, twisting and bowing, can occur in this region, which hinder the stability of the forming operation and can lead to unacceptable product irregularities. The most difficult problem confronting designers of roll schedules today is how to optimize the forming process without resorting to trial and error solutions. The total strain is determined by the final product profile, so the number of roll passes needed to complete the forming operation decreases, as the roll angle in each forming stage is increased. Now, because the strip is continuously deforming as it passes through the deformation zone, the strain-rate in the material depends on the strain gradient along the deformation length and the velocity of the strip. In order to produce a component in the minimum time, the maximum rate of work must be done in the deformation zone. However, continuous fiber reinforced sheets behave like constrained fluids in their molten states, so the forming stresses are proportional to the strain-rate. Therefore, roll schedule optimization means maximizing the sheet velocity and minimizing the number of roll stations, while keeping the strain-rate, and hence the stresses, below a critical failure level.

ROLL FORMING OF CFRT SHEETS

Figure 2 shows a schematic diagram of a possible set-up for roll forming CFRT sheets. The composite sheets are pulled off a number of rolls and fed into a consolidation process from which the consolidated sheet moves directly into the first set of rollers in the roll-forming mill. At each roll station, the strip is progressively formed into its appropriate shape and emerges in its finished form on the right-hand side of Figure 2. During this process, it is possible to put the strip into tension by increasing the roller rotational speed at successive roll stations. Extra roll

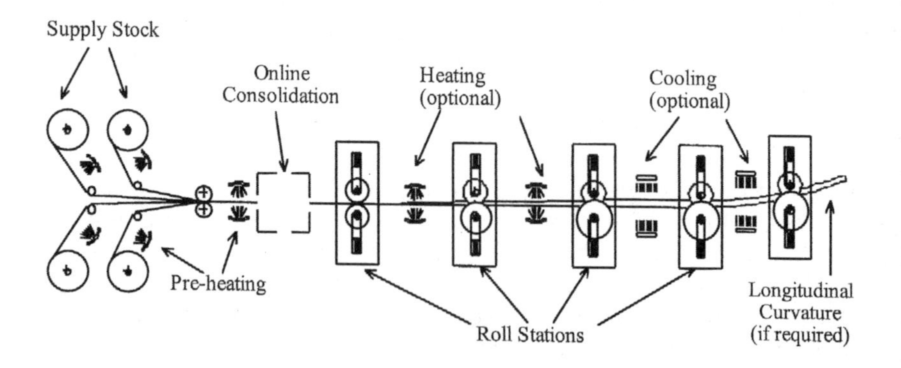

Figure 2 A schematic diagram of a possible roll-forming arrangement for CFRT sheets.

stations may also be positioned after the last forming roll to hold the strip cross-section constant, while it is being cooled, or to introduce longitudinal curvature into the section, if required. Different types of materials can be used as feed stock, namely (1) plies containing unidirectional fiber reinforcement, (2) woven cloths, with two or three directions of fiber reinforcement, and (3) knitted fabrics, with a multitude of knitted patterns to suit many different structural applications. In addition to the three basic material types, there is also a range of methods for combining the resin with the fibers. The fully impregnated plies can be consolidated into high-quality laminates much faster than the other sheets, because the fibers are already fully wet and the matrix does not need to flow through the fiber bed. The slowest consolidation process is needed when a completely unimpregnated sheet is being used. The preforms can all be wound onto feeder rolls and fed into the production process in a continuous manner. Instead of using preform materials on a continuous basis, another starting point could be to use a fully consolidated laminate. The advantage of using a preconsolidated laminate is that the preforms can be cut and aligned at any desired angle to form a laminate with an infinite number of fiber orientations and architecture. While this option alleviates the need for an on-line consolidation process, it is more or less impossible to wind stiff laminates onto feeder rolls. Thus the length of each formed section is limited by the length that can be preconsolidated and conveniently handled. However, a discontinuous material supply procedure eliminates some of the advantages of employing a continuous roll-forming process. Also, particular problems, such as aligning the strip with the forming rolls to eliminate sideways tracking and feeding the leading edge of the strip into roll stations with high roll angles, may arise. On the other hand, online consolidation requires special equipment, such as additional heating supplemented by compressing rollers or a more expensive double-belt press. Roll-forming speed may also get affected by the rate of consolidation, which normally is not very fast. Another important aspect is to keep the temperature of the material within its processing window, while it is being formed. A convenient way to achieve this is to have individual heating units between the roll stations, as illustrated in Figure 2. However, one major advantage of using polypropylene (PP) is that it has a very large processing window (during cooling it can be deformed even around 125°C) and intermediate heating may be completely eliminated, enabling a high forming speed to be achieved at a minimal equipment cost. Polypropylene (PP) based composites such as Plytron® (unidirectional prepregs) and Twintex® (commingled PP/glass fibers) have been reported to be roll formed at a speed higher than 10 m/min and it has been indicated that a higher speed is achievable.

The next consideration is how to retain the desired shape of the produced component. A dummy roll, which draws sufficient heat out of the laminate to cause solidification, is a possibility. Alternatively, the strip can be passed through a water bath or a spray of cold air/water before it exits the final roll station. The positioning of the exit rolls or placement of a former can also be used to generate desired longitudinal curvature in a formed section. However, the form-fixing procedure must be performed with the thermorheological properties of the composite in mind. When the produced sections are cooled, a *spring-forward* effect is commonly observed [7]. Instead of trying to spring-back like a sheet metal material, CFRT sheets tend to reduce the internal angle in a bend upon cooling. The amount of spring-forward depends on the ratio of the in-plane to out-of-plane thermal contraction during cooling and also on the temperature distribution in the part [8]. With a correct combination of roll design and controlled cooling, the desired final cross-section can be obtained. Unlike a material, such as PEEK, which has a relatively small postmelting processing window, transient thermorheological properties become very important in the case of polypropylene-based composites for correct shape production [9].

REFERENCES

1. Mander, S.J., Panton, S.M., Dykes, R.J. and Bhattacharyya, D. (1997) Roll forming of sheet materials, in *Composite Sheet Forming*, Chap. 12, (ed. D. Bhattacharyya), Elsevier, Amsterdam, pp. 473–516.
2. Mander, S.J., Bhattacharyya, D. and Collins, I.F. (1995) Roll forming of fiber reinforced composite sheet, in *Proceeding of the 10th International Conference on Composite Materials*, III (eds. A. Poursartip and K. Street), Woodhead Publishing, Cambridge, pp. 413–420.
3. Cogswell, F.N. (1992) *Thermoplastic Aromatic Polymer Composites*, Butterworth–Heinemann, Oxford.
4. Wang, E.L. and Gutowski, T.G. (1991) Laps and gaps in thermoplastic composites processing, *Composites Manuf.*, **2**, 69–78.
5. Martin, T.A., Christie, G.R., Bhattacharyya, D. (1997) Grid strain analysis and its applications in thermoplastic sheet forming, in *Composite Sheet Forming*, (ed. D. Bhattacharyya), Elsevier, Amsterdam, pp. 217–246.
6. Bhattacharyya, D., Smith, P.D., Yee, C.H. and Collins, I.F. (1984), The prediction of deformation length in cold roll forming, *J. Mech. Working Tech.*, **9**, 181–191.
7. Zahlan, N. and O'Neill, J.M. (1989) Design and fabrication of composite components: The spring-forward phenomenon, *Composites*, **20**(1), 77–81.
8. Ruddock, G.J. and Spencer, A.J.M. (1996). Effect of cooling rate on the solidification of a cylindrical channel section, *Proceedings of the 4th International Conference on Flow Processes in Composite Materials*, The University of Wales, Aberystwyth, UK.
9. Dykes, R.J., Horrigan, D.P.W. and Bhattacharyya, D. (1997) Numerical analysis of shape fixability of continuous fiber reinforced thermoplastics, Proceedings of the *Eleventh International Conference on Composite Materials*, ICCM-II, Vol. 9,

Gold Coast, Australia (14–18 July 1997), (ed. M.L. Scott), Woodhead Publishing, Cambridge, UK, pp. 352–359.

Keywords: roll forming, continuous fiber reinforced polypropylene, on-line consolidation, trellis mechanism, deformation length, spring-back/forward, fabric reinforced polypropylene, consolidation, Plytron,® Twintex®.

Rolltrusion processing of polypropylene for property improvement

J.H. Magill

INTRODUCTION

Property enhancement through mechanical deformation has figured prominently in material property improvements for more than half a century. For several decades now, there has been considerable diversity in the techniques employed to create stronger and improved polymers. Many of these have been reviewed in the literature [1] and will not be mentioned further in this article. The goal of much of the experimental work has been to produce polymers with mechanical properties approaching their respective theoretical tensile strength and tensile modulus. Although there has not been complete agreement on these parameters, our knowledge has advanced considerably. Some literature claims have placed the measured tensile modulus of polymers close to theoretical predictions. While this situation may be realized in thin fibers, it is unlikely to be attained in much larger polymer sections. This is particularly true where the materials possess significant polydispersity and where heterogeneities, such as molecular chain length, microfibrils, macrofibrils, chain ends, various defects and kink bands, are present in the crystalline phase. All of these and other features impair, rather than improve, polymer property enhancement. Rolltrusion was introduced initially to facilitate significant overall property enhancement in an affordable manner without emphasis on ultrahigh goals. A rolltruded

Polypropylene: An A–Z Reference
Edited by J. Karger-Kocsis
Published in 1999 by Kluwer Publishers, Dordrecht. ISBN 0 412 80200 7

morphology is commensurate with a triaxial crystallite morphology where the crystallites are interconnected by variously stretched 'amorphous' tie molecules that bridge them in three directions [2]. This significant texture arises because of the considerable 'take-off' tension that is simultaneously coupled with a very high compressive load applied to the workpiece in the roller nip region (Figure 1). As a result of this deformation mode the processed polymer, whatever its initial state, has experienced a triaxial stress field that is responsible for the three-dimensional (3D) morphology induced and verified through testing. For each polymer, the processing conditions must be established beforehand. The resulting workpiece (if additives are absent) is transparent, as the inhomogeneities within it are too small to scatter white light randomly. Consequently, commercial plastics of moderate molecular weight have been converted into quality materials with much better properties, that are sometimes comparable with some engineering plastics, except for their ability to withstand high temperatures. From published results on rolltruded polypropylene, the 3D texture that is created is unique and can be used to advantage to produce strong selective membranes [3] for vapor and liquid transmission. Relatively thin transparent films of many polymers may be produced now without the use of solvents (noxious or otherwise), which is a considerable process advantage in membrane/film preparation.

EXPERIMENTAL PROCEDURE

This will only be described schematically (Figure 1) since we are merely concerned here with the basic features of rolltrusion [1, 2]. Variously sized polymer billets of all kinds, and strips/rolls of different widths and thicknesses can be converted into very strong clear sheets. About a dozen different polymers, copolymers and composites have been rolltruded successfully. During processing, the workpiece only becomes thinned in the thickness (compression) direction, there being little or no change in the width of the billet/specimen entering and leaving the rolls; hence there is distinction from regular twin-mill rolling *per se* in the process and the morphology that is produced. Sheet-like polymer extrudates can also be converted into processed articles with improved 3D properties, without loss of transparency, and need for conventional cross-plying techniques that are often used to improve properties.

MORPHOLOGY OF ROLLTRUDED POLYMERS

The improvements in properties arises as a result of well-oriented crystallites (in essence crystallographically disposed) within the workpiece interconnected essentially in three mutually perpendicular directions by tie molecules as illustrated in Figure 1(b). Mechanical properties

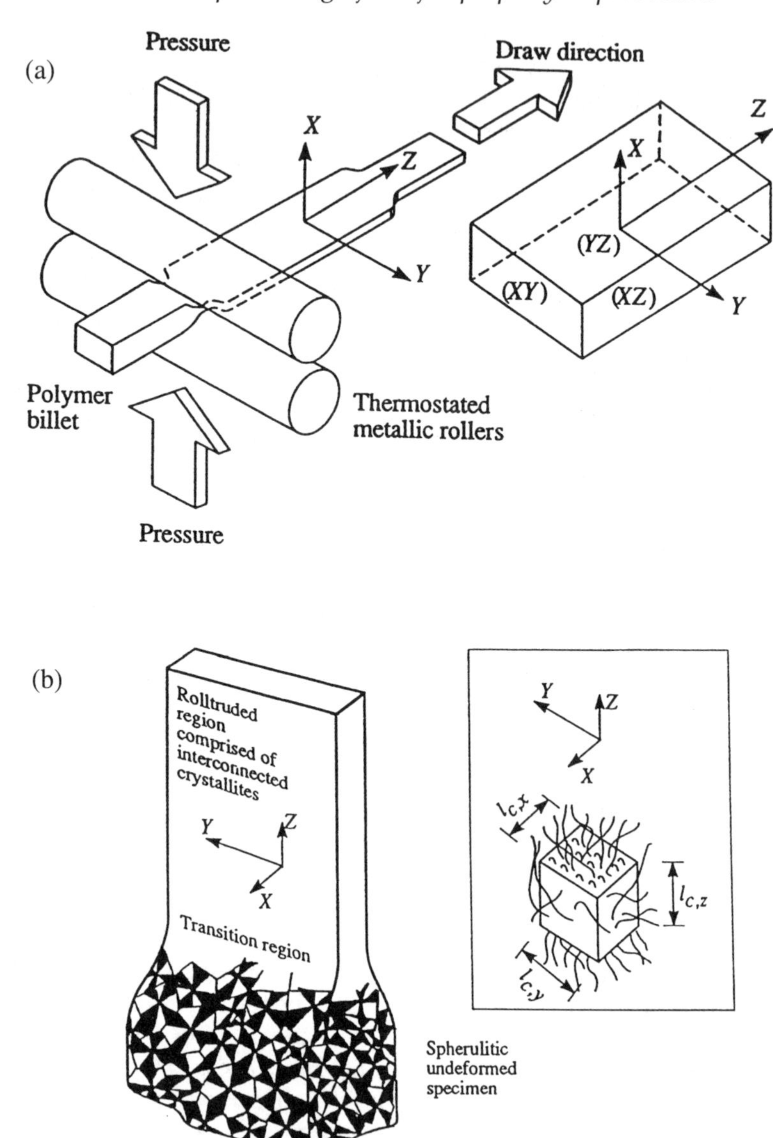

Figure 1 Rolltrusion processing schematic: (a) triaxial orientation process; (b) morphological schematic of the unprocessed and processed polymer with an enlarged insert depicting a crystallite with tie-molecule connections.

of all rolltruded materials follows a general pattern for all polymers of moderate to high molecular weights. It has been established that the relative improvements in properties that occur with respect to the coordinates of the original workpiece such that $Z >> Y > X$ direction (see insert in Figure 1). This pattern and resultant properties have been substantiated using conventional testing techniques that are commonly used for polymer characterization [1, 4].

MECHANICAL PROPERTIES

Rolltruded PP specimens were tested using a MTS 800 machine according to ASTM D-38 procedures from 25°C up to the highest rolltrusion temperatures for preparation of the workpiece [2, 4]. Unfortunately, mechanical measurements in the *X*-direction can only be made in compression because of the thinness of the specimens. Still, high-frequency sonic moduli tests were made on selected PP when an improvement in modulus with draw ratio (DR) was noted. Sonic measurements were made in other specimen directions (*Y* and *Z*) and were found to be in agreement with tensile measurements within 5–10% for deformation ratios as high as $\times 25$. This provided an important cross-check on properties. Material strengthening tie molecule connections, as well as increasing their numbers, upgrade the mechanical properties, such that $Z >> Y > X$ in the three directions indicated in Figure 1(a). It is likely that there are some incorporated fully extended chains but all the experimental evidence indicates that model Figure 1(b) is preferred since significant necking and elongation occurs when the rolltruded workpiece is subsequently drawn in the *Y* direction. It is instructive that crystallites rotate and transform from their rolltruded 3D morphology into a one-D (uniaxial state) according to microbeam X-ray diffraction measurements conducted at different stages of this transverse drawing process. Besides, necking occurs and this, too, is consistent with the fact that tie molecules must also interconnect crystallites in the *Y* direction.

THERMAL MEASUREMENTS

These were made using a Perkin-Elmer DSC2 calorimeter fitted with computer integration accessories. Baseline corrections were made for all thermograms whatever the test run conditions. The significance of these measurements has already been addressed in the literature [5] and other aspects of the work will be addressed elsewhere.

X-RAY MEASUREMENTS

Wide angle X-ray measurements (WAXS) were made in three directions *X*, *Y* and *Z* for each specimen using monochromatic CuK(α)-radiation with an XRD5 GE diffractometer fitted with a graphite monochromator

[1, 4]. A Laue-type film camera (temperature regulated) using CuK(α)-radiation was employed for photographic recording purposes for PP [2].

Small angle X-ray diffraction (SAXS) measurements were carried out using: (1) a Rigaku–Denki, 6 kW generator fitted with a small angle camera with a vacuum path; (2) selected measurements were also performed on the 10 m SAXS facility at the National Center for Small Angle Scattering, Oakridge, Tennessee. These diffraction patterns verified that there was a measurable long period in the three mutually perpendicular directions of the rolltruded specimens, in line with the 3D model proposed [6], and the evidence cited in the mechanical properties section of this manuscript.

BIREFRINGENCE MEASUREMENTS

Thin sections were cut from selected specimens of PP with deformation ratios in the range from × 1 to almost × 40 and for regular polyethylene. Their birefringence, perpendicular to the *X*, *Y* and *Z* coordinates of selected specimens, was measured with a calibrated Berek compensator. WAXS measurements were also made on similar samples. It was found that the crystalline and amorphous orientation factors, f_c and f_a respectively, were determined assuming that rolltruded specimens were of a two-phase nature – a questionable assumption – because tie molecules in the intercrystallite regions of a well-oriented amorphous phase are not randomly arrayed. Many are taut due to the constraints imposed during processing. A linear correlation of tensile strengths is found only with f_a, not f_c, over the entire deformation range [7]. The crystallite orientation function is only linear up to deformation ratios below × 10 deformation ratio, after which it is essentially invariant at higher ratios [7]. The implications of this result are clear and they will be mentioned shortly.

WEAR MEASUREMENTS

Wear rates were measured for rolltruded PP investigated for planes perpendicular to *X*, *Y* and *Z* directions. Anisotropic properties have been reported [8].

COMPUTER MODELLING

Using a modified commercial software package termed ABAQUS (initially formulated for metalworking technology), an attempt was made to simulate some features of the rolltrusion process at several deformation ratios $< \times 5$. Simulated deformed grid patterns compare favorably with experimentally observed workpiece deformations. Besides, useful computer plots for trends in strain and stresses with processing time (equivalent to distance processed along workpiece) were obtained for a few processing temperature and deformation ratios (9).

STATE OF KNOWLEDGE

Property improvements for rolltruded polypropylene and other plastics have been reasonably well documented, but there is still much unpublished. A self-consistent model of the processed workpiece (Figure 1) embodies all the experimental features that are consistent with a 3D morphology comprised of crystalline, oriented amorphous and unoriented amorphous polypropylene.

The high stresses which the polymer experiences in the roller nip region transform spherulites into well-organized lamellae that are connected by tie molecules to provide strength enhancement according to their populations expressed as $Z >> Y > X > XYZ$ (isotropic condition) of initial workpiece. *XYZ* denotes the starting sample. The special thermotropic network that is produced via rolltrusion is one in which the crystallites function in a sense as the 'crosslinks', akin to a conventional crosslinked system in most of its mechanical functions.

Table 1 illustrates selected improvements in bulk tensile modulus and tensile strength for PP made on bulk specimens and indicates that there is considerable improvement in mechanical properties in the *Z* and *Y* directions with corresponding enhancement for compression in the *X* direction of these rolltruded specimens. High-frequency sonic measurements show that there is a moderate increase in tensile modulus (up to $\times$ 1.5–2) for deformation ratios up to $\times$ 20. Other polymers have been found to behave similarly.

These property changes (enhancements in strength) are accomplished without complete chain extension or even reverting to using ultra-high

Table 1 Mechanical property enhancement for rolltruded polypropylene

Polymer sample	*Test direction*	*Maximum modulus* (GPa)	*Maximum strength* (MPa)	*Modulus enhancement factor*	*Strength enhancement factor*
Original iPP	Z or Y[a]	0.7	27.6	1.0	1.0
	X	0.7	38.6	1.0	1.0
Oriented	Z	21.3	524.0	28.1	19.0
	Y	2.3	44.8	3.0	1.6
	X	2.1	104.0	3.0[b]	2.7

[a] The original sample was unoriented so tensile properties show no directional dependence.
[b] Compression tested.

molecular weight materials. Physical evidence from X-ray measurement (WAXS and SAXS), birefringence and mechanical property measurements, demonstrate that there is a rapid improvement in f_c as the crystallite orientation tends to $\times 10$ ratios. All of the evidence so far supports the condition that the crystallites are aligned crystallographically with the molecular chain axis in a uniaxial–uniplanar fashion, i.e. in the (Z) direction of the highest strength and modulus. The other two axes that are oriented mutually perpendicular to each other at right angles to the *c* direction also display substantial mechanical improvements. This 'single crystal-like' morphology is unique in deformation processing. Ordered crystallites are interconnected in three directions via tie-molecules and amorphous regions.

Within a rolltruded PP and other polymers, the long period in the *X*, *Y*, and *Z* directions of a rolltruded specimen depends upon processing conditions, but more on the processing temperature. The intercrystallite tie molecular region is typically about 25–30% of the overall long period in the *Z* direction for homopolymers and even greater in the case of copolymers. Quality mechanical properties can be obtained without using ultra-high molecular weight polymer(s).

Again, significant support for the 3D or triaxial model (Figure 1) is obtained from the fact that tensile strength for large rolltrusion deformations (up to $\times 30$ or more) increases almost linearly with f_a, not with f_c (see Figure 3(b), ref. [7]). This may be interpreted (notwithstanding the simple two-phase model) that the load bearing crystallites are essentially fully aligned at orientations of $\times 10$ and beyond this point further strengthening comes about from tie molecules. Note that f_c under similar test conditions is far from linear for up to large deformations and this factor tends to unity at DR values well below $\times 10$). The 'amorphous' f_a versus tensile strength is almost linear from low to large deformations. The tie molecule (amorphous /oriented)–tensile strength trends suggests an enhancement with DR up to high values.

Correspondingly, the wear behavior (essentially a surface property) of processed polymer shows 3D anisotropic behavior [8].

REFERENCES

1. Shankernarayanan, M.J., Sun, D.C., Kojima, M. and Magill, J.H. (1983) *J. Int. Polymer Proc. Soc.*, **1**, 66.
2. Berg, E.M., Sun, D.C. and Magill, J.H. *SPE ANTEC*, Tech. Papers, **35**, 625 (1989); *Polymer Engng Sci.*, **29**, 715 (1989); *Polymer Engng Sci.*, **30**, 635 (1990).
3. Ciora, Jr, R.J. and Magill, J.H. *J. Polymer Sci., Polymer Phys. Ed.*, **32**, 1035 (1992); *Polymer*, **35**, 949 (1994); *J. Appl. Polymer Sci.*, **58**, 1021 (1995).
4. Lin, C.C., Kojima, M. and Magill, J.H. (1992) *J. Mat. Sci.*, **27**, 5849.
5. Sun, D.C. and Magill, J.H. (1989) *Polymer Engng Sci.*, **29**, 1503.
6. Sun, D.C., Magill, J.H. and Lin, J.S. (1988) *Bull. Am. Phys. Soc.*, **33**, 717.

7. Magill, J.H., Sun, D.C. and Shankernaryanan, M.J. (1987) *J. Appl. Polymer Sci.*, **34**, 2337.
8. Voss, H., Magill, J.H. and Friedrich, K. (1987) *J. Appl. Polymer Sci.*, **33**, 1745.
9. Sun, D.C., Berg, E.M. and Magill, J.H. (1989) *Proc. Am. Chem. Soc., Div. Polymer Materials Eng. Sci.*, **61**, 555.

Keywords: crystallite draw ratio, long period, mechanical properties, membrane, three-dimensional morphology, orientation factor, roll-trusion, SAXS, tie molecules, WAXS.

Size exclusion chromatography

K. Lederer

INTRODUCTION

Size exclusion chromatography (SEC) separates macromolecules according to their hydrodynamic volume and has become the dominant method for the determination of molecular weight or mass distribution (MWD or MMD) of synthetic polymers [1].

Crystalline polyolefins as polypropylene (PP) with high tacticity and copolymers of propylene with ethylene and/or higher α-olefins containing sufficiently long isotactic PP blocks can only be dissolved under conditions (solvent/temperature) which cause complete melting of the crystalline domains. Therefore, SEC of PP and crystalline copolymers of propylene must be carried out at elevated temperatures which requires the special equipment of 'high-temperature SEC (HT-SEC)' which is commercially available from several sources, e.g. Millipore-Waters Corp. (Milford, MA, USA) and Polymer Laboratories Ltd (Church Stretton, Shropshire, UK).

EXPERIMENTAL CONDITIONS

HT-SEC of PP is usually carried out at a temperature ranging from 135°C up to 150°C. This temperature must be maintained in the SEC apparatus from the point of injection of the sample solution throughout the chromatographic columns and the concentration detector or the multidetection system at the end of the SEC line. 1,2,4-trichlorobenzene (TCB) is preferably used as solvent and eluent. The concentration of the sample solution is usually in the range of 0.1–0.3 g/l. SEC columns based on

Polypropylene: An A–Z Reference
Edited by J. Karger-Kocsis
Published in 1999 by Kluwer Publishers, Dordrecht. ISBN 0 412 80200 7

polystyrene (PS) gel made by copolymerization of styrene with 1,4-divinylbenzene as a crosslinking agent are dominant. Sometimes, normal phase silica-gel columns are used; in this case, o-dichlorobenzene is the preferred solvent to prevent adsorption of polymer, e.g. PS, on the columns.

As PP is sensitive to thermooxidative degradation, an antioxidant such as 2,6-di-tert.butyl-4-methylphenol is usually added to the solvent in a concentration of about 0.5 g/l. The complete dissolution of PP pellets usually requires at 150°C about 4 h, and at 170°C about 1 h, respectively, and should be carried out in a nitrogen atmosphere with occasional stirring; too intensive stirring may cause molecular degradation of the high molar mass fraction.

DATA EVALUATION

As in any kind of column chromatography, the raw data of SEC (Figure 1(a)) is the trace of the signal from the concentration detector over the retention volume, $S(V)$, which is proportional to the concentration, $C(V)$:

$$C(V) = k(V)S(V). \tag{1}$$

If the factor k is independent of the retention volume, V, as in the case of a refractive index detector applied to sample solutions of propylene homopolymer, the normalized concentration signal, $e(V) = S(V)/\int S(V)\,dv$, often called the eluogram, is a close approximation to the differential mass distribution of the retention volume, $w(V)$. The conversion of $w(V)$ to the differential mass distribution of molar mass, $w(M)$, requires a knowledge of the dependence of molar mass, M, on the retention volume, V (Figure 1(b)). This dependence must be determined by molar mass calibration of SEC.

In the case of propylene homopolymer, the universal calibration procedure established by Grubisic *et al.* [2] can be applied using anionically polymerized polystyrene molar mass standards with narrow MMD; for TCB at 135°C, the following values of the parameters K and a in the equation:

$$[\eta] = K\,M_r, \tag{2}$$

where $[\eta]$ = limiting viscosity number (dl/g) and M_r = relative molecular mass, are recommended:

$K = 1.75 \times 10^{-4}$ dl/g, $a = 0.670$ for polystyrene [3];
$K = 1.90 \times 10^{-4}$ dl/g, $a = 0.725$ for propylene homopolymer [4].

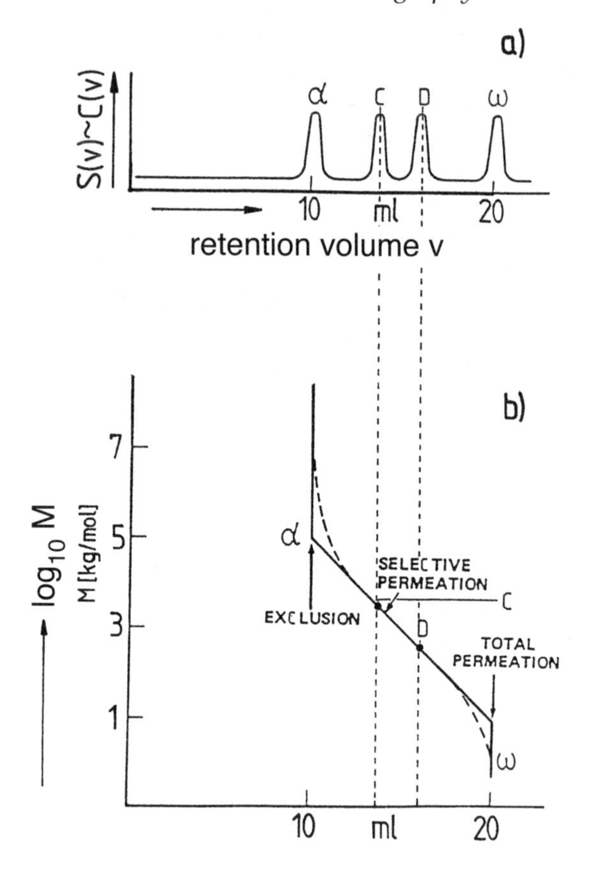

Figure 1 SEC data evaluation: (a) chromatogram, S (V), of monodisperse samples α, C,D, ω with molar mass decreasing from α to ω; (b) typical dependence of molar mass M on retention volume, V; $\log_{10} M$ is plotted versus V.

In the case of copolymers of propylene with ethylene and/or higher α-olefins, the value of the parameter a will be close to $a = 0.725$ (for propylene homopolymer) and the parameter K can be estimated by the expression [4]:

$$\frac{K}{1.9 \times 10^{-4}} = \left(\frac{M^*}{M}\right)^{0.725} \tag{3}$$

where M = molar mass of a propylene homopolymer, and M^* = molar mass of a copolymer of propylene with ethylene and/or higher α-olefins with the same number of carbon atoms in the backbone chain as the propylene homopolymer of molar mass M.

Keeping in mind that, according to Grubisic *et al.* [2], macromolecules eluting at the same retention volume exhibit the same molar hydrodynamic volume, described by the product $[\eta] \times M = K\, M^{1+a}$, the measured dependence of M and $[\eta] \times M$, respectively, on the retention volume V of polystyrene standards with known M and narrow MMD can be transformed to the dependence of molar mass on the retention volume, V, for the eluted fractions of propylene homopolymer or of the types of propylene copolymer specified above. Usually, this dependence is presented in the semilogarithmic plot $\log M$ versus V (Figure 1(b)); the knowledge of the function $\log M(V)$ then allows the straightforward calculation of the logarithmic differential mass distribution of molar mass $w^*(\log M)$ according to:

$$w^*(\log M) = w(V)/(\mathrm{d} \log M/\mathrm{d}V). \tag{4}$$

As $w(V)$ fulfills the normalization condition $\int_{v=0}^{\infty} w(V)\, \mathrm{d}V = 1$, $w^*(\log M)$ is normalized to fulfill the condition:

$$\int_{\log M=0}^{\log M_{max}} w^*(\log M)\, \mathrm{d} \log M = 1. \tag{5}$$

PEAK BROADENING

As shown in Figure 1(a), even a monodisperse sample will give a chromatogram which can usually be approximated by a Gaussian function $D(V)$:

$$D(V) = \frac{1}{\sigma\sqrt{2\pi}} \exp\left[-\frac{(V - V_0)^2}{2\,\sigma^2}\right] \tag{6}$$

where V_0 = the retention volume at the peak maximum. The parameter σ describes the extent of peak broadening and is called the peak broadening parameter.

As a first approximation, σ can be considered to be independent of the retention volume V; then the influence of σ on the eluogram $e(V)$ is given by the equation:

$$e(V) = \int_{V_0=0}^{\infty} w(V_0) \frac{1}{\sigma\,\sqrt{2\pi}} \exp\left[-\frac{(V - V_0)^2}{2\,\sigma^2}\right] \mathrm{d}V_0 \tag{7}$$

The resulting difference between $e(V)$ and $w(V_0)$ in equation (7) depends both on σ and on $w(V_0)$, i.e. on the MMD, w^* $(\log M)$, of the analyzed grade of PP and on the calibration curve, $\log M(V_0)$, of the column set.

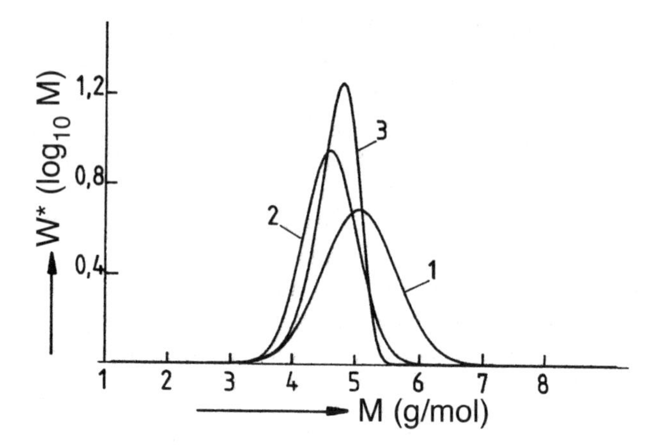

Figure 2 Typical cases of molar mass distribution of propylene homopolymer: curve 1, high molar mass grade ($M_w = 882\,kg/mol$, $M_w/M_n = 6.0$); curve 2, controlled rheology grade ($M_w = 200\,kg/mol$, $M_w/M_n = 2.57$); curve 3, typical metallocene grade ($M_w = 200\,kg/mol$, $M_w/M_n = 2.0$).

Typical cases of the differential mass distribution of molar mass, $w^*(\log M)$, of propylene homopolymer are shown in Figure 2. Curve 1 is a logarithmic normal distribution ('Wesslau' distribution) of a conventional Ziegler–Natta grade with very high molar mass and broad MMD ($M_w = 882\,kg/mol$, $M_w/M_n = 6.0$; M_w = weight average molar mass, M_n = number average molar mass). Curve 2 shows the MMD of a controlled rheology grade for injection molding (Wesslau distribution with $M_w = 200\,kg/mol$, $M_w/M_n = 2.57$) and curve 3 is a 'most probable (Schulz–Flory)' distribution of a metallocene grade with $M_w = 200\,kg/mol$ and $M_w/M_n = 2.00$.

The peak broadening parameter, σ, can be experimentally determined by coupling of SEC with a light scattering detector [5]. A rough estimate of σ can be made from the number N of theoretical plates ($N = V_0^2/\sigma^2$) determined for the same SEC system with low molar mass compounds, e.g. toluene.

For a typical column set used for SEC of PP, the calibration curve, $\log_{10} M(V)$, in its linear range (Figure 1(a)) may be approximated, for example, by $\log_{10} M = 12.84 - 0.566 \times V$, where M = molar mass in kg/mol, and V = retention volume in ml. For this example, $N = 21\,000$ determined with toluene corresponds to $\sigma = 0.17$ ml, and $N = 3\,000$ to $\sigma = 0.45$ ml, respectively. Usually, a newly delivered modern high-quality SEC column set with a linear calibration range similar to the example given above fulfills the criterion $N > 20\,000$. After being used in HT-SEC

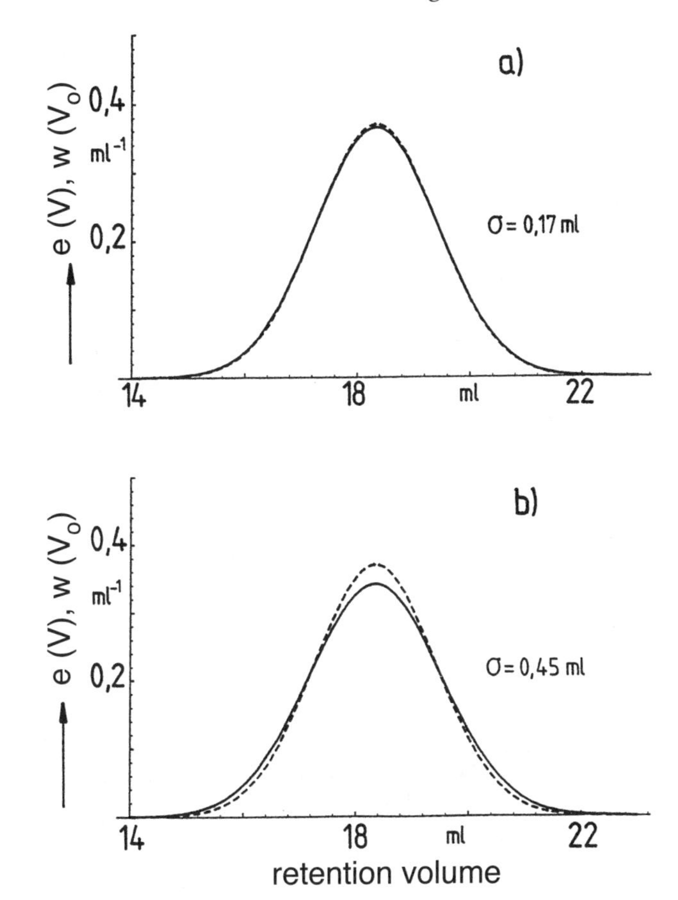

Figure 3 Calculated peak broadening in SEC of Ziegler–Natta PP blow-molding and thermoforming grade (Wesslau – MMD; $M_w = 467$ kg/mol, $M_w/M_n = 5.6$); ——— e (V), ----- w (V_0). (a) $\sigma = 0.17$ ml, (b) $\sigma = 0.45$ m.

over longer periods of time, such column sets will however gradually deteriorate leading to a considerably lower value of N, e.g. $N < 5000$.

Figure 3 and Figure 4 illustrate the effect of peak broadening on the eluogram in the case of a broad Wesslau molar mass distribution and a narrow Schulz–Flory distribution assuming $\sigma = 0.17$ ml (Figures 3(a) and 4(a)) and $\sigma = 0.45$ ml (Figures 3(b) and 4(b)), respectively.

In the case of a narrow MMD with $M_w/M_n < 3$, there can be a considerable influence of the peak broadening effect on SEC results, if column sets with a lower number of theoretical plates are used; then, the

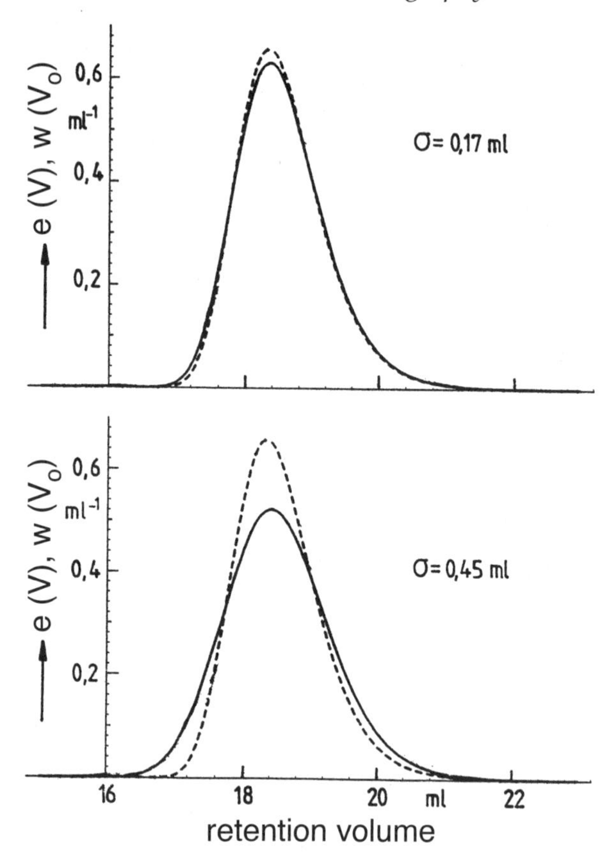

Figure 4 Calculated peak broadening in SEC of a metallocene polypropylene grade (Schulz–Flory MMD; $M_w = 200$ kg/mol, $M_w/M_n = 2.0$); ——— e (V). ----- w (V_0). (a) with $\sigma = 0.17$ ml, (b) with $\sigma = 0.45$ ml.

comparison of the MMD of the same sample of polypropylene determined in different laboratories may lead to greater discrepancies [6].

REFERENCES

1. Yau, W.W., Kirkland, J.J and Bly, D.D. (1979) *Modern Size Exclusion Chromatography*, J. Wiley, New York.
2. Grubisic, Z., Rempp, P. and Benoit, H. (1967) A universal calibration for gel permeation chromatography. *J. Polymer Sci., Polymer Lett. Ed.*, **5**, 753–759.
3. Rudin, A., Grinshpun, V. and O'Driscoll, K.F. (1984) Long chain branching in polyethylene. *J. Liquid Chromat.*, **7**, 1809–1821.
4. Scholte, Th.G., Meijerink, N.L.J., Schoffeleers, H.M. and Brands, A.M.G. (1984) Mark–Houwink equation and GPC calibration for linear short branched

polyolefins, including polypropylene and ethylene-propylene copolymers. *J. Appl. Polymer Sci.*, **29**, 3763–3782.
5. Lederer, K., Beytollahi-Amtmann, I. and Billiani, J. (1994) Determination and correction of peak broadening in SEC of controlled rheology polypropylene. *J. Appl. Polymer Sci.*, **54**, 47–55.
6. Lederer, K., and Mingozzi, I. (1997) Molecular characterization of commercial polypropylene with narrow and broad distribution of molar mass. *Pure Appl. Chem.*, **69**, 993–1006.

Keywords: size exclusion chromatography (SEC), molecular weight distribution (MWD), universal calibration, homopolymer, copolymer, Schulz-Flory distribution, Wesslau distribution, controlled rheology grade, metallocene grade, peak broadening, weight average molecular mass, number average molecular mass.

Solid-state forming of polypropylene

A. Ajji and N. Legros

INTRODUCTION

The replacement of some high-performance structural materials, such as fiber reinforced composites, by more conventional materials forces scientists and engineers to look at new processes to enhance the properties of the latter. In fact, the utilization of composites has some shortcomings among which weight, recyclability and fiber to matrix adhesion should be mentioned. Hence, the development of ultrahigh modulus polymer products is of paramount importance, in view of their significantly lower density. The carbon–carbon bond being the strongest one, a full alignment of these bonds in the same direction would lead to a material with a very high modulus and strength. Polypropylene (PP) has a theoretical ultimate modulus of 50 GPa and an ultimate tensile strength of 16 GPa. These values are considered unlikely to be achieved because the polymer assumes a random entangled and twisted configuration which has a low bearing capacity. In recent years, it was realized that the greatest modulus and strength would result from an anisotropic structure of highly oriented, extended and densely packed chains [1–3].

Solid-state forming processes can induce a permanent deformation of the internal structure, namely, the conversion of an initially isotropic and spherulitic structure to a fibrillar structure. The fibrils are made of oriented and extended molecular chains which ensure mechanical connection between crystals and load transfer. It can be realized that, for maximum mechanical performance, all polymeric chains should be

Polypropylene: An A–Z Reference
Edited by J. Karger-Kocsis
Published in 1999 by Kluwer Publishers, Dordrecht. ISBN 0 412 80200 7

extended along the deformation direction. This macroscopic deformation should not be confused with the conventional melt extrusion process which may involve some molecular orientation, nor with the fiber and film-forming processes treated elsewhere in this handbook. Indeed, during any melt processing operation, some molecular orientation occurs because of the cooling history and the viscoelastic nature of polymeric materials, but the fraction of extended chains is exceedingly small. This plastic deformation should be carried out below the melting temperature of the polymer, which is around 165°C for PP (depending on its crystalline structure). For some semicrystalline polymers, there exist an α-transition temperature corresponding to a secondary transition and above which lamellar slip can occur, thus facilitating crystal subunits' motion within the larger crystal unit, therefore allowing the formation of extended chain crystals. This temperature is around 100°C for PP [1, 3].

SOLID-STATE DEFORMATION PROCESSES FOR PP

Many solid-state orientation processes, among which are drawing, extrusion, compression and rolling, have been used to produce high-modulus polymers. All these techniques rely on aligning the existing crystalline structure into a highly oriented fibrillar structure by an extensional deformation process. Each of these techniques also has several variations and unique features associated with it. For example, all of them can be isothermal or adiabatic, drawing can be free or constrained deformation, extrusion can be direct or hydrostatic, with or without haul-off tension and rolling can be with or without tension. Here, we will particularly discuss the extrusion and roll-drawing processes.

In the ram extrusion process, an isotropic polymer billet at a temperature below the melting temperature is forced by a piston through a die of reducing cross-section, as shown in Figure 1(a). Depending on the die geometry, extrusion can be used to produce rods, tubes, flat or shape profiles. The extrusion ratio is defined as the ratio of the cross-sectional area from the billet to the extrudate. The molecular rearrangement that occurs is similar to that obtained in the drawing process, namely, elongational deformation of the polymer. However, extrusion processes have two main limitations. First, the process is slow; an attempt to extrude at higher rates results in process instabilities, resulting in highly distorted extrudate. Nevertheless, hydrostatic extrusion provides lubrication in the deformation zones, reducing the pressure required to extrude the polymer. Secondly, they are discontinuous; preformed billets have to be used. One possibility is die-drawing which can be operated continuously and consists of pulling the polymer through a shaped die. In this process, the polymer does not maintain a large contact area with the die during the deformation.

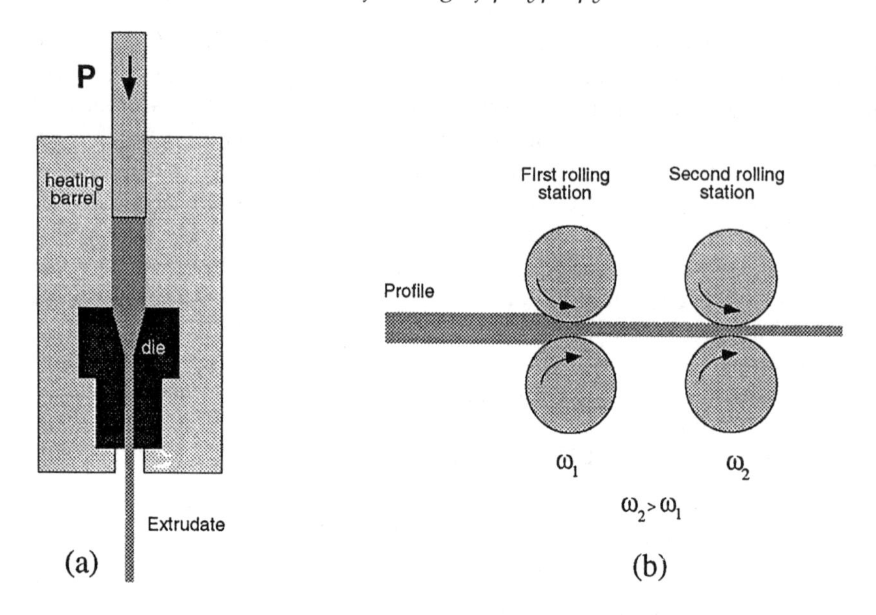

Figure 1 Sketches of the ram extrusion (a) and roll-drawing (b) processes.

Rolling processes, on the other hand, have been used for a long time to produce PVC and rubber sheet stock. Others used it for room temperature rolling of polymers, primarily polyethylene (PE) and PP, claiming thickness reduction ratios of 5:1. Solid-state rolling of PP at elevated temperatures yielded thickness reduction ratios up to 11:1, in which a fibrillar morphology was created. They also showed that speeds as high as 20 m/min can be achieved by rolling adiabatically and that tension influences the extent of orientation in the amorphous phase, which affects the chemical and thermal stability of the polymer. An illustration of the roll-drawing process is shown in Figure 1(b). The roll-drawing assembly consists generally of a series of pairs of rolls (heated or not) preceded and followed by temperature-conditioning steps. The process parameters are profile and rolls temperatures, rolls speeds, gaps between the rolls and tension. In this process, draw ratio is expressed as an increase in length or as the thickness reduction [4].

EFFECT OF PROCESS PARAMETERS ON DEFORMATION

The most important parameters in solid-state deformation are the polymer forming temperature, the rate of deformation, the deformation and the tension or pressure. The effect of the polymer temperature is illustrated in Figure 2 in the case of the extrusion. An increase in the extrusion temperature leads to a decrease of the total pressure required

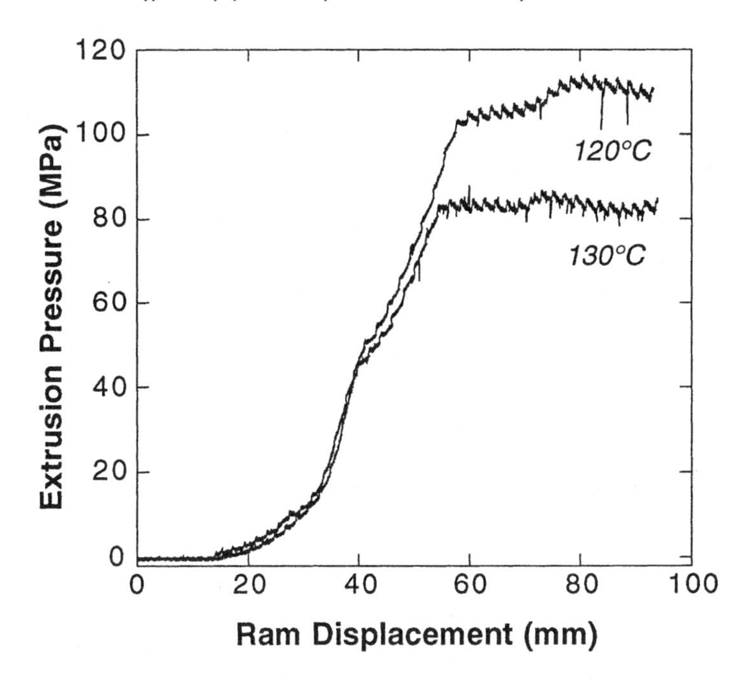

Figure 2 Extrusion pressure versus ram displacement for PP.

to extrude the material, due to the reduction in the material flow stress with temperature. The total pressure is the sum of the die pressure and friction stresses between the polymer and the tool. Lubricating reduces the pressure necessary to deform the polymer by limiting friction forces. The rate of deformation is governed by the deformation speed, i.e. extrusion speed or roll speed, and the draw ratio, i.e. die geometry or gap between the rolls. If the extrusion speed is high, the deformation-induced heating in the PP cannot be properly conducted away through the tooling, resulting in a rise in the actual material temperature, and thus in a decrease of final mechanical properties due to the molecular relaxation. In roll-drawing, tension allows for the control of the elastic recoil and relaxation of the polymer at the exit from the roll gap. This tension has to be high enough to minimize these effects and to avoid surface roughness and fracture but below the break stress of the polymer [4].

Mechanical calculations for the extrusion process showed that PP can prematurely leave contact with the die wall, particularly in decreasing strain rate dies at high drawing speeds. Increased friction of the die wall and increased drawing speed cause this 'loss of contact' point to move progressively downstream in all die cases. Theoretically predicted drawing loads show acceptable agreement with experimental test results when low values of friction coefficient are assumed. The loads required to extrude PP are much lower when lubrication is used.

Studies of the uniplanar deformation of isotactic PP (iPP) in uniaxial compression by forging at several constant temperatures from 20 to 140°C showed that at a forging temperature below the smectic-phase stability limit temperature (T_s), the efficiency of draw, parallel to the uniplanar direction, was high (97%) and independent of the compression ratio, pressure and draw temperature (T_d). Above T_s, the achievable compression ratio (CR) increased steadily with T_d. At a given T_d, the CR increased with increasing applied pressure, and this tendency was more prominent above T_s. The T_s moved higher as the compression rate was increased. Figure 2 illustrates the effect of temperature on load required to solid state deform PP. It is clearly seen that this load decreases with increasing temperature.

STRUCTURE DEVELOPED IN SOLID-STATE DEFORMED PP

Deformation in the solid state will affect PP in two aspects: crystalline structure (and content) and molecular orientation. Crystallinity as a function of draw ratio are shown in Figure 3 for a PP homopolymer (PP1) and PP copolymer (PP2) using solid-state extrusion and rolling respectively. Crystallinity, determined by differential scanning calorimetry (DSC) was observed to increase with deformation. Polarized-light microscopy showed that the replacement of spherulites by an oriented structure is complete at a draw ratio of about 4. Distinct fibrils are visible in scanning electron micrographs of fracture surfaces. Above this deformation, molecular orientation still takes place but mainly by a more and more important alignment of bridging molecules between the amorphous and crystalline phases (tie-molecules).

The response of spherulites to a deformation appeared to result from a number of mechanisms including slip, tilt (rotation in the plane of orientation) and twinning of the crystalline lamellae. Besides this complexity, which is inherent in all spherulitic polymers, the plastic deformation of PP is affected by the early occurrence of a necking process. Ultrahigh molecular weight PP has also been studied and its cold-drawing orientational behavior was observed to be very different from that of conventional PP.

The drawing behavior of several syndiotactic polypropylenes (sPP) was also examined as a function of molecular weight and initial morphology. It was found that these materials can be drawn to a much lesser extent than comparable isotactic polypropylene. This limited drawability was attributed to the absence of an effective crystalline c-slip process.

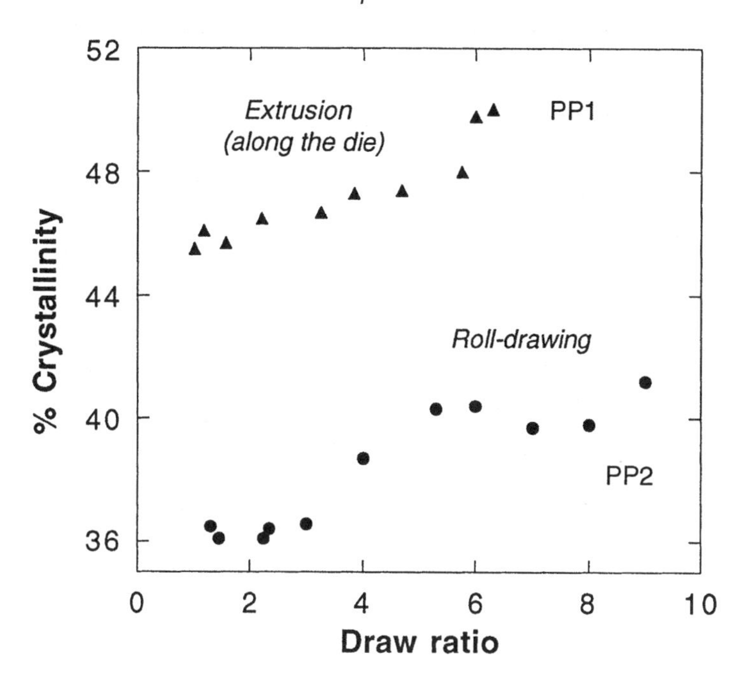

Figure 3 Crystallinity as a function of draw ratio for extruded and roll-drawn PP.

PROPERTIES

In terms of performance, many researchers studied the structure–property relationships in oriented polymers and particularly PP. It was found that mechanical strength is enhanced triaxially for roll-drawn PP. An example of such property enhancement is shown in Figures 4 and 5 in terms of modulus and strength respectively as a function of draw ratio for roll-drawn or solid-state extruded PP. For biaxially rolled PP, the tensile strength of a sheet, initially elongated by 5.0 in longitudinal direction (L) and then elongated by 1.5 in the transverse direction (T), reached almost 100 MPa, and the value was three times as large as that of the stock in all directions in the plane. This combination of the first rolling elongation of 5.0 and second rolling elongation of 1.5 was the condition to get substantially the same rolling elongation in both L and T-directions because the uniaxially rolled sheets become thicker by shrinking when reheated for biaxial rolling and the substantial second rolling elongation is larger than 1.5 [5].

Oriented PP multidirectional (OPP) laminates (with $(0)_2$, $(0/90)_s$ and $(0/\pm 45)_s$ lay-ups) with improved properties in more than one direction in the plane obtained using a hot-plate welding technique were also

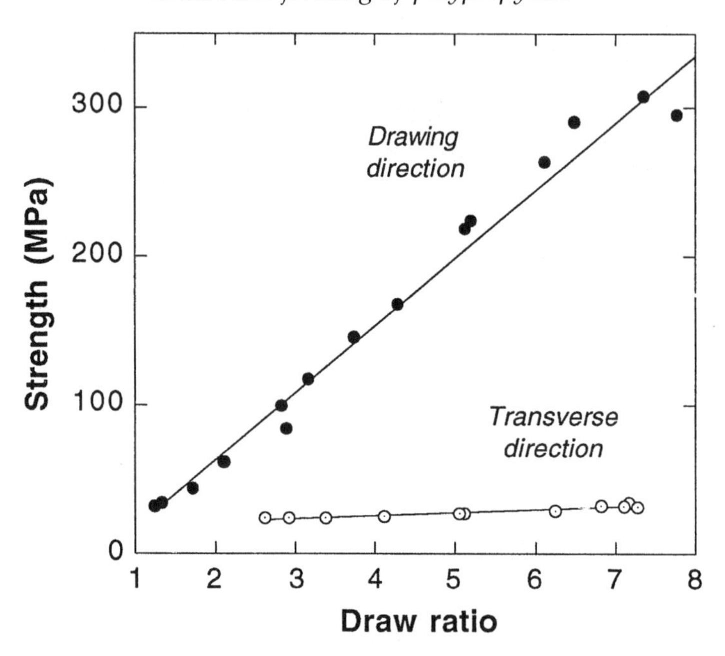

Figure 4 Strength of roll-drawn PP as a function of draw ratio.

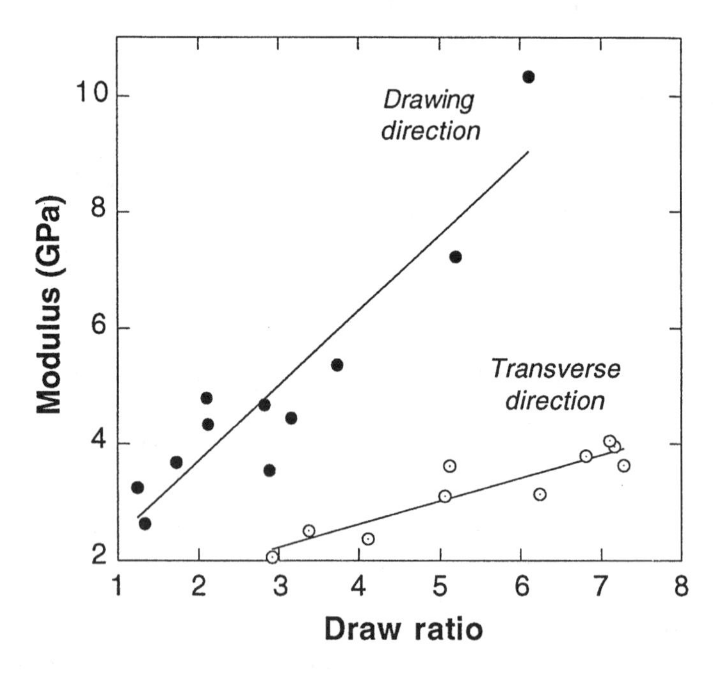

Figure 5 Modulus of roll-drawn PP as a function of draw ratio.

studied. The in-plane properties of the laminates were successfully predicted with classical laminate theory (CLT), which is commonly used to predict the properties of fiber reinforced materials. These laminates can be quasi-isotropic and were found to have improved modulus (up to 6 GPa for $(0/90)_s$ laminates) and strength (up to 150 MPa for $(0/90)_s$ laminates) as well as exceptionally good impact toughness.

For sPP, at a given draw ratio, the modulus was remarkably lower for sPP than for iPP, yet the tensile strengths were not significantly different. The maximum tensile modulus and strength of sPP achieved were 3.0 and 0.33 GPa, respectively. These values were remarkably lower than those (20 and 0.60 GPa, respectively) achieved for an iPP having comparable molecular characteristics, reflecting the low crystal modulus, drawability and crystallinity of sPP compared to those of iPP.

Cold rolling led also to an increase in the creep strain and secondary creep strain-rate. The creep activation energy was found to increase with increasing rolling reduction. Within the secondary creep stage, the creep process in polypropylene is mainly due to the α-relaxation process and most of the creep strain was recoverable.

REFERENCES

1. Ward, I.M. (1985) The preparation, structure and properties of ultra-high modulus flexible polymers. *Adv. Polymer Sci.*, **70**, 1–70.
2. Proceedings of SPE RETEC meeting, *New Advances in Oriented Plastics*, September, 1987, Atlantic City, NS, USA, SPE, Brookfield, CT, USA.
3. Zachariades, A.E. and Porter, R.S. (eds) (1988) *High Modulus Polymers*, Dekker, New York.
4. Ajji, A., Legros, N., Dufour, J. and Dumoulin, M.M. (1996) High performance materials obtained by solid state forming of polymers. *J. of Reinforced Plast. Composites*, **15**, 652–662.
5. Higashida, Y., Watanabe, K. and Kukuma, T. (1991) Mechanical properties of uniaxially and biaxially rolled polymer sheet. *ISIJ Int.*, **31**, 655–660.

Keywords: solid-state extrusion, roll-drawing, orientation, crystallinity, mechanical properties, α-relaxation, solid-state deformation, compression ratio, morphology, tie molecules, lamellar orientation, laminate theory, fibrillar structure, rolling.

Special polypropylene fibers

Martin Jambrich and Anton Marcinčin

INTRODUCTION

Special types of polypropylene (PP) fibers were developed in order to widen their application. The special fibers may be ranged into the following main groups:

- multicomponent fibers containing at least two different polymers;
- fibers with changing cross-section geometry (profiled and partly filled);
- bioactive fibers.

MULTICOMPONENT POLYPROPYLENE FIBERS

Blending of polymers is one of the common methods used to prepare new polymer materials with 'new' properties. Nowadays this approach is of paramount interest from the viewpoint of both science and industry. Blending of polymers has the following technical and economic advantages:

- 'tailoring' of new polymers with unique properties;
- fast development associated with rapid commercialization;
- property diversification of existing grades.

Blending of polymers is practiced also in the production of synthetic (chemical) fibers. Multicomponent fibers are usually prepared by mixing more than two polymers. In this case, the ordinary spinning equipment

Polypropylene: An A–Z Reference
Edited by J. Karger-Kocsis
Published in 1999 by Kluwer Publishers, Dordrecht. ISBN 0 412 80200 7

with the usual shape of nozzles or special spinning equipment with nozzles of a different geometry may be used.

Generally, two different polymers in the melt can form a homogenous or heterogeneous blend. This depends on the thermodynamics of polymer–polymer interaction and/or kinetics of the mixing process. The majority of polymer pairs are immiscible and form systems with distinct separated phases. A typical example of spinning thermodynamically immiscible polymer blends using the classical circle-shaped nozzles results in a polyfibrillar fiber of the blend matrix/fibrillar (M/F) type (Figure 1). In this case, the phenomena called 'specific spinning' can be observed [1]. The expression 'specific' means that the formation of fine fibers of one polymer in the matrix of another polymer occurs before the blend enters the nozzle channel, not after passing the outlet. In this case, the number of fine fibers is a result of the microrheology of the blend dosed to the spinning nozzle. A large number of ultrafine fibers is created in the matrix in each orifice during this process. Thus, the dispersed phase polymer will be transformed into fine fibers. The final properties of fibers produced in this way are determined mainly by the weight ratio of the components, by their viscosity ratio, as well as by the interfacial tension of the components in the melt.

Using both spinning nozzles with changed geometry of orifices and combined nozzles, further fiber variations such as side by side (S/S), core and side (C/S) and others can be produced [2] – see Figure 1.

The most common combinations of PP are those with polyethylene (PE), polyamide (PA) and polyethyleneterephthalate (PET). The fibers spun out of these blends find numerous practical applications, especially [1, 3]:

- preparation of microfibers (PP/PA, PP/PET);
- production of thermobonding types (PP/PE);
- production of surface dyeable fibers (PP/PA,PP/PET,PP/polyvinyl-alcohol);
- preparation of fibers of high elastic properties (PP/PA, PP/PET);
- M/F fibers for exclusive and sports garments.

Figure 1 shows a SEM picture of a fiber on the basis of PP/PA/-interfacial agent polymers (polypropylene/polyamide/PP grafted with maleic anhydride).

The polyfibrillous fibers are suitable for the preparation of microfibers, with fineness 0.1–0.5 μm, which are easily dyeable by dye solution and possess improved elastic properties compared to PP fibers [1].

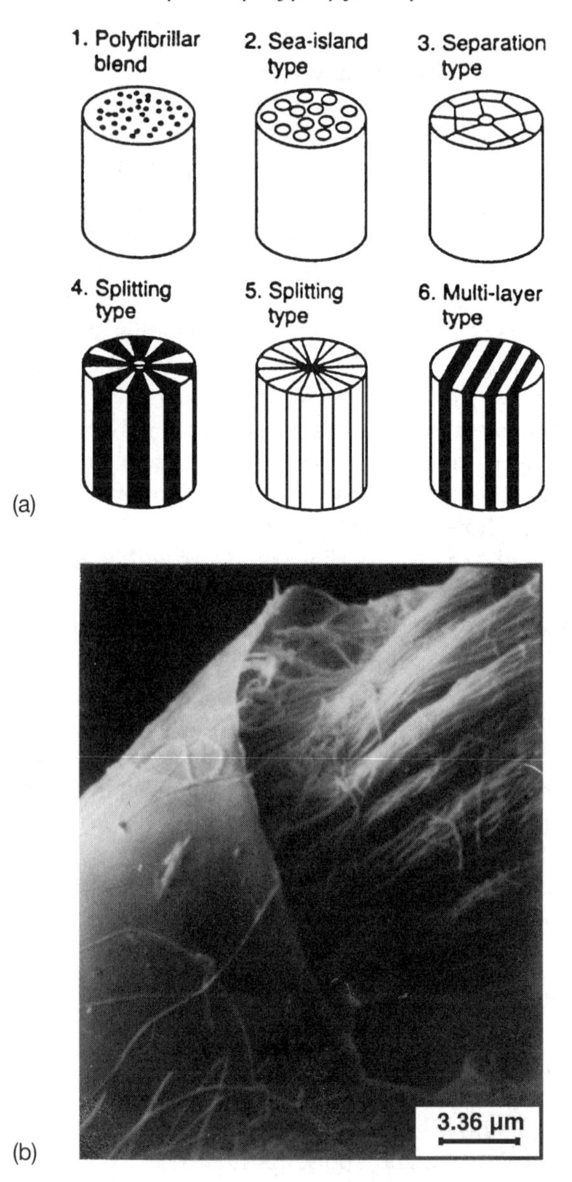

Figure 1 Scheme of various types of fibers from two polymers (a) [2] and SEM photograph of PP/PA6 blend fiber polyfibrillar type after partial extraction of PP matrix (b) [1].

Multicomponent fibers produced from polymer blends have interesting possibilities for many kinds textile goods including technical textiles [3].

FIBERS WITH CHANGED CROSS-SECTION AND PARTLY FILLED

Fibers with changed cross-section (geometry-profiled) represent an interesting assortment. The survey of the most common PP profiled fibers produced is given in Figure 2.

These fibers are produced using noncircular shaped nozzles. By changing the cross-section geometry, i.e. the profile of the fibers, the specific surface and parameters of supermolecular and morphological structure are changed. The specific surface and structure of PP fibers significantly influence their physicochemical reactivity and thus, some physical properties, e.g. separation activity for nonpolar organic solvents (crude oil products) are improved [4]. The physical properties can also be varied by adding inorganic fillers as well ($CaCO_3$, $BaSO_4$, $Mg(OH)_2$). Fillers affect the separation efficiency (Table 1); in addition they enhance the adhesion in composite materials. Such fibers are recommended for textile materials suitable for cleaning waste water and removing the soil from organic pollutants.

The profiled PP fibers significantly influence the geometry and volume of textiles and, in combination with other chemical or natural fibers, they improve the physiological parameters of textile goods for children's clothing, sport and winter garments. In addition, the fibers have higher adhesion in composites, and higher filtration efficiency for both solid and gaseous media. The profiled PP fibers exhibit higher capillary elevation and sorption properties also, which may be advantageously exploited in production of hygienic textiles. There is a possibility to construct three-dimensional textiles for 'anchoring' microorganisms which are able to decompose water pollutants lipolytically.

PP fibers with changed cross-section are very promising and they should have a good future based on their application possibilities.

BIOACTIVE POLYPROPYLENE FIBERS

Antimicrobial finishing of fibers is frequently used to protect textiles against various pathogenic microorganisms. Microbial contamination of textile goods depends mainly on the air and its activity on the material. The critical value of relative humidity (RH) for growing of microorganisms lies in the range of 70–80% RH in case of fungi, while for bacteria the value is above 98%. Antibacterial textiles find their applications in the following sectors:

- common textile goods, such as garments, underwear, socks, linen, carpets, towels, mattresses, blankets, decorative textiles;
- medical textiles, e.g. linen, coverings, clothes for doctors and patients;

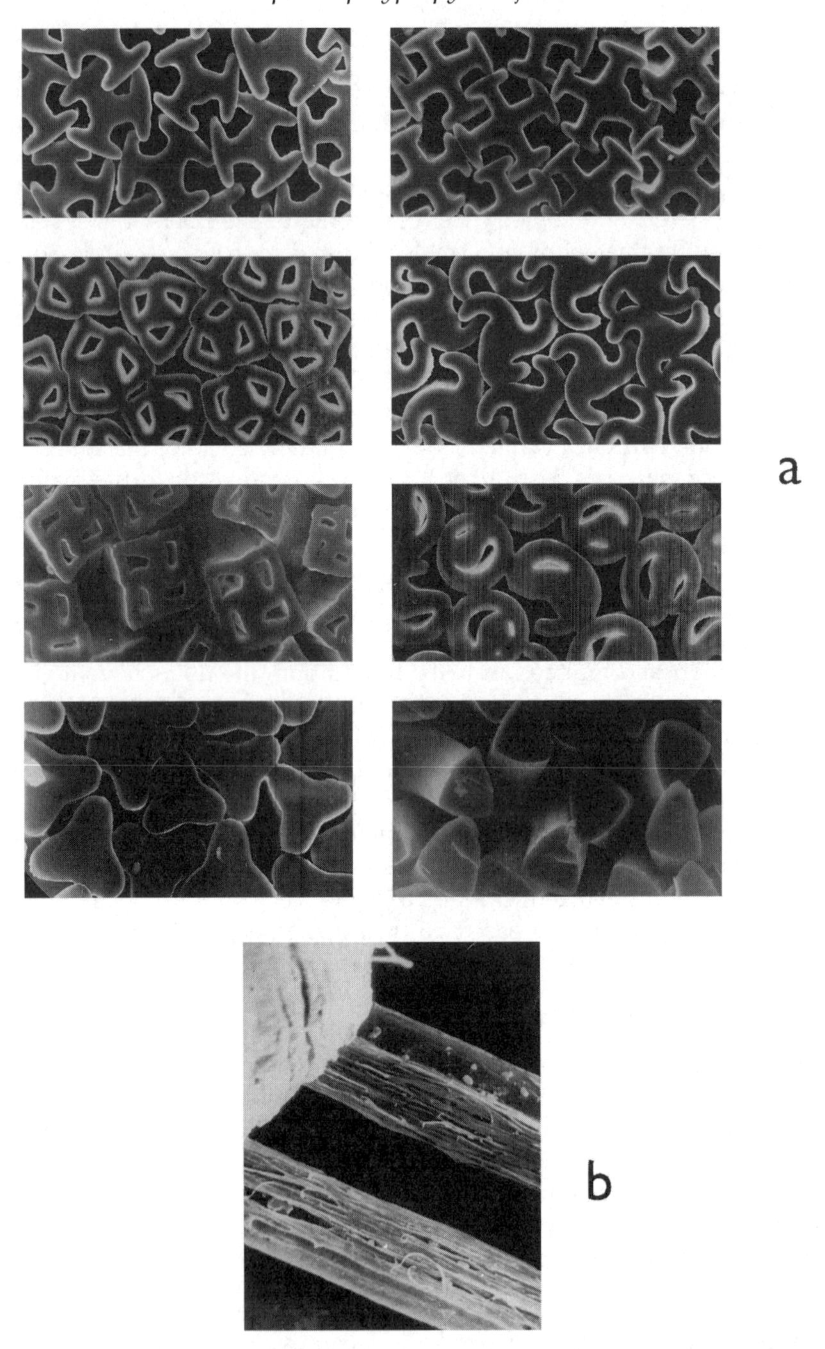

Figure 2 (a) SEM photographs of cross-sections of PP fibers with a linear density of 27 dtex. (b) SEM photograph of hollow-additivated PP fibers with linear density of 30 dtex.

Table 1 Geometry parameters of PP fibers and their separation efficiency for crude oil components from waste water [DIN 38409; STN 83052]

Cross section geometry	*Additive and concentration* (%)	*Linear density* (dtex)	*Separation efficiency* (%)
○	$Mg(OH)_2$ 50	~30	73
○	$BaSO_4$ 50	~30	75
○	$CaCO_3$ 50	~30	60
○	–	1.3	90
○	–	27	61
✠	–	27	80
⅄ (curved)	–	27	74
Y	–	27	65

- special textiles for bandages and swabs;
- technical textiles designed for air conditioning equipment, filtering materials and geotextiles.

Bioactive materials are usually produced, depending on the end-use, in the following ways:

- by chemical modification (the biological active agent is chemically bonded to the fiber);
- by physical modification (sorption of the antibacterial agents by the supermolecular structure of the fiber);
- by deposition of hardly soluble active substances which are 'fixed' to the surface of the fiber by means of binders.

PP fibers of hydrophobic paraffin character have a special position among bioactive materials. One usually takes advantage of the limited solubility and migration in PP when solid particles or natural polymeric derivatives are incorporated. Additives based on inorganic fillers, such as Ag^+, Cu^+, Zn^{2+} ions, were developed in addition to additives from the class of antibacterial organic substances. Inorganic fillers are introduced in the PP in form of highly concentrated dispersions. In case of inorganic fillers, the concentration varies from 0.5–1.5 wt.% and that of the organic ones is about 0.2%. The antimicrobial agents have to fulfil requirements of permanence (washing and cleaning) without migration and environmental contamination. The antibacterial 'finishing' should not reduce the mechanical–physical properties of the fiber [5] (Table 2).

Table 2 Properties of PP bacteriostatic fibers modified by Ag (ISO R 846, DIN 53930, AATCC 100–1988, 90–1970)

Filler C, (wt.%)	*Tenacity* (cN.dtex^{-1})	*Elongation* (%)	*Antibacterial efficiency* (%)	
			Escherichia coli	*Staphylococcus aureus*
0.7	3.40	30.0	0.0	18.3
1.0	3.35	29.0	60.6	52.0
1.5	3.20	28.6	67.4	71.4
2.0	3.13	28.7	68.9	77.5

C = agent.

REFERENCES

1. Jambrich, M. and Ďurčová, O. (1993) Fibres from polypropylene-polyamid 6 blend. *Fibres and Textiles East. Europe*, **1**(1), 34–37.
2. Hongu, T. and Phillips, G.O. (1990) *New Fibers*. Ellis Horwood, New York.
3. Krištofič, M. (1994) Modified polypropylene fibres. *Fibers and Textiles East. Europe*, **2**(5), 38–39.
4. Revil'áková, J., Jambrich, M. and Staškovanová, A. (1993) Fibre properties with changed cross-sectional geometry. *Fibres and Textiles East. Europe*, **1**(2), 17–20.
5. Marcinčin, A., Ujhelyiová, A., Legéň, J. *et al.* (1997) Bioactive polypropylene fibers, *Vlakna a Textil (Fibres and Textiles)*, **4**(2), 38–43.

Keywords: bicomponent fiber, multicomponent fiber, fibers of non-circular cross-section, sorption properties, separation, bioactive fibers, filled fibers, antibacterial textiles.

Spherulitic crystallization and structure

József Varga

In a quiescent melt, isotactic polypropylene (iPP) crystallizes into spherulitic structures [1, 2]. Spherulite is a spheriform aggregate of chain-folded fibrillar or lamellar primary crystallites. According to a simplified model, crystallites start from a central, pin-point type nucleus and grow uniformly in all spatial directions radially, with noncrystallographic small angle branching in between. The branching of growing crystallites provides complete space filling. The phenomenological theory of spherulitic crystallization of polymers was developed and later partially revised by Keith and Padden [3] and Vaughan [4]. According to this theory, the contaminating components (e.g. fractions with low, if any, tendency to crystallize) segregated from the melt in the vicinity of the growing crystal surface are responsible for the branching of the fibrils and so for the space filling. Based on trasmission electron microscopic (TEM) studies on spherulites Vaughan [4] drew attention to some contradictions in the phenomenological theory and proposed a 'dominant/subsidiary growth model' for the morphology development in semicrystalline polymers.

Morphological features, kinetics of growth, formation of structure and melting behavior of iPP spherulites were discovered [1, 2 and references therein]. During the crystallization of iPP, being a polymorphic material with several modifications [5], different types of spherulites may develop, which imply crystallites of the α-, β- and γ-modification [1, 2, 5–7]. All these polymorphs consist of right-, and/or left-handed three-fold helices with 0.65 nm chain axis repeat distance. The molecular

Polypropylene: An A–Z Reference
Edited by J. Karger-Kocsis
Published in 1999 by Kluwer Publishers, Dordrecht. ISBN 0 412 80200 7

aspects of crystal structure and morphology of different iPP polymorphs were recently reviewed by Lotz *et al.* [7]. The types and structural features of spherulites formed are markedly influenced by the thermal conditions of crystallization, by the thermal prehistory of melt, by the mechanical load applied to the crystallizing melt (pressure, shear or tensile stresses), and by the presence of extraneous materials (α- and β-nucleating agents, additives, etc.) [1, 2]. The spherulitic crystallization and structure of iPP may be readily visualized by polarized light microscopy (PLM). Their constituent elements, the primary crystallites, may be usually observed by electron and atomic force microscopy (AFM) [4]. Following two-dimensional crystallization (apart from its very early stage) in a thin film of the melt, PLM reveals the development of birefringent disc-like objects on randomly formed nuclei, radially growing at a constant rate (linear growth rate) under isothermal conditions. Due to the impingement of growing spherulitic fronts, the texture formed will consist of polygonal formations confined by straight or curved lines after the complete crystallization. The possible types of boundary lines developed between two adjacent spherulites and their mathematical description were analyzed by Varga [1, 2]. The average size of spherulites might be reduced by elevating the density of nuclei by adding different nucleating agents or by decreasing the crystallization temperature (T_c) [1].

α-SPHERULITE

During the crystallization of commercial iPP grades, essentially the α-modification is formed, sometimes accompanied by a lower or higher amount of β-modification. The α-modification of iPP (α-iPP) seems to be the thermodynamically stable form (but, see a critical analysis about the types of iPP polymorphism in [1]). The crystal cell of α-iPP is monoclinic with parameters $a = 0.665$ nm, $b = 2.096$ nm, $c = 0.65$ nm and $\beta = 99°80'$, which consists of alternating right- and left-handed helices [5, 7].

Three different types of α-spherulites might grow from the melt depending on the T_c: positive radial (α_I) below ≈ 134°C, negative radial (α_{II}) above ≈ 138°C, and mixed-type spherulites (α_m) in the intermediate temperature range [2]. The last have no inevitable birefringence and show no Maltese cross in the optical micrographs. Figure 1 shows SEM micrograph of an α_{II}-type spherulite. It can be seen that the radial spherulite has a compact crystal array and contains straight lamellae. The formation of α-spherulites has a lower threshold temperature. At high supercooling (i.e. at $T_c < \approx 90$°C) a mesomorphic (smectic) structure forms instead of a crystalline (spherulitic) one [5]. According to the structural model of spherulites based on the radial growth and small-angle branching of primary crystals, polymers are expected to produce

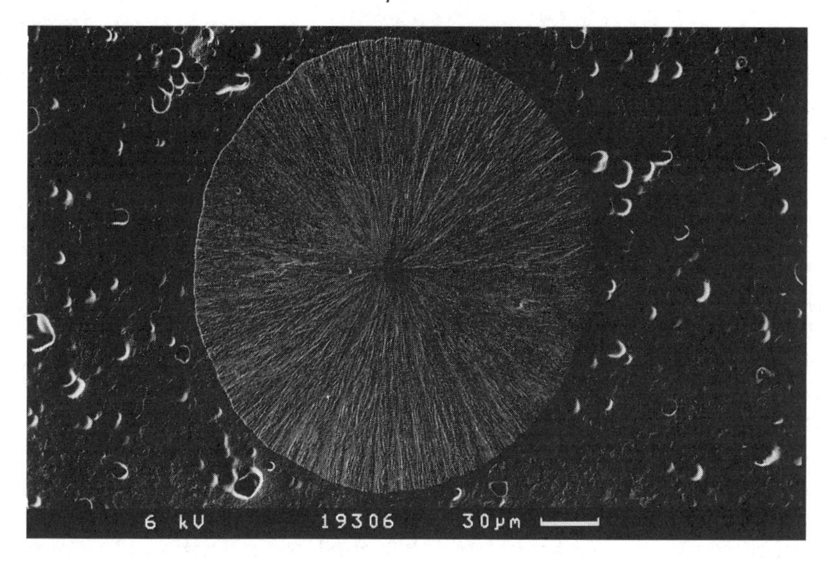

Figure 1 Scanning electron micrograph showing an α-spherulite of iPP crystallized at 143°C and than etched.

only spherulites with negative birefringence. The positive birefringence and the mixed optical character of some α-spherulites were interpreted by the 'cross-hatched' structure consisting of radial (R) and tangential (T) lamellae [1, 7, 8]. T-lamellae meet the radius of spherulite at an angle of about 80°. T-lamellae form in consequence of the high-angle αα-lamellar branching. The molecular origin of this branching and the consequent cross-hatched structure were clarified by Lotz *et al.* [7]. They revealed that the formation of wide angle αα-branching was a unique case of autoepitaxy (rotation twinning). The sign of birefringence of α-spherulites is controlled by the proportion of T-lamellae. The optical character of α-spherulites converts into positive if the proportion of T-lamellae exceeds one-third [1]. Changes in the fraction of T-lamellae also play a role in the dependence of type and birefringence of α-spherulites on T_c. It was detected by TEM that increasing T_c led to a reduced amount of T-lamellae. In fact, their formation ceased above T_c of about 160°C [8]. Consequently, the optical character of spherulites shifts in the positive to mixed to negative direction with increasing T_c. On heating, the positive and mixed α-spherulites gradually transformed into the negative type. This observation can be explained by the partial melting of T-lamellae during heating. The lower melting temperature range of T-lamellae may be associated with their smaller thickness as compared to the radial ones [1].

The experimental melting point of α-spherulites crystallized under usual thermal condition is ~165°C. Literature data for the equilibrium melting temperature of α-iPP ($T^0_m(\alpha)$) are very contradictory. The reported values fall into the range 174–220°C [1], the author accepts $T^0_m(\alpha)) = 208$°C as the most reliable one [2].

β-SPHERULITE

β-phase rich or even pure β-phase with spherulitic structure can be prepared by using a selective β-nucleating agent [1]. The experimental data indicated a possible upper ($T(\beta\alpha) = 140$°C) and lower temperature ($T(\alpha\beta) = 100$–110°C) limit of the formation of pure β-iPP [1, 2, 9]. The β-modification of iPP (β-iPP) has a trigonal cell with parameters $a = b = 1.101$ nm, $c = 0.65$ nm containing three isochiral helices. This cell is frustrated; the three helices do not have similar orientation [7].

Depending on T_c, negative radial (β_{III}) or negative banded β-spherulites (β_{IV}) might form during the melt crystallization [2]. According to our observation, undoubtedly β_{III}-type forms at $T_c < 120$°C, while the appearence of β_{IV}-type is favored by higher T_c. In Figure 2, a banded spherulite formed at 130°C and some characteristics of its lamellar structure are shown [9]. The banded β-spherulites contain lamellae twisted around their longitudinal axes. In the dark bands, lamellae lying flat-on are dominating while the clear bands represent the region where the lamellae stand on their edges. Twisting of lamellae in the intermediate regions between the dark and clear bands is illustrated in Figure 2(b), where it is perceptible that lamellae standing on edges lean to the plane of the sample. In the PLM pattern of the banded spherulites, more or less regular concentric dark rings appear corresponding to the region where the lamellae are lying flat-on. This implies that the polymer chains are almost or exactly perpendicular to the plane of the lamellae.

An interesting feature of the β-spherulitic growth is the β to α modification transition (βα-growth transition). This was observed on the growing crystal front under appropriate thermal conditions. This transition takes place on the surface of growing β-spherulites crystallized between $T(\alpha\beta)$ and $T(\beta\alpha)$ after heating above $T(\beta\alpha)$ or cooling below $T(\alpha\beta)$ during step-wise isothermal crystallization [1, 2]. Recently it was found [9] that the β-growth transition may occur even under isothermal conditions above $T(\beta\alpha)$ in the presence of a highly active β-nucleant which induces the formation of β-phase even above $T(\beta\alpha)$ (at least not too far from it). Similarly, a βα-growth transition could be observed during the isothermal crystallization of β-nucleated random propylene copolymers with low comonomer content [1]. The βα-growth transition was interpreted on the basis of secondary βα-nucleation [2]. The βα-nuclei formed on the growing fronts of β-spherulites propagated into

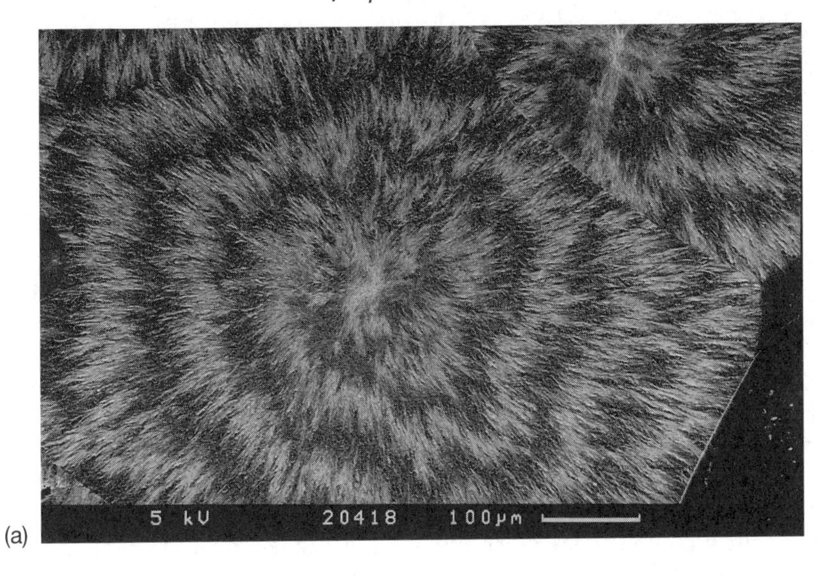

(a)

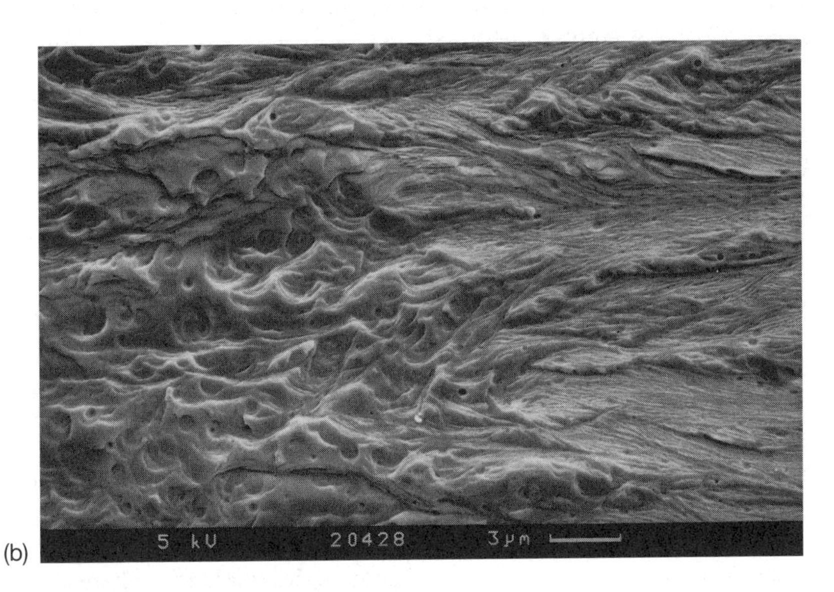

(b)

Figure 2 SEM micrographs of an etched banded β-spherulite (a) crystallized at $T_c = 130°C$ and its lamellar structure (b) [9].

α-spherulitic segments. These successively block up the space for the further growth of the β-spherulite [1, 2, 9], resulting in characteristic βαtwin spherulites. The frequent occurrence of α-nuclei formation increases considerably with T_c elevation [2]. The kinetic condition for the modification transitions is a faster growth rate of the new phase than that of the basic crystal. The βα-growth transition observed above $T(\beta\alpha)$ and below $T(\alpha\beta)$ clearly suggests that here the growth rate of the α-phase (G_α) exceeds that of the β-phase (G_β). The fact that G_α is higher than G_β above $T(\beta\alpha)$, has been proved by growth kinetic measurements [2]. In the case of propylene random copolymers, the kinetic requirement of the βα-transition of growth is also fulfilled because G_α is higher than G_β in a wide temperature range [1].

β-iPP is a unique polymer with respect to its melting and recrystallization characteristics [1]. They are highly dependent on the thermal posthistory of the crystalline sample (melting memory effect). β-spherulites cooled below a critical temperature before heating recrystallize in the α-modification during partial melting. The α-spherulites, recrystallized from β-modification, preserve their former substructure (e.g. their banded character) and have a higher melting point than α-spherulites originally present in the sample formed isothermally. The critical recooling temperature range is $T_R^* = 100$–$110°C$ [1, 2]. On the other hand, β-iPP samples that were not cooled below T_R^* are not susceptible to βα-recrystallization. They melt separately like thermodynamically stable modifications. A similar phenomenon could be observed during annealing of β-iPP (annealing memory effect) [1]. Consequently, the tendency to βα-recrystallization is not a general feature of β-iPP, it is associated only with a given thermal history, namely, cooling below T_R^*. The experimental melting point of nonrecooled β-phase is ~ 155°C. The most reliable value of the equilibrium melting point of the β-phase is $T_m^0(\beta) = 184 \pm 4°C$, which was obtained with pure β-iPP when the disturbance of βα-recrystallization was eliminated [1].

γ-SPHERULITE

The γ-modification of iPP (γ-iPP) may form in degraded, low molecular weight iPP or in samples crystallized under high pressure [5, 6]. Certain propylene copolymers with low comonomer content (4–10 wt.%) crystallize preferentially in γ-form, as well. γ-iPP has a face-centred orthorombic unit cell with parameters $a = 0.85$ nm, $b = 0.993$ nm and $c = 4.241$ nm containing isochiral helices. The cell structure proposed by Brückner *et al.* [5] is unique in polymer crystallography: the chain axes in adjacent crystal layers are not parallel. The angle between the chain stems is about 80°. γ-iPP is not usually observed as an independent phase, but crystallizes with and within the α-spherulites. According to Lotz *et al.* [7], the positive spherulites observed in samples with mixed polymorphic composition of α- and γ-iPP are probably made of a framework of radiating α-lamellae on which γ-lamellae branch. The branching angle between the α- and γ-lamellae is about 40°, in contrary to wide angle αα-lamellar branching (~80°C) giving rise to the formation of cross-hatched structure in α-spherulites [5, 7]. Recently, Phillips and Mezghani [6] showed that the pure γ-iPP crystallized at 200 MPa and 187.5°C forms negative radial spherulites with clear Maltese cross. No cross-hatching occurs when more than 60% γ-iPP is present in the sample with mixed polymorphic composition. The high molecular weight iPP samples crystallized at high pressure melt above 150°C. Contrary to the former observation, no γ to α recrystallization during the partial melting of γ-iPP was mentioned [6]. The equilibrium melting temperature of γ-iPP at atmospheric pressure has a value of $T_m^0(\gamma) = 187.6°C$ [6].

THE EARLY STAGE OF SPHERULITIC CRYSTALLIZATION

In the very early stage of α-spherulitic growth, the formation of square-like hedrites, i.e. quadrites, were observed [8, and references therein]. The α-quadrites are two-dimensional arrays of intercrossed lath-like lamellae with an approximately rectangular outline. Quadrites seen flat-on in the polymer melt show very low birefringence. On the other hand, quadrites seen edge-on are rod-like or elongated particles with strong positive birefringence [8, 9].

According to optical microscopic, scanning electron microscopic (SEM) and AFM studies on the high-temperature isothermal crystallization of polypropylene in the presence of active β-nucleating agents, a hedritic structure was formed [9]. β-hedrites lying flat-on are characteristic hexagonal formations (hexagonites) with low birefringence. They are multilayer clusters of lamellar crystallites (Figure 3(a)). On the surface of the lamellae hexagonal formations (etch pits) appear which refer to a screw dislocation (Figure 3(b)). These are responsible for branching and proliferation of the lamellae. β-hedrites standing edge-on were rod-like formations with strong negative birefringence [9].

The α- and β-hedrites might be transformed into spherulites in the later stage of growth because of lamellar branching, giving rise to randomization of the original structure. Therefore, they could be regarded as the precursors of spherulites. The rod-like precursors transform into radially symmetric spherulitic form through sheaf-like then oval (ovalite) arrangements. The lower T_c is, the earlier the stage of growth at which hedritic crystallization turns to spherulitic. Since the precursors of α- and β-spherulites are hedrites (quadrites or hexagonites, respectively), no spherical symmetry exists in the vicinity of the centre of

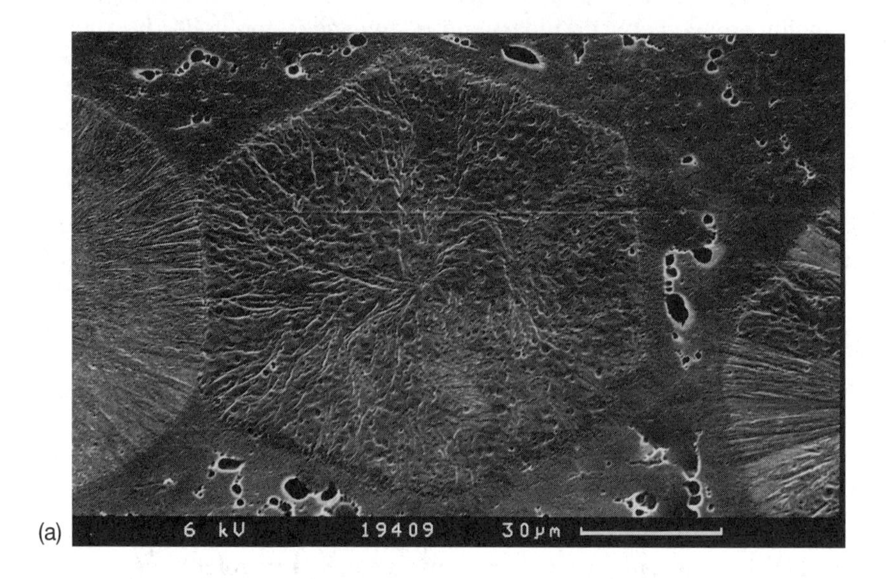

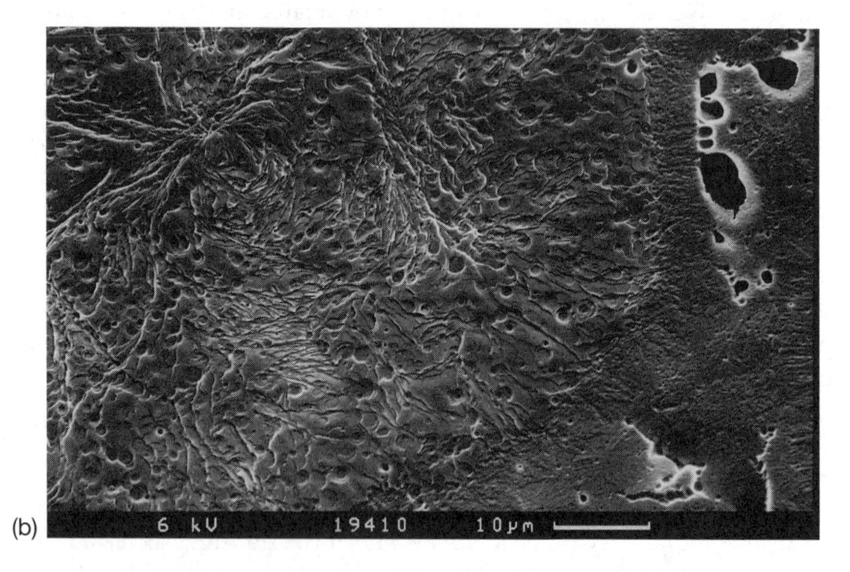

Figure 3 SEM micrograph of a β-hexagonite (a) and some details of its lamellar morphology (b). After partial crystallization at $T_c = 138°C$, the sample was quenched and than etched.

the spherulites in contrast to the spherulite as a whole. This non-homogeneous central region in α- and β-spherulites was revealed experimentally by SEM and TEM [1].

THE LATE STAGE OF SPHERULITIC CRYSTALLIZATION [10, and references therein]

In the late stage of the crystallization, melt inclusions are encapsulated by the crystallized phase. The contraction caused by the proceeding crystallization within the inclusion leads to a reduced ('negative') pressure. It has been discovered that the growth rate of spherulites within the inclusion was lower and the melting point of the portion of crystals, formed from the melt enclosed by the spherulites, was higher than that of the surrounding α-spherulites. These observations can be attributed to the depression in the equilibrium melting point (T_m^0) due to the reduced pressure. At a constant T_c, a reduced equilibrium melting point is equal with a decrease in the actual supercooling ($\Delta T = T_m^0 - T_c$). This reduced supercooling explains both the elevated melting point and the reduced growth rate of the spherulites formed inside the inclusions. Another consequence of the formation of melt inclusions is the appearance of crack-like boundaries between the spherulites crystallized within the encapsulated melt in high temperature ranges ($T_c > \sim 135$°C). At lower T_c values, however, the 'negative' pressure is released with formation of vacuum bubbles. On the surface of vacuum bubbles a crystal front of the β-modification could be developed. The formation of β-iPP was confirmed by the strong negative birefringence of the related structure and its melting and recrystallization behavior and by SEM. The formation of β-iPP on the vacuum bubbles can be attributed to the extension-induced crystallization of the melt [10].

REFERENCES

1. Varga, J. (1995) Crystallization, melting and supermolecular structure of isotactic polypropylene, in *Polypropylene: Structure, Blends and Composites, Vol. I., Structure and Morphology*, (eds J. Karger-Kocsis), Chapman & Hall, London, pp. 56–115.
2. Varga, J. (1992) Supermolecular structure of isotactic polypropylene. *J. Mat. Sci.*, **27**, 2557–79.
3. Keith, H.D., and Padden F.J. (1963) A phenomenological theory of spherulitic crystallization. *J. Appl. Phys.*, **34**, 2400–19.
4. Vaughan, A.S. (1992) The morphology of semicrystalline polymers. *Sci. Prog. Oxford*, **76**, 1–65.
5. Brückner, S., Meille, S.V., Petraccone, V. and Pirozzi, B. (1991) Polymorphism in isotactic polypropylene. *Prog. Polymer Sci.*, **16**, 361–404.
6. Phillips, P.J. and Mezghani, K. (1996) Polypropylene, isotactic (polymorphism), in *The Polymeric Materials Encyclopedia*, Vol. 9, (ed. J.C. Salamone), CRC Press, Boca Raton. FL, pp. 6637–49.

7. Lotz, B., Wittmann, J.C. and Lovinger, A.J. (1996) Structure and morphology of poly(propylenes): A molecular analysis. *Polymer*, **37**, 4979–92.
8. Olley, R.H. and Bassett D.C. (1989) On the development of polypropylene spherulites. *Polymer*, **30**, 399–409.
9. Varga, J. and Ehrenstein, G.W. (1997) High-temperature hedritic crystallization of the β-modification of isotactic polypropylene. *Colloid Polymer Sci.*, **275**, 511–519.
10. Varga, J. and Ehrenstein, G.W. (1996) Formation of β-modification in its late stage of crystallization. *Polymer*, **37**, 5959–63.

Keywords: annealing memory effect, α-, β- and γ-modification, birefringence, cross-hatching, crystallization, equilibrium melting temperature, extension-induced crystallization, growth rate, growth transition, hedrite, hexagonite, melting memory effect, morphology, nucleating agent, polarized light microscopy (PLM), polymorphism, precursor, quadrite, recrystallization, spherulite, scanning electron microscopy (SEM) transmission electron microscopy (TEM).

Split fiber production

B.L. Deopura

INTRODUCTION

New forms of polypropylene (PP) fibrous material have emerged to supplement the established forms of staple fiber and continuous filaments [1]. These are of two distinct types. The first is stretched tape used extensively for woven sack fabrics. The second type relate to 'split fiber' tape yarns produced by splitting or fibrillation of highly oriented stretched tapes.

FIBRILLATION

The fibrillation or axial splittability of a drawn tape relates to longitudinal splitting tendency under the influence of a given transverse force. A scanning electron micrograph of a fibrillated tape is shown in Figure 1. The ratio of tensile strength in the length direction to that in the transverse direction is a good measure of axial splittability of drawn tapes. In other words, the smaller is the transverse tensile strength of tape the greater is the fibrillation tendency.

A test procedure has been developed to estimate fibrillation tendency of flat tapes. Pretwisted tapes were studied in tensile testing mode [2]. Typical parameters of testing are (a) length of the tape as 200 mm, (b) twist angle 1080° (3 × 360°). Care is needed during the mounting of twisted tape in the jaws of tensile testing system to ensure alignment of sample along the tensile stress direction. The kink in the force–elongation curve indicates axial splitting of the tape during the test. The tensile stress at the kink is inversely proportional to the fibrillation tendency.

Polypropylene: An A–Z Reference
Edited by J. Karger-Kocsis
Published in 1999 by Kluwer Publishers, Dordrecht. ISBN 0 412 80200 7

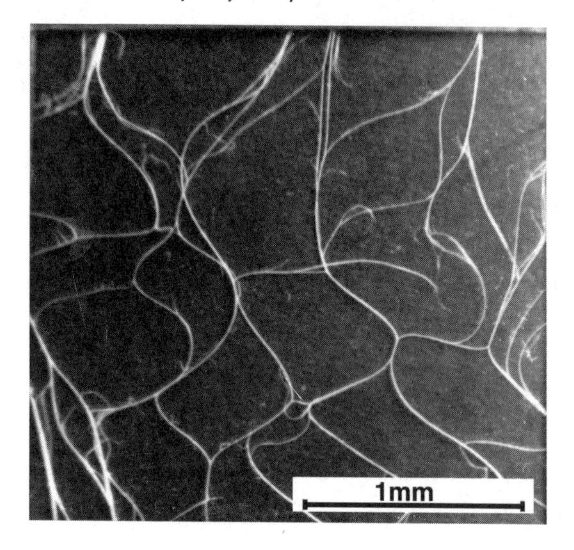

Figure 1 Scanning electron micrograph of fibrillated tape.

FILM FIBER PROCESSING

Polymer characteristics

PP is the material of choice for split fiber production, due to the ease of processing and excellent mechanical properties. The main characteristics of PP film fiber systems are high tensile strength, stiffness and softening point, lower elongation and very high fibrillation tendency. PP with a melt flow index (MFI) of 4–6 g/10 min is used for split film fibers, whereas, a MFI of 0.3–0.5 g/10 min is favored for strappings. A relatively high MFI PP is preferred for split film products due to ease of processing, high fibrillation and increased draw ratio. PP has also been used in the form of blends with other olefins for increased fibrillation tendency.

Tape production

In order to make film for converting into fibrous products, the techniques used are similar to those of film tape manufacturing [1]. The processing comprises the following steps: (1) extrusion of the polymer melt; (2) film solidification by cooling; (3) uniaxial stretching at elevated temperature; (4) heat relaxation setting; and (5) roll up. PP chips are fed to the extruder equipped with a flat die. The quenching of the film is typically done through water, although chilled rolls can be used for the purpose. Efficient quenching process through liquid media allows higher stretching in subsequent processing steps. Flat film is cut into flat tapes of width in the range of 1–20 mm. The cutting operation can be performed before or after stretching, although former is the preferred approach. Cutting followed by stretching leads to a reduction in tape width and thickness, with more pronounced monoaxial orientation and higher tendency for splitting.

The stretching process

Stretching of the PP film is used to reach the desired mechanical properties. Conditions of stretching, i.e. the stretching system, stretch ratio and draw temperature, can be varied to impart the strength and elongation characteristics needed. Stretching improves the orientation of polymer molecules and increases the crystallinity of PP. Beyond a certain stretch ratio, minifibrils (fiber-like fine structure) are formed. The fibrillation tendency increases with increasing stretch ratio. Due to lower intermolecular cohesion, relative ease of splitting is observed between the fibrils in highly oriented PP tapes. Thus, PP is the preferred material for film–fiber products. Typically, stretching temperatures vary between 120–170°C with stretch ratio up to 1:11 and feed line speed of 25–30 m/min. Tapes produced in the stretching step need heat setting to remove built-in stresses, due to molecular orientation and to guarantee dimensional stability. Heat setting is done in line through a combination of hot-chill rolls. Typically, the hot roll temperature is in the range of 130–140°C and the shrinkage is about $\sim 5\%$.

The splitting process

The critical step in the production of fibrous products from films is the splitting process. This process can be performed before, during or after stretching. In a typical production process, film is sliced into tapes, stretched and followed by splitting process. A typical splitting process by needle roller fibrillation unit is shown in Figure 2.

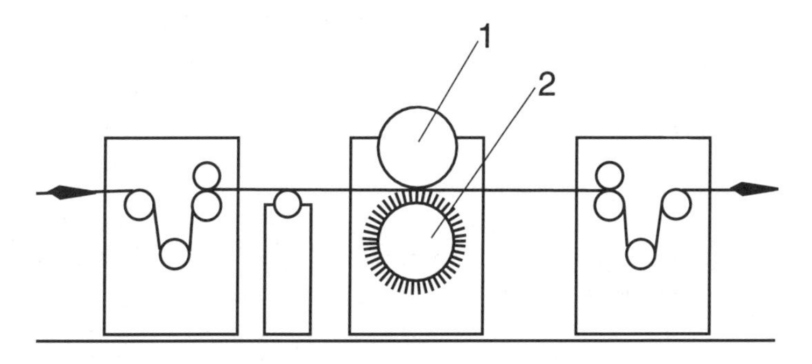

Figure 2 Needle roller fibrillation unit: 1, rubber coated pressure roll; 2, pinned fibrillated roll.

Film is pressed against the needles of the fibrillator roller. A rubber coated press roller mounted above the fibrillation roller forces the film into the needles. The fibrillation roller speed (V_r) is much greater than the line speed (V_f). The ratio of these speeds is termed as fibrillation ratio. The mesh width (L) is given by:

$$L = a\{1 - V_f/V_r\} \quad (1)$$

where a is the contact length (cm). The mesh structure is well discernible in Figure 1.

The average fineness (titre) of single fiber segments T_{sf} can be calculated by the expression:

$$T_{sf} = T_{tf}/a\{V_r/V_f - 1\}\{N\ M\}. \quad (2)$$

where N is the number of needles or blades per cm along the width of the roller, M is the number of needle rows per unit length of roller circumference and T_{tf} is total titre of the film. Some typical data of tape characteristics and processing conditions are given in Table 1.

The actual degree of fibrillation varies widely from film to film. It depends mainly on the splitting tendency of the PP film. The resulting parameters of the fibrillation process are fibrillation ratio, fibrillator dimensions, contact length and titre (fineness or linear density). The calculated values of titre of the fiber segment are normally higher than the actual values. This is due to an additional self-splitting due to transverse mechanical stresses during fibrillation. Additives, such as pigments, fillers and polymer blending activate additional splitting. It is crucial that needles of the fibrillator roller penetrate sufficiently into the film.

The geometry of network texture of the fibrillated film can be controlled by the arrangement of needles on the roller. Regular meshing

Table 1 Tape characteristics with respect to the resulting split-film fineness

Tape characteristics	*Parameter*
Tape width (before stretching)	2 cm
Tape thickness (before stretching)	100 μm
Stretch ratio	1:8
Tape thickness (after stretching)	35 μm
Tape titre (after stretching)	2200 dtex
Tape titre per unit film width	3150 dtex
Contact length (a)	2.5 cm
V_r/V_f	3
N	13
M	1.1
Film fiber titre	44 dtex

results from needles arranged in staggered lines. The pattern can be varied by changing the distance between the needles. Regular networks give high strength as compared to film products with irregular network texture. The distance between needle rows is also of importance. Too short a distance causes the 'fakir effect'. In this case, too many needles touch the film at the same time and cannot penetrate but 'carry' it. A fibrillator tube with sinusoidal placement of pins in the rows counteracts the fakir effect.

A combination of small neeedle roll diameter with short contact length between film and pinroll gives a highly irregular network with random small meshes with coarse interlinking fiber segments. On the other hand, needle rolls with large diameter and long contact length result in weak fibrillates having an irregular mesh network.

Profiling of films with special dies

An alternative approach to splitting the film is to use profile dies. Profiling of the film can be carried out before, during or after the extrusion. Typically, the melt stream is divided into, let us say, 200 individual streams between filter screen block and the die slot. The melt is thereafter extruded through a normal die under reunification of separated streams resulting in lengthwise 'grooved' film. On stretching such a film, it splits into single filaments equal in number to the number of streams formed before the die slot. Relatively fine split fibers (low titre) are produced by means of profile die.

By lengthwise profiling the film during extrusion, the splitting process can be predetermined and directed. The splitting tendency depends on the ratio of thick and thin sections of the film. During stretching, the film starts to split due to necking. Additional mechanical stresses directed tangentially to the film, such as twisting, complete the splitting process. The process leads to highly uniform splitting, softer feel and good handling.

PROPERTIES AND APPLICATIONS

The mechanical properties of split film yarns are comparable to that of filament yarn or staple yarn. Melt spun filaments exhibit a circular and highly uniform cross-section. Film-based fibers have rectangular, lemon-type or a banana-type cross-section and larger diameter variation. These differences result in changes of softness, handling, resilience, bulk and fiber to fiber friction. The main applications of split fibers are summarized in Table 2.

Fabrics made from fibrillated film have a somewhat rough touch and film-like lustre. In the processing of film through a needle roller, there is

Table 2 Areas of applications of split PP fibers. Dimensions in mm, fineness (titre) in dtex, tenacity in cN/dtex, elongation and shrinkage in % units.

Application area	*Quality requirement*	*Draw ratio*	*Dimensions; mechanical properties*	*mm. (dtex)*
Carpet backing fabric	Low heat shrinkage	1:1	Tape width	1.2/2.4
	High strength	1:5	Thickness	0.05
	Temperature resistance		Titre	550/1100
	Splitting tendency		Tenacity	4
	Dull appearance		Elongation	20
			Shrinkage	1 (150°C)
Canvas over tarpaulin	High strength	1:7	Tape width	2.4
			Thickness	0.04
			Titre	850
			Tenacity	5
			Elongation	20
Sacks, bags	High strength	1:7	Tape width	3.0
	High abrasion resistance		Thickness	0.03
	Controlled elongation		Titre	800
			Tenacity	4.2
			Elongation	30
Ropes	High tensile strength	1:9 to 1:11	Tape width	20–60
	Controlled elongation	(1:15)	Thickness	0.04–0.10
	Good fibrillation tendency		Titre	15 000–50 000
Twine	High tensile strength	1:9 to 1:11	Tape width	30–80
	High knot strength		Thickness	0.03–0.06
			Titre	14 000–30 000
Release fabrics	High strength	1:7	Tape width	2.1
			Thickness	0.04
			Titre	750
Filter fabrics	Low heat shrinkage	1:7	Tape width	1.0/2.0
	High abrasion resistance	1:5	Thickness	0.04
			Titre	350/700
Reinforcing fabrics	Low heat shrinkage	1:7	Tape width	2.0
	Controlled elongation	1:5	Thickness	0.03
	Heat resistance		Titre	550
Wallpapers, household textiles	Ultraviolet resistance	1:7	Tape width	1.2–3.0
	Low static charging		Thickness	0.035
	Uniform dyeing		Titre	350–900
	Textile handle		Tenacity	5.5
			Elongation	15
			Shrinkage	5 (150°C)

a decrease in tenacity and elongation of the yarn strand. The drop in these properties increases with decreasing titre of the split strand. Needle (or blade) roll fibrillation is the most extensively used process.

From market point of view and from the structure of the fiber industry, the future for film fiber products may lie in the careful choice of application area, choosing applications where cheap fibrous materials or small-volume specialities are involved. In the tex range >1, there is a huge application potential for split film yarns. Table 2 gives a selection of various applications.

REFERENCES

1. Krassig, H.A., Lenz J. and Mark, H.F. (1984) *Fiber Technology,* Dekker, New York.
2. Mahajan, S.J., Deopura, B.L. and Wang, Y. (1996) Fibrillation behavior of oriented tapes of polyethylene and its blends IV, *J. Appl. Polymer Sci.,* **60,** 1551–59.

Keywords: split fiber, fibrillation, fibrillation tendency, processing, stretching, tape splitting, mesh structure, titre, film 'profiling', properties, use.

Squeeze flow in thermoplastic composites

T.A. Martin and D. Bhattacharyya

INTRODUCTION

In order to permanently alter the shape of a thermoplastic composite, without degrading its material properties, it is necessary to heat the polymer matrix prior to forming. Crystalline and semicrystalline polymers must be completely melted to allow the long polymer chains to slide past one another. In this fluid state, thermoplastic composites behave like incompressible viscoelastic liquids, which convect the embedded fibers as they deform. For composite sheets containing unidirectional continuous-fiber reinforcement, the flow behavior is highly anisotropic and the extensional flow is severely limited along the fiber direction. It is therefore convenient to simplify the kinematics of forming by assuming that the fibers are inextensible and aligned in the plane of the sheet. This idealization has been utilized for linearly elastic solids, elastic/plastic solids, viscous fluids and linear viscoelastic solids [1], because it reduces the number of deformation modes to two: shear along the fibers and shear transverse to the fibers.

Squeeze flow is the type of planar flow that occurs when an incompressible resin is compressed between two rigid platens. If the resin contains a single family of fibers, the body spreads out perpendicular to the fibers as the thickness of the sheet diminishes. An alternative name for the process is therefore 'transverse squeeze flow'. Forming operations, such as diaphragm forming, matched die forming, roll forming, tape laying and many more, initiate squeeze flow because they apply

Polypropylene: An A–Z Reference
Edited by J. Karger-Kocsis
Published in 1999 by Kluwer Publishers, Dordrecht. ISBN 0 412 80200 7

surface pressures normal to the composite sheet. Often this flow is undesirable as it can significantly alter the fiber distribution in various regions of the finished part; however, it is essential for consolidation and bonding processes, since it helps heal the laps and gaps in adjacent layers of a laminate [2]. Sometimes the thickness of the surface layer is dramatically affected by the interaction between the tool and the composite sheet and, in other cases, both sheet thickening and sheet thinning occur simultaneously within the laminate; for example, in a single curvature reverse bend [3]. These thickness variations occur as a direct result of transverse squeeze flow. Since this mode of deformation is an important flow mechanism, it is necessary to consider it in isolation. In order to investigate the thermorheological squeezing response of thermoplastic composite materials, a parallel plate plastometer is typically used.

RHEOLOGICAL MODELS

In this chapter, the deformation of continuous-fiber reinforced thermoplastic (CFRT) sheets is considered. Experimental work carried out by Barnes and Cogswell [4] on the squeeze flow of a unidirectional thermoplastic composite, subjected to a constant load, identified three distinct phases of deformation: (1) viscous transverse flow, (2) limiting transverse flow and (3) elastic recovery. By studying a plan view of the specimens after forming, it was noticed that the fibers did not remain perfectly straight and aligned during squeezing. Instead the fibers curved in the plane near the free edges, so that the plan view of the initially square samples took up the shape of a barrel. While the fibers in the central region remained straight, the fibers which flowed the furthest experienced the greatest distortion. It was proposed that this distortion led to fiber locking, which limited the transverse flow of the composite and dramatically increased the apparent viscosity of the laminate with time. With a further increase in the load, the thicker samples (1 mm) exhibited fiber jetting, whereby fibers in the central region of the specimen flowed axially out of it. Neither the barrelling effect nor the fiber jetting effect have been reported elsewhere and may be a function of the initial width/length or thickness/length ratio of the sheet. Finally, the elastic recovery of the specimens was associated with the relaxation of the stress in the twisted fibers at the edges.

The simplest approach to modelling the behavior of an idealized (incompressible and inextensible) thermoplastic composite under squeeze flow conditions has been proposed by Rogers [5] using a Newtonian fluid continuum model. Provided the fibers do not bend or rotate during forming, the inextensibility constraint only allows a plane strain two-dimensional velocity field. Full-stick and full-slip are considered as the two possible interface conditions between the platen and the fluid. In the full-slip case, the flow mechanism is purely extensional in the transverse fiber direction and this deformation is equivalent to pure shear across the fibers. When the full-stick condition is satisfied, an inhomogeneous flow field is generated which causes the free edges of the sheet to bulge. These edges remain traction free, but tend to spread onto the platens during forming. In both cases, the width of the strip, l, increases as the platen gap, h, decreases. Employing the incompressibility constraint, the pressure distribution is integrated over the contact surface area, to determine the squeeze force, F, as a function of the cross-head speed, $\dot{h}$.

$$F\ (\text{Full-slip}) = -\frac{8\eta_T l\dot{h}}{h} \qquad F\ (\text{Full-stick}) = \frac{2\eta_T l(l^2 + 3h^2)\dot{h}}{h^3} \tag{1}$$

where η_T is the Newtonian transverse viscosity.

For a constant cross-head speed, the platen gap decreases in a nonlinear manner with time and the predicted compression load is highly dependent on the interface condition. In reality, a slip-stick boundary condition applies at the contact surface, unless special care is taken to lubricate the interface to ensure full-slip.

Balasubramanyam *et al.* [6] generalized Rogers's Newtonian fluid model by incorporating a sliding friction condition at the platen/sheet interface and a slightly different traction condition at the free edges. The large variation in their experimental results suggests that the contact condition is difficult to repeat in practice and corresponds to neither a full-stick nor a full-slip situation. The results from their theoretical model also compared poorly with the constant load squeeze flow behavior of APC-2 (unidirectional carbon fiber containing polyether etherketone – (CF/PEEK)). The authors attribute this lack of agreement to a stick-slip platen interface condition and a dramatic increase in the transverse viscosity with decreasing sheet thickness due to fiber locking. It is most probable that a Newtonian fluid model is inadequate to capture the essential features of the flow.

Wang and Gutowski [2] investigated the squeeze flow problem by considering an idealized continuum under constant loading. However, instead of using a Newtonian fluid model, they employed a power-law shear thinning fluid model ($\tau = m\dot{\gamma}^n$) and a full-stick boundary condition at the platen/sheet interface. The average width of the strip was calculated by assuming volume constancy. The pressure distribution across the contact surface was then derived as a function of the platen gap and the nonlinear viscous behavior of the fluid. From their experimental work on Radel 8230 and APC-2 laminates, a nondimensional transverse spreading parameter was plotted against a nondimensional time variable and compared favorably with the model predictions. The results indicate that the degree of deformation is highly dependent on the consistency index, *n*, the pseudo-Newtonian viscosity, *m*, and the geometry of the laminate.

Mander [7] conducted experimental squeeze flow tests on Plytron® (unidirectional glass fiber reinforced polypropylene – GF/PP) to study the effects of forming speed and temperature on the material response. In all cases an increase in the forming temperature resulted in a decrease in the shear viscosity, as expected. Figure 1 shows the viscosity of Plytron at 170°C for experiments satisfying the full-stick interface condition. These values were back-calculated from the experimental load data using equation (1) for the full-stick interface condition. In each test, the apparent viscosity initially rises in a nonlinear manner before reaching a constant value at a strain of approximately 0.1. Furthermore, an increase in the cross-head speed from 0.5 mm/min to 5.0 mm/min causes a decrease in the steady-state viscosity from approximately 5000 Pa s to 1000 Pa s. This behavior indicates a shear thinning response in the material, which cannot be accounted for using a Newtonian fluid model. In the full-stick case, a series of square grids marked on the surface of the specimens normal to the fiber direction demonstrated an inhomogeneous velocity field during forming. Therefore, the material at the edges of the strip bulged and flowed onto the platens, causing the surface contact area to increase. No resin percolation along the length of the fibers was observed during the tests, confirming the suitability of the inextensible fiber model. Also, the fibers remained in straight parallel lines as the laminate spread, so no barrelling was seen. For the case of full-slip squeeze flow, the response of the polypropylene composite is seen to be quite different in Figure 2. The viscosity of the Plytron increases nonlinearly as a function of strain and continues to rise without reaching a plateau. These values were back-calculated from the experimental load data using equation (1) for the full-slip condition. An increase in the squeeze flow rate causes a decrease in the viscosity, as would be expected for a shear thinning polymer. Apart from the extreme difference in shape between the curves in Figure 1 and Figure 2, the full-slip viscosities are much higher than those measured for the full-stick flow. This may be explained by the inhomogeneous velocity flow field, in the full-stick case, which results in a higher effective strain-rate and hence greater shear thinning.

Using the concept of a shear thinning fluid, Shuler and Advani [8] investigated the use of a Carreau fluid model to fit the squeeze flow data for a clay/nylon and an APC-2 composite material. Their results were calculated numerically assuming a full-stick flow condition and seem to match their experimental data quite well. In particular, they studied the

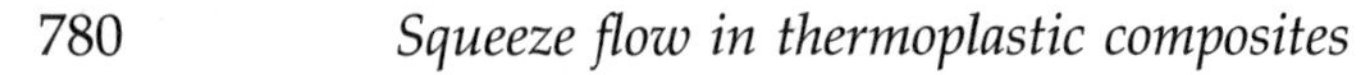

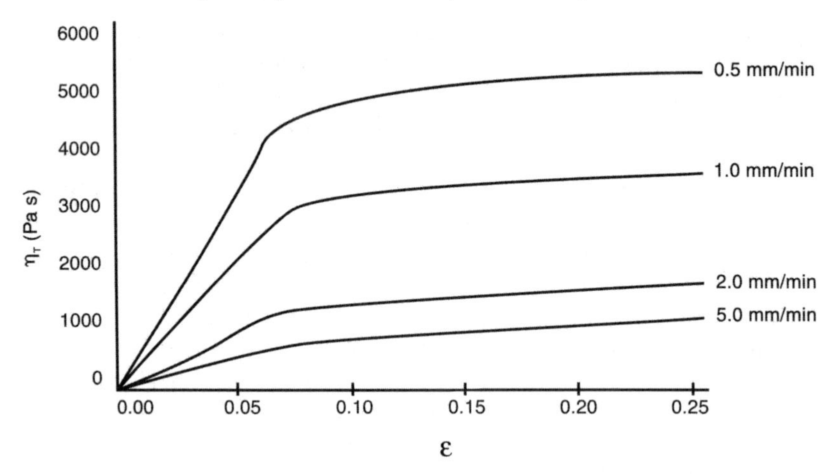

Figure 1 Transverse shear viscosity versus engineering strain for various cross-head speeds (full-stick condition).

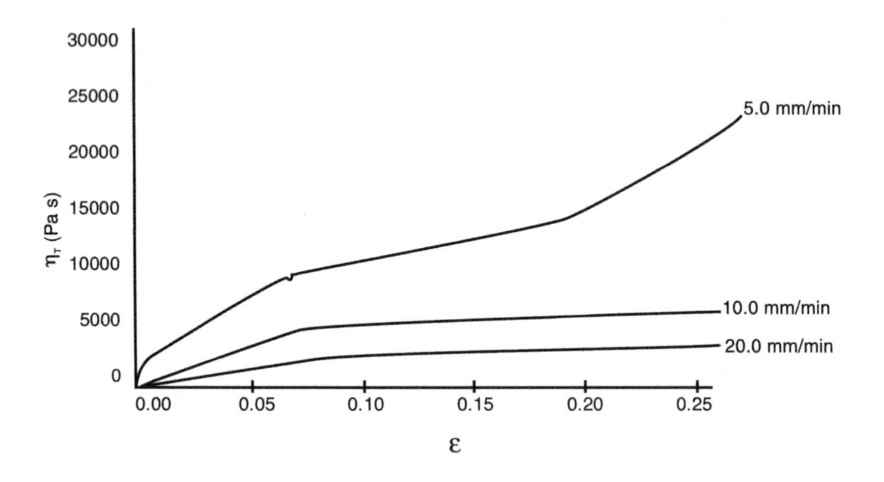

Figure 2 Transverse shear viscosity versus engineering strain for various cross-head speeds (full-slip condition).

effects of the fiber volume fraction and the fiber diameter on the transverse shear viscosity using a parallel plate plastometer. A feature of their work, which was different from the previously mentioned authors, was the use of equal sized platens and samples. As each specimen was compressed, the laminate flowed outside the platens so that the contact surface area remained constant. The boundary traction effect of the material outside the platens was ignored. It is not clear how this assumption affects the accuracy of the theoretical model. The behavior of the clay/nylon composite reveals that increasing the fiber volume fraction significantly increases the shear viscosity, while the shear thinning index remains equal to that for the unfilled matrix material. A higher fiber volume fraction also causes shear thinning to begin at lower shear rates. Using the same fiber volume fraction but smaller diameter fibers causes the shear viscosity to increase dramatically, while slightly retarding the inception of shear thinning. The Carreau fluid model adequately describes the behavior of APC-2 at small strain rates and fits the experimental data quite well at higher strains. Using the concept of a lubricated fiber bundle with aligned fibers in a Carreau fluid, the effect of fiber volume fraction on the squeeze force was shown to agree qualitatively with the experimental results. However, the possibility that anything other than a fluid model should be needed to quantify the results was not discussed.

Results obtained by Mander [7] indicate that only observing the loading phase of a squeeze flow test does not give sufficient information about the material behavior to fully characterize the transverse shear viscosity. When the cross-head is stopped, the load demonstrates a definite stress relaxation phenomenon consistent with viscoelastic behavior. It is tempting to use Rogers's [1] linear viscoelastic model for idealized materials to quantify this behavior. However, it should be remembered that the strains in squeeze flow experiments are not small and therefore small strain viscoelastic theory should not be applied. In this case, since the plane strain deformation is essentially isotropic, it is possible to use one of the many finite strain viscoelastic material models discussed by Bird *et al.* [9], such as the Lodge rubber-like liquid model, the convected Jeffrey's model and the Giesekus model. However, these models cannot describe the full three-dimensional anisotropic behavior of the continuum in general flow.

A final note to mention concerns work carried out by Bland *et al.* [10] on the squeeze flow behavior of a glass mat thermoplastic PP (GMT-PP) (see the related chapter in this book). Instead of containing continuous aligned fiber reinforcement, this material consisted of short glass fibers randomly distributed in an incompressible PP matrix. Because the material behaved in a quasi-isotropic manner, it was possible to use a circular blank for the experimental work. In this way a planar axisymmetric flow field was generated and a power-law shear thinning model was used to describe the fluid properties. The viscoelastic effects were ignored. The experimental squeezing force curves for the full-stick and the full-slip conditions were almost identical in shape, although the full-stick loads were slightly higher. The full-slip results compared reasonably well with the theoretical model; however, no comparison was made for the full-stick case and no details were tabulated for the calculated pseudo-Newtonian viscosity or the consistency index.

REFERENCES

1. Rogers, T.G. (1989) Rheological characterisation of anisotropic materials, *Composites*, **20**(1), 21–27.
2. Wang, E.L. and Gutowski, T.G. (1991) Laps and gaps in thermoplastic composites processing, *Composites Manufact.*, **2**, 69–78.
3. Cogswell, F.N. (1987) The processing science of thermoplastic structural composites, *Int. Polymer Process.*, **1**, 157–165.
4. Barnes, J.A. and Cogswell, F.N. (1989) Transverse flow processes in continuous fibre-reinforced thermoplastic composites, *Composites*, **20**(1), 38–42.
5. Rogers, T.G. (1989) Squeezing flow of fibre-reinforced viscous fluids, *J. Engng Math.*, **23**, 81–89.
6. Balasubramanyam, R., Jones, R.S. and Wheeler, A.B. (1989) Modelling transverse flows of reinforced thermoplastic materials, *Composites*, **20**, 33–37.
7. Mander, S.J. (1997) Roll forming of thermoplastic composite materials, Ph.D. Thesis, The University of Auckland, Auckland, New Zealand.
8. Shuler, S.F. and Advani, S.G. (1996) Transverse squeeze flow of concentrated aligned fibres in viscous fluids, *J. Non-Newtonian Fluid Mech.*, **65**, 47–74.
9. Bird, R.C., Armstrong, R.C. and Hassager, O. (1987) Dynamics of Polymeric Liquids, vol. 1, Fluid Mechanics, 2nd edn, Wiley, New York.
10. Bland, J.H., Bersee, H.E.N. and Gibson, A.G. (1997) Squeeze flow testing of polypropylene and glass mat thermoplastics at compression moulding strain rates, in *Proceedings of the 11th International Conference on Composite Materials*, Gold Coast, Queensland, Australia, (ed. M.L. Scott), Woodhead Publishing, Cambridge, UK, vol. 4, pp. 400–410.

Keywords: squeeze flow, Plytron®, full-slip condition, full-stick condition, unidirectional continuous-fiber reinforced composites, shear thinning fluid, Carreau fluid, APC-2, composite laminate, UD laminate.

Structure–property relationships in polypropylene fibers

Ismail Karacan

INTRODUCTION

Polypropylene (PP) is one of the most successful commodity synthetic polymers, with world production capacity reaching 4 million tonnes a year including continuous and staple yarn. In volume terms, polypropylene fiber takes the fourth place after polyester, polyamide and acrylic fibers. PP is widely used in many applications because of its low density (0.905 g/cm^3), high crystallinity as well as its high stiffness and hardness. Of course, the resulting high strength-to-weight ratio is undoubtedly the main advantage of PP in many industrial end-uses. The main applications of PP range from carpet industry, domestic textiles, clothing to industrial textiles.

Because of the thermoplastic nature of PP, the melt spinning technique is normally used for fiber formation. The physical properties of PP fiber are shown to depend on the initial material characteristics, such as melt viscosity, and the fiber formation conditions, such as extrusion temperature, cooling rate and spinline stress [1]. In terms of fiber mechanical properties, there is now a general agreement that high tenacity fibers can be produced using a polymer of low melt viscosity. In order to improve the mechanical properties, a multistage drawing procedure seems to lead to improved fiber performance [2, 3].

Polypropylene: An A–Z Reference
Edited by J. Karger-Kocsis
Published in 1999 by Kluwer Publishers, Dordrecht. ISBN 0 412 80200 7

STRUCTURAL CHARACTERIZATION

Isotactic polypropylene (iPP) is generally regarded as a semicrystalline polymer since its wide angle X-ray diffraction patterns show the characteristics of both a crystalline phase (as indicated by the sharp reflections) and an amorphous phase (as indicated by a diffuse halo).

The crystalline phase contains individual crystallites which are formed from molecular chains folded back on one another and connected by tie molecules. So far, four different crystal forms (α, β, γ and 'smectic') have been identified mainly from X-ray diffraction studies. The majority of studies suggest that monoclinic (α-phase) of iPP is the most common crystalline form with helical chain conformations in the unit cell. The chains in the crystalline regions adopt a 3_1 helix with a symmetry class of C_3. The helical chains of PP stretch rather easier than the extended chains of polyethylene. The presence of bulky —CH_3 groups increase the chain cross-sectional area and reduce the possibility of attaining highly extended chain conformation. Due to the helical chains, the calculated chain modulus of polypropylene (about 50 GPa) is much lower than the calculated chain modulus of polyethylene (about 300 GPa). Furthermore, the —CH_3 groups are believed to lower the packing density, thereby reducing the modulus. The amorphous phase, on the other hand, includes complex chain molecules whose tacticity is believed to exclude them from crystallization process. These is strong evidence, however, that these molecules are still largely in the helical conformation.

CHARACTERIZATION OF ORIENTATION

In the literature it has been shown conclusively that there is a direct relationship between the degree of orientation of molecular chains and the mechanical properties, such as Young's modulus.

The data obtained from birefringence measurements, X-ray diffraction and infrared and Raman spectroscopy could be used to estimate the degree of orientation averages. These techniques, of course, provide structural information at different levels, with their own limitations. For example, the use of birefringence can be used to estimate the average orientation in terms of second-order orientation averages, whereas the use of X-ray diffraction only provides the crystalline orientation averages. However, the use of infrared spectroscopy [4, 5] provides the orientation averages of both crystalline and amorphous phases.

Next it will be shown, using two sets of as-spun iPP fibers produced from two different polymers, how the processing conditions and the quality of the raw material affect the structure–property relationships. The sample details are presented in Table 1. The polymers used were Finaprop 10080S and 7039S with melt flow index (MFI) values of 35 and

Table 1 Sample details of as-spun isotactic polypropylene fibers

Polymer	T_s (°C)	*MFI* (g/10 min)	$\Delta n \times 10^3$	M (cN/tex)
10080S	200	35	4.0	36.5
	220	36	3.0	30.7
	240	37	2.8	25.2
	260	39	2.4	26.8
	280	48	2.0	21.7
7039S	200	11	15.0	96.1
	220	12	13.0	71.9
	240	12	12.0	61.3
	260	14	11.0	58.0
	280	28	9.0	37.1

T_s = extrusion temperature; MFI = melt flow index; Δn = birefringence, M = 2–5% secant modulus [1].

12 g/10 min, respectively. The as-spun fibers were produced with extrusion temperatures between 200 and 280°C.

OPTICAL BIREFRINGENCE

Optical birefringence can be defined as the difference between the refractive index for light polarized parallel to the fiber axis ($n_{//}$) and that for light polarized perpendicular to the fiber axis ($n_{\perp}$) and is a measure of overall molecular orientation. The refractive indices of the fiber samples were measured, with the aim of determining birefringence values, using an image splitting Carl Zeiss interference microscope. The refractive indices were measured by matching the refractive index of commercially available liquids. This method is far more accurate than the direct measurements.

The fibers produced from the polymer with low melt viscosity (MFI = 35 g/10 min) possess a paracrystalline structure with low orientation. The X-ray diffraction patterns of these fibers show halos indicating poor orientation. The overall molecular orientation as shown by the birefringence values decreases with increasing extrusion temperature.

The as-spun fibers produced from the polymer with a high melt viscosity of MFI = 12 g/10 min show a highly oriented α-monoclinic structure. As indicated by the X-ray diffraction patterns and equatorial traces, these fibers are more oriented than the fibers produced with MFI = 35 g/10 min. In fact, the subsequent drawing of the fiber produced from the polymer with low melt viscosity leads to high-tenacity fibers. The overall molecular orientation of as-spun fibers from polymer with high melt viscosity is found to be higher than the fibers produced from

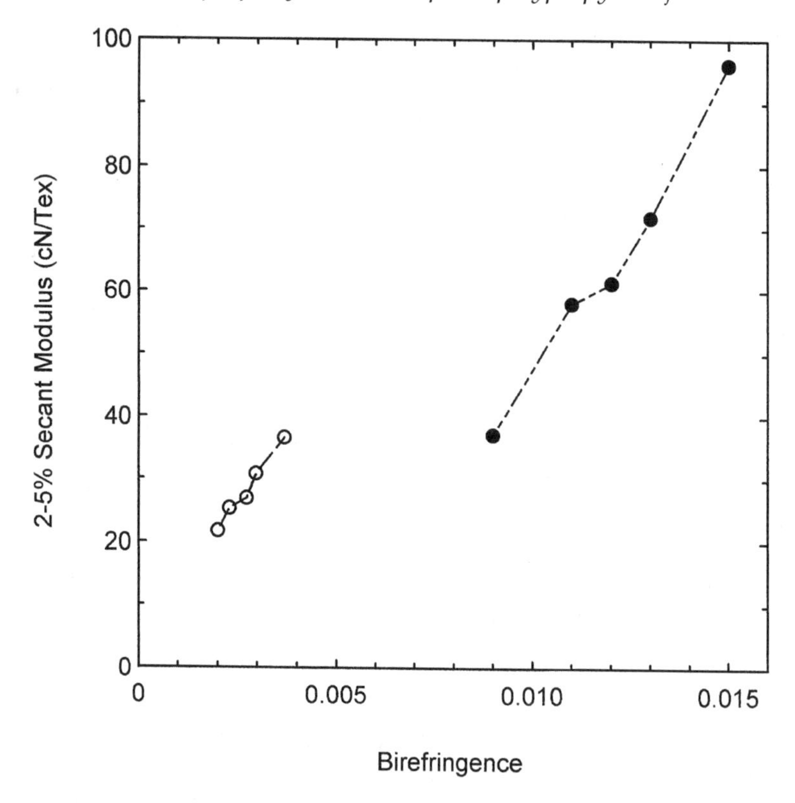

Figure 1 Comparison of 2–5% secant modulus with the optical birefringence: (○) 10080S polymer fibers; (●) 7069S polymer fibers.

polymer with low melt viscosity. In all cases, the overall molecular orientation as measured by the optical birefringence decreases with increasing extrusion temperature (Table 1).

INFRARED SPECTROSCOPY

The orientation averages obtained from the infrared spectroscopy is only limited to second order averages as described by the second-order Legendre polynomial of $\cos\theta$:

$$\langle P_2 \rangle = \frac{1}{2}(3\langle \cos^2\theta \rangle - 1) = \frac{D-1}{D+2}\frac{D_0+2}{D_0-1} \tag{1}$$

where the angular brackets indicate the average values and θ is the angle between the molecular axis and the fiber axis direction. $D_0 = \cot^2\alpha$, α being the angle between the transition moment associated with the vibrational mode and the molecular axis. D is known as the dichroic ratio and is defined as follows:

$$D = A_{//}/A_{\perp} \quad (2)$$

where $A_{//}$ and $A_{\perp}$ are the measured absorbances for radiation polarized parallel and perpendicular to the fiber axis direction, respectively. For samples with very small birefringences such as the present as-spun fiber samples shown in Table 1, equations (1) and (2) are sufficiently accurate. In the literature the second orientation average, $\langle P_2 \rangle$, is also known as Herman's orientation function.

The polarized infrared spectrum of isotactic polypropylene in the 750–1080 cm^{-1} region shows at least seven well-defined peaks. Of these peaks, those at 841, 973 and 998 cm^{-1} can be described as being very strong in intensity and show parallel polarization characteristics, whereas the peaks at 809, 900, and 940 cm^{-1} can be described as being medium in intensity and show perpendicular polarization characteristics. These is also a doublet with peaks at 1036 and 1045 cm^{-1}, both components of which are medium in intensity and show parallel polarization characteristics.

Infrared results: parallel modes

841 cm^{-1}

This peak has been assigned to a combination of CH_2 rocking, C—C stretching, CH_3 rocking and C—CH_3 stretching vibrations and is regarded as a 'regularity' band because of its narrowest half-height width in the spectral region studied. In a previous study [4], the Herman's orientation parameter determined from this peak has been found to correlate well with the orientation parameter determined from X-ray diffraction. In the absence of X-ray diffraction, this peak may be used for determining the chain axis orientation of crystalline material. As shown in Figure 2(a), the Herman's orientation parameter determined from this peak is higher than the orientation averages determined from the 998 cm^{-1} and 973 cm^{-1} peaks.

998 cm^{-1}

This peak has also been assigned to a combination of C—C stretching, CH_3 rocking, CH_2 rocking, CH_2 wagging, CH bending, CH_2 twisting and C—CH_3 stretching vibrations. The half-height width of this peak is greater than that of 841 cm^{-1} and is usually regarded as being due to the crystalline or highly ordered material. As shown in Figure 2(b), the Herman's orientation averages determined from this peak are always

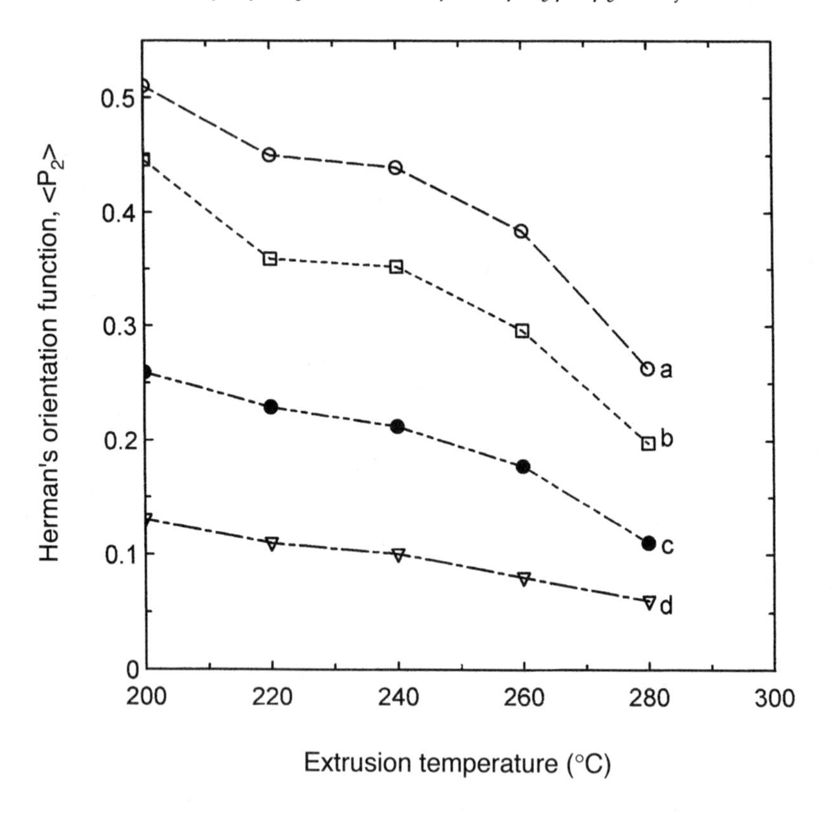

Figure 2 The effect of spinning temperature on the orientation parameter $\langle P_2 \rangle$ of as-spun fibres produced from polymer 7069S. $\langle P_2 \rangle$ values are obtained from the (a) 841 cm^{-1}; (b) 998 cm^{-1}; (c) 973 cm^{-1}; (d) 974 cm^{-1} infrared peak.

lower than those obtained from the 841 cm^{-1}. This result suggests that this peak is slightly sensitive to less ordered material. Hot-stage infrared studies show that, after melting the polymer above its melting temperature, the infrared spectrum still contains this peak with much reduced intensity.

973 cm^{-1}

The half-height of this peak is higher than the peak at 998 cm^{-1}. The transition moment angle of this peak is estimated to be 23° with the chain axis [4]. As shown in Figure 2(c), the Herman's orientation averages determined from this peak is always lower than the peaks at 841 and 981 cm^{-1}, respectively, and this suggests that this peak is even more sensitive to the disordered material. In order to improve the curve fitting, this peak is fitted with two peaks at 972 and 974 cm^{-1}, respectively. The behavior of the peak at 972 cm^{-1} is found to be similar to that of the peak at 998 cm^{-1}. In an earlier paper [4], the behavior of the peak at 974 cm^{-1} was found to correspond to the amorphous material. As shown in Figure 2(d), this peak has been used for the determination of the orientation averages of amorphous phase.

CONCLUSIONS

It has been demonstrated that there is a direct relationship between the fiber mechanical properties and the molecular orientation (Figure 1). A similar relationship exists between the processing conditions, molecular orientation and mechanical properties. It has been shown that a combination of optical birefringence and the infrared spectroscopy techniques can be used for the measurement of molecular orientation. The results show that for a set of fiber samples, the 841 cm^{-1} infrared peak can be confidently used for the estimation of crystalline orientation averages. The 973 cm^{-1} peak can be separated using curve fitting procedures into 972 and 974 cm^{-1} components and the latter component can be used to estimate the amorphous orientation averages.

REFERENCES

1. Wang, I.-C., Dobb, M.G. and Tomka, J.G. (1995) Polypropylene fibres: An industrially feasible pathway to high tenacity. *J. Text. Inst*, **86**, 383–392.
2. Wang, I.-C., Dobb, M.G. and Tomka, J.G. (1996) Polypropylene fibres: Exploration of conditions resulting in high tenacity. *J. Text. Inst*, **87**, 1–12.
3. Abo El Haaty, M.I., Bassett, D.C., Olley, R.H., Dobb, M.G., Tomka, J.G. and Wang, I.-C. (1996) On the formation of defects in drawn polypropylene fibres, *Polymer*, **37**, 213–218.
4. Karacan, I., Taraiya, A.K., Bower, D.I. and Ward, I.M. (1993) Characterisation of one-way and two-way drawn isotactic polypropylene films. *Polymer*, **34**, 2691–2701.
5. Karacan, I., Bower, D.I. and Ward, I.M. (1994) Molecular orientation in uniaxial, one-way and two-way drawn poly(vinyl chloride) films. *Polymer*, **35**, 3411–3422.

Keywords: amorphous orientation, crystalline orientation, Herman's orientation factor, infrared spectroscopy (IR), IR band assignment, optical birefringence, orientation, structure–property relationships in fibers, X-ray diffraction

Surface modification of polypropylene by additives

I. Chodák and I. Novak

Polypropylene (PP) is highly apolar material, therefore a modification aimed at the creation of a more polar surface is an important issue considering, for example, wettability, adhesion, barrier properties or dyeability.

The modification of PP surfaces can be done by various means, such as grafting more polar monomers onto the surface, controlled oxidation or plasma treatment. Incorporation of a suitable component into the bulk polymer is one of the possibilities commonly considered [1, 2].

Various low molecular organic compounds are used for the modification of PP in bulk. The list of effective modifiers includes fatty acids, their salts, unsaturated dicarboxylic acids and their anhydrides, plasticizers, surfactants, organic peroxides and ketones. Most of the relevant data can be found in patent literature [e.g. 3, 4].

Fatty acids and their salts have been well studied regarding the effect on adhesion properties of PP. The optimal concentration of the additive depends on its solubility in the polymer. When investigating the addition of fatty acids or their salts, the highest adhesion values are usually observed at a rather low content of the additive (quite often only at 0.1–0.2 wt.%). This concentration is around the solubility threshold of the fatty acids in PP. The addition of higher amounts of the modifier results in an extensive diffusion of the additive on the polymer surface leading to a formation of a layer with low cohesive strength. When compounds with higher solubility in PP, such as the esters of higher fatty acids, e.g. ethyl palmitate, are added, the optimal concentration for achieving the

Polypropylene: An A–Z Reference
Edited by J. Karger-Kocsis
Published in 1999 by Kluwer Publishers, Dordrecht. ISBN 0 412 80200 7

highest adhesion was substantially higher, around 5 wt.%. Similar results were obtained when investigating the effect of some ketones, e.g. diheptadecyl ketone.

The operative mechanism is the physical adsorption of the long aliphatic chain in the hydrophobic PP phase, while the polar part of the additive penetrates the surface and increases its polarity. Hence, suitable polar functional moieties can be bound onto the surface by chemisorption. A significant increase in adhesion is observed so that even cohesive destruction of the adhesive joints has been reported in some cases.

When studying the compounds with the same length of aliphatic chain, the values of mechanical work of adhesion of modified PP towards several more polar polymers was found to decrease in the order: sodium stearate > oleic acid > stearic acid.

The strength of an adhesive joint between PP and polar polymers can be also increased by an addition of surfactants. Anionic as well as nonionic surfactants are used, such as dodecyl benzene sulphonate, derivatives of oxyethylated esters of fatty acids, or polyglycol esters of alkyl phenols. The highest values of mechanical work of adhesion were observed for PP containing 0.2 wt.% of the additive. The positive effect of the surfactants is explained by the adsorption on the interface leading to the increase of surface wettability and of the contact area between the adhering substrates. Further, the decrease of internal stresses in PP as a result of the presence of the modifier can contribute to an increase of the strength of the adhesive joint. The optimal concentration of the additive in the polymer depends on its solubility. A formation of monomolecular layer on the polymer–substrate interface leads to a maximal adhesion. Further increase in the additive concentration results in formation of a thicker interlayer with low cohesive strength.

A positive effect regarding the adhesion was observed when plasticizers were added. Their effect is based on the increase of molecular mobility in the amorphous phase. Exceeding the threshold additive concentration leads to a decrease in adhesion, as with other additives discussed above. This effect is caused by a reduction of the contact area. The importance of threshold solubility is demonstrated by the addition of nonpolar vaseline oil which is miscible with PP in the whole concentration range. The increase in the vaseline oil concentration results in a monotonous rise of the adhesion parameters until cohesive fracture is reached.

Addition of maleic anhydride (MA) by simple physical mixing is also beneficial to the adhesion properties of PP. The threshold concentration in this case is rather high, being around 5 wt.%. It is more straightforward, however, to graft MA onto the PP chain via free radical initiation, e.g. either by thermal decomposition of peroxides or photochemically. The amount of grafted maleic anhydride is usually below

1 wt.% and leads to a significant improvement of the adhesion [5]. Besides MA, other unsaturated acids, anhydrides or imides can be grafted onto the PP surface resulting in an increase of adhesion properties. Organic peroxides are common initiators of the grafting reaction.

Peroxides or some ketones are effective even in the absence of any other component. Thermal or ultraviolet decomposition of these compounds generates PP peroxides and hydroperoxides, leading to increased polarity and enhanced adhesive properties.

The grafting can be also initiated by exposure of the PP surface to plasma, high/energy irradiation or ultraviolet radiation (see the chapters entitled Surface modification of polypropylene by plasmas, and Resistance to high-energy radiation). The monomer is introduced after irradiation, or simultaneously with the initiation. Abrasion of polymer surfaces can also serve for the generation of grafting sites. Abrasion of PP in liquid epoxy resin leads to a grafting of epoxy onto the PP surface. A dramatic increase in wettability and bondability is achieved in this way.

The adhesion properties of PP can be also modified by addition of a more polar polymer. The effect of the polymer additive is similar to that of a low molecular plasticizer. The major advantage is a negligible migration of the polymeric additive to the PP surface, compared to low molecular plasticizers. The efficiency of a particular polymer depends on the morphology of the blend, especially regarding the dispersion of the additive in the PP matrix. Different polymers have been investigated as additives for a modification of surface properties of PP. Besides grafted copolymers of PP with maleic anhydride, acrylic or methacrylic acid, the list of effective modifiers includes block copolymers 1,3-butadiene-co-styrene and isoprene-styrene, or chlorinated and chlorosulphonated PP.

A linear change of free surface energy on the composition of the blend was observed in some cases but usually the dependence on the composition of the blend is more complicated. The free surface energy of the polymer blends depends primarily on the amount of polymeric components and on their free surface energies. A rather common feature is a phenomenon of excess free surface energy resulting from a preferential accumulation of one polymeric component in the surface layer of the matrix polymer. This effect was observed, for example, for the blends of PP with polyethylene-co-vinyl acetate or polyethylene-co-acrylic acid.

All the methods mentioned above are rather effective regarding the surface modification of PP, demonstrated, for example, by a significant increase in the strength of the adhesive joint between PP and the polar surface. However, the modification always leads to certain decrease in other, especially mechanical, properties. From this point of view, the addition of triblock copolymers or grafting with maleic anhydride can be considered as the most suitable procedures.

REFERENCES

1. Wool, R.P. (ed.) (1995) *Polymer Interfaces, Structure and Strength*, Hanser, Munich.
2. Wu, S. (1982) Modification of polymer surfaces: Mechanisms of wettability and bondability improvements, Chap. 9, in *Polymer Interface and Adhesion*, Dekker, New York, pp. 279–336.
3. Ishii, H., Takamatsa, H., Hirayama, Y. and Kono, N. (1989) *Jpn. Kokai Tokkyo Koho*, JP 01 240 542.
4. Murukami, H., Suda, T., Machida, M. and Umemura, S. (1988) *Jpn. Kokai Tokkyo Koho*, JP 01 279 939.
5. Sclavons, M., Carlier, V., de Roover, B., Franquinet, P., Devaux, J. and Legras, R. (1996) The anhydride content of some commercial PP-g-MA: FTIR and titration, *J. Appl. Polymer Sci.*, **62**, 1205.

Keywords: polarity, adhesion, wettability, solubility, diffusion, physical separation, chemical sorption, grafting, free surface energy.

Surface modification of polypropylene by plasmas

Chi-Ming Chan

Polypropylene (PP) is a polymer with good mechanical properties which has been used in a wide range of applications. However, due to its inertness and low surface energy, PP has to be surface treated for applications in coatings, bonding, printing and metallization. The frequently used surface treatments for PP include plasma, corona and flame treatments. Plasma treatment of polymer surfaces has been studied extensively for many years [1]; however, it has not been used as extensively as corona and flame treatment in industry. Plasma treatments must be carried out in a vacuum and the cost of industry-scale equipment for such treatment is very high. The process parameters are highly system-dependent; the optimal processing conditions developed for one system usually cannot be adopted by another system.

In spite of these disadvantages, plasma treatment of polymers is an attractive process to produce the required surface modification. By using different types of gas, various chemical functionalities can be introduced on the surface. In general, more uniform surfaces are produced by plasmas than by flame and corona treatments. The modification is typically confined to the surface without changing the bulk physical and chemical properties of the polymer.

A plasma can be broadly defined as a gas containing charged and neutral species, including some or all of the following: electrons, positive ions, negative ions, radicals, atoms and molecules. Reactions between gas-phase species and surface species, and reactions among surface species, produce functional groups and crosslinking, respectively at the

Polypropylene: An A–Z Reference
Edited by J. Karger-Kocsis
Published in 1999 by Kluwer Publishers, Dordrecht. ISBN 0 412 80200 7

surface. Examples of these reactions include plasma treatment by argon, ammonia, carbon monoxide, carbon dioxide, fluorine, hydrogen, nitrogen, nitrogen dioxide, oxygen and water.

A typical system consists of a gas inlet, reactor vessel, vacuum pump, matching network and power source. Various reactors have been used in plasma processing. For DC and low-frequency glow discharges, internal electrodes are necessary. As the frequency increases, electrodes may be placed outside the reactor vessel.

To understand the relationship between the performance of the modified surfaces and their physical and chemical properties, the surfaces have to be analyzed by surface-sensitive techniques that have a very small sampling depths (in the order of 5 nm). The characterization of plasma-treated surfaces usually involves surface-sensitive techniques, such as X-ray photoelectron spectroscopy (XPS), secondary ion mass spectrometry (SIMS), contact angle measurement and atomic force microscopy [1]. XPS, which is also known as electron spectroscopy for chemical analysis (ESCA), is probably the most commonly used technique in the characterization of polymer surfaces. XPS provides quantitative chemical composition and identification of functional groups with the aid of chemical derivatization. Angular-resolved XPS is frequently used to yield chemical composition in depth. The sampling depth of 3–5 nm is typical for XPS. SIMS can provide structural information of the modified surfaces. It complements XPS because SIMS can differentiate among polymers that have very similar XPS spectra. In addition, it can provide an elemental map of the surface. Contact angle measurements are often used to study the surface dynamics of functional polar and apolar groups. When water is used as the testing liquid, hydrophilicity and hydrophobicity of polymer surfaces can be determined. Atomic force microscopy has been very useful in determining polymer surface morphology because it can image nonconducting samples, such as polymers and ceramics.

Air, oxygen, nitrogen, argon, water, carbon dioxide, CF_4, CF_3H, CF_3Cl, CF_3Br, SF_6 and SOF_2 plasmas have been used to modify polypropylene surfaces to provide the required physical and chemical surface properties for various applications. Oxygen and oxygen-containing plasmas can react with polypropylene to produce a variety of oxygen functional groups, including C—O, C=O, O—C=O, C—O—O, and CO_3 at the surface. Table 1 is a summary of the surface chemical composition of the polypropylene surfaces treated by several oxygen-containing plasmas. The surface energy of the treated samples is substantially higher than that of the control sample. The concentration of different functional groups present on the plasma-treated surfaces can be determined by curve fitting of the XPS C1s spectrum. Table 2 shows the relative percentage amounts of the oxygen functional groups detected. The

Table 1 Surface chemical composition of the plasma-treated PP surfaces detected by XPS [2]

*Treatment (pressure=*10 Pa; *time=*7 min; *current density =*40 μA cm^{-2})	*Atomic ratio*		*Surface energy (mJ m^{-2})*
	O/C	*N/C*	
Untreated	0.06	–	29.08
80% Ar–20% O_2 plasma	0.28	–	60.01
80% H_2O–20% Ar plasma	0.44	0.22	54.54
H_2O plasma	0.43	0.03	47.10
Air plasma	0.21	0.02	65.71

Table 2 Relative percentage amounts of the oxygen functional groups on the plasma-treated PP surfaces [2]

Treatment (pressure = 10 Pa; *time* = 8 min; *current density* = 40 μA cm^{-2})	*C—H, C—C*	*C—O—C C—O—H*	*C=O O—C—O*	*O=C—OH*
80% Ar–20% O_2 plasma	82	9	5	4
80% H_2O–20% Ar plasma	74	21	2	3
H_2O plasma	78	13	5	4
Air plasma	82	9	5	4

oxygen functional groups are responsible for the improved bondability and printability of the PP films. Figure 1 shows the lap shear strengths of plasma-treated PP against steel. All these plasma-treated PP samples, except the water-plasma-treated sample, shows significant improvement in adhesion with mild steel. Besides the chemical changes caused by the plasma, it has been shown that surface roughness of PP increases after air-plasma treatment. The increase in the surface roughness is possibly caused by the preferential attack on the amorphous phase of PP by the plasma.

Surface fluorination of polymers is an important industrial process because it can provide PTFE-like (PTFE: polytetrafluoroethylene) surfaces on relatively inexpensive polymers. Fluorine-containing plasmas have been used to impart hydrophobic properties to polymer surfaces. In a fluorine-containing plasma, surface reactions, etching and plasma polymerization can occur simultaneously. Strobel *et al.* [3] studied surface modification of polypropylene with CF_4, CF_3H, CF_3Cl and CF_3Br plasmas. For example, CF_4-plasma treatment on polypropylene generated predominantly CF and CF_2 functional groups and a smaller amount of CF_3 group, as shown in Figure 2(a). However, when CF_3Cl

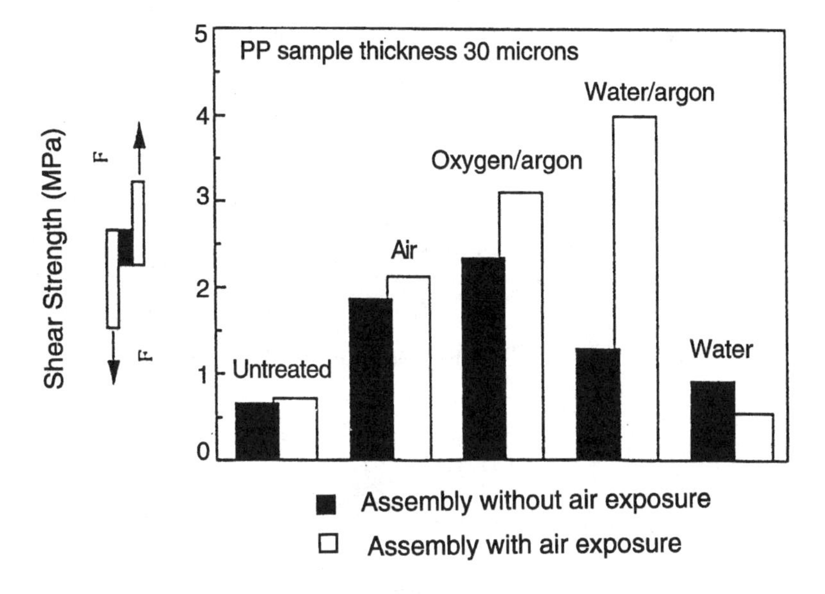

Figure 1 Shear strength of plasma-treated PP against steel.

and CF_3Br plasmas are used, chlorination and bromination of the PP surface occur, respectively and fluorination of the surface occurs only to a small extent. In a CF_3H plasma, polymerization is the dominant mechanism, leading to formation of a polymer layer on PP samples. The C1s spectrum of this polymer layer is shown in Figure 2(b). Comparing the CF_4 and CF_3H plasma-treated PP surfaces, similar fluorine functionalities were found, but their relative concentrations are different. The formation of a polymer layer when fluorine-containing plasmas were used to treat PP was determined to be the reason for the observed reduction in the permeation rates for several organic solvents [4].

The performance of the plasma-treated surfaces are related to their storage time in various environments. The aging of the plasma-treated PP has been also a subject of many studies [5]. When a polymer is exposed to an oxygen-containing plasma, the surface changes to a high-energy state (increase in surface tension) due to the formation of polar groups. Various surface studies indicate that the decrease in the surface energy when the treated surface is placed in a low-energy medium, such as air or a vacuum, is caused by the rotation of the polar groups in the bulk or the migration of low molecular weight fragment to the surface to reduce the interfacial energy. When a low-energy surface formed by treating a polymer in a fluorine-containing plasma is placed in a high-energy medium, such as water, the apolar groups will move away from the surface into the bulk in order to minimize the interfacial energy. This

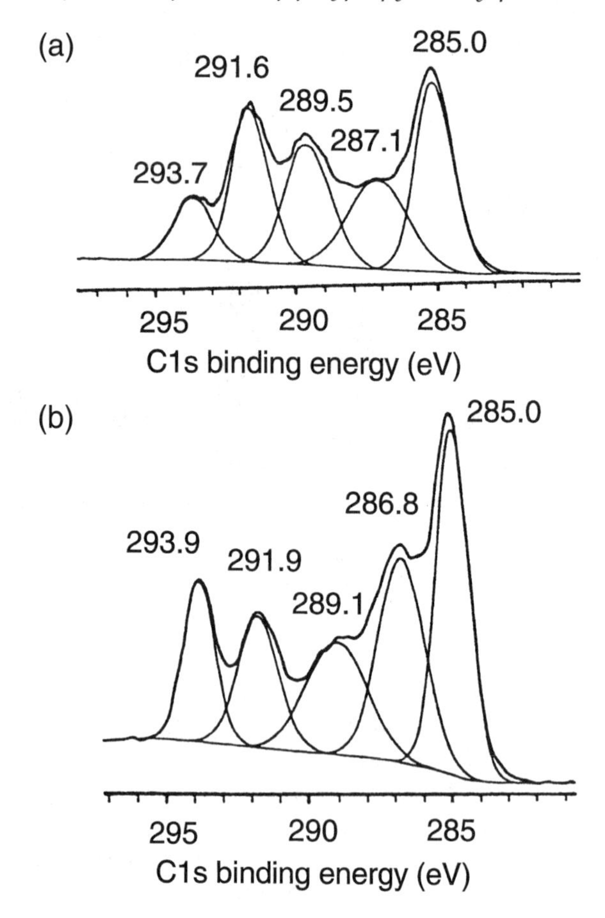

Figure 2 C1s spectra of PP exposed to (a) a CF_4 plasma and (b) a CF_3H plasma [3].

phenomenon is usually described as the aging of a treated surface. The aging of plasma-treated polymer surfaces is a very complex phenomenon that is strongly affected by treatment parameters, the nature of the polymer and storage conditions. Contact angle measurement, which is a very surface-sensitive technique, has been successfully used to study the dynamic characteristics of polymer surfaces in various environments.

REFERENCES

1. Chan, C.-M. (1994) *Polymer Surface Modification and Characterization*, Hanser, New York.
2. Greenwood, O.D., Boyd, R.D., Hopkins, J. and Badyal, J.P.S. (1995) *J. Adhesion Sci.*, **9**, 311.
3. Strobel, M., Corn, S., Christopher, C., Lyons, S. and Korba, G.A. (1985) *J. Polymer Sci.*, **23**, 1125.
4. Friedrich, J.F., Wigant, L., Unger, W., Wittrich, A.H., Prescher, D., Erdmann, J., Gorsler, H.-V. and Nick, L. (1995) *J. Adhesion Sci. Technol.*, **9**, 1165.
5. Occhiello, E., Morra, M., Morini, G., Garbassi, F. and Humphrey, P. (1991) *J. Appl. Polymer Sci.*, **42**, 551.

Keywords: electron spectroscopy for chemical analysis (ESCA), functionality, lap shear strength, plasma treatment, secondary ion mass spectrometry (SIMS), surface modification, surface energy, X-ray photoelectron spectroscopy (XPS).

Surface treatment of polypropylene by corona discharge and flame

Chi-Ming Chan

Polymers such as polyethylene, polypropylene and polystyrene are very inert and have quite low surface energies. Therefore it is necessary to treat the surfaces of these polymers prior to processes such as printing or bonding [1]. Corona and flame treatments are commonly used in the industry to improve bondability and printability of polyolefin films and large-size objects because of the speed and ease of processing. The set-up for such an operation is very simple and cost-effective. A corona treatment system consists of an electrode connected to a high-voltage source and a grounded roll as shown in Figure 1. The grounded roll is usually covered with an insulating material such as polyester, ceramic, epoxy or silicon rubber. The corona occurs when a high voltage is applied across the electrodes to cause ionization of air. A plasma is formed and a light blue color can be observed in the air gap. This atmospheric pressure plasma is called a corona discharge. The insulating material, which covers the grounded roll, prevents a direct arc between the two electrodes. However, the roll covering material is subjected to damage due to heat and the corona. One of the solutions to this problem is resolved by using a quartz tube filled with irregularly shaped metal pellets as the electrode. Quartz, being impervious to damage by heat or corona, is a good material for this application. To control the environment in which the corona is generated, the corona equipment can be placed in a chamber that is continuously flushed with dried or humidified air.

Polypropylene: An A–Z Reference
Edited by J. Karger-Kocsis
Published in 1999 by Kluwer Publishers, Dordrecht. ISBN 0 412 80200 7

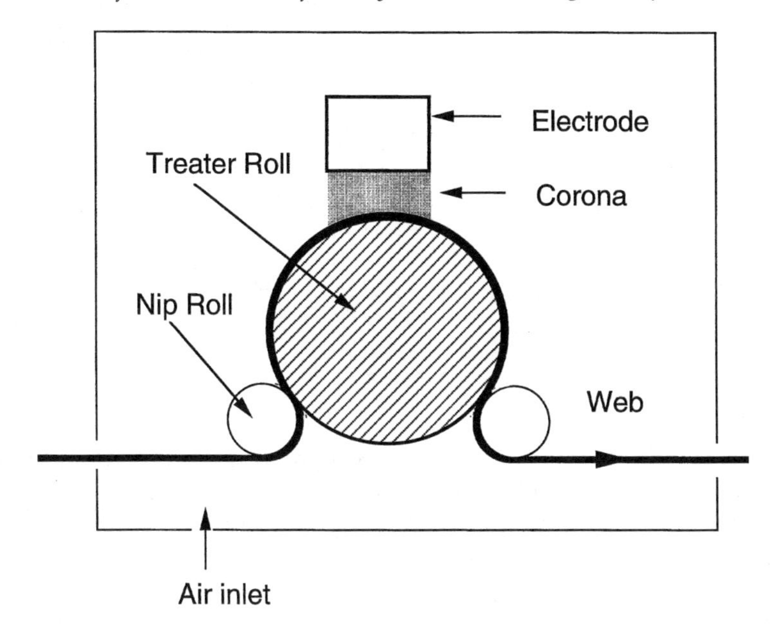

Figure 1 A schematic diagram showing a corona discharge apparatus.

The surface properties of the treated films were found to be related to the normalized energy (E), which can be determined from the net power and the film velocity [2]:

$$E = \frac{P}{wv} \tag{1}$$

where P is the net power, w is the electrode width, and v is the film velocity. The typical unit for E is J/cm^2. As the corona is in contact with a polymer surface, it can cause the surface to oxidize. Electrons, ions, excited neutrals and photons which are present in a discharge can react with a polymer surface to form radicals:

$$\mathrm{RH} \longrightarrow \mathrm{R}^\bullet + \mathrm{H}^\bullet$$
$$\mathrm{RR'} \longrightarrow \mathrm{R}^\bullet + \mathrm{R'}^\bullet$$

These radicals react rapidly with atmospheric oxygen [2]:

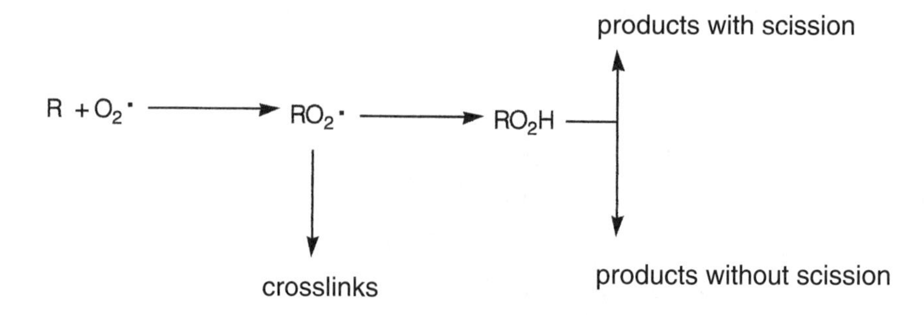

Scheme 1

These reactions make possible crosslinking and functionalization of the polymer surface with and without chain scission. The decomposition of the hydroperoxide groups produce oxygen functional groups on the surface, including C—OH, C=O and COOH.

Chain scission produces low molecular weight oxidized materials. The creation of low molecular weight oxidized material on the surface of polyolefins during air-corona treatment has been reported by a number of researchers. This material is very likely to be the product of free-radical oxidation processes. Experimental data obtained from X-ray photoelectron spectroscopy (XPS) and attenuated total reflectance infra-red (ATR-IR) spectroscopy have shown that the low molecular weight products formed on the polypropylene are soluble by polar solvents, such as water, ethanol and acetone, suggesting that they are not strongly bonded to the substrate. Strobel *et al.* reported that $0.05\,J/cm^2$ is the threshold energy for generation of these materials on polypropylene [3]. The amount of oxidized materials created on corona-treated polymer surfaces increases as the energy input to the corona increases. Humidity is another important factor which affects the formation of low molecular weight oxidized materials.

Roughening, which occurs on severely corona-treated samples, is attributed to the agglomeration of low molecular weight oxidized materials at high relative humidities. The extent of roughening is a function of the humidity and energy. At low relative humidities, no roughening was detected on a polypropylene surface, regardless of the energy input to the discharge. However, roughening was readily observed at high relative humidities (80%), even at low energies ($1.7\,J/cm^2$).

Flames have been used in various industries for surface treatment of plastics. This technique has been used to treat telephone cables to promote adhesion between the cable surface and a hot-melt adhesive,

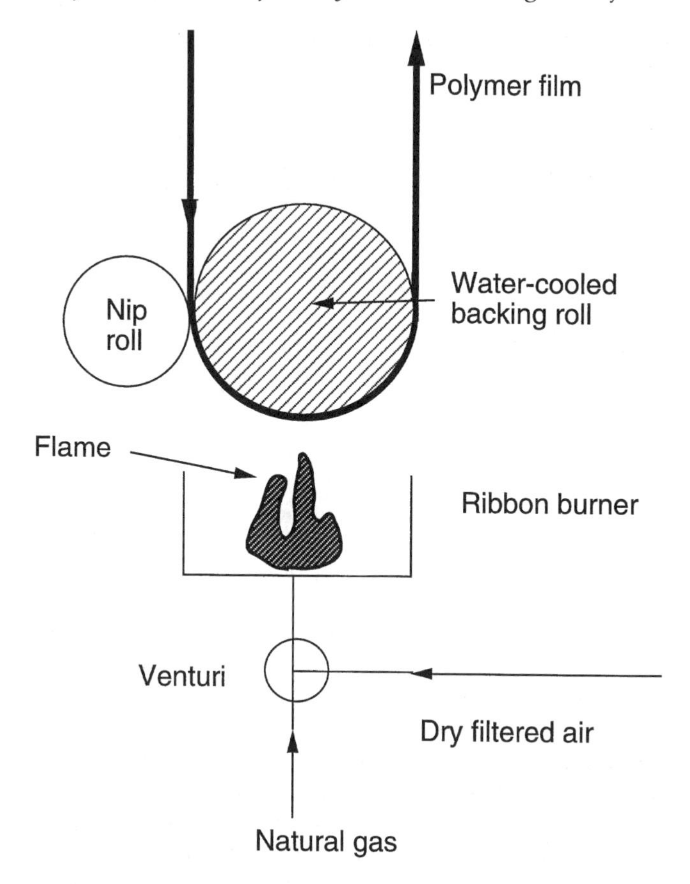

Figure 2 A schematic diagram showing a flame treatment apparatus [3].

because a torch is very portable. Flame treatment has also been commonly employed to enhance ink permanence on polymer surfaces. A schematic diagram for a flame treatment system is show in Figure 2.

Important variables for flame treatment are the air-to-gas ratio, air and gas flow rates, the distance between the tip of the flame and the object to be treated, the nature of the gas and the treatment time. The effect of these variables for flame treatment of polypropylene was studied by Sutherland *et al.* [4]. Oxidation at the polymer surface caused by flame treatment could be attributed to the high flame temperature (1000–2000°C) or reactions with many excited species in the flame. Flame treatment of polypropylene was studied by Garbassi *et al.* [5] using XPS and secondary ion mass spectrometry. XPS analyses revealed that oxygen functional groups such as OH, C=O, COOR and COOH were detected on flame-treated polypropylene surfaces. Garbassi *et al.* [5]

concluded that the site of the oxidative attack is the methyl pendant group:

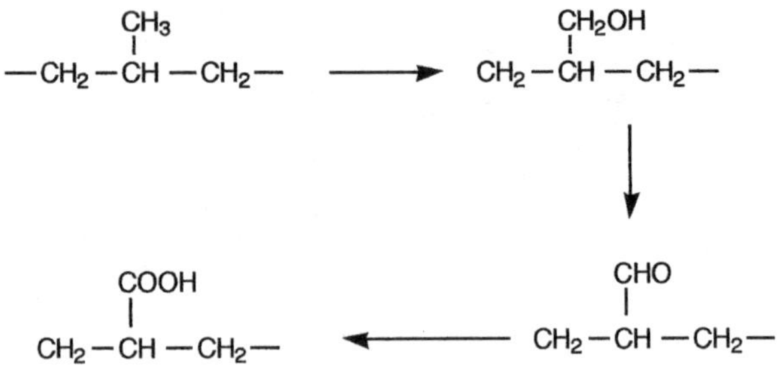

$$-CH_2-\underset{|}{\overset{CH_3}{}}CH-CH_2- \longrightarrow CH_2-\overset{CH_2OH}{\overset{|}{CH}}-CH_2-$$

$$\downarrow$$

$$CH_2-\overset{COOH}{\overset{|}{CH}}-CH_2- \longleftarrow CH_2-\overset{CHO}{\overset{|}{CH}}-CH_2-$$

Scheme 2

After the formation of the hydroxyl group, the oxidation proceeds with the formation of formyl and carboxyl groups.

The surface properties of treated polypropylene films by different techniques are summarized in Table 1. The results indicate that corona, flame and plasma treatments produce approximately the same O/C ratios on the surface. However, the wettabilities of the treated surfaces are different, as determined by the contact angle measurements. The corona- and flame-treated surfaces clearly have the lowest advancing contact angle (θ_a) and the flame-treated surface has the lowest receding contact angle (θ_r). Even though the O/C ratios for these treated surfaces are similar, the flame-treated surface has a higher wettability. One of the possible explanations is that flame treatment may produce different concentrations of various functional groups on the polypropylene surface.

Table 1 Surface properties of treated polypropylene films [3]

Treatment	*Exposure time* (s)	*XPS O/C atomic ratio*	θ_a(°)	θ_r(°)
None	–	0.0	117	95
Corona (1.7 J/cm^2)	0.5	0.12	71	52
Corona (0.17 J/cm^2)	0.05	0.07	74	50
Flame	0.04	0.12	73	24
Remote air plasma	0.1	0.12	83	33
Ozone only	1800	0.13	85	63
Ultraviolet/air	600	0.085	83	51
Ultraviolet/air plus O_3	500	0.14	75	53

Corona and flame treatments are very effective in oxidizing polypropylene surfaces. Many studies have shown that both bondability and printability of polypropylene and other polymers can be improved substantially by these two techniques.

REFERENCES

1. Chan, C.-M. (1994) *Polymer Surface Modification and Characterization*, Hanser, New York.
2. Strobel, M., Dunatov, C., Strobel, J.M., Lyons, C.S., Perron, S.J. and Morgen, M.C. (1989) *J. Adhesion Sci. Technol.*, **3**, 321.
3. Strobel, M., Walzak, M.J., Hill, J.M., Lin, A., Karbashewski, E. and Lyons, C.S. (1995) *J. Adhesion Sci. Technol.*, **9**, 365.
4. Sutherland, I., Brewis, D.M., Heath, R.J. and Sheng, E. (1991) *Surface Interface Anal.* **17**, 507.
5. Garbassi, F., Occiello, E., Polato, F. and Brown, A. (1987) *J. Mat. Sci.*, **22**, 1450.

Keywords: contact angle, corona treatment, flame treatment, functionality, surface energy, surface modification, wettability, XPS.

Textile applications of polypropylene fibers

Martin Jambrich and Pavol Hodul

INTRODUCTION

Polypropylene (PP) fibers belong to the youngest generation of large-scale produced chemical fibers. Their production is the second largest after polyethylene terephthalate (PET) [1].

The development of PP fibers is favored by a sufficient amount of the basic raw material, low price, easy fiber forming as well as by their physicomechanical properties which enable the wide application in fields of home textiles, technical textiles and clothing.

The main categories of PP fibers, fibrous and related materials are the following: monofilaments, multifilaments, staple fibers, nonwoven textiles (spunbonds, melt blowns), tapes, split films and others [1, 2, 3, 5].

FIELDS OF APPLICATION OF PP FIBRES AND FIBROUS MATERIALS

The main fields of application of fibers, fibrous and other PP-based textile materials are the following: carpets, underlays, rugs, hygiene textile products, tapes, ropes, clothing (home, sport, children's protective), geotextiles and agrotextiles, wraps and big bags, technical textiles (filter and separation materials for the car industry, for composites manufacturing, for enviromental-ECO textiles), textiles for medicine and others.

The traditional basic classification of PP staple fibers is carpet, woollen, cotton types and microfibers. These fiber types differ not only in

Polypropylene: An A–Z Reference
Edited by J. Karger-Kocsis
Published in 1999 by Kluwer Publishers, Dordrecht. ISBN 0 412 80200 7

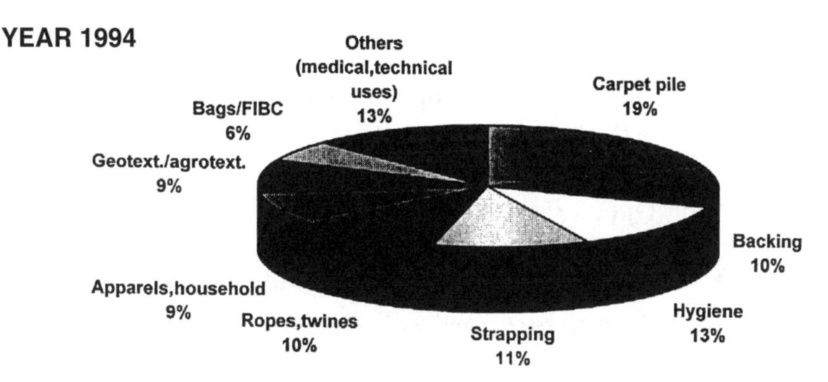

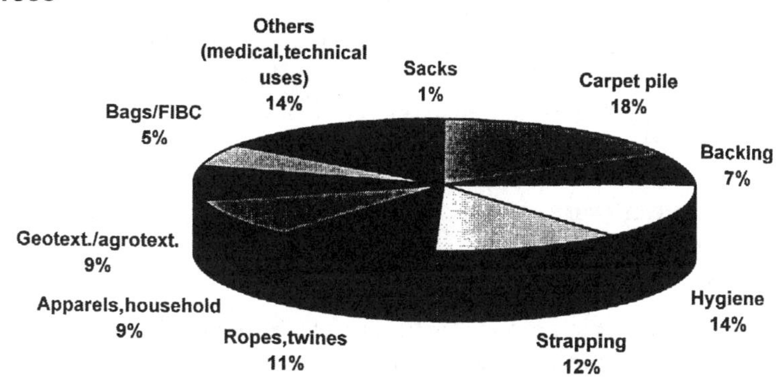

Figure 1 Total consumption of textile polypropylene in Western Europe.

their softness, but also in their morphological structure, (some fibers even in the molecular composition of the polymeric as well). The structure, or properties change according to the field of application (Figure 1).

A fundamental portion of PP staple fibers is used in home textiles. PP staple fibers of the carpet type are used in:

- piles or underfelts for tufted carpets;
- needle felts and nonwoven floor coverings;
- nonwoven decoration textiles and wall linings;
- dewatering felts and filter fabrics.

Woollen types of PP staple fibers are used either alone or in blends with viscose staple fibers and natural fibers, particularly in:

- protective and work clothing for various environments and climatic conditions;
- environmental ecosystems;

- nonwoven fabrics for hygiene products and worsted blankets;
- knitted winter or sport socks and hand and machine knitting yarns;
- nonwoven fabrics for warming interlining, filtration of gases and liquids, e.g. deoiling filters;
- textiles for building industry and transport.

The cotton type of PP staple fibers has also found wide application, either in pure yarns or in blended ones, particularly with viscose staple fibers and cotton. Their main applications are:

- fabrics for alkali-resistant uniforms and overalls, but also for usual work clothing;
- base fabrics for special technologies (needle punching, tufting); technical filter fabrics for agriculture, building industry, transportation, etc.;
- fabrics for bedspreads, bandages, shoemaker's ducks;
- interlock knitted fabrics for underwear and one-ribbed knitted fabrics for bedspreads, sportswear, etc.;
- shrinkage fibers for blended fleece for the production of synthetic leather;
- nonwoven fabrics from multicomponent fibers for filtering gases or liquid media;
- preparation of ECO-materials.

A considerable amount of PP is used in form of nonwoven materials (prepared by the melt blown technique and reinforced by mechanical, chemical or thermal treatment):

- filter materials;
- underfelts for tufted carpets;
- wraps and hygiene products;

The main application field, however, is in the form of geotextiles and agrotextiles for:

- building of roads, highways, forest and field driveways;
- building of sport and recreational areas;
- stabilization of embankments, dikes, water reservoirs and channels;
- building of railroads and tunnels;
- afforestation and grassing.

Endless PP fibers are available in a broad assortment and different softness, various supermolecular structure and physicomechanical properties and cross-section geometry.

The main production volume is given by transversally and longitudinally textured BCF (bulked continuous filament), which are used in the sector of home textiles, especially as:

- tufted carpets with loop and cut pile;
- upholstery fabrics and covers, decorative textiles, textile wall coverings, etc.

To this field, the so-called multicomponent PP/PE (polyethylene) fibers can be added. They are used as cigarette filters, but showing specific properties they can also be used for the filtration of gaseous media.

Endless fiber-textile yarns, smooth or textured, represent a relatively large capacity of production which is increasing all over the world. They are used in:

- textiles for needle punched blankets;
- technical textiles for filtration, leakproof sheets, elastic belts and straps;
- warps of upholsteries and coverings in cars;
- needle punched fabrics for the manufacture of nonwoven textiles.

In the clothing industry, PP fibers are used for underwear, children's and ladies' clothes, socks, gloves, sport and leisure wear, special protective working clothes, e.g. for chemical plants and laboratories. PP-yarn textured and of various geometry enables preparation of the so-called integrated textile materials which represent a suitable product for clothing, in terms of physiology of clothing and hygiene properties (Figure 2; Table 1).

The application of special types of PP fibers is growing in medicine, particularly as surgical threads, bandages, pads, prostheses, membranes for ultrafiltration of blood, wraps for diapers, bedsore protection mats, mulls, surgical uniforms, etc.

PP fibers and their products have broad fields of application, and they are further being developed to meet new requirements by combining them with natural and other chemical fibers.

For further market penetration, new PP fibers are developed by innovation in the manufacturing processes and by physicochemical modifications such as:

- modified high-shrinkage types for new types of carpets, covering cloths, upholstery textiles and blankets, quilts (nonallergic), prepared by new combination of PP stereoisomers;
- special types for floor coverings, wall linings and wall coverings, bed sheets, etc.;
- in the field of textile fabrics, modified types in blends with other fibers are being developed for protective and working clothes for demanding applications in the industry, medicine and biolaboratories;
- fibers with changing transversal and longitudinal geometry for winter and sport clothing and for warming interlinings;

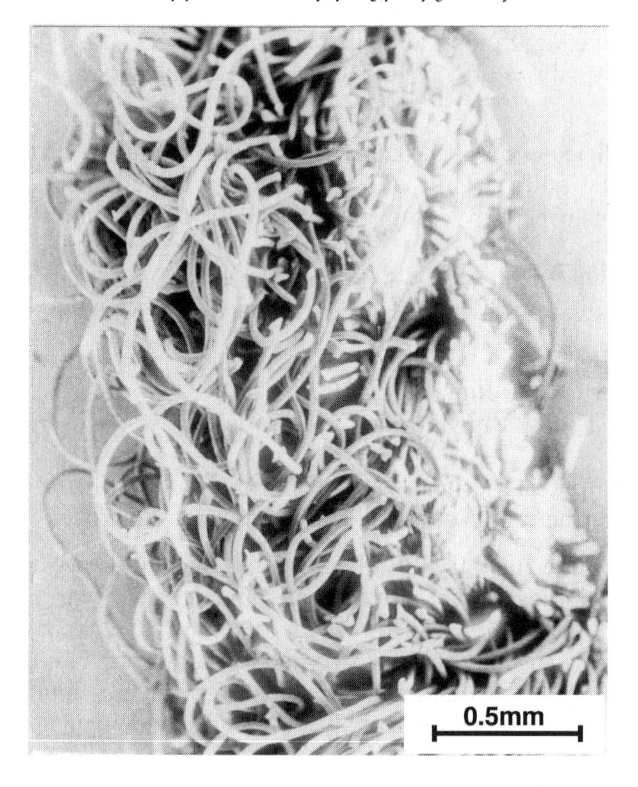

Figure 2 Cross-section SEM photograph of PP/cotton integrated fabric.

Table 1 Some basic and mechanical properties of integrated textiles of PP/cotton

Fabrics	*Property*									
	a	*b*	*c*	*d*	*e*	*f*	*g*	*h*	*i*	*j*
Integrated cotton/PP loop	1.6	220	1.8	25	115 85	34	0.35	2.6	12	4.5
Pure cotton	0.7	140	1.3	23	115 110	14	0.1	1.6	16	3.5

a, thickness (10^{-3} m); b, unit weight ($g.m^{-2}$); c, air permeability ($m^3\ m^{-2}$); d, water vapor permeability (s); e, thermal absorptivity ($J\ m^{-2}\ s^{-0.5}\ K^{-1}$); f, thermal resistance ($10^{-3}\ Km^2\ W^{-1}$); g, electrical potential (kV); h, half-time of discharge (s); i, time of obtainment of 100% relative humidity underneath a clothing moisture (min); j, thermal-insulating values Δt, (°C) in a rest condition.

- fibers for integrated textiles with good properties of humidity transportation and comfort of clothing (underwear, children's diapers, sportswear);
- blended fibers for a new generation of synthetic leather;
- microfibers prepared by classical procedures, as multicomponent M/F (matrix/fibrillar) blend and by melt-blown procedures. They are used for fashioned and sports clothing and for technical applications;
- fibers for technical textiles, such as geotextiles for building of roads, highways, for stabilization of embankments, dikes, water reservoirs, for afforestation, etc.;
- modified and high tenacity fibers for reinforcement of concrete for building industry, and for thermal insulation of buildings, special fibers for sport playgrounds, ropes and packages;
- modified types of fibers for drainage felts, filter materials, air conditioning;
- fibers for solving eco-problems, particularly for separation of crude oil from water and gas substances and for protection of soil against contamination by harmful agents;
- special fibers for transport, agriculture, food-processing and chemical industries;
- multicomponent fibers on the basis of PP of M/F type (polyfibrous and multicore) with changed mechanical and chemical properties suitable for preparation of microfibers;
- special fibers and three-dimensional textile constructions, such as carriers of bacteria (lipolytic microorganisms) for biological cleansing of waste water;
- bicomponent fibers of S/S (side/side) and C/S (core/sheet) types, especially heterophyllous and thermobonding ones;
- bioactive PP fibers with broad application in textile and medicine products [4];
- special modified PP fibers and fibrous materials for the car industry (exhibiting high ultraviolet stability, flame retardancy and high elasticity). PP textiles are being tested for seats, doors, storage rooms and roof textile coverings. PP nonwoven materials are substituting PET nonwoven materials step by step.

REFERENCES

1. Jambrich, M., Štupák, A., Jambrich, P., Budzák, D., Javorek, M. and Lučiviansky, J. (1997) Developments in the production of polyolefine fibres in the world and in Slovakia. *Fibres Text. East. Europe*, **5**(1), 14–20, 25.
2. An. (1996) Unterschätzte PP-Fasern. *Technische Textilien*, **39**, September, 117.
3. *Polypropylene in Europe in 1995*. A Newsletter from European Association for Textile Polyolefins, No. 12, March, 6.

4. Marcinčin, A., Šesták, J., Pašková, E. and Marcinčinová, T. (1996) Bioactive chemical fibres. *Fibres and Textiles*, **3**(1), pp. 24–27.
5. Ahmed, M. (1982) *Polypropylene Fibers – Science and Technology*, Elsevier, Amsterdam.

Keywords: use, application, chemical fiber, polypropylene fiber, monofilament, multifilament, staple fiber, tapes, spunbond, melt blown, split film, textile yarn, knitted fabrics, clothing, nonwoven fabric, home textiles, upholstery, geotextiles, agrotextiles, composites, medical textiles, automotive textiles, bulk continuous filaments (BCF), Eco-textiles, integrated fabric.

Textile polypropylene fibers: fundamentals

Anton Marcinčin and Martin Jambrich

INTRODUCTION

Polypropylene (PP) fibers contain, according to ISO 2076, minimally an 85% portion of macromolecular polypropylene chains and maximally 15% of another fiber-forming polymer, the content of nonfiber-forming substance being unlimited. Of the stereoregular isomers, only isotactic PP is used for fiber preparation. Nowadays, new types of PP (syndiotactic, metallocene based) have been developed for fiber and film production. The first industrial company that launched the production of PP fibers was Montecatini Co. (1959), then after 1960 I.C.I., Celanese, Hercules, etc. Japanese companies started the production of bicomponent fibers. Currently, the global production amounts to 4200 kt of PP fibers in wide assortment of textile, industrial and special types. Table 1 presents a survey of the world production of PP fibers in 1992–1995.

BASIC TYPES OF PP FIBERS

The diameter of fiber or the surface of the cross-section are generally expressed in term of linear density or titre (T) of the fiber in tex units ($10^{-6}\,\mathrm{kg\,m^{-1}}$) respectively in derived units ktex, dtex, mtex. One tex means that a fiber 1000 m long has the weight of 1 g. The expression $T_{\mathrm{dt}} = 167\,\mathrm{dtex} \times f\ 32 \times 2$, for example, means that it is a multifilament 167 g/10 000 m with 32 monofilaments multiplied by two. The individual titre $T_{\mathrm{i}} = 167{:}32 = 5.2\,\mathrm{dtex}$. PP fibers can be divided into four basic groups according to technology and the use [1–3]:

Polypropylene: An A–Z Reference
Edited by J. Karger-Kocsis
Published in 1999 by Kluwer Publishers, Dordrecht. ISBN 0 412 80200 7

Table 1 World production of polyolefin fibers in kt. (Approx. 85% PP and 15% PE fibers)

Parts of the world	*1992*	*1994*	*1995*
Western Europe	1088	1225	1298
Eastern Europe	161	161	180
North America	958	1149	1157
South America	170	216	251
Asia	1392	1670	1859
Africa	32	34	33
Australia (New Zealand)	37	39	38
Total	3838	4494	4816

1. *mono- and multifilaments.* Monofilaments are infinite individual fibers with the linear density of $T_i \cong 150$ dtex and higher; multifilaments are infinite fibers formed by a certain number of monofibers:
   - textile silk, flat, twisted, interlaced, textured, high elastic, low elastic. $T_{dt} \approx 50$–600 dtex, $f \approx 16$–150, $T_i \approx 1.0$–10.0 dtex;
   - microfibers, textile $T_i \approx 0.3$–1.0 dtex, technical types $T_i \approx 10^{-3}$–0.3 dtex;
   - industrial high-tenacity fibers, $T_{dt} \approx 100$–500 dtex, $T_i \approx 3.0$–20.0 dtex;
   - BCF (bulk continuous fibers), $T_{dt} \approx 1000$–4000 dtex, special types up to 10 000 dtex, $T_i \approx 12$–27 dtex,
2. *staple fibers.* These are fibers determined for yarn and nonwovens production with the characteristic cutting length and purpose of use: cotton type $T_i \approx 3.0$–6.0 dtex; wool type $T_i \approx 6.0$–9.0 dtex, cutting length 30–90 mm, carpet types $T_i \approx 11$–30 dtex; special types $T_i \approx 100$–200 dtex, cutting length 60–90–150 mm. Fineness range of staple color is $T_i \approx 1.0$–3.0 dtex.
3. *spunbonded types.* $T_i \approx 2$–15 dtex and melt-blown color $T_i \approx 0.01$–10 dtex for nonwoven materials.
4. *fibers and tapes from films.* Fibrillated or cut $T_i \approx 10$–200 dtex and melt-blown technology from films $T_i \approx$ over 50 000 dtex.

REQUIREMENTS ON FIBER-FORMING POLYPROPYLENE

Molecular weight

In contrast to linear polyamides and polyesters, PP may be processed to form fibers over a wide range of molecular weight ($\bar{M}$), approximately between 80 000 and 300 000 g mol^{-1} which corresponds to melt flow index (MFI) approximately from 30–35 g/10 min to 5.0–6.0 g/10 min. For melt-blown technology, the MFI requirement is 100–1000 g/10 min. The molecular weight distribution expressed as the ratio of weight average molecular weight (M_w) and number average molecular weight (M_n) is in the range 3–5 for degraded PP types and 5–8 for the others. During thermal degradation by organic peroxides the distribution of $\bar{M}$ of PP is narrowed. The relation between intrinsic viscosity (IV = [η]) measured in tetraline at 135°C, molecular weight ($\bar{M}$) and MFI is approximately $\log[\eta] = \log k + \alpha \log \bar{M} = A + B \log(\mathrm{MFI})$, where $k = 1.04\ 10^{-4}$, $\alpha = 0.8$, $A = 0.50$, $B = 197$ [1, 2].

Rheological properties

Fiber-forming PP melts behave like non-Newtonian viscoelastic liquids having drop of viscosity with increasing share rate. The declination from a Newtonian flow and elasticity of PP melt are essentially higher comparing with polyamides and polyesters. Typical values of melt viscosity at processing conditions are 150–450 Pa s [4].

Isotacticity index

This is required to be maximum, i.e. 98–99% of isotactic stereoisomer. A small amount of atactic portion 2.0–5.0% operates as plasticizer.

Thermo- and photooxidation stability

PP for textile applications must be stabilized. Sterically hindered amines and phenol derivatives are most often used as light and thermal stabilizers.

TECHNOLOGY OF POLYPROPYLENE FIBERS

Polypropylene fibers are industrially prepared by spinning from the melt by the conventional low oriented yarn (LOY) or partially oriented yarn (POY) methods, further by aerodynamic procedures, such as spun bonded yarn or using the melt-blown technology [2, 3]. A certain portion of fibers is prepared by the technology of cutting and splitting of films (about 23% in Europe and 27% in USA). The polymer is melted exclusively in extruders placed vertically or horizontally and having a screw diameter of ∅ = 60–150 mm. In standard extruders, the ratio of the length and diameter (L/D) of the screw is 28–30. The melt temperature varies from 220–240°C (for resins of MFI 25–35 g/10 min) to 290–300°C (for resins of MFI 6–8 g/10 min). The mass coloration is usually connected with spinning process. During melting of the polymer, its molecular weight decreases by about 20% on the average without consequences for its fiber-forming properties. The first melt filtration of the PP occurs at the head of the extruder and the last one at the spinning head.

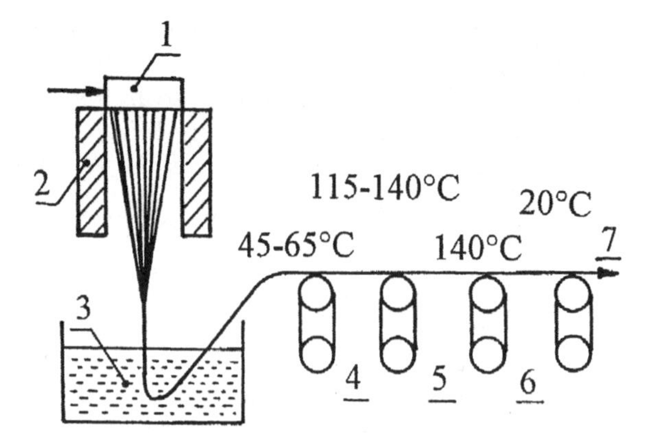

Figure 1 Spinning plant for high-tenacity polypropylene yarns: 1, spinning head; 2, heating zone; 3, water cooling bath; 4, 5, drawing; 6, relaxing zone; 7, take up.

PP silk

The molecular weight of the polymer is 180–200 000 g mol^{-1}, MFI 25–30 g/10 min, $M_w/M_n \approx 4.0$. Spinning process is conventional LOY and POY. Circular spinnerets with the number of orifices of 50–150 and the capillary diameter from 0.3–0.6 mm are used for spinning. Take-up speeds are in the range of 500–1200 m/min for the LOY process and 3000–4500 m/min for the POY fibers. After extrusion, the fibers are air-cooled and lubricated by a suitable finishing agent. Modified LOY technology is also used for preparing PP monofilaments and fibers with high titres (air and water cooling, dry-wet spinning with take-up speeds of 25–50 m/min). Using POY method and rarely high oriented yarn (HOY) for spinning, the silk filaments are joined on a lubricating head, positioned 0.6–1.5 m from spinneret, in order to reduce the air resistance. Microfibers with 0.6–1.0 dtex and high-tenacity technical fibers (6.7–7.5 cNdtex^{-1}) are prepared by a procedure similar to the preparation of silk types. Textile flat fibers are produced by stretching of the LOY and POY fibers, textured fibers (DTY, drawing texturing yarn) are made by texturing (see Figure 1).

PP staple fibers

The older discontinuous method with separate spinning, uses circular spinnerets with the number of orifices up to 150. The spinning speed is 400–600 m/min. New procedures use the so-called compact plant with short spinning line, ring spinnerets and central cooling under the spinneret. The spinning speed is low, 20–50 m/min. The number of spinneret holes is 40 000 for fine fibers to 3 dtex, 22 000 for 3 dtex, 15 000 for 6 dtex, and 7000 for 16 dtex. Extruder, filtration block and spinning head represent a compact aggregate which is usually equipped with volumetric or gravimetric automatic dosers for the PP granules and additives in the form of concentrates. The spun tows are pulled off at constant speed by

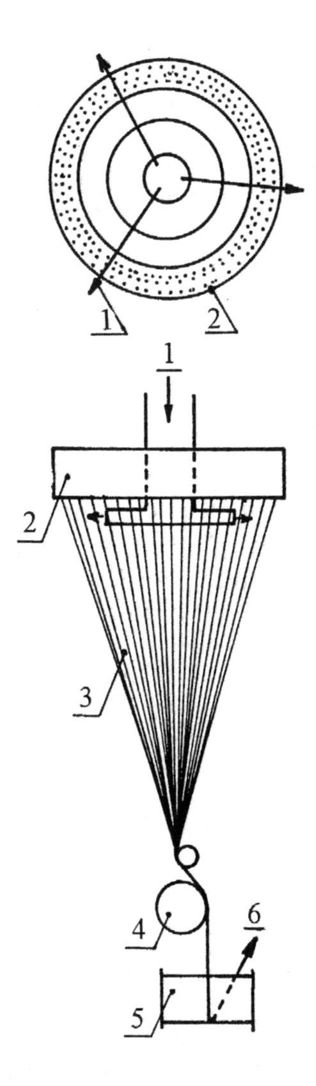

Figure 2 Spinning section of a compact spinning plant for polypropylene staple fibers: 1, cooling air; 2, ring spinneret; 3, threads; 4, 5, rollers; 6, take up.

the godets of the first draw unit. The drawing process is continuous on horizontal plant with heated channels (90–100°C) between godets. Texturing is done by packing. The cable is stabilized at 100°C and cut to the required length and pressed into a bale (Figure 2).

Newly developed technology for PP staple fiber is similar to BCF technology with modified cutting process at higher spinning and take-up speed.

BCF (bulk continuous fibers)

These are textured fibers with the titre from 1000 to 3500 dtex, mainly for upholstery and carpets. They are produced in a continuous way: extruder melting-spinning-lubrication-drawing-texturing-setting-interlacing-take-up. The spinning speed is 600–1000 m/min, its take-up speed 2200–2500 m/min (on new equipment it is above 3000 m/min). The combination of extruders and melt distribution to the heads with three spinnerets allows the so-called three-color spinning (multicolor effect).

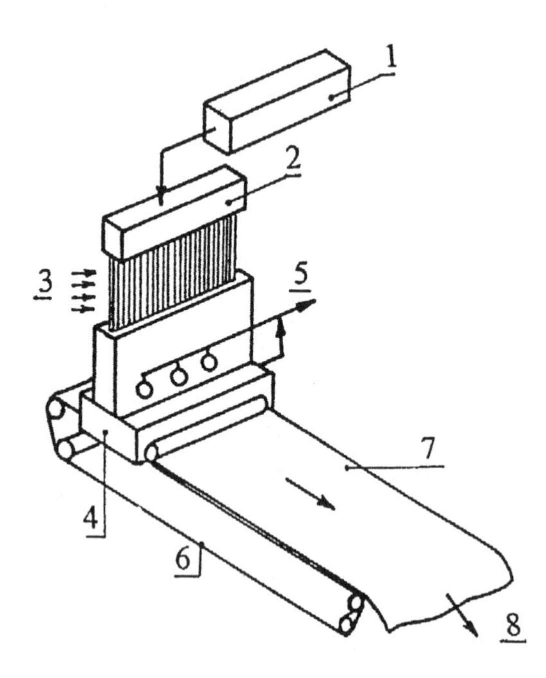

Figure 3 Spunbond line: 1, extruder; 2, spinning head; 3, cooling air; 4, web-formation chamber; 5, suction blower; 6, conveyor screen; 7, nonwoven material; 8, calander, winder.

Spunbonded fibers

These are other typical polypropylene fibers which are produced when, after cooling under the spinneret, the fiber is taken up in an aerodynamic way (ejectors) and is placed on an transport belt as nonwoven material. Take-up speed is 1000–1500 m/min (Figure 3).

Melt blowing

The melt-blown spinning process 'stretches' the polymer melt leaving the capillaries of the spinneret using compressed air stream. The fibers, randomly deposited on a conveyor belt, form a non-woven textile fabric – see the chapter entitled Melt blowing technology in this book.

STRUCTURE AND PROPERTIES OF POLYPROPYLENE FIBERS

The degree of crystallinity (β_c) of polypropylene fibers is 0.50–0.80. From the four crystalline modifications of PP two appear in the process of fiber formation: the less stable modification (smectic-hexagonal) showing a melting temperature (T_m) 142–144°C and stable modification (monoclinic) with $T_m = 158$–167°C. Spherulitic structure is typical of isotropic fibers with low orientation. Lamellar structure is formed at low stretching ratios and long setting times at temperatures above 100°C. Standard, oriented fibers have a fibrillar microstructure. The orientation of macromolecules and supermolecular formations is expressed using the factors of average orientation (f_o), orientation of crystalline regions (f_c) and orientations of amorphous regions (f_a). From the point of view of orientation the bimodal structure with *a* and *c* axially oriented particles is characteristic for PP fibers. With increasing f_o the *c*-axially oriented formations prevail. Basic mechanical and physical properties, such as tenacity (T), elongation (ε) and elasticity (E), depend on the degree of crystallinity and orientation of polypropylene fibers. Some semi-empirical relations indicate a direct relation between the mechanical properties and microstructure, e.g. $T = a + b\, f_a\, (1 - \beta_c)^{-1}$; $\varepsilon = a + b\, (1/\beta_c)\, f_a^{-1}$; $E = a + b\, f_a$, where a, b are coefficients, E = Young's modulus. Table 2 contains some structure parameters and physical properties of PP fibers [5].

Some other typical properties of PP fibers are: density at 20°C, 900–905 kgm^{-3}; tenacity (in cN/dtex) of fibers 1.3–3.0, filaments 1.5–6.0, high-tenacity fibers 7.0–10.0; elongation for fibers 30–200%, for filaments 20–150%, for high-tenacity fibers 8–10%; elastic deformation for fibers 10%, for filaments 80%, for high tenacity color 85–90%; sorption of water is 0%, shrinking in water at 95°C is 0–5%. Important temperatures are: glass transition temperature (T_g) from −10 to −20°C; fixation/ironing 110–130°C; softening 140–150°C; $T_m = 160$–172°C, self-ignition 430–

Table 2 Typical structure parameters and mechanical and physical properties of PP fibers (PP characteristics: MFI = 26.7 g/10 min, M_w/M_n = 3.4)

Parameter	*Conventional LOY + drawing*		*Conventional POY 250°C take up (m. min^{-1})*		
	λ = 3.5				
	T = 20°C	*T = 120°C*	*2000*	*3000*	*5000*
β_c	0.6	0.66	0.55	0.58	0.58
f_0	0.64	0.62	0.50	0.57	0.54
f_a	0.33	0.10	0.29	0.25	0.20
f_c	0.98	0.98	0.89	0.95	0.97
T(cN/dtex)	3.55	3.36	2.23	2.27	2.26
ε (%)	119	89	261	240	202
E(cN/dtex)	25.1	28.4	15.2	21.9	30.8

450°C. The specific heat is 1.70–2.10 kJ/(kg K), melting enthalpy is 60–90 kJ/g, specific electric resistance is 10^{-10}–10^{15} Ω/m. PP fibers are soluble only at temperatures above 100°C in tetraline, decaline, xylene, chlorobenzene. No toxic products are formed at ignition. Fibers are oleophilic, resistant to acids, alkali and they are thermostable to 100–120°C.

REFERENCES

1. Ahmed, M. (1982) *Polypropylene Fibres – Science and Technology*, Elsevier, Amsterdam.
2. Fourne, F. (1993) Polypropylen-Fasern: Herstellung, Eigenschaften, Einsatzgebiete. *Chemiefasern/Textilindustrie* **43/95**, Oct., 811–822.
3. Schäfer, K. (1995) Machines, lines and processes for the production of PP fibres and filaments. *Chemical Fibres Int. (CFI)*, **45**, April, 116–123.
4. Marcinčin, A., Legéň, J., Zemanová, E. *et al.* (1995) Rheological properties of polypropylene – Viscoelasticity. *Vlákna a textil – Fibres and Textiles*, **2**(1), 7–14.
5. Jambrich, M., Diačik, I. and Kanitra, E. (1985) Einstufige Herstellung von PP Filamentgarnen. *Chemiefasern/Textilindustrie*, January, 31–33.

Keywords: bulk continuous fibers (BCF), crystallinity, drawing, elasticity, elongation, fiber-forming polymer, fiber monofilament, low oriented yarn (LOY), melt blown process, melting, monofilament, multifilament silk, orientation, partially oriented yarn (POY), staple fibers, spinning, spunbonded fibers, structure, technical fiber, tenacity, textile fiber, texturing.

Thermal antioxidants

S. Al-Malaika

INTRODUCTION

The performance of polyolefin-based artifacts is adversely affected by the various stages of their lifecycle involving polymer manufacture, storage, processing/fabrication and the environment. Polyolefins, like other hydrocarbon polymers, are highly susceptible to attack by molecular oxygen; and the auto-oxidation process (i.e. oxidation which starts slowly but its rate increases gradually in a catalytic manner) is the major cause of irreversible deterioration leading to loss of useful properties and ultimate mechanical failure of the polymer artifacts.

Hydrocarbon polymers vary in their inherent resistance to oxidation, depending on their chemical structures and their physical and morphological characteristics. Increasing chain branching, for example, leads to more rapid auto-oxidation, and indeed polypropylene (PP) is the most oxidizable polymer out of the other major commercial polyolefins, e.g. low and high density polyethylenes (PE). The oxidative degradation processes of PP, however, are normally accelerated under the influence of temperature, ultraviolet light and other factors, such as mechanical stress, atmospheric pollutants, adventitious metal ion contaminants, though the effect of the first two parameters is, by far, the most important. Both thermal- and photo-oxidation (see the chapter 'Photo-stabilizers' in this book) processes are responsible for the 'aging' and deterioration of properties of PP.

The adverse effects of oxidation, however, can be minimized by the use of oxidation inhibitors (referred to as antioxidants). Indeed, the commercialization of PP, and its use for outdoor applications, would

Polypropylene: An A–Z Reference
Edited by J. Karger-Kocsis
Published in 1999 by Kluwer Publishers, Dordrecht. ISBN 0 412 80200 7

certainly have not been possible without the successful development of effective antioxidants. The effects of thermally-initiated oxidation processes and the role of thermal antioxidants will be discussed in this chapter (photo-initiated oxidation and photoantioxidants are treated in the chapter entitled Photostabilizers).

THERMAL OXIDATION OF POLYPROPYLENE

Thermooxidative degradation may occur at all stages of the lifecycle of PP (polymerization, storage, processing, fabrication and in-service) but its effect is most pronounced during the high-temperature, high-shear conversion processes (e.g. extrusion, injection molding, blow molding, internal mixing) used to produce the final fabricated article, including reprocessing and recycling operations. The high temperatures required for these polymer conversion processes are deterimental to the stability of the macromolecular structure and especially in the presence of molecular oxygen.

The thermal oxidation of PP is essentially a free radical chain process (Figure 1) characterized by the three basic steps of:

1. initiation (leading to the production of the first free radicals in the chain sequence);
2. propagation (giving the most important molecular product of oxidation, the hydroperoxide); and
3. termination (eliminating the radical species).

The nature of the primary initiation reaction of molecular oxygen with PP is complicated by interference from other reactions, such as hydroperoxide initiation or metal catalysis (see reactions 1 and 4, Figure 1). The propagation reaction (reaction 3, Figure 1) is 20% faster at a tertiary carbon atom of PP than at a secondary carbon atom (e.g. in PE). Furthermore, in PP this reaction, which is the most important in the auto-oxidation process and is the rate determining step (RDS) is particularly facilitated by intramolecular hydrogen abstraction leading to the formation of adjacent hydroperoxides along the polymer chain (Figure 2); these are less stable than isolated hydroperoxides and lead to an increased rate of initiation. Consequently, the kinetic chain length (i.e. the average number of oxidation cycles, reactions 3 and 4, before termination occurs) for PP auto-oxidation is quite high, approximately 100 (compared to only about 10 for the more oxidatively stable PE).

The deterioration of physical properties of the polymer results mainly from the thermal breakdown of hydroperoxides (which is accelerated by the presence of metal ion impurities (reaction 4, Figure 1, reaction 2,

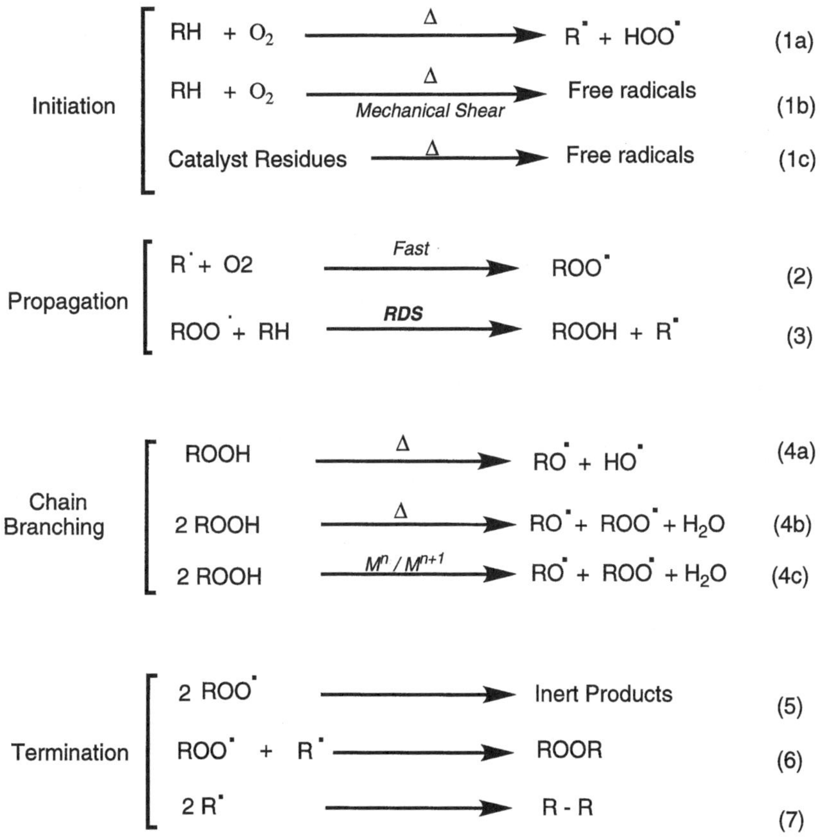

Initiation	$RH + O_2$	$\xrightarrow{\Delta}$	$R^{\bullet} + HOO^{\bullet}$	(1a)
	$RH + O_2$	$\xrightarrow[\text{Mechanical Shear}]{\Delta}$	Free radicals	(1b)
	Catalyst Residues	$\xrightarrow{\Delta}$	Free radicals	(1c)
Propagation	$R^{\bullet} + O2$	$\xrightarrow{\text{Fast}}$	$ROO^{\bullet}$	(2)
	$ROO^{\bullet} + RH$	$\xrightarrow{\text{RDS}}$	$ROOH + R^{\bullet}$	(3)
Chain Branching	ROOH	$\xrightarrow{\Delta}$	$RO^{\bullet} + HO^{\bullet}$	(4a)
	2 ROOH	$\xrightarrow{\Delta}$	$RO^{\bullet} + ROO^{\bullet} + H_2O$	(4b)
	2 ROOH	$\xrightarrow{M^n / M^{n+1}}$	$RO^{\bullet} + ROO^{\bullet} + H_2O$	(4c)
Termination	$2\ ROO^{\bullet}$	$\longrightarrow$	Inert Products	(5)
	$ROO^{\bullet} + R^{\bullet}$	$\longrightarrow$	ROOR	(6)
	$2\ R^{\bullet}$	$\longrightarrow$	R - R	(7)

Figure 1 Free radical chain reaction involved in PP (RH) thermal oxidation.

Figure 2)). The free radicals which are formed during the initiation and propagation steps may undergo further reactions leading to insertion of various oxygen-containing groups in the oxidized polymer, which will lead not only to changes in its molecular structure but also to deterioration of properties. For example, the breakdown of PP-hydroperoxides leads to the formation of alkoxyl radicals which undergo unimolecular decomposition (β-scission) leading to chain scission and reduction in the polymer molecular weight, see Figure 3 (see [1] for a review on oxidation of polymers).

It is important to mention that oxidation occurs only in the amorphous region of the polymer and since PP is substantially crystalline, a relatively small degree of oxidation will cause rapid impairment of the physical and mechanical properties of the polymer.

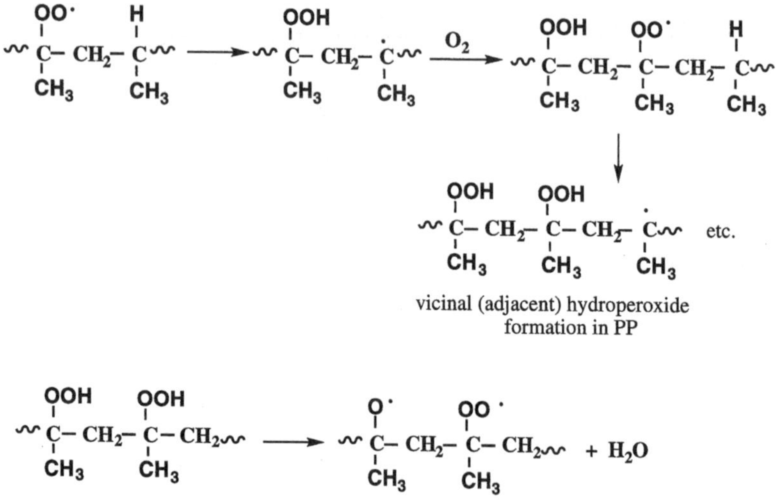

$$\sim C(OO^{\bullet})(CH_3)-CH_2-CH(CH_3)\sim \longrightarrow \sim C(OOH)(CH_3)-CH_2-\dot{C}(CH_3)\sim \xrightarrow{O_2} \sim C(OOH)(CH_3)-CH_2-C(OO^{\bullet})(CH_3)-CH_2-CH(CH_3)\sim$$

$$\downarrow$$

$$\sim C(OOH)(CH_3)-CH_2-C(OOH)(CH_3)-CH_2-\dot{C}(CH_3)\sim \text{ etc.} \quad (1)$$

vicinal (adjacent) hydroperoxide formation in PP

$$\sim C(OOH)(CH_3)-CH_2-C(OOH)(CH_3)-CH_2\sim \longrightarrow \sim C(O^{\bullet})(CH_3)-CH_2-C(OO^{\bullet})(CH_3)-CH_2\sim + H_2O \quad (2)$$

Figure 2 Intramolecular propagation and formation of vicinal hydroperoxides in PP and their decomposition.

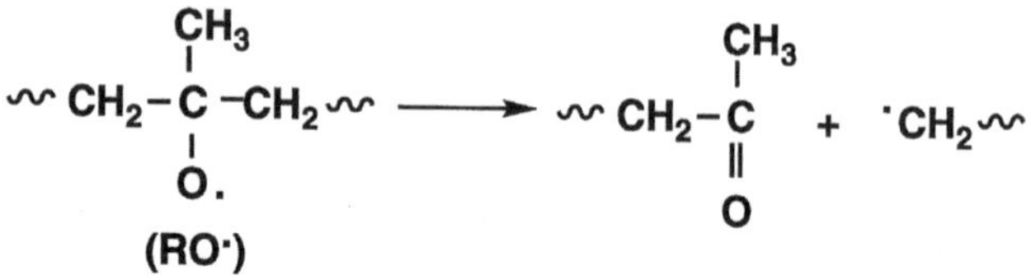

$$\sim CH_2-C(CH_3)(O^{\bullet})-CH_2\sim \;(RO^{\bullet}) \longrightarrow \sim CH_2-C(CH_3)=O + {}^{\bullet}CH_2\sim$$

Figure 3 β-scission in PP.

THERMAL ANTIOXIDANTS: STABILIZATION OF PP AND MECHANISMS OF ANTIOXIDANT ACTION

Chemical compounds which can inhibit or retard the oxidative degradation of polymers are referred to as 'antioxidants'. Thermal antioxidants for PP are chemical agents which stabilize the polymer against the effects of high temperature. Inhibition of thermal oxidation can be achieved by the use of low levels (usually 0.05–0.25 wt.%) of antioxidants which are incorporated during the fabrication process. Antioxidants are usually classified according to their mode of action when interrupting the overall oxidation process: chain breaking (CB) antioxidants (sometimes referred to as primary antioxidants) and preventive antioxidants. CB antioxidants act by removing the propagating radicals, $ROO^{\bullet}$ and $R^{\bullet}$, whereas preventive antioxidants (e.g. peroxide decomposers, PD) act by preventing or inhibiting the generation of free radicals in the auto-oxidation cycle (for detailed discussion on antioxidant mechanisms, see [2]).

Melt processing antioxidants

The effectiveness of melt processing antioxidants is normally assessed by subjecting the antioxidant-containing polymer to processing conditions similar to those used in practice, e.g. multiple extrusion, and measuring the changes in melt flow index (MFI) of the polymer, and comparing these values with control samples processed similarly but in the absence of the antioxidant. Sterically hindered phenols (chain breaking donor, (CB-D) antioxidants) are amongst the most important processing antioxidants for PP. Important commercial examples are based on alkylated hindered phenols, e.g. BHT, Irganox 1076 and 1010 (see structures I) which act by donating a hydrogen atom to the reactive peroxyl radical (formed in propagation reaction 2, Figure 1), hence reducing it to a hydroperoxide and forming a much less reactive antioxidant radical (a hindered phenoxyl radical). The latter can terminate by further reaction with a second peroxyl radical, hence does not continue the kinetic chain (Figure 4). Synthetic hindered phenols, e.g. Irganox 1010, are very effective processing antioxidants for PP (Table 1). Unfortunately, due to the formation of quinonoid transformation products during further oxidation reactions of the 'primary' phenoxyl radicals, many hindered phenols impart various degree of color to the polymer, the extent of which depends on their chemical structure. However, polymer discoloration can be minimized, by incorporating peroxidolytic antioxidants, e.g. phosphites, with hindered phenols during processing.

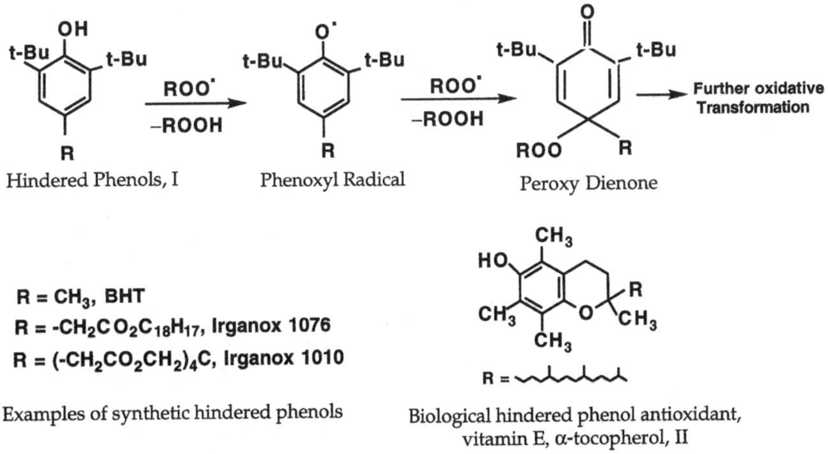

Hindered Phenols, I

Phenoxyl Radical

Peroxy Dienone

R = CH_3, BHT
R = $-CH_2CO_2C_{18}H_{17}$, Irganox 1076
R = $(-CH_2CO_2CH_2)_4C$, Irganox 1010

Examples of synthetic hindered phenols

Biological hindered phenol antioxidant, vitamin E, α-tocopherol, II

Figure 4 Chain breaking antioxidant reactions of hindered phenols.

Table 1 Melt stabilizing efficiency of antioxidants in PP (processed in an internal mixer at 190°C). Melt flow index (MFI) measured at 230°C and 2.16 kg

Antioxidant	*% w/w*	*MFI* (g/10 min)
Control (no antioxidant)	0	11.7
Irganox 1010 (AO–4)	0.05	5.9
Irganox 1010 (AO–4)	0.10	4.6
Irganox 1010 (AO–4)	0.20	3.9
Irganox 1076 (AO–3)	0.20	3.7
BHT (AO–1)	0.20	3.7
Irgafos 168 (AO–14)	0.05	7.7
Irgafos 168 (AO–14)	0.10	7.2
Irganox PS 800 (AO–17)	0.10	8.4
α-tocopherol (AO–9)	0.05	3.8
α-tocopherol (AO–9)	0.10	3.6
Irganox 1010+Irgafos 168	0.05+0.05	4.3
Irganox 1010+Irganox PS 800	0.05+0.10	4.9

The commercially available, hydrocarbon soluble, biological antioxidant vitamin E has highly effective antioxidant properties in living organisms giving rise to nontoxic transformation products in the body. The antioxidant mechanism and the melt stabilizing role of α-tocopherol the most active form of vitamin E (see Figure 4 for structure), in PP has been studied recently [3]. Table 1 shows the superior performance of α-tocopherol as melt stabilizer for PP compared to one of the best commercial synthetic hindered phenols, Irganox 1010, especially at very low concentrations. Further, α-tocopherol shows also good color stability in the polymer.

Amongst the simplest peroxidolytic antioxidants are the alkyl and aryl phosphites which also act as good melt stabilizers in PP (Table 1). Phosphites reduce hydroperoxides to alcohols with a 1 : 1 stoichiometry and are therefore referred to as 'stoichiometric peroxide decomposers (PD-s)', Figure 5(a). In addition to their stoichiometric peroxidolytic activity, some phosphite esters also behave as catalytic peroxide decomposers (PD-c) in addition to having some chain-breaking (CB) activity;

$$(RO)_3P + ROOH \xrightarrow[(PD\text{-}s)]{a} ROH + (RO)_3P{=}O$$

$$(RO)_3P + H_2O \xrightarrow{b} ROH + H_3PO_3$$

Figure 5 Reactions of phosphites with hydroperoxide and water.

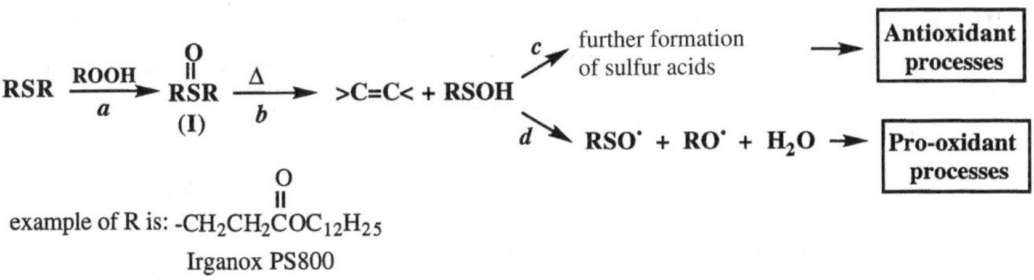

$$\mathrm{RSR} \xrightarrow[a]{\mathrm{ROOH}} \mathrm{R\overset{O}{\overset{\|}{S}}R}\ (\mathrm{I}) \xrightarrow[b]{\Delta} \mathrm{{>}C{=}C{<}} + \mathrm{RSOH}$$

c: further formation of sulfur acids → **Antioxidant processes**

d: $RSO^{\cdot} + RO^{\cdot} + H_2O$ → **Pro-oxidant processes**

example of R is: $-CH_2CH_2\overset{O}{\overset{\|}{C}}OC_{12}H_{25}$

Irganox PS800

Figure 6 Simplified antioxidant mechanism of thiodipropionate esters as peroxidolytic antioxidants.

the contribution of each of these modes to the overall mechanism depends on the structure of the phosphite, oxidizability of the substrate and the reaction conditions. A major problem associated with the use of phosphites in PP stabilization, however, is the fact that most phosphites suffer from hydrolytic instability leading to lower molar mass products. Figure 5(b) shows the formation of an inorganic acid and a low molar mass phenol. The consequence of this reaction is the reduction of flow properties of the additive system (undesirable change in handling characteristics), increase in corrosion of metal surfaces, formation of dark colored spots and gel formation.

A second class of peroxidolytic antioxidants are based on sulfur-containing compounds which are more complex in their behavior than the phosphites. They react with hydroperoxides in a catalytic reaction which leads to the formation of various sulfur acids that are responsible for the catalytic nonradical destruction of hydroperoxides. In the case of the 'simple' dialkyl sulphides, e.g. thiodipropionate esters such as the antioxidant Irganox PS800, although they are good melt stabilizers (Table 1), their main disadvantage is that their catalytic peroxide decomposition to sulfur acids (Figure 6) is always preceded by an observed pro-oxidant stage, due to the formation of free radicals (Figure 6, reaction d), see [4] for a review on processing antioxidants.

Thermooxidative antioxidants for stabilization during service

It is important to point out here that stabilizing PP for service under elevated temperature conditions is very different from stabilizing it against the high temperatures experienced during processing and fabrication. Processing temperatures are very much higher (200–300°C) than that experienced by the polymer under service aging conditions which is generally below the polymer melting point (rarely above 100°C). The efficiency of thermooxidative antioxidants is assessed following accelerated aging of the polymer containing the antioxidant in circulating air ovens at elevated temperatures. The thermally aged PP samples are then tested for changes in physical (e.g. changes in tensile strength, elongation to break and time to embrittlement) and chemical (e.g. monitoring carbonyl concentration by spectroscopic measurements) characteristics.

Under normal conditions of thermal oxidation, the ratio of alkylperoxyl radicals to alkyl radicals is very high. The most effective antioxidants for the above conditions of high temperature aging are CB-D antioxidants, e.g. hindered phenols. Effective processing antioxidants with high chemical intrinsic activity, however, may not necessarily serve as good antioxidants under the lower temperatures experienced during service unless they possess suitable physical characteristics, e.g, low volatility. Therefore, antioxidants with higher molar mass and lower volatility, e.g. Irganox 1010, are potentially more effective than those which have the same antioxidant function but with lower molar mass, such as BHT (see Figure 4 for antioxidant structures).

Peroxide decomposers, e.g. sulfur-containing compounds, enhance the performance of high molar mass phenols under high temperature service conditions. For example, in polyolefins, dialkyl sulfides (e.g. PS 800) are often used as a peroxide decomposer synergists.

Synergistic effects of thermal antioxidants

A cooperative interaction between more than one antioxidant (or antioxidant function) which leads to a greater overall antioxidant effect than the sum of the separate individual antioxidants is referred to as synergism. Synergism can arise from the combined action of two chemically similar antioxidants, e.g. two hindered phenols (homosynergism), or when two different antioxidant functions are present in the same molecule (autosynergism) or due to the cooperative effects between mechanistically different classes of antioxidants, e.g. the combined effects of chain breaking antioxidants and peroxide decomposers (heterosynergism), [5].

Powerful synergism is achieved, for example, in the melt and thermal stabilisation of PP by using combinations of hindered phenols (CB-D) and phosphites and/or phosphonites (PD) and sulfur compounds (PD); in fact phosph(on)ites are seldom used alone. The latter enhance the melt stabilizing effect of hindered phenols (Table 1) and reduce discoloration of the polymer caused by phenol transformation products. The cooperative effect of hindered phenols (PhOH) and phosphites (P) occurs through two steps, whereby phenols scavenge alkylperoxyl radicals and phosphites decompose peroxides in a nonradical reaction which leads to enhanced melt stability of the polymer (Figure 7). Further interaction between the colored transformation products of phenol and the phosphite antioxidant, or its product(s), results in noncolored products, hence

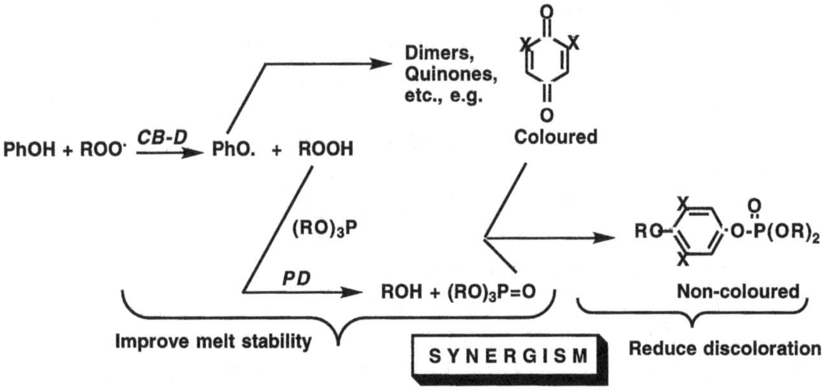

Figure 7 Synergism between phosphites (PD) and hindered phenols (CB-D).

a decrease in the extent of polymer discoloration (Figure 7). Similarly, synergism between hindered phenols and sulfur compounds (e.g. Irganox 1010 + Irganox PS 800, Table 1) in PP has been shown to occur and results from several cooperative reactions involving the oxidation products (e.g. phenoxyl) of the parent hindered phenol and the sulfur (e.g. sulfenic acid and sulfinyl radical) antioxidants leading to the removal of the PP propagating radicals (e.g. alkyl peroxyl and alkyl radicals) from the system, in addition to the peroxidolytic action of the sulfur compound which contributes also to the overall heterosynergism observed.

FACTORS AFFECTING THE EFFICIENCY OF THERMAL ANTIOXIDANTS

In order to inhibit polymer oxidation, the antioxidants have to be present in sufficient concentration at the various oxidation sites. Antioxidants must, therefore, be physically retained in polymers during the high-temperature processing/fabrication operations and under aggressive service conditions; their performance can sometimes be dominated by physical phenomena such as compatibility with, or volatility and extractibility from, the polymer matrix.

Safety of antioxidants is a major issue in polymer stabilization and stringent regulations are already in place, in most countries, on the use of antioxidants in applications involving the human environment, e.g. food packaging and medical implants. In such cases, physical loss of antioxidants into the contact media can have severe toxicological consequences in addition to risks associated with premature failure of the polymer product. Although all antioxidants which are licensed for use in polymers for food contact and medical applications have to undergo strict toxicity testing regimes, their approval, however, does not necessarily mean that their oxidation products (derived from the parent antioxidant during processing or as a result of its antioxidant action in the substrate) are also nontoxic. Potentially toxic transformation products derived from antioxidants during polymer processing, e.g. for food packaging, are generally not tested for toxicity.

This presents a particularly difficult problem in view of the immense amount of research that would be involved in evaluating toxicity of all possible transformation products that can be formed from the 'approved' commercial antioxidants. Recent research has highlighted new directions:

1. the use of oligomeric (high molar mass) antioxidants which show great persistence in the polymer [6];
2. tieing (grafting) antioxidants chemically on polymer backbones so that they cannot physically migrate from the polymer during processing or in service [7], and
3. the use of biological antioxidants which are already in use in foods or as diet supplement. In this context, the use of vitamin E, a natural antioxidant, has been found [3] to impart superior melt stabilization to PP at very low concentrations (Table 1) and would be more acceptable as a 'safe' antioxidant for applications such as food packaging and in human contact environment.

REFERENCES

1. Al-Malaika, S. (1993) Autoxidation, in *Atmospheric Oxidation and Antioxidants*, Vol. 1, (ed. G. Scott), Elsevier Applied Science, London, pp. 45–82.
2. Scott, G. (1993) Antioxidants: chain breaking, in *Atmospheric Oxidation and Antioxidants*, Vol. 1, (ed. G. Scott), Elsevier Applied Science, London, pp. 121–160; Al-Malaika, S. (1993) Antioxidants: Preventive mechanisms, *ibid*, pp. 161–224.
3. Al-Malaika, S. and Issenhuth, S. (1996) Processing effects on antioxidant transformation and solutions to the problem of antioxidant migration, in *Advances in Chemistry Series-249*, (eds. R.L. Clough, K.T. Gillen and N.C. Billingham), American Chemical Society, Washington, DC, pp. 425–440; Al-Malaika, S., Ashley, H. and Issenhuth, S. (1994) *J. App. Polymer Sci, Part A, Polymer Chem*, **32**, 3099–3113.
4. Scott, G. (1993) Oxidation and stabilisation of polymers during processing, in *Atmospheric Oxidation and Antioxidants*, Vol. 2, (ed. G. Scott), Elsevier Applied Science, London, pp. 141–218.
5. Scott, G. (1993) Synergism and antagonism, in *Atmospheric Oxidation and Antioxidants*, Vol. 2, (ed. G. Scott), Elsevier Applied Science, London, pp. 431–460.

6. Pospisil, J. (1990) Macromolecular stabilisers, *Oxidation Inhibition of Organic Materials*, Vol. 2, (ed. P. Klemchuk and J. Pospisil), CRC Press, Boca Raton, FL, pp. 193–224.
7. Al-Malaika, S. (1993) New trends in stabilisation of polyolefins, *Macromolecules 1992*, (ed. J. Kahovec), VSP, Zeist, Netherlands, pp. 501–515.

Keywords: antioxidants, thermooxidative degradation, thermal oxidation, chain breaking antioxidants, preventive antioxidants, peroxide decomposers, vitamin E, phosphites, phosphonites, hindered phenols, sulfur-containing compounds, synergism, migration, toxicity.

Thermally stimulated currents of polypropylene and its composites

György Bánhegyi

THEORETICAL BACKGROUND

The theory of dielectric relaxation is briefly summarized in another chapter 'Dielectric relaxation and dielectric strength of polypropylene and its composites' in this book. The thermally stimulated current (TSC) technique is a special subclass of dielectric measuring methods. Nevertheless, because of its versatility and specific features, it is worth treating separately.

Thermally stimulated currents are measured using a direct current (DC) field and a linear heating program, although other polarization techniques, such as corona charging, irradiation, tribo-electrification, etc., are also known. Two basic modes are applied: (1) thermally stimulated polarization (TSP) and (2) thermally stimulated depolarization (TSD). TSP is usually measured under constant DC voltage and increasing temperature, while in the case of TSD the sample is polarized at an elevated temperature (under isothermal or nonisothermal conditions), then the sample is cooled back to a lower temperature, where the DC field is removed and the short circuit discharging current is measured under a linearly increasing heating program.

TSP current is the sum of two components: one is of dipolar origin, the other is ascribed to conduction. The dipolar part is proportional to the time derivative of the polarization (dipole moment density), which can be described as follows, assuming a single relaxation time (τ) and an exponential dielectric response function [1]:

Polypropylene: An A–Z Reference
Edited by J. Karger-Kocsis
Published in 1999 by Kluwer Publishers, Dordrecht. ISBN 0 412 80200 7

$$j_{\text{pol}} = \frac{\mathrm{d}P}{\mathrm{d}t} = -\frac{P}{\tau} + \varepsilon_0 \frac{\varepsilon_R - \varepsilon_U}{\tau} E \tag{1}$$

where j is the current density, P is the polarization, ε_0 is the vacuum permittivity, ε_R and ε_U denote the relaxed and unrelaxed permittivities respectively, E is the field strength. Under nonisothermal conditions, the temperature dependence of the relaxation time must be taken into account. In this case the time dependence of polarization can be described as [1]:

$$P = \varepsilon_0(\varepsilon_R - \varepsilon_U)\, E\{1 - \exp\,(-\,\alpha_r \int_{t_r}^{t} a_T \mathrm{d}t)\} \tag{2}$$

where α is the reciprocal of the relaxation time, index r denotes a reference value, where the polarization process begun, and a_T is the so called shift factor, which can be described by the Arrhenius or by the Williams–Landel–Ferry (WLF) equation. Introducing the concept of reduced time [1]:

$$\xi = s \int_{T_R}^{T} a_T \mathrm{d}T \tag{3}$$

the nonisothermal formalism can be converted to the isothermal one. The distribution of relaxation times has to be taken, of course. Under nonisothermal conditions the distribution of preexponential frequency factors should be distinguished from the distribution of activation energies. If only the former is present, the system is called thermorheologically simple, which means that a single, temperature-dependent shift factor is enough to convert isothermal and nonisothermal polarization curves into each other. In the latter case, when the material exhibits thermorheologically complex behavior, the transformation is much more complicated.

Under TSD conditions, the second part of equation (1) becomes zero, and the equation is simpler. It can be shown that the TSD current exhibits a maximum in a relatively narrow temperature range for a single relaxation time. The position and width of this peak depends on the preexponential factor of the relaxation time and on the activation energy. The activation energy can also be determined from the initial slope of the TSD current curve. If a distribution of relaxation times is present, the so-called thermal sampling technique is used. This involves polarization during cooling in a very narrow temperature range, followed by depolarization. This technique makes possible the sampling of relaxation time distribution and the analysis of overlapping relaxation mechanisms.

It has been shown that the TSD curves of dipolar dielectrics can be well correlated with the temperature-dependent low frequency loss spectra. In the case of TSD the effective frequency is, however, in the order of 10^{-3}–10^{-4} Hz, which explains its great resolving power and its sensitivity to phase heterogeneity.

The thermally stimulated relaxation in heterogeneous dielectrics, consisting of relaxing components and exhibiting interfacial relaxation, presents a special problem, which has been solved exactly only for bilayer systems [2]. Here, the accumulation of space charge at the interfaces may cause anomalous depolarization effects (currents of reversed polarity). The same is true for samples measured with air gap or with one-sided electrodes (this latter is frequently used for corona charged samples).

In the case of low polarity samples, such as polypropylene, another mechanism, namely space charge relaxation, is at least as important as dipolar relaxation. Space charges can be intrinsic or injected. Intrinsic charges may originate from ionizable impurities, defects, etc., while injected charges (electrons or holes) come from the electrodes. Both intrinsic or injected charges may be mobile or immobile. In the presence of mobile space charges (where the mobility can be determined either by ionic mobility, i.e. diffusion or by scattering or trapping-detrapping processes), the homogeneous field distribution becomes distorted. If only intrinsic charges are present, which cannot be neutralized at the electrodes (nonohmic electrodes), the so-called heterocharge distribution is present (the electrode and the space charges are of opposite polarity), which results in a higher than average field at the electrodes. In the case of charge injection, *homocharges* are present, i.e. electrons are injected from the cathode, holes from the anode. In this case, the local field at the electrode is reduced. Heterocharges are usually observed in samples containing additives at lower fields, while homocharges are normally present in very clean samples at high fields. In the presence of space charge relaxation, the ratio of the current maximum and the polarizing field becomes nonlinear. Charge injection is influenced by the electrode material as well. Trapping and detrapping are governed by the relaxation processes in the polymer.

TSC IN POLYPROPYLENE HOMO- AND COPOLYMERS

The main relaxation processes detected by TSC are in good agreement with those revealed by other techniques (see, for example, [3, 4]), although the large resolution and the thermal sampling method allows a more refined analysis. Figure 1 shows some TSD curves measured in a wide temperature range, taken from [3] and [4], while Figure 2 shows some other data (taken from [5–7]) measured in the high-temperature

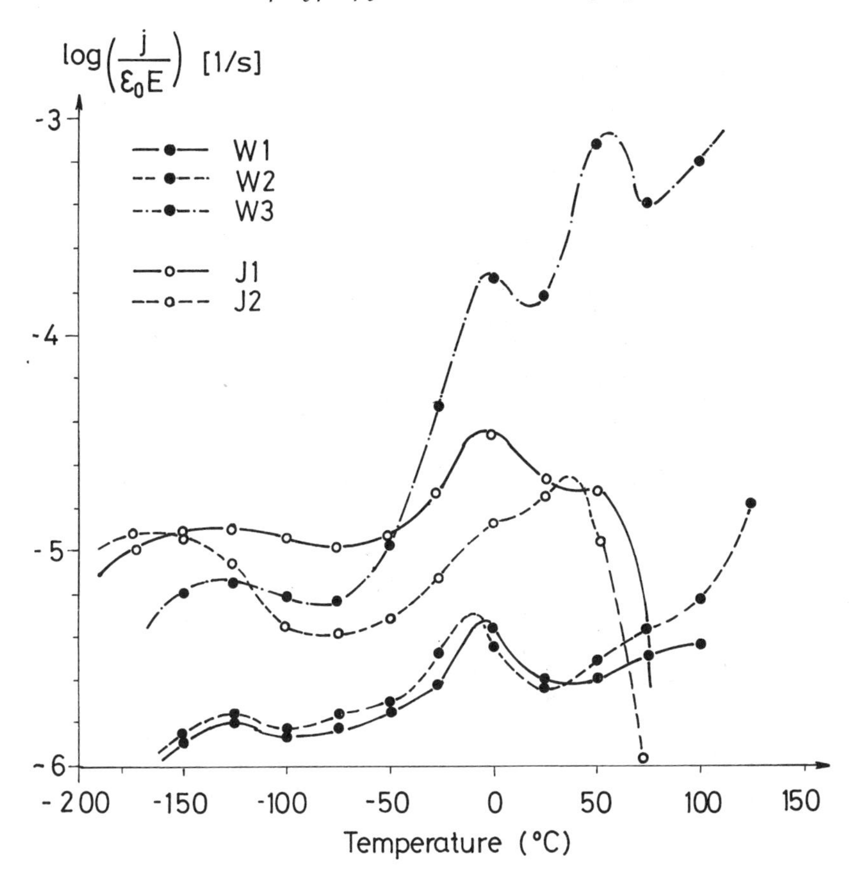

Figure 1 Thermally stimulated depolarization current curves in reduced current density units, measured on PP homopolymer samples in a wide temperature range. W1, reference; W2, annealed; W3, oxidized (data taken from [3]); J1, nondrawn film; J2, drawn film (data taken from [4]).

region. The figures present the data in reduced current density units ($j/\varepsilon_0 E$), as this is independent of the field and geometrical conditions if the relaxation is of dipolar origin. In the presence of space charge relaxation, even reduced current densities cannot be directly compared. Different heating rates, polarization times and temperatures also may give rise to differences between data measured in different laboratories.

Curves W1 and W2, taken from [3], represent very clean samples, the only difference is that W2 was measured after annealing for 16.5 h at 140°C in vacuum. This causes mainly recrystallization, which is reflected in a minor modification of the β peak, ascribed to the glass transition of

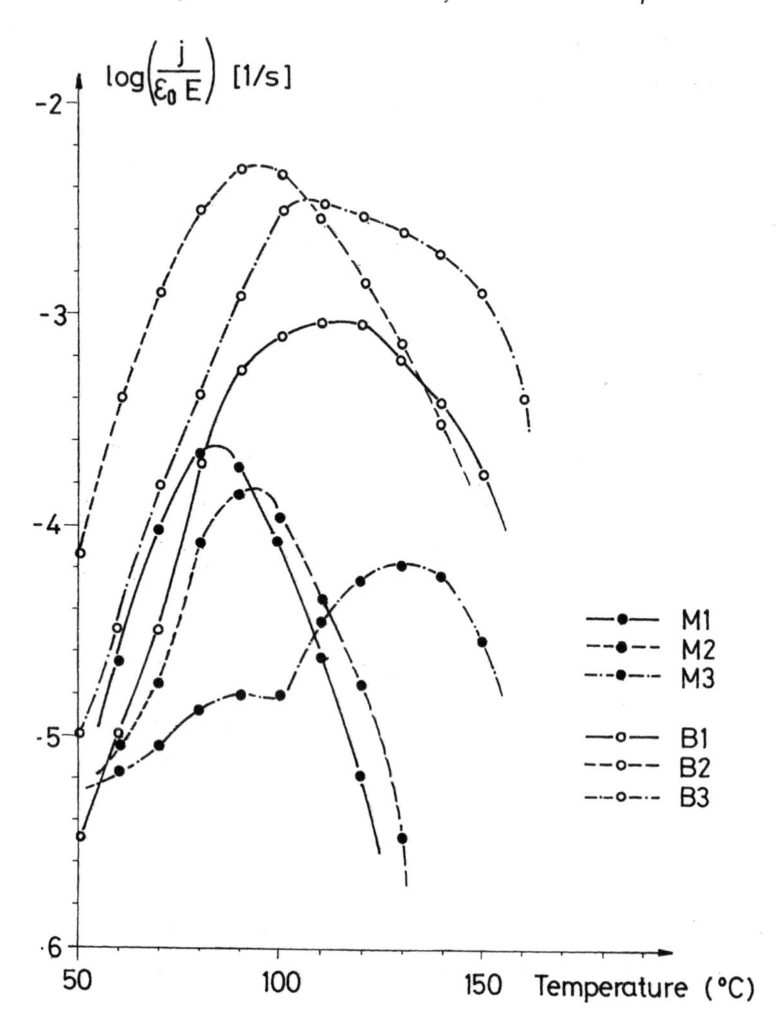

Figure 2 Thermally stimulated depolarization current curves in reduced current density units, measured on homopolymer samples in the pre-melting range. M1, not annealed; M2, annealed at 120°C; M3, annealed at 130°C; (data taken from [5]); B1, nonfilled homopolymer sample (data taken from [6]); B2 and B3, nonfilled homopolymer sample (data taken from [7]).

the amorphous phase. Curve W3 was measured after 15.5 h annealing but in air, which causes serious oxidation. This results in a very significant increase of the depolarization current in the whole temperature range, especially above the glass transition temperature, and a strong premelting peak is observed around 50–60°C.

Curves J1 and J2, taken from [4], were measured on thin films, J1 on nondrawn, J2 on drawn samples (draw ratio 6 at 130°C). In this case, the polarization temperature is very low, so relaxation peaks in the premelting region are strongly suppressed. Nevertheless, the effect of drawing is clearly present even in the low-temperature region. In the case of the drawn sample, mechanical stress hinders the polarization, thus a lower current is observed for the glass transition. Elementary spectra (with a single relaxation time) were investigated between −40 and +70°C by the thermal sampling technique. From the Arrhenius plots, it turned out that the processes observed with a polarization temperature lower than 0°C the preexponential factors and the activation energies are correlated to each other ('compensation effect'):

$$\log \tau_0 = \log \tau_c - \frac{E_a}{RT_c} \tag{4}$$

where τ_0 is the preexponential factor, E_a is the activation energy, τ_c (0.12 s) and T_c (26°C) are the critical relaxation time and temperature respectively. For the drawn sample, τ_c was 10^{-4} s and T_c shifted to 90°C. In the higher-temperature region, two relaxation peaks out of seven did not follow the Arrhenius behavior.

Curves M1–M3 of Figure 2, taken from [5], indicate the effect of annealing on the TSD curves. Sample M1 was cooled from 190°C and was not annealed, while M2 and M3 were annealed at 120 and 130°C respectively. Annealing results in increasing crystallite size, decreasing inter-crystalline domain size, which causes decreased trap density (lower TSD current) and decreased activation energy. The activation energies determined for curves M1, M2 and M3 are 191, 175 and 112 kJ/mole respectively. If, however, the sample is cooled from 220°C, the effect of annealing is much less on both the morphological parameters and on the TSD currents. The maximum currents and the activation energy (156 kJ/mole) are practically the same for both annealed and for un-annealed samples, only the peak temperatures shifts somewhat upwards.

Curves B1, B2 and B3 of Figure 2, taken from [6] and [7], represent three nonfilled industrial PP grades. In these references, the effect of fillers and other additives are studied (see the next section of this chapter), here only the curves of pure polymers are shown for comparison, which exhibit much higher maximum currents than the rest of the cited samples. In [7], we have shown that, although the general trends and the orders of magnitude are clear, the reproducibility of TSD curves measured on semicrystalline systems is somewhat limited, since batch-to-batch variations exist, and blocking conditions have a great influence on the TSD current values.

A TSC study of isotactic and atactic polypropylene, high density polyethylene, ethylene–propylene block copolymers, ethylene–propylene rubbers and blends thereof [8] in the γ and β relaxation range helped to identify the molecular origin of the relaxation peaks. Two β processes were observed in the copolymer samples, one (β_1) around $-5°C$, the other one (β_2) at about $-50°C$. The former is attributed to the PP chains, while the latter is ascribed to microscopically random ethylene–propylene rubber segments.

TSC IN FILLED POLYPROPYLENE COMPOSITES

The dielectric behavior of mineral filled PP composites is complicated by at least two factors: one is the presence of phase heterogeneity, the other is the presence of adsorbed water. The former leads to the appearance of a new relaxation mode, interfacial relaxation and modifies the trap distribution. As to the latter, water is itself dielectrically active (dipolar relaxation) and its presence can modify the interfacial relaxation process, especially if the water layer adsorbed on the filler/matrix interface can dissolve some mobile ions.

Investigations on $CaCO_3$ filled PP [6, 7] and on clay filled ethylene–propylene rubber (EPR) [9, 10] revealed a fairly complex behavior, summarized below. The presence of a small amount of additive (filler [6, 7, 9], or surfactant [7]) tends to decrease the polarization current measured in the thermally stimulated [6, 7] or in the isothermal [9] mode. At higher concentrations, the polarization current increases again. The same is true for the depolarization current maxima in the premelting range. Reference [9] explains this by changes in the filler morphology and connectivity, while [6] and [7] assume the superposition of two effects: the creation of new traps at low additive concentration, which is overcompensated by the interfacial polarization at higher concentrations. In case of clay filled EPR [10], the authors could clearly detect a new depolarization current peak, not present in pure EPR, which showed a correlation with filler concentration, which could be ascribed to the interfacial polarization mechanism. The detection of this peak is easier in EPR than in pure PP, as in the latter the effect is masked by the presence of a strong space charge relaxation ascribed to the crystalline/amorphous interface, which is not present in the amorphous rubber.

The presence of stearate surface treatment [6] or other surfactants [7] leads to the appearance of new depolarization peaks and increases the TSP and TSD currents. Adsorbed water causes further complications in the TSP curves, as it is gradually desorbed between 30 and 80°C. Some representative TSD curves of filled PP composites are collected in Figure 3.

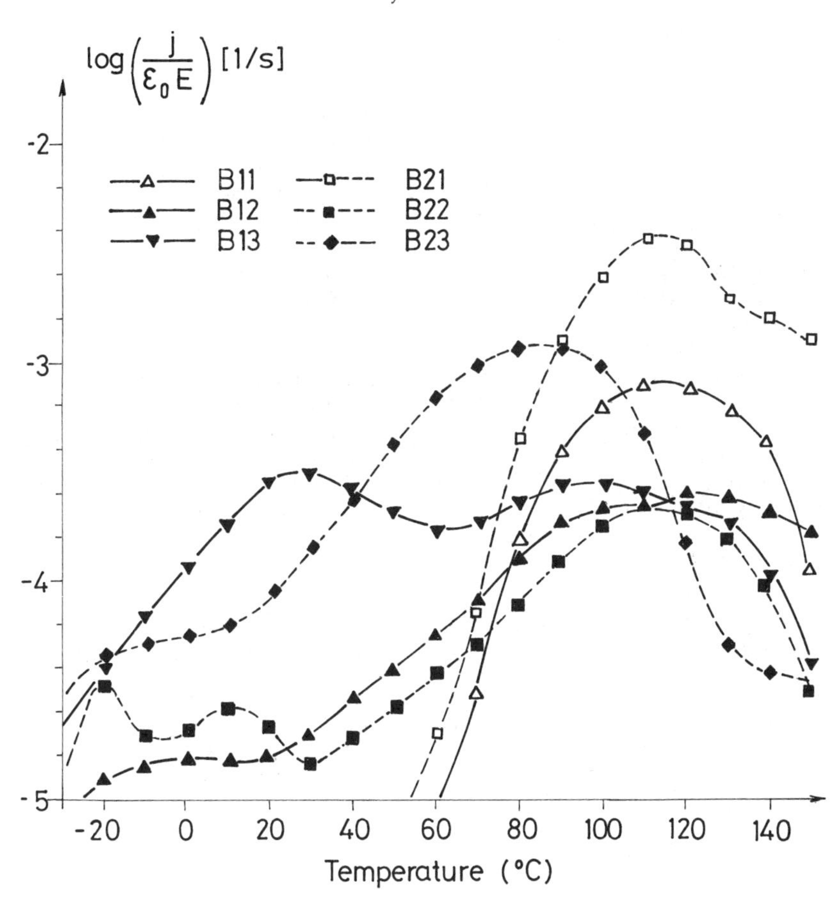

Figure 3 Thermally stimulated depolarization current curves in reduced current density units, measured on filled and nonfilled PP samples. B11, homopolymer; B12, 30 wt.% $CaCO_3$; B13, 30 wt.% stearate treated $CaCO_3$; (data taken from [6], measurement conditions are identical to B1 of Figure 2); B21, homopolymer; B22, 30 wt.% $CaCO_3$; B23, 30 wt.% $CaCO_3$ and 2 wt.% nonionic surfactant, (data taken from [7]), measurement conditions are identical to B2 and B3 of Figure 2).

REFERENCES

1. van Turnhout, J. (1971) Thermally stimulated discharge of polymer electrets, *Polymer J.*, **2**, 173–191.
2. Bánhegyi, G. (1984) Computer simulation of interfacial polarization in stratified dielectric systems, II. Bilayer dielectric, temperature domain, *Colloid Polymer Sci.*, **262**, 967–977.
3. Weber, G. (1979) Thermisch stimulierte Entladung als Methode zur Untersuchung der Dynamischen Eigenschaften von Polymeren, *Prog. Colloid Polymer Sci.*, **66**, 125–133.

4. Jarrigeon, M., Chabert, B. *et al.* (1980) Multiple transitions in isotactic polypropylene around and above the glass transition, *J. Macromol. Sci. Phys.*, **B17**, 1–24.
5. Myslinski, P. and Kryszewski, M. (1987) The effect of crystalline structure on the thermally stimulated discharge current in isotactic polypropylene, *Acta Polymerica*, **38**, 253–258.
6. Bánhegyi, G., Karasz, F.E. *et al.* (1990) AC and DC Dielectric properties of some polypropylene/calcium carbonate composites, *Polymer Engng Sci.*, **30**, 374–383.
7. Bánhegyi, G., Marosi, G. *et al.* (1992) Studies of thermally stimulated current in polypropylene/calcium carbonate/surfactant systems, *Colloid Polymer Sci.*, **270**, 113–127.
8. Ronarc'h, D., Audren, P. *et al.* (1985) Amorphous phase separation in propylene block copolymers as revealed by thermostimulated depolarization measurements, I. Complex spectra study, *J. Appl. Phys.*, **58**, 466–473.
9. Yamanaka, S., Fukuda, T. *et al.* (1991) Electrical properties of EPR with filler, *Proceedings of the 3rd International Conference on Properties and Applications of Dielectric*, IEEE, New York. *Materials*, pp. 1003–1006.
10. Jeffrey, A.-M. and Damon, D.H. (1995) Dielectric relaxation properties of filled ethylene propylene rubber, *IEEE Trans. on Dielectrics and El. Insulation*, **2**, 394–408.

Keywords: thermally stimulated current (TSC), thermally stimulated polarization (TSP), thermally stimulated depolarization (TSD), Williams–Landel–Ferry (WLF) equation, relaxation, thermorheology, activation energy, ionic mobility, interfacial relaxation.

Thermoforming of fiber reinforced composite sheets

D. Bhattacharyya

Thermoplastic polymers can undergo a reversible phase change from solid to liquid, thereby enabling the development of shaping and joining methods analogous to those for conventional metallic materials. In addition, high toughness, corrosion resistance, long shelf life and the ability to incorporate various types of fiber reinforcement to suit the end use of the formed product are added advantages. In the case of fiber reinforced thermoplastics (FRTP) in sheet form, these potential processing advantages have led to the development of thermoforming processes and to the adoption of sheet forming techniques typical of those used with metallic materials [1]. The FRTP material is heated above the melting point of the matrix and formed into a desired shape. While not fully realized, this feature provides the potential to break from the labor-intensive, hand lay-up techniques and lengthy curing periods prevalent with thermosetting matrix materials, and move to efficient mass production of composite parts. This section gives a general overview of thermoforming techniques and discusses the mechanisms associated with them.

FUNDAMENTAL ISSUES IN THERMOFORMING

The potential advantages of rapid component manufacturing using FRTP sheets and associated reduction in production costs can be of tremendous benefit to various industries. These advantages arise from (1) the

Polypropylene: An A–Z Reference
Edited by J. Karger-Kocsis
Published in 1999 by Kluwer Publishers, Dordrecht. ISBN 0 412 80200 7

possibility of implementing efficient forming processes in order to convert composite sheets into three-dimensional geometries, and (2) the fact that, with component redesign and the use of a suitable heating/cooling system, a rapid manufacturing cycle can be achieved. Furthermore, if processing is done on existing metal forming equipment with minimal modifications, the manufacturers will adopt the thermoplastics more readily as replacements for traditional metallic sheets or more advanced thermosetting materials. However, the understanding of the deformation mechanisms for these materials is much more demanding. This is due to a number of factors including: (1) the time-dependent viscoelastic behavior of the matrix at high temperatures; (2) the possibility of hyper-anisotropic material behavior due to fiber collimation; (3) fiber movement during forming; and (4) the complicated fiber/matrix and tool/workpiece interactions.

These factors lead to the necessity of understanding a few critical issues for successful implementation of any thermoforming process using thermoplastic composites. These forming issues include: (1) a good knowledge of fiber reorientation during a thermoforming process; (2) an understanding of forming conditions that might lead to local instabilities, such as out-of-plane sheet buckling, in-plane fiber wrinkling and necking or thinning; (3) a macro-understanding of the effects of various factors, such as forming speed, temperature, tooling and fiber architecture; (4) the thermorheological behavior of such materials in the molten condition at various shear rates.

The key limitations of current technology can be attributed to the lack of predictive capabilities for sheet forming behavior resulting in high tooling costs and long cycle-times at high temperatures. The final acceptability of any manufacturing process is determined by its ability to produce parts of acceptable quality and cost with reasonable ease. Though the actual mechanism of a sheet forming process can be quite complex, a general understanding of the process characteristics can be gained through some simple experimental techniques, such as dome forming, vee-bending and squeeze testing [2, 3] and the operating conditions can be optimized for actual manufacturing applications. A fair amount of information regarding fiber reorientation and local instabilities can also be obtained from these simulated tests which are widely used in the sheet metal industry.

MATERIALS

Under the general description of thermoplastic composite sheet, a number of material options exist with various forms of fibers (e.g. unidirectional, mat, knitted, woven, chopped strand, etc.) reinforcing a thermoplastic polymer matrix. Further to the differences in their mechanical properties, each form of reinforcement has its own merits and demerits from the manufacturing point of view. A disadvantage with the unidirectional plies is that they can only be used with the fibers aligned along the strip length, unless the fibers are stitched together, as they cannot support their own weight when they are molten. If off-axis fibers are required, the stitching can interfere with the sheet deformation. However, unidirectional plies can be successfully combined to produce bidirectional or multidirectional (quasi-isotropic) laminates for secondary forming operations. The fiber reinforcement in commingled, powder/melt coated fiber tows or knitted fabric form renders the material more drapable and often more complex shapes are realized. On the other hand, the board-like fully impregnated plies can be consolidated into high quality laminates much faster because the fibers are already fully wet and the necessary resin percolation can be kept to a minimum. The slowest consolidation rate is achieved when a completely unimpregnated sheet is used. A laminate (a stack of plies with desirable orientation and sequence) can be consolidated *in situ* while forming or may be preconsolidated. For a more complete discussion on this topic the readers are referred to Cogswell [1]. Here attention will be paid to the composites made from semicrystalline polypropylene (PP) matrix, which due to their melting range of 160–180°C and relatively cheap prices, offer a good scope of usage to designers and manufacturers. Furthermore, it has been shown [4, 5] that during the cooling phase, before PP recrystallizes, these composites retain their formability until around 120°C leaving a fairly large processing window. The most widely used PP-composite sheet, especially in the automotive industry, is available in the glass mat reinforced form (see the chapter entitled Polypropylene in automotive applications). However, more recently other types of material, such as Plytron® (unidirectional fiber reinforced preimpregnated sheets) and Twintex® (woven glass/PP co-mingled rovings), have become available in the market increasing the options to the users. Random chopped fiber reinforced materials such as Azdel® are also quite suitable when the directionality is not so important.

THERMOFORMING PROCESSES

In a typical thermoforming process, the feed material in sheet form is heated in a separate heating system or within the forming equipment. The FRTP sheet is subsequently forced to conform to a predetermined shape. After forming, the product is allowed to cool below the crystallization temperature of the matrix material before being removed from the constraint of the tool. Many different sheet forming processes have been devised and tested since the inception of composite sheet materials in the early 1980s. Some of these, such as matched-die forming, rubber-block forming and hydroforming are adapted directly from similar sheet metal processes while others, most notably diaphragm forming, have been devised specifically for FRTP materials and make use of the low forces to mold molten thermoplastics. For composites where high melting temperature/high pressure forming is required, autoclave forming proves to be attractive. Okine [6] and Cogswell [1] cover these descriptions well and this section will only attempt to briefly introduce some of the techniques which are particularly suitable for materials such as polypropylene composites.

In rubber-block forming, one half of the tool is replaced by a thick pad of rubber which, under pressure, conforms to the contoured tool half. The rubber pad may or may not be profiled to the tool geometry and is permanently attached to the platen. In hydroforming the pressure is applied through a fluid medium behind a flexible medium and the sheet preform is deformed against a male or female tool. However, with all these various options available, the matched-die and diaphragm forming techniques are considered to very promising with respect to versatility, cost and product quality. The matched-die forming process, Figure 1(a), utilizes conventional sheet metal forming presses with heated platens. High-quality equipment might be required if fine control over processing parameters is necessary. Moreover, the heating of the work material may be performed outside the press system and then transferred to forming machine without losing too much heat or the heaters can be moved away after the heating phase is complete. For diaphragm forming, Figure 1(b), the sheet preform is held between two deformable diaphragms, which are then clamped, heated with sheet preform to the processing temperature and deformed over a profiled tool. Originally superplastic aluminium alloys were used as the diaphragm material but more recently the use of polymeric materials [7] has become predominant. For isothermal forming, obviously the whole operation has to be carried out within a temperature-controlled environment.

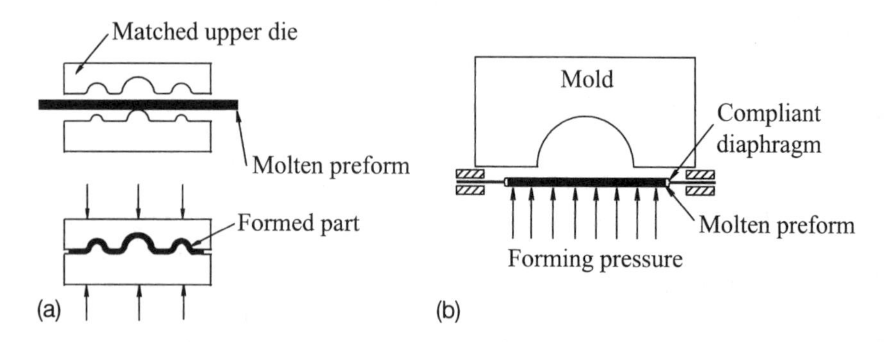

Figure 1 Schematic views of (a) matched-die forming, (b) diaphragm forming.

Another completely different type of thermoforming method, which may be broadly categorized as a continuous thermoforming process, is roll forming [4]. In this process, a composite sheet is preheated and passed through a set of contoured rolls, normally arranged in tandem. Roll forming is one of the most versatile sheet forming processes and, if used wisely, can go a long way to fully realizing the manufacturing potential of FRTP sheets (see the chapter entitled Roll forming of composite sheets).

FLOW MECHANISMS AND THEIR EFFECTS

In their molten states, thermoplastic polymers behave like incompressible viscous fluids. As mentioned earlier, PP remains deformable over a wide temperature window while cooling from its melt temperature. In this thermal region, it exhibits nonlinear viscoelastic fluid properties which are not easily quantified. In addition to the complicated response of the matrix, the fibers introduce a high degree of anisotropy as they severely limit any extensional flow along their paths. To categorize the analysis of a forming process, a hierarchy of four basic flow mechanisms are normally considered, namely resin percolation, transverse squeeze flow, interply shear and intraply shear. These flow mechanisms are commonly present in combination depending on the forming operation and pertain to sheets with unidirectional plies. The first flow type, resin percolation, is typically associated with the consolidation process, because the matrix polymer flows through the fiber bed to wet the fibers. Transverse squeeze flow is essential in welding and compression-molding processes, as it enables the laps and gaps to be 'healed' between adjacent layers [8]. The third and fourth flow mechanisms, interply and intraply shear, are required in shaping processes which induce single and double curvature in a sheet. Consolidated laminates are often constructed from unidirectional prepreg materials with the fibers aligned in two or more directions, while other laminates are constructed from bidirectional cloths with the fibers woven together. The part quality depends not only on the processing conditions but also on the fiber architecture and their positioning with respect to the tool reference system [2]. Furthermore, the way the matrix and fibers are allowed to move during deformation influences the onset of instability. Under the same processing and preform conditions, drawing and draping (associated with diaphragm forming and matched-die forming respectively) can produce quite different part qualities. Due to surface traction created by the diaphragms, compressive instability in the form of fiber wrinkling or out-of-plane buckling is minimized in diaphragm forming whereas matched-die forming tends to produce a better surface finish [9]. Attention has also to be given to the shape conformance of the products.

Unlike metallic materials, where spring-back is monotonically common, the FRTP sheets might display both spring-back and spring-forward depending on the processing conditions and the materials used [10]. The thermomechanical behavior of the composite becomes even more important for PP-based materials for which a relatively large processing window is available. In the final analysis, the decision of choosing a particular process is guided by the achievable part quality and more importantly, by the cost competitiveness of the product. Using 'activity based costing', it has been clearly shown that the thermoforming of PP-based composite products, under many situations, could provide required properties and be quite cost competitive at the same time [5]. Therefore, thermoplastic composite materials deserve serious consideration as alternatives to thermosetting polymers and metals.

REFERENCES

1. Cogswell, F.N. (1992) *Thermoplastic Aromatic Polymer Composites*, Butterworth–Heinemann, Oxford.
2. Martin, T.A., Bhattacharyya, D. and Pipes, R.B. (1992) Deformation characteristics and formability of fiber-reinforced thermoplastic sheets, *Composites Manufact.*, **3**, 165–172.
3. Martin, T.A., Bhattacharyya, D. and Collins, I.F. (1995) Bending of fibre-reinforced thermoplastic sheets, *Composites Manufact.*, **6**, 177–187.
4. Mander, S.J., Panton, S.M., Dykes, R.J. and Bhattacharyya, D. (1997) Roll forming of sheet materials, in *Composite Sheet Forming*, Chap. 12, (ed. D. Bhattacharyya), Elsevier, Amsterdam, pp. 473–516.
5. Krebs, J., Bhattacharyya, D. and Friedrich, K. (1997) Production and evaluation of secondary composite aircraft components – A comprehensive study, *Composites, Part A*, **28A**, 481–489.
6. Okine, R.K. (1990) Composites sheet forming – Thermoplastic, in *Polymer Composites*, (ed. S.H. Munson-McGee), NIST, GCR 90–577–1, US Dept. of Commerce, pp. 129–149.
7. Mallon, P.J., O'Brádaigh, C.M. and Pipes, R.B. (1989) Polymeric diaphragm forming of continuous fibre reinforced thermoplastic matrix composites, *Composites*, **20**, 48–56.
8. Wang, E.L. and Gutowski, T.G. (1991) Laps and gaps in thermoplastic composites processing, *Composites Manufact.*, **2**, 69–78.
9. Krebs, J., Friedrich, K. and Bhattacharyya, D. (1997) A direct comparison of matched-die versus diaphragm forming, *Composites, Part A*, **29A**, 183–188.
10. Zahlan, N. and O'Neill, J.M. (1989) Design and fabrication of composite components: The spring-forward phenomenon, *Composites*, **20**, 77–81.

Keywords: thermoforming, fiber reinforced thermoplastic, hyperanisotropy, rapid manufacturing, matched-die forming, diaphragm forming, roll forming, interply shear, intraply shear, wrinkling, out-of-plane buckling, economics, hydroforming, roll forming, Plytron®, Twintex®, composite, laminate, fabric-reinforcement, unidirectional (UD), fiber reinforcement.

Thermoforming of polypropylene

E. Harkin-Jones, N.J. Macauley and W.R. Murphy

THE THERMOFORMING PROCESS

Thermoforming is a generic name for shaping a sheet of plastic which has been heated until soft [1]. When soft and pliable, the sheet is stretched over or into a cool mold. Once the material has adopted the shape of the mold it is cooled, so that it retains its new shape. This is why the process is called thermoforming – forming by means of heat. This sequence of events, heating, forming (or stretching) and cooling, forms the process cycle, and is at the heart of all thermoforming.

There are two main types of thermoforming. (1) vacuum forming and (2) pressure forming. In vacuum forming, the plastic sheet is heated until it becomes soft and pliable, the heater is removed and the mold is rapidly evacuated. The reduction in pressure on the mold side allows deformation of the sheet, by atmospheric pressure, into the mold cavity. On contacting the mold, the sheet cools rapidly and retains the shape of the mold. The force available to deform the sheet is limited to that provided at atmospheric pressure and this type of thermoforming is thus limited to thin gauge, low-modulus polymers.

Pressure forming can be used to form high-modulus polymers in a wide range of sheet gauges. In this process, the heated, softened sheet is forced into the mold by means of a positive pressure (up to 0.3 MPa) applied above the sheet.

Both vacuum forming and pressure forming may be used in conjunction with a mechanical plugging device which is used to enhance the material deformation capabilities of the process and to improve the

Polypropylene: An A–Z Reference
Edited by J. Karger-Kocsis
Published in 1999 by Kluwer Publishers, Dordrecht. ISBN 0 412 80200 7

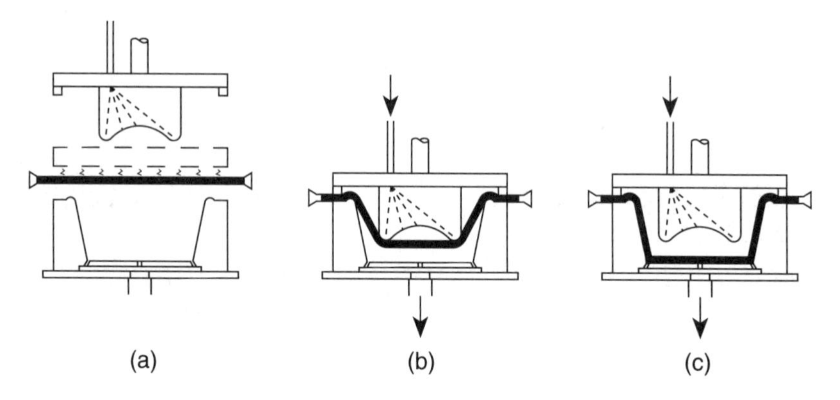

Figure 1 Plug assist pressure forming (a) sheet is raised to formable temperature, (b) plug assist is driven into sheet clamping it against the mold rim, (c) air pressure forces sheet into intimate contact with mold wall and the mold cavity is allowed to vent.

material distribution in the finished product. This is known as plug assist thermoforming (Figure 1).

THERMOFORMING OF POLYPROPYLENE

A number of polymers may be formed by the thermoforming process, some more successfully than others. Amorphous polymers, such as polystyrene (PS), upon which the industry has traditionally relied, are by far the easiest to thermoform. These materials are, however, being replaced at a rapid rate, particularly in the packaging field, by semicrystalline polymers such as polypropylene (PP).

PP's semicrystalline nature imparts to it an excellent combination of physical, thermal and chemical properties which makes it a premium material in packaging applications. Its commercial exploitation in this sector has, however, been impeded by several material-related properties which make it difficult to thermoform. The main problems are (1) low melt strength leading to sagging of the sheet during heating and which in turn results in parts with poor material distribution, and (2) a narrow temperature processing window within which the sheet may be formed.

The melting behavior and thermal characteristics of PP contributing to the aforementioned problems have been extensively studied and reported upon. They are largely due to the semicrystalline nature of PP which results in a sharp melting point at around 165°C. Amorphous polymers, such as PS and polyvinyl chloride (PVC), gradually soften on heating above their glass transition temperatures. Conversely, PP melts very sharply on reaching its melting point and changes abruptly from a crystalline solid to a fluid with low melt strength.

A melt-phase thermoformable material requires a substantial viscous component to permit flow under stress as the material is stretched and, to ensure good mold conformation. An elastic component is also required to resist flow in the absence of the deforming stress and to impart rigidity and strength to the final article. The specific combination of these two material properties defines the thermoformability of a material. Hylton and Cheng [2] quantified these parameters using creep and creep recovery tests. They concluded that a material possessing an elastic modulus of 10^6 Pa and an extensional viscosity of 10^7 Pa s at the required melt forming temperature would produce the best results. Successful thermoforming also demands that the viscoelastic properties remain fairly constant over a wide temperature range. The material should therefore possess a wide rubbery plateau region such that visco-elastic properties are maintained in the presence of sheet temperature fluctuations. This is illustrated in Figure 2. A wide plateau exists for amorphous polymers such as high impact PS, but is absent in crystalline polymers such as PP [3]. Upon heating, PP passes through its viscoelastic region very rapidly, resulting in poor melt strength and sag. This presents a very narrow processing window to the manufacturer.

In response to the problems encountered in the melt-phase thermo-forming of PP, the Shell Chemical Co. developed a process known as

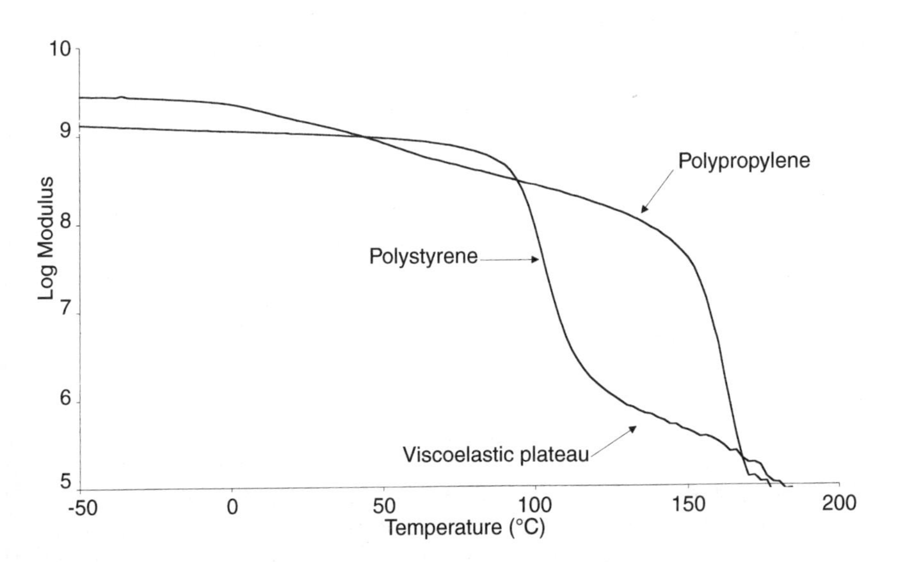

Figure 2 Log modulus versus temperature for PP and PS showing the wide viscoelastic plateau region for the amorphous polymer.

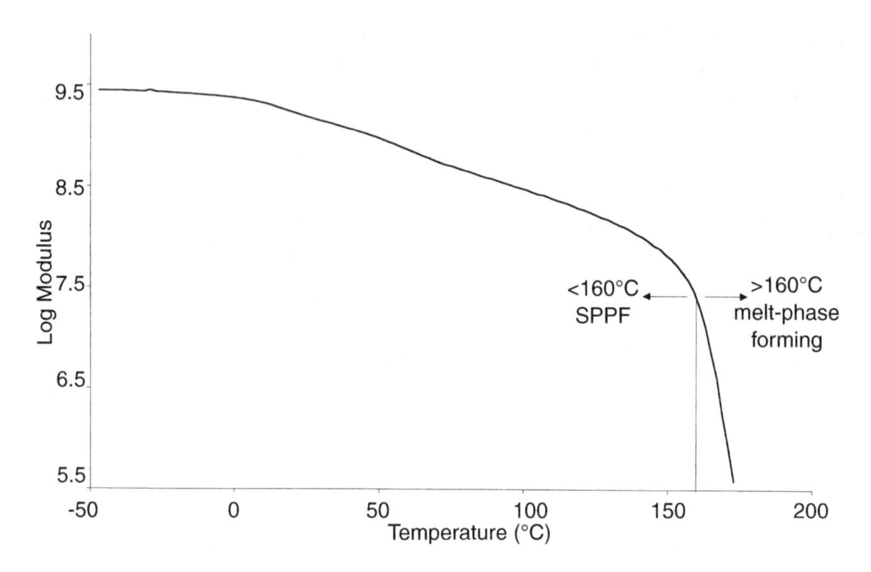

Figure 3 Log modulus versus temperature for PP sheet. The temperature ranges for melt-phase and SPPF are shown.

solid-phase pressure forming (SPPF) in which PP is formed prior to its melting point. In SPPF the sheet is heated to 160°C (compared with > 160°C for conventional melt-phase thermoforming). Since this is below the crystalline melting point the sheet can support its own weight without sagging during the heating stage of thermoforming and is ductile enough to be formed into parts with the aid of mechanical plugs and pressure. The fundamental difference between conventional thermoforming and SPPF is apparent from Figure 3 which illustrates how the modulus of PP decreases very rapidly upon heating above 160°C. SPPF therefore requires greater pressure to deform the sheet into the mold compared with melt-phase thermoforming, which utilizes sheet with a lower modulus offering less resistance to the stretching deformation.

Although SPPF can overcome some of the problems associated with melt-phase forming of PP there are a number of disadvantages inherent in the method.

One disadvantage of SPPF is the level of residual stress and orientation remaining in the final product. This arises from stretching the sheet prior to its softening point. The presence of greater orientation enhances the clarity and stiffness of the formed article. One undesirable consequence of this, however, is the unsuitability of SPPF in the production of containers for high temperature applications. If used at elevated temperatures, e.g. in microwave oven trays, the frozen-in orientation and stresses have an opportunity to relax, causing distortion of the container.

Melt-phase forming is therefore required in the production of such items. The manufacture of 'deep draw' products (large depth to diameter ratio) by SPPF is also difficult because of the high forming pressures required to push the sheet into the mold extremities and also because of the high levels of unidirectional orientation imparted to the end product. This orientation may lead to dimensional instability at high temperatures and also to a tendency for the product to split easily along the lines of orientation. With amorphous materials, the problem of excessive orientation in deep draw products may be alleviated by increasing mold temperature or cooling time. This is not an option for PP since such conditions will lead to an increase in end product crystallinity with a corresponding decrease in optical properties and impact strength.

Another disadvantage is that SPPF requires a large capital outlay to either replace or modify existing production equipment in order to meet the high-pressure requirements of the process.

Because of the disadvantages of SPPF which have been outlined above, efforts are continually being directed towards improving PP resins for melt-phase processing. Initially, high molecular weight resins with excellent sag resistance were developed. These materials, however, proved to be extremely difficult to extrude into the high-quality sheet required for subsequent thermoforming. Efforts are currently being directed towards producing a resin which has balanced extrusion and thermoforming capabilities. Metallocene resins, which are now being produced in commercial quantities, may provide the answer but it is likely that their narrow molecular weight distribution will also give rise to extrusion problems. High molecular weight resins with a broad molecular weight distribution appear to present the best solution to obtaining the required balance [4]. Other methods of improving sag resistance without compromising extrudability of the sheet material include the use of acrylic-modified PP and the incorporation of mineral fillers into the PP resin.

Although advances are being made in materials to overcome the deficiencies of PP in relation to melt strength, inherent problems still exist with end product shrinkage which is a consequence of the semicrystalline nature of this material. Mold design and selection of cooling conditions are thus two areas where caution must be exercised. Mold designers generally allow 1.8% for shrinkage with PP although the range of shrinkage values observed in practice may be 1.2–12.2%. (For amorphous materials such as PS, upon which the industry has traditionally relied, shrinkage figures are considerably lower, ranging from 0.3% to 0.8%). In order to obtain *consistent* shrinkage with PP, it is essential to maintain tight control over mold temperature.

Minimum shrinkage will be obtained at the lowest mold temperatures but this condition will lead to maximum retention of process-induced orientation. A comprise may therefore have to be reached in selection of mold temperature. No compromises should however be made in relation to maintaining tight control of the selected mold temperature if product consistency is to be achieved.

REFERENCES

1. Throne, J.L. (1987) *Thermoforming*, Hanser, Munich.
2. Hylton, D.C. and Cheng, C.Y. (1988) Thermoforming polypropylene on conventional equipment. *Plast. Engng*, **44**(4), 55–57.
3. Macauley, N. (1996) *Extrusion and Thermoforming of Polypropylenes*, PhD. Thesis, The Queen's University of Belfast.
4. Drickman, R. and McHugh, K.E. (1993) Balancing extrusion and thermoforming capability for polypropylene. *J. Plast. Film Sheeting*, **9**, 22–34.

Keywords: thermoforming, solid-phase pressure forming (SPPF), vacuum forming, pressure forming, processing window, viscoelasticity, residual stress, sagging, shrinkage, quality control.

Thermoplastic dynamic vulcanizates

József Karger-Kocsis

DEFINITION, CLASSIFICATION, PRODUCTION

Thermoplastic dynamic vulcanizates (TDV) belong to a novel family of thermoplastic rubbers. They are blends composed of thermoplastic resins and elastomers in which the fully or partially cured elastomer component is finely dispersed (mean particle size: 1–2 μm) in the thermoplastic matrix. If both the matrix and dispersion-giving polymer are of polyolefin nature, they are called thermoplastic olefin elastomers (TPO). The matrix is mostly polypropylene (PP), whereas the elastomer component is either a saturated (EPM) or nonsaturated (EPDM) ethylene/propylene (E/P) rubber (in the abbreviations M and D stay for monomer and diene, respectively). Among the TPO grades, a further distinction is made depending whether the elastomeric phase is crosslinked (vulcanized, cured; TPO-V) or not (blend or elastomeric alloy; TPO-O). TDVs are produced via dynamic vulcanization, credited to Gessler, that was later intensively studied, explored and brought for commercialization by Fischer (Uniroyal) and Coran and coworkers (Monsanto, later Advanced Elastomer Systems) [1, 2]. Dynamic crosslinking means an intensive kneading, mixing operation of a blend composed of a crosslinkable elastomer and a noncrosslinkable thermoplastic resin, performed above the melting temperature of the latter. The vulcanizing agents and related additives should result in a selective crosslinking of the rubber (i.e. not affecting the molecular characteristics of the matrix polymer during melt blending). As usual rubber vulcanizing systems (such as sulfur,

Polypropylene: An A–Z Reference
Edited by J. Karger-Kocsis
Published in 1999 by Kluwer Publishers, Dordrecht. ISBN 0 412 80200 7

sulfur-donating, phenolic curatives) with the exception of peroxides (see the chapter 'Crosslinking of polypropylene' in this book) do not react with PP, which on the other hand outperforms other polyolefins in respect to strength and melting temperature (affecting the related mechanical and thermal behavior), so the preferred thermoplastic resin is PP.

Manufacturing of TPO-V occurs either discontinuously (Banbury or other type of inner mixers) or continuously (mixing extruders of single- or twin-screw design). During mixing the blends having a thermoplastic to rubber ratio in the range of 30:70 and 70:30, the initially formed continuous rubbery phase disintegrates and breaks up into a fine dispersion. The resulting mean particle size depends on several compositional and processing-related parameters. It is obvious that the mixing parameters (residence time, locally accommodated shear and elongational stresses) should be matched to the vulcanization rate of the rubber component. Simultaneously with the development of a rubber dispersion (becoming finer and finer due to mixing) the stress–strain behavior of the compound reaches higher and higher stress and strain values. For a TDV with fine rubber dispersion, the stress–strain response resembles that of a conventional static vulcanized rubber [1, 2].

THERMOPLASTIC AND ELASTIC CHARACTER

Since PP forms the matrix, the thermoplastic feature of the resulting TDV is obvious. What is the reason, however, for the rubbery performance and especially for the elastic recovery? Interestingly, this aspect is less studied in spite of its paramount importance. In a recent pioneering work of Kikuchi *et al.* [3], a two-dimensional finite-element modelling (2D FEM) was applied to shed light on the elastic recovery. The authors demonstrated by this model that the plastic flow of the matrix ligaments between the cured rubbery particles is highly strain dependent. At lower strains the equatorial matrix ligaments (viz transverse to the loading direction) whereas, at much higher strains, the polar matrix ligaments (viz along the loading direction) deform plastically, hampering the elastic recovery. The matrix ligaments remaining still within the elastic limits at high deformations guarantee the recovery of TPO-V. This explanation, though quite unexpected, is very straightforward in order to get a deeper understanding on the morphology dependence of the elastic behavior. This finding may affect the research and development (R&D) activity following, until recently, a trial and error approach. The target is to develop a suitable morphology that is controlled by material (composition, composition range, surface energy, crosslink density, curatives etc.) and processing (e.g. residence time, shear rate) parameters, which are partly also interrelated. It is worth noting, that the understanding of the structure–property relationships of melt-blended impact-modified thermoplastics contributed to the development of TDVs ('knowledge transfer').

PROPERTIES

Comparing the properties of TPO-O and TPO-V of the same constituents at the same hardness, the benefits given by the fully cured rubber in TPO-V become clear (improved set properties, fluid and chemical resistance, service temperature and tensile strength). These are achieved, however, at the cost of some other properties (e.g. reduced melt flow, tear

Table 1 Comparison of the properties of TPO-V grades of the same components for the lowest and highest hardness range, respectively. Data based on Sarlink® types

Properties	*Unit*	*Standard*	*Types*	
Shore hardness (5 s)	°	ISO 868	45–50A	50–55D
Density	gcm^{-3}	ISO 1183	0.95	0.96
In flow direction:		ISO 37 (II)		
Tensile strength	MPa		2.8	20.7
Modulus at 100% elongation	MPa		2.7	17.6
Elongation at break	%		≈150	≈500
Transverse to flow direction:				
Tensile strength	MPa		3.9	22.0
Modulus at 100% elongation	MPa		1.2	13.5
Elongation at break	%		≈600	≈650
Tear strength (trouser)	kNm^{-1}	ISO 34A	5	64
Compression set		ISO 815		
22h, 70°C	%		44	75
22h, 100°C	%		57	81
Volume swell		ISO 1817		
72h, 100°C water	%		+5	+2
72h, 100°C ASTM oil 1	%		+60	+9
Hot air aging (28d, 125°C)		ISO 188		
Hardness change	°		+2	−5
Retention tensile strength	%		123	85
Retention elongation at break	%		118	90

strength and elongation at break). The limits of the service temperature of TPO-V are linked to the rubber (about −60°C) and PP phase (about 135°C), respectively. The properties vary in a broad range depending on the hardness and thus on the composition range of rubber and PP (Table 1). It is worthwhile to note that, although the set (both tensile and compression) values of TDVs are good, they are always inferior to those of traditionally vulcanized rubbers. The most widespread TPO-V grades are offered under the tradename Santoprene® (Advanced Elastomers Systems, Ohio, USA) and Sarlink® (DSM, Holland). It should be emphasized here that TDVs can be produced by using other elastomers than EPM and EPDM. So, other diene rubbers, such as natural rubber (NR), butadiene rubber (BR), styrene-butadiene rubber (SBR) and nitrile rubber (NBR), can also be used as ingredients [2, 4, 5]. In case of the NBR of polar character, the PP/NBR blend should be compatibilized accordingly. A further option for property modification is to use multicomponent TDVs [2]. Interested readers are referred for further information to references [1, 2, 4, 5].

PROCESSING

Most commercial TDVs are fully formulated and tailored for given molding operations. The melt viscosity of TDVs at low shear rate are considerably beyond the uncured blends. Their viscosity decays with increasing shear rate (shear thinning or pseudoplastic feature) but this behavior is less temperature sensitive than in case of uncured blends [1, 2, 4, 5]. Due to this beneficial shear thinning effect, TDV is shaped mostly by injection molding which represents a molding operation at high shear rates. In order to get uniform cavity filling, the mold layout should meet the following criteria: short sprues and runners, runners of identical length, cavities arranged in a balanced pattern. It is also essential to position the weld lines (see the chapter 'Weldlines' in this book) for nonfunctional regions of the parts. Molds produced according to these guidelines yield an optimum melt flow, flash-free molding, short production cycles and, last but not least, products of uniform quality. TPO-V is succesfully molded on reciprocating screw injection-molding machines, provided they have a proper clamping force (in respect of the projected surface of the parts a clamping pressure of 50–80 MPa is required). The mass of the product should not exceed 70% of the maximum shot size of the injection-molding machine. Co-injection, two-component molding or overmolding (both by thermoplastics and a similar TDV of different hardness) can also be practiced. Now Santoprene® grades are available with good adhesion to polyamides and ABS resins. The hot runner system which avoids the use of sprues and runners is gaining also in acceptance.

For extrusion and extrusion blow molding, the machinery designed for PP processing (i.e. length to diameter ratio of the screw >20, compression ratio ≈ 3) is suitable. Tubes and hoses can be produced without and with sizing. In case of extrusion blow molding, external vacuum calibration should be adopted. It is noteworthy that both injection and extrusion molding may result in some flow-induced mechanical anistropy (Table 1). The extrusion foaming by means of physical blowing agents (with water, too) of TPO-V is also viable.

Irrespective of the high melt viscosity at low shear rates, TDV are available also for processing at low shear rates, such as calendering (see the chapter 'Calendering of polypropylene' in this book).

TPO-V scrap can be easily reprocessed by adding to the virgin material; up to 10 wt.% generally no deterioration in the mechanical and rheological properties can be observed. Usually no change in the mechanical properties of TPO-V occurs until about 5 reprocessing cycles. This is in harmony with the thumbnail rules known for PP.

MARKET, APPLICATIONS

The total world demand for thermoplastic elastomers, including all types, will reach about 1.5 million tons in the year 2000. The annual growth rate of polyolefin based thermoplastic elastomers was 9.1% between 1985 and 1995, and for the next five years growth of about 7.6%/year was forecast (which may be optimistic) by the Freedonia Group (Cleveland, Ohio, USA). The overall market share of the thermoplastic elastomers is about 10% in the nontyre market.

The majority of TPO-V (about 40%) is applied in the automotive sector (airbag cover, axel sleeves, bumper fascia, underhood cables and hoses, sealings) where they replace mostly thermoset rubbers and PVC. In case of bumpers, TPO-O grades are preferentially used (see the chapter 'Bumper recycling technology' in this book). The second big market (about 30%) for TPO-V is building and construction (window glazing, weather seal, expansion joint, roofing membranes, etc.), followed by electric, electronic applications (about 15%; covering wire and cable jacketing, electric plugs; see also the chapter 'Polypropylene in cable applications' in this book). TPO-V is gaining acceptance also among medical goods in form of tubes (dialysis, blood collection) and sealing (safety needle, stoppers). In case of technical rubber goods (seals, gaskets, bushings, etc.), TDVs are replacing conventional rubbers. Since their set properties are, however, inferior to the traditional rubbers, the replacement is associated with some 'redesign' of the product.

OUTLOOK

The application fields of TPO grades will be diversified next due to the commercialization of metallocene polymerized polyolefinic rubbers (see the chapter 'Elastomeric polypropylene homopolymers using metallocene catalysts' in this book). These soft-touch grades of segmented or block copolymer nature (in which the physical network structure is given by crystalline domains built up of segments of several macromolecular chains; the crystalline domains can thus be approached by the 'fringed-micelle' model) will compete and thus penetrate into markets of TPO-O (melt blends or reactor blends) and TPO-V grades (dynamic vulcanized types). At present such polyolefin elastomers (Engage®) are marketed by DuPont Dow Elastomers.

REFERENCES

1. Rader, C.P. (1988) Elastomeric alloy thermoplastic vulcanizates, in *Handbook of Thermoplastic Elastomers*, 2nd edn, Chap. 4, (eds B.M. Walker and C.P. Rader), Van Nostrand Reinhold, New York, pp. 85–140.
2. Coran, A.Y. and Patel, R.P. (1995) Thermoplastic elastomers by blending and dynamic vulcanization, in *Polypropylene: Structure, Blends and Composites*, Vol. 2, Chap. 6, (ed. J. Karger-Kocsis), Chapman & Hall, London, pp. 162–201.
3. Kikuchi, Y., Fukui, T., Okada, T. and Inoue, T. (1991) Elastic–plastic analysis of the deformation mechanism of PP-EPDM thermoplastic elastomer: Origin of rubber elasticity, *Polymer Engng Sci.*, **31**, 1029–1032.
4. De, S.K. and Bhowmick, A.K. (eds) (1990) *Thermoplastic Elastomers from Rubber–Plastic Blends*, Ellis Horwood, Chichester.
5. Coran, A.Y. (1987) Thermoplastic elastomers based on elastomer–thermoplastic blends dynamically vulcanized, in *Thermoplastic Elastomers*, Chap. 7, (eds N.R. Legge, G. Holden and H.E. Schroeder), Hanser, Munich, pp. 133–161.

Keywords: thermoplastic polyolefin rubber, thermoplastic dynamic vulcanizate, dynamic vulcanization, properties, processing, market, applications, morphology, Santoprene®, Sarlink®, Engage®, elastic recovery.

Warpage and its prediction in injection-molded parts

T. Matsuoka

INTRODUCTION

Warpage during injection molding impedes the part assembly and decreases the output of finished products. In general, warpage is caused by thermal shrinkage due to unbalanced filling and cooling, and poor mold and part design. It involves many factors related to part geometry, material, mold, machine and molding conditions. These factors are combined in a complex way. Computer-aided engineering (CAE) techniques are used to predict warpage for its reduction or control.

MOLD SHRINKAGE

In injection molding, the polymer is cooled from the melt temperature to a temperature below the solidification, which is accompanied by a reduction of pressure in a mold before the molded item is ejected from the mold. Since the ejected molded part is still hot, its cooling to room temperature occurs under atmospheric pressure. As a result of temperature and pressure changes, the cooled polymer shrinks according to the *PVT* (pressure-specific volume–temperature) relationship or the thermal expansion coefficient of the polymer. The linear mold shrinkage is 0.1–0.8% for amorphous polymers and 1.0–2.0% for crystalline polymers. For polypropylene, it is about 1.7%. Shrinkage affects the dimensional accuracy and dimensional stability of molded items.

Polypropylene: An A–Z Reference
Edited by J. Karger-Kocsis
Published in 1999 by Kluwer Publishers, Dordrecht. ISBN 0 412 80200 7

FACTORS OF WARPAGE

Causes of warpage are summarized in Figure 1. In injection molding, polymer melt is injected into a narrow cavity at a high flow rate and then is subjected to high shear stresses during filling. Since the polymer is usually stretched in the flow direction, the shrinkage is larger in the flow direction than in the transverse direction. Molecular orientation induces different shrinkage and causes warpage and residual stresses in molded parts. For fiber reinforced thermoplastics, fiber orientation is a main reason of warpage. Fiber orientation induces a strong anisotropy in properties of molded parts. For example, thermal expansion in the orientation direction is smaller than that in the perpendicular direction. Even if the temperature change is uniform, the shrinkage is direction-dependent. The problem caused by fiber orientation can be solved by designing proper gate locations and setting adequate molding conditions. The shrinkage is affected by the degree of crystallization, which increases with decreasing cooling rate. In the packing stage, an additional polymer is forced in to compensate for shrinkage. Inadequate packing, e.g. over-packing, changes shrinkage and becomes a further reason for warpage. Undesirable temperature distribution occurs in molded parts owing to insufficient mold cooling and poor design of a part's thickness. In this way, different shrinkage due to orientation,

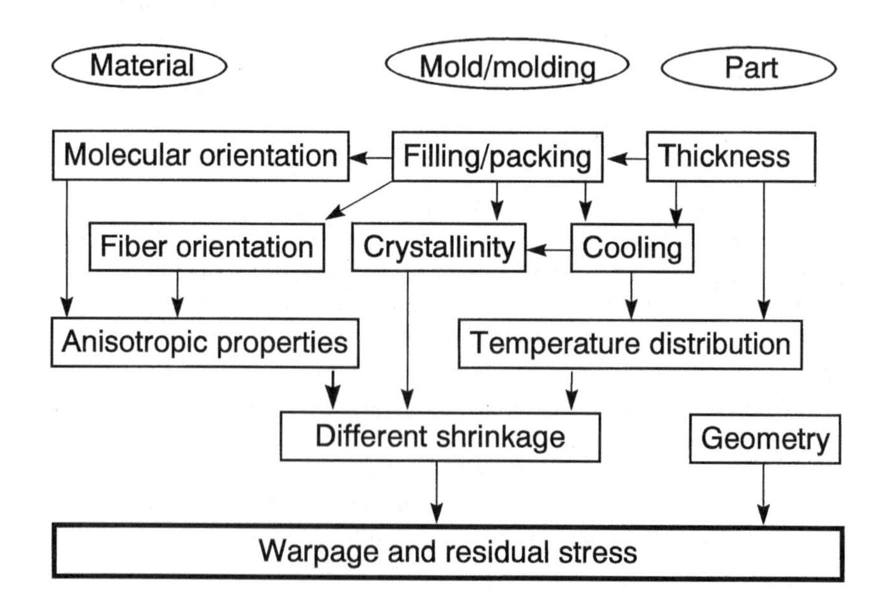

Figure 1 Factors of warpage of molded parts in injection molding.

temperature distribution, thickness change and differential crystallization cause warpage during molding.

MECHANISM OF WARPAGE

Figure 2 shows two illustrations of warpage of simple molded parts. A strip is molded under conditions of different mold temperatures between upper and lower molds. Assuming that the upper mold temperature is higher than that of the lower mold and the properties of the strip are uniform, the strip will be warped like a dish because the thermal shrinkage is larger at the top than at the bottom after ejecting. The mechanism of warpage is similar to a bimetallic strip, which is constructed by laminating two metals with different thermal expansion coefficients. The bimetallic strip will bend during a temperature change because of a difference in thermal expansions between its constituting components. Ribs are designed into molded parts to prevent warping. The rib is, however, a cause of warpage itself. As shown in Figure 2, the ribbed plate, in which the plate is thicker than the rib, will be deformed toward the side without rib, even if the mold temperature is uniform. Since the temperature at the plate is higher than that at the rib at the ejection time, the thermal shrinkage is larger at the plate than at the rib. Therefore, the warpage mentioned above is induced also in a ribbed plate. This warpage is due to thermal stresses and strains. The thermal stress and strain field is the reason for the warpage of injection molded parts.

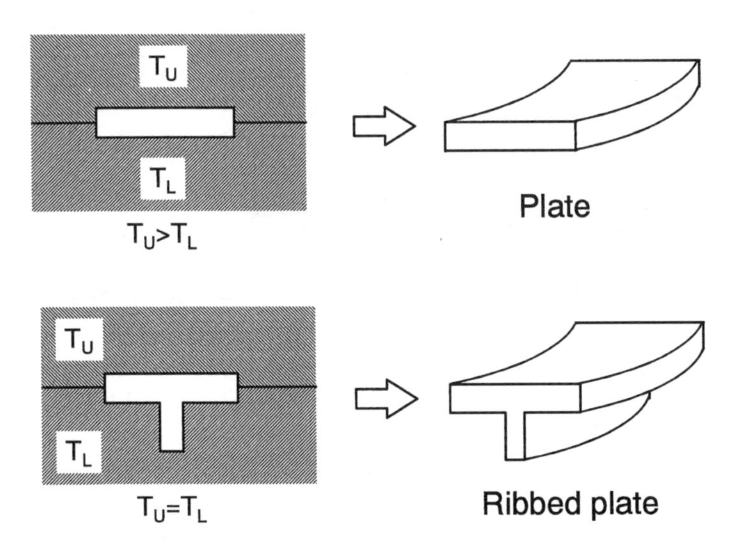

Figure 2 Warpage of molded parts due to different thermal shrinkage.

For an L-shaped molded part, its flanks come closer to each other because of the local shrinkage at the corner. This is called the corner effect. The inner warpage of box-shaped parts is also due to this corner effect. For a center-gated disk, if the inner shrinkage is larger than the outer one, the disk will twist, otherwise it will dome. These phenomena are called buckling.

PREDICTION OF WARPAGE

The outline of a computer-aided engineering (CAE) system, IMAP, is shown in Figure 3 [1, 2]. Analysis programs for mold cooling, mold filling, fiber orientation, and thermal stress are integrated into the system for predicting warpage of three-dimensional thin-walled molded parts. They are linked to data files. The mold data are defined according to a boundary element model considering the geometry of the mold and its cooling system. The cavity data are taken from a finite-element model (the geometry of the molded part is sectioned by a mesh commonly used in every program). This meshing is made by using common mesh generators. In the polymer data files, the rheological, thermal and

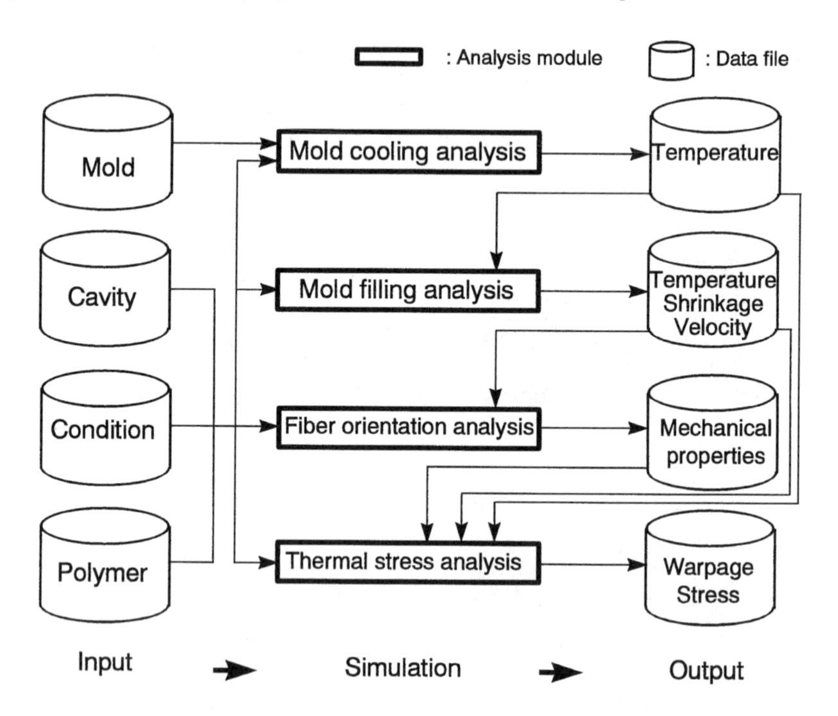

Figure 3 Integrated injection molding analysis programs (IMAP) to predict warpage of thin-molded parts.

mechanical properties of various polymers are stored. The polymer properties required for the analysis are extracted from the polymer data files by specifying the polymer code in a process condition data file. The process condition data file contains a set of molding conditions, computation parameters and the polymer code. The output data file, in which computed results are written, is individually generated by each analysis program.

Mold cooling analysis by CAE was originally developed to achieve balanced cooling of the moldings. Moreover, it is now applied for the integrated analysis of warpage, too. The mold cooling analysis is carried out to estimate the distribution of the mold surface temperature by considering the heat transfer from polymer to coolant and air through the mold wall. The calculated mold surface temperatures are used as input data for the boundary conditions of the mold filling and thermal stress analyses. The heat and flow behavior of the polymer in the mold during filling, packing and cooling, is simulated by mold filling analysis. The outcomes related to warpage are the distributions of velocity, temperature and shrinkage of the polymer. For fiber reinforced thermoplastics, fiber orientation is predicted from the velocity distributions by using the Folgar–Tucker model [3] orientation. The fiber orientation analysis includes the estimation of the anisotropy in the mechanical

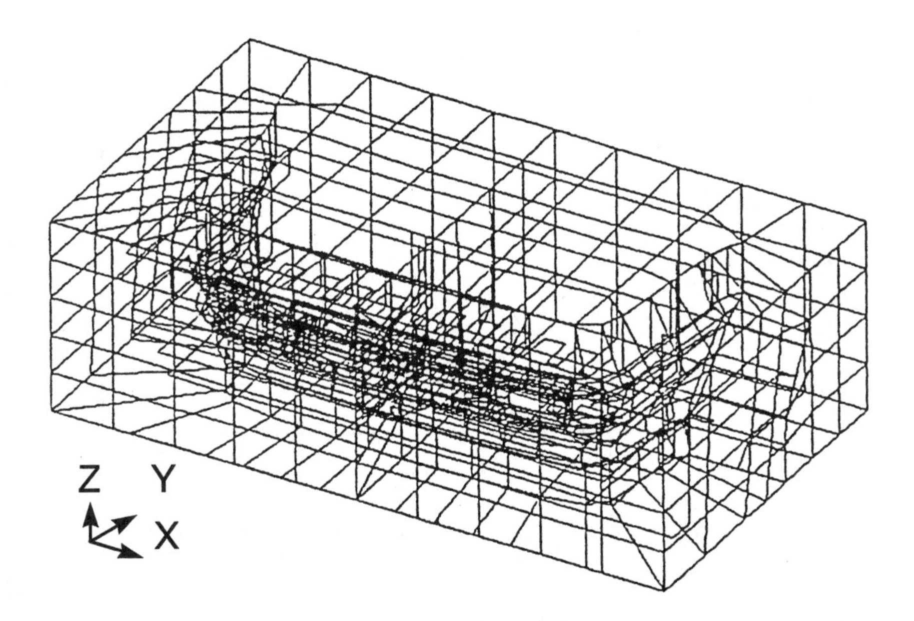

Figure 4 Meshed model for the mold of an automotive front bumper.

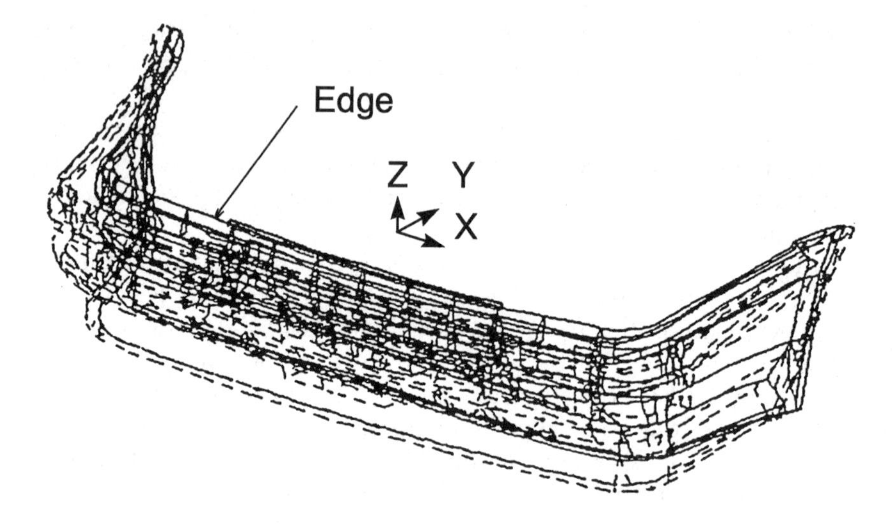

Figure 5 Predicted warpage of the bumper (solid lines, warpage; dashed lines, original shape).

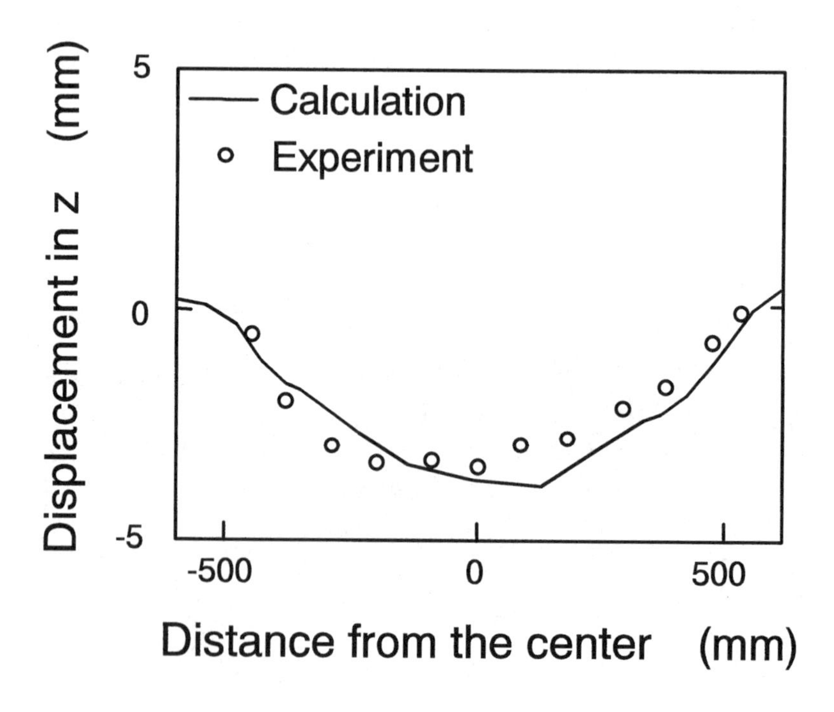

Figure 6 Comparison between the predicted and experimental displacements of the upper edge of the bumper.

properties, elastic moduli, shear moduli, and thermal expansion coefficients in the molded parts. The modified Halpin–Tsai's equation, Schapery's equation, and the classical laminate theory are adopted for the assessment of the material properties [4, 5]. Finally, warpage is predicted by the thermal stress analysis in which the temperature difference between upper and lower molds, temperature distribution in the molded parts at the ejection, difference of shrinkage, anisotropic material properties due to fiber orientation are considered.

Figure 4 shows a meshed model for predicting the warpage of an automotive front bumper molded of polypropylene (PP). The model consists of triangular elements for the cavity, quadrangular elements for the mold surface, and linear elements for the cooling lines and the runner system. Figure 5 shows the predicted warpage of the bumper. The solid line indicates the warped shape, whereas the dashed line means the original. The predicted z-directional displacement of the upper edge of the bumper is in good agreement with the experimental result as shown in Figure 6.

REFERENCES

1. Matsuoka, T., Takabatake, J., Koiwai, A., Inoue, Y., Yamamoto, S., and Takahashi, H. (1991) Integrated simulation to predict warpage of injection molded parts. *Polym. Eng. Sci.*, **31**, 1043–1050.
2. Matsuoka, T. (1995) Fiber orientation prediction in injection molding, in *Polypropylene: Structure, Blends and Composites*, Chap. 3, (ed. J. Karger-Kocsis), Chapman & Hall, London, pp. 113–141.
3. Folgar, F. and Tucker III, C.L. (1984) Orientation behavior of fibers in concentrated suspensions. *J. Reinforced Plast. Composites*, **3**, 98–119.
4. Lewis, T.B. and Nielsen, L.E. (1970) Dynamic mechanical properties of particulate-filled composites. *J. Appl. Polymer Sci.*, **14**, 1449–1471.
5. Schapery, R.A. (1968) Thermal expansion coefficients of composite materials based on energy principles. *J. Composites Mat.*, **2**, 380–404.

Keywords: computer-aided engineering (CAE), fiber orientation analysis, finite element method, injection molding, integrated injection-molding analysis programs (IMAP), mechanical properties, mold filling analysis, mold cooling analysis, *PVT*-diagram, residual stress, shrinkage, thermal stress analysis, warpage.

Weathering

J.R. White

INTRODUCTION

The use of polypropylene (PP) outdoors is limited by its weatherability. The main cause of degradation is photooxidation, promoted by ultraviolet (UV) irradiation, and in hot sunny climates components made from PP may fail within a few days' outdoor service unless they are suitably protected. Failure is usually by brittle fracture resulting from surface embrittlement. Chemical stabilizers added to the polymer can increase the lifetime very considerably and should be used even if a component is to be exposed to sunlight only occasionally. Much of the research into polymer weathering and stabilization has been conducted on PP because of its importance and its sensitivity to outdoor conditions.

PHOTOOXIDATION AND STABILIZATION

Photooxidation reactions

PP weathering involves oxidative degradation, proceeding as a radical reaction with branched kinetic chains [1–4]. This chain reaction is composed of four main stages, beginning with initiation in which the polymer molecule, PH, is converted into a radical:

(1) *Initiation*:

$$PH + O_2 \rightarrow P^\bullet + HO_2^\bullet \quad (1)$$

Polypropylene: An A–Z Reference
Edited by J. Karger-Kocsis
Published in 1999 by Kluwer Publishers, Dordrecht. ISBN 0 412 80200 7

(2) *Propagation*:

$$P^{\bullet} + O_2 \rightarrow POO^{\bullet} \quad (2)$$

$$POO^{\bullet} + PH \rightarrow POOH + P^{\bullet} \quad (3)$$

(3) *Chain branching*:

$$POOH \rightarrow PO^{\bullet} + HO^{\bullet} \quad (4)$$

$$2POOH \rightarrow POO^{\bullet} + PO^{\bullet} + H_2O \quad (5)$$

Radicals formed in (4) and (5) then re-enter the propagation phase via the following reactions:

$$PO^{\bullet} + PH \rightarrow POH + P^{\bullet} \quad (6)$$

$$HO^{\bullet} + PH \rightarrow P^{\bullet} + H_2O \quad (7)$$

(4) *Termination*:

$$P^{\bullet} + P^{\bullet} \rightarrow \text{inactive products} \quad (8)$$

$$P^{\bullet} + POO^{\bullet} \rightarrow \text{inactive products} \quad (9)$$

$$POO^{\bullet} + POO^{\bullet} \rightarrow \text{inactive products} \quad (10)$$

In the case of photooxidation, initiation is promoted by a photon in the UV part of the solar spectrum. Initiation is caused by collision of a photon with an impurity, such as a polymerization catalyst residue, not by direct interaction with the polymer. Further molecular degradation occurs because the radicals are unstable and may undergo scission or crosslinking reactions. Scission dominates, but both reactions lead to deterioration of the engineering properties of the material.

Hydroperoxides produced by reaction (3) or by other means can be decomposed by UV radiation with wavelength below 360 nm giving a $PO^{\bullet}$ radical (reaction (4)). Hydroperoxide decomposition is central in the degradation of polypropylene. Carbonyl groups are produced during PP oxidation and are themselves photolabile; aldehyde and ketone groups produced during processing may influence the subsequent photodegradation behavior.

Photostabilization

Several additives are available that improve photostability. Ultraviolet screening can be provided by pigments, including carbon black (or can be provided by a reflective coating). Ultraviolet absorption can be provided by additives that are transparent to visible light and do not alter the appearance of the product, but the energy that they absorb must not be available for transfer to and breakage of nearby polymer bonds. Excited state deactivation (or quenching) provides a method for the dissipation of energy absorbed by a polymer chromophore in place of the damaging scission reaction. The tendency for carbonyl photolysis to lead to backbone cleavage can be reduced by deactivating excited carbonyl species using transition metal chelates. As with UV absorbers, the quencher must be able to dissipate the acquired energy. Radical scavengers break the oxidation chain, so limiting the damage. The scavenger is often a free radical that is relatively stable and does not itself initiate reactions with the undamaged polymer, reacting with alkyl radicals when they are produced by photooxidation. Examples are nitroxyls and phenoxyls.

Some stabilizers act via cyclic reaction paths and are regenerated, remaining effective at lower concentrations or for longer periods of time than would otherwise occur. This is the case with hindered amine light stabilizers (HALS), an example of which is bis(2,2,6,6-tetramethyl-4-piperidinyl) sebacate (often known by a commercial name: Tinuvin 770), which are especially effective in improving weatherability in polypropylene and other polyolefins.

Hydroperoxide decomposition into inactive products is an important method of stabilization for it prevents generation of radicals by reactions such as (4) and (5). Hydroperoxides are more potent photo-initiators than carbonyl groups and may be produced during processing as well as by photooxidation. Thermal stabilizers are thus required to limit degradation during processing, otherwise the polypropylene may be sensitized to subsequent photodegradation. Decomposition of hydroperoxides can be achieved by reaction with phosphite esters or nickel chelates, or by a catalytic action by a range of compounds including mercaptobenzothiazoles.

Some stabilizers appear to provide protection through more than one mechanism. For example, stabilization by HALS is generally attributed to radical scavenging but it has been suggested that in addition HALS may promote hydroperoxide decomposition, though this has been disputed by some researchers.

Stabilizer molecules often migrate within a polymer, which is an advantage when they diffuse to replenish those lost by reaction in the surface region but is a disadvantage if they diffuse to the surface and are lost by processes other than those in which they provide sacrificial protection. HALS (and other stabilizers) may be copolymerized with PP to reduce mobility, as may be required in films. A mixture of low molecular weight stabilizer and polymer-bound stabilizer sometimes appears to be the best option.

MECHANISMS OF FAILURE

The major failure mode with weathered PP is fracture nucleated in an embrittled surface layer. Photodegradation is concentrated strongly in the surface zone partly because the UV intensity is greatest there but mainly because of the ready supply of oxygen. The UV transmittivity of PP is such that the intensity in the interior is sufficient to promote photochemical degradation, and any absorbed oxygen located there is quickly consumed. The reaction near to the surface is so rapid in sunny conditions that oxygen diffusing in from the surface (to replace that used by reaction) is consumed before it can penetrate far into the polymer and no more oxidation is possible in the interior as long as UV irradiation continues. Chain scission releases molecule segments from entanglements and assists secondary crystallization (chemicrystallization). Weathered PP often develops tensile residual stresses in the surface as a consequence of the volumetric contraction caused by crystallization, further enhancing its vulnerability to cracking.

When the surface is put into tension (by bending, or by the application of a tensile test) cracks form readily in the embrittled layer and often propagate into the interior, leading to failure (Figure 1). Even though the interior may be relatively undegraded chemically, a crack in the embrittled layer may provide sufficient stress concentration to cause brittle fracture or a ductile tearing failure instead of deformation and drawing.

If UV exposure is prolonged the surface may become so degraded that it is no longer effective in supporting a critical crack that propagates into the interior. This may be because of the formation of multiple cracks which mutually unload or because stress transfer from the highly degraded surface into the relatively undamaged interior becomes ineffective. It is believed that this is why some studies have shown that the strength of polypropylene may recover somewhat after a period of exposure (Figure 2), an observation also recorded with polycarbonate and poly(vinyl chloride). It has been observed that failure may nucleate at the surface facing away from the UV source after prolonged exposure. This is because similar damage develops at this surface, but more slowly. Eventually the surface facing away from the UV source becomes ineffective also and failure may occur from an internal site. Such observations are possible only in components that are weathered for extensive periods before being exposed to a high mechanical stress, whereas in service it is common for a component to be subjected to stress application regularly or even continuously, in which case it fails when the strength falls below the applied stress level for the first time.

If the UV source is applied only intermittently the oxygen level may recover (partially) during the dark periods. This occurs during nighttime, though in sunny climates the dominant effect is the daytime surface degradation. The presence of stabilizers has an important effect on the depth-distribution of degradation. An effective stabilizing system slows the reaction near the surface so much that oxygen diffusion into

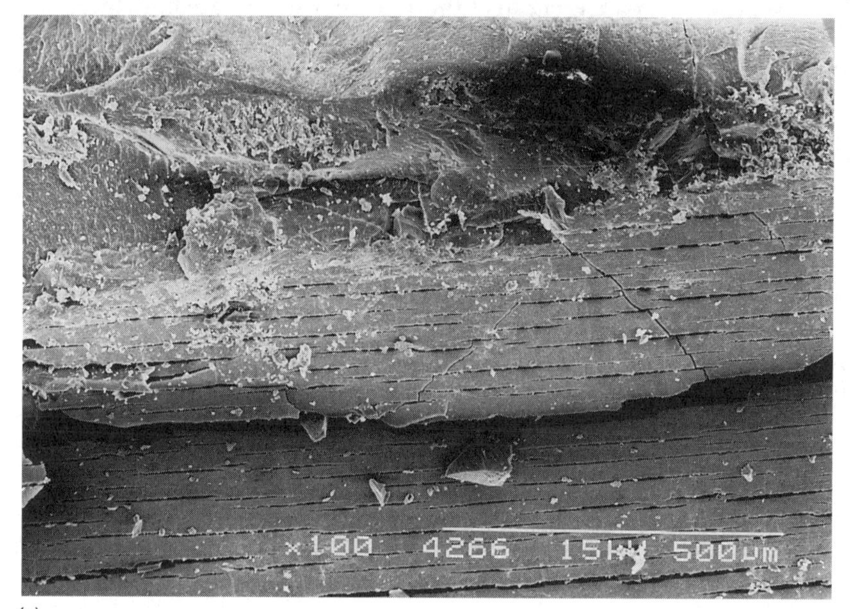

(a)

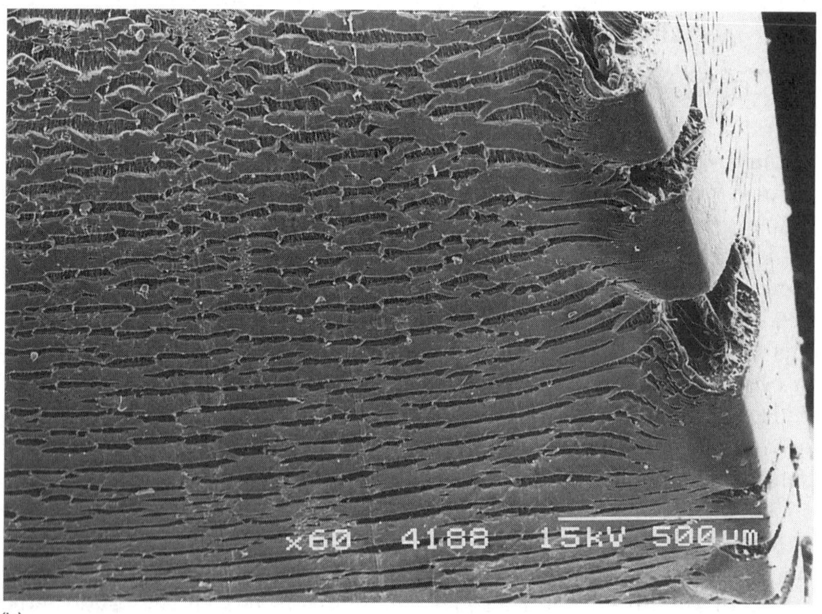

(b)

the interior is possible and the level of degradation there may be higher than in an unstabilized grade.

TESTING AND LIFETIME PREDICTION

The standard tests for assessing the durability of polymers in outdoor exposure (and other degradative environments) have been listed and discussed by Brown [5]. He concluded that the standard tests are of little value in predicting service lifetime and discussed some of the reasons for this. He also noted the lack of correlation between the results of natural weathering and laboratory weathering, a topic also discussed by other authors [2, 4, 5].

The usual strategy for testing the weatherability of polymers is to expose samples outdoors or in the laboratory for various periods of time and then to characterize them chemically or by mechanical tests. The strength and toughness are key properties even if the material is not to be used explicitly for load bearing. The separation of the exposure and of the test represents a serious departure from service conditions. This is because in service the component is likely to be subjected to stress continuously or intermittently throughout its life and it fails the first time a critical stress is applied. In the case of PP, a minimum in strength has been observed to occur after 3–6 weeks under one set of laboratory exposure conditions. If a large period is chosen between times at which samples are withdrawn for testing such a minimum might be missed yet in service the component would fail if a critical stress were applied while it was in this state and the apparent recovery would be of no use. A further objection to the separation of exposure and stress application is that tensile stress has been found to accelerate photooxidation.

Artificial laboratory weathering conditions are usually chosen to accelerate degradation. Elevated temperature is almost always used. This accelerates all of the degradation reactions and diffusion rates but in doing so it may distort the balance between the various reactions. Many reactions involve the products of other reactions and depend on diffusion. These factors will also affect the degradation depth and, consequently, the fracture mechanics.

Figure 1 Cracks formed in the surface of polypropylene samples during tensile testing after (a) weathering for 12 weeks in Jeddah, and (b) after 4 weeks laboratory ultraviolet exposure while held in bending. (See Qayyum, M.M. and White, J.R. (1993) *Polym. Degrad. Stab.* **41**, 163 and Tong, L. and White, J.R. (1996) *Polym. Degrad. Stab.* **53**, 381).

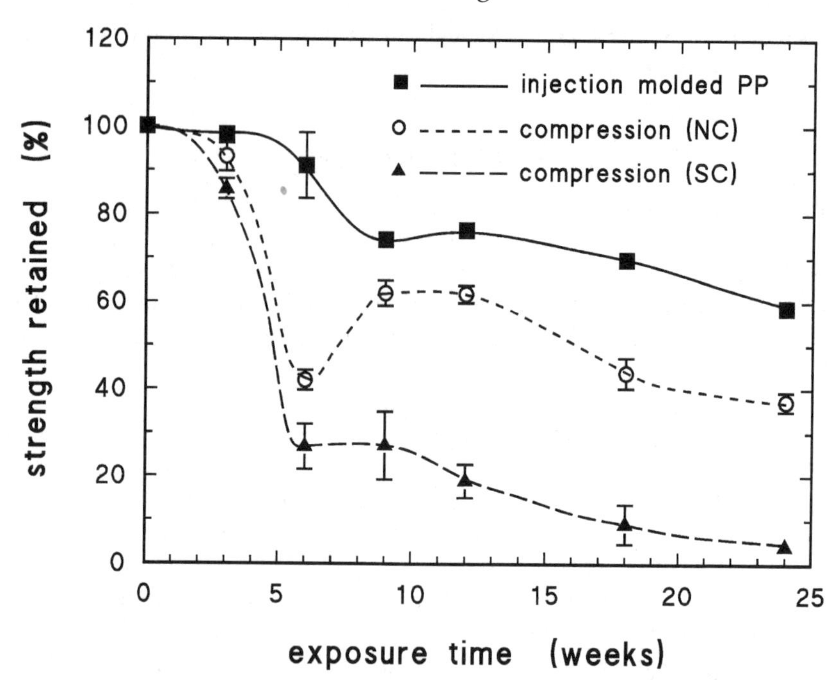

Figure 2 Strength of injection-molded and compression-molded polypropylene bars after various periods of laboratory exposure. The strength retained is the percentage strength compared with that of the unexposed sample. Compression-molded samples were cut from plaques that were cooled in a cold press ('normally cooled': NC) or left to cool slowly in the hot press after turning the power off ('slow cooled': SC). Injection-molded polypropylene containing talc as nucleator gave results similar to those for the NC compression-molded samples. (See Rabello, M.S. and White, J.R. (1997) *Polym. Degrad. Stab.* **56**, 55).

In view of all of these weaknesses in test procedures it is hardly surprising that there is poor correlation between the results for laboratory and natural weathering. Similarly, it is difficult to predict service lifetime from the data gathered. Accurate lifetime modelling would require a knowledge of all reaction rate constants and the diffusion coefficients of oxygen and all reaction products involved in the reactions. Even if all of this information were available, it would require a most complex analysis to model degradation accurately, though it might be possible to identify the most important reactions and to develop a simpler yet satisfactory analysis based on fewer parameters.

CONCLUSIONS

PP is very sensitive to weathering. The chemical degradation process involves many reactions, some of which involve the products of other reactions. Changing the conditions may change the balance of the reactions and artificial weathering tests are rarely successful in predicting outdoor performance. The lifetime can be enhanced very significantly by the inclusion of a suitable stabilizing system. The mechanism of failure may be influenced by the conditions and by the nature and concentration of the stabilizer.

REFERENCES

1. Rabek, J.F. (1996) *Photodegradation of Polymers (Physical Characteristics and Applications)*, Springer, Berlin.
2. White, J.R. and Turnbull, A. (1994) *J. Mat. Sci.*, **29**, 584.
3. Allen, N.S. and Edge, M. (1992) *Fundamentals of Polymer Degradation and Stabilisation*, Elsevier Applied Science, London.
4. Davis, A. and Sims, D. (1983) *Weathering of Polymers*, Applied Science, London.
5. Brown, R.P. (1991) Survey of status of test methods for accelerated durability testing, *Polymer Testing*, **10**, 3.

Keywords: artificial weathering, excited state deactivation, HALS, indoor exposure, outdoor exposure, photodegradation, photooxidation, radical scavenging, residual strength, residual stress, mechanical properties, surface cracking, surface embrittlement, UV exposure, UV screening, weathering.

Weldlines

B. Fisa and A. Meddad

INTRODUCTION

In polymer processing, the term 'weldline' is used to designate the interface created when two streams of polymer, flowing in the mold, are brought into contact. Initially devised to describe surface imperfections in injection-molded parts, the term is in today's context somewhat misleading since what we call a weldline is, in fact, a zone often several millimeters wide extending throughout the entire part thickness. In the weldline region, many material structural features differ from those away from the weldline [1]. To part designers, weldlines present a concern in terms of appearance and strength. It is interesting to note that the development of many exciting plastics, some of them polypropylene (PP) based, has been hampered by weak weldlines. Plastics reinforced with long glass fibers or high aspect ratio platelets and multiphase polymer alloys are examples of such materials. Weldlines have been called, in this context, the 'Achilles' heel' of these materials.

This article addresses the issue of weldlines in PP-based plastics. It is divided into three parts. We begin with a brief description of phenomena leading to weldlines in injection-molded parts. The following section considers the structure and properties. In the last section, methods used to reduce the detrimental effects of weldlines are mentioned.

FORMATION

There are numerous situations that can lead to weldlines in injection-molded parts. In the literature it is often stated that the weldlines can be divided into two categories. A 'cold weldline' is formed when two melt

Polypropylene: An A–Z Reference
Edited by J. Karger-Kocsis
Published in 1999 by Kluwer Publishers, Dordrecht. ISBN 0 412 80200 7

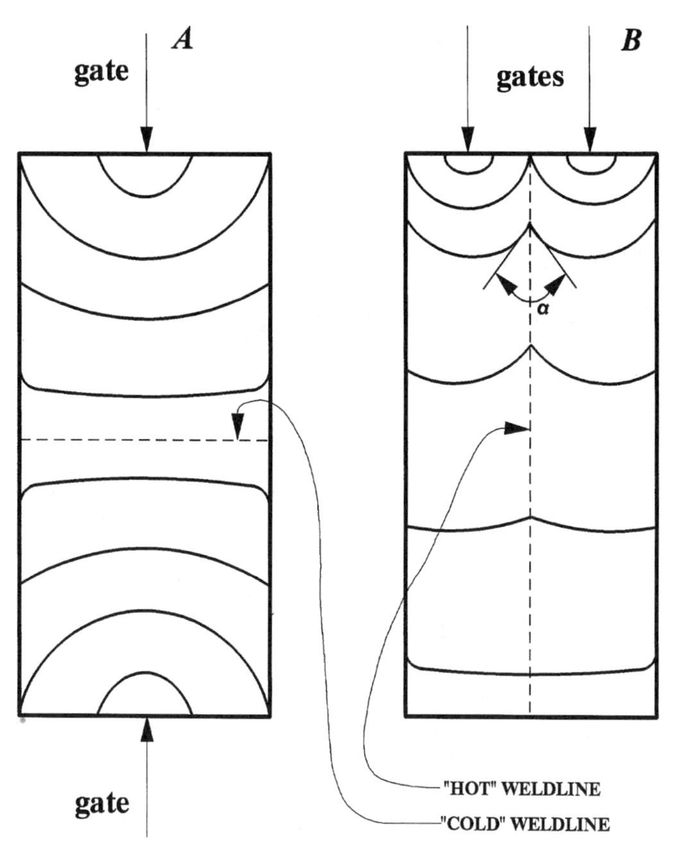

Figure 1 Two types of weldlines in injection-molded parts (mold-filling patterns in double-gated rectangular cavities). (a) Melt fronts flowing in opposite directions. (b) Mold with two gates side by side. Weldlines are indicated by dashed lines.

fronts, flowing in two opposite directions, collide and become rapidly immobilized (Figure 1(a)). The last area to be filled is close to the mold surface leading occasionally to a V-shaped ridge on the part surface. In some plastics, this 'V-notch' may contain poorly bonded layers but in PP its effect is usually limited to being a source of stress concentration in the part. Two key aspects of the mold filling determine the structure of injection molded parts: the cold mold walls and the so-called fountain flow (Figure 2). As the plastic flows into the mold, the first melt to contact the cold mold wall solidifies instantaneously forming an immobile 'frozen layer'. Additional flow then takes place within the envelope created by the frozen layer. The melt velocity at the frozen layer–melt boundary being nil, the fluid elements in the core move faster

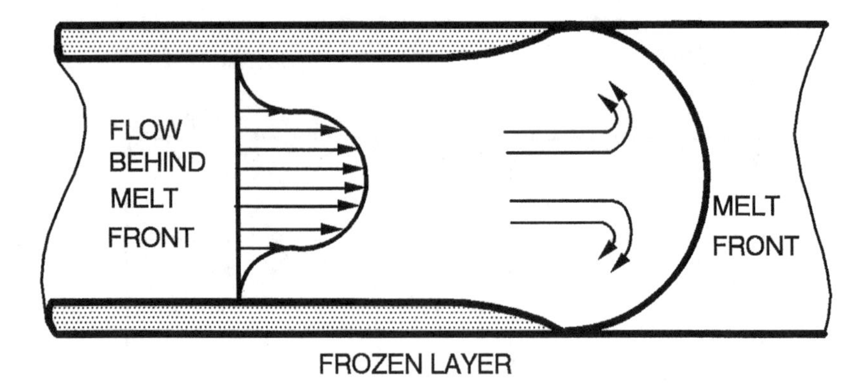

Figure 2 Fountain flow. Formation of the frozen layer from the expanding melt front, showing also idealized shape of the velocity profile in the core behind the melt front.

until they reach the melt front. There they expand and flow to the wall. As a result of the stretching which takes place in the melt front area, the fluid that has stretched most is the one which finds itself in the vicinity of the cold mold wall. Depending on the nature of the plastic and on the process parameters, some flow-generated orientation by the stretching is frozen. Polymer chains, filler and reinforcement particles in the weldline area then assume an orientation different from the bulk of the material. The local material anisotropy often leads to the weakening of the part.

The second type of weldline is illustrated in Figure 1(b). This 'hot weldline' is produced for example when the plastic penetrates into the mold via two gates located side by side (Figure 1(b)). In the beginning, the melt spreads around in a semicircular fashion until the melt fronts meet halfway between the gates forming a weldline which, in this area, is similar to the one resulting from the collision of two opposing melt fronts. As the recombined front advances, one expects the polymer melt to 'forget' the initial division and the weldline to disappear. The recovery of the without weldline structure in neat polymers has been related to the value of the melt front angle α: when the angle reaches a value of the 'weldline vanishing angle', the weldline has effectively disappeared. It is interesting to note that the weldline types shown in Figure 1 have been translated into standard mold-cavity shapes for weldline-strength evaluation (ASTM-D647 method).

Weldlines are also formed in molds of variable depth [2]. In such cases, the melt advances faster in thicker sections (Figure 3). For example, when the gate is located in the thinner section, the melt will flow into the thicker section, accelerate compared the thinner one and flow partly back leading

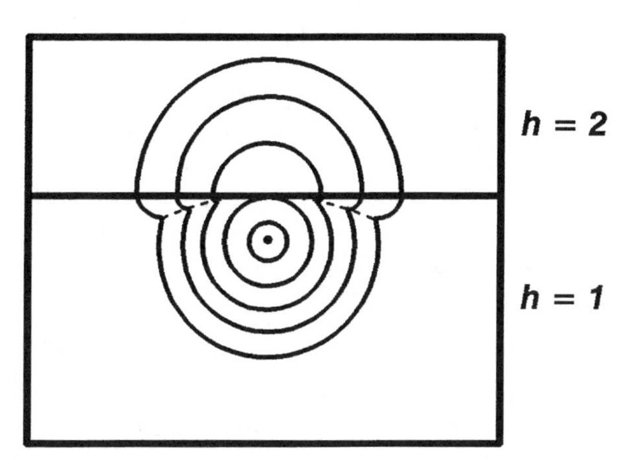

Figure 3 Mold-filling patterns in a cavity of variable thickness, *h*. The dashed line indicates weld line.

to a weldline. In moldings with holes, weldlines are formed when the melt stream, divided as it flows around the hole-forming insert, recombines.

When the flow in the mold is perfectly symmetrical, as is the case in Figure 1, the weldline will be planar across the part thickness. However, in most 'real' parts, the weldline topography will not be planar. Consider a relatively simple molding shown in Figure 4. When the left side of the

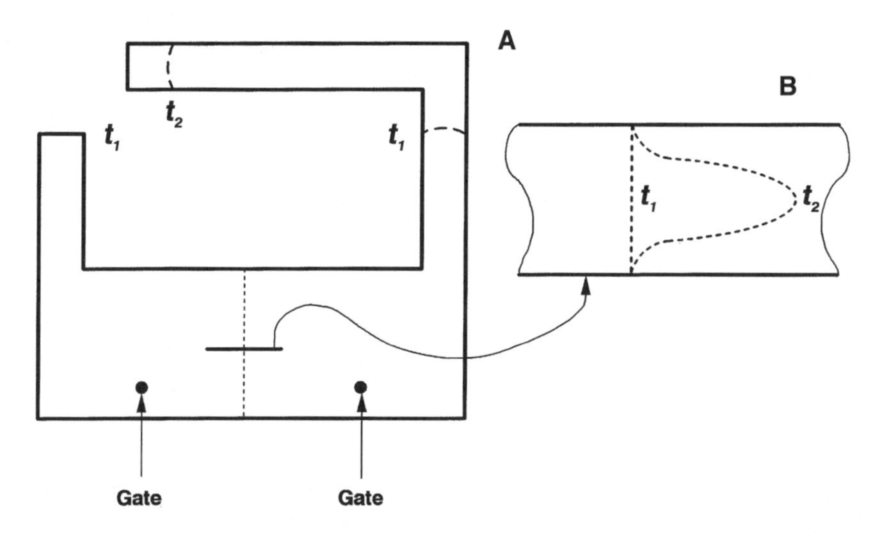

Figure 4 Weldine topography (b) during mold filling of a panel shown on the left (a). At time t_1, the weldline is planar. At time t_2, the weldline surface has been deformed by flow from both gates.

mold is filled (at time t_1), the weldline has already formed and is perpendicular to the panel surface. The polymer in contact with the mold wall has immobilized while the core is still molten. The melt required to fill the rest of the mold will be supplied from both gates and, as a result, the initially planar weldline will assume a bell-like shape. The control of the weldline topography, across the thickness during the injection-molding cycle, has become in recent years the preferred method of dealing with the deleterious effect of weldlines [3].

STRUCTURE AND PROPERTIES

In neat polymers, the loss of strength due to the weldline depends on a variety of factors. The incomplete diffusion across the interface has been recognized as the principal source of weakness of weldlines in some neat amorphous polymers, such as polystyrene (PS), polymethylmetacrylate (PMMA) and others. Since these plastics have to be molded with mold temperature below the glass transition temperature (T_g) of the polymer, chains, particularly those near the mold wall, become immobilized at T_g before a strong bond can develop. With semicrystalline polymers, such as PP ($T_g \approx -10°C$), the mold temperature is above the T_g. The polymer chains cannot be 'frozen' and maintain their mobility until they become incorporated into a crystallite. This suggests that in neat PP, processed using normal parameters, the strength loss due to an incomplete diffusion across the weldline is limited. In fact, even in the most unfavorable case of the weldline resulting from a head-on collision of two melt fronts, the weldline factor (strength of sample with a weldline divided by that of an equivalent weldline-free specimen) is typically between 0.9 and 1. Differences of local molecular orientation in and away from the weldline are thought to account for this limited strength loss.

The mechanical properties of the weldline zone can be influenced by the material characteristics and by the processing conditions. In practical terms, the freedom one has to improve the weldline strength by changing processing parameters, such as the melt and mold temperatures, to favor chain diffusion is limited (polymer degradation and longer cycle time) [4].

Most of the results published on weldlines in reinforced plastics were obtained using some sort of a double-gated dog bone shaped tensile bar mold in which the gates are located in the end of the dog bone (e.g. ASTM-D647). With this type of mold, the fibers in the weldline posses a random-in-plane orientation, the plane of orientation being parallel to the plane of the weldline. There is no evidence that the fibers cross the weldline plane. Away from the weldline, fibers are oriented parallel to flow. With the weldlines formed by flow around an insert, the fibers acquire a parallel-to-flow orientation throughout the entire thickness and

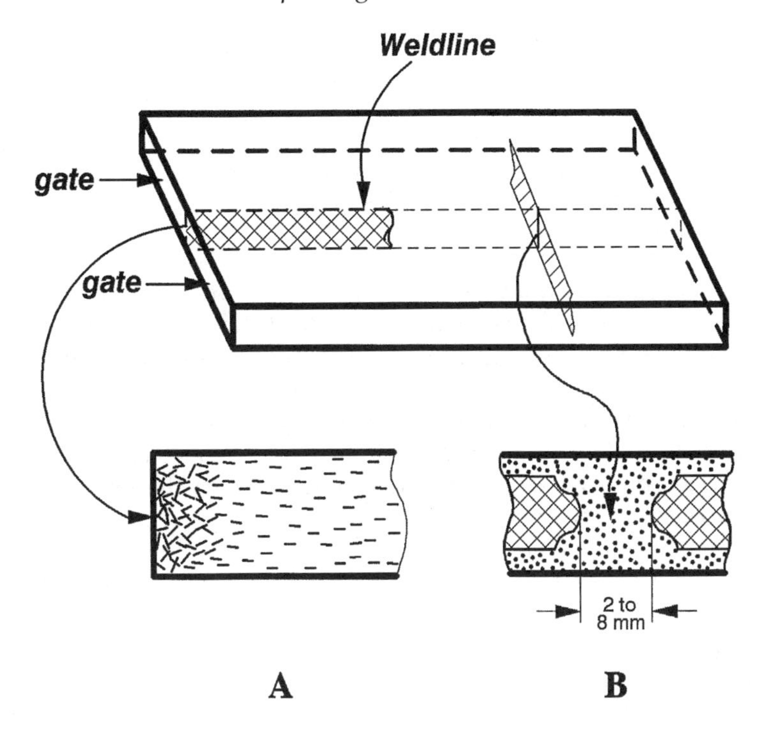

Figure 5 Fiber orientation in and away from the weldline in glass reinforced polypropylene. (a) Parallel to flow view. (b) Perpendicular to flow view.

keep it for a considerable distance from the insert. Figure 5 shows the fiber orientation (Figure 5(a) parallel to flow and Figure 5(b) perpendicular to flow) observed in a rectangular mold cavity with short glass fiber reinforced thermoplastic. The weldline width depends on the type of material and cavity depth but is essentially independent of the distance from the gate. (It is between 2 and 8 mm). Away from the weldline zone, the well known skin-core structure is observed [5]. In this case, the fibers in the skin are oriented parallel to flow and perpendicular to flow in the core. In glass fiber reinforced PP, the weldline strength across the weldline is approximately equal to that of the unreinforced matrix regardless of the material strength away from the weldline (provided there is a good adhesion at the fiber matrix interface). With poor interfacial adhesion, the weldline strength is further reduced.

IMPROVING THE WELDLINE

The early literature on weldlines, aimed at alleviating the strength loss via the optimization of processing parameters, has shown that the room for improvement was rather limited. Various approaches are now available; many of them can be found in the patent literature. Recently efforts at solving the 'weldline problem' concentrated on the weldline topography. There are several processes in which the weldline is made to be nonplanar. By various means the weldline, instead of appearing like a butt joint, is made to be analogous to a tongue-and-groove joint. This can be achieved by:

- Influencing the weldline topography by an appropriate control of the shut-off system of the hot runner nozzles (sequential filling). At the start of the mold filling operation, the melt first flows uniformly through the two nozzles until such time as one of the nozzles is shut. Following this, the remainder of the cavity is filled via the still open second nozzle, causing the weldline to be displaced. A similar approach employs sequential filling during which the injection gates are opened at different times leading to a relocation or complete elimination of the weldline.
- Influencing the weldline topography through appropriate gating: the principle of the method is shown in Figure 4.
- The 'push–pull' injection-molding process exploits the machine technology of multimaterial color injection molding with modified control system. The machine is made up of two injection units which are controlled completely independently of each other. The cavity is filled simultaneously by both units via two or more gates (Figure 6). After the flow fronts meet, the cavity is full and the weldline is formed, the control electronics permits one unit at a time to be retracted, whilst the second continues to inject the molten material and reverses the flow of the still molten core to move it in the reverse direction, pushing it through the more viscous barrier of material at the flow joint, thus providing an enlarged area and an improved strength in the joint.

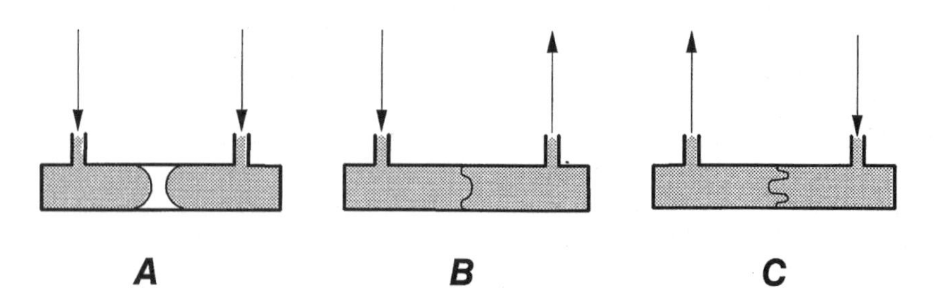

Figure 6 The principle of the push-pull molding process. (a) Left unit forward, right unit forward. (b) Left unit forward, right unit backward. (c) Left unit backward, right unit forward.

Of course, the outermost layers of the weldline zone, having solidified before the onset of pressure cycling, may remain weak. However, it is conceivable that emerging technologies, such as the one in which the mold surface in contact with the polymer is briefly heated to reduce the chain orientation, will bring further progress.

REFERENCES

1. Fellahi, S., Meddad, A., Fisa, B. and Favis, B.D. (1995) Weldlines in injection-molded parts: A review. *Adv. in Polymer Technol.*, **14**, 169–195.
2. Bangert, H. (1985) Development and design of injection molds. *Kunststoffe German Plast.*, **75**, 542–549.
3. Lanvers, A., Galuschka, S. and Tietz, W. (1994) Injection molding made transparent – new development in process simulation. *Engng Plast.*, **7**, 223–291.
4. Mennig, G. (1992) Mechanical characteristic data of injection molding weld lines. *Kunststoffe German Plast.*, **82**(3), 235–238.
5. Akay, M. and Barkley, D. (1985) Processing-structure-property interaction in injection molded glass-fiber-reinforced polypropylene. *Composite Structures*, **3**, 269–293.

Keywords: weldline, injection molding, cold weldline, hot weldline, flow, skin-core structure, weldline strength, fiber orientation, fountain flow, gate.

Wood–polypropylene composites

T. Czvikovszky

MANUFACTURING METHODS

Wood, being a fibrous polymer composite itself, and an abundant renewable raw material, offers its cellulosic fiber as a reinforcement for synthetic polymers. Indeed, the tensile strength of the cellulosic fiber is similar to that of an E-glass fiber (up to 1 GPa), and its modulus is also high enough (up to 70 GPa) [1].

PP is very suitable for filling, reinforcing and blending. Compounding PP with fibrous natural polymers of biomass origin is one of the most promising routes to create natural–synthetic polymer composites. The key question is whether they are cheap filled compounds or high value, reinforced composites. It depends on the cohesion, cooperation between the synthetic PP matrix and the natural fiber.

There has been much work done in the last two decades to bring together wood fiber, which is an irregular, hygroscopic, thermally sensitive, polar polymer, with polypropylene, which is an apolar, hydrophobic, highly crystalline synthetic, requiring relatively high temperature to melt together with a compounding partner [1, 2, 3].

To compatibilize such incompatible partners, there have been several physical, chemical and technological solutions proposed, such as:

- intensive, intimate melt mixing without additives;
- mixing with special compatibilizers and processing aids;
- mixing in presence of reactive additives;
- reactive processing (reactive extrusion, injection molding) applying special simultaneous treatments (electron-beam, ultraviolet, etc.).

Polypropylene: An A–Z Reference
Edited by J. Karger-Kocsis
Published in 1999 by Kluwer Publishers, Dordrecht. ISBN 0 412 80200 7

To mix the fluffy, low-bulk-density, almost sawdust-like (flour-like) woodfiber into the PP is the most difficult step of technology. High-speed, high-shear mixers can bring such mixtures up to >200°C in seconds into a very intimately mixed melt [1]. There are no data, however, about the large scale feasibility of high-shear mixing.

The technology of polymer blends and alloys offers nowadays a whole range of chemical compatibilizers and processing aids [4, 5]. The effect of those processing aids includes changing surface tensions, promoting fiber wettability, lubricating, facilitating the fusion of individual resin particles, so lowering the thermal load (temperature and heating time) during processing, and creating possible physical (or chemical) bonds between components. For the present purpose, maleic anhydride modified PP (maleated PP, MPP) seems to be the most suitable. Besides MPP, there are many other acrylic modified, vinyl-functional or silicon-functional, vinyl-silane, etc., types of additives (from DuPont, 3M, BF Goodrich, Dow, Eastman Kodak, etc.) on the market [5]. The trouble is that applying those high value (>15 US$/kg) additives, added in 1–2% to the composite, we may lose more than 20% of economic benefit of the wood-PP composites.

Reactive additives may also be applied to bond the wood fiber into the PP matrix [3, 4]. These are commercial multifunctional (unsaturated) monomers and oligomers, such as acrylates, polyesters, urethane-acrylates, epoxy-acrylates etc, which can be 'activated' to bond wood fiber and PP through several initiation methods (peroxides, heat, electron-beam, etc) [3, 4]. In fact, some of the earlier group of chemical compatibilizers (e.g. vinyl-silanes) are promising similar effects providing bridge-forming (partial 'crosslinking') chemical reactions between the composite components.

Reactive processing as a polymer modification technique is of growing importance in the field of compounding, alloying and recycling [1, 5]. The MPP itself is made by reactive extrusion. Crosslinking of thermoplastics can be performed through adequate initiators acting during the extrusion or injection molding. Reactive extrusion and reactive injection molding (RIM) as modern plastics-processing technologies have long since outgrown the original polyurethane technologies. In the contemporary high-power mixing extruders, e.g., where the intermeshing twin-screw construction is assuring intimate melt mixing with a typical 1–3 min dwelling time of the melt, the chemical reaction required for bonding can be performed 'just in time' during the plastic processing.

Special simultaneous treatments during the conventional or reactive processing can also provide better control of the required bonding of wood fiber and PP. Electron-beam (EB) treatment with industrial scale electron machines, or plasma treatment assuring better connections of the interface of wood fiber–PP composites are nowadays cheap enough

(< 0.15 US$/kg of end product) to be applied in, for example, automobile applications of PP and its composites [3].

PROPERTY PROFILE

In spite of the great number (> 200) of publications on wood fiber–PP (WF-PP) composites in the last two decades [1], not too many details of the commercially produced WF-PP products have been released. On the other hand, the growing number of relevant patents (actually more than 60) show the industrial potential of such products, elaborated in most case by big, multinational companies, such as BASF, Hoechst, Imperial Chemical Industries (ICI), DuPont, SOLVAY, Mitsubishi, Ford Motor Co. and others. It is clear from the applications of WF-PP composites in the American, European and Japanese automobile industry, that their production in the late 1990s surpassed 100 000 tons/year [1].

What are those attractive properties of WF-PP composites, to justify such a large production? In Table 1, which contains the most important properties (comparing them with the raw PP serving as matrix for the last Canadian WFRP product), it becomes clear that flexural strength and modulus of elasticity as well as the heat tolerance is increased compared

Table 1 Properties of wood-fiber PP composites

	HM-PP (BASF, D)	*WF-PP (ICI, UK)*	*Woodstock (GOR, Italy, USA)*	*WFRP-S (AECL, Canada)*	*PP*
Composition					
Polypropylene (wt.%)	60	60	60	65	100
Wood fiber (wt.%)	40	40	40	35	0
Properties					
Flexural strength (MPa)		31.3	43.5	62.5	37.8
Flexural modulus (GPa)		2.17	2.86	3.46	1.41
Tensile strength (MPa)	34.0	19.3	24.2	48.8	37.3
Tensile modulus (GPa)	4.0	2.68	3.38	3.89	1.87
Impact strength/notched Izod (J/m)				11	12
Heat distortion temperature					
at 1.85 N/mm (°C)			100	116	
at 0.45 N/mm (°C)	106			146	
Thermal tolerance[a] (°C)		52		142	51
Melt flow index (g/10 min) (at 230°C, 2.16 kg)		0.3	1.0	4.4	20

[a] Temperature at 1 GPa flexural modulus.

to the PP. Those properties are all important in designing automobile interior trims.

So, by reinforcing PP with wood fiber, we obtain a strong, stiff PP compound with increased heat tolerance, and by using a significant amount of fibrous natural polymer, almost an order of magnitude

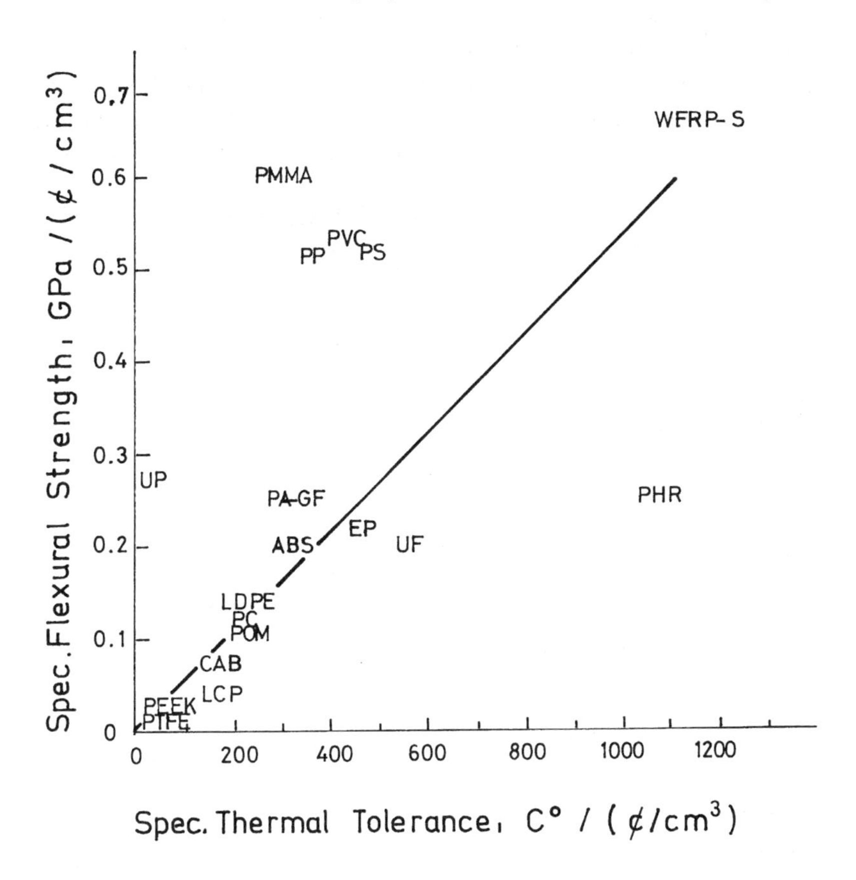

Figure 1 Cost-related (specific) flexural strength of major thermoplastics, versus cost-related (specific) thermal tolerance. The unit cost is the market price in US cents (1992) of 1 cm^3 plastics. The thermal tolerance is the temperature difference (ΔT) over room temperature ($\Delta T = T^* -$ room T), by which temperature (T^*) the flexural modulus is equal to 1 GPa. Designations, abbreviations: WFRP-S, wood fiber reinforced PP (S type) of AECL, Canada (See Table 1); PMMA, polymethylmethacrylate; PVC, polyvinyl chloride; PS, polystyrene; PP, polypropylene; UP, unsaturated polyesters; PA-GF, glass fiber (35%) reinforced polyamide; PHR, phenolic resin; EP, epoxy resin; ABS, acrylonitrile/butadiene/styrene copolymer; UF, urea/formaldehyde; LDPE, low density polyethylene; PC, polycarbonate; POM, polyoxymethylene; CAB, cellulose acetate butyrate; LCP, liquid crystal polymers; PEEK, polyether-etherketone; PTFE, polytetrafluorethylene.

cheaper than PP. On the other hand, we pay for that benefit with a decrease of impact strength and with a more difficult processability, reflected in lowered melt flow index (MFI), increased melt viscosity.

The benefits are more pronounced if we relate the flexural strength and thermal tolerance to the unit cost (US cents/cm^3). Figure 1 shows those cost-related properties, mapping all the major thermoplastics types. The linear regression line shows that there is a weak correlation between cost-normalized flexural strength and heat tolerance, and the WFRP-S type wood fiber reinforced PP composite has a top position, being the cheapest solution when strong, heat-tolerant thermoplastic plates are required.

Similar cost-normalized data are compiled in Figure 2, comparing the flexural modulus as a function of the impact strength. Here again, we

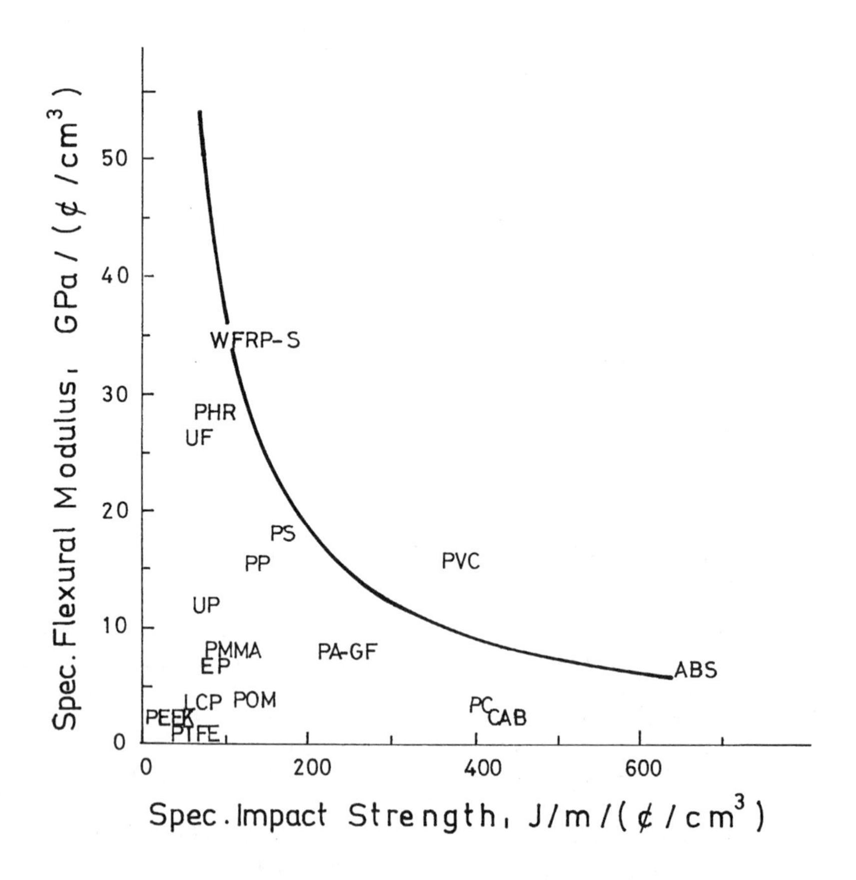

Figure 2 Cost-related (specific) flexural modulus of major thermoplastics versus cost-related (specific) impact strength. (For definitions and abbreviations, see Figure 1). Impact strength: notched, Izod.

plot the cost-related features of the major plastics including engineering plastics as well.

Figure 2 reflects the well-known fact that flexural modulus and impact strength, as well as rigidity and flexibility, are inversely related to each other. WFRP-S shows a modest impact strength, however it is on the 'cutting edge' of that hyperbolic function. For its modest price, WFRP-S offers the best compromise between stiffness and flexibility. The major advantage of the WF-PP composite that it is extremely cost-effective, similar to most conventional thermosets (PHR, UF, etc.) However, it belongs to the family of thermoplastics, suitable for extrusion, thermoforming and, in its most recent version [3], can be injection molded as well.

PROCESSING TECHNOLOGY

Cellulose fiber reinforced PP composites are typically made with short (0.5–5 mm) fibers, but there are reinforced PP composites on the market with long cellulosic fibers (> 5 mm) as well. The latter variants, being similar to the glass fiber reinforced composite prefabricates (sheet and bulk molding compounds, and glass mat reinforced thermoplastics; SMC, BMC, GMT, respectively) are limited in their processing technology. The manufacturing of those long fiber reinforced composites is performed typically by mixing cellulosic fibers and PP fibers applying conventional nonwoven textile technologies (carding, mat-forming, etc.). The long fiber PP composite prefabricates (mats) are suitable for press molding or thermoforming, but are unsuitable for extrusion or injection molding.

The main stream of wood fiber PP composites is short fiber reinforced, suitable for typical thermoplastic processing. The main steps of this technology are: (1) pre-mixing, (2) plastic transformation of the melt (typically sheet extrusion), and (3) thermoforming (vacuum- or pressure-forming) of the sheet.

To make premix of short wood fibers and PP, several proven technologies and compounding machines can be used. From the most simple tumbling barrel, all kinds of ribbon blender, vertical screw silo mixer or high speed 'fluid-bed' mixer can serve. The main difficulty in this step is the difference in the components' bulk density. It is difficult to mix the fluffy, low-bulk-density sawdust with the solid polymer. PP powder or small granules (< 1 mm) of 'Spheripol' type PP is more suitable for mixing with wood fiber. The additives described earlier help the wetting in the mixing step, too. Proprietary details of the technology of commercial products have not been released. For most advanced wood fiber–PP composite (see WFRP-S in Table 1), fluid mixer has been used to mix together 35% fluffy 'wood pulp' (dry fibrous raw material of paper making) with 63% PP (in powder form), in presence of about 2% viscous-liquid reactive additive (acrylic type oligomer). Those high-speed mixers are applied to make typical 'dry-blend' premixes in powdery form for rigid, or semirigid PVC compounds used in extrusion or injection molding. The high-speed (up to 2000 rev/min) mixer of close to 1 m^3 capacity (with a typical 300 kg load) brings the powder-like components to 'fluid-state', the friction causing a temperature increase (from room temperature to 60–80°C in 3–6 min) which helps the homogenization.

The difficulty in compounding and processing of WF-PP composites arises from the significantly increased melt viscosity (related to the PP) because of the high amount of solid (nonfusible) wood fiber (up to 50 wt.%). This high viscosity, characterized by the low value of MFI, held back the industrial application of the composite until the arrival of high-power, high-shear melt-processing machines, such as the intermeshing twin-screw extruder. In fact the Woodstock type wood fiber–PP composite (Table 1) was elaborated and succesfully commercialized in Europe, America and Japan by the producers of an Italian manufacturer (G.O.R, ICMA San Giorgio, Milano) of high-power twin-screw extruders. Those extruders with vented (vacuum-assisted) mid-zone transform the 'dry-blend' type pre-mix powder into typically 1500 mm wide, 2.0 mm thick sheets. The melt leaving the 'coathanger' type die is fed into an assembly of three roll calender. The sheets can be thermoformed with or without textile coating (typically nonwoven fabrics of PP or polyethylene terephthalate, PET, fibers) afterwards. The new types of WFRP can be injection molded as well.

WF-reinforced PP is an excellent, cost-effective, recyclable thermoplastic raw material which is made partially from synthetic (PP) and partially from natural polymer of biomass origin (WF), reusing a byproduct of the woodworking industry (a fraction of sawdust). Its application is steadily increasing as a backing panel of textile covered automobile interior trims, such as door panel, rear quarter cover, rear shelf, roof liner, spare wheel covering and package compartment covering, etc.

REFERENCES

1. Wolcott, M.P. (ed.) (1993) *Wood-Fiber/Polymer Composites*, Forest Product Society, Madison, WI.
2. Singh, A. and Silverman, J. (eds) (1991) *Radiation Processing of Polymers*, Hanser–Oxford University Press, Munich, Oxford.
3. Czvikovszky, T. (1995) Reactive recycling of multiphase polymer systems. *Nucl. Instr. Methods Phys. Res., B.*, **105**, 233–237.
4. Gächter, R. and Müller, H. (eds) (1990) *Plastics Additives*, Hanser–Oxford University Press, Munich, Oxford.

5. Brandrup, J., Bittner, M., Michaeli, W. and Menges, G. (1996) *Recycling and Recovery of Plastics*, Hanser–Gardner, Munich, Cincinnati.

Keywords: wood fiber, reinforcing, composite, reactive extrusion, automotive engineering applications, compounding, compatibilization, economics, sawdust.

X-ray scattering

S.V. Meille and S. Brückner

X-ray scattering methods with polymers generally require a source of monochromated radiation, which for convenience is usually Cu-Kα with a wavelength of 0.15418 nm, giving optimum compromise between resolution and scattered intensity. Scattering studies can give at wide angles (WAXS) information about the crystal structure, the crystallinity and the morphology of samples which may be either uniaxially oriented fibers, bulk specimens or films. Caution is necessary with WAXS patterns because they may vary substantially depending on: (1) degree of crystallinity; (2) crystallite size and perfection; (3) copresence of different crystalline modifications; and (4) preferred orientation. The crystallinity of unoriented samples can be readily determined using standard procedures with reference to the pattern of the amorphous polymer. For oriented samples, appropriate averaging procedures are necessary to obtain reliable crystallinity values. The crystallite size influences essentially the full width at half-maximum (FWHM) of the diffraction maxima while crystallite defects may have more subtle effects.

Small angle X-ray scattering (SAXS) requires more sophisticated experimental set ups and supplies complementary morphological information. High-brilliance X-rays available from synchrotrons allow for time-resolved structural and morphological studies, high spatial resolution and access to very small scattering angles.

ISOTACTIC POLYPROPYLENE (iPP) [1–3]

Since the β and γ phases of iPP are unstable under elongational stress, fiber diffraction patterns can be obtained only for the α-form and for

Polypropylene: An A–Z Reference
Edited by J. Karger-Kocsis
Published in 1999 by Kluwer Publishers, Dordrecht. ISBN 0 412 80200 7

mesomorphic samples quenched from the melt. The relatively high intensity near the meridian on the third layer of these patterns clearly points to the three-fold helical conformation of iPP.

The α, β and γ crystalline modifications (see the chapter 'Polymorphism in crystalline polypropylene' in this book) of iPP can be readily identified from their wide angle X-ray diffraction patterns (WAXD) shown in Figure 1. The pattern of a quenched sample in the semiordered mesophase, showing two broad maxima at $d = 0.6$ and 0.42 nm are also displayed in Figure 2. Crystallinity as determined from WAXD measurements, for properly crystallized highly isotactic samples, normally ranges between 60 and 70% but can be as high as 85%. Since α-iPP is readily obtained from melts, from solution or under tensile stress, this phase is normally prevalent while the other polymorphs are often only impurities. Diffraction patterns of α-iPP are characterized by five strong reflections at $d = 0.626$, 0.519, 0.477, 0.419 and 0.404 nm. The distinction between α_1 structure (space group C2/c) and the more ordered α_2 structure ($P2_1/c$ symmetry) is based on the fact that in the former the hkl reflections with (h + k) odd are forbidden, this does not apply in the latter. Specifically, as shown by Corradini *et al.*, reflections 231 and 052, characteristic of the α_2 structure, cause a measurable relative increment of the diffraction maximum located between 34.4 and 36.0° 2θ (Cu-Kα). The β-form is identified by the strong maximum at $d = 0.55$ nm corresponding to the interchain projection distance of its hexagonal packing. It is hard to obtain macroscopic samples of this phase without some α-phase impurities. Four of the five main reflections of the γ-phase, at $d = 0.64$, 0.53, 0.419 and 0.406 nm are very close to α-form maxima, due to the substantial structural similarities of the two structures which share a common building block (the 'bilayer', see the chapter 'Polymorphism in crystalline polypropylene' in this book) and only the fifth maximum at $d = 0.442$ nm is specific.

The coexistence of different phases can be established with reference to the 0.477 and the 0.442 nm peaks specific respectively to the α- and to the γ-form. These maxima together with the 0.55 nm reflection of the β-form can be used to determine empirically the proportions of the three crystalline phases present in a given sample with due consideration of preferred orientation. Additional care must be taken with samples crystallized at pressures of a few hundred atmospheres because of the possibility that α- and γ-iPP cocrystallize within the same lamella giving rise to diffraction spectra which are not the sum of the patterns due to the two separate phases.

Aside from normal fiber patterns with the fiber axis coinciding with the molecular axis, bi-oriented morphologies are frequent in α-iPP films due to selfepitaxy related to the quadritic morphology. Patterns may result evidencing two distinct crystallite orientations, for which the a^*

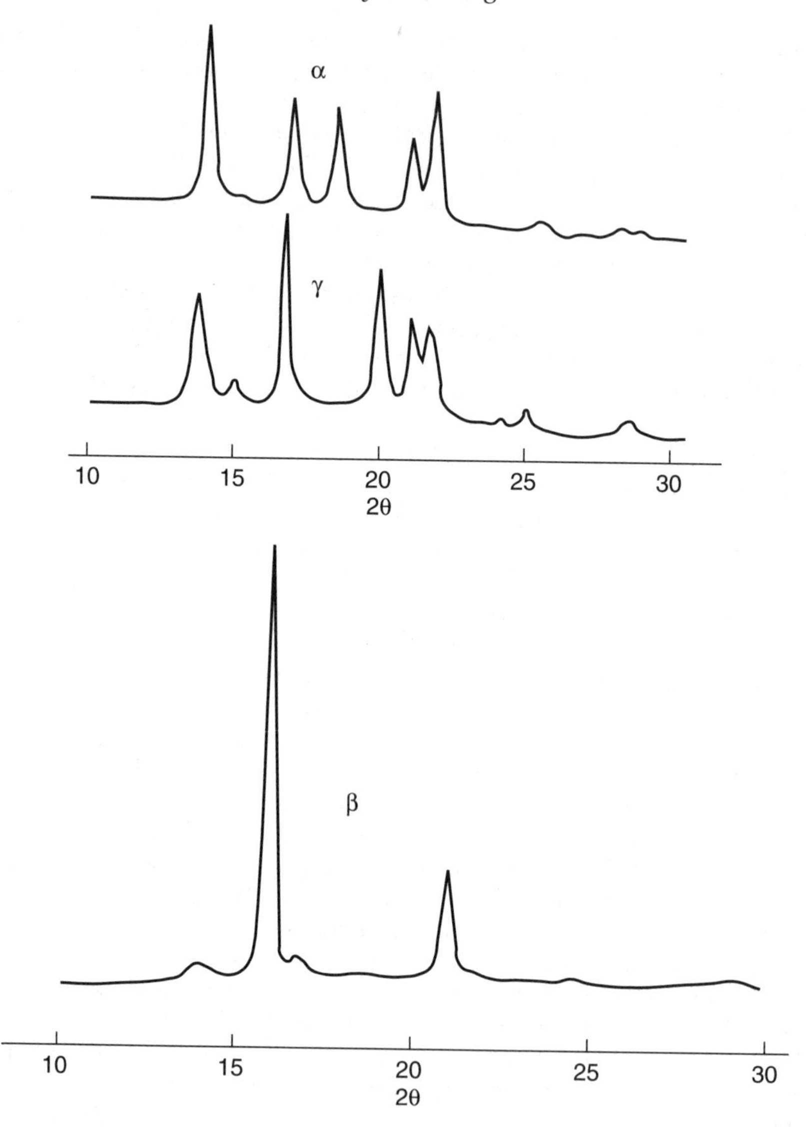

Figure 1 The three powder diffraction profiles of α-iPP, β-iPP and γ-iPP.

direction of the former corresponds to the c^* axes direction of the latter. The b axis in such cases is often perpendicular to the film plane and as a result in diffractometric scans in reflection geometry the 040 reflection at 0.52 nm may be substantially enhanced. Oriented patterns (but not along the molecular axis), result with β-iPP from portions of individual spherulites, by crystallization in temperature gradients or in any case due to the

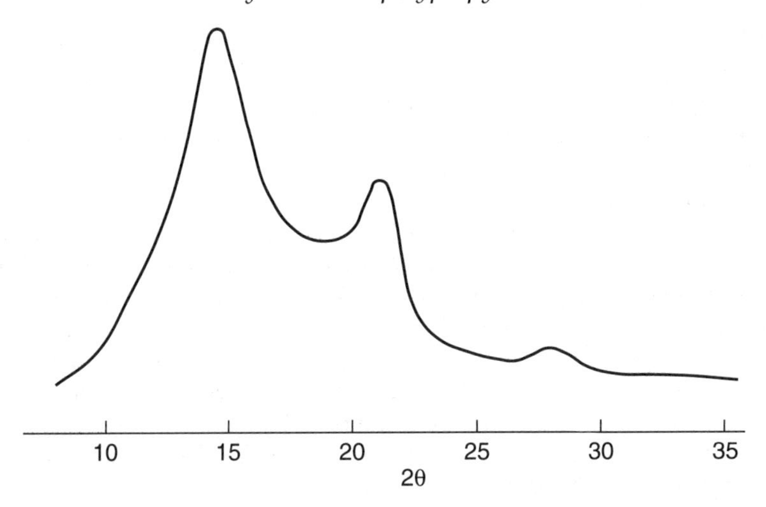

Figure 2 The power diffraction profile of the mesomorphic form of iPP.

preferred growth of this phase along the a^* axis. In the case of γ-iPP, the reported oriented patterns arise in the presence of oriented α-form, probably because of the tendency of γ-iPP to crystallize by an epitaxial mechanism also on oriented α-form crystals. In all such cases, estimation of relative proportions of coexisting phases is problematic.

Lamellar thicknesses have been measured with the SAXS technique for a number of different α- and γ- form samples. In the case of α-iPP, Bragg periodicities typically vary between 12 nm for water-quenched samples to 40 nm and more for highly isotactic samples crystallized and annealed at high temperatures. The coexisting α- and γ- forms in samples crystallized in the 0–2000 bar range give rise to a single low angle diffraction maximum, suggesting that under these conditions the two modifications may be copresent within individual lamellae whose thickness surprisingly decreases with increasing pressure.

SYNDIOTACTIC POLYPROPYLENE [4, 5]

Syndiotactic polypropylene (sPP) is characterized by the existence of three regular conformers with different fiber repeats. Three crystalline modifications result, all giving fiber patterns as opposed to what is found in iPP.

Un-annealed fibers drawn at room temperature contain the planar zigzag low-temperature orthorhombic form characterized by a 0.505 nm c axis, obviously corresponding to the layer line spacing. The strongest equatorial reflections of this form are the 020, 110 and 130 at 0.558, 0.473 and 0.303 nm respectively (with Cu-Kα radiation, $2\theta = 15.9$, 18.8 and

29.5°). Quite intense are also the 021, 111 and 112 upper layer reflections ($d = 0.376$, 0.346 and 0.223 nm respectively). The planar zig-zag conformer upon annealing above 100°C readily transforms into the high-temperature form characterized by the $(T_2G_2)_n$ conformation (where T and G stand for trans and gauche, respectively) and a helical periodicity of 0.74 nm.

This second modification can be obtained also by crystallization from solution and from the melt and, although it presents always the same subcell with $a = 1.45$, $b = 0.56$ and $c = 0.74$ nm, it may present different experimental diffraction patterns. The fiber repeat corresponding to the c axis and the features of fiber photographs are consistent with twofold helical symmetry. In more detail this form shows basically two limit different diffraction patterns (and combinations thereof) indicating different packing arrangements on a similar lattice. The C-centered structure (all isochiral helices, space group $C222_1$) originally proposed by Corradini *et al.* is characterized experimentally by strong diffraction maxima at $d = 0.725$, 0.522 and 0.431 nm ($2\theta = 12.2$, 17.0 and 20.6°, Cu-Kα) corresponding to reflections 200, 110 and 111 (see the chapter 'Polymorphism in crystalline polypropylene' in this book). The other

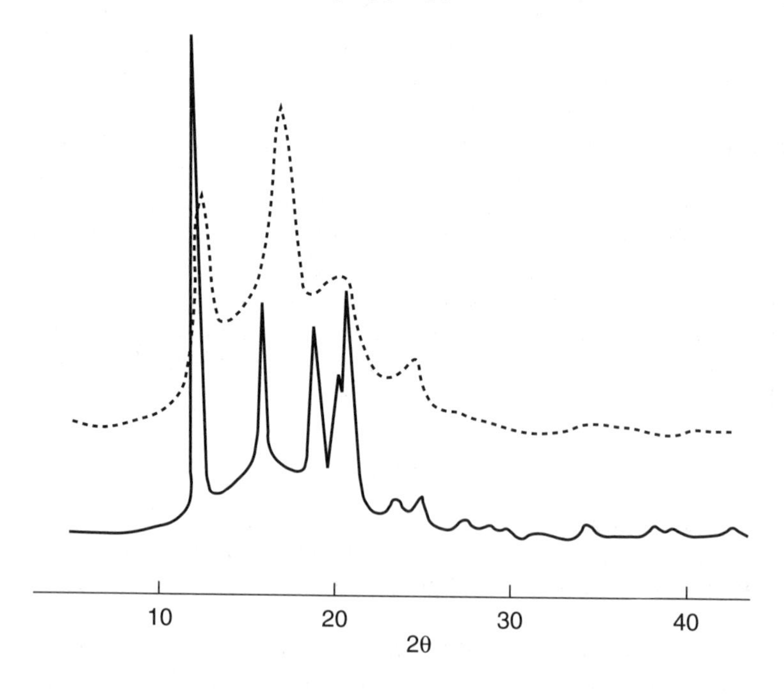

Figure 3 The two limit powder diffraction profiles observed in the high temperature orthorhombic form of sPP. The dashed line refers to the $C222_1$ model whereas the full line refers to Ibca model.

limiting structure obtained by crystallization at higher temperature or by prolonged annealing is apparently a perturbation of the Ibca model proposed by Lovinger *et al.* It shows in powder X-diffraction patterns peaks at $d = 0.725$, 0.56, 0.47 and 0.431 nm ($2\theta = 12.2$, 15.8, 18.8 and 20.6°, Cu-Kα). Considering the doubling of the *b* axis these maxima correspond to the 200, the 020, the 211 and the 121 reflections. The two limit powder diffraction profiles are shown in Figure 3. It is notable that the interpretation, the proposed structural models and the diffraction patterns of sPP have all been substantially evolving with the improved control of the stereoregularity recently achieved with the new metallocene catalysts.

The last polymorph of sPP involves a $(T_6G_2T_2G_2)_n$ conformation and is obtained by exposition to solvents of fibers of the all *trans* polymorph. The helical periodicity, corresponding to the *c* axis is 1.16 nm. The value is readily obtained from the layer line spacing in diffraction patterns which are also consistent with only translational periodicity along the chain axis. The main characteristic maxima of this form (triclinic) are at $d = 0.67$ nm (010 and 011), 0.51 nm and 0.45 nm ($2\theta = 13.2$, 17.2 and 19.5°, Cu-Kα) with a number of different reflections contributing to the last two diffraction peaks. SAXS studies on solution crystallized samples of sPP appear to indicate that the lamellar thickness correlates both with the annealing temperature and with the degree of syndiotacticity.

REFERENCES

1. Brückner, S., Meille, S.V., Petraccone, V. and Pirozzi, B. (1991) Polymorphism in isotactic polypropylene. *Prog. Polymer. Sci.*, **16,** 361–403.
2. Lotz, B., Wittmann, J.C. and Lovinger, A.J. (1996) Structure and morphology of poly(propylenes): A molecular analysis. *Polymer*, **37**, 4979–4992.
3. Natta, G. and Corradini, P. (1960) Structure and properties of isotactic polypropylene, *Nuovo Cimento Suppl.*, **15,** 40–51.
4. Rodriguez-Arnold, J., Zhengzheng, B. and Cheng, S.Z.D. (1995) Crystal structure, morphology, and phase transitions in syndiotactic polypropylene. *JMS Rev. Macromol. Chem. Phys.*, **C35(1),** 117–154.
5. De Rosa, C., Auriemma, F. and Corradini, P., (1996) Crystal structure of form I of syndiotactic polypropylene. *Macromolecules*, **29,** 7452–7459.

Keywords: X-ray, powder pattern, fiber pattern, WAXS, SAXS, isotactic, α-form, β-form, γ-form, mesophase, lamellar thickness, crystallinity, syndiotactic, orthorhombic form (high temperature, low temperature), triclinic form.

Ziegler–Natta catalysis and propylene polymerization

R. Mülhaupt

INTRODUCTION

Transition metal catalyzed olefin polymerization is playing a key role in the development of novel versatile and environmentally friendly polymeric materials incluing both commodity and specialty polymers. Polyolefins, such as polypropylene (PP), combine low price and heat distortion temperature $> 100°C$ with attractive mechanical proprties, e.g. strength, stiffness, impact resistance, low weight, corrosion resistance, and versatility in applications, ranging from automotive moldings to packaging and textile fibers. Moreover, polyolefins are compatible with most modern recycling processes. Upon heating above 400°C, polyolefins are readily degraded to form oil and gas which can be used as raw materials in petrochemical industry. Both crude oil and polyolefins are hydrocarbons with very similar molecular architectures, similar high energy content but different molar mass. In fact, PP and other polyolefins represent a solid modification of crude oil. Polyolefins meet the demand for sustainable development and help preserve valuable resources. As shown in Figure 1, crude oil is thermally cracked to form olefin monomers, which are polymerized using transition metal catalysts at low pressure in energy-efficient processes. Polymer properties can be tailored by means of catalysis and processing to meet the demands of highly diversified applications. After completing their product life cycle, polyolefins are recycled by remolding or by recovering the oil feedstock, which can be used to produce new olefin monomers, to substitute oil in

Polypropylene: An A–Z Reference
Edited by J. Karger-Kocsis
Published in 1999 by Kluwer Publishers, Dordrecht. ISBN 0 412 80200 7

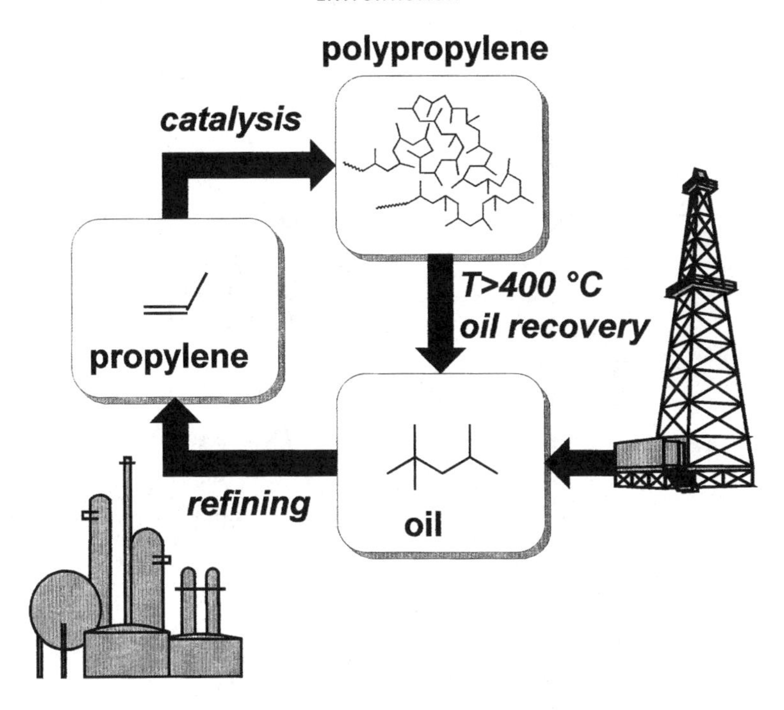

Figure 1 Polyolefin life cycle: energy-saving catalytic low pressure polymerization affords polyolefins which are readily recycled by recovery of oil and gas feedstocks upon thermal degradation.

the petrochemical industry or to serve as an oil substitute in steel mills or incinerators.

Progress in olefin polymerization technology, including catalyst as well as process technology, has greatly simplified polyolefin production. During the 1960s PP production was associated with byproduct formation and extensive polymer purification, e.g. removal of corrosive catalyst residues and low molecular wax-like polymer fraction. Modern gas phase processes convert propene gas into PP in quantitative yields, thus preventing environmental pollution. Modern PP is competing very successfully with materials such as glass, wood, and metals, and is substituting other, less environmentally friendly and expensive polymers. As illustrated in Figure 2, soon PP production in Western Europe is expected to surpass that of PVC. At the beginning of the 21st century, forty years after its discovery, catalytic olefin polymerization still represents one of the most innovative fields in chemistry and technology. A major challenge in catalyst and process development is to design highly active catalysts which control both molecular and supermolecular polyolefin architectures.

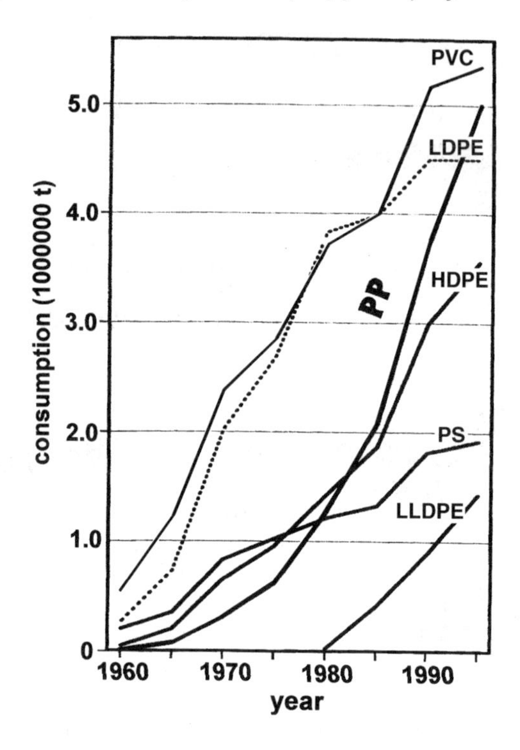

Figure 2 Consumption of commodity thermoplastics in Western Europe: polyvinylchloride (PVC), low density polyethylene (LDPE), polypropylene (PP), high density polyethylene (HDPE), polystyrene (PS) and linear low density polyethylene (LLDPE). Data was obtained from BASF AG.

HISTORY AND BASIC PRINCIPLES OF CATALYTIC OLEFIN POLYMERIZATION

During the mid-1930s, free radical ethylene polymerization was developed by ICI. At temperatures above 150°C and pressures exceeding 1000 atm, radical initiators produce radicals which initiate free radical ethylene polymerization. Due to inter- and intramolecular chain transfer reaction via H atom transfer with the free radical at the polymer chain end, high pressure polyethylene contains short- and long-chain branches which reduce polyethylene (PE) crystallinity and consequently also PE density. Therefore this class of polyolefins became known as low density polyethylene (LDPE). The history of polyolefins was compiled by Seymour and Cheng [1].

During the 1950s, several research groups explored the potential of novel organometallic compounds. It was Karl Ziegler at the Max-Planck Institut für Kohlenforschung in Mülheim, Germany who discovered that

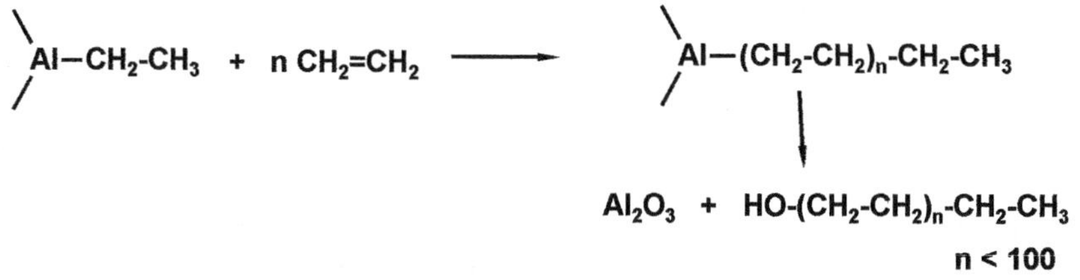

$$>Al{-}CH_2{-}CH_3 + n\,CH_2{=}CH_2 \longrightarrow >Al{-}(CH_2{-}CH_2)_n{-}CH_2{-}CH_3$$

$$\downarrow$$

$$Al_2O_3 + HO{-}(CH_2{-}CH_2)_n{-}CH_2{-}CH_3$$

$$n < 100$$

Figure 3 Ziegler's 'Aufbaureaktion' produces alcohols and alumina.

aluminum alkyls can insert ethylene into the metal–carbon bond to form paraffins. Ziegler's 'Aufbaureaktion', shown in Figure 3, is still being used to produce high purity alumina and linear 1-alcohols, which are of interest in production of ionic and nonionic surfactants. In the presence of nickel compounds, only 1-butene, the ethylene dimer, was found. When Ziegler and his coworkers investigated the role of transition metal compounds, such as zirconium and titanium halides, they discovered that group IV transition metal compounds activated with main group metal alkyls, especially aluminum alkyls, catalyzed the polymerization of ethylene at low pressure ('Mülheimer Niederdruckverfahren'). The catalytic cycle is displayed in Figure 4. The aluminum alkyl activator

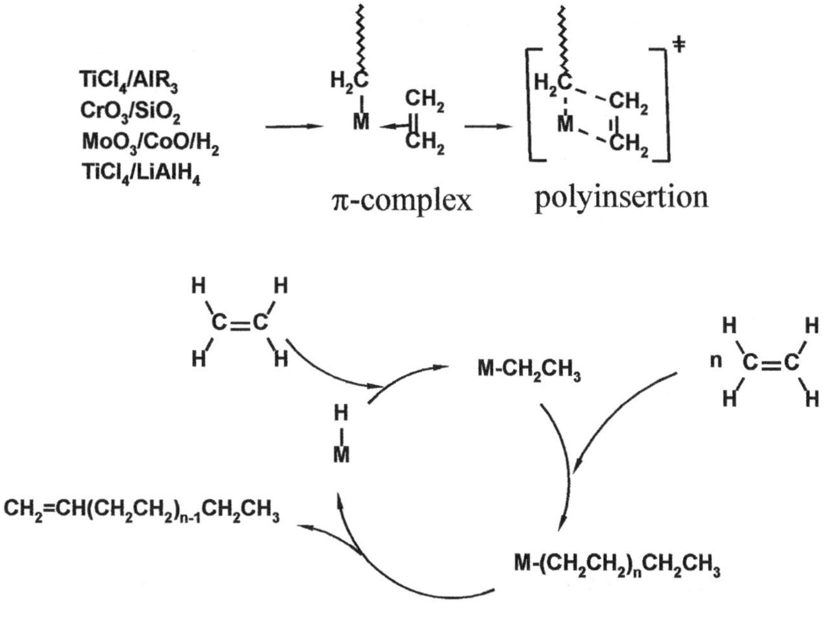

$TiCl_4/AlR_3$
CrO_3/SiO_2
$MoO_3/CoO/H_2$
$TiCl_4/LiAlH_4$

$\longrightarrow$ π-complex (H_2C–M ← $CH_2{=}CH_2$) $\longrightarrow$ polyinsertion ($[H_2C \cdots CH_2 / M \cdots CH_2]^{\ddagger}$)

$H_2C{=}CH_2$ + H–M → $M{-}CH_2CH_3$

$M{-}CH_2CH_3$ + $n\,H_2C{=}CH_2$ → $M{-}(CH_2CH_2)_nCH_2CH_3$

$M{-}(CH_2CH_2)_nCH_2CH_3$ → $CH_2{=}CH(CH_2CH_2)_{n-1}CH_2CH_3$ + H–M

Figure 4 Catalytic cycle of ethylene polymerization.

alkylates the transition metal compounds which can be reduced to form lower valent transition metal compounds. During chain propagation, ethylene is inserted into the transition metal alkyl bond. The combination of transition metal compounds with main group metal alkyls became known as Ziegler catalysts. Within a few years, the highly flammable aluminum alkyls, originally thought to be exotic laboratory chemicals, and linear high density polyethylene (HDPE), were produced in industry. The pioneering advances of Ziegler's group were reviewed by Wilke [2] and Martin [3].

Also during the 1950s, Phillips Petroleum Co. disclosed activator-free catalyst systems consisting of CrO_3 on silica gel. Standard Oil of Indiana used MoO_3/CoO mixed metal oxide catalysts which were pretreated with hydrogen. Ziegler and Phillips catalysts, as well as other catalyst systems, function according to the same principle displayed in Figure 4. Transition metal alkyls, formed either by alkylation with aluminum alkyls or by reducing chromium (VI) with ethylene, activate ethylene by π-complex formation, thus promoting ethylene insertion into the transition metal alkyl bond at low pressure. Today, it is well established that chain growth occurs exclusively at the transition metal center. Due to the absence of intermolecular chain transfer, the resulting polyethylene is linear and exhibits high density (HDPE).

Chain termination results predominantly from either β-hydride elimination or cleavage of the transition metal alkyls by means of hydrogen, which is used to control molecular weight in industrial processes (Figure 5). Other mechanisms, such as methyl transfer to yield vinyl end groups, are much less common. Vinyl groups are also formed upon β-hydride transfer following a '2,1' insertion. Both hydride transfer and hydrogenation yield transition metal hydrides which are considered to be the the key intermediates of the catalytic cycle displayed in Figure 4.

β-hydride elimination

$$\sim\!\sim\!\sim\text{CH(H)}-\text{CH}_2\text{M} \longrightarrow \sim\!\sim\!\sim\text{CH}=\text{CH}_2 + \text{H}-\text{M}$$

molecular weight control with hydrogen

$$\sim\!\sim\!\sim\text{H}_2\text{C}-\text{CH}_2\text{M} + \text{H}-\text{H} \longrightarrow \sim\!\sim\!\sim\text{H}_2\text{C}-\text{CH}_2-\text{H} + \text{M}-\text{H}$$

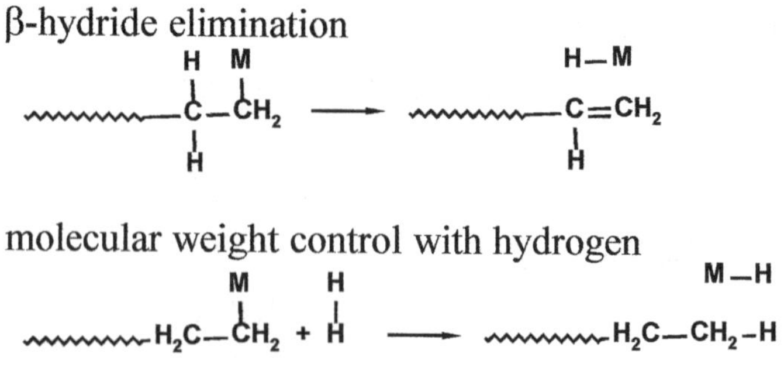

Figure 5 Chain terminating reactions.

Ziegler catalysts based upon group IV transition metal alkyls can be used to polymerize ethylene, 1-olefins, such as propylene, 1-butene, 1-hexene, 1-octene, 4-methyl-1-pentene, dienes, such as 1,3-butadiene and isoprene, and also cycloolefins, such as norbornene and ethylidene norbornene. They fail to incorporate olefins with internal double bonds, e.g. butene-2, and polar olefins, such as methylmethacrylate. Frequently, isobutene and polar monomers form homopolymers via free radical or ionic polymerization, initiated by means of catalyst components. Since conventional Ziegler catalysts and their components represent strong Lewis acids, addition of polar additives can cause severe catalyst poisoning. Therefore, high monomer purities are required to achieve polymerization with highly active catalyst systems. Ethylene was copolymerized with a few mol.% of 1-olefins to incorporate n-alkyl side chains which account for lower density (LLDPE for linear low density polyethylene). At propene and 1-olefin contents ranging between 20 and 60 mol.%, the resulting copolymers (referred to as EPM for ethylene/propylene copolymers and EPDM for ethylene/propylene terpolymers containing small amounts of diene to incorporate olefinic side chains as cure sites for sulfur-based rubber vulcanization) are rendered amorphous and rubbery. Typical dienes are ethylidenenorbornene, 1,4-hexadiene, and 4-vinyl-1-cyclohexene.

In 1954 Giulio Natta at the Milan Polytechnic Institute, sponsored by Montecatini Company, applied Ziegler's catalyst system successfully to produce isotactic PP (iPP). Propylene is prochiral and produces polymers where the repeating unit contains a stereogenic carbon atom. Such diastereoisomers exhibit different physical properties. It was Natta and his group who discovered that fractionation by means of boiling solvents represents the method of choice to achieve separation of different diastereoisomers. Stereoirregular amorphous polypropylene – referred to by Natta as atactic polypropylene (aPP; see the chapter entitled Amorphous or atactic polypropylene) – is soluble in boiling diethyl ether. The aPP exhibits random sequences of the two possible configurations of the stereogenic carbon atom of the propylene repeat unit. In contrast, highly stereoregular iPP is crystalline, insoluble in boiling ethers as well as boiling n-heptane, and melts at 165°C. According to Natta's X-ray structure analysis, iPP is composed of PP sequences with the identical configuration of the stereogenic carbon atoms of adjacent repeat units. Crystalline iPP identified by using Natta's analytical methods, was also present in small amounts in solid polypropylene obtained at Phillips. In 1981, after almost thirty years of legal battles, the US patent on iPP was finally assigned to Phillips [4].

Today the stereoregularity is readily analyzed by means of ^{13}C nuclear magnetic resonance (NMR) spectroscopy on PP solutions. The methyl group signal of aPP and the assignment of pentad sequences is shown in

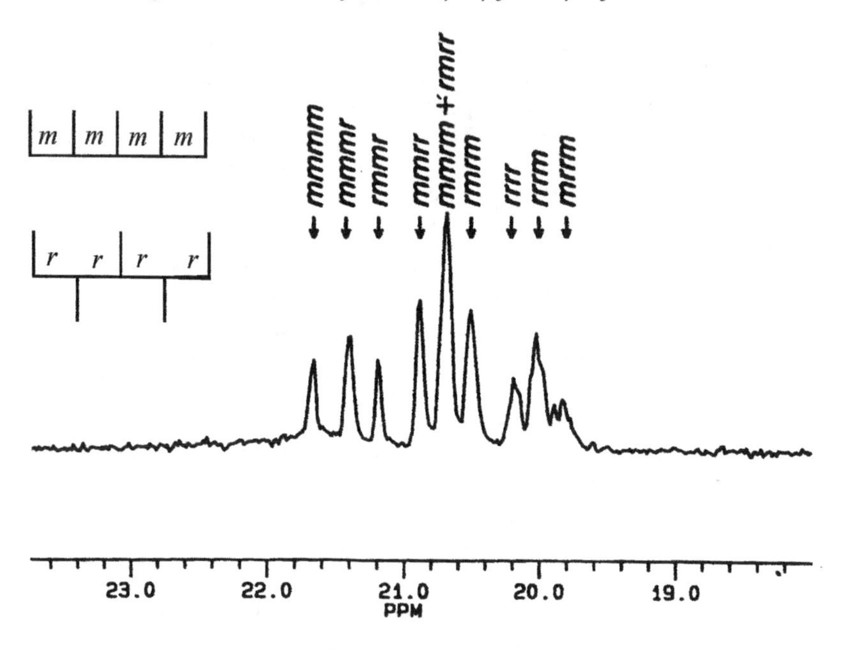

Figure 6 ^{13}C NMR spectrum of atactic polypropylene assignment of the pentad sequences of the methyl group signal.

Figure 6. The spectrum of iPP shows only one methyl group signal typical for the meso (mmmm) pentad, whereas the racemic (rrrr) pentad is typical for syndiotactic polypropylene (sPP). The remarkable history of polyolefins and the development of the stereoselective 1-olefin polymerization is described in reviews by Pino and Mülhaupt [5], covering the first 25 years, and by Brintzinger *et al.* [6, 7], covering modern aspects of metallocene-based propylene polymerizations. PP technology is the subject of several books [8–11] including the chapter 'Industrial polymerization processes' in this book.

During the 1980s, Fink reported a new mechanism for 1-olefin polymerization which became known as 2,ω-polymerization, migratory polyinsertion, or chain walking mechanisms, respectively [12–14]. While polymerization of 1-pentene using conventional titanium catalysts produced poly(1-pentene), nickel-based catalysts were found to yield poly(ethylene-alt-propylene), which is equivalent to methyl-branched polyethylene (Figure 7). With higher 1-olefins, the spacer length in between two methyl branches can be controlled. According to Fink, multiple β-hydride eliminations followed by repeated reinsertion of Ni–H allow migration of Ni-alkyl along the 1-olefin chain. Most likely for steric reasons, only the terminal Ni-alkyl is able to continue inserting 1-olefin.

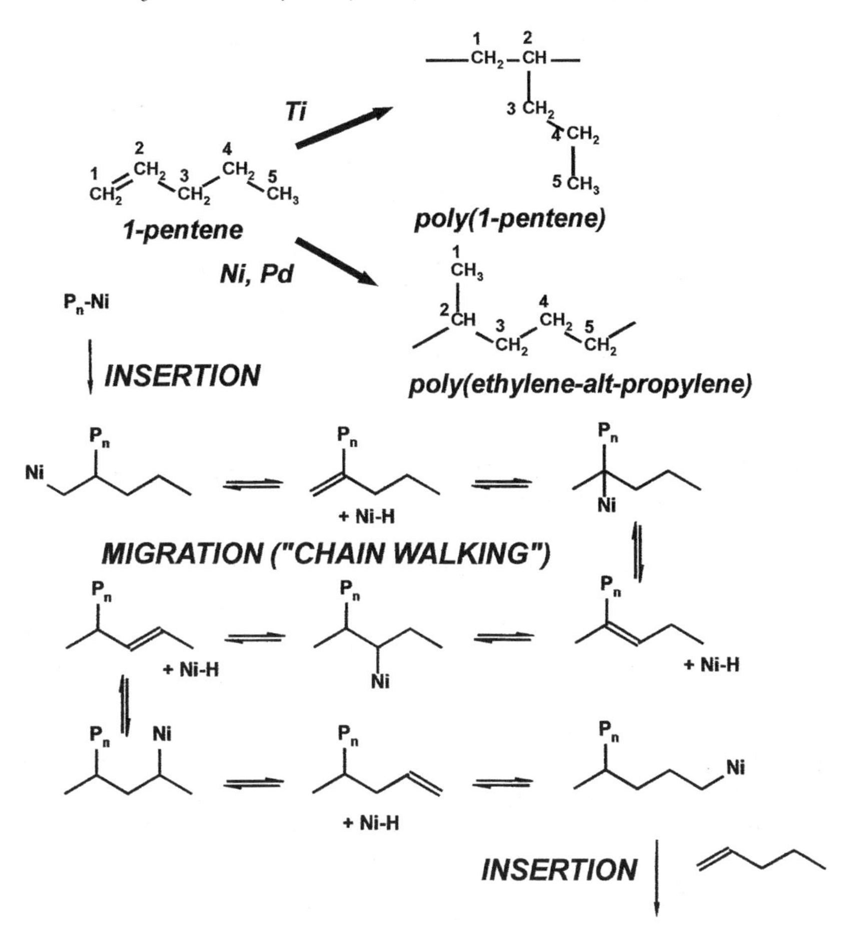

Figure 7 Fink's 2,ω polymerization of 1-olefins produces methyl-branched polyethylene.

This migratory polyinsertion process is also typical for bisimine-based Ni- and Pd-catalysts reported by Brookhart *et al.* [69]. In contrast to Ni, Pd inserts ethylene during migration, thus producing highly branched PEs. Formation of linear, methyl-branched and alkyl-branched PE is shown in Figure 8. In contrast to Ti catalysts, Ni- and Pd-based catalyst systems are known to tolerate functional groups.

While most Ziegler catalysts involve chain termination and chain transfer, only a few catalyst systems have been reported to afford living propylene polymerization where all polyolefin chains contain active transition metal alkyl end groups. A characteristic feature of living olefin polymerization is the increase of degree of polymerization proportional

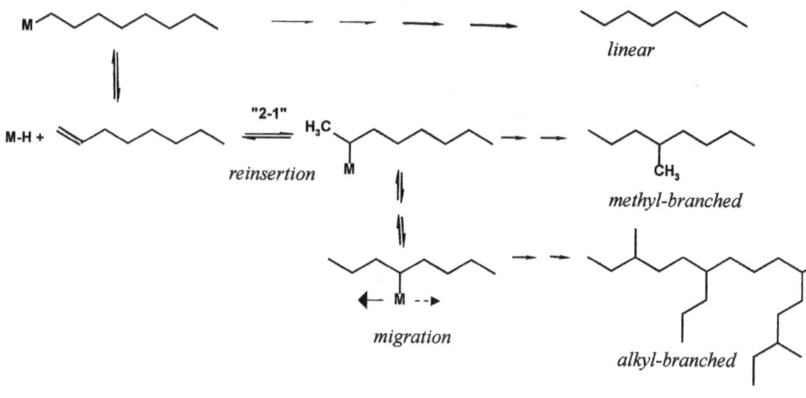

M
linear
"2-1"
M-H +
H_3C
reinsertion
M
CH_3
methyl-branched
M
migration
alkyl-branched

Figure 8 Mechanisms leading to linear, methyl and alkyl-branched polyethylene in ethylene homopolymerization.

to the increase of conversion. The first example of living polymerization was described by Doi *et al.* [15] who polymerized propylene at temperatures of −78°C using VCl_4 activated with $AlEt_2Cl$. Recently Van der Linden *et al.* [16] and Scollard and McConville [17] used novel diamide complexes of titanium, e.g. $[RN(CH_2)_3NR']TiMe_2$ with $R=2,6\text{-}{}^{i}Pr_2C_6H_3$ $R'=2,6\text{-}Me_2C_6H_3$, activated with equimolar amounts of $B(C_6F_5)_3$, to initiate living nonstereoselective polymerization such as hexene-1, octene-1, and decene-1 at room temperature in toluene or methylenechloride solution. Molar masses of living polyolefins varied between 4300 and 148 100 g/mol and polydispersities M_w/M_n varied between 1.05 and 1.11. This development of novel initiators for living olefin polymerization could lead to the preparation of novel families of block and star copolymers, impact-resistant polymer blends, thermoplastic elastomers, and rubbers.

REGIO- AND STEREOSELECTIVE 1-OLEFIN POLYMERIZATION

An important feature of Ziegler–Natta and related catalytic olefin polymerization is the opportunity of controlling poly(1-olefin) molecular architecture via regio- and stereoselectivities of the catalyst systems. Regioselectivity depends upon the insertion type of 1-olefin insertion. Insertion where the transition metal alkyl bond is formed to the carbon atom in 1-position of the 1-olefin is referred to as '1,2' insertion. Accordingly, metal alkyl bond formation involving the carbon atom in 2-position corresponds to '2,1' insertion. In the case of 100% regioselectivity, as shown in Figure 9, the resulting polymers are composed of head-to-tail polyolefins. Lower regioselectivity, i.e. simultaneous '1,2' and '2,1'

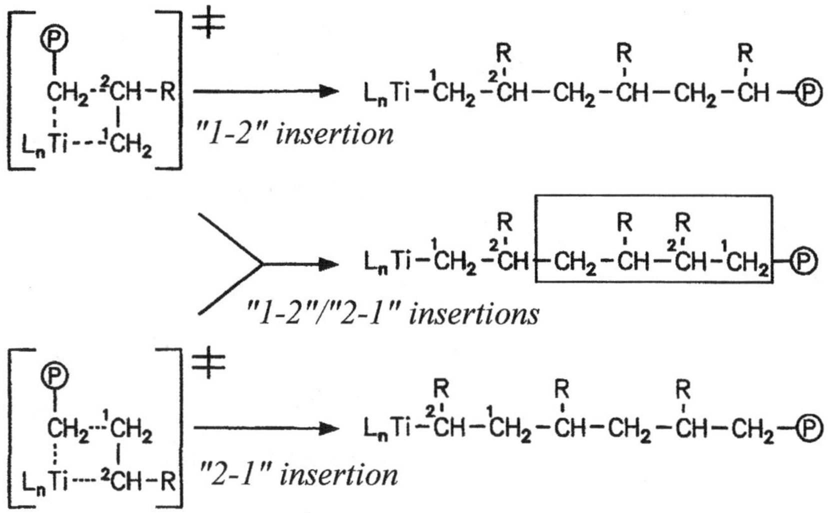

Figure 9 Regioselectivity and insertion type.

insertions, yield regioirregular polypropylenes with 'head-to-head' enchainment as evidenced by $—CH_2—CH(CH_3)—CH(CH_3)—CH_2—$ ('head-to-head') structural units which can be detected by means of NMR spectroscopy. According to Zambelli, the insertion type can be identified by NMR spectroscopy investigation of end groups [18]. For instance, n-propyl end groups reflect the presence of '1,2'-type insertion, while isopropyl endgroups are typical for '2,1' insertions.

Polymerization of *cis*-1-deuterio-propylene gave exclusively *erythro*-diisotactic PP, whereas the *trans*-1-deuterio-propylene afforded exclusively the *threo* PP stereoisomer (Figure 10). This investigation clearly demonstrates that insertion takes place via *cis* addition of the transition metal alkyl to the double bond [19, 20].

The steric control depends upon the capability of the growing chain or the catalytically active site, respectively, to distinguish between the two enantiofaces of the prochiral 1-olefin. As shown in Figure 11, the chirality of catalytically active complex or the stereogenic carbon atom of the propylene unit at the chain end play a key role in steric control.

In the case of propylene polymerization using $VCl_4/AlEt_2Cl$ at −78°C, the steric interaction between the methyl group of the last inserted propylene unit and the '2-1'-inserting propylene monomer accounts for formation of sPP (chain-end control). However, at elevated temperatures, this catalyst system produces mainly isotactic polymers with poor regioselectivity, as evidenced by formation of head-to-head units in the polymer backbone [21]. At temperatures below −30°C, the homogeneous

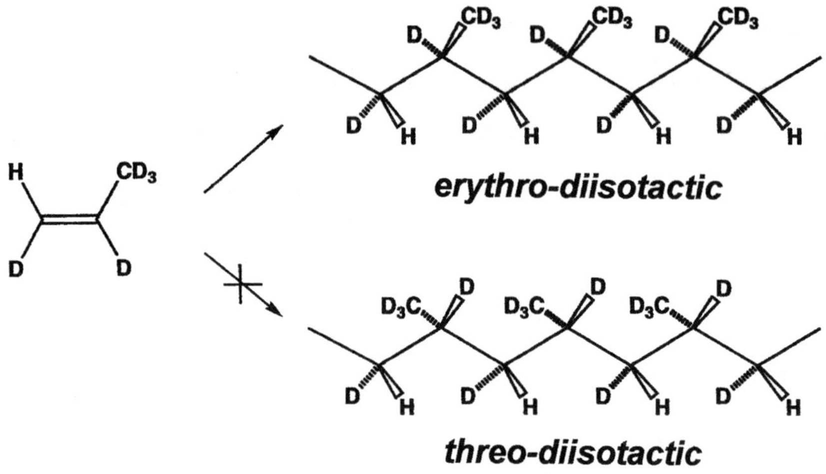

Figure 10 *Cis* addition of propylene was verified by microstructure analysis of poly(*cis*- and *trans*-1-deuterio-propylene).

methylaluminoxane (MAO)-activated Cp_2TiPh_2 catalyst, which was first described by Ewen [22], produced isotactic stereoblock PP containing alternating segments of propylene repeat units with opposite chirality of the stereogenic carbon atom. Upon heating above room temperature, Ewen's homogeneous catalyst, which polymerizes propylene via '1–2' insertion, was rendered completely nonstereoselective, producing aPP without losing high regioselectivity. While Doi's low-temperature, vanadium-based syndioselective catalysts could not meet the high standards of industrial production, Ewen *et al.* [23] developed metallocene-based catalyst systems, such as MAO-activated isopropylidene(fluorenyl) (cyclopentadienyl)zirconiumdichloride, which forms sPP in very high yields according to catalytic-site control mechanism (see the chapter entitled Metallocene catalysts and tailor-made polyolefins).

In catalytic-site control, the chirality of the metal center is responsible for stereoselectivity. For more than two decades, it was thought that only heterogeneous catalysts could produce such chiral transition metal sites. However, during the early 1980s, Brintzinger *et al.* [24] introduced asymmetric ansa-metallocenes such as racemic ethylenebisindenylzirconiumdichloride. Upon activation of racemic ansa metallocenes with MAO, Ewen [25], and Kaminsky *et al.* [26], obtained homogeneous isospecific catalysts, whereas the corresponding meso metallocene catalyst was much less active and gave completely atactic PP. This was the origin of highly versatile new families of metallocenes (see the chapter

chain-end control

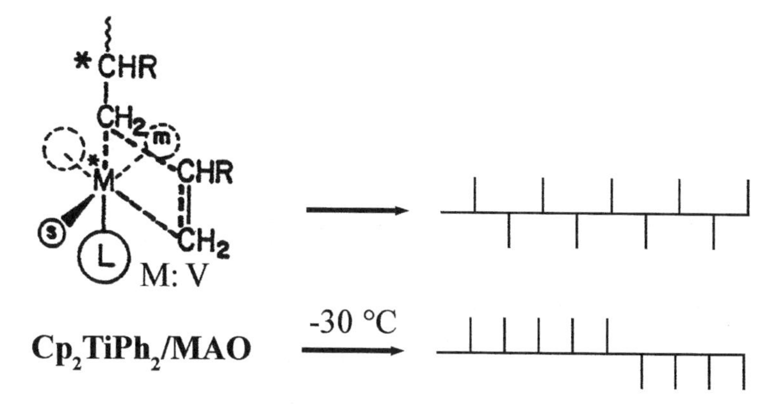

catalytic-site control

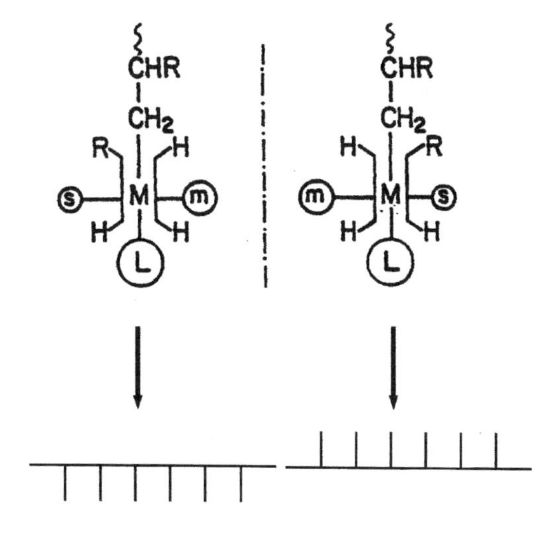

Figure 11 Stereoselectivity via chain-end and catalytic site control.

concerning metallocene catalysts in this book) for the production of iPP, sPP and stereoblock PP.

Again, ^{13}C NMR spectroscopy represents the method of choice to distinguish between chain-end and catalytic-site control mechanisms using the PP microstructure as fingerprint. Recently, Randall reported the characterization of steric defects in PP by means of ^{13}C NMR spectroscopy for iPP as well as sPP [27]. As outlined in Figure 12, false insertion affords a change in chirality of the stereogenic carbon atom of

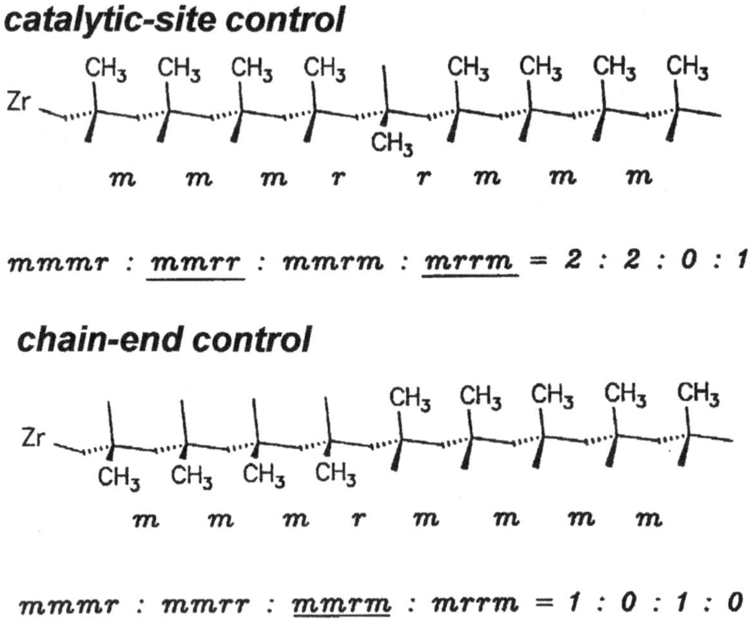

catalytic-site control

CH_3 CH_3 CH_3 CH_3 CH_3 CH_3 CH_3 CH_3

Zr

CH_3

m m m r r m m m

mmmr : *<u>mmrr</u>* : *mmrm* : *<u>mrrm</u>* = 2 : 2 : 0 : 1

chain-end control

CH_3 CH_3 CH_3 CH_3 CH_3

Zr

CH_3 CH_3 CH_3 CH_3

m m m r m m m m

mmmr : *mmrr* : *<u>mmrm</u>* : *mrrm* = 1 : 0 : 1 : 0

Figure 12 Polyolefin microstructures serve as finger prints to identify the origins of steric control.

the inserted propylene which is perpetuated in the case of chain end control, whereas the change of chirality remains isolated in the case of enantiomorphic site control. According to statistics, it is possible to identify typical sequence distributions reflected by characteristic ratio of pentad sequences.

The molecular architectures of PP are the key to controlled superstructure formation and PP properties, e.g. stiffness, strength, impact resistance and optical clarity. Only stereoregular PP is able to crystallize. During the late 1950s Natta [28] showed that iPP forms 3_1 helices which fold to form lamella-type structures. Three different arrangements of helical iPP chains are known as α-, β- and γ-modification (see the chapter entitled Polymorphisms in crystalline polypropylene along with those on the β- and γ-phase PP), which can be identified by means of wide angle X-ray diffraction (WAXS). Polymorphism of isotactic polypropylene was reviewed by Phillips and Mezghani [29]. Chain folding accounts for formation of lamellar structures, which are displayed in Figure 13 for the α-modification of iPP. Three-dimensional growth, known as crosshatching, yields spherulitic superstructures which are visible as Maltese cross-like structures in polarized light. Variation of crystallization conditions and PP microstructure permit control of the crystallization process.

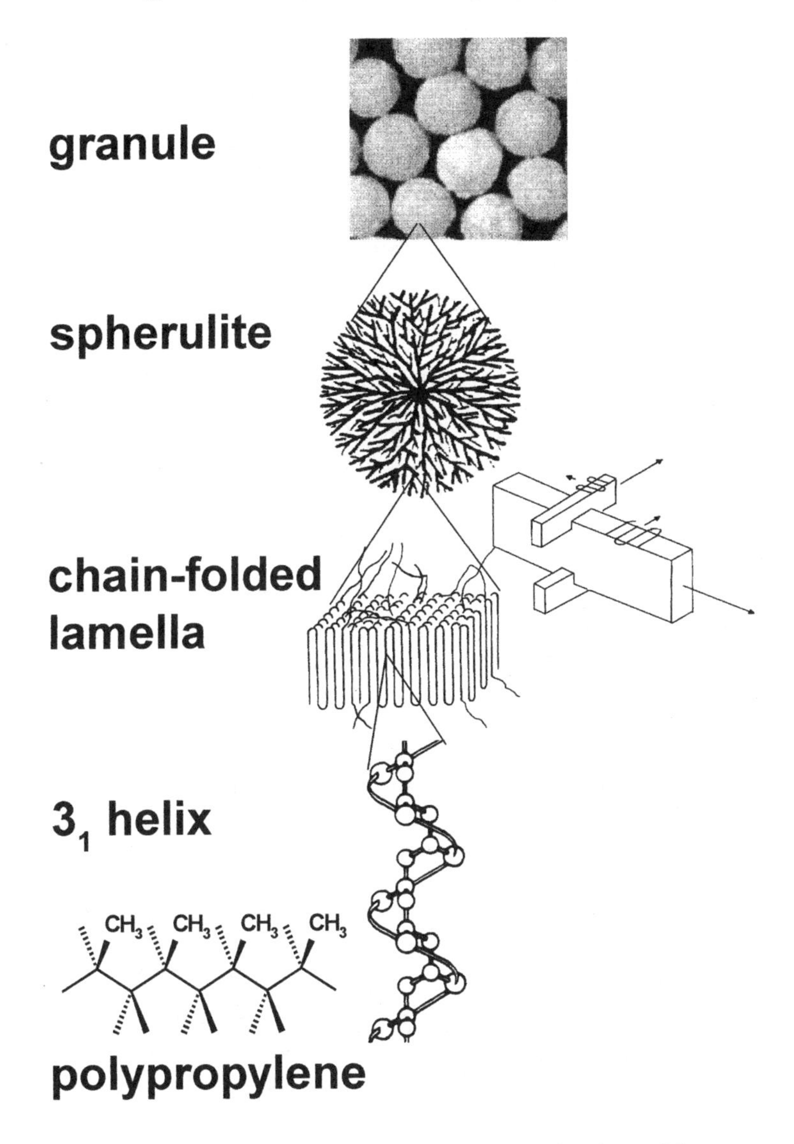

Figure 13 Polypropylene molecular and supermolecular architectures.

Recently, metallocene catalysts were used to incorporate randomly distributed stereo- and regioirregularities into the iPP chain. Decreasing isotactic segment length promotes formation of the γ-modification [30, 31]. Optical clarity of γ-iPP is significantly improved with respect to α-iPP. Special nucleating agents have been introduced to favor formation of the β-modification [32]. Crystallization of sPP, which forms 2_1 helices, have been reported by Lovinger *et al.* [33, 34] and Thomann *et al.* [35] to form nanoscaled bundle-like structures and no large spherulitic three-dimensional superstructures. Modern catalyst technology gives excellent control on molecular polyolefin architectures, which is the key to control superstructure formation during processing.

PROGRESS IN ZIEGLER–NATTA CATALYSIS AND OLEFIN POLYMERIZATION PROCESSES

Since Ziegler's and Natta's pioneering advances during the early 1950s, PP production was revolutionized in remarkably regular intervals of approximately fifteen years. Important development steps and new catalyst and process generations are shown in Figure 14. Innovations have stimulated rapid growth of PP production and enhanced PP's competitiveness with respect to other more expensive or less environmentally friendly polymers. The assignment of numbers for catalyst generations is somewhat arbitrary because of the large number of innovative catalyst systems. In fact, there is considerable confusion to be found in the literature with respect to the numbers of catalyst generations. In Figure 14 we used generation numbers to characterize quantum-leap progress in catalyst and process technology, instead of measuring significant advances within one specific catalyst family. Advances in Ziegler–Natta catalysis were reviewed by Pino [5], Brintzinger [6], Tait *et al.* [36–38], and Corradini *et al.* [39].

The first catalyst generation exhibited rather poor catalyst activities and stereoselectivities with respect to iPP production. Extensive purification was required to remove color-forming, corrosive catalyst residues

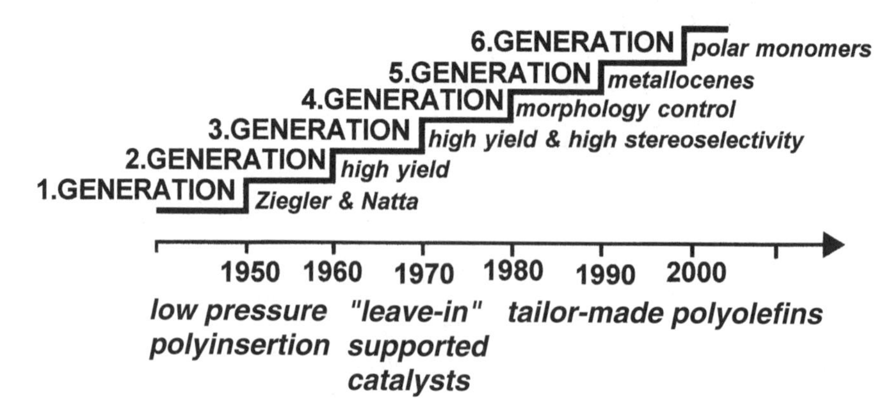

Figure 14 Polypropylene catalyst generations.

and polymeric byproducts such as low molecular weight and aPP. Therefore, a long-standing objective of iPP development has been to improve both catalyst activity and stereoselectivity simultaneously. Since catalytically active transition metal alkyls require at least one vacant coordination site, bulk titanium cannot participate in the catalytic reaction. It is not surprising that many conventional titanium halide-based catalysts, prepared according to Ziegler by alkylation and reduction of $TiCl_4$ with aluminum alkyls, gave rather poor catalyst activities with less than 1% of the transition metal involved in polymerization because most titanium halides were located in the bulk of the $TiCl_3$. Moreover, it was recognized by Natta and his group that crystal modification played an important role in improving stereoselectivities. Therefore, catalyst development was aimed at preparation of supported catalysts where active titanium catalytically active sites are located at the surface and possess adequate configuration to afford high stereoselectivity. The main directions of catalyst developments are shown in Figure 15. Successful strategies and progress in catalyst design were reviewed comprehensively by Albizzati and coworkers [40].

The first catalyst generation, developed by Natta and his group, was composed of γ-$TiCl_3$ activated with $AlEt_2Cl$. Although the stereoselectivity was improved from 40 to 90% iPP with respect to Ziegler's $TiCl_4$/AlR_3 catalyst systems, extensive PP purification was required to produce commercial iPP. Moreover, propylene polymerization was performed as slurry polymerization in an inert hydrocarbon medium, which required special solvent recycling steps. It was soon recognized that improving the surface area and the presence of anhydrous $AlCl_3$, which can substitute inactive bulk $TiCl_3$, as well as the presence of weak electron donors, such as sterically hindered dialkylethers, during catalyst preparation promoted catalyst performance. The first successful improvements led to the development of 'Solvay'-type δ-$TiCl_3$/$AlCl_3$/isoamylether/$AlEt_2Cl$ catalyst system with tenfold catalyst activities and stereoselectivity exceeding 95%. This type of catalyst system was investigated by Nielsen [41]. Some of these second generation catalysts and modified systems are still being applied today in conventional iPP slurry processes.

During the late 1980s, third generation supported catalysts were introduced exhibiting significantly improved stereoselectivities without sacrificing high catalyst activities. At Montedison and Shell, it was discovered that super highly active catalysts were obtained when $TiCl_4$ was supported on anhydrous high-surface-area magnesum chloride in the presence of electron-donating Lewis bases [42–46]. The crystal structure of $MgCl_2$ is isotype to that of γ-$TiCl_3$. Therefore, $MgCl_2$ can substitute inactive bulk $TiCl_3$ and offers the equivalent coordination site for immobilizing titanium alkyl halide complexes at the $MgCl_2$ surface.

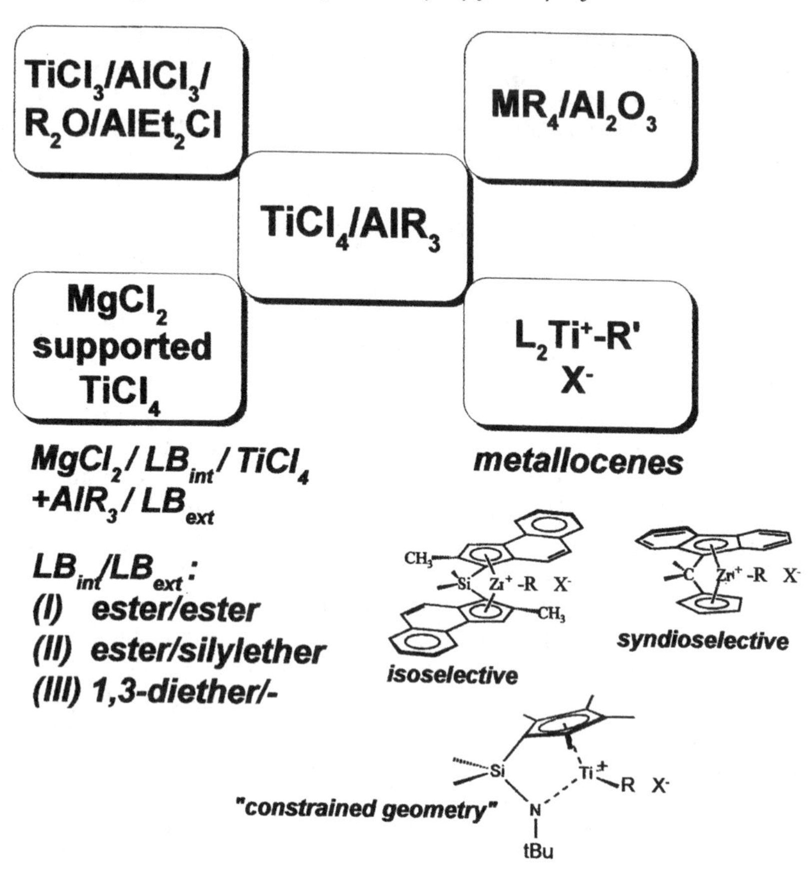

Figure 15 Development of group IV catalysts.

Activation of $MgCl_2$ is required to afford high specific surface areas $> 40\,m^2/g$. This can be achieved by means of mechanical grinding or by *in situ* preparation of $MgCl_2$, e.g. by chlorinating magnesium alkyls.

A typical process for catalyst preparation is shown in Figure 16. After dehydration, $MgCl_2$ is ground together with $TiCl_4$ and a Lewis base, such as diethylphthalate, as internal electron donor. This internal Lewis base facilitates deagglomeration of primary catalyst particles during propylene polymerization. While conventional $TiCl_3/AlEt_2Cl$ catalysts are encapsulated in a iPP shell, which is highly impermeable to propylene, deagglomeration during polymerization is responsible for very high concentration of active sites and less diffusion limitations. In comparison to first generation catalysts, the number of active sites is increased by one or two orders of magnitude. The external Lewis base, which is added

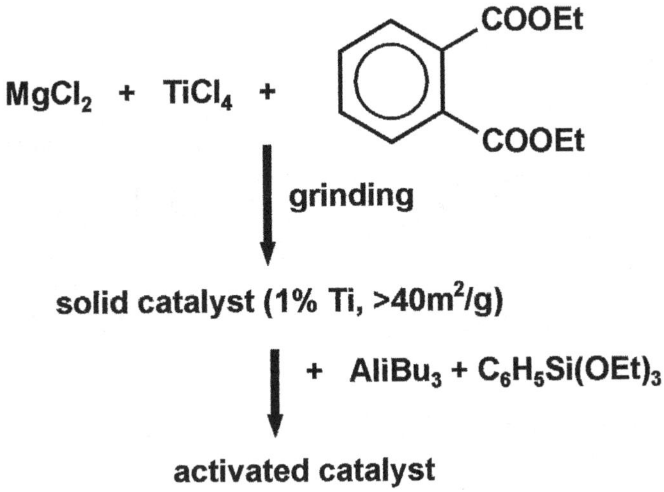

$MgCl_2$ + $TiCl_4$ + (diethyl phthalate: C_6H_4(COOEt)(COOEt))

↓ grinding

solid catalyst (1% Ti, >40m^2/g)

↓ + $AliBu_3$ + $C_6H_5Si(OEt)_3$

activated catalyst

Figure 16 Preparation of $MgCl_2$-supported catalysts.

together with the activator aluminum alkyl, poisons selectively non-stereospecific sites. Most likely for steric reasons, highly isoselective sites are much less Lewis acidic with respect to the sterically less hindered non-selective sites. In a series of equilibria involving complex formation of Lewis bases with Ti-alkyl sites as well as with the aluminum activator alkyl, the complex formation with highly Lewis acidic nonstereoselective sites is favored. The selectivity of Lewis bases modifiers with respect to catalytically active sites appears to follow the 'lock–key' principle which was proposed by several groups in analogy to enzyme/substrate or enzyme/coenzyme complexes, respectively.

Originally esters of aromatic carboxylic acids, such as ethyl benzoate or dialkylphthalate, were used as preferred external and internal Lewis bases. Although esters poisoned nonstereospecific sites selectively, the total catalyst activity decreased markedly with increasing external ester/Al molar ratio and with increasing stereoselectivity [47]. In a subsequent improvement during the 1980s, weaker silylether Lewis bases, e.g. phenyltriethoxysilane, were used as external Lewis bases in conjunction with diesters as internal Lewis bases in order to achieve high stereo-selectivities without encountering such pronounced losses of catalyst activity [48–51]. During the late 1980s and early 1990s, 1,3-diethers were introduced as a new class of electron-donating Lewis bases [52]. When using 1,3-diethers as internal Lewis bases, both the resulting catalyst activities and stereoselectivities were extraordinarily high. Such catalyst systems do not require the use of additional external Lewis bases. In fact,

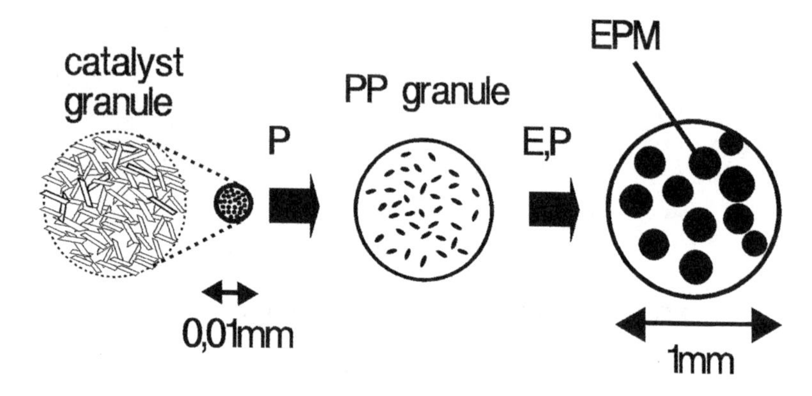

Figure 17 Reactor granule technology uses spherical catalyst particles as templates to control morphology.

the aluminum alkyl activator does not appear to be able to remove the diether from the solid catalyst component [53]. Development of this 'third generation' catalyst (Figure 14) is still in progress and illustrates the very attractive potential of Lewis-base-modified $MgCl_2$-supported catalysts for improving polypropylene properties.

Based upon the insight in the correlation between catalyst structure and morphology development, Galli and coworkers observed that the catalyst particles can act as templates for the formation of PP polymer particles (Figure 17), when using spherical catalyst particles, which are composed of a large number of agglomerated much smaller primary particles. As a function of the catalyst morphology, dense and microporous PP granules can be obtained. This new catalyst generation, referred to as fourth generation catalysts in Figure 14, became known as reactor granule technology or Spheripol® iPP developed by Montell. Preferably PP particle formation is performed in a liquid propylene using a loop reactor. As outlined in Figure 17, the iPP granules contain uniformly dispersed active primary catalyst particles. In a second stage in a gas-phase reactor, ethylene is copolymerized with propylene inside iPP particles to produce impact-modified iPP/EPM reactor blends (Catalloy® technology of Montell). The microporous iPP granules can be used as reactors to polymerize monomers, such as styrene, acrylics or styrene/maleic anhydride, inside of the iPP particles to afford multiphase iPP with improved balance of stiffness/toughness/heat distortion temperature (Montell's Hivalloy® technology). Reactor granule technology has been reviewed by Galli and coworkers [54, 55].

During the 1980s, another catalyst generation based upon metallocenes – referred to as fifth generation in Figure 14 – became available. In contrast to multicenter first generation catalysts, being composed of numerous catalytically active sites with different stereoselectivities, catalyst activities, and reactivities with respect to comonomer incorporation, metallocenes are composed of a single type of catalytically active center. Single-site metallocenes, activated with MAO or boranes, give very uniform homo- and copolymers with polydispersities M_w/M_n of approximately 2 and molecular weight independent comonomer incorporation. Moreover, metallocenes give excellent control of molecular weight, end groups, long-chain branching, comonomer incorporation of 1-olefins as well as styrene and cycloolefins, and polymer morphology. Cyclopolymerization and copolymerization of olefins with cycloolefins expand the range of polyolefins in the range of engineering resins with high heat distortion temperature. Reviews by Kaminsky [56] and Hamielec [57] cover opportunities for polyolefin synthesis. More detailed description of metallocene technology is presented in the chapters on metallocene catalysis in this book. The development of well-defined metallocene catalysts is setting the stage for the development of tailor-made polyolefins with new polymer architectures and property combinations.

As shown in Figure 15, the search for activator-free catalysts systems led to the development of MR_4/Al_2O_3 with M=Ti,Zr and R=CH_2Ph, $CH_2C(CH_3)_2Ph$. In contrast to highly stereoselective catalysts, these catalyst systems are much less stereoselective. However, they are able to produce both highly isotactic and high molecular weight low stereoregular PP simultaneously. The very flexible low stereoregular PP represent stereoblock polymers containing both flexible atactic and rigid isotactic segments. Due to their high molecular weight, there exist more than two isotactic segments which can cocrystallize to form temporary thermoreversible crosslinks via crystallite formation. This development by Setterquist and Collette at Du Pont has led to a wide range of flexible PP materials including thermoplastic elastomers [58, 59]. Rather soft stereoblock PPs were reported by Wisseroth [60] who used gas phase polymerization of propylene. Today special metallocene catalysts are available to tailor stereoblock PP and thermoplastic elastomers [61, 62]. Aluminium-alkyl-mediated chain transfer between stereoselective and nonstereoselective sites of metallocene catalysts were applied by Chien and coworkers to prepare thermoplastic elastomers consisting of alternating flexible atactic and stereoregular segments [63]. Also high molecular weight aPP, produced by means of nonstereoselective metallocenes, appear to be interesting elastomer blend components for PP [64].

Recently, Ni- and Pd-based catalysts have been developed in order to copolymerize nonpolar ethylene, 1-olefins or cycloolefins with polar comonomers, such as carbon monoxide and methyl acrylate or 1-olefins, containing polar groups which are separated by at least two methylene group spacers from the polymerizable vinyl group. While CO is a very potent catalyst poison for most Ti-based catalysts system, Pd-based catalysts afford strictly alternating olefin/CO copolymers in very high yields. Carilon® from Shell [65] and Ketonex® from BP [66] are semi-crystalline thermoplastic polyketones with a melting temperature of 220°C. Incorporation of a few percent of propylene into alternating ethylene/CO copolymers lowers the melting temperature and facilitates processing. High content of propylene renders polyketones elastomeric [67]. A wide range of new polyketone and polyspiroketals have been prepared. The development of alternating olefin/CO polymerization was reviewed by Drent and Budzelaar [68]. With bisimine complexes of nickel and palladium, described by Brookhart *et al.* [69], ethylene can be copolymerized with methylacrylate to form highly branched polyethylene with pendant n-alkyl and esteralkyl chains. In future, novel generations of transition metal catalysts will exploit the potential of both nonpolar and polar petrochemical feedstocks ranging from ethylene and propylene to CO, acrylics, acrylonitrile and vinylacetate.

Propylene polymerization processes, including slurry, gas-phase and liquid pool polymerization, have been reviewed by Lieberman and Barbe [70]. Progress in catalyst development is reflected by significantly simplified polymerization processes and markedly reduced environmental pollution. As is apparent from Figure 18, which displays the general scheme of an olefin polymerization process, in gas-phase and liquid pool processes hydrocarbon diluents and deactivations as well as polymer purifications steps are eliminated. Reactor granule technology forms

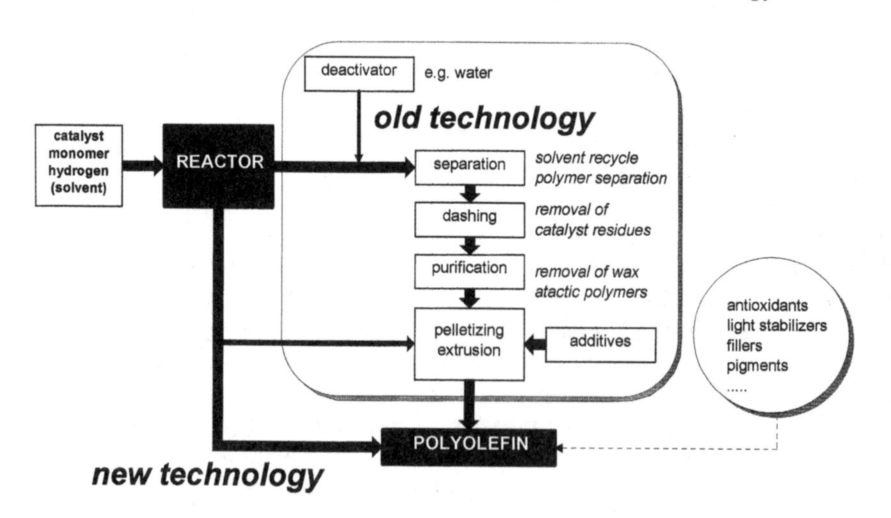

Figure 18 Olefin polymerization processes using first generation and modern catalyst systems.

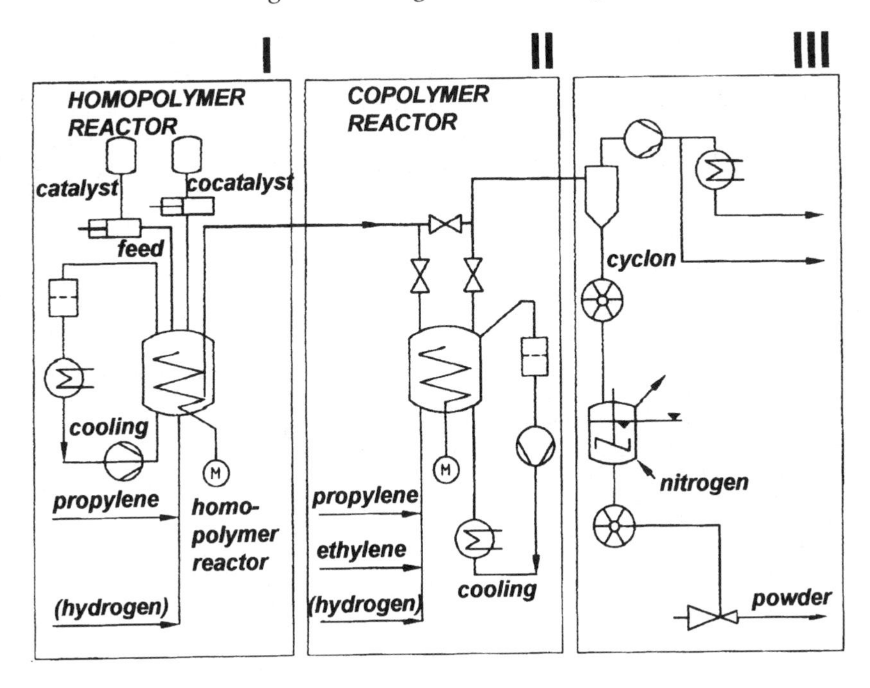

Figure 19 Novolen gas-phase process (courtesy of BASF AG).

pellet-sized iPP particles during polymerization, thus eliminating pelletizing extrusion.

Figure 19 displays the scheme of the Novolen® gas-phase polymerization [71, 72] to illustrate modern environmentally friendly propylene polymerization and the potential of cascade reactor system for production of reactor blends. This technology is exploiting the potential of new generations of supported catalysts which give high yield, high selectivity, controlled comonomer incorporation and controlled PP morphology development. In the first stage (I), catalyst and activator are fed into the stirred tank reactor together with propylene gas. Hydrogen can be used to control molecular weight. In the second stage (II), ethylene/propylene are copolymerized in polypropylene particles to incorporate up to 40% rubber. Depending on the catalyst system and polymerization reaction condition, it is possible to vary the sequence of the EPM. Both random and block copolymers can be formed. The resulting impact-modified iPP reactor blend exhibits improved balance of stiffness and low-temperature impact resistance. In the disengagement zone (III), residual propylene gas is recycled and the iPP powder recovered. Also, the molecular weight distribution of iPP can be varied in sequenced gas-phase polymerization reactors to improve processability and mechanical properties.

Gas-phase polymerization technology benefits from the development of single-site metallocenes [73] which enhance control of PP properties in order to meet the demands of highly diversified applications. Simplified solvent-free processes combined with catalyst design represents the key to the development of novel polyolefin families with tailor-made properties.

REFERENCES

1. Seymour, R.B. and Cheng, T. (eds) (1986) *History of Polyolefins*, Reidel, Dordrecht.
2. Wilke, G. (1996) in *Ziegler Catalysts*, (eds G. Fink, R. Mülhaupt and H.H. Brintzinger), Springer-Verlag, Berlin, p. 1.
3. Martin, H. (1996) in *Ziegler Catalysts*, (eds G. Fink, R. Mülhaupt and H.H. Brintzinger), Springer-Verlag, Berlin, p. 15.
4. Hogan, J.P. and Banks, R.L. (1986) in *History of Polyolefins*, (eds R.B. Seymour and T. Cheng), Reidel, Dordrecht, p. 103.
5. Pino, P. and Mülhaupt, R. (1980) *Angew. Chem. Int. Ed.*, **19**, 857.
6. Brintzinger, H.H., Fischer, D., Mülhaupt, R., Rieger, B. and Waymouth, R.M. (1995) *Angew. Chem. Int. Ed. Engl.*, **34**, 1143.
7. Fink, G., Mülhaupt, R. and Brintzinger, H.H. (eds) (1996) *Ziegler Catalysts*, Springer-Verlag, Berlin.
8. Moore, E.P. (ed.) (1996) *Polypropylene Handbook*, Hanser, Munich.
9. Van der Ven, S. (1990) *Polypropylene and other Polyolefins*, Elsevier, Amsterdam.
10. Karger-Kocsis, J. (ed.) (1995) *Polypropylene: Structure, Blends and Composites*, Vols 1–3, Chapman & Hall, London.
11. Vasile, C. and Seymour, R.B. (eds) (1993) *Handbook of Polyolefins*, Dekker, New York.
12. Fink, G., Möhring, V., Heinrichs, A. and Denger, Ch. (1992) *ACS Symp. Ser.*, **496**, 88.
13. Schubbe, R., Angermund, K., Fink, G. and Goddard, R. (1995) *Macromol. Chem. Phys.*, **196**, 467.
14. Fink, G., Möhring, V.M., Heinrichs, A., Denger, C., Schubbe, R.H. and Mühlenbrock, P.H. (1996) *Polymer Mat. Encycl.*, **6**, 4720.
15. Doi, Y., Ueki, S. and Keii, T. (1979) *Macromolecules*, **12**, 814.
16. Van der Linden, A., Schaverien, C.J., Meijboom, N., Ganter, C. and Orpen, A.G. (1995) *J. Am. Chem. Soc.*, **117**, 3008.
17. Scollard, J.D. and McConville, D.H. (1996) *J. Am. Chem. Soc.*, **118**, 10008.
18. Zambelli, A., Sacchi, M.C. and Locatelli, P. (1983) in *Transition Metal Catalyzed Polymerizations Alkenes and Dienes Part* A, (eds R.P. Quirk, H.L. Hsieh, P.J.T. Tait), MMI Press, Harwood, New York, p. 83.
19. Natta, G., Farina, M. and Peraldo, M. (1958) *Atti Acad. Lincei Rend.*, **25** (Ser. 8), 424.
20. Zambelli, A., Giongo, M. and Natta, G. (1968) *Makromol. Chem.*, **112**, 183.
21. Zambelli, A., Bajo, E. and Rigamonti, E. (1978) *Makromol. Chem.*, **179**, 1249.
22. Ewen, J.A. (1984) *J. Am. Chem. Soc.*, **106**, 6355.
23. Ewen, J.A., Jones, R.L., Razavi, A. and Ferrara, J.D. (1988) *J. Am. Chem. Soc.*, **110**, 339.
24. Wild, F.R.P.W., Zsolani, L., Huttner, G. and Brintzinger, H.H. (1982) *J. Organomet. Chem.*, **232**, 233.

25. Ewen, J.A. (1984) *J. Am. Chem. Soc.*, **106**, 6355.
26. Kaminsky, W., Külper, K., Brintzinger, H.H. and Wild, F.R.P.W. (1985) *Angew. Chem. Int. Ed. Engl.*, **24**, 507.
27. Randall, J.C. (1997) *Macromolecules*, **30**, 803.
28. Natta, G. (1955) *J. Polymer Sci.*, **16**, 143.
29. Phillips, P.J. and Mezghani, K. (1996) *Polymeric Materials Encyclopedia*, Vol. 9, (ed. J.C. Salamone), CRC Press, Boca Raton, FL, p. 6637.
30. Fischer, D. and Mülhaupt, R. (1994) *Macromol. Chem. Phys.*, **195**, 1433.
31. Thomann, R., Wang, C., Kressler J. and Mülhaupt, R. (1996) *Macromolecules*, **29**, 8425.
32. Binsbergen, F.L. and Lange, B.G.M. (1968) *Polymer*, **9**, 23.
33. Lovinger, A.J., Lotz, B., Davis, D.D. and Schumacher, M. (1994) *Macromolecules*, **27**, 6603.
34. Schumacher, M., Lovinger, A.J., Agarwal, P., Wittmann, J.C. and Lotz, B. (1994) *Macromolecules*, **27**, 6956.
35. Thomann, R., Wang, C., Kressler, J., Jüngling S. and Mülhaupt, R. (1995) *Polymer*, **36**, 3795.
36. Tait, P.J.T. (1989) in *Comprehensive Polymer Science*, Vol. 7, (eds G. Allen, J.C. Bevington, G.C. Eastmond, A. Ledwith, S. Russo and P. Sigwalt), Pergamon, Oxford, p. 1.
37. Tait, P.J.T. and Watkins, N.D. (1989) in *Comprehensive Polymer Science*, Vol. 7, (eds G. Allen, J.C. Bevington, G.C. Eastmond, A. Ledwith, S. Russo and P. Sigwalt), Pergamon, Oxford, p. 533.
38. Tait, P.J.T. and Berry, I.G. (1989) in *Comprehensive Polymer Science*, Vol. 7, (eds G. Allen, J.C. Bevington, G.C. Eastmond, A. Ledwith, S. Russo and P. Sigwalt), Pergamon, Oxford, p. 575.
39. Corradini, P., Busico, V., Guerra, G. (1989) in *Comprehensive Polymer Science*, Vol. 7, (eds G. Allen, J.C. Bevington, G.C. Eastmond, A. Ledwith, S. Russo and P. Sigwalt), Pergamon, Oxford, p. 29.
40. Albizzati, E., Giannini, U., Collina, G., Noristi, L. and Resconi, L. (1996) in *Polypropylene Handbook*, (ed. E. P. Moore, Jr.), Hanser, Munich, p. 11.
41. Nielsen, R.P. (1983) in *Transition Metal Catalyzed Polymerization*, (ed. R.P. Quirk), Harwood, New York, p. 47.
42. Chadwick, J.C. (1995) in *Ziegler Catalysts*, (eds G. Fink, R. Mülhaupt and H.H. Brintzinger), Springer-Verlag, Berlin, p. 428.
43. Galli, P., Barbe, P.C. and Noristi, L. (1984) *Angew. Makromol. Chem.*, **120**, 73.
44. Barbe, P.C., Cecchin, G. and Noristi, L. (1987) *Adv. Polymer Sci.*, **81**, 1.
45. Tait, P.J.T., Zohuri, G.H., Kells, A.M. and McKenzie, I.D. (1995) in *Ziegler Catalysts*, (eds G. Fink, R. Mülhaupt and H.H. Brintzinger), Springer-Verlag, Berlin, p. 344.
46. Goodall, B.L. (1983) in *Transition metal catalyzed polymerizations, Part A*, (ed. T.P. Quirk), Harwood, New York, p. 355.
47. Kashiwa, N. and Kojoh, S. (1995) *Macromol. Symp.*, **89**, 27.
48. Soga, K., Shiono, T. and Doi, Y. (1988) *Makromol. Chem.*, **189**, 1531.
49. Spitz, R., Bobichon, C. and Guyot, A. (1989) *Makromol. Chem.*, **190**, 707.
50. Hu, Y. and Chien, J.C.W. (1988) *J. Polymer Sci. Polymer Chem.*, **26**, 2003.
51. Seppälä, J.V., Härkönen, M. and Luciani, L. (1989) *Makromol. Chem.*, **190**, 2535.
52. Albizzati, E., Giannini, U., Morini, G., Smith, C.A. and Zeigler, R.C. (1995) in *Ziegler Catalysts*, (eds G. Fink, R. Mülhaupt and H.H. Brintzinger), Springer-Verlag, Berlin, p. 413.

53. Iiskola, E., Pelkonen, A., Kakkonen, H.J., Pursiainen, J. and Pakkanen, T.A. (1993) *Makromol. Chem. Rapid Commun.*, **14**, 133.
54. Galli, P., Haylock, J.C. and Simonazzi, T. (1995) in *Polypropylene, Structure, Blends and Composites*, (ed. J. Karger-Kocsis), Chapman & Hall, London, Vol. 2, p. 1.
55. Galli, P. and Haylock, J.C. (1992) *Macromol. Symp.*, **63**, 19.
56. Kaminsky, W. (1997) *Macromol. Chem. Phys.*, **197**, 3907.
57. Hamielec, A.C. (1996) *Prog. Polymer Sci.*, **21**, 651.
58. Shih, C.K. and Su, A.C.L. (1987) in *Thermoplastic Elastomers*, (eds N.R. Legge, G. Holden and H.E. Schroeder), Hanser, Munich, p. 91.
59. Collette, J.W., Tullock, C.W., MacDonald, R.N., Buck, W.H., Su, A.C.L., Harrell, J.R., Mülhaupt, R. and Anderson, B.C. (1989) *Macromolecules*, **22**, 3851.
60. Wisseroth, K. (1997) *Chemiker Zeitung*, **101**, 271.
61. Llinas, G.H., Dong, S.H., Mallin, D.T., Rausch, M.D., Lin, Y.G., Winter, H.H. and Chien, J.C.W. (1992) *Macromolecules*, **25**, 1242.
62. Coates, G.W. and Waymouth, R.M. (1995) *Science*, **267**, 217.
63. Chien, J.C.W., *Macromolecules*,in print.
64. Resconi, L. and Silvestri, R. (1996) in *Polymeric Materials Encyclopedia*, (ed. J.C. Salamone), CRC Press, Boca Raton, FL, p. 6609.
65. Ash, C.E. and Flood, J.E. (1997) *Polymer Mat. Sci. Engng*, **76**, 110.
66. Bonner, J.G. and Powell, A.K. (1997) *Polymer Mat. Sci. Engng*, **76**, 108.
67. Abu-Surrah, A.S., Eckert, G., Pechhold, W., Wilke, W. and Rieger, B. (1996) *Macromol. Rapid Commun.*, **17**, 559.
68. Drent, E. and Budzelaar, P.H.M. (1998) *Chem. Rev.*, **96**, 663.
69. Johnson, L.K., Mecking, S. and Brookhart, M. (1996) *J. Am. Chem. Soc.*, **118**, 267.
70. Lieberman, R.B. and Barbe, P.C. in *Encyclopedia of Polymer Science and Engineering*, Vol. 13, (eds H.F. Mark, N.M. Bikales, C.G. Overberger and G. Menges), Wiley, New York, p. 464.
71. Rümpler, K.-D., Jaggard, J.F.R. and Werner, R.A. (1988) *Kunststoffe*, **78**, 602.
72. Langhauser, F., Kerth, J., Kersting, M., Kölle, P., Lilge, D. and Müller, P. (1994) *Angew. Makromol. Chem.*, **223**, 155.
73. Hungenberg, K.D., Kerth, J., Langhauser, F., Marcinke, B. and Schulund, R. (1995) in *Ziegler Catalysts*, (eds G. Fink, R. Mülhaupt and H.H. Brintzinger), Springer-Verlag, Berlin, p. 363.

Keywords: PP production, Ziegler–Natta catalysts, catalytic polymerization, chain branching, high density polyethylene (HDPE), low density polyethylene (LDPE), linear low density polyethylene (LLDPE), ethylene/propylene copolymer (EPM), ethylene/propylene/diene terpolymer (EPDM), living polymerization, stereoselectivity, regioselectivity, selective catalysts, chirality, microstructure control, polymorphism, crosshatching, spherulite, polidispersity, metallocene catalysts, single-site catalysts, atactic PP (aPP), syndiotactic PP (sPP), Hivalloy®, Novolen®, Catalloy®, Spheripol®, supported catalysts, industrial polymerization, history, gas-phase polymerization, slurry polymerization, polymerization.

Index